STP 1074

Fracture Mechanics: Twenty-First Symposium

J. P. Gudas, J. A. Joyce, and E. M. Hackett, editors

ASTM
1916 Race Street
Philadelphia, PA 19103

ASTM Publication Code Number (PCN): 04-010740-30
ISBN: 0-8031-1299-8
ISSN: 1040-3094

NOTE

The Society is not responsible, as a body,
for the statements and opinions
advanced in this publication.

Peer Review Policy

Each paper published in this volume was evaluated by three peer reviewers. The authors addressed all of the reviewers' comments to the satisfaction of both the technical editor(s) and the ASTM Committee on Publications.

The quality of the papers in this publication reflects not only the obvious efforts of the authors and the technical editor(s), but also the work of these peer reviewers. The ASTM Committee on Publications acknowledges with appreciation their dedication and contribution of time and effort on behalf of ASTM.

Printed in Baltimore, Md.
August 1990

Foreword

The ASTM Twenty-First National Symposium on Fracture Mechanics was held in Annapolis, Maryland, on 28–30 June 1988. Its sponsor was Committee E-24 on Fracture Testing.

The co-chairmen for this symposium were John P. Gudas, David Taylor Research Center; James A. Joyce, United States Naval Academy; and Edwin M. Hackett, David Taylor Research Center. They have also served as editors of this volume.

Contents

ELASTIC-PLASTIC FRACTURE MECHANICS (II)

MICROMECHANICS OF FRACTURE

COMPUTATIONAL MECHANICS

Introduction

The success of the Twenty-First National Symposium on Fracture Mechanics, held on 28-30 June 1988 in Annapolis, Maryland, and sponsored by ASTM Committee E-24 on Fracture Testing, demonstrated the continued rapid development occurring in this field. Papers were solicited from all areas of fracture mechanics and its applications. Contributions representing a wide range of topics came from the United States and six foreign countries. New work is presented in elastic-plastic fracture, dynamic fracture, transition fracture in steels, micromechanical aspects of the fracture process, computational mechanics, fracture mechanics testing, and applications of this technology. Each area poses its own challenges, and developments proceed somewhat independently. This volume aids the researcher in keeping abreast of these varied aspects of the discipline of fracture mechanics.

The diligent work of the Symposium Organizing Committee, the authors, and the reviewers is gratefully appreciated. We would particularly like to recognize the efforts of the ASTM staff including Mr. Hans Greene, Ms. Kathy Friend, Ms. Wendy Dyer, Ms. Kathy Greene, Ms. Monica Armata, Ms. Rita Harhut, and Mr. Allan Kleinberg. Finally, the assistance of Mrs. Mary Cropley and Ms. Amanda Ewen of the David Taylor Research Center is gratefully acknowledged.

J. P. Gudas
J. A. Joyce
E. M. Hackett

Elastic-Plastic Fracture Mechanics (I)

W. A. Sorem,[1] R. H. Dodds, Jr.,[2] and S. T. Rolfe[3]

An Analytical Comparison of Short Crack and Deep Crack CTOD Fracture Specimens of an A36 Steel

REFERENCE: Sorem, W. A., Dodds, R. H., Jr., and Rolfe, S. T., **"An Analytical Comparison of Short Crack and Deep Crack CTOD Fracture Specimens of an A36 Steel,"** *Fracture Mechanics: Twenty-First Symposium, ASTM STP 1074,* J. P. Gudas, J. A. Joyce, and E. M. Hackett, Eds., American Society for Testing and Materials, Philadelphia, 1990, pp. 3–23.

ABSTRACT: The effect of crack-depth to specimen-width ratio on crack tip opening displacement (CTOD) fracture toughness is an important consideration in relating the results of laboratory tests to the behavior of actual structures. Deeply cracked three-point bend specimens with crack-depth to specimen-width ratios (a/W) of 0.50 are most often used in laboratory tests. However, to evaluate specific weld microstructures or the behavior of structures with shallow surface cracks, specimens with a/W ratios much less than 0.50 often are required. Laboratory tests reveal that three-point bend specimens with short cracks ($a/W = 0.15$) exhibit significantly larger critical CTOD values than specimens with deep cracks ($a/W = 0.5$) up to the point of ductile initiation.

In this study, finite element analyses are employed to compare the elastic-plastic behavior of square (cross-section) three-point bend specimens with crack-depth to specimen-width ratios (a/W) ranging between 0.50 and 0.05. The two-dimensional analysis of the specimen with an a/W ratio of 0.15 reveals a fundamental change in the deformation pattern from the deep crack deformation pattern. The plastic zone extends to the free surface behind the crack concurrent with the development of a plastic hinge. For shorter cracks ($a/W = 0.10$ and 0.05), the plastic zone extends to the free surface behind the crack prior to the development of a plastic hinge. For longer cracks ($a/W > 0.20$), a plastic hinge develops before the plastic zone extends to the free surface behind the crack.

These results prompted further study of specimens with an a/W ratio of 0.15 using three-dimensional, elastic-plastic finite element analyses. Results of the short crack ($a/W = 0.15$) analysis are compared to the results of the deep crack ($a/W = 0.50$) analysis reported previously by the authors. In the linear-elastic regime (characterized by small-scale plastic deformation) the relationship of stress ahead of the crack tip to CTOD is identical for the short crack and the deep crack specimens. At identical CTOD levels in the elastic-plastic regime (large-scale plasticity, hinge formation), the crack tip stress is significantly lower for specimens with $a/W = 0.15$ than for specimens with $a/W = 0.50$. Correspondingly, at equivalent stress levels, the CTOD for the short crack is approximately 2.5 times the CTOD for the deep crack. This observation has considerable significance in the application of CTOD results to failure analysis or specification development where the fracture mechanism is cleavage preceded by significant crack tip plasticity.

KEY WORDS: elastic-plastic fracture mechanics, CTOD, crack depth, short crack, toughness, finite element, constraint

[1] University of Kansas, Lawrence, KS 66045; currently at Exxon Production Research Company, Houston, TX 77252.
[2] University of Illinois, Urbana, IL 61801.
[3] University of Kansas, Lawrence, KS 66045.

The crack-depth to specimen-width ratio (a/W) has a significant effect on the fracture toughness results in the elastic-plastic regime where brittle fracture is preceded by significant crack tip blunting but no ductile tearing. Typically, deep cracks ($a/W = 0.5$) are tested in laboratory specimens to develop maximum constraint (stress triaxiality) at the crack tip and therefore provide conservative estimates of material toughness. While this approach may be appropriate for the general characterization of material fracture toughness, in the evaluation of existing flaws in structures it is more appropriate to model the actual degree of constraint present in the structural component. In weldments, for example, the testing of short cracks becomes particularly important, since the region of lowest toughness does not necessarily occur at a crack depth halfway through the specimen. With the variation of microstructures through the heat-affected zone (HAZ) and local brittle zones (LBZs), it is probable that a deeply cracked laboratory specimen would not produce the conservative fracture toughness values expected.

Several methods of laboratory testing for fracture toughness within the elastic-plastic regime are standardized. The two most widely used are the J-integral and the crack tip opening displacement (CTOD) test procedures. Both procedures could be extended to include short crack specimens. Currently, the CTOD test procedure (BS 5762, "Methods for Crack Opening Displacement Testing") has a major advantage in that it is the only standard which allows testing throughout the entire realm of fracture toughness from linear-elastic to fully-plastic behavior. J-integral standards (e.g., ASTM E 813, "J_{Ic}, A Measure of Fracture Toughness") are currently limited to determining the initiation of ductile tearing (J_{Ic}), but have often been extended to quantify brittle fracture (J_c) prior to initiation of ductile tearing.

Recent experimental investigations [1–7] have examined the effect of a/W ratio on the fracture behavior of three-point bend specimens. Both critical values and ductile initiation values of CTOD and J-integral were reported. A variety of materials were tested including low-strength steels, high-strength steels, and weldments. The experimental studies [1–7] have demonstrated CTOD and J-integral values for short crack specimens ($a/W < 0.20$) to be significantly larger than for deep crack specimens ($a/W = 0.50$).

Most CTOD-based studies have used the BS 5762 CTOD versus crack mouth opening displacement (CMOD) relation to analyze the behavior of short crack specimens. This relation is based on a small-scale yielding component and a plastic rotation component. A major difficulty for CTOD testing is assessing the plastic rotation factor for specimens with a/W ratios less than 0.2. This rotation factor has been experimentally determined with dual clip-gage techniques [3,5] and rubber crack replication techniques [3,4,6,7]. The rotation factor for specimens with a/W ratios of approximately 0.15 has been reported as low as 0.20 to as high as 0.45.

This investigation focuses on three objectives. The first is to determine the a/W ratio at which the plastic zone extends from the crack tip to the free surface behind the crack for a material with significant strain hardening. This a/W ratio is expected to define the boundary between short crack behavior and deep crack behavior. The second objective is to establish a relationship to calculate CTOD from the measured load-CMOD record or alternative measurement of specimen response. Ideally, this relationship would merely extend the current BS 5762 equation. The third objective is to determine the effects of the crack depth on stresses near the crack tip in relation to the CTOD levels.

Finite element analyses are conducted to study the effect of crack depth on the nonlinear behavior of CTOD fracture toughness test specimens. Two-dimensional (plane-stress and plane-strain) analyses are conducted using the uniaxial stress-strain properties of an A36 steel. Square cross-section, three-point bend specimens with crack-depth to specimen-width ratios (a/W) of 0.50, 0.20, 0.15, 0.10, and 0.05 are analyzed. The stress distributions are compared to determine the effect of crack depth on specimen behavior and crack tip constraint. A three-dimensional, elastic-plastic analysis is performed on full-size (31.8 by 31.8 by 127 mm (1.25 by 1.25 by 5.0 in.)) and sub-size (12.7 by 12.7 by 50.8 mm (0.50 by 0.50 by 2.0 in.)) three-point bend specimens with a/W ratios of 0.15. Comparisons of the numerical results demonstrate the effect of crack depth relative to specimen size and also the effect of absolute crack depth. Results of the short crack (a/W = 0.15) CTOD specimens are compared to numerical results of the deep crack (a/W = 0.50) CTOD specimens previously analyzed by Sorem et al. [8].

Material Properties

The material properties for the finite element analysis were taken from the corresponding experimental study on a 31.8 mm (1.25 in.) thick A36 steel plate in its as-rolled condition. The engineering stress-strain curve obtained from a standard 12.8 mm (0.505 in.) diameter longitudinal tensile test conducted at a slow loading rate is shown in Fig. 1. The A36 steel had an ultimate stress to yield stress ratio of 1.86 and a strain hardening exponent of 0.23. Tensile tests were conducted at room temperature and thus typify the stress-strain properties over a temperature range of 0°C (32°F) to 21°C (70°F). These temperatures corresponded to the transition region between brittle and ductile behavior of the A36 steel.

Finite Element Analysis Procedure

Elastic-plastic finite element analyses were conducted on square three-point bend specimens with a/W ratios ranging from 0.05 to 0.50. Finite element meshes for two of these specimens are shown in Fig. 2. The analyses predicted the deformation for both small- and large-scale plasticity with no simulation of crack growth. The finite element solutions employed the conventional, linear strain-displacement relations based on small geometry

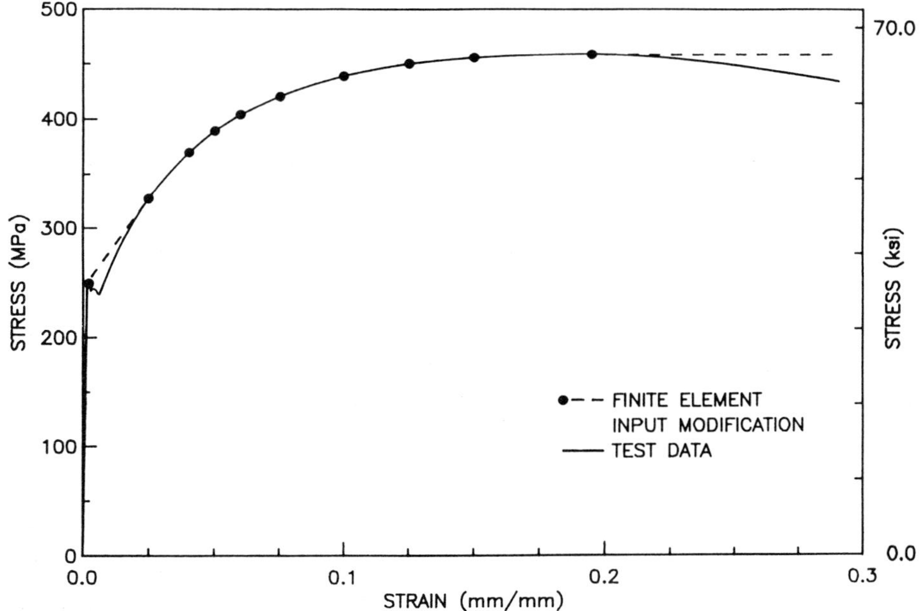

FIG. 1—*A36 steel tensile test showing modification for finite element analysis.*

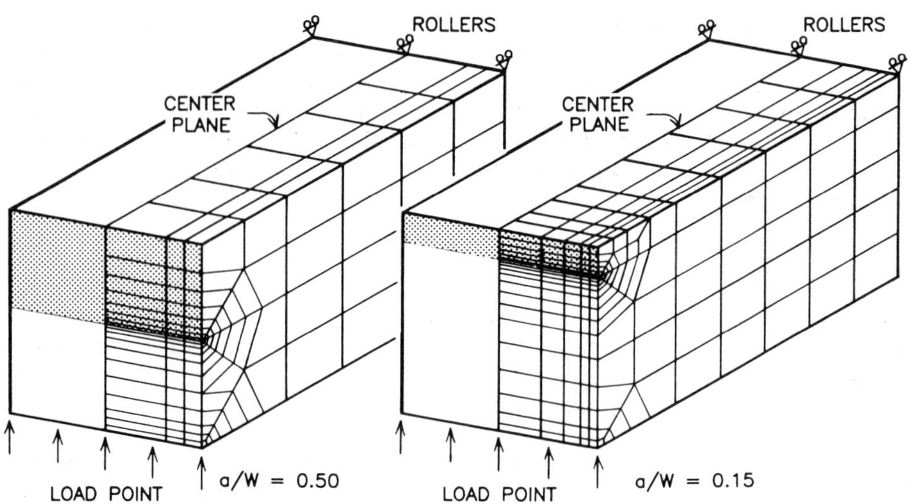

FIG. 2—*Three-dimensional finite element analysis mesh for the three-point bend specimens.*

change assumptions. Numerical computations were performed with the POLO-FINITE structural mechanics system [9,10]. The analytical procedure was identical to that adopted for the study of deeply cracked square specimens considered by Sorem et al. [8]. Details of the finite element procedure are provided in the Appendix.

Finite Element Results

Plastic Zone Distributions

Two-dimensional finite-element analyses (FEA) were performed on the full-size, square CTOD specimen geometry with crack-depth ratios (a/W) of 0.50, 0.20, 0.15, 0.10, and 0.05. Plastic zone sizes obtained from the 2-D FEA were compared at various linear-elastic and elastic-plastic CTOD levels. The plastic zones developed based on the von Mises equivalent stress for the five plane-strain models at applied CTOD levels of 0.0254, 0.0533, and 0.109 mm (1.00, 2.10, and 4.30 mils) are shown in Fig. 3. Plastic zones for the specimens with crack-depth ratios of 0.05 and 0.10 extended from the crack tip to the free surface behind the crack before a plastic hinge formed. For the specimen with a crack-depth ratio of 0.15, the formation of a plastic hinge coincided with the plastic zone extending back to the free surface. The specimen with an a/W ratio of 0.20 developed a plastic hinge well before the plastic zone extended back to the free surface. The plastic zone of the deep crack specimen ($a/W = 0.50$) was contained completely in the plastic hinge region and never reached the free surface behind the crack. The boundary between short crack and deep crack specimens apparently occurs at an a/W ratio of about 0.15 for this structural steel which undergoes significant strain hardening.

This definition of short and deep crack behavior agrees with the theoretical slip-line field discussed by Matsoukas et al. [6] and the results of several experimental studies [3–6]. The crack depth at which the slip-line field first extends to the free surface was found to be $0.177W$ ($a/W = 0.177$) for a rigid-plastic material. This does not imply that specimens with

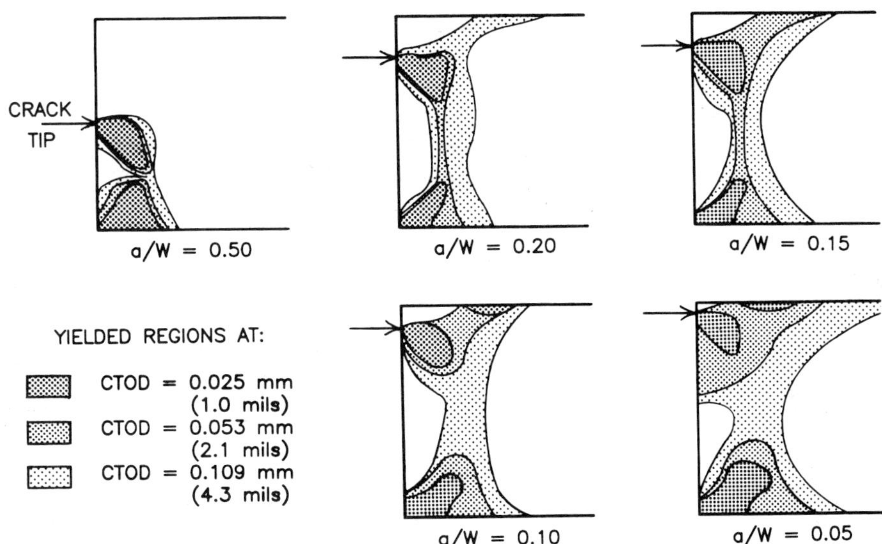

FIG. 3—*von Mises stress distributions for two-dimensional plane-strain A36 steel specimens.*

a/W ratios greater than 0.177 behave the same as deep crack ($a/W = 0.50$) specimens. The plastic zone for the deep crack specimen was always confined to the hinge region and did not extend to the back surface. Even though the plastic hinge formed before the plastic zone reached the back surface in the specimen with $a/W = 0.20$, continued strain hardening of the material caused the plastic zone to eventually reach the back surface.

Comparison of Numerical and Experimental Results

Load versus crack mouth opening displacement (CMOD) records from the plane-strain, plane-stress, and three-dimensional finite element analyses for the square ($W = 31.8$ mm (1.25 in.)) CTOD model are compared to a measured load versus CMOD record in Fig. 4. The measured load versus CMOD record is typical of specimens tested in a temperature range between 0°C (32°F) and 21°C (70°F). At these temperatures, the stress-strain properties of the A36 material are similar to the room temperature tensile properties input to the finite element model.

At equivalent CMOD levels, the plane-strain analysis provides an upper bound and the plane-stress analysis provides a lower bound to the experimentally measured load. The CMOD at the center plane of the three-dimensional model is plotted versus load, since it matches the location of the CMOD measurement in the test specimen. The load-CMOD record of the 3-D finite element model accurately predicts the experimental load-CMOD record.

Although very good agreement is achieved between the finite element and the experimental load-CMOD records, this "global" agreement does not necessarily verify the accuracy of predicted response in the crack tip region, specifically the CTOD. A procedure was developed to provide a comparison between the crack opening profile predicted by the 3-D finite element analysis and the actual crack profile.

A short crack ($a/W = 0.15$) test specimen was loaded well beyond plastic hinge development to a CMOD of approximately 0.427 mm (16.8 mils) and then the load was removed.

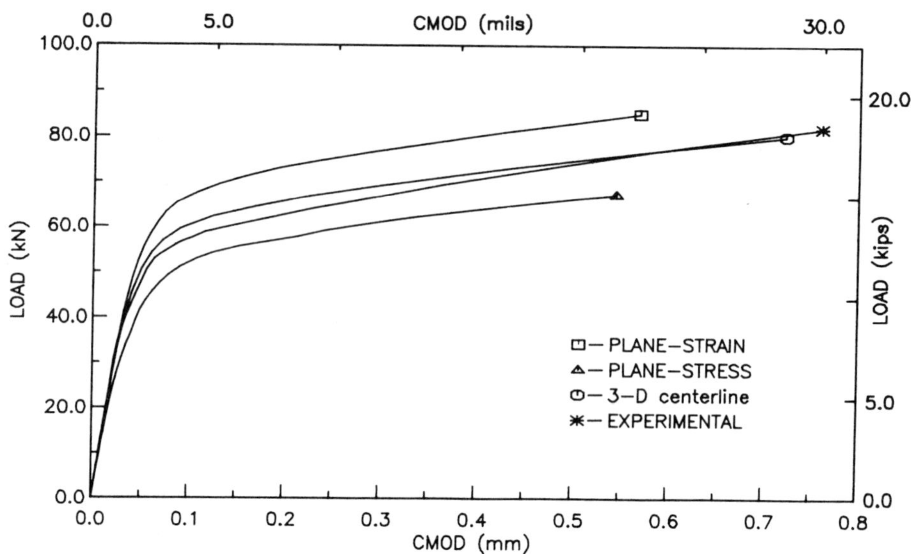

FIG. 4—*Load versus CMOD for square (31.8 by 31.8 mm) A36 steel specimens with* a/W
= 0.15.

The residual plastic component of CMOD was approximately 0.366 mm (14.4 mils) as illustrated in Fig. 5. The specimen was cut in half longitudinally through the center plane, polished, and etched. Micrographs were taken of the crack profile at the center plane of the unloaded specimen (Fig. 5a). These micrographs show the residual plastic displacement of the crack profile in the unloaded specimen. The specimen was then reloaded in a special fixture to the original CMOD of 0.427 mm (16.8 mils). Micrographs of the crack profile of the reloaded specimen were taken and compared to the predicted crack profile (superimposed) at the equivalent CMOD level (Fig. 5b). The crack profile of the test specimen is predicted very accurately by the finite element model. This comparison provides the needed validation of the 3-D finite element modeling procedures.

CMOD-CTOD Relation

The CMOD-CTOD relation for the short crack specimen is developed from the finite element results. Variations of the CTOD and CMOD through the thickness clearly demonstrate the three-dimensional character of the short crack specimen. The CTOD values are taken directly from the deformed finite element mesh using the 90° intercept method [11] at six positions through the half-thickness of the specimen.

The CMOD-CTOD results for the full-size square specimen are shown in Fig. 6. Load step 20 and load step 24 designated on the figure show the relationship of CTOD levels and CMOD levels through the thickness of the specimen. The CTOD remains nearly constant over the center 70% of the specimen and then decreases significantly near the outside free surface. The CMOD exhibits the opposite behavior. Unlike the deep crack specimen, which has no variation of CMOD over the thickness of the specimen, the CMOD for the short crack specimen is smallest at the center plane and increases as the outside free surface is approached. Similar CMOD variations are observed in the laboratory specimens.

The CMOD-CTOD results of the sub-size square specimen are shown in Fig. 7. Again, load steps 20 and 24 designated on the figure show the through-thickness relationships of

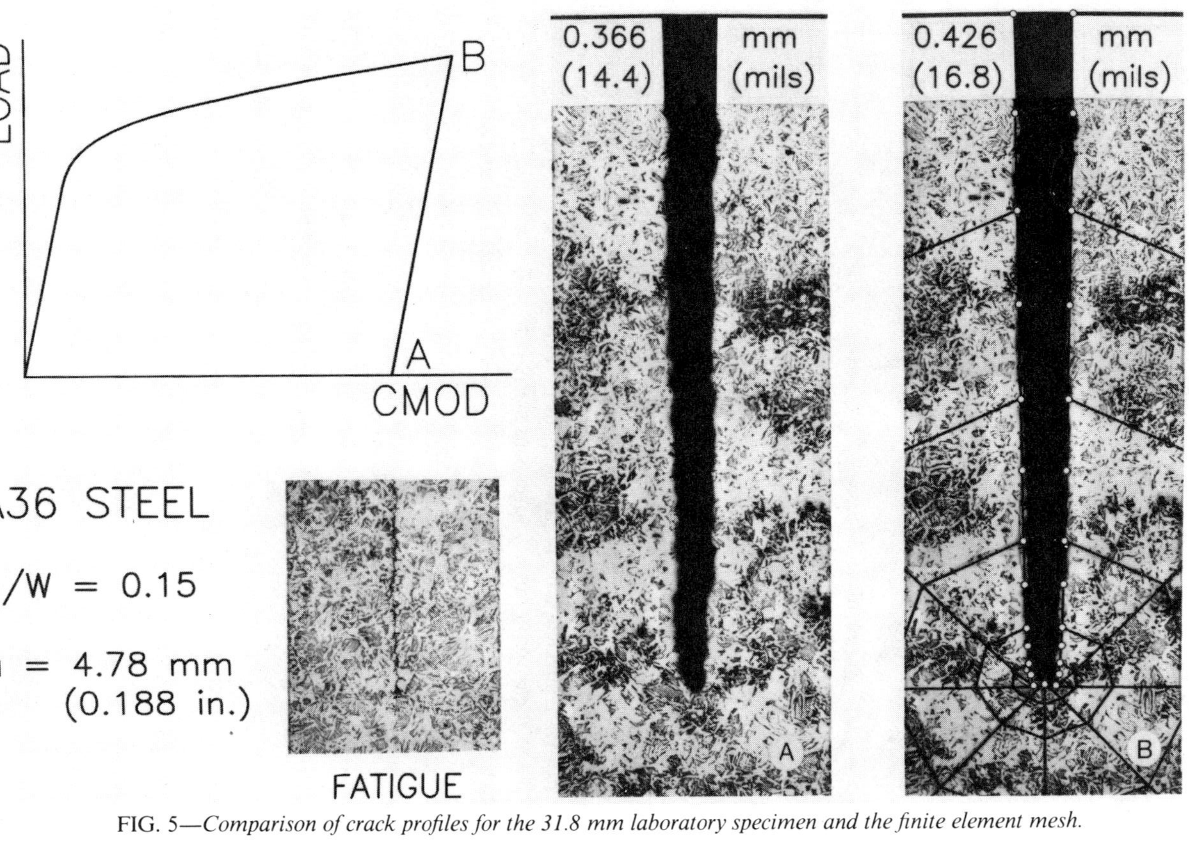

FIG. 5—Comparison of crack profiles for the 31.8 mm laboratory specimen and the finite element mesh.

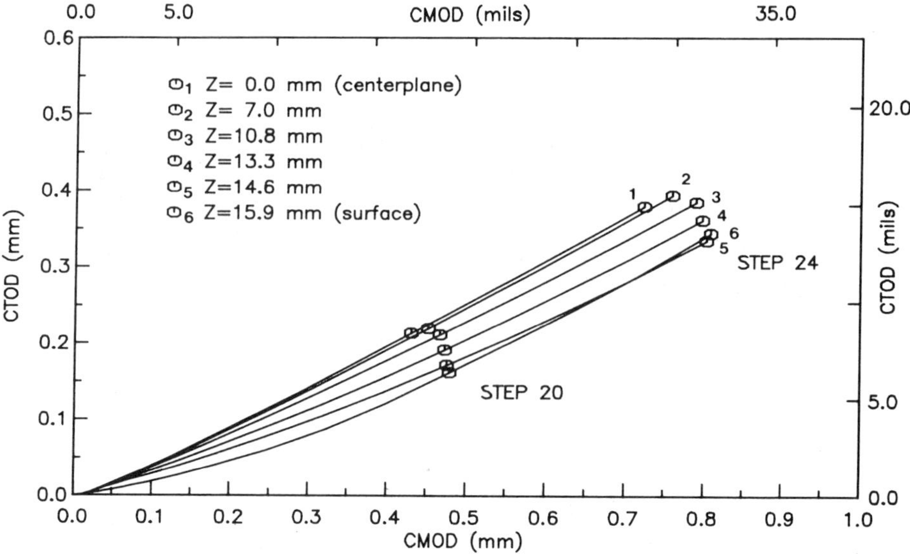

FIG. 6—*CTOD versus CMOD for square (31.8 by 31.8 mm) A36 steel specimens with a/W = 0.15.*

CTOD and CMOD levels. At step 20 (CMOD of 0.32 mm (13 mils)), response of the sub-size specimen is similar to that of the full-size specimen. The CTOD remains nearly constant over the center 70% of the specimen and decreases near the outside surface. The CMOD is largest at the outside surface. The relative deviation between outside surface measurements and center plane measurements is more consistent with step 24 of the full-size specimen

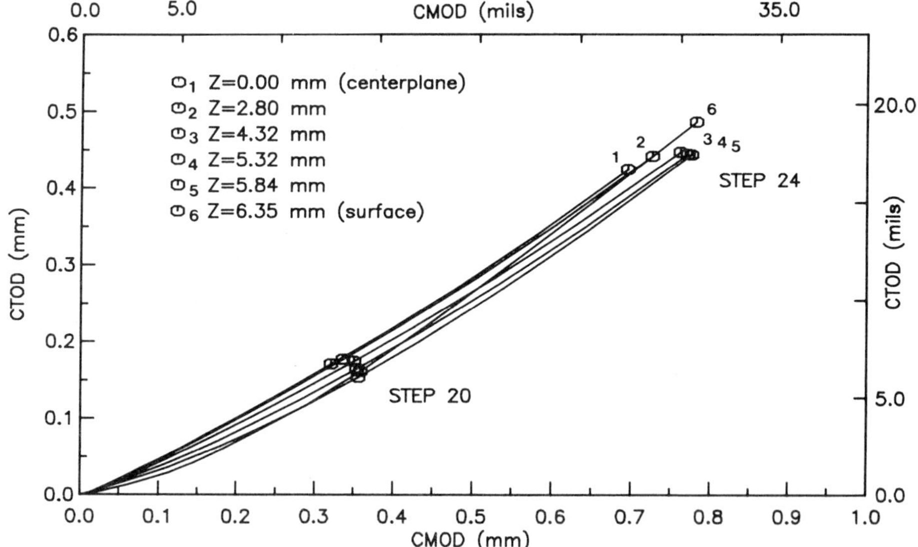

FIG. 7—*CTOD versus CMOD for square (12.7 by 12.7 mm) A36 steel specimens with a/W = 0.15.*

(CMOD of 0.72 mm (28 mils)). At step 24 of the sub-size specimen analysis (CMOD of 0.70 mm (27 mils)), the CMOD relationships through the thickness are consistent with previous steps, but the CTOD relationships are reversed. The CTOD is smallest at the center plane, increases slightly over the intermediate layers, and then increases to a maximum at the outside surface. The cause of this reversal in trends is not yet established. Experimental data are insufficient to distinguish this response as the actual response or an anomaly of the finite element modeling and solution (e.g., large displacement effects and mesh refinement near the free surface).

Analysis of BS 5762 Equation

A CMOD-CTOD relation for the short crack specimen is required to determine the CTOD from the measured load-CMOD record. Conducting a finite element analysis for each specimen type and each material is impractical. An empirical relation similar to the existing BS 5762 equation is desirable.

The BS 5762 equation relating CTOD to applied load and CMOD consists of two parts: (1) a small-scale yielding (SSY) contribution, which is often referred to as the elastic contribution, and (2) a fully plastic contribution:

$$\delta = \frac{K^2(1 - \nu^2)}{2\sigma_{ys}E} + \frac{RF(W - a)V_p}{RF(W - a) + a}$$

where

K = theoretical stress intensity factor = $YP/B\sqrt{W}$,

Y = stress intensity coefficient for a three-point bend specimen having span $(S) = 4W$:

$$= \frac{6(a/W)^{1/2}(1.99 - a/W[1 - a/W][2.15 - 3.93\,a/W + 2.7(a/W)^2])}{(1 + 2a/W)(1 - a/W)^{3/2}}$$

P = applied force,
B = specimen thickness,
W = specimen depth,
ν = Poisson's ratio,
σ_{ys} = 0.2% offset yield strength at temperature of interest,
E = Young's modulus at temperature of interest,
RF = plastic rotation factor,
a = physical crack length (initial length + stable crack growth), and
V_p = plastic component of clip-gage displacement.

The elastic (SSY) component of CTOD is a function of the stress intensity factor, K_I. Three methods are used to calculate the stress intensity factor from the linear-elastic finite element analysis. First, K_I is calculated from the displacement of the quarter-point crack tip elements [12]. Second, J-integral calculations [13] are correlated to K_I. Third, K_I is calculated from the stress at the Gauss points ahead of the crack tip [14]. The K_I values obtained from the three methods are nearly identical. The K_I value is constant over the center 80% of the specimen and then gradually decreases near the outside surface. The calculated K_I value at the center plane agrees within 2% of the theoretical equation (ASTM E 399, "Plane-Strain Fracture Toughness of Metallic Materials"). The free surface value of K_I is approximately 15% less than the center plane value.

The contribution to CTOD from the SSY component is determined from the elastic-plastic finite element analysis. For the first few load steps, the plastic component is essentially zero; CTOD is simply a function of the SSY component K_1. The CTOD of the displaced finite element mesh is compared with the theoretical K_1 value to determine the value of the constraint factor (m). The constraint factor is 1.7, which is 15% less than the BS 5762 equation but within the bounds of the 1.0 to 2.0 value described by Dawes [15]. The effect of changing the constraint factor from 2.0 to 1.7 is negligible in the elastic-plastic regime because the plastic component is the dominant contribution to CTOD. Since using 2.0 yields a conservative estimate of elastic CTOD, it is unchanged in the BS 5762 equation.

The fully-plastic component of CTOD is based on the rigid body rotation of the specimen about a point ahead of the crack tip (Fig. 8). For deeply cracked specimens, the plastic rotation factor has been shown to be approximately 0.45 [16]. The plastic rotation factor for the short crack specimens is determined from the FEA following two independent procedures. First, the outer four nodes on the crack profile at the center plane (closest to the crack mouth) are used to locate the plastic center of rotation. The linear-elastic displacements of these nodes are subtracted from the total displacements to determine the location of the nodes due only to the plastic deformation. Performing a simple linear regression on these node locations provides the location of the plastic rotation point. The second method is a calculation of the rotation point using the FEA load-CMOD record and the measured CTOD value. Each term in the equation is known (K, v, σ_{ys}, E, W, a, V_p) with the exception of the rotation factor (RF). The rotation factor is simply back calculated from the 90° intercept value of CTOD.

Both methods predict rotation factors between 0.2 and 0.3 for the short crack ($a/W = 0.15$) specimen. Therefore the apparent center of plastic rotation for the short crack specimen is closer to the crack tip than for the deep crack ($a/W = 0.50$) specimen. This value of the rotation factor is for a low strength (A36) steel and has not been shown to be material independent.

The CTOD is calculated from the 3-D finite element load-CMOD record and the BS 5762 equation with the adjusted rotation factor. A comparison of the calculated CTOD and the measured CTOD for the full-size square specimen is shown in Fig. 9. Use of a plastic rotation

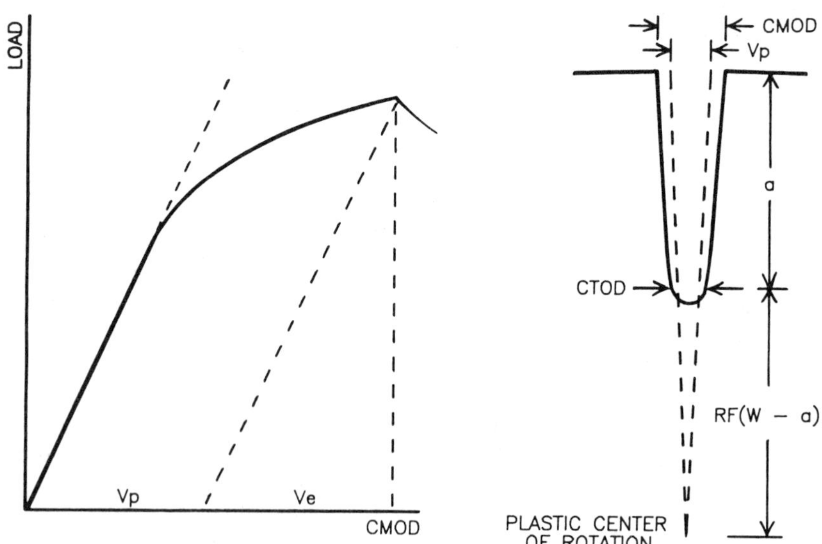

FIG. 8—*Schematic derivation of elastic and plastic CMOD components and plastic rotation factor.*

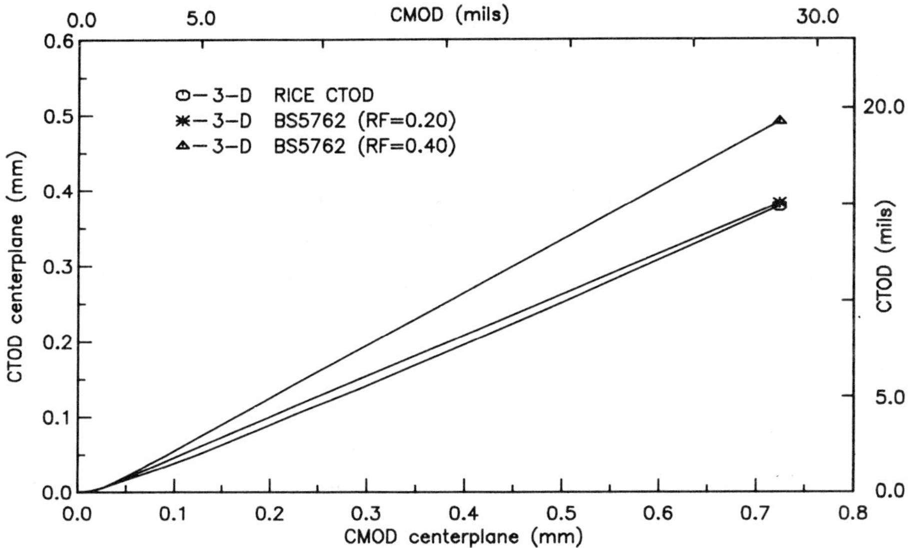

FIG. 9—*CTOD versus CMOD for square (31.8 by 31.8 mm) A36 steel specimens with a/W = 0.15.*

factor of 0.4 considerably overestimates the CTOD and therefore produces unconservative estimates of CTOD. The plastic rotation factor of 0.2 predicts a very close approximation to the CTOD determined analytically from the FEA. Similar trends in the results are obtained for the subsize specimen (Fig. 10). The last load steps of the FEA show a nonlinear relationship between the measured CTOD and the CMOD. This is reflected by an apparent increase

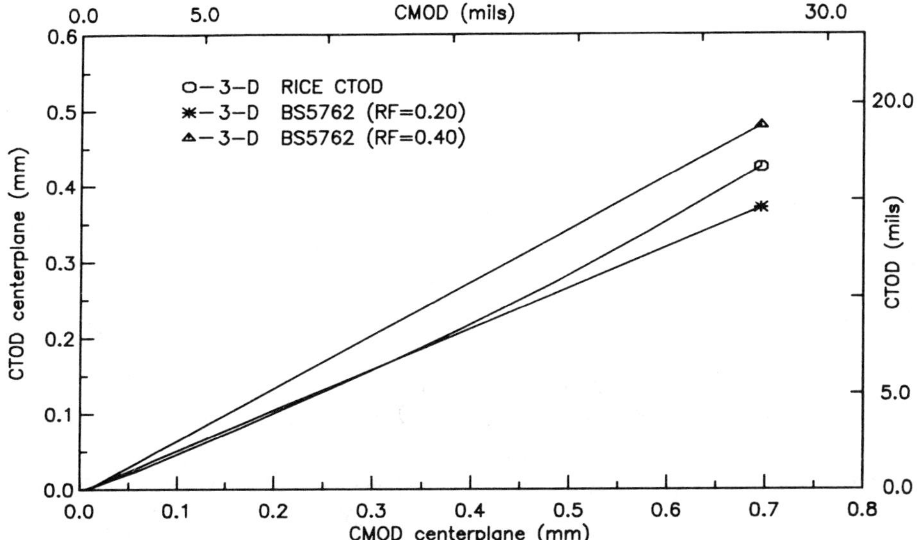

FIG. 10—*CTOD versus CMOD for square (12.7 by 12.7 mm) A36 steel specimens with a/W = 0.15.*

in the back-calculated rotation factor. The rotation factor calculated from the slope of the crack flanks does not indicate a similar apparent change of rotation factor.

Stress Distributions

Stresses near the crack tip at locations through the thickness are compared for the two short crack specimens at the same absolute position from the crack tip ($r = 0.36$ mm (0.0142 in.) and $\theta = 55°$). This location corresponds to the first Gauss quadrature point to yield near the crack tip outside of the degenerated isoparametric elements and the zone of large strain effects. Since this location corresponds to that considered by Sorem et al. [8] for the deep crack specimens, a comparison of the short crack square specimens with the deep crack square specimens is possible.

The through-thickness variation in crack opening stress (σ_y) is shown in Figs. 11, 12, 14, and 16 for a number of CTOD levels throughout the linear-elastic and elastic-plastic regions. The axes are nondimensionalized to accommodate the different specimen thicknesses. The through-thickness dimension (Z) is divided by one half the specimen thickness (t); on the horizontal axis, 0.0 corresponds to the specimen center plane and 1.0 corresponds to the outside free surface. The opening mode stress is divided by the room-temperature static yield stress (248 MPa (36 ksi)).

The transverse stress (σ_x) and through-thickness stress (σ_z) are not shown in these figures. The through-thickness variation of σ_x is similar to that of σ_y but has approximately one half the magnitude. The magnitude of σ_z is approximately one half of σ_y at the specimen center plane, but decreases to zero at the outside surface.

The distribution of the opening mode stress for essentially linear-elastic behavior is shown in Fig. 11. The CTOD is 0.0002 mm (0.0125 mils), which corresponds to a stress intensity factor of 5.5 MPa $\sqrt{\text{m}}$ (5 ksi $\sqrt{\text{in}}$). The stress distribution for the full-size short crack specimen is nearly identical to that of the full-size deep crack specimen. The opening mode stress

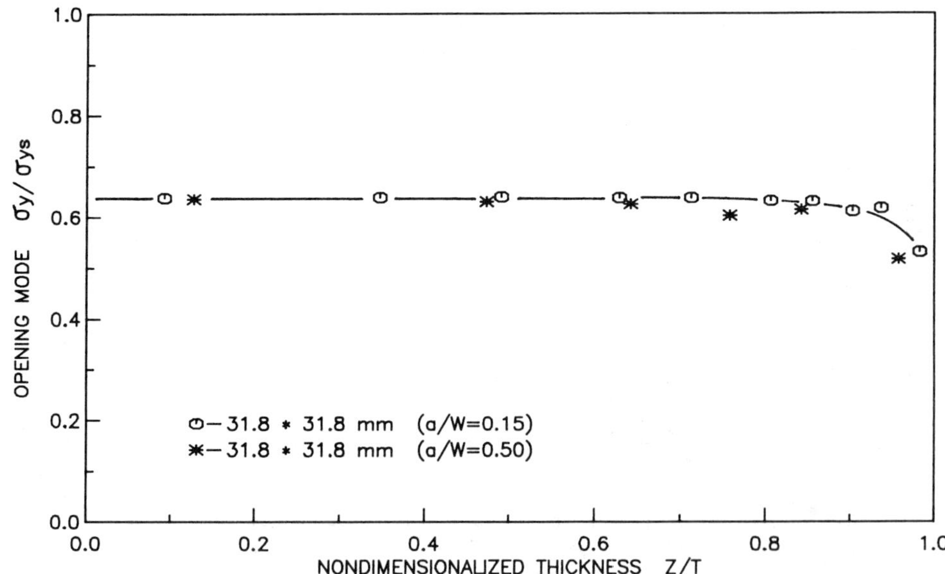

FIG. 11—*Crack opening stress distributions at centerplane CTOD = 0.0002 mm ($K_I =$ 5.5 MPa $\sqrt{m}$).*

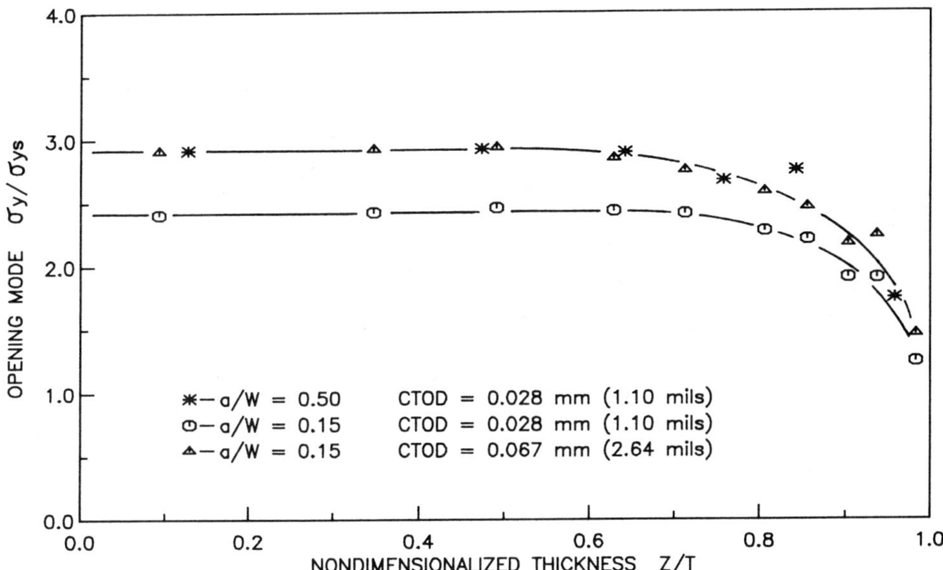

FIG. 12—*Crack opening stress distributions for 31.8 by 31.8 mm A36 steel specimens.*

is constant over the center 70% of the specimen and then decreases slightly at the outside (free) surface. Very small plastic zones exist at the crack tip for this level of loading.

The opening mode stress distributions for a CTOD of 0.028 mm (1.1 mils) are shown in Fig. 12 for full-size specimens with a/W ratios of 0.15 and 0.50. The maximum stress occurs at the center plane and is nearly constant over the center 60% of the specimen. Similar through-thickness variations are exhibited by the short crack and deep crack specimens, but the maximum stress developed in the short crack specimen is approximately 20% less than for the deep crack specimen. At a CTOD of 0.067 mm (2.6 mils), the short crack specimen develops an opening mode stress equivalent to the deep crack specimen at a CTOD of 0.028 mm (1.1 mils). Thus the CTOD of the short crack specimen is approximately 2.4 times the CTOD of the deep crack specimen at equivalent values of opening mode stress.

The corresponding von Mises stresses are shown in Fig. 13. At a CTOD of 0.028 mm (1.1 mils), the yielded regions are contained at the crack tip and at the point of load application for both the short crack and the deep crack specimens and are similar in absolute size. Significant yielding occurs behind the crack tip of the short crack specimen reaching toward the free surface. At a CTOD of 0.067 mm (2.6 mils), a plastic hinge develops and the yielded region extends from the crack tip to the surface behind the crack tip. Development of a plastic hinge does not correspond to the attainment of a limit load, since the low-strength steel undergoes significant strain hardening. Nevertheless, there is a significant increase in specimen compliance.

The opening mode stress distributions for a CTOD of 0.102 mm (4.0 mils) are shown in Fig. 14 for full-size specimens with a/W ratios of 0.15 and 0.50. Similar stress distributions are exhibited for the short and deep crack specimens. The maximum stress occurs over the center 50% of the specimen and then decreases at the outside surface. The maximum stress developed in the short crack specimen is approximately 85% of the value in the deep crack specimen at the same CTOD. At a CTOD of 0.251 mm (9.9 mils), the short crack specimen develops an opening mode stress equivalent to the deep crack specimen at a CTOD of 0.102

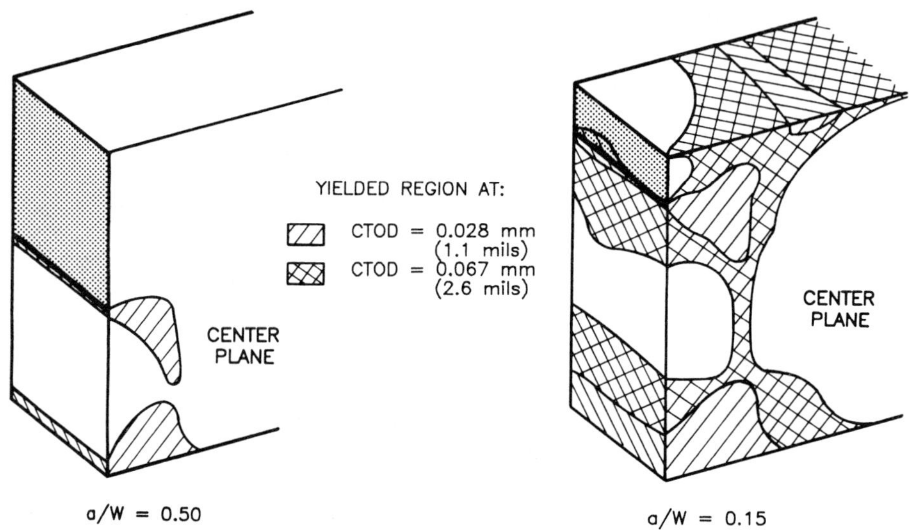

FIG. 13—*von Mises stress distributions for 31.8 by 31.8 mm A36 steel specimens.*

mm (4.0 mils). Thus, to obtain a maximum opening mode stress equivalent to that of the deep crack specimen at a CTOD of 0.102 mm (4.0 mils), the CTOD of the short crack specimen must be approximately 2.5 times larger than the CTOD of the deep crack specimen.

The corresponding von Mises stress at the center plane of each specimen is shown in Fig. 15. At a CTOD of 0.102 mm (4.0 mils), a plastic hinge develops in the deep crack specimen. Further yielding occurs in the short crack specimen at CTOD levels of 0.102 mm (4.0 mils) and 0.251 mm (9.9 mils) and extensive yielding occurs on the free surface behind the crack tip.

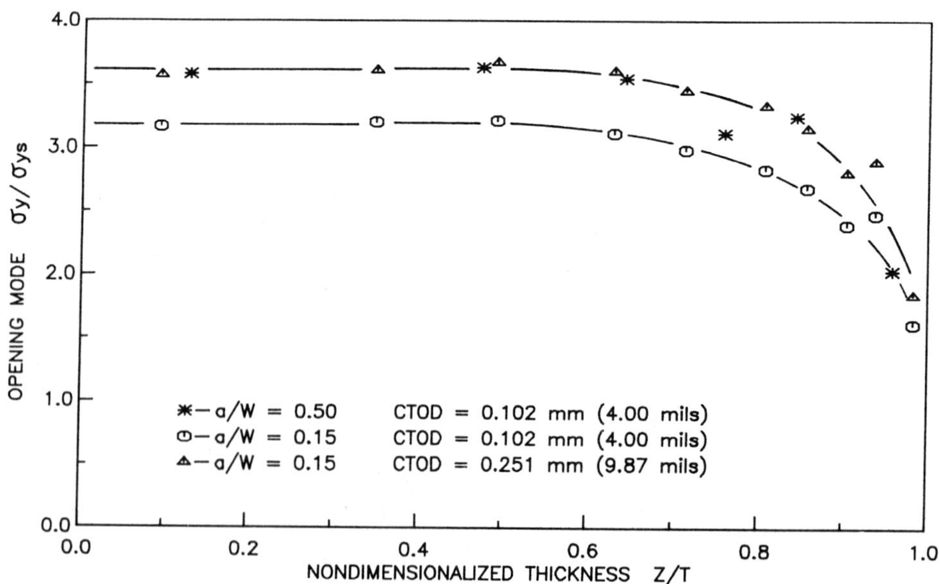

FIG. 14—*Crack opening stress distributions for 31.8 by 31.8 mm A36 steel specimens.*

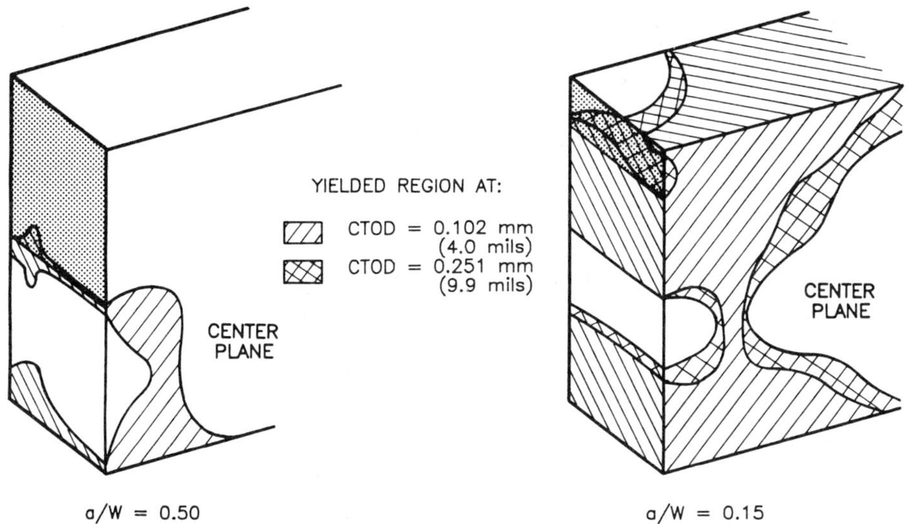

FIG. 15—*von Mises stress distributions for 31.8 by 31.8 mm A36 steel specimens.*

A comparison of the opening mode stress distribution for the subsize specimen with a/W = 0.15 to the stress distribution for the full-size specimen with a/W = 0.15 is shown in Fig. 16. The stress distributions are compared at the same CTOD levels as previously discussed. For nearly linear-elastic behavior (K_I = 5.5 MPa$\sqrt{m}$ (5 ksi$\sqrt{in.}$)) the opening mode stress of the sub-size specimen is identical to that of the full-size specimen. At a CTOD level of 0.028 mm (1.1 mils) the behavior of the sub-size specimen is similar to that of the full-size

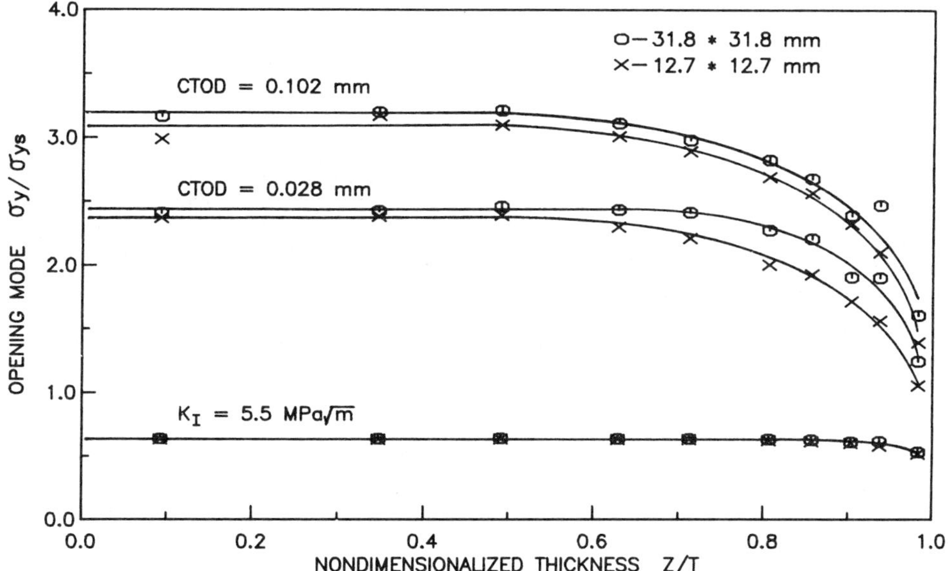

FIG. 16—*Crack opening stress distributions for 31.8 mm and 12.7 mm A36 steel specimens with a/W = 0.15.*

specimen, but the maximum opening mode stress is slightly less (2%) and the stress decreases more quickly near the outside free surface on the sub-size specimen than on the full-size specimen. Similar behavior occurs at a CTOD of 0.102 mm (4.0 mils). The opening mode stress is consistently 4% less for the sub-size specimen than for the full-size specimen at the same relative location through the thickness of the specimen.

At the (center plane) CTOD of 0.380 mm (14.9 mils), the CTOD is significantly higher at the outside surface than at the center plane of the sub-size specimen (Fig. 7). The stress is constant over the center 50% of the specimen and then decreases significantly near the outside free surface. Therefore the increased surface CTOD is not reflected in the opening mode stress distribution through the thickness of the specimen. The opening mode stress is approximately 4% less for the sub-size specimen than for the full-size specimen.

A comparison of the plastic zones developed based on the von Mises stresses for the full-size and the sub-size short crack specimens is shown in Fig. 17. The performance of the sub-size specimen at a CTOD of 0.170 mm (6.7 mils) is very similar to the full-size specimen at a CTOD of 0.378 mm (14.9 mils). The extent of elastic-plastic behavior appears proportional to the specimen dimensions. The yielded region of each specimen is extensive and undergoes considerable strain hardening. The region of fully-plastic behavior (on the plateau of the stress-strain curve) is very small and limited to the crack tip elements. At a CTOD of 0.425 mm (16.7 mils), the sub-size specimen exhibits more extensive elastic-plastic behavior, but the fully plastic zone is still small.

To summarize the effect of crack depth on the near crack tip stress field, the opening mode stress at the center plane of the short crack and deep crack specimens is shown in Fig. 18 for the entire CTOD range. Throughout the linear-elastic regime, the short ($a/W = 0.15$) and the deep ($a/W = 0.50$) crack specimens exhibit the same level of stress at the same CTOD. Throughout the elastic-plastic regime, the stress in the specimen with $a/W = 0.50$ is approximately 20% higher than the specimen with $a/W = 0.15$ at the same CTOD level. Therefore, at equivalent stress levels, the CTOD of the short crack specimen is approximately 2.5 times as large as the CTOD of the deep crack specimen.

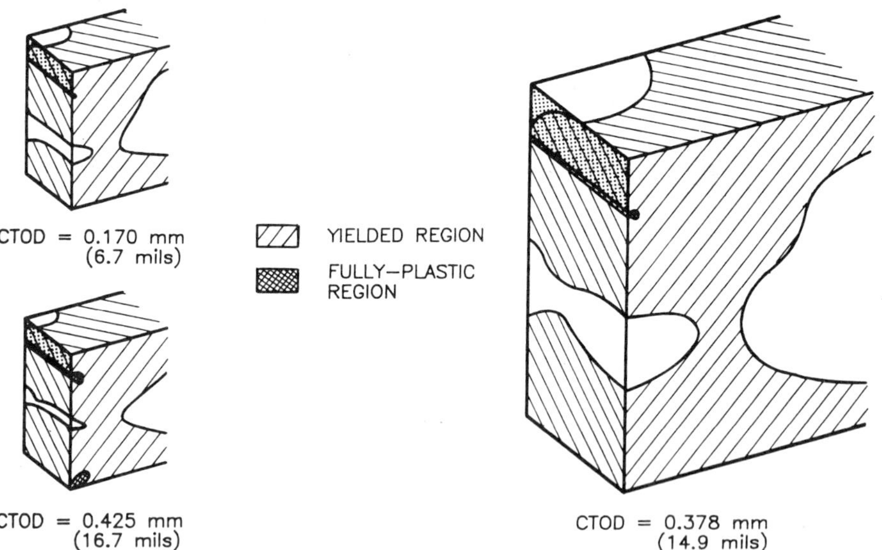

FIG. 17—*von Mises stress distributions for 31.8 mm and 12.7 mm A36 steel specimens with* a/W = 0.15.

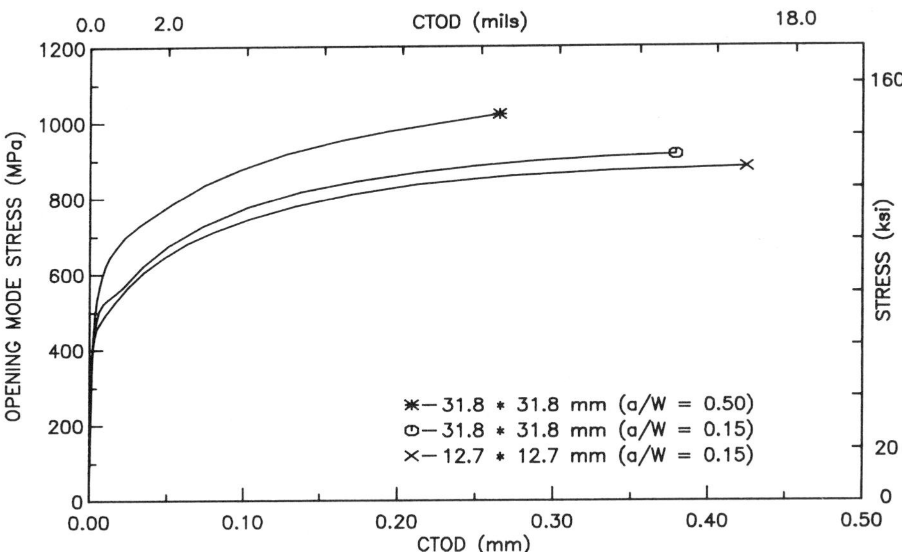

FIG. 18—*Maximum crack opening stress versus CTOD for A36 steel specimens with* a/W *= 0.15 and 0.50.*

Discussion

Investigations of the short crack three-point bend specimen [3–6] show that the boundary separating short crack and deep crack specimens occurs between the crack-depth to specimen-width ratios (a/W) of 0.15 and 0.20, depending on the degree of strain hardening. Between these a/W ratios, the yielded region at the crack tip also reaches the free surface behind the crack. When this occurs, the stress triaxiality near the crack tip is relaxed and a loss of constraint occurs. Milne and Chell [17] and Pisarski [18] have used the RKR model proposed by Ritchie, Knott, and Rice [19] to describe the effects of constraint on cleavage fracture. In the RKR model, cleavage fracture occurs when the maximum principal stress ahead of a sharp crack exceeds a critical value (σ_f) over a characteristic distance. Cleavage fracture generally occurs if there is a large degree of triaxiality because yielding is restricted and the maximum principal stress ahead of the crack tip is elevated.

In the region of linear-elastic (SSY) fracture behavior, the plastic zone is contained at the crack tip and negligible loss of constraint occurs. Similar stress states are developed at the crack tip irrespective of the specimen crack depth. Therefore the short crack and deep crack specimens produce equivalent CTOD or K_I results, reaffirming the applicability of a single parameter characterization.

In the region of elastic-plastic fracture behavior, the plastic zone is no longer limited to small-scale yielding at the crack tip. Plasticity in the short crack specimen extends to the free surface behind the crack and a loss of constraint occurs at the crack tip. This loss of constraint (or loss of stress triaxiality) requires that the short crack specimen undergo considerably larger strains than the deep crack specimen to develop equivalent opening mode stresses at the crack tip. Larger strains imply increased crack tip blunting and higher CTOD values. Therefore, if fracture occurs when the maximum principal stress ahead of the crack exceeds the critical value, then short crack specimens should exhibit significantly larger critical CTOD values than deep crack specimens.

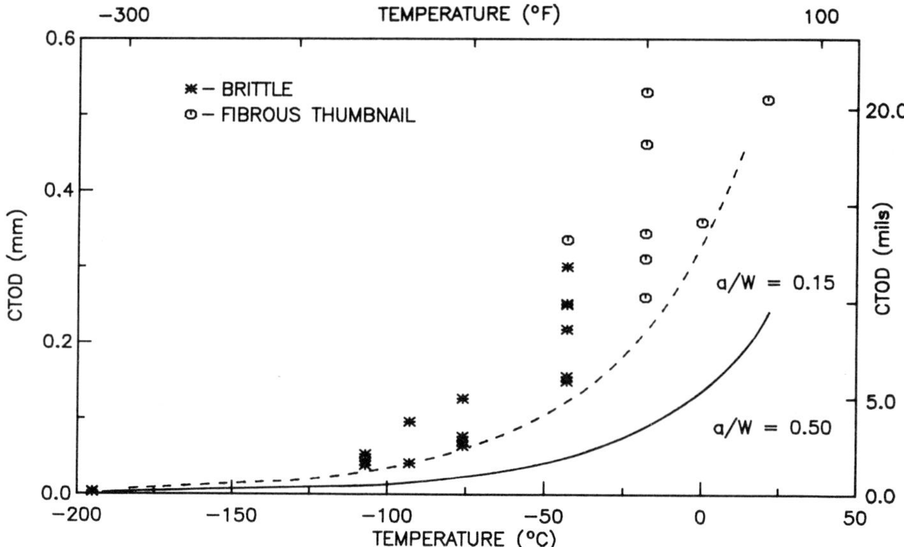

FIG. 19—*CTOD versus temperature for A36 steel specimens (31.8 by 31.8 mm) with* a/W *ratios of 0.15.*

These observations developed from numerical analyses are substantiated by the preliminary results of the corresponding experimental comparison of short crack and deep crack CTOD specimens [20]. The experimental test results of the short crack ($a/W = 0.15$) specimens are shown in Fig. 19. A lower bound curve was constructed through the lowest experimental values of the deep crack specimens. The deep crack lower bound curve was magnified by 2.5 in the lower transition region and agreed extremely well with the lowest experimental values from the short crack specimens.

Conclusions

Elastic-plastic finite element analyses were conducted on a low-strength structural steel (A36) to determine the effect of crack depth on the behavior of square three-point bend specimens. Two specimen sizes were modeled ($W = 31.8$ mm (1.25 in.) and $W = 12.7$ mm (0.50 in.)) using the standard geometry of W by W by $4W$. The analyses were employed to predict the deformation up to the point of a cleavage-type failure with no stable crack growth. The short crack results were compared to results for a deep crack ($a/W = 0.50$) specimen previously analyzed by Sorem et al. [8]. The results of this study on the short crack specimen and the comparisons with the deep crack specimen may be summarized as follows:

1. A comparison of the 2-D finite element results of various crack-depth to specimen-width ratios (a/W) showed a fundamental change in the nonlinear stress field at an a/W value of approximately 0.15. Specimens with shorter cracks ($a/W = 0.10$ and 0.05) showed yielding to the free surface behind the crack (back surface) well before the formation of a plastic hinge. Specimens with deeper cracks ($a/W = 0.20$) developed a plastic hinge before the plastic zone extended from the crack tip to the back surface.

2. From the 3-D finite element analysis conducted on the specimen with $a/W = 0.15$, an analytical CTOD-CMOD relationship was developed. With this relationship, the BS 5762

procedure for calculating CTOD from the load-CMOD record worked very well if the plastic rotation factor was decreased from 0.40 to 0.20.

3. A comparison of square specimens with a/W ratios of 0.15 and 0.50 showed that at the same linear-elastic K_1 level, short crack specimens and deep crack specimens developed equivalent opening mode stresses near the crack tip consistent with a single parameter characterization of fracture driving force. At equivalent CTOD levels in the elastic-plastic regime, the short crack specimens exhibited significantly lower opening mode stresses near the crack tip compared to the deep crack specimens. Correspondingly, at equivalent stress levels, the short crack specimens exhibited CTOD values approximately 2.5 times larger than the deep crack CTOD values.

In summary, the results of this study suggest that the short crack CTOD specimen loses crack tip constraint when the yielded region at the crack tip extends to the free surface behind the crack. This loss of constraint requires that the short crack specimens undergo considerably more crack tip blunting and plastic zone development than the deep crack specimen to develop the same opening mode stress near the crack tip. These results have considerable significance in the application of CTOD values to failure analysis or specification development where the fracture mechanism is cleavage preceded by significant crack tip plasticity.

APPENDIX

The finite element solutions employed conventional, linear strain-displacement relations based on small geometry change assumptions. Crack growth was not simulated in the finite element analyses. Numerical computations were performed with the POLO-FINITE structural mechanics system [9,10].

The 2-D model for both plane-strain and plane-stress analyses employed the 8-node isoparametric element. Due to symmetry of the 3-point bend specimen, only one (right) half of the specimen was modeled. A total of 71 elements and 262 nodes were incorporated in the model. The crack tip was represented with 8-node isoparametric elements collapsed into triangles. This is a common technique to develop a strain singularity of the order $1/\sqrt{r}$ for linear analysis [21] and $1/r$ for elastic-plastic analysis [22]. The size of the crack tip elements was adjusted for each geometry to be <5% of the crack depth. The collapsed 8-node elements provided a convenient means to extract CTOD from the displacements of the initially coincident nodes at the crack tip using the 90° intercept method.

The 3-D finite element model consisted of 20-node isoparametric elements. The element mesh was constructed by replicating the 2-D grid pattern in the through-thickness dimension to define five element layers through the half thickness (Fig. 2). Element layers varied in thickness, with thinner elements on the free surface to accommodate the sharp transition to a traction-free external surface from essentially plane-strain conditions on the interior near the crack front. Symmetry constraints were applied over the remaining ligament on the crack plane and the longitudinal center plane; only one quarter of the specimen was modelled given the two symmetry planes. The nonlinear model contained 355 elements and 2052 nodes.

The 3-D, 20-node isoparametric elements were collapsed into triangular prisms at the crack tip. Their behavior is analogous to that of the collapsed 2-D, 8-node isoparametric element [23]. The coincident corner nodes of the collapsed element develop a strain singularity of the order $1/\sqrt{r}$ for linear analysis and $1/r$ for elastic-plastic analysis. The coincident mid-side nodes of the collapsed element do not develop the same extent of singular region. The CTOD was obtained from the displacements of the initially coincident corner nodes of the collapsed elements using the same procedure previously outlined for the 2-D analysis.

Thus six CTOD values were available through the specimen half thickness for the 3-D analysis.

The finite element model for the sub-size square specimen was generated by simply scaling the coordinates of the full-size square specimen. The models were scaled to insure identical locations for comparison of stresses and strains (radius and angle from the crack tip) independently of the overall specimen size. Identical element aspect ratios were also maintained.

Reduced integration (2 by 2 for 2-D and 2 by 2 by 2 for 3-D) was used in the analysis to eliminate artificial locking under the incompressibility conditions imposed by plastic deformation [24]. Normal integration ($3 \times 3 \times 3$ for 3-D) was used for the degenerated crack tip elements of the 3-D model to prevent spurious zero-energy modes. Plasticity was modeled using incremental theory with a von Mises yield surface, associated flow rule, and isotropic hardening [25].

The stress-strain properties of the material for the finite element analysis were derived from a standard 12.8 mm (0.505 in.) diameter uniaxial tensile test (Fig. 1). A piecewise linear approximation provided a better match with the experimental curve than did a power law approximation. However, due to the Newton-Raphson iteration technique employed to solve the equilibrium equations, a "stiffening" material caused the solution to diverge when large load steps were taken [26]. Consequently, the small yield plateau was replaced with a straight line from the yield point to a point of tangency on the strain hardening curve to produce a stress-strain curve with a monotonically decreasing slope.

The finite element models were loaded directly below the uncracked ligament. By imposing load increments, rather than displacement increments, the solutions converged much faster. Approximately 25 load steps were imposed on each finite element model. A convergence tolerance of 0.5% on the ratio of the norm of the residual load vector to the norm of the applied load increment was maintained throughout the analyses. Initially, the load was imposed directly below the uncracked ligament at a single node for 2-D analysis and at a line of nodes through the specimen thickness for 3-D analysis. As the size of the plastic zone increased, it was necessary to distribute the load on the adjacent nodes to be consistent with the local deformation observed at the loading roller in the laboratory tests.

References

[1] de Castro, P. M. S. T., Spurrier, J., and Hancock, P., "An Experimental Study of the Crack Length/Specimen Width (a/W) Ratio Dependence on the Crack Opening Displacement (COD) Test Using Small-Scale Specimens," in *Fracture Mechanics (Eleventh Conference), ASTM STP 677,* C. W. Smith, Ed., American Society for Testing and Materials, Philadelphia, 1979, pp. 486–497.

[2] Sumpter, J. D. G., "The Effect of Notch Depth and Orientation on the Fracture Toughness of Multi-Pass Weldments," *International Journal of Pressure Vessels and Piping,* Vol. 10, 1982, pp. 169–180.

[3] Cotterell, B., Li, Q.-F., Zhang, D.-Z., and Mai, Y.-W., "On the Effect of Plastic Constraint on Ductile Tearing in a Structural Steel," *Engineering Fracture Mechanics,* Vol. 21, No. 2, 1985, pp. 239–244.

[4] Li, Q.-F., "A Study About J_i and δ_i in Three-Point Bend Specimens With Deep and Shallow Notches," *Engineering Fracture Mechanics,* Vol. 22, No. 1, 1985, pp. 9–15.

[5] Li, Q.-F., Zhou, L., and Li, S., "The Effect of a/W Ratio on Crack Initiation Values of COD and J-integral," *Engineering Fracture Mechanics,* Vol. 23, No. 5, 1986, pp. 925–928.

[6] Matsoukas, G., Cotterell, B., and Mai, Y.-W., "Hydrostatic Stress and Crack Opening Displacement in Three-Point Bend Specimens with Shallow Cracks," *Journal of the Mechanics and Physics of Solids,* Vol. 34, No. 5, 1986, pp. 499–510.

[7] Zhang, D. Z. and Wang, H., "On the Effect of the Ratio a/W on the Values of δ_i and J_i in a Structural Steel," *Engineering Fracture Mechanics,* Vol. 26, No. 2, 1987, pp. 247–250.

[8] Sorem, W. A., Dodds, R. H., Jr., and Rolfe, S. T., "An Analytical and Experimental Comparison of Rectangular and Square Crack-Tip Opening Displacement Fracture Specimens of an A36 Steel," *Nonlinear Fracture Mechanics: Vol. II—Elastic-Plastic Fractures, ASTM STP 995,* American Society for Testing and Materials, Philadelphia, 1989, pp. 470–494.

[9] Lopez, L. A., "Finite: An Approach to Structural Mechanics Systems," *International Journal for Numerical Methods in Engineering,* Vol. 11, No. 5, 1977, pp. 851–866.

[*10*] Dodds, R. H. and Lopez, L. A., "A Generalized Software System for Nonlinear Analysis," *International Journal for Advances in Engineering Software,* Vol. 2, No. 4, 1980.

[*11*] Rice, J. R., "A Path Independent Integral and the Approximate Analysis of Strain Concentration by Notches and Cracks," *Journal of Applied Mechanics, Transactions of the American Society of Mechanical Engineers,* Vol. 35, June 1968, pp. 379–386.

[*12*] Shih, C. F., de Lorenzi, H. G., and German, M. D., "Crack Extension Modeling with Singular Quadratic Isoparametric Elements," *International Journal of Fracture,* Vol. 12, No. 4, 1976, pp. 647–651.

[*13*] Dodds, R. H., Jr., Carpenter, W. C., and Sorem, W. A., "Numerical Evaluation of a 3-D J-Integral and Comparison with Experimental Results for a 3-Point Bend Specimen," *Engineering Fracture Mechanics,* Vol. 29, No. 3, 1988, pp. 275–285.

[*14*] Westergaard, H. M., "Bearing Pressures and Cracks," *Transactions of the American Society of Mechanical Engineers, Journal of Applied Mechanics,* 1939.

[*15*] Dawes, M. G., "Elastic-Plastic Fracture Toughness Based on the COD and *J*-Contour Integral Concepts," in *Elastic-Plastic Fracture, ASTM STP 668,* J. D. Landes, J. A. Begley and G. A. Clarke, Eds., American Society for Testing and Materials, Philadelphia, 1979, pp. 307–333.

[*16*] Matsoukas, G., Cotterell, B., and Mai, Y.-W., "On the Plastic Rotation Constant Used in Standard COD Tests," *International Journal of Fracture,* Vol. 26, No. 2, 1984, pp. R49–R53.

[*17*] Milne, I. and Chell, G. G., "Effect of Size on the *J* Fracture Criterion," in *Elastic-Plastic Fracture, ASTM STP 668,* American Society for Testing and Materials, Philadelphia, 1979, pp. 358–377.

[*18*] Pisarski, H. G., "Influence of Thickness on Critical Crack Opening Displacement (COD) and *J* Values," *International Journal of Fracture,* Vol. 17, No. 4, Aug. 1981, pp. 427–440.

[*19*] Ritchie, R. O., Knott, J. F., and Rice, J. R., "On the Relationship Between Critical Tensile Stress and Fracture Toughness in Mild Steel," *Journal of the Mechanics and Physics of Solids,* Vol. 21, 1973, pp. 395–410.

[*20*] Sorem, W. A., Rolfe, S. T., and Dodds, R. H., Jr., "An Experimental Comparison of Short Crack and Deep Crack CTOD Fracture Specimens of an A36 Steel," to be published.

[*21*] Barsoum, R. S., "Application of Quadratic Isoparametric Finite Elements in Linear Fracture Mechanics," *International Journal of Fracture,* Vol. 10, No. 4, Dec. 1974, pp. 603–605.

[*22*] Barsoum, R. S., "Triangular Quarter-Point Elements as Elastic and Perfectly-Plastic Crack Tip Elements," *International Journal for Numerical Methods in Engineering,* Vol. 11, No. 1, 1977, pp. 85–98.

[*23*] Bloom, J. M. and Van Fossen, D. B., "An Evaluation of the 20-Node Quadratic Isoparametric Singularity Brick Element," *International Journal of Fracture,* Vol. 12, No. 1, Feb. 1976, pp. 161–163.

[*24*] Dodds, R. H., "Effect of Reduced Integration on the 2-D Quadratic Isoparametric Element in Plane Strain Plasticity," *International Journal of Fracture,* Vol. 19, No. 3, 1982, pp. R75–R78.

[*25*] Nayak, G. C. and Zienkiewicz, O. C., "Elasto-Plastic Stress Analysis: A Generalization for Various Constitutive Relations Including Strain Softening," *International Journal for Numerical Methods in Engineering,* Vol. 5, No. 1, 1972, pp. 113–135.

[*26*] Wellman, G. W., Rolfe, S. T. and Dodds, R. H., Jr., "Three-Dimensional Elastic-Plastic Finite Element Analysis of Three-Point Bend Specimens," in *Fracture Mechanics: Sixteenth Symposium, ASTM STP 868,* American Society for Testing and Materials, Philadelphia, 1985, pp. 214–237.

R. Herrera[1] *and J. D. Landes*[2]

Direct *J-R* Curve Analysis: A Guide to the Methodology

REFERENCE: Herrera, R. and Landes, J. D., **"Direct *J-R* Curve Analysis: A Guide to the Methodology,"** *Fracture Mechanics: Twenty-First Symposium, ASTM STP 1074,* J. P. Gudas, J. A. Joyce, and E. M. Hackett, Eds., Americian Society for Testing and Materials, Philadelphia, 1990, pp. 24–43.

ABSTRACT: A direct method for evaluating *J-R* curves from load displacement data is proposed. This method eliminates a need for automatic crack length monitoring equipment. The method uses the geometric normalization suggested by Ernst for deeply cracked bend specimens to develop calibration curves which relate load, displacement, and crack length. Given two of these parameters, the third can be determined from the calibration curves. These curves are developed by assuming a functional form with unknown constants and evaluating the unknowns at calibration points in the test. This method was previously proposed using a power law functional form. In this paper a more general functional form is proposed which combines a power law and straight line. The method is studied for a variety of materials and specimen sizes. From the results a recommendation is made on how to generally apply the method.

KEY WORDS: elastic-plastic fracture, *J* integral, fracture toughness, *J-R* curve, normalization method, plastic deformation pattern

J-R curves represent a basic fracture toughness property for materials which fail in a ductile manner [*1*]. Standard tests for developing *J-R* curves rely on automatic methods for measuring crack advance [*2*]; the elastic unloading compliance method is most frequently used [*3,4*]. The automatic crack measuring methods require sophisticated equipment and techniques; sometimes a good deal of effort is required to achieve good results.

Landes and Herrera have recently proposed a method for directly determining a *J-R* curve from a load versus displacement record. This method does not require automatic crack length measuring equipment [*5,6*]. The method is based on the principle of normalization of deformation properties of a material for which load, displacement, and crack length can be functionally related [*7,8*]. If an appropriate functional form is assumed, the crack length can be determined directly from corresponding load and displacement values.

The functional form previously assumed was a power law [*6*]. The method was applied to a single material, A508 steel, where compact specimens of different sizes were analyzed to develop *J-R* curves by this direct method of normalization. These *R* curves were compared with ones determined by elastic compliance measurements. The direct method worked well for this material, possibly because a power law gave a good representation of the material stress-strain properties. A power law is not a general representation for deformation and therefore may not generally apply for the direct analysis of the *J-R* curve.

The method, however, only requires that a known functional form be used; this functional

---

[1] National University of Mar Del Plata, Mar Del Plata, Argentina.
[2] Department of Engineering Science and Mechanics, The University of Tennessee, Knoxville, TN 37996.

form does not necessarily have to be a power law. Therefore to make the method more applicable, a general representation of deformation is needed. This paper proposes a general functional form which can be used in the direct method. Using this general function the method is evaluated by applying it to several steels and specimen sizes. The results are used to suggest general guidelines for analyzing the load versus displacement record of any metal undergoing ductile fracture to develop the *J-R* curve without automatic crack length monitoring equipment.

Review of Normalization

The principle of normalization follows from the work of Ernst et al. [7,8], who showed that load, P, can be written as separable multiplicative functions of crack length, a, and displacement, v, when the plastic component of displacement is used:

$$P = G\left(\frac{a}{W}\right) H\left(\frac{v_{\mathrm{pl}}}{W}\right) \tag{1}$$

where a and v_{pl} are normalized by a width dimension, W, and displacement is separated into elastic and plastic components:

$$v = v_{\mathrm{el}} + v_{\mathrm{pl}} \tag{2}$$
$$= PC\left(\frac{a}{W}\right) + v_{\mathrm{pl}}$$

where $C(a/W)$ is a compliance function relating load and elastic displacement.

For deeply cracked bend type specimens the function form for the crack length dependence, $G(a/W)$, is known [8]. The load can be normalized by this function:

$$P_N = \frac{P}{G\left(\dfrac{a}{W}\right)} = H\left(\frac{v_{\mathrm{pl}}}{W}\right) \tag{3}$$

so that normalized load, P_N, is only a function of v_{pl}/W, that is, the deformation character of the material. The functional form of $H(v_{\mathrm{pl}}/W)$ should then reflect the form of the material stress-strain relationship. This stress-strain relationship can often be a power law [9]. Therefore it seems reasonable to assume that a power law functional form might work for $H(v_{\mathrm{pl}}/W)$.

A power law function given by

$$\frac{v_{\mathrm{pl}}}{W} = AP_N^m \tag{4}$$

where A and m are unknown constants, was used to analyze an A508 steel where compact specimens varied in size from $W = 25$ mm to $W = 500$ mm [6]. The unknown constants were evaluated by taking load and displacement values at the initial and final crack lengths, where crack lengths were measured from the surface of the broken specimen. The method for choosing the appropriate values for m and A is described in Ref 6. The resulting *J-R* curve from this analysis was compared with the *J-R* curve generated by the elastic unloading compliance method. When the compliance method was successful in determining the physically measured crack length, the two *J-R* curves agreed very well. When compliance method predicted the crack length poorly, the method of normalization was judged to improve the quality of the *J-R* curve. An example of a successful use of both is given in Fig. 1, where Fig. 1*a* shows the normalized load P_N as a function of v_{pl}/W and Fig. 1*b* shows the resulting *J-R* curve.

The success with which the method was applied to the A508 steel could be directly related to the stress and strain properties which can be well represented by a power law relationship [10]. There are other materials where a power law is not a good description of the stress-strain relationship. A different functional form should be postulated for the $H(v_{pl}/W)$ values

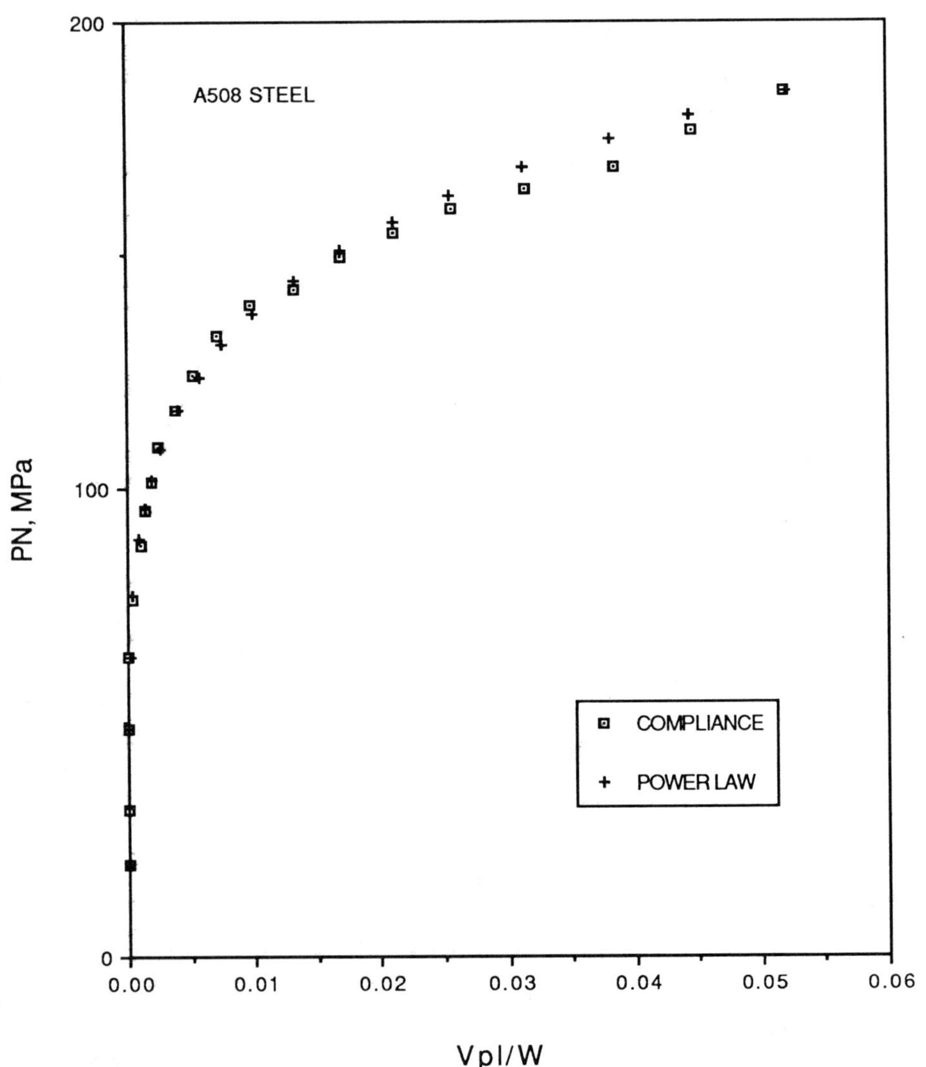

FIG. 1*a*—P_N *versus* V_{pl}/W *for A508 steel, W = 50 mm, comparing power law with compliance.*

of these materials. An important consideration in choosing a functional form is that unknown constants can only be evaluated when load, crack length, and displacement are known for a common point. Only initial and final crack lengths can be determined from a specimen fracture surface, in principle leaving only two points for determining unknowns. The choice of the functional form for $H(v_{pl}/W)$ must then be one which is compatible with the stress and strain relationship and allows unknown parameters to be evaluated at the known values of crack length.

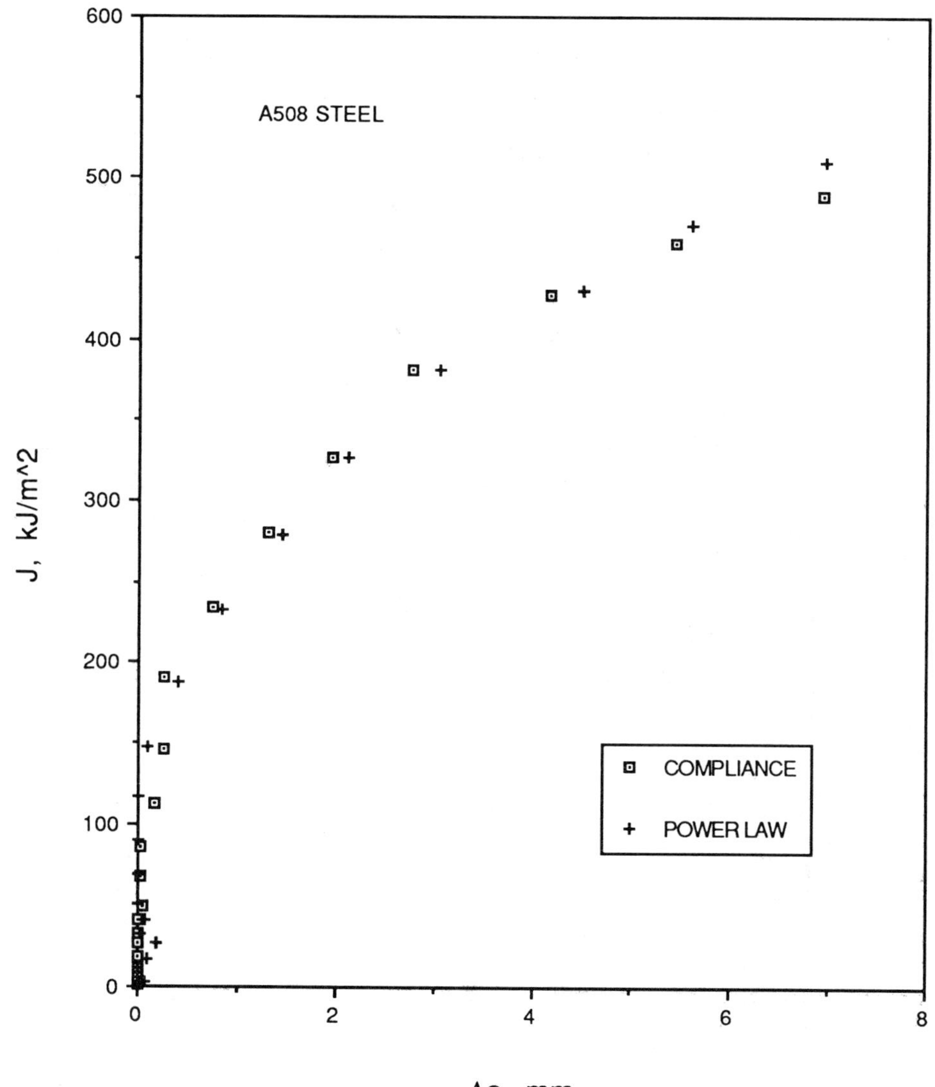

FIG. 1*b*—J *versus* Δa R-*curve for A508 steel,* W = 50 mm, *comparing power law with compliance.*

Deformation Relationships

Austenitic stainless steel is a material whose stress-strain relationship does not follow a power law. A previous study had been conducted to develop J-R curves for stainless steel specimens using the elastic compliance procedure [11]. An example test was re-analyzed using the normalization procedure with a power law fit for P_N versus v_{pl}/W following the method described in Ref 6. The result (Fig. 2) was not particularly good. This result reinforces the idea that a power law stress-strain relationship is necessary for the method to work with a power law $H(v_{pl}/W)$ function.

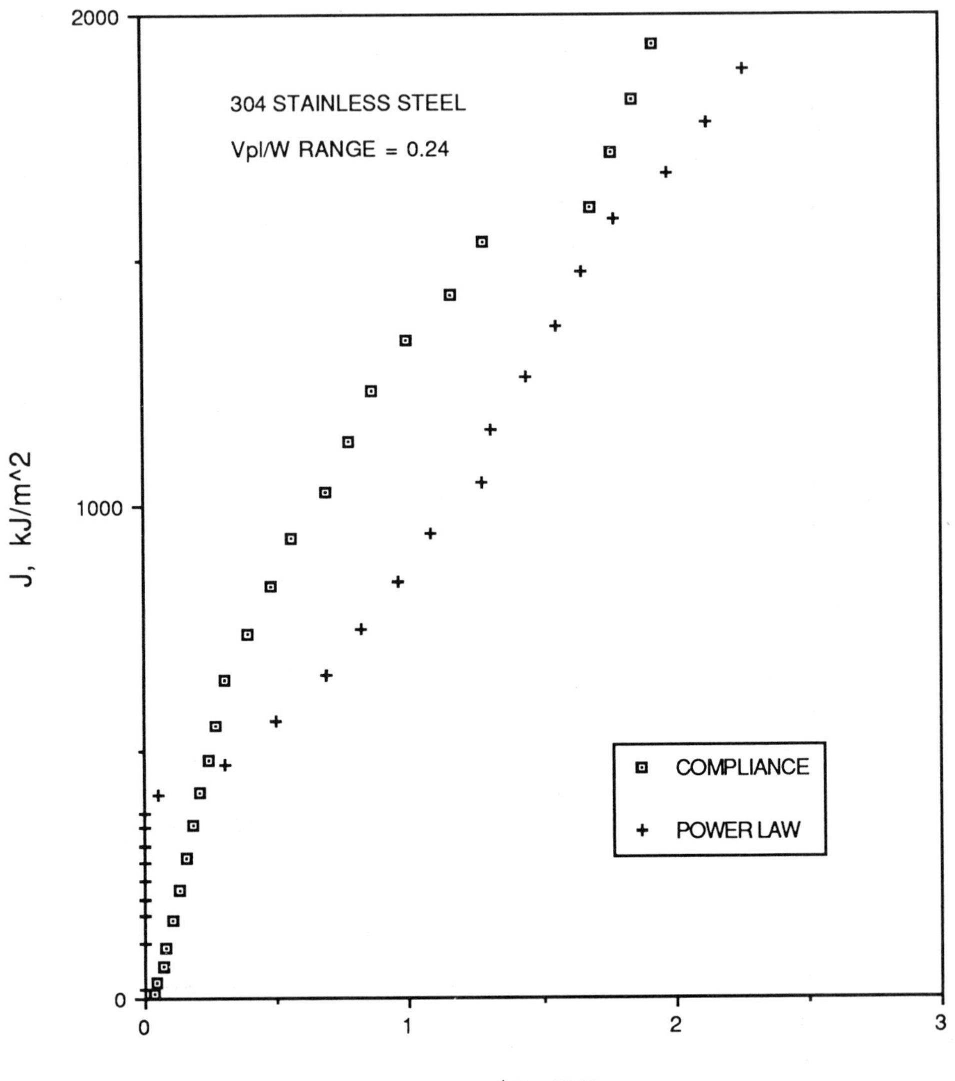

FIG. 2—J *versus* Δa R-*curve for Type 304 stainless steel,* W = 50 mm, *comparing power law with compliance.*

The study in Ref *11* had suggested that series of power laws could be used to model the stress-strain behavior of the stainless steel with a different exponent being used over different ranges of strain (Fig. *3a*). If the stress-strain relationship is to be used as a model of the P_N versus v_{pl}/W relationship, a similar relationship could be used. However, this does not fit the criteria suggested in the previous section of keeping the relationship simple enough to be described by two unknown constants.

A careful examination of the stress-strain relationship for this material shows that a straight line gives a better fit of the data over a larger range of strain (Fig. *3b*). Using this as

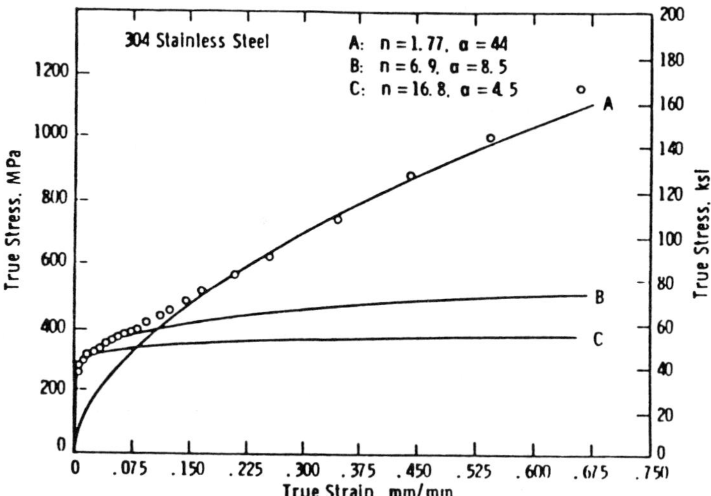

FIG. 3a—*True stress versus true strain for Type 304 stainless steel showing power law fits.*

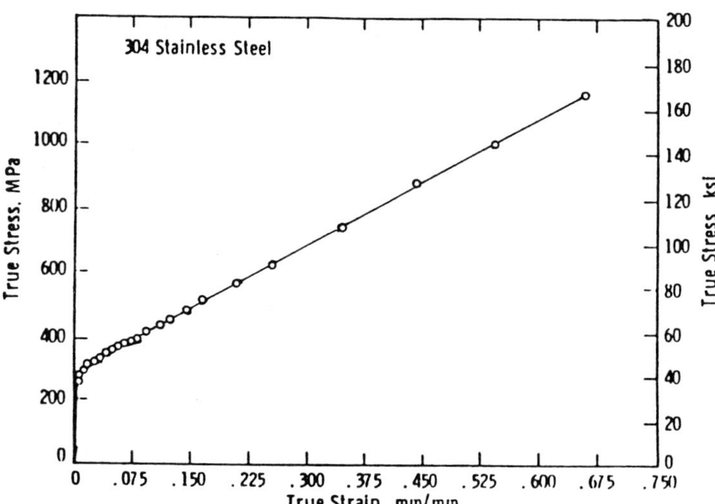

FIG. 3b—*True stress versus true strain for Type 304 stainless steel adding straight line fit.*

a model for the P_N versus v_{pl}/W relationship would suggest that the $H(v_{pl}/W)$ function in Eq 4 could be replaced by the straight line relationship

$$P_N = D_1 + D_2 \left(\frac{v_{pl}}{W} \right) \tag{5}$$

where D_1 and D_2 are unknown constants. This has only two unknowns which can be determined at the initial and final crack lengths.

Using the straight line relationship in Eq 5 and a procedure similar to one described previously (Eq 6), the normalization method was applied to the stainless steel specimen of Fig.

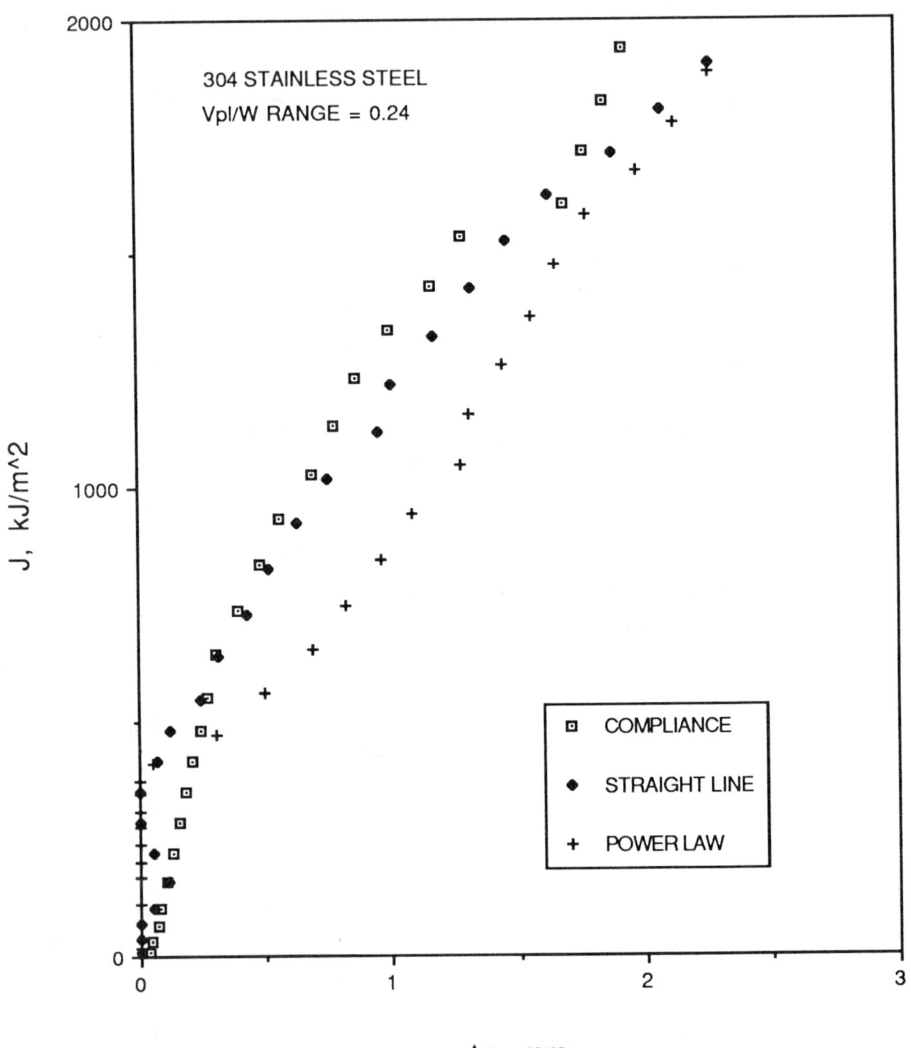

FIG. 4—*J versus Δa R-curve for Type 304 stainless steel comparing compliance, straight line, and power law (W = 50mm).*

2. The resulting *J-R* curve is shown in Fig. 4, where all three methods—compliance, power law $H(v_{pl}/W)$, and straight line $H(v_{pl}/W)$—are compared. The straight line function gives a much better match to the one developed from the compliance procedure. The three methods are also compared in terms of P_N versus v_{pl}/W in Fig. 5. This shows that the deformation behavior is better described by a straight line.

This result suggests that the normalization procedure can work well for determining *J-R* curves if the functional form of $H(v_{pl}/W)$ is known. That functional form could be similar to the one from stress and strain relationship. However, given a record of load and displacement for an arbitrary material that has no stress-strain data available, a proper choice of the $H(v_{pl}/W)$ functional relationship may be difficult. It would be helpful to develop some general guidelines for the arbitrary choice of an $H(v_{pl}/W)$ function.

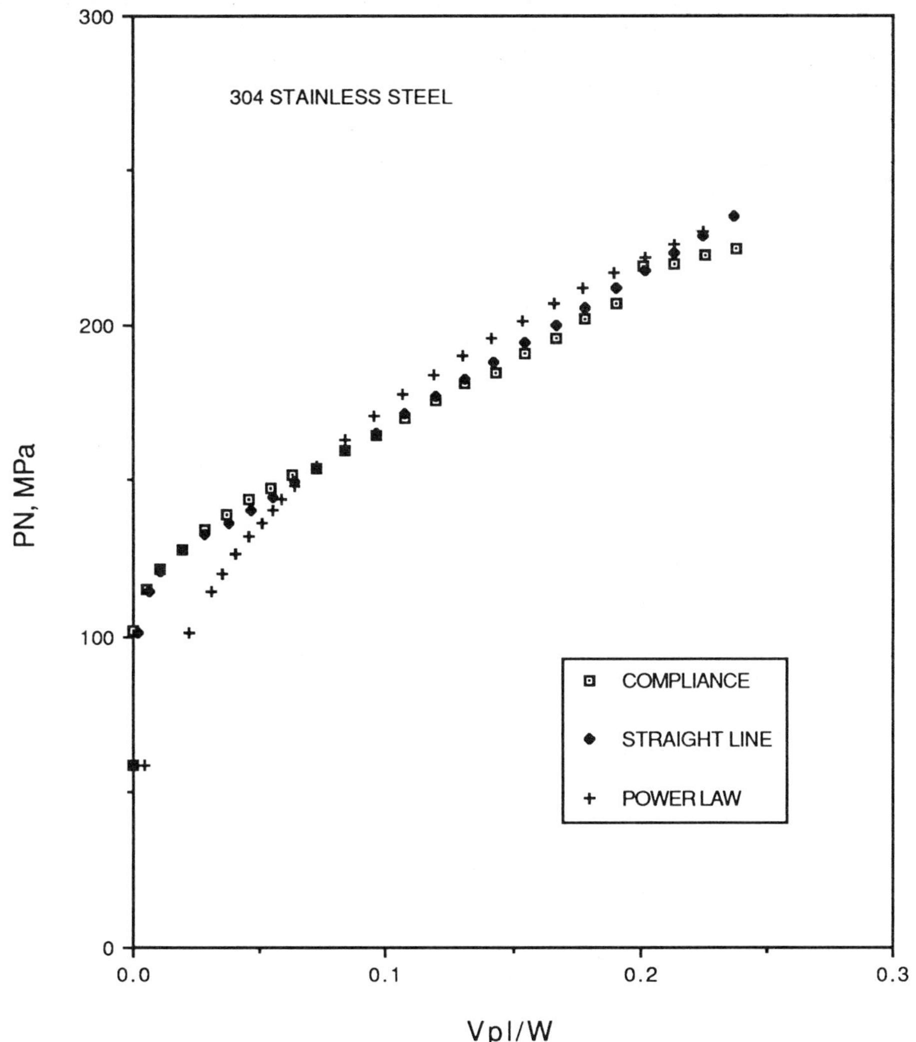

FIG. 5—P_N *versus* V_{pl}/W *for Type 304 stainless steel* (W = 50 mm).

Comparing the normalized load and plastic displacement for the stainless steel specimen in Fig. 5 with that for the A508 steel specimen in Fig. 1a an obvious difference is in the range of v_{pl}/W needed to develop the J-R curve. For the A508, the maximum v_{pl}/W value is about 0.05, but for the stainless steel this value is greater than 0.20. Evaluation of other materials showed that generally specimens which had a v_{pl}/W range of 0.05 or less worked well with a power law fit to the normalized curve whereas a range of v_{pl}/W greater than 0.1 worked well with a straight line fit.

The conclusion from this is that the material deformation pattern seems to describe a power law for the short v_{pl}/W range and approaches a straight line at larger v_{pl}/W. A more accurate way to approximate the functional form of $H(v_{pl}/W)$ would be to start with a power law for low v_{pl}/W and then change to a straight line at some value of v_{pl}/W around 0.05. Although this appears to be the ideal way to develop a functional form of $H(v_{pl}/W)$, it presents a procedural difficulty in that an intermediate calibration point cannot be determined from the information available. Only an initial and a final value of crack length can be used with the appropriate load and displacement pairs as calibration values for determining unknown constants. However, for many of the specimens crack growth begins after some

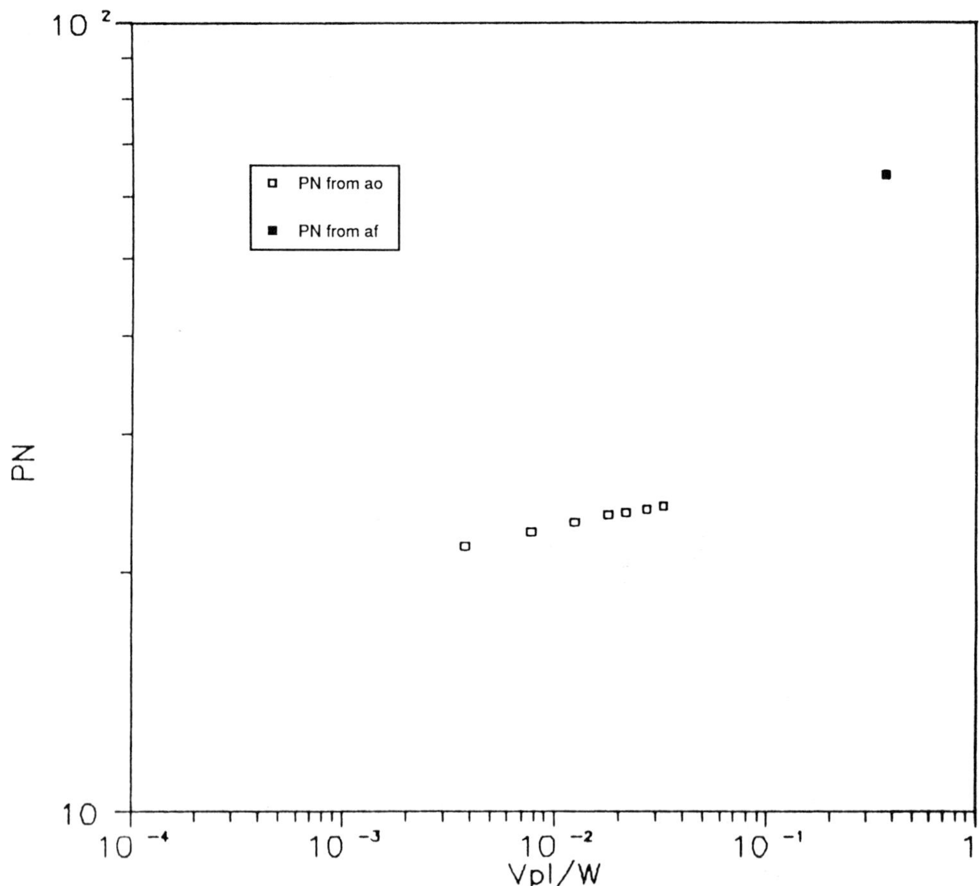

FIG. 6—*Log* P$_N$ *versus log* V$_{pl}$/W *comparing normalization at the initial and final crack lengths.*

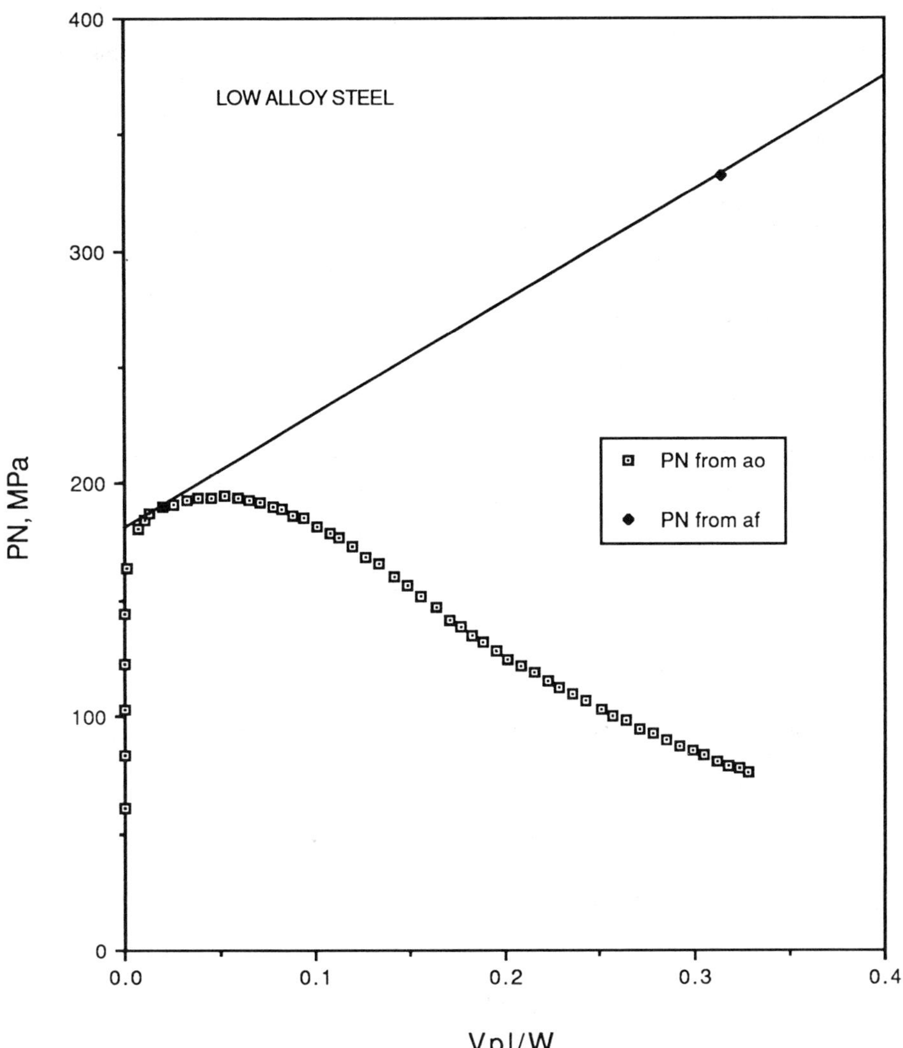

FIG. 7—P_N *versus* V_{pl}/W *showing the tangent point of transition between power law and straight line (*W = 50 mm*).*

amount of v_{pl}/W has been developed. Therefore several pairs of P_N and v_{pl}/W could be associated with the initial crack length and used to determine the exponent in the power law. An example of this is given in Fig. 6, where the two axes are plotted in a log-log format so that the exponent can be identified from the slope of the resulting line. This figure shows that initially a power law relationship is established. However, when the point associated with the final crack length is plotted, it is obviously not part of the same power law relationship. Therefore, using the combined two functional forms, power law and straight line, appears appropriate but the point of transition between the two must be determined from some additional information. Looking at the data from Fig. 6 replotted on linear axes (Fig. 7), a simple transition point is available. A line drawn from the final calibration point tangent to the top of a curve of P_N versus v_{pl}/W, where P_N has been developed using the initial crack length for normalizing, gives the simplest method for developing J-R curves. Although this results in a simple procedure, there is no technical justification for choosing this transitional point. Therefore the procedure can only be evaluated empirically, preferably by developing J-R curves for a variety of specimen sizes and material types where automatic crack length measurement has been used to develop the J-R curve. The resulting J-R curves developed by the normalization procedure can be compared with J-R curves developed with the automatic crack length monitoring system. This was done for many cases and the results will be presented in a later section. First a more detailed description of the procedure will be presented.

General Normalization Procedure

To develop a J-R curve directly from load and displacement pairs the following steps were used:

1. Initial and final crack lengths are measured from the specimen fracture surface.
2. Using the initial crack length only, normalized values of P_N are calculated from

$$P_N = \frac{P}{G(a/W)} = \frac{PW}{b^2 B g\left(\dfrac{b}{W}\right)} \tag{6}$$

where b is $W - a$ and $g(b/W) = \exp(0.522\, b/W)$ for the compact specimen ($g = 1$ for the bend specimen).

3. Plastic displacement is determined by subtracting an elastic component from the total displacement following Eq 1:

$$v_{pl} = v - PC\left(\frac{a}{W}\right) \tag{7}$$

where the compliance is determined at the initial crack length.

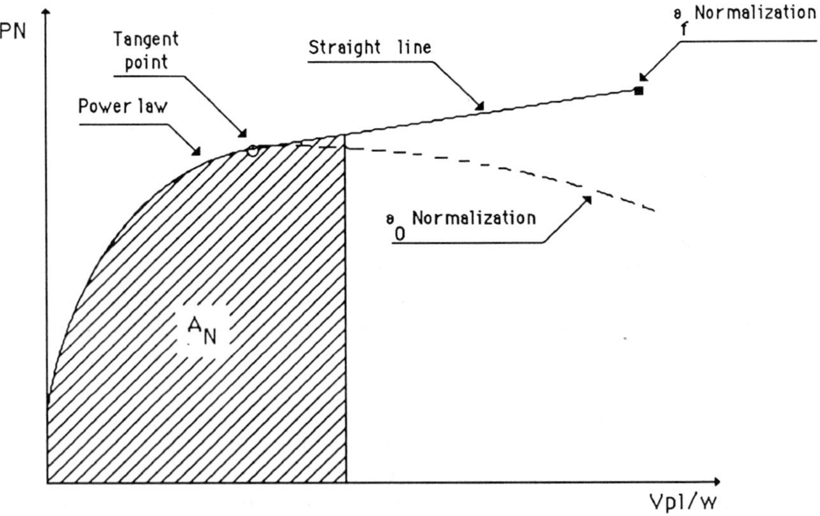

FIG. 8—*Schematic showing the method for applying a combination law normalization.*

4. The values of P_N and v_{pl} are determined for the final load and displacement point using the final crack length.

5. Values of P_N versus v_{pl}/W are plotted on linear axes as in Fig. 8. The point corresponding to the final crack length, a_f, forms one end of the straight line; the other is the tangent point to the curve of P_N versus v_{pl}/W calculated from the initial crack length, a_0. If a discrete number of points is used, the tangent point is one which forms the smallest slope when taken in pair with the final point.

6. Using these two points as end points for the straight line, unknown constants D_1 and D_2 from Eq 5 are determined.

7. J and Δa pairs are then determined from the straight line normalized load and displacement starting at the point after tangency. Crack extension is $\Delta a = a_i - a_0$ where a_0 is the initial crack length and a_i is the current crack length determined from the solution to the equation

$$v = v_{el} + v_{pl} \tag{8}$$

$$= PC\left(\frac{a_i}{W}\right) + (P_N - D_1)/D_2$$

where

$$P_N = \frac{PW}{b_i^2 Bg\left(\dfrac{b_i}{W}\right)} \tag{9}$$

where $b_i = W - a_i$.

Values of J are determined by evaluating an elastic and a plastic component

$$J = J_{el} + J_{pl} \tag{10}$$

where

$$J_{el} = \frac{K^2}{E'} \tag{11}$$

as in the standard test methods and J_{pl} is evaluated directly from the area under the normalized load and displacement curve, A_N, including both the power law and straight line portions (Fig. 8) by

$$J_{pl} = \eta_i b_i g\left(\frac{b_i}{W}\right) A_N \tag{12}$$

where η_i is the eta used in standard test methods. For a known $H(v_{pl}/W)$ function the area A_N can be written as a function of v_{pl} or P_N by integrating the appropriate functional forms.

Results

This method was applied to a number of specimen sizes and material types where the elastic compliance method was also used to evaluate the crack extension. The materials are listed in Table 1 along with their tensile properties. All specimens used in the study were of the compact geometry. Along with this method, the original normalization method of Ref

TABLE 1—*Properties of materials.*

Material	Yield Strength, MPa	U.T.S., MPa
A106	324	558
A508	386	545
A533B	441	600
304SS	294	614
HSLA	510	600
3Ni steel	614	731

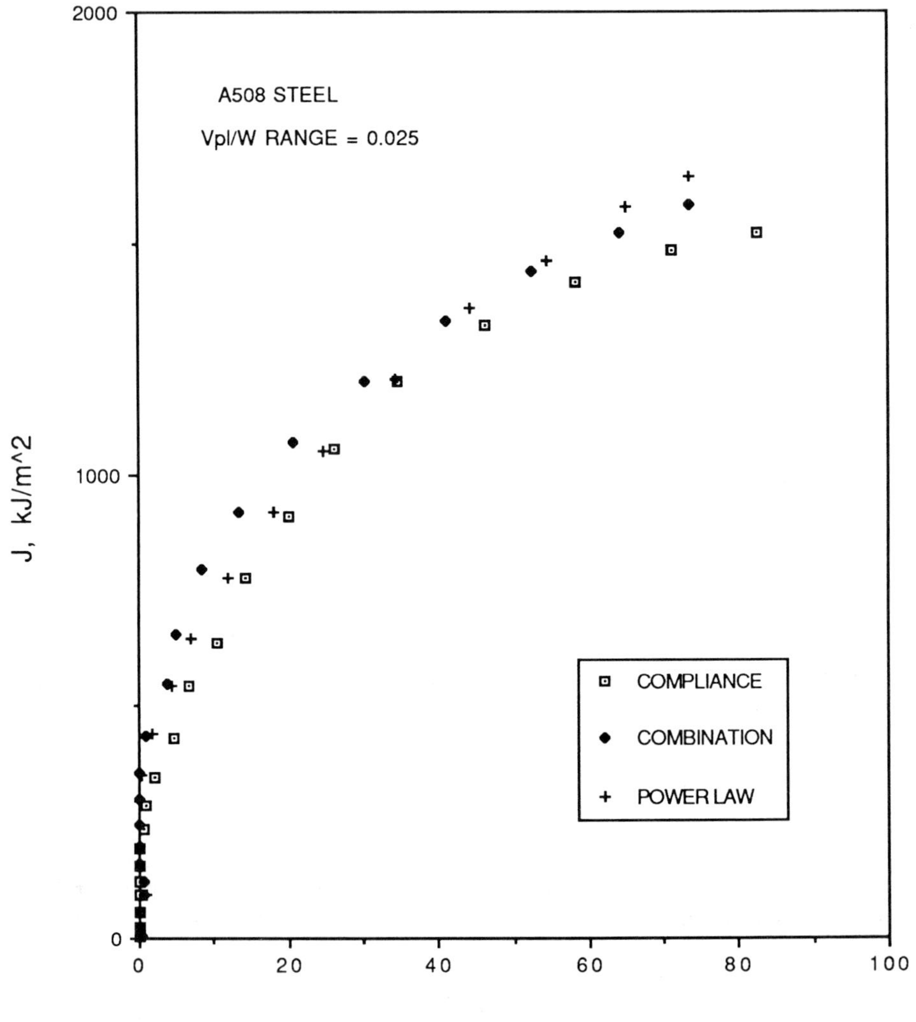

FIG. 9—J *versus* Δa R-*curve A508 steel comparing compliance with power law and combination fits (W = 50 mm).*

6, where a power law was used to represent the entire normalized curve P_N versus v_{pl}/W, was also used to evaluate the *J-R* curve. One objective was to compare the *J-R* curves resulting from the present method of combining a power law and straight line for the P_N versus v_{pl}/W curve with the previous one which used only the power law. The comparison of the two cases will be made relative to the range of v_{pl}/W covered in the test.

For short ranges of v_{pl}/W the A508 data previously analyzed can be considered. Figure 9 shows the analysis of the largest specimen, $W = 500$ mm, where the *J* versus Δa *R*-curve is plotted for compliance, power law, and the combination of power law and straight line. The maximum range of v_{pl}/W is about 0.025, which is less than the 0.05 limit for a power law

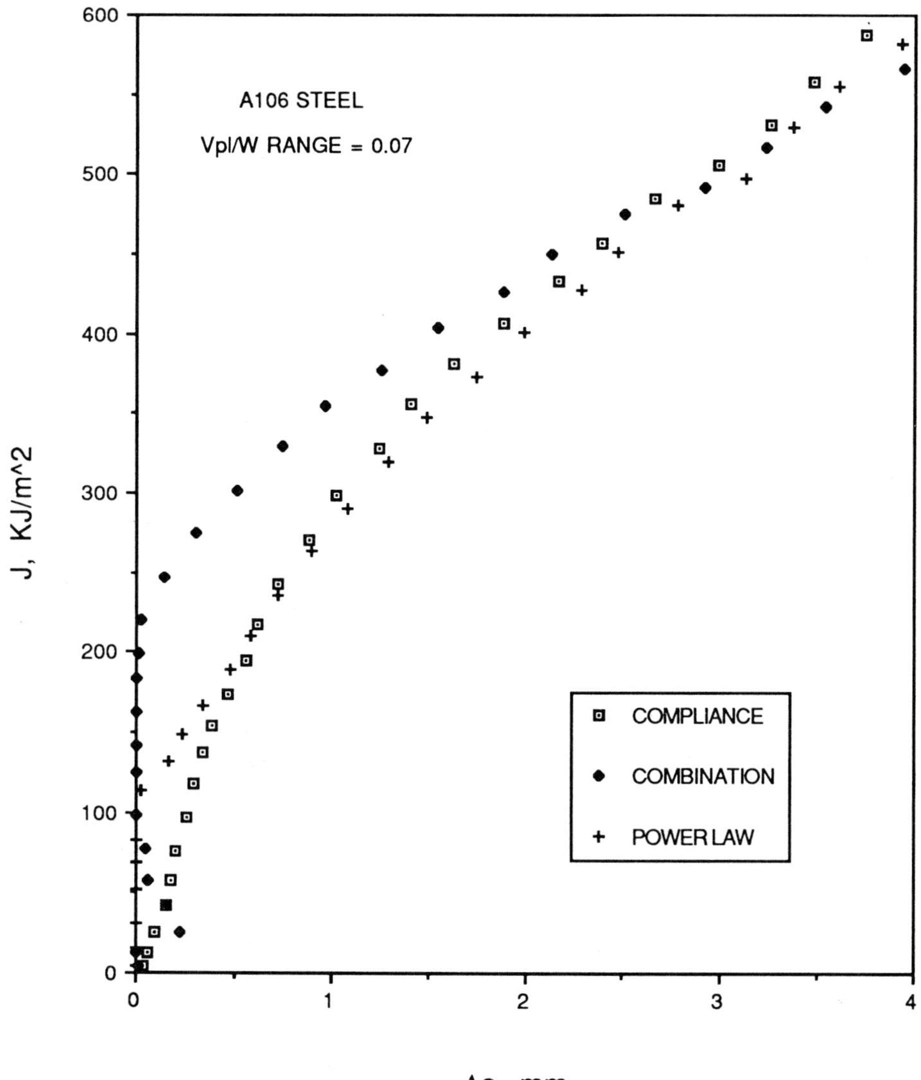

FIG. 10—J *versus* Δa R-*curve for an A106 steel comparing compliance with power law and combination fits (*W* = 50 mm).

function. The results in Fig. 9 show that the power law normalization gives an R-curve more compatible with the compliance generated one than the power law-straight line combination.

Figures 10, 11, and 12 show three different test results for an A106 steel originally tested by Hackett and Joyce [*12*], where the range of v_{pl}/W changed from 0.07 to 0.15. These results show that the shortest range v_{pl}/W works best with only power law normalization, the longest works best with the combined normalization, and the intermediate range could use either.

Discussion

The results presented in Figs. 9 to 12 show a consistent pattern. That is, for using a normalization procedure the fit of P_N versus v_{pl}/W should be only a power law (the method of

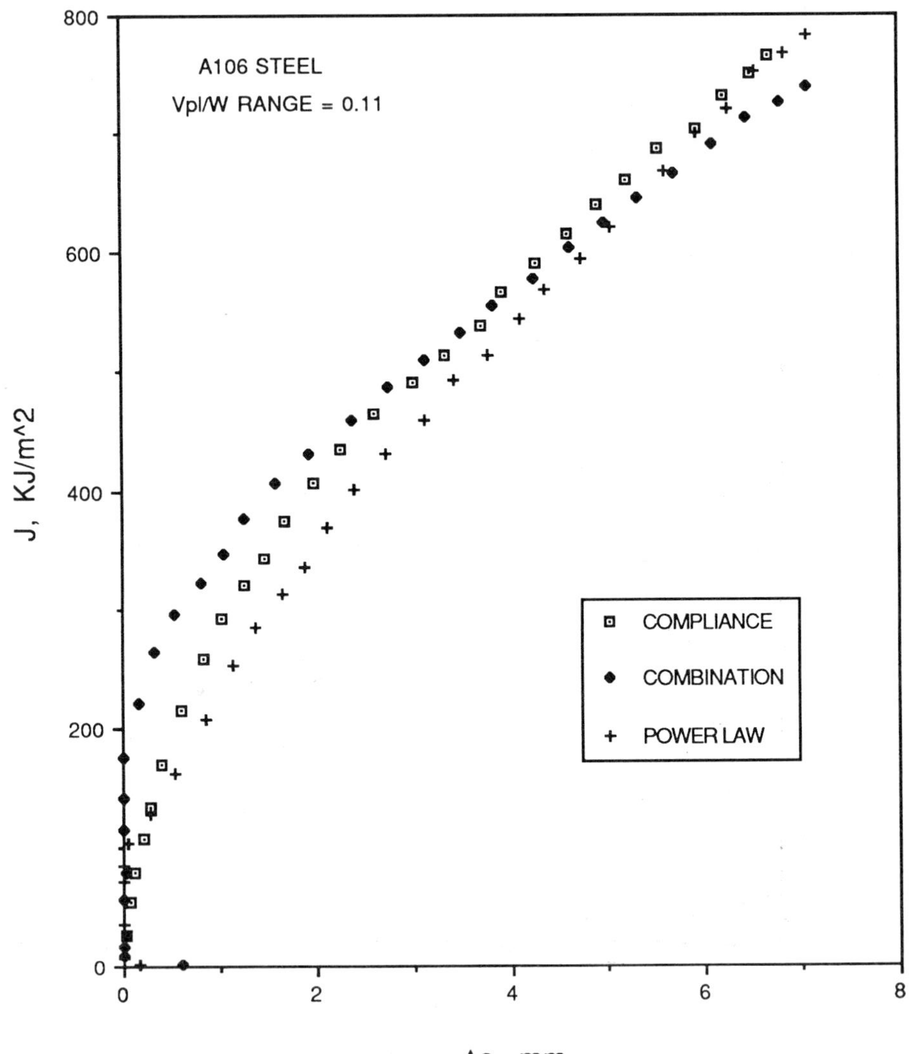

FIG. 11—J *versus* Δa R-*curve for an A106 steel comparing compliance with power law and combination fits (W = 50 mm).

Ref *6*) when the total range of v_{pl}/W is small, perhaps less than 0.05. However, when v_{pl}/W is greater than 0.10, the combination of a power law and straight line gives the best normalized curve for determining the *J-R* curve. This conclusion is not completely general. Figure 13 shows a result for a 3Ni steel with a v_{pl}/W range of 0.035 where both methods work well; and Fig. 14 shows a result for a pressure vessel steel with a v_{pl}/W range of 0.12 where both methods work well. Although these exceptions occurred, applying the two methods as suggested gave good results for the cases examined.

The method was applied to a large number of cases all of which could not be shown here. The result was generally that the *J-R* curves successfully generated by the elastic compliance procedure could be as well be determined by a normalization procedure. The two procedures

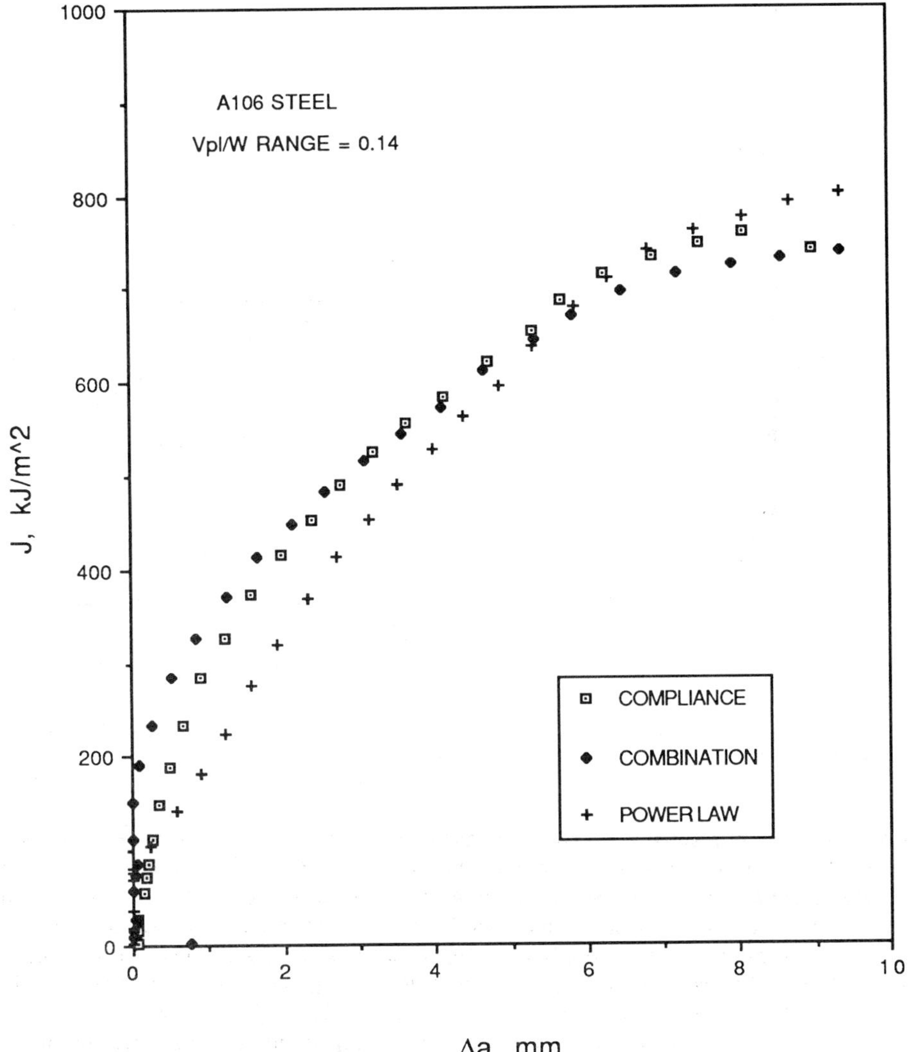

FIG. 12—J *versus* Δa R-*curve for an A106 steel comparing compliance with power law and combination fits (W = 50 mm).

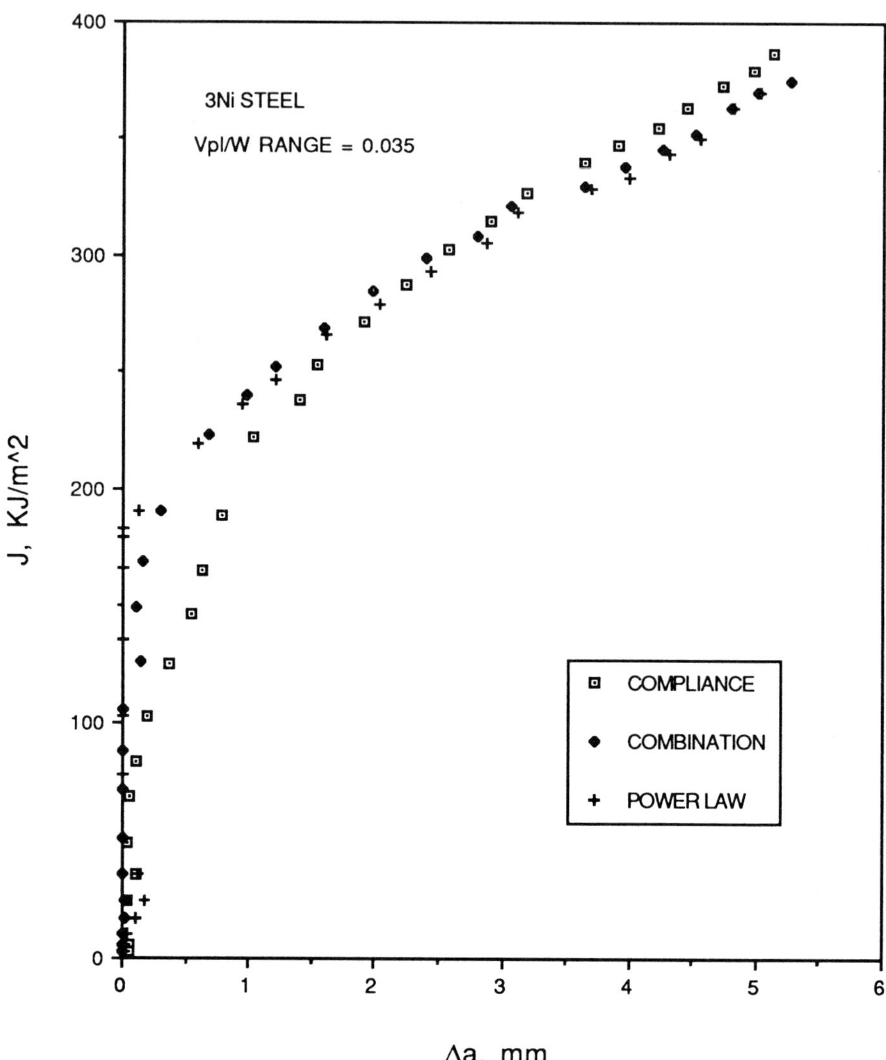

FIG. 13—J versus Δa R-curve for a 3Ni steel comparing compliance with power law and combination fits at low V_{pl}/W range (W = 50 mm).

give nearly identical results when the compliance result gives a good estimate of measured physical crack length. When the measured physical crack length is different from the one predicted by compliance, the normalization method appears to work better. One example can be taken from results in Fig. 9, where the final measured crack length was different from the compliance generated one. In this case, the final region of the J-R curve is corrected to end at the final measured crack length. An additional example is shown in Fig. 15, where the final compliance prediction of crack length is about 50% of the measured value. Here the normalization procedure gives a much better J-R curve, one that matches the physical measurements.

The normalization procedure, then, can be used to accurately develop J-R curves. Since

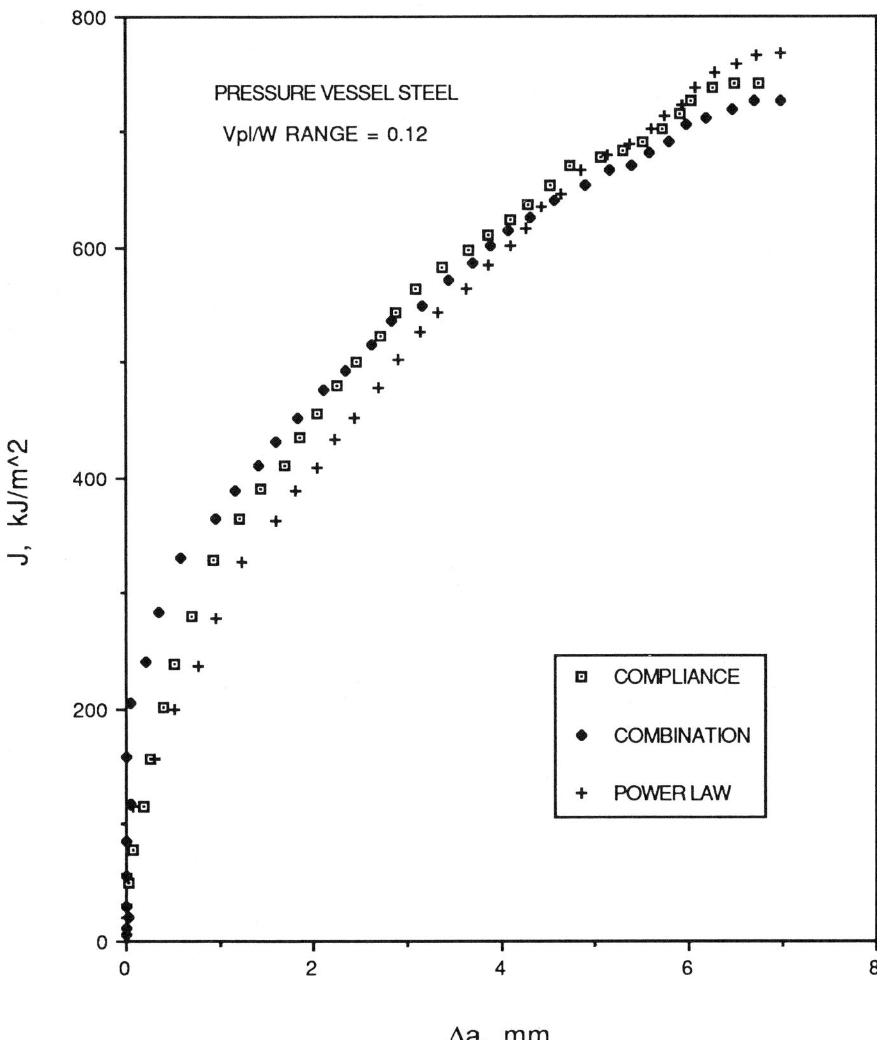

FIG. 14—J *versus* Δa *for a pressure vessel steel (A533B) comparing compliance with power law and straight line fits at a high* V_{pl}/W *range (W = 50 mm).*

it requries only computations for a given load versus displacement record and does not need sophisticated automatic crack length measuring equipment, it has the potential of being more generally usable. Laboratories which do not have the required equipment or the technique to perform compliance testing can use a normalization procedure to develop *J-R* curves. The method could be incorporated into the standard test methods used in elastic-plastic fracture testing.

The procedure works well for the range of v_{pl}/W less than 0.05 where only a power law normalization fit is needed. For the case where a power law and straight line combination is used, there are some obvious differences between the normalization and compliance methods. As can be seen in Figs. 9 to 15, the normalization method does not develop a blunting

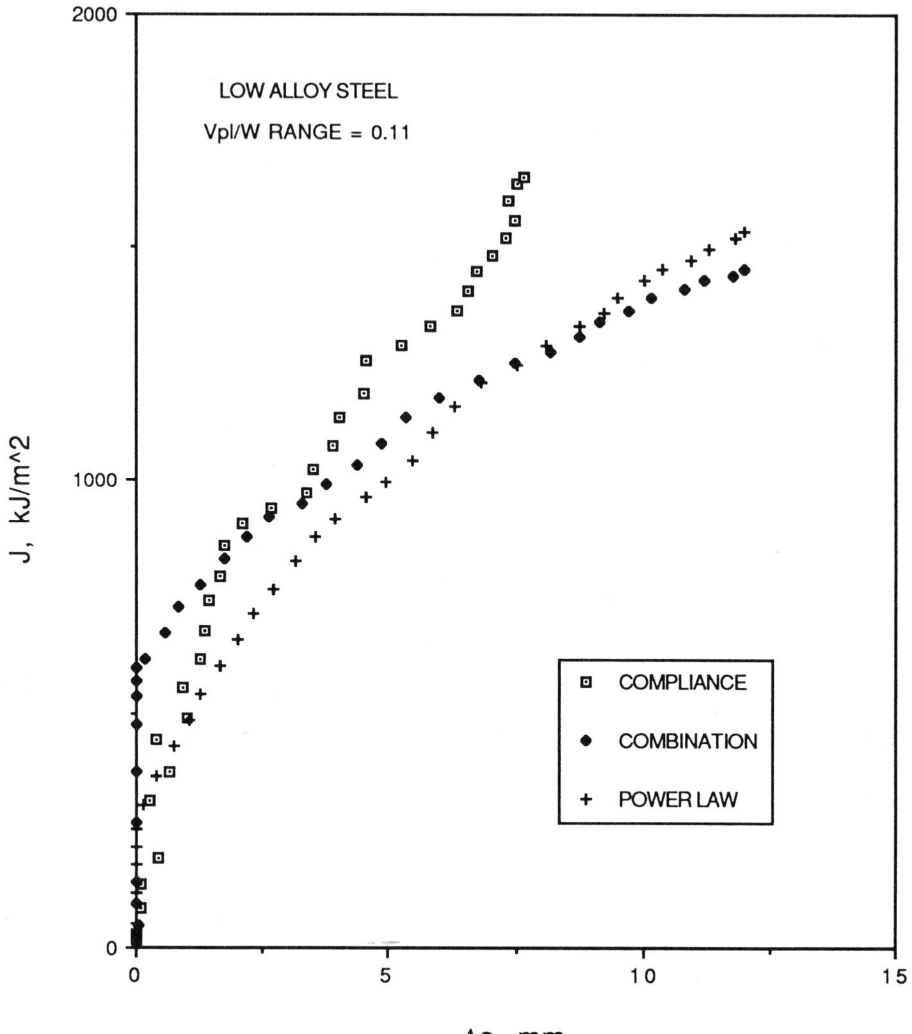

FIG. 15—J *versus* Δa R-*curve for a high-strength low-alloy steel (HSLA) showing the results for a bad compliance prediction of measured crack length* (W = *100 mm*, B = *25 mm).*

line as was historically seen in *J-R* curve and J_{Ic} testing. This occurs because crack growth is not predicted until there is a difference between the functionally fitted normalized curve and a normalized curve developed with initial crack length. In the combined method this does not occur over the power law region, only over the straight power law region. Therefore the power law serves only to provide an anchoring point for the straight line.

Although the method of choosing the power law-straight line transition by picking a tangent point works adequately, there is no good technical justification for this. A better method of choosing this transition point could be found and is the subject of further investigation.

Summary

The method of directly developing *J-R* curves from load versus displacement records represents an advantage over the commonly used elastic compliance method. The results from this work show that the method can be applied by choosing the appropriate form of the calibration function $H(v_{pl}/W)$. This form should be a power law for v_{pl}/W less than 0.05 and a combination of a power law and straight line for v_{pl}/W greater than 0.1. The method was applied to a variety of steels and a range of specimen sizes. The *J-R* curve generated by the method agrees well with compliance generated results whenever the compliance prediction of crack length is accurate. When the compliance prediction is not so accurate, the normalization procedure appears to give better results.

To gain more confidence in this procedure it would be well to submit it to further examination. It is recommended that it be applied as a second method of *J-R* curve evaluation whenever automatic crack length equipment is being used. If after a reasonable amount of evaluation the method proves to work well, it is suggested that this method be considered in a standard test procedure to develop *J-R* curves.

References

[1] Landes, J. D. and Begley, J. A., "Recent Developments in J_{Ic} Testing," in *Developments in Fracture Mechanics Test Methods Standardization, ASTM STP 632*, W. F. Brown, Jr., and J. G. Kaufman, Eds., American Society for Testing and Materials, Philadelphia, 1977, pp. 57–81.

[2] "Standard Test Method for Determining *J-R* Curve," ASTM Method E 1152, *1987 ASTM Book of Standards*, Vol. 03.01.

[3] Clarke, G. A., Andrews, W. R., Paris, P. C., and Schmidt, D. W., "Single Specimen Tests for J_{IC} Determination" in *Mechanics of Crack Growth STP 590*, American Society for Testing and Materials, Philadelphia, 1976, pp. 24–42.

[4] Joyce, J. A. and Gudas, J. P., "Computer Interactive J_{Ic} Testing of Navy Alloys," in *Elastic-Plastic Fracture, ASTM STP 668*, J. D. Landes, J. A. Begley, and G. A. Clarke, Eds., American Society for Testing and Materials, Philadelphia, 1979, pp. 451–468.

[5] Landes, J. D. and Herrera, R., "A New Look at *J-R* Curve Analysis," *International Journal of Fracture*, Vol. 36, 1988, pp. R9-R14.

[6] Herrera, R. and Landes, J. D., "A Direct *J-R* Curve Analysis of Fracture Toughness Tests," *Journal of Testing and Evaluation*, Vol. 16, No. 5, Sept. 1988, pp. 427–449.

[7] Ernst, H. A., Paris, P. C., Rossow, M., and Hutchinson, J. W., "Analysis of Load Displacement Relationship to Determine *J-R* Curve and Tearing Instability Material Properties," in *Fracture Mechanics, ASTM STP 677*, C. W. Smith, Ed., American Society for Testing and Materials, Philadelphia, 1979, pp. 581–599.

[8] Ernst, H. A., Paris, P. C., and Landes, J. D., "Estimations on *J*-Integral and Tearing Modulus *T* from a Single Specimen Test Record," in *Fracture Mechanics: Thirteenth Conference, ASTM STP 743*, Richard Roberts, Ed., American Society for Testing and Materials, Philadelphia, 1981, pp. 476–502.

[9] Kumar, V., German, M. D., and Shih, C. F., "An Engineering Approach for Elastic Plastic Fracture Analysis," Electric Power Research Institute, NP 1931, Topical Report, July 1981.

[10] Landes, J. D., McCabe, D. E., and Ernst, H. A., "Fracture Testing of Ductile Steels, Electric Power Research Institute, NP-5014, Final Report of Research Project 1238–2, Jan. 1987.

[11] Landes, J. D. and McCabe, D. E., "Toughness of Austenitic Stainless Steel Pipe Welds," Electric Power Research Institute, NP-4768, Topical Report on Research Project 1238-2, Oct. 1986.

[12] Hackett, E. M. and Joyce, J. A., unpublished results.

R. J. Sanford[1] and R. W. Judy, Jr.[2]

Application of the Method of Caustics to *J*-Testing with Standard Specimen Geometries

REFERENCE: Sanford, R. J. and Judy, R. W., Jr., **"Application of the Method of Caustics to *J*-Testing with Standard Specimen Geometries,"** *Fracture Mechanics: Twenty-First Symposium, ASTM STP 1074,* J. P. Gudas, J. A. Joyce, and E. M. Hackett, Eds., American Society for Testing and Materials, Philadelphia, 1990, pp. 44–55.

ABSTRACT: The optical caustic technique offers unique opportunities for experimental measurement of fracture parameters under conditions where conventional instrumentation either can not be used or is not available. The apparent simplicity of the method has attracted the interest of metallurgists and materials scientists. However, there are subtle problems with the method which complicate the interpretation of results. Some of the problems have not been fully explored and others need clarification before the method can be routinely applied. Of particular concern is the fact that existing theories for caustic behavior apply only in special cases which are rarely, if ever, attained in test specimens of practical dimensions for fracture testing. Nonetheless, the potential promise of the method justifies a systematic study of the influence of various parameters, both physical and optical, on the observed behavior of the caustic.

In previous work, the authors demonstrated that there is a correlation between the measured caustic diameter and the value of the *J*-integral parameter obtained from mechanical measurements for planar 1T compact tension specimens in various steels, subject to some restrictions. As might be expected, this correlation did not agree with existing theoretical relations since the theories are all based on plane stress assumptions. In the present paper, these correlations are studied further and the influence of imaging distance and specimen thickness examined in detail. In addition, the characterization of caustic behavior in standard bend specimens is investigated.

For these studies, a new four-lens caustic camera was used which permits the simultaneous determination of four caustics with varying image distances at each increment of applied load. With the aid of this camera, sufficient data were obtained to determine empirical relations for the calculation of the *J*-integral values from optical measurement of the caustic.

KEY WORDS: fracture, caustics, high-strength steels, *J*-integral, fracture mechanics

Methods based on the principles of optical caustics have been used in experimental programs to infer the fracture properties of materials, principally in studies of dynamic fracture in steels; in these steels, fracture resistance levels are such that cracks extend under essentially brittle conditions [*1–7*]. The major advantage of this technique is that it is based on optics and does not require any sensors or tranducers to be attached to test specimens, thus obviating the usual concerns of interference of the instrumentation with the progression of the fracture or the response time of the electronic equipment used to make many measurements in a very short time. For use in dynamic fracture studies, it is only necessary to have a light source and high-speed photographic equipment, provided there is confidence in the estab-

---

[1] Professor of Mechanical Engineering, University of Maryland, College Park, MD 20742.
[2] Supervisory Mechanical Engineer, Naval Research Laboratory, Washington, DC 20375.

lished relations between caustic dimensions and actual, instantaneous loading parameters. Relations between caustic size and applied loading parameters have been rigorously derived for the conditions of plane stress constraint levels and elastic loading conditions; these circumstances exist rarely in real metallic materials and thicknesses. The extent to which the analysis can be applied to include plasticity and crack growth is being explored by several investigators [7,8].

Past work described in the literature on applications of caustics to measure fracture resistance properties shows that experiments involving metals typical of naval alloys in standard sample sizes would violate the fundamental assumptions regarding crack tip deformation and general plasticity of the far-field regions, remote from the crack tip. This is because caustics result from surface deformation in specimens, in particular, the out-of-plane deformations. These factors do not, however, rule out the use of caustics to measure fracture properties of ductile metals; they do indicate that the existing body of technology cannot be simply transferred to compute relationships between caustic dimensions and aspect ratios for fracture under plane strain or mixed mode constraint. An analytical approach to the problem would be extraordinarily complex, since it would involve three-dimensional elastic and plastic deformation fields at crack tips in finite bodies. A logical first step in determining whether caustics can be used reliably, therefore, is an experimental investigation of the behavior of caustics under a variety of conditions of constraint in different metal samples of varying inherent fracture resistance. Accordingly, the subject of this paper is the experimental observation of crack-tip caustic behavior in three steel samples. This behavior was evaluated using two standard fracture test techniques under constraint conditions which thoroughly violated the basic assumptions of the existing technology.

Background

A review of the principles involved in the formation of caustics in reflected light is helpful at this point in order to properly interpret the experimental results presented later. To observe the caustic, a highly reflective specimen containing a sharp crack is illuminated with light from a point source while being loaded. As illustrated in Fig.1 [5], the caustic is a band of bright light surrounding a shadow zone around the crack tip. In many cases, the small band of bright light is not readily observable in reflected light, particularly when surface reflectivity is less than perfect, and the dark shadow spot is used to determine the caustic size. The caustic results from concentration of the light due to the distortion of the specimen surface by strains associated with crack tip stress fields. The size and shape of the caustic are determined by the characteristics of the strain field at the crack tip and by the distance z_0 of the virtual image plane from the plane of the deformation (i.e., the specimen surface). A camera is focussed at the virtual image plane to photograph the caustic. Note from the figure that the caustic diameter varies in non-linear fashion with image-plane distance and that at locations near the specimen surface the caustic cannot be observed.

The transverse diameter of the caustic is the dimension of principal interest for developing relationships with the applied loading parameters K and J. The shape of the caustic is influenced by the anisotropy and by the strain hardening characteristics of the material. The caustic size varies with the z_0 dimension; the region on the specimen surface which contains the deformations giving rise to the caustic is called the generating circle or, more properly, the initial curve. The location of the initial curve is not fixed but is proportional to z_0 and the applied load, such that its diameter is a fixed fraction of the caustic diameter. Thus the experimental determination of caustic behavior requires that the location of the initial curve be known. The initial curve can be very close to the crack tip, inside the plastic zone or outside the plastic zone, depending on z_0 distance. As can be readily imagined, the location

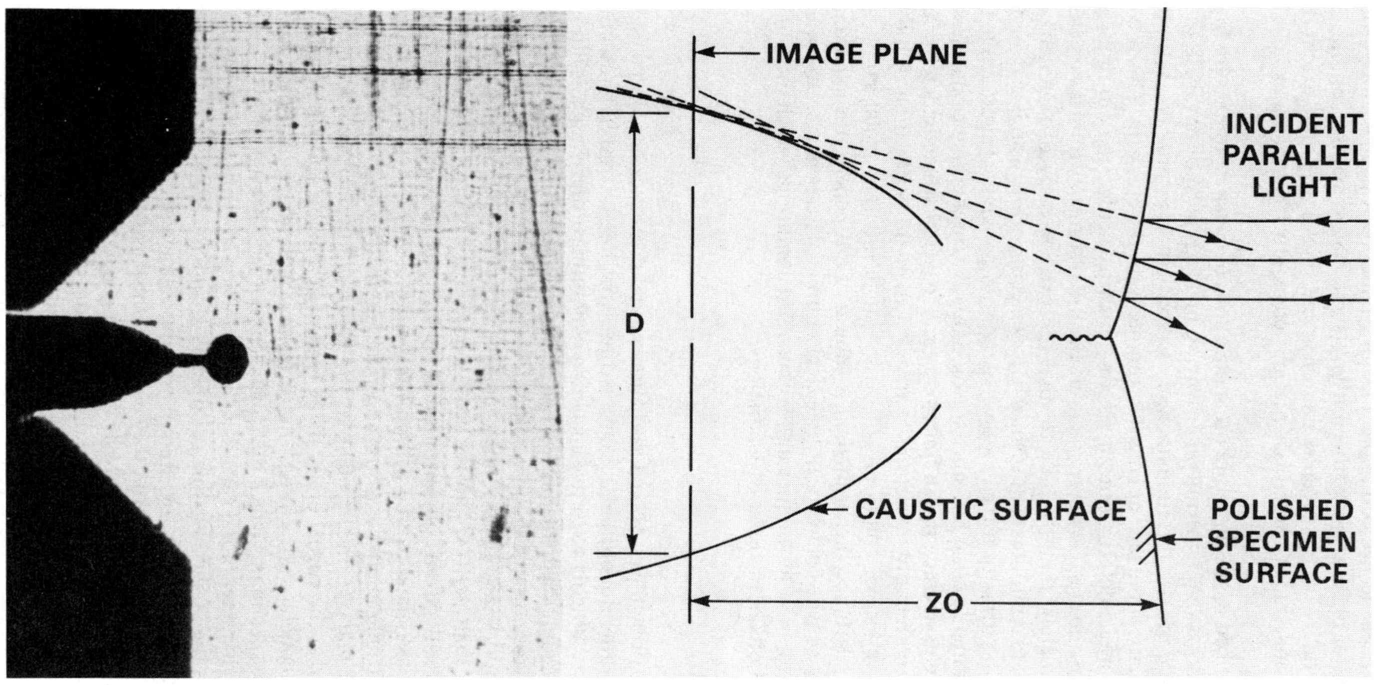

FIG. 1—*Formation of a caustic in reflection* (right) *by deformation of the surface of a metallic test specimen at the tip of a sharp crack* [5]. (left) *Typical caustic recorded in experiments reported previously* [8].

of this curve has a profound effect on the behavior of the caustic, since the caustic is a man-ifestation of the local strain fields at the site of the initial curve. It is pointed out that in the course of experimental determinations of fracture properties, the increase in load causes increases in both the caustic diameter and the plastic zone size, provided the z_0 distance is held constant; thus it is possible for the initial curve to move from the elastic field surround-ing the plastic zone into the region inside the plastic zone during the test.

Formulations for the relation between caustic diameter and loading parameters have been reported in the form of power law equations for several special cases [1–6]:

$$K = C_1 \frac{D^{2.5}}{z_0 T} \qquad \text{for elastic behavior} \tag{1}$$

$$J = C_2 \frac{D^3}{z_0 T} \qquad \text{for elastic-perfectly plastic materials} \tag{2}$$

where D is caustic diameter and T is specimen thickness. These equations assume plane stress constraint and idealized material response.

Materials and Procedures

Three steels representing varying strength-toughness combinations from high-strength/low-toughness to low-strength/high-toughness were used for evaluation of caustic behavior. These steels were 4340 heat-treated to 1200 MPa (170 ksi) yield strength, a 5-Ni steel and an HSLA steel, A710. Details of the material properties and thicknesses are given in Table 1. The 4340 plate was machined into specimens of varying thickness to provide plane strain constraint (25 mm thick), near plane stress constraint (2.5 mm thick), and a mixed-mode constraint condition (10 mm thick). The other two materials were tested in the full thickness condition only, which provided mixed-mode constraint for each steel. Throughout the exper-iments, an effort was made to observe caustics on the same plate surface for each material sample; the purpose of this was to minimize any effects of gradients in material properties.

The specimen configurations used for the experiments were a standard 1T compact ten-sion (CT) design, with varying thickness, and standard bend bars (ASTM E 813). Before testing, one surface of each CT specimen was polished using standard metallographic tech-niques; in most cases, the polishing operation was performed before fatigue precracking. The surfaces of the bend specimens were prepared using a transfer-mirror technique developed specifically for these tests. This technique, in which an aluminized mirror coating is first deposited on an optical quality surface and then transferred to the specimen by bonding with a thin epoxy coating [9], produces high quality surfaces that are superior to mechanically polished specimens. The coating and the adhesive layer were sufficiently thin that surface displacements were faithfully reproduced for measurement of the caustic. Load-shedding techniques were used in precracking to minimize residual surface deformation. The experi-

TABLE 1—*Properties of materials used in study of caustic behavior.*

Alloy	Plate Thickness, mm	Yield Strength, MPa	Ultimate Strength, MPa	J_{Ic}, kJ/m^2
4340 steel	25	1207	1282	28
5-Ni steel	50	965	1014	93
A710 steel	13	717	778	...

ments consisted of standard fracture mechanics tests to measure K and J integral conducted at static load rates, while at the same time photographs were taken of the caustics at predetermined intervals. Experiments were performed in a closed-loop servohydraulic test machine using a potential drop method to monitor crack extension. A personal computer and appropriate A-D converters and signal conditioning equipment were used for data collection. Computation of the K and J parameters required that load, load-line displacement, and the potential drop signal be recorded; these data were taken at preset increments of load-line displacement (4×10^{-4} mm or 0.0004 mm). The computer program allowed momentary pauses in the test to photograph the caustic and to record the data point number at which the photograph was taken; this operation was performed with only 1 or 2 s interruptions in most cases and had only minor, if any, influence on the test results. All experiments were conducted in the displacement control mode. At the conclusion of the tests, the specimens were fatigued to mark the crack extension and were then fatigued at higher cyclic loads to failure. Computations of the K and J parameters were performed in accordance with ASTM E 813. Specific parameters to be computed were stress-intensity factor (K) and the J-integral (J). As prescribed in ASTM E 813, the elastic and plastic components of J were computed separately and combined; the values of J reported in this paper are the combined, or total, value of J.

The experiments were conducted using a recently developed four-lens camera system [*10*]. The camera, shown schematically in Fig. 2, was configured so that each lens was focussed on a different image plane. This was accomplished by carefully selecting lens of different focal lengths while maintaining the bellows length fixed. The image planes for the experiments were approximately 55, 291, 512, and 657 mm. Illumination of the specimen was provided by a single white light source and a system of four fiber optic cables terminated with needle point probes, one for each camera lens. In order to provide image separation on the film plane, a field lens was used for collimation of the light (Fig. 3). The photographs of the caustics were taken with positive-negative Polaroid film. Measurements of the caustic diameters were made by projecting an enlargement of the negatives onto a digitizing tablet. Each measurement was repeated three times and the average value used in all subsequent analysis. The four-lens camera permitted four caustics to be recorded at a single value of applied J or K

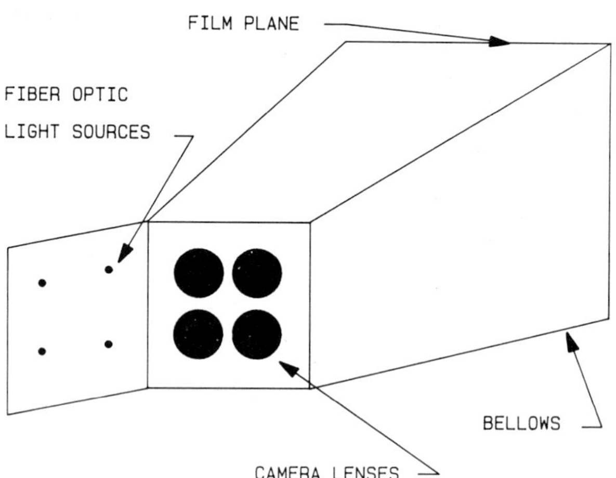

FIG. 2—*Schematic illustration of four-lens caustic camera and fiber optic light source array used to record multiple caustics at each load increment. Each of the four camera lenses is of a different focal length to produce varying z_0 distances for the same bellows length.*

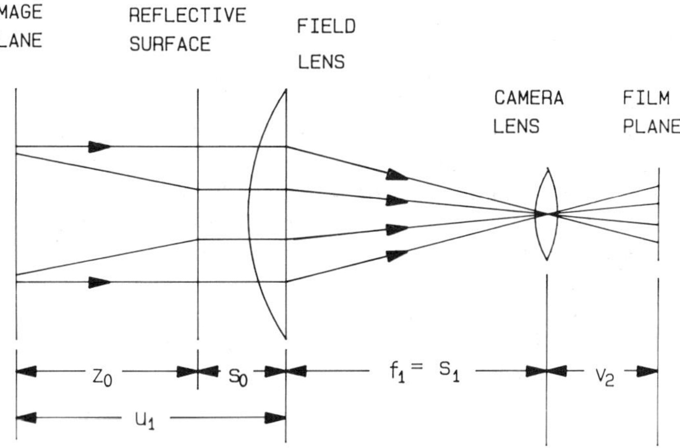

FIG. 3—*Optical ray diagram for the optical system used to record multiple caustics in collimated light. As shown in Fig. 2, only the reflected light is illustrated; the light from the fiber optic source also passes through the field lens.*

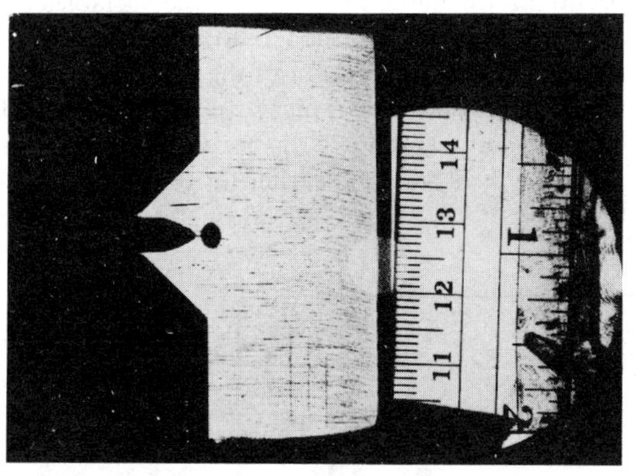
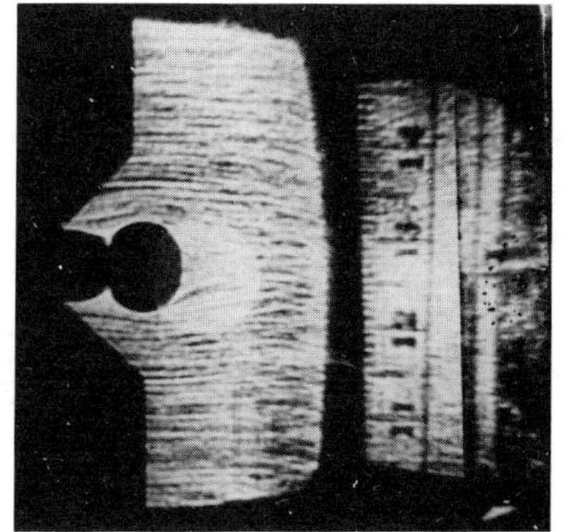

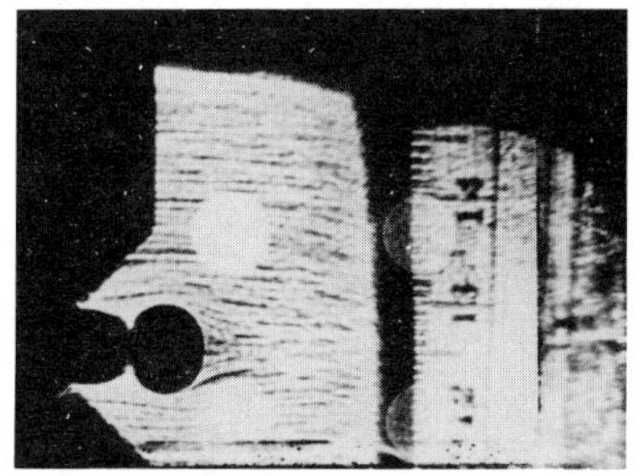

FIG. 4—*A set of caustics photographed with the four-lens camera system. The z_0 distances are 55, 291, 512, and 627 mm respectively, moving counterclockwise from upper left. Since the diameter of each caustic depends on z_0, an increase in caustic dimensions is observed as the z_0 distance is lengthened.*

without pausing to refocus the camera. A typical set of caustics recorded with this camera is shown in Fig. 4. The scale, shown in Fig. 4, was mounted on a front-surface mirror located at one of the image planes and provides a direct measurement of the image plane magnification for each image in the set.

Samples of 4340 steel were tested at thicknesses of 25, 10, and 2.5 mm in the CT configuration and at thicknesses of 25 and 10 mm in the bend configuration to investigate the influence of constraint mode on the relation of caustic diameter to applied load. Full thickness samples of 5-Ni steel and A710 steel were tested in both specimen configurations to study the influence of strength and flow properties on this relation. Measurements of the caustic diameter were made at predetermined increments of applied J as described above. A computerized data acquisition system was employed to record load, load-line displacement, and the signal from the potential drop crack length measurement system; these data were sufficient to compute crack length, K and J, as described above.

Discussion

The data set of caustic diameter versus J recorded by the methods described above included variations of materials, specimen type, thickness, and z_0; these data will be presented in summary form. For each experiment, the data show, as expected, a strong influence on z_0 distance. This is illustrated in Fig. 5, which shows the results of a typical experiment. The general goals of the project were to experimentally define the effects of the variables on the J-D relation, to use this information to establish ranges where the relations are consistent, and to construct a master curve from the data at hand. Examination of the published theoretical relationships between the factors, of which Eqs 1 and 2 are typical, shows that J should be inversely proportional to both z_0 and thickness T, and should be proportional to D raised to some power. The exponential relation of J and D was reported earlier for the same materials used in the present experiments, and was observed in plots of all of the data in the

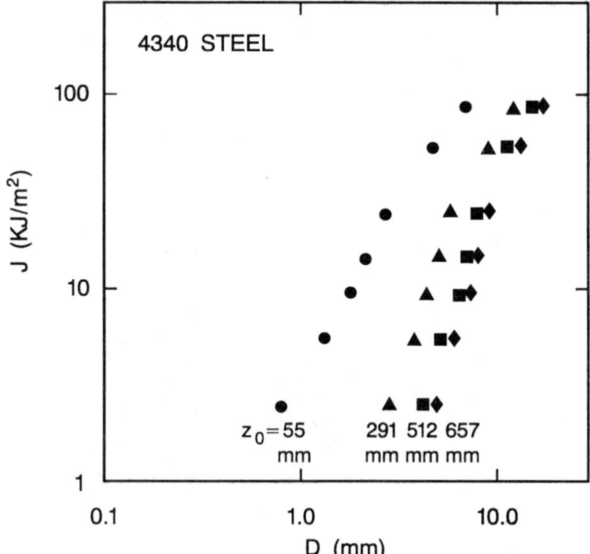

FIG. 5—*Data taken from an experiment with a compact tension specimen of 4340 steel. The effect of* z_0 *is consistent for all the experimental data.*

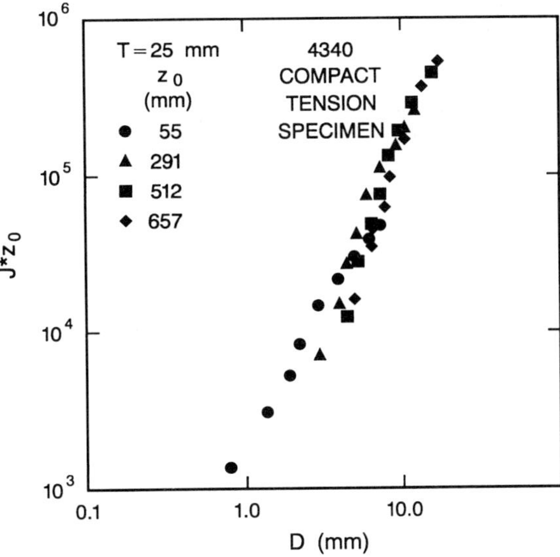

FIG. 6—*Relation of product of* J*z_0 *to* D *for* T = 25 mm *shows that at given values of* D, J *is directly proportional to* z_0. *Also,* J *is an exponential function of* D.

format of Fig. 5. By plotting the product J*z_0 versus D (using only data recorded in tests with compact tension specimens) on logarithmic scales at a constant thickness, the relation of z_0 can be illustrated; this is done for 4340 steel in three different thicknesses in Figs. 6 to 8. For clarity not all the recorded data are shown in these figures; the data plotted were selected to illustrate the range and fit with regard to z_0 and thickness. In each figure, the good

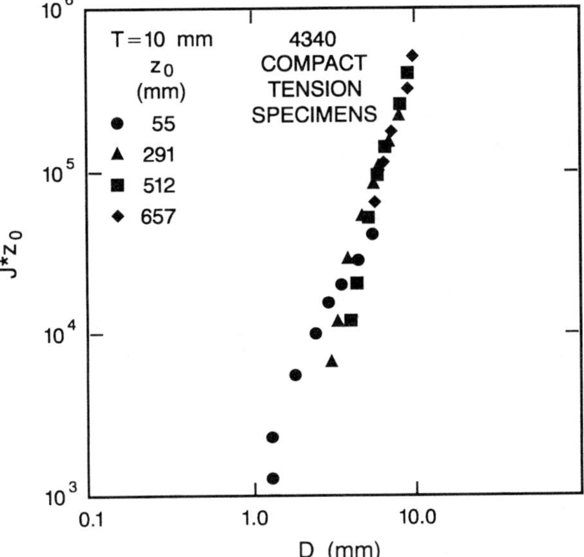

FIG. 7—J*z_0 *versus* D *for 4340 steel at 10 mm thickness.*

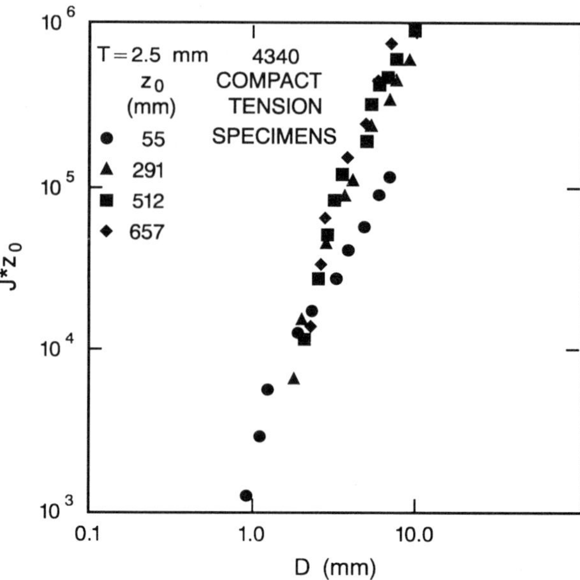

FIG. 8—$J*z_0$ *versus* D *for 4340 steel at 2.5 mm thickness. Note that the values recorded at* z_0 *= 55 mm do not follow the trend set by the values recorded at longer* z_0 *values. This trend is also present in Figs. 6 and 7 but to a lesser extent.*

fit to a straight line shows that the effect of z_0 can be taken in this manner. In each of these figures, the data taken at z_0 = 55 mm deviate from the main body of data; this is due to the location of the initial curve well within the plastic zone, particularly at the higher values of applied J. This effect, described in the earlier work [8], is caused by intense local strains, which cause local inhomogeneities that distort the caustic. Inspection of the data in Figs. 6 to 8 shows that they do not fit a single line, so that thickness has a significant effect which must be separately determined. All the data shown in Figs. 5 to 8 were taken from experiments with compact tension specimens. The influence of specimen configuration will be discussed later.

In the analytical relations between caustic diameter and applied J for plane stress conditions, it was shown that the effect of thickness should be one of direct proportionality. As a first approximation, a fit of all the present data (which, clearly, do not conform to plane stress conditions) in the form of the product $J*z_0^*T/2$ versus z_0 is illustrated in Fig. 9. This simple approach collapses the data to a very tight band for the 4340 material and is the basic master curve for relating J and D. The limits on the use of the master curve are that the initial circle must be confined to the singularity-dominated region or the elastic region (i.e., the z_0 distance must be large enough to exclude the region of intense plastic deformation). It is also noted that in all the data, conditions of crack propagation were excluded. Using these conditions, the data recorded with compact tension specimens for 5-Ni steel and A710 steel are plotted in Fig. 10, with the data trend line for 4340 steel being shown for comparison. The slight deviation of data for the more ductile steels from the trend line indicates that the simplified approach to accounting for thickness may not be exact and could be improved by a more rigorous analysis of the displacements which cause the caustic in the first place. Such an endeavor is well beyond the scope of the present investigation. It is also observed that the 5-Ni steel data conform to the trend line established for 4340 steel much more closely than

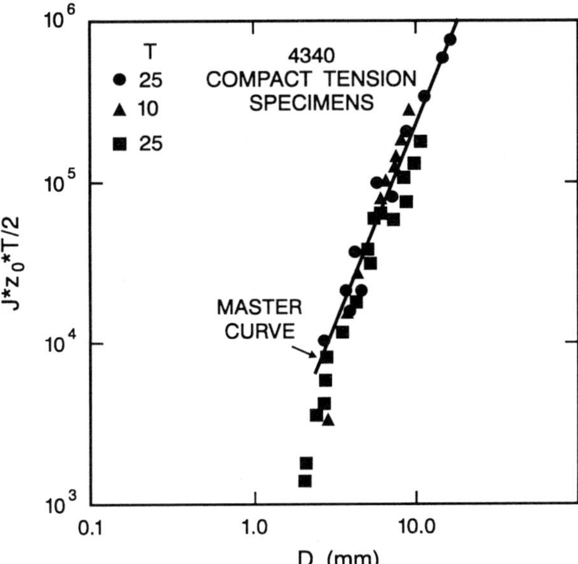

FIG. 9—*The product of* $J*z_0*T/2$ *collapses the data to a small scatter band when plotted versus* D, *which shows that the effect of thickness can be approximated as a direct dependence. This is the master curve for compact tension specimens of any thickness.*

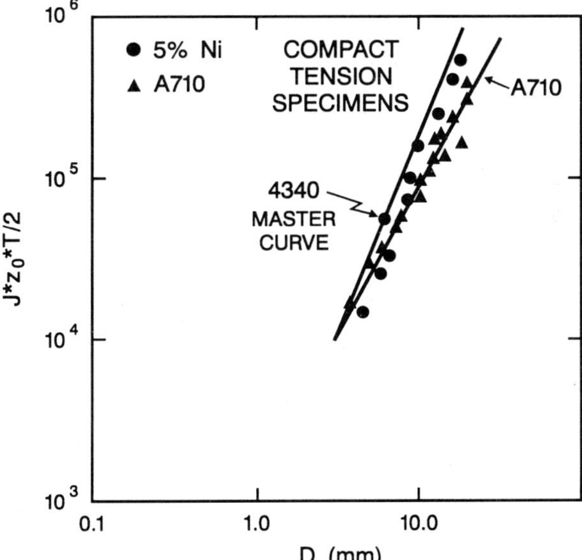

FIG. 10—*Compact tension data for 5-Ni and A710 steels. Note that the A710 data deviate from the 4340 master curve trend line due to its unusual elastic-plastic behavior.*

did the data for A710 steel. Approximations of the relative sizes of plastic zones to initial circle diameter and their locations with regard to the crack tips led to the conclusion that the discrepancy is due to the location of the initial circle in the elastic or singularity dominated region for the 4340 and 5-Ni steels and inside the plastic zone for the A710 steel. In previous work with these steels [8], it was shown that the A710 steel exhibited upper and lower yield points and had a very flat tensile flow curve, which led to the formation of a closely confined zone of intense yielding near the crack tip. It is concluded that the same phenomena led to the deviation of the A710 data from the main trends in this investigation.

Data from experiments with bend specimens are compared to the results shown above for compact tension specimens in Fig. 11. Bend specimen data were subjected to the same restrictions on z_0 distance that were applied to the compact tension data. The same trends were observed for bend specimens in that 4340 steel and 5-Ni steel data followed one relation and the data for A710 steel followed another. In Fig. 11, the trend lines shown are taken from the earlier figures. The good comparison, particularly at the higher values of applied J, for both trend lines shows that there is no significant influence of specimen geometry (i.e., the master curve which relates the caustic diameter to J is, within limits, independent of specimen type). One would expect considerable difference in the behavior of the caustics observed during crack propagation in the two specimen types; however, under purely static conditions before crack extension, there should be no difference, as was observed within experimental error.

Conclusions

Experimental determinations of the relations between caustic diameter and the level of applied J-integral were made, including variables of material strength and flow properties, constraint mode, specimen type, and image plane distance. All of these variables were

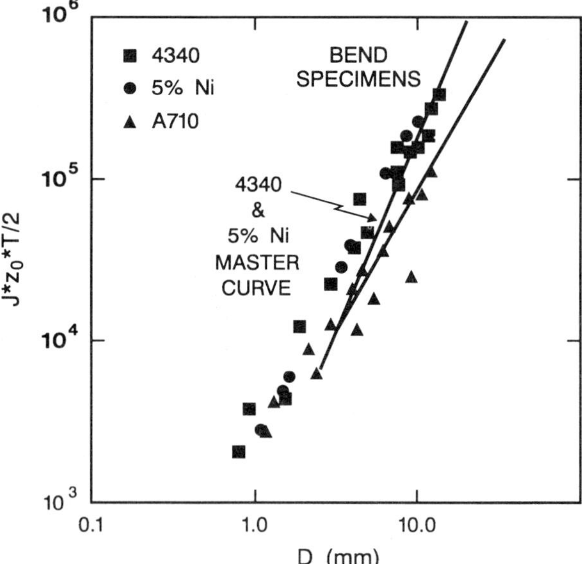

FIG. 11—*Comparison of bend bar data for all three steels to the trend established for compact tension specimens. The good fit illustrates that there is no dependence on specimen type for the relation between caustic diameter and J-integral.*

resolved, within limits, to produce a preliminary master curve using very simplified assumptions. Use of image plane distances such that the initial circle is in the singularity-dominated zone or the region of elastic loading is essential to obtaining a well-defined caustic that can be related to externally applied loading parameters. The intense local strains that are produced inside the plastic zone cause distortion of the caustic and the results are unpredictable. As expected, the relation between caustic diameter and applied J is not as predicted by plane stress theory. Resolution of the data into a master curve format was accomplished by assumptions that the influences of z_0 and material thickness could be modeled as direct proportionalities; the data fit was sufficiently good to support that assumption. The variations of tensile flow properties as represented by three different steels were not a significant factor, unless, as demonstrated by the A710 data, the material exhibits unusual flow characteristics. Finally, there was no distinct difference in results attributable to specimen geometry (compact tension versus bend specimens). All of these findings support the premise that, within the boundaries of the experiments, the caustic diameter has a singular relation to applied loading parameters, as expressed by the J-integral.

References

[1] Kalthoff, J. F., Beinert, J., and Winkler, S. "Measurements of Dynamic Stress Intensity Factors for Fast Running and Arresting Cracks in Double-Cantilever-Beam Specimens," in *Fast Fracture and Crack Arrest, ASTM STP 627,* American Society for Testing and Materials, Philadelphia, 1977, pp. 161–176.

[2] Kalthoff, J. F., "Stress Intensity Factor Determination by Caustics," presented to International Conference on Experimental Mechanics, Society for Experimental Stress Analysis and Japan Society of Mechanical Engineers, Honolulu, Maui, Hawaii, 1982.

[3] Manogg, P., "Anwendung der Schattenoptik zur 2 Untersuchung des Zerreibvorgangs von Platten," Dissertation, Universitat Freiburg, West Germany, 1964.

[4] Rosakis, A. J., "Experimental Determination of the Fracture Initiation and Dynamic Crack Propagation Resistance of Structural Steels by the Optical Method of Caustics," Ph.D. thesis, Brown University, Providence, R.I., 1982.

[5] Rosakis, A. J. and Freund, L. B., "Optical Measurement of the Plastic Strain Concentration at a Crack Tip in a Ductile Steel Plate," *Engineering Materials and Technology,* Vol. 104, April 1982, p. 115.

[6] Rosakis, A. J., Ma, C. C., and Freund, L. B., "Analysis of the Optical Shadow Spot Method for a Tensile Crack in a Power-Law Hardening Material," *Applied Mechanics,* Vol. 50, Dec. 1983.

[7] Zhender, A. T. and Rosakis, A. J., "Basic Experiments on the Dynamic Fracture Initiation and Unstable Creep Growth of Cracks in Metals Using Reflected Shadowgraphy," presented to Third International ASTM Symposium on Nonlinear Fracture Mechanics, 6–8 Oct. 1986, Knoxville, Tenn.

[8] Judy, R. W., Jr., and Sanford, R. J., "Correlation of Optical Caustic Behavior with Fracture Properties of High Strength Steels," in *Nonlinear Fracture Mechanics: Volume I—Time-Dependent Fracture, ASTM STP 995,* American Society for Testing and Materials, 1989, pp. 340–357.

[9] Sanford, R. J. and Collison, P., "Preparing Transfer Mirrors for Reflective Caustic Studies," *Experimental Techniques,* in press.

[10] Sanford, R. J. and Gerstein, J. S., "Design and Calibration of a Multi-Lens Caustic Camera," in *Proceedings,* Sixth International Conference on Experimental Mechanics, Portland, Ore., June 1988.

G. M. Wilkowski,[1] C. W. Marschall,[1] and M. P. Landow[1]

Extrapolation of C(T) Specimen *J-R* Curves

REFERENCE: Wilkowski, G. M., Marschall, C. W, and Landow, M. P., "**Extrapolation of C(T) Specimen *J-R* Curves,**" *Fracture Mechanics: Twenty-First Symposium, ASTM STP 1074*, J. P. Gudas, J. A. Joyce, and E. M. Hackett, Eds., American Society for Testing and Materials, Philadelphia, 1990, pp. 56–84.

ABSTRACT: In many structural flaw evaluations, frequently it is necessary to predict the maximum load where a ductile tearing failure mode is expected. In such a calculation, large ductile crack growth needs to be accounted for to predict the maximum load, especially for large structures and/or lower toughness materials. One difficulty in such an evaluation is that the amount of crack growth obtainable in C(T) specimens that can be machined from the material (e.g., from a pipe) is relatively small. Because of this, it is necessary to extrapolate the *J-R* curves to the larger amount of crack growth. No generally accepted method exists for such extrapolations.

This paper discusses methods used to evaluate *J-R* curves extrapolation techniques. As part of this study, several materials were evaluated using C(T) specimens of various sizes but with the same thickness. This is frequently the case for piping, whereas for a pressure vessel or other heavy-wall structure the specimens will frequently not even be the same thickness as the structure. The results in this study showed more geometry effects with the J_M-R curve than expected.

KEY WORDS: fracture, *J-R* curve, crack, crack growth, elastic-plastic fracture

In the evaluation of many structures such as cracked pipes, it is necessary to predict the fracture behavior using small fracture specimens machined from the structure. For example, in the case of a pipe, the curvature of the pipe dictates that the specimen be relatively small compared to the pipe. On occasions it becomes necessary to be able to extrapolate the crack growth resistance curve from the small fracture specimens to predict the maximum load or fracture stability under displacement-controlled stresses.

To evaluate the extrapolation schemes for *J-R* curves, a series of nine tests using C(T) specimens have been conducted as part of the Nuclear Regulatory Commission's (NRC) Degraded Piping Program conducted at Battelle [*1,2*]. These tests were conducted on specimens machined from plate materials and weldments in the same plates. Standard C(T) specimens were tested as well as planform C(T) specimens of larger size. The planform specimens had the same thickness as the smaller specimens, but the other specimen dimensions were larger. Paris et al. [*3*] first used large planform C(T) specimens to obtain *J-R* curves for large amounts of crack growth.

In this study, J_D-R curves, analyzed in accordance with ASTM E 813-81 and E 1152-87, and J_M-R curves were evaluated. The crack growth in the C(T) experiments was well beyond the ASTM validity limits for crack growth. For the cases of C(T) specimens on weldments, circumferential throughwall cracked pipe bending experiments were in one case conducted and in another case are currently being planned to assess the predictions from the C(T) specimens.

---

[1] Battelle, Columbus, OH 43201.

C(T) Specimen Test Matrix and Test Procedures

This section contains a description of the experiments performed in the present study. The aim of the experimental effort was to provide the necessary data that could be analyzed to obtain *J-R* curves for different amounts of crack growth. In all experiments, the C(T) specimen geometry was used. To obtain data corresponding to increasing amounts of crack growth, the planar dimensions of the specimens, where the material was 25.4 mm (1 in.) thick, were chosen to be 1T, 3T, and either 9.5T or 10T.[2] Figure 1 shows the three typical specimen sizes. For materials that were 9.6 mm (0.38 in.) thick, the planform specimen sizes were 0.5T, 1.5T, and 3T.

Eight series of tests were conducted in this study (Table 1). This included Type 304 austenitic steel purchased to ASTM Specification A240 along with GTAW and submerged arc welds in the stainless steel, and ASTM A516 Grade 70 pressure-vessel quality ferritic plate steel and a submerged arc weld in that plate. (Note: A516 Grade 70 plate is also used in seam welded nuclear piping.) Table 1 indicates which specimens were side grooved. The side grooves were 20% of the thickness (10% on each side). C(T) specimens of the three planform sizes were machined from each material. Plan dimensions were in accordance with ASTM Test for J_{Ic}, A Measure of Fracture Toughness (E 813-81). Each specimen was fatigue precracked using the guidelines given in ASTM E 813-81 to produce a starter crack length of $0.5W$, where W is the specimen width.

All tests were instrumented to measure load and displacement. Crack extension was determined by the direct-current electric potential (d-c EP) method and by periodic unloadings during the test to mark the crack front.

In all tests, displacement rates were selected to cause crack initiation in 10 to 20 min. Tests were terminated when the crack had extended by an amount equal to 40 to 60% of the original uncracked ligament. The specimens were then broken into two pieces, and initial and final crack lengths were measured. In the case of Type 304 stainless steel, specimens were heat tinted at 540°C (1000°F) before breaking to clearly mark the extent of crack growth.

Typical load-versus-displacement and crack-extension records are shown in Fig. 2, in this case for a 1T specimen. Note that in the initial 10T specimen test, the d-c EP data were found to be unsatisfactory, probably because of inadequate electrical insulation of the anti-buckling plates. Crack growth records for these tests were obtained from markings on the fracture surface produced by periodic unloadings. The point of crack initiation was then determined by extrapolating the crack extension versus displacement curve to zero crack extension. In subsequent large planform C(T) tests, the buckling plates had sufficient electrical insulation so that good d-c EP data were obtained. For those tests in which satisfactory d-c EP values (U) were obtained as a function of displacement (δ), the initiation point U_0 was taken to be the point of departure from linearity of the U versus δ curve. Crack growth beyond initiation was calculated from U/U_0 using the expression given by Johnson [4]. The term $2Y$ in this expression, which represents spacing of the voltage probe connections, was allowed to vary in proportion to δ as the test progressed [5]. ASTM E 813-81 states that the calculated and actual crack growth values should agree within 15% at the end of the test. This value was met in all but two tests, where it was exceeded slightly. It might be argued that a relation, other than that of Ref *4*, could provide a better agreement. However, the fact that the calculation sometimes overpredicted the actual crack growth by as much as 15%, and at other times underpredicted by as much as 18%, suggests that the Johnson expression is reasonably good on the average, but a nonlinear correction may be worth pursuing.

---

[2] Original measurements made in English units.

FIG. 1—*Photographs of planform C(T) specimen tests. (a) 1T, 3T, and 10T stainless steel nonside-grooved C(T) specimens. (b) 10T C(T) with antibuckling plates.*

TABLE 1—*Listing of a series of planform C(T) specimen tests.*

Material	C(T) Specimen Sizes	Thickness, mm (in.)	Sidegrooved
304 Stainless Base	1T, 3T, 10T	25.4 (1.0)	No
304 Stainless Base	1T, 3T, 10T	25.4 (1.0)	Yes
304 Stainless Steel	0.5T, 1.5T, 3T	9.65 (0.38)	No
304 Stainless Base GTAW	0.5T, 1.5T, 3T	9.65 (0.38)	No
304 Stainless SAW	1T, 3T, 9.5T	25.4 (1.0)	No
A516 Gr. 70 Base	1T, 3T, 10T	25.4 (1.0)	No
A516 Gr. 70 Base	1T, 3T, 10T	25.4 (1.0)	Yes
A516 Gr. 70 SAW	1T, 3T, 9.5T	25.4 (1.0)	No

The experimental load, displacement, and crack extension records were used to generate J_D- and J_M-resistance curves by appropriate estimation schemes. The estimation scheme results are discussed in the following section.

Calculated *J-R* Curves from Experimental Data

In elastic-plastic fracture mechanics practice, *J-R* curves are usually developed using experimental data of the kind shown in Fig. 2 together with a *J*-estimation method. For C(T) specimens, ASTM E 1152 provides a method for calculating *J*, called J_D, as a function of Δa. The method, originally proposed by Ernst and Paris [6], includes a correction term to account for crack extension. This method was used to develop *J-R* curves for each of the specimens.

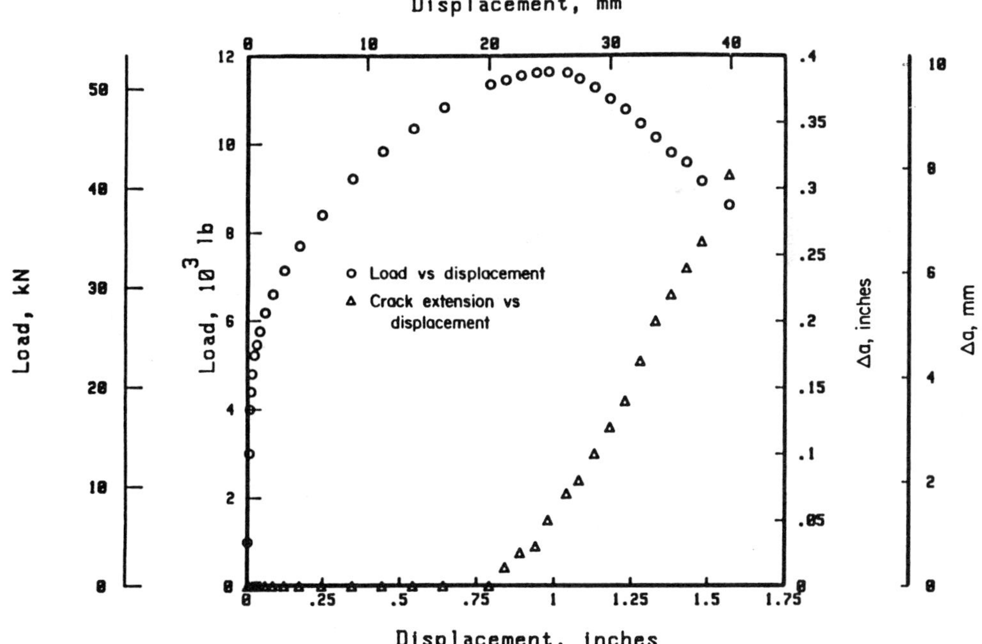

FIG. 2—*Typical load-displacement-crack growth data used in analyses. (Data are from nonside-grooved 1T C(T) stainless steel specimen.)*

The expression for J_D used in the calculation is

$$J_{D(n+1)} = [J_{D(n)} + (\eta/b)_n(A_{n,n+1})/B_N][1 - (\gamma/b)_n(a_{n+1} - a_n)] \tag{1}$$

where A is the area under the load-versus-displacement curve, and B_N is the net thickness. This is the ASTM E 813-81 relation. The current ASTM E 1152 J-R curve relation was also used. For the materials in this effort, there was insignificant difference between the two equations.

The subscripts n and $n + 1$ relate to the test record increments, and the parameters η, γ, and b (defined below) are updated at each increment:

$$\eta = 2 + 0.522\, b/w$$
$$\gamma = 1 + 0.76\, b/w$$
$$b = [w - (a_0 + \Delta a)]$$

The modified J, or J_M [7], was the result of an investigation aimed at finding a parameter which would characterize crack growth beyond that characterized by J_D. J_M is defined as

$$J_M = J_D - \int_{a_0}^{a} [\partial(J_D - G)/\partial a]_{\delta_{\text{pl}}}\, da \tag{2}$$

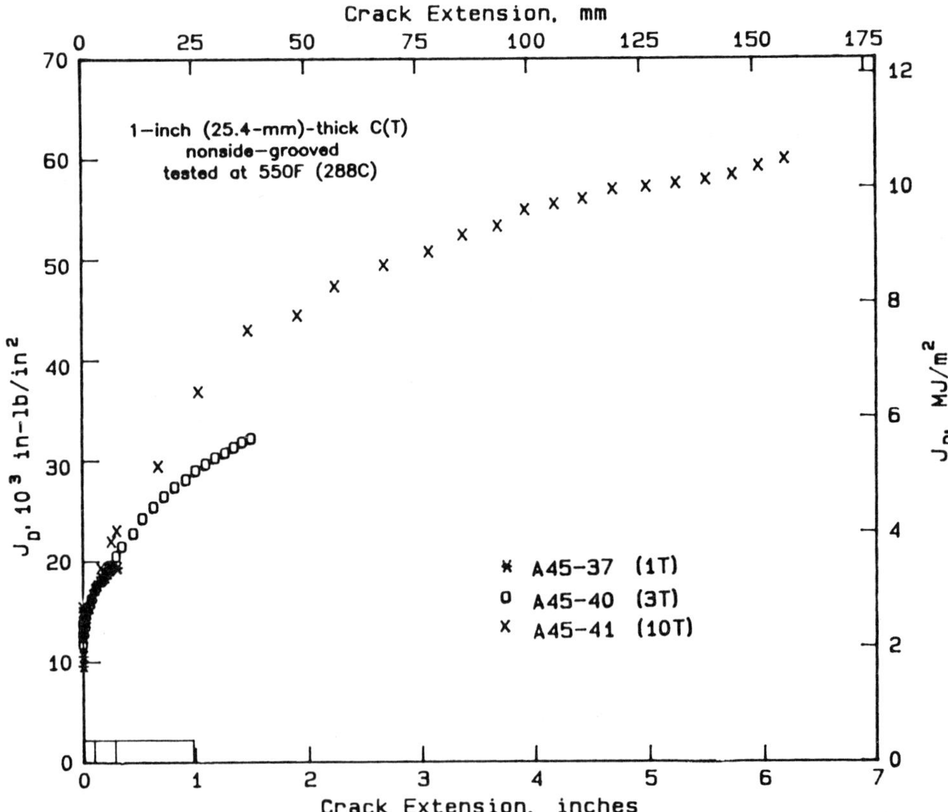

FIG. 3—J_D-R *curves from 25.4-mm (1-in.)-thick nonside-grooved TP304 stainless steel C(T) tests. (Boxes in lower left corner are ASTM validity limits.)*

where G is the Griffith's linear elastic energy release rate. Therefore the quantity differentiated with respect to a is the plastic component of J_D, and δ_{pl} is the plastic component of the displacement. In order to ensure that these calculations were correct, the J_D- and J_M-resistance curves were calculated by several organizations using data generated at Battelle Columbus Division. A miniature round-robin was conducted with Westinghouse, Oak Ridge National Laboratory, Materials Engineering Associates, and David Taylor Research Center. In this round-robin the same load-displacement crack-growth data were given to each group for the 10T nonside-grooved stainless steel compact specimen. The calculated *J-R* curves were found to be in close agreement [*1*].

Nonside-Grooved 25.4-mm-Thick TP304 Stainless Steel Results

These efforts involved tests on 25.4-mm (1 in.)-thick Type 304 stainless steel base metal. The tests were conducted at 288°C (550°F). The orientation of these specimens from the plate material was the same as if the plate was seam welded into pipe, with the rolling direction in the pipe axis, and the crack was growing in the circumferential direction. In all C(T) tests the crack was grown to 40 to 60% of the ligament.

For the nonside-grooved Type 304 stainless steel specimens, the J_D-R curves showed slight specimen size dependence (Fig. 3). The 1T C(T) specimen J_D-R curve was very close to the 3T C(T) specimen J_D-R curve. The J_M-R curves indicate that the smaller specimens tend to

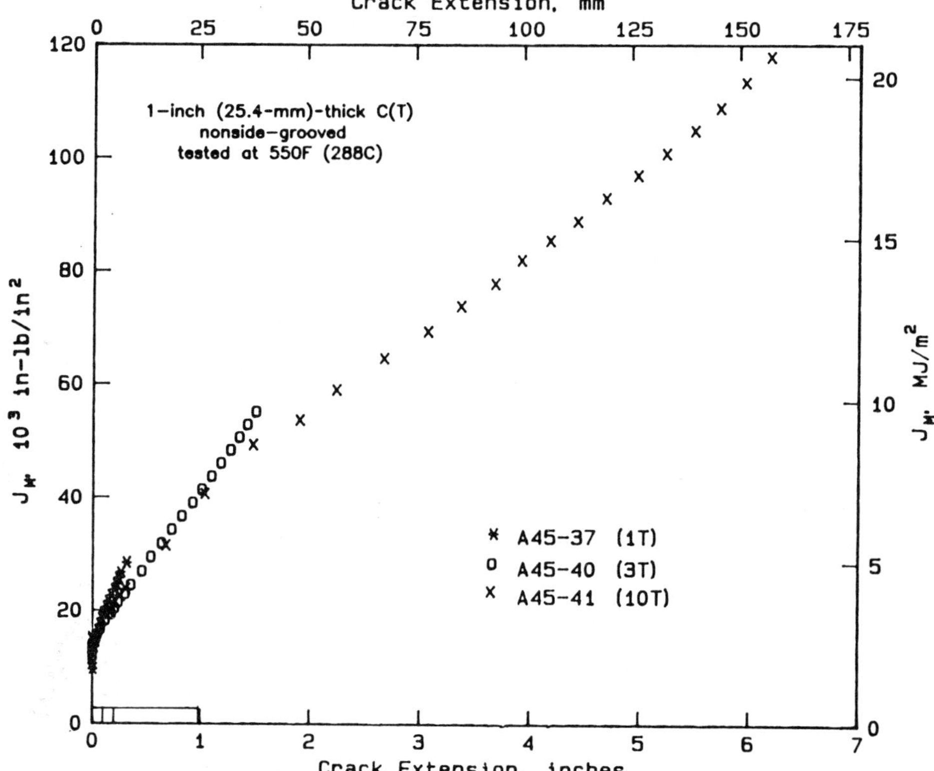

FIG. 4—J_M-R *curves from 25.4-mm (1-in.)-thick nonside-grooved TP304 stainless steel C(T) tests. (Boxes in lower left corner are ASTM validity limits.)*

display higher slopes than for the larger specimens (Fig. 4). This could result in a nonconservative extrapolation of the J-R curve from the smaller specimens.

Side-Grooved 25.4-mm-Thick TP304 Stainless Steel Results

These specimens came from the same plate as the nonside-grooved specimens described above. The side-grooved Type 304 stainless steel C(T) specimen J_D- and J_M-R curves are shown in Figs. 5 and 6. With large crack growth the 1T specimen J_D-R curve reaches a maximum and then decreases. This is frequently seen when the crack growth data are well beyond the ASTM limits. The 3T specimen agreed with the 10T specimen until the crack growth exceeded 30% of the ligament.

The side-grooved J_M-R curves are shown in Fig. 6. As with the nonside-grooved specimens, the 1T specimen shows a steeper slope than the larger specimens. The 10T specimen slope is significantly less than even the 3T specimen in this test series.

Nonside-Grooved 9.5-mm-Thick TP304 Stainless Steel Results

The next series of planform C(T) specimen J-R curves was developed in an evaluation of stainless steel TIG weld specimens. Base metal TP304 stainless steel specimens were also

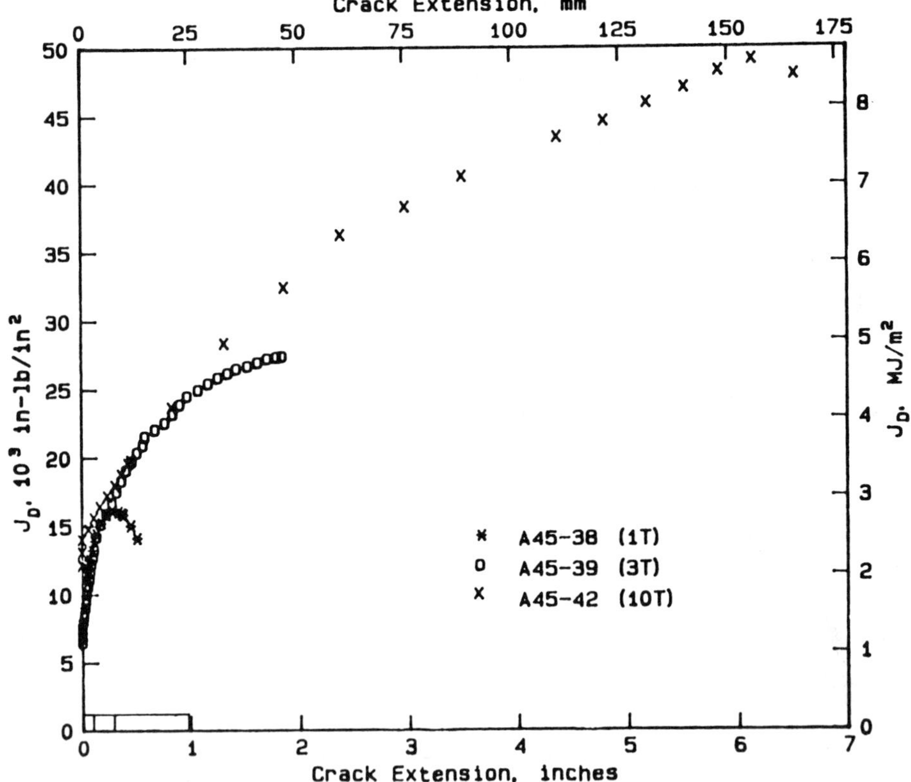

FIG. 5—J_D-R *curves from 25.4-mm (1-in.)-thick 20% side-grooved TP304 stainless steel C(T) tests. (Boxes in lower left corner are ASTM validity limits.)*

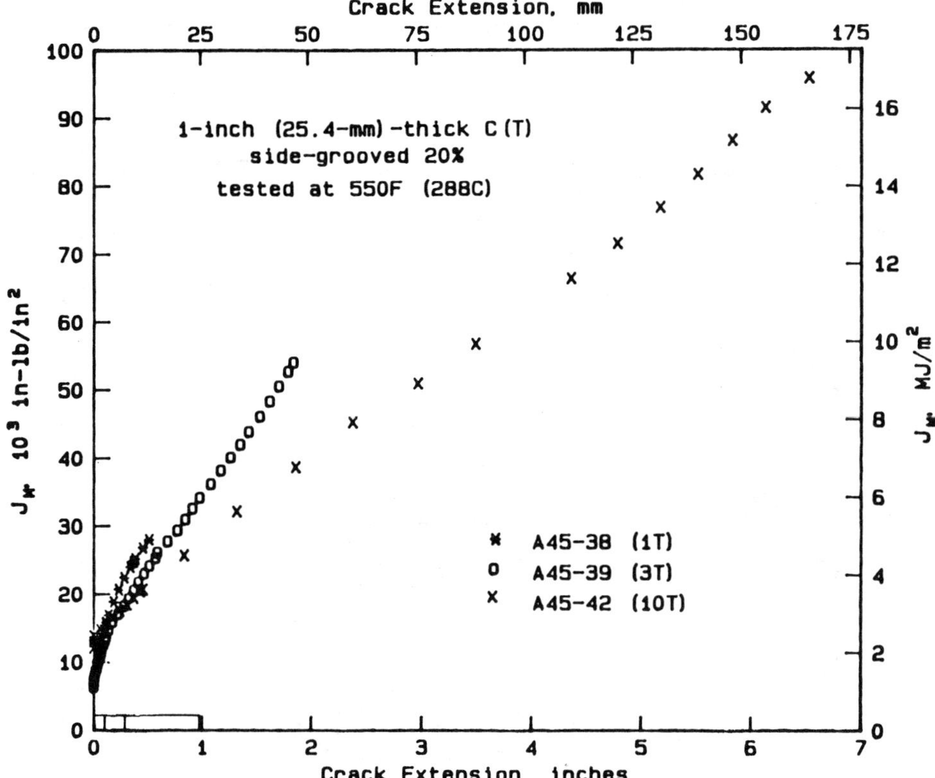

FIG. 6—J_M-R *curves from 25.4-mm (1-in.)-thick 20% side-grooved TP304 stainless steel C(T) tests. (Boxes in lower left corner are ASTM validity limits.)*

tested for comparison to the weld specimens. All specimens were nonside-grooved. The thickness of these specimens was 9.5 mm (0.38 in.). The planform specimen sizes were 0.5T, 1.5T, and 3T C(T) specimens. All specimens were tested at 288°C (550°F). Finite element analyses were conducted on the 0.5T and 3T C(T) specimens.

For the base metal specimens, the 0.5T specimens had a lower J_i than the larger specimens.[3] These J_i values were significantly lower than the J_i values of the 25.4-mm (1-in.)-thick specimens cited in the first series of planform C(T) specimen testing. It is not known whether this difference is due to thickness or normal variations in toughness for Type 304 stainless steel.

For the base metal specimens, the J_D-R curves showed significant geometry dependence (Fig. 7). The geometry dependence in this case was much greater than for the 25.4-mm (1-in.)-thick Type 304 stainless steel specimens.

The J_M-R curves showed less goemetry dependence than the J_D-R curves (Fig. 8). The smallest specimen showed a greater slope than the larger specimens, which is a nonconservative trend.

---

[3] The term J_i is used to define J at initiation which may not be a valid ASTM J_{Ic} value.

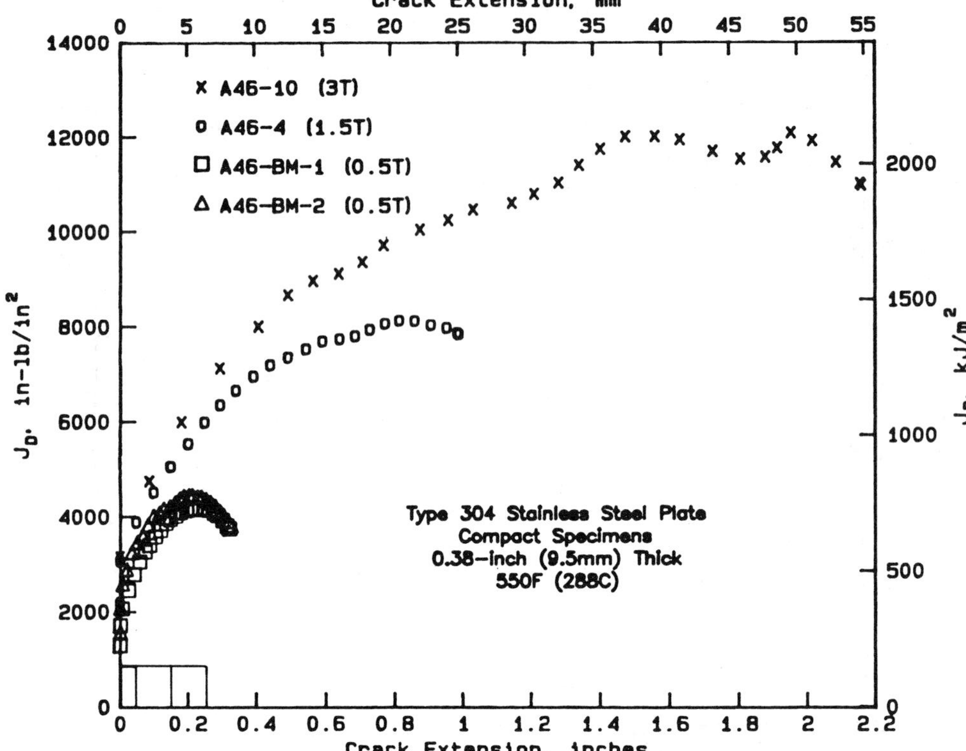

FIG. 7—J_D-R *curves from 9.5-mm (0.38-in.)-thick nonside-grooved TP304 stainless steel C(T) tests. (Boxes in lower left corner are ASTM validity limits.)*

Nonside-Grooved 9.5-mm-Thick TP304 Stainless Steel GTA Weld

These specimens were made from the same plate material used in the specimens described in the immediately preceding section. For the gas tungsten arc weld (GTAW) specimens, the 0.5T C(T) specimens had a lower J_i than the larger specimens, with the 1.5T C(T) specimen having the highest value. The 0.5T C(T) specimen had the weld crown machined off, whereas the larger specimens had the weld crown left on. In the case of the specimens with the weld crown left on, it was assumed that the thickness (B) of the specimen was that of the weld. Since J-R curves from these specimens were to be compared to welded pipe test J-R curves with a weld crown left on, it was desired to have the weld crown left on the larger specimens.

For the GTA weld specimens, the general shape of the entire J_D-R curves were not as well ordered with the specimen size, perhaps because of the influence of the weld and the weld crown left on the larger specimens (Fig. 9). The initial slope of the 0.5T C(T) specimens was steeper than that for the larger specimens, which is consistent with the previously discussed base metal J-R curves.

For the GTA weld specimens, the J_M-R curves were much steeper for the 0.5T C(T) specimens (Fig. 10). The 0.5T specimens also showed an upward hooking, which was a concern raised at a recent J_M-R curve workshop [8].

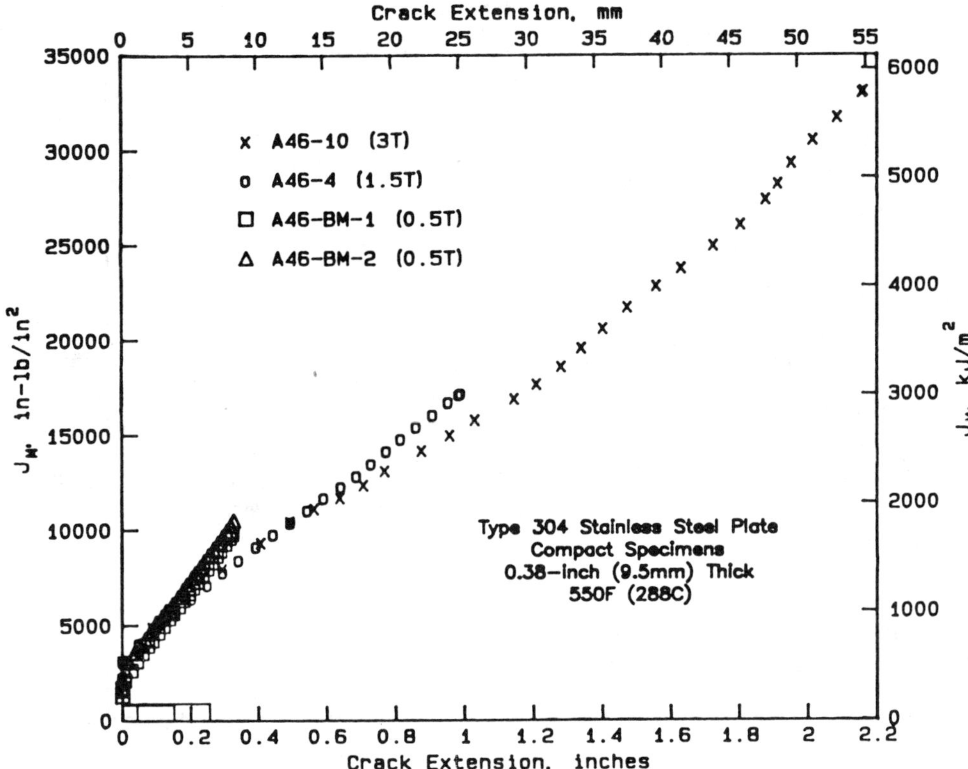

FIG. 8—J_M-R *curves from 9.5-mm (0.38-in.)-thick nonside-grooved TP304 stainless steel C(T) tests. (Boxes in lower left corner are ASTM validity limits.)*

Nonside-Grooved 25.4-mm-Thick TP304 Stainless Steel SAW Results

The next series of planform C(T) specimen tests were on a stainless steel SAW. The weld procedure used came from a U.S. BWR specification. The plate material in the welded specimens was the same as used in the first series of tests described on 25.4-mm (1-in.)-thick Type 304 stainless steel. Only nonside-grooved specimens of 1T, 3T, and 9.5T planform size were tested. The weld crown was left on 3T and 9.5T specimens. All specimens were tested at 288°C (550°F).

The larger planform specimens had a larger J_i. This is the general trend of the other data. With large amounts of crack growth, there was significant geometry dependence of the J_D-R curves (Fig. 11).

Once the crack growth was greater that 6.3 mm (0.25 in.), the J_M-R curves were in close agreement (Fig. 12). The 1T C(T) specimens in this case had a lower slope, which is not consistent with the general trend of the other platform specimen test series.

Nonside-Grooved 25.4-mm Thick A516 Grade 70 Results

These efforts involved tests on 25.4-mm (1-inch)-thick A516 Grade 70 carbon steel. The tests were conducted at 288°C (550°F). The orientation of these specimens from the plate

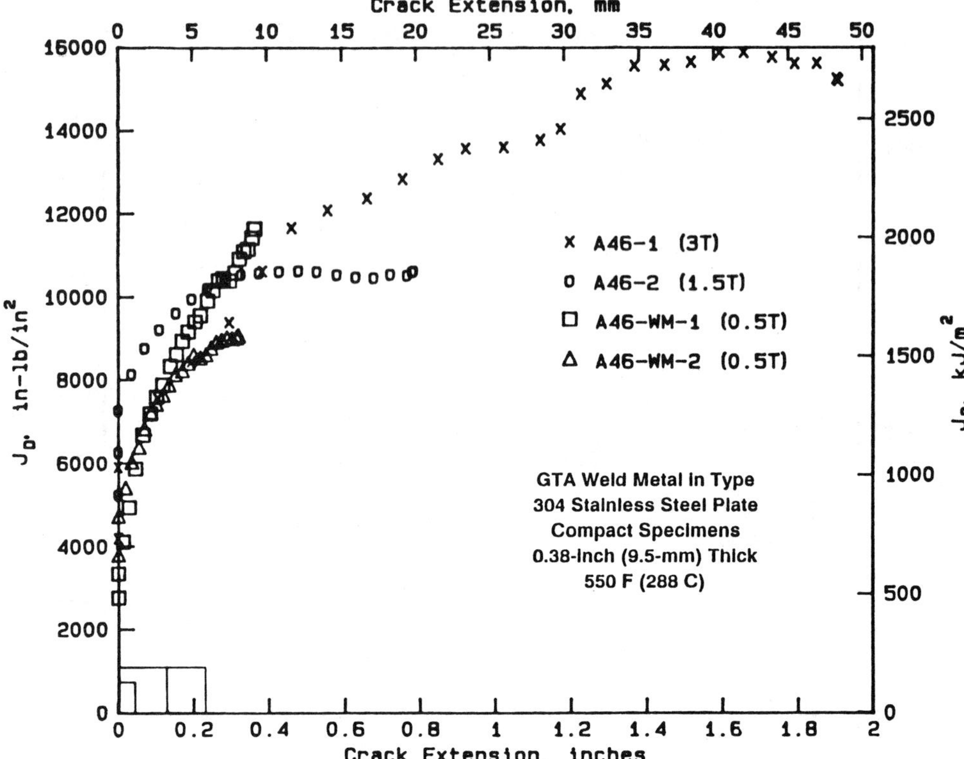

FIG. 9—J_D-R *curves from 9.5-mm (0.38-in.)-thick nonside-grooved TP304 GTA welded stainless steel C(T) tests. (Boxes in lower left corner are ASTM validity limits.)*

material was the same as if the plate was seam welded into pipe, with the rolling direction in the pipe axis, and the crack was growing in the circumferential direction.

The nonside-grooved 3T and 10T C(T) A516 Grade 70 ferritic steel specimens had the crack quickly turn 90 deg from the intended crack plane. This was the low toughness direction in the plate. This was similar to the fracture behavior observed in a cold-leg (also an A516 Grade 70 ferritic steel) circumferential through-wall-cracked pipe test reported in Ref *1*. This crack turning made it impossible to calculate the *J-R* curve from the C(T) experiments. Only the J_i values could be determined from this series of planform C(T) tests.

For the carbon steel specimens, the J_i values were clearly dependent on the specimen size. The stainless steel specimens showed less effect of specimen size on J_i, but generally were more sensitive to side-grooving. With the exception of one specimen, all of the side-grooved specimens from both materials had a lower J_i than the nonside-grooved specimens. These data are shown in Fig. 13.

Side-Grooved 25.4-mm-Thick A516 Grade 70 Results

For the side-grooved A-516 Grade 70 ferritic steel C(T) specimens, both J_D and the modified J, J_M resistance curves were calculated. The specimens have a total side-grooving depth

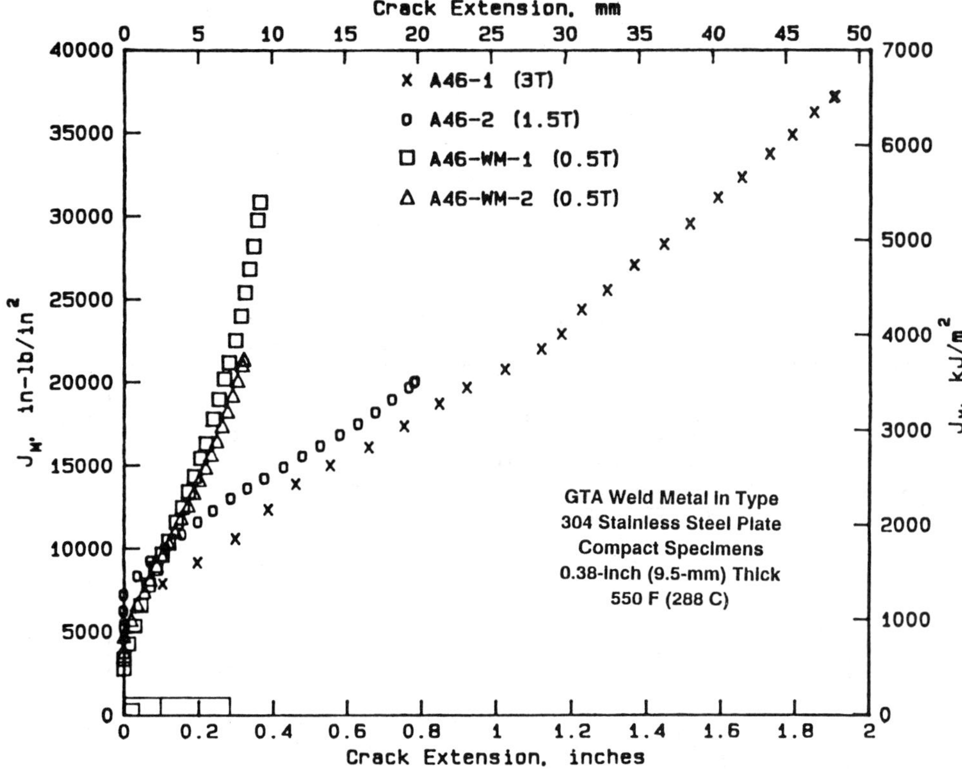

FIG. 10—J_M-R *curves from 9.5-mm (0.38-in.)-thick nonside-grooved TP304 stainless steel C(T) tests. (Boxes in lower left corner are ASTM validity limits.)*

of 20% of the thickness (10% per side). The J_D-R curves showed significant size dependency, the trend being that the smaller the specimen the lower the J_D-R curve (Fig. 14). The J_M-R curves showed much less size dependency and after 6.3 mm (0.25 in.) of crack growth all of the different specimen J_M-R curves were virtually identical (Fig. 15). Note that for the 1T C(T) specimens, this amount of crack growth is about 25% of the ligament, which far exceeds the ASTM limit of 10% of the ligament.

Nonside-Grooved 25.4-mm-Thick A516 Grade 70 SAW Results

This weld was made using a procedure obtained from a U.S. PWR vendor. The weldments were made in the same plate used in the first series of planform C(T) specimens described in this section. The 3T and 9.5T specimens were tested with the weld crown left on. All specimen sizes were tested without side grooves. The test temperature was 288°C (550°F). The results of these tests are summarized below.

The J_D-R curves are shown in Fig. 16. This was the lowest toughness material tested in the Degraded Piping Program [*1*]. The initiation toughness of the larger specimens was higher than for the smaller specimens. This is consistent with the other planform C(T) series. The 1T C(T) specimen J_D-R curves agreed well with the 9.5T C(T) specimen *J-R* curve.

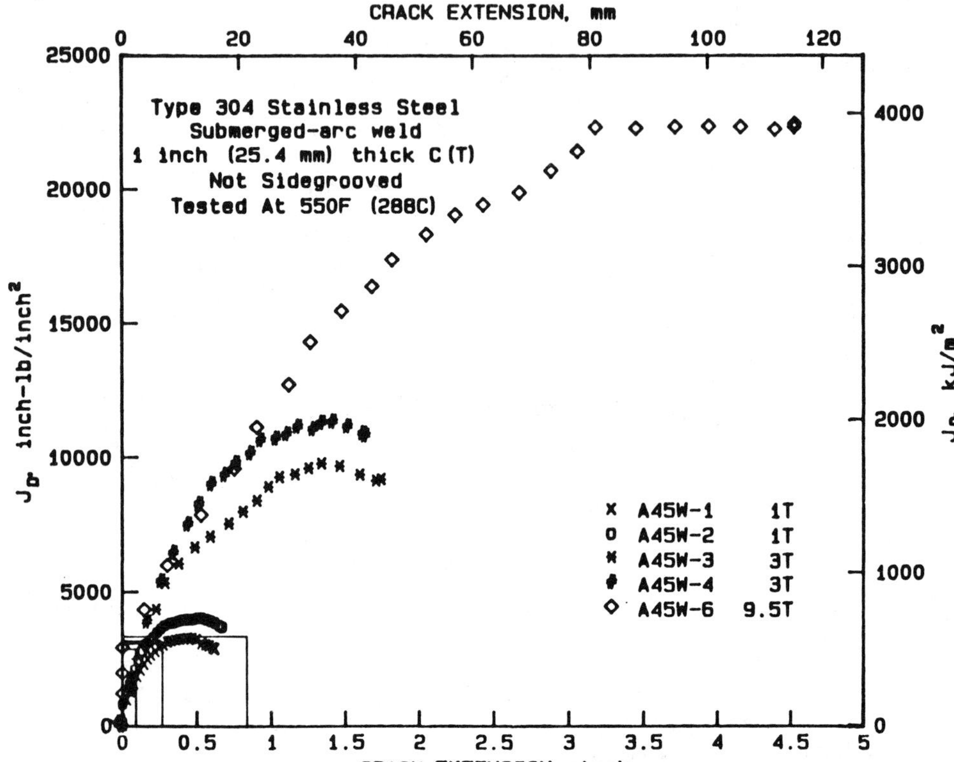

FIG. 11—J_D-R *curves from 25.4-mm (1-in.)-thick nonside-grooved TP304 stainless steel SAW C(T) tests. (Boxes in lower left corner are ASTM validity limits.)*

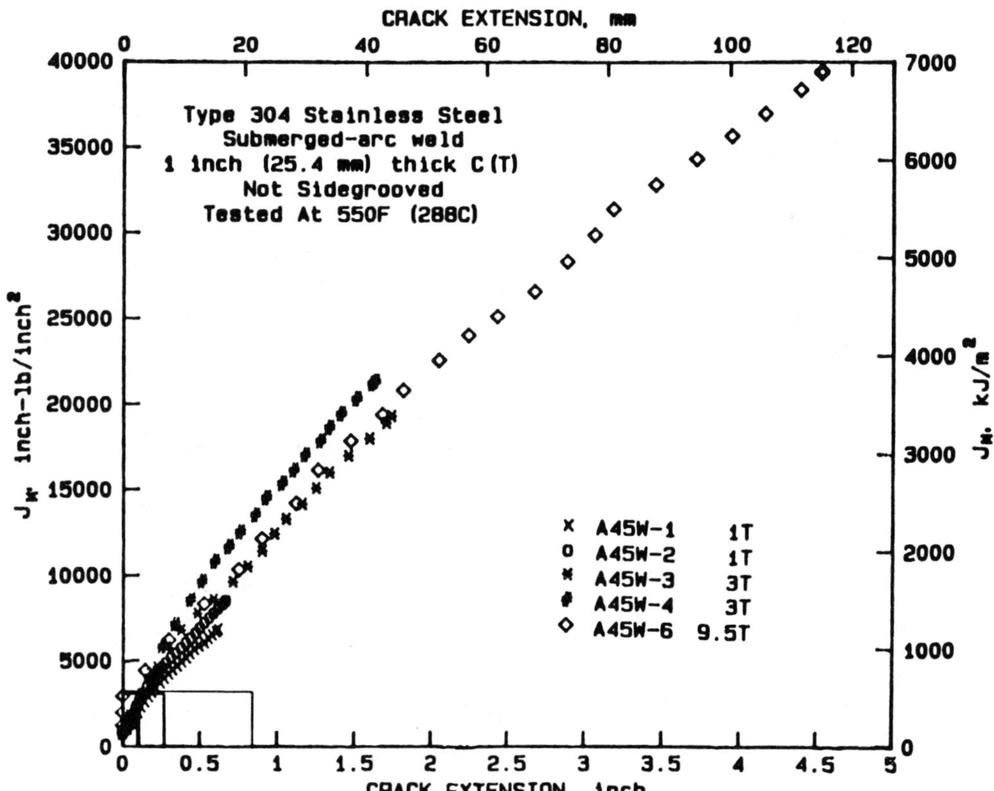

FIG. 12—J_M-R *curves from 25.4-mm (1-in.)-thick nonside-grooved TP304 stainless steel SAW C(T) tests. (Boxes in lower left corner are ASTM validity limits.)*

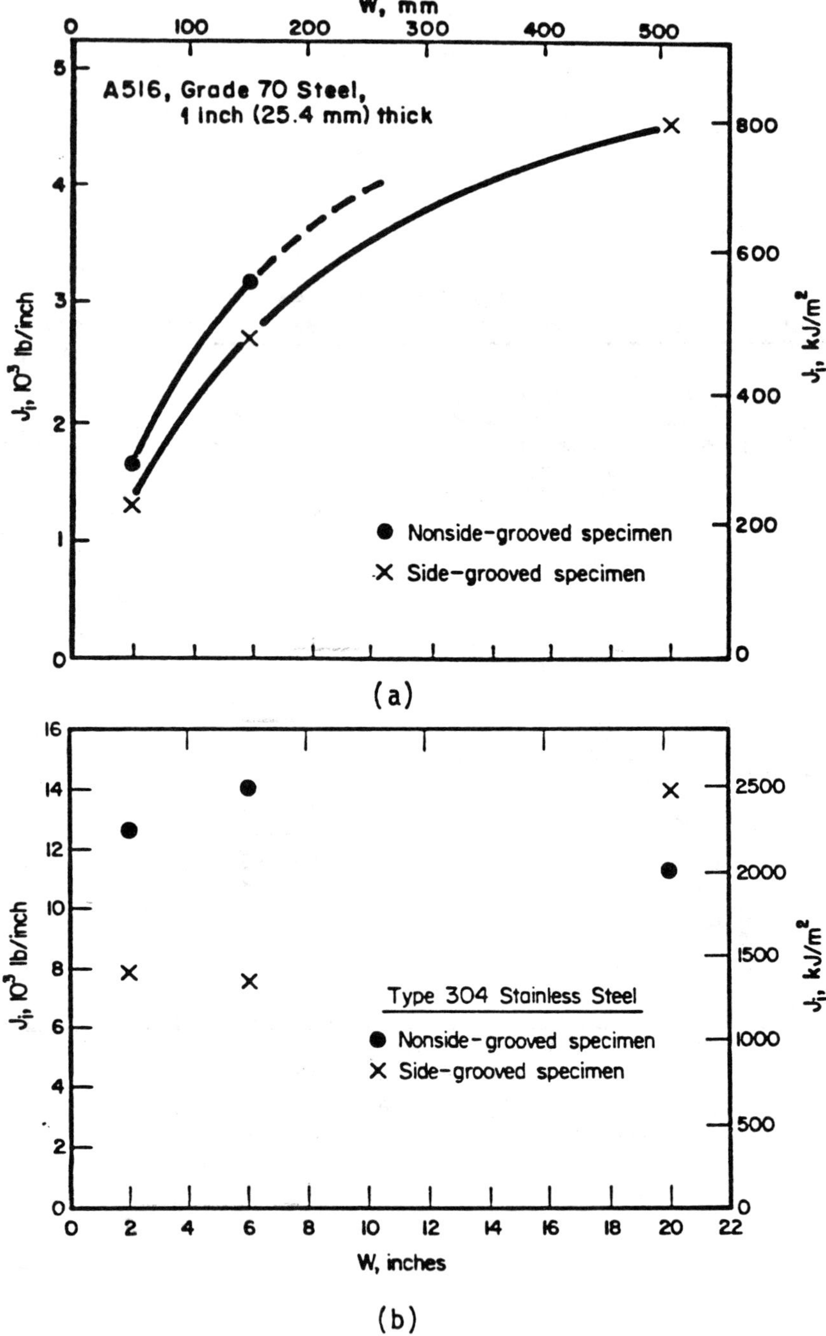

FIG. 13—J_i *versus specimen size from 25.4-mm (1-in.)-thick A516 Grade 70 C(T) tests.*

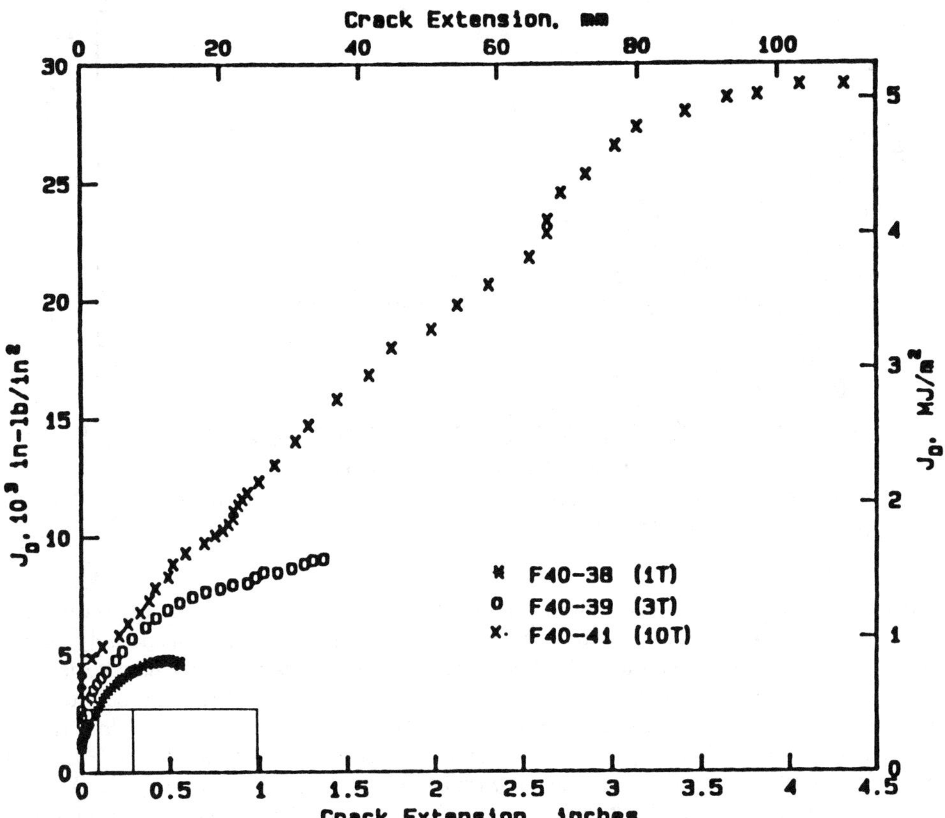

FIG. 14—J_D-R *curves from 25.4-mm (1-in.)-thick side-grooved A516 Grade 70 carbon steel C(T) tests. (Boxes in lower left corner are ASTM validity limits.)*

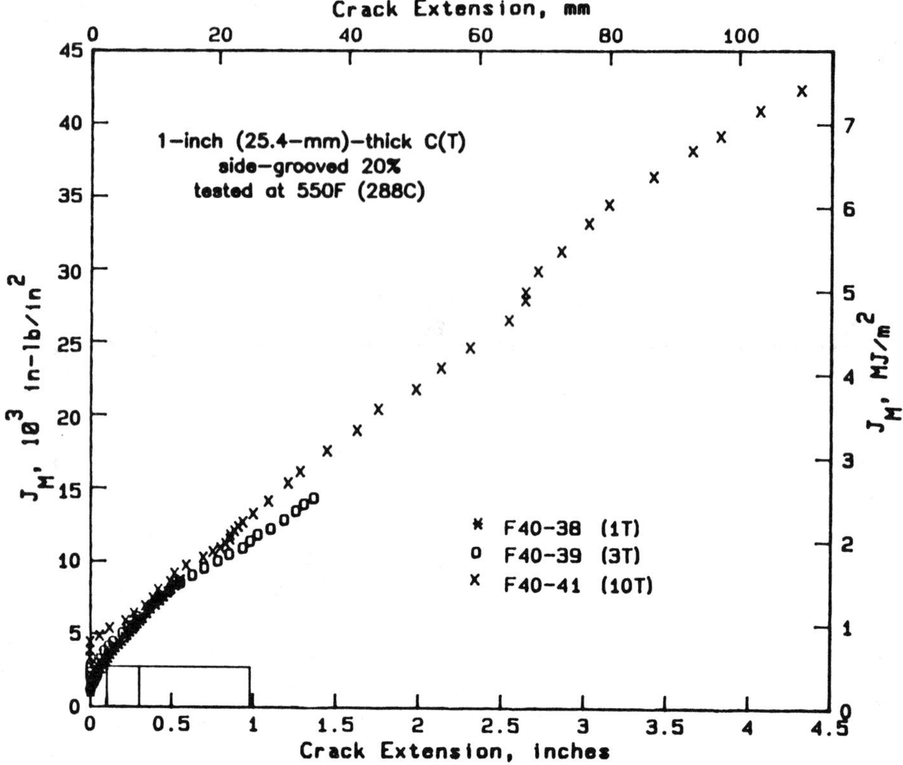

FIG. 15—J_M-R *curves from 25.4-mm (1-in.)-thick nonside-grooved A516 Grade 70 carbon steel C(T) tests. (Boxes in lower left corner are ASTM validity limits.)*

For the J_M-R curves, the initial slopes of the *J-R* curves were in agreement. With more than 5.1 mm (0.2 in.) of crack growth, the 1T specimen J_M-R curve was considerably steeper than for the larger specimens (Fig. 17). This is potentially nonconservative behavior, depending on how the through-wall-cracked pipe behaves. A pipe test is currently planned to assess this effect in the Degraded Piping Program [1]. It can also be noted in Fig. 17 that the 1T specimen J_M-R curve is hooking upward, which is a concern with other J_M-R curve data [8]. This upward hooking of J_M-R curves occurs with large crack growth, above the ASTM limits. It was observed in low and high toughness materials (e.g., see the high toughness TIG weld data).

Evaluation of Extrapolation Procedures for *J-R* Curves

Three different extrapolation procedures are evaluated in this paper. Generally, these involve using the smallest specimen data to make an extrapolation, then comparing the extrapolated small specimen data to the larger specimen data.

NUREG-1061 Extrapolation Method

This extrapolation method was developed in a report generated by the NRC for LBB assessment of through-wall cracked pipe [9]. The steps in the procedure are as follows:

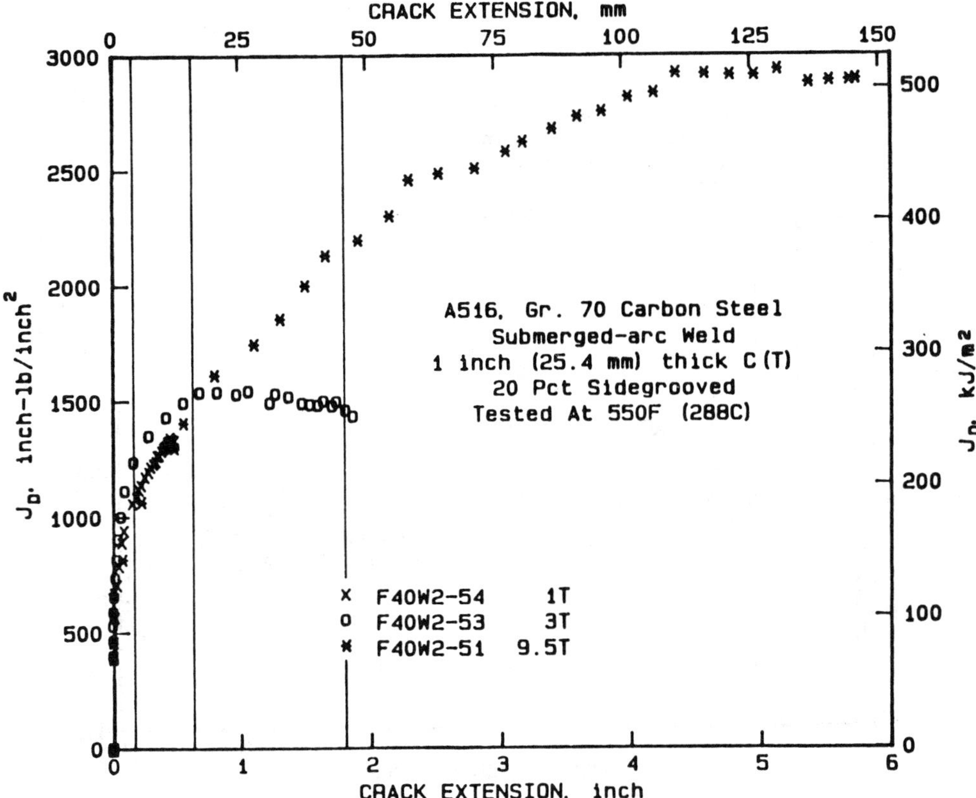

FIG. 16—J_D-R *curves from 25.4-mm (1-in.)-thick nonside-grooved A516 Grade 70 carbon steel SAW C(T) tests. (Lines on left side of figure represent ASTM validity limits.)*

1. Use the ASTM E 813 analysis procedure to calculate J_D-R curve.
2. The maximum allowable J, J_{max} value is when ω is greater than five, where

$$\omega = b \, dJ/(J \, da) \tag{3}$$

3. Plot the *J-R* curve data on a *J-dJ/da* or *J-T* diagram where

$$T = E \, dJ/(\sigma_f^2 \, da) \tag{4}$$

4. When $J = J_{max}$, linearly extrapolate the *J-T* curve as a tangent.
5. The linear extrapolation of the *J-T* curve is allowed up to a *J* value of $2J_{max}$.

This procedure is illustrated in Fig. 18. To evaluate this procedure, the planform specimen data were used. Since the stainless steel data were not valid by ASTM E 813, only the ferritic A516 base metal and SAW data could be used. To do this, the 1T C(T) specimen data were extrapolated by this procedure and compared to the larger specimen data.

The A516 Grade 70 base metal data and extrapolation are shown in Fig. 19, which shows that the NUREG 1061 Extrapolation Procedure is conservative relative to the 3T and 10T

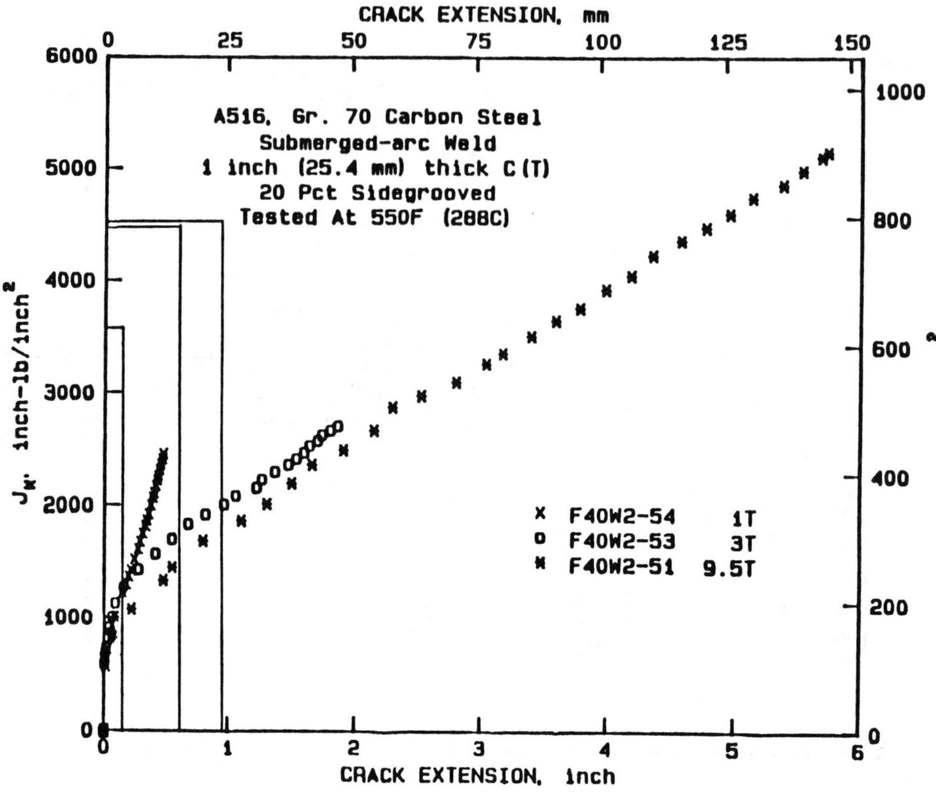

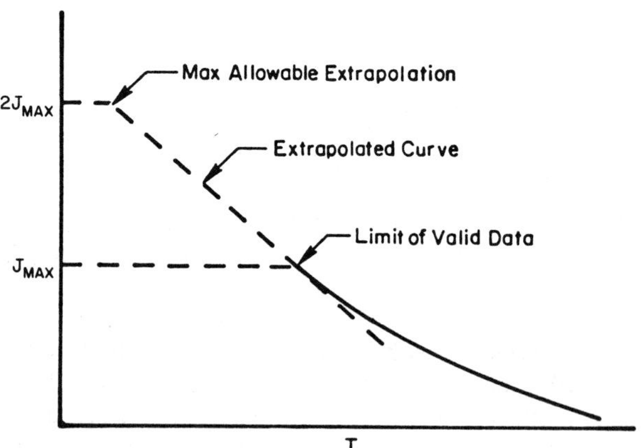

FIG. 17—J_M-R *curves from 25-4-mm (1-in.)-thick nonside-grooved A516 Grade 70 carbon steel SAW C(T) tests. (Boxes in left corner are ASTM validity limits.)*

FIG. 18—*Schematic of NUREG 1061 J-R curve extrapolation method.*

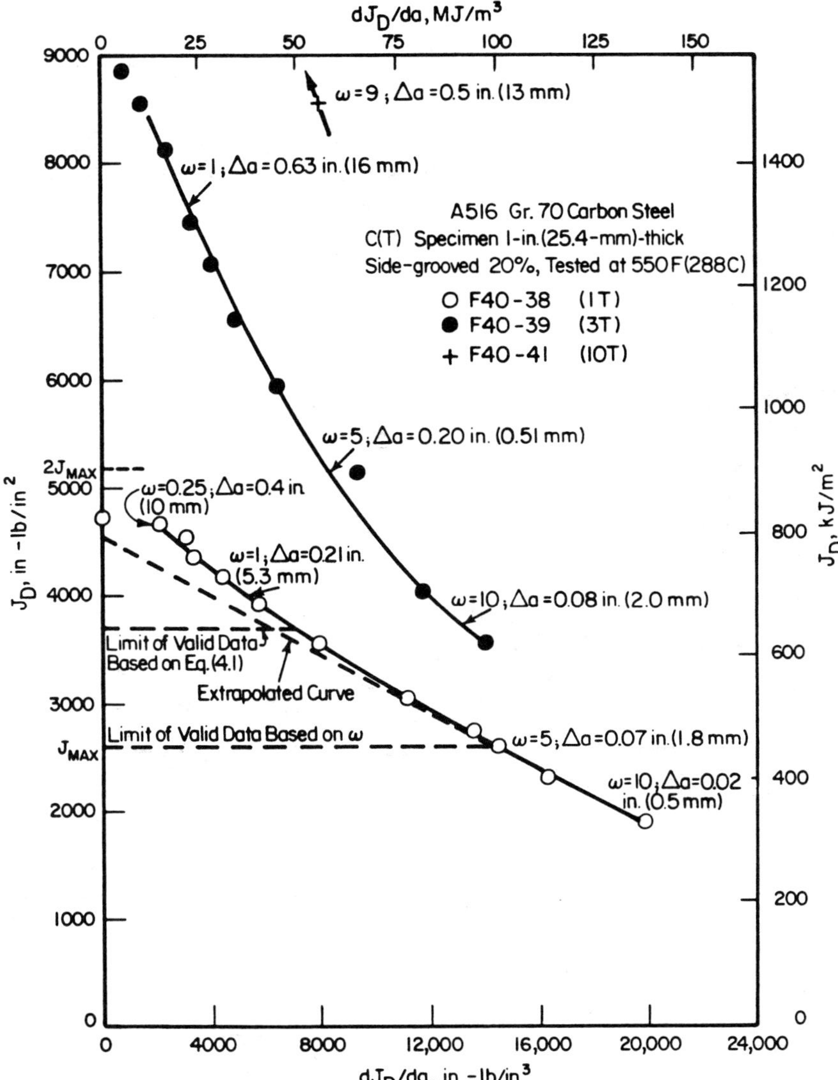

FIG. 19—*Evaluation of NUREG 1061 extrapolation method using A516 Grade 70 20% side-grooved C(T) data.*

specimen data. Note that the extrapolated curve reaches a tearing modulus value of zero before it reaches a J value of $2J_{max}$. Also note that the extrapolated $2J_{max}$ value only corresponds to a Δa of 10 mm (0.4 in.).

The A516 SAW data and extrapolated curves are shown in Fig. 20. The extrapolated J-T curve in this case was above the 10T specimen curve for small amounts of crack growth in the 10T specimen. This was an unexpected trend when looking at the data in J-T or J-dJ/da space. Also note that the $2J_{max}$ value corresponds to a Δa value of only 3.6 mm (0.14 in.).

In summary, this method appears to be too restrictive in that the extrapolated Δa values

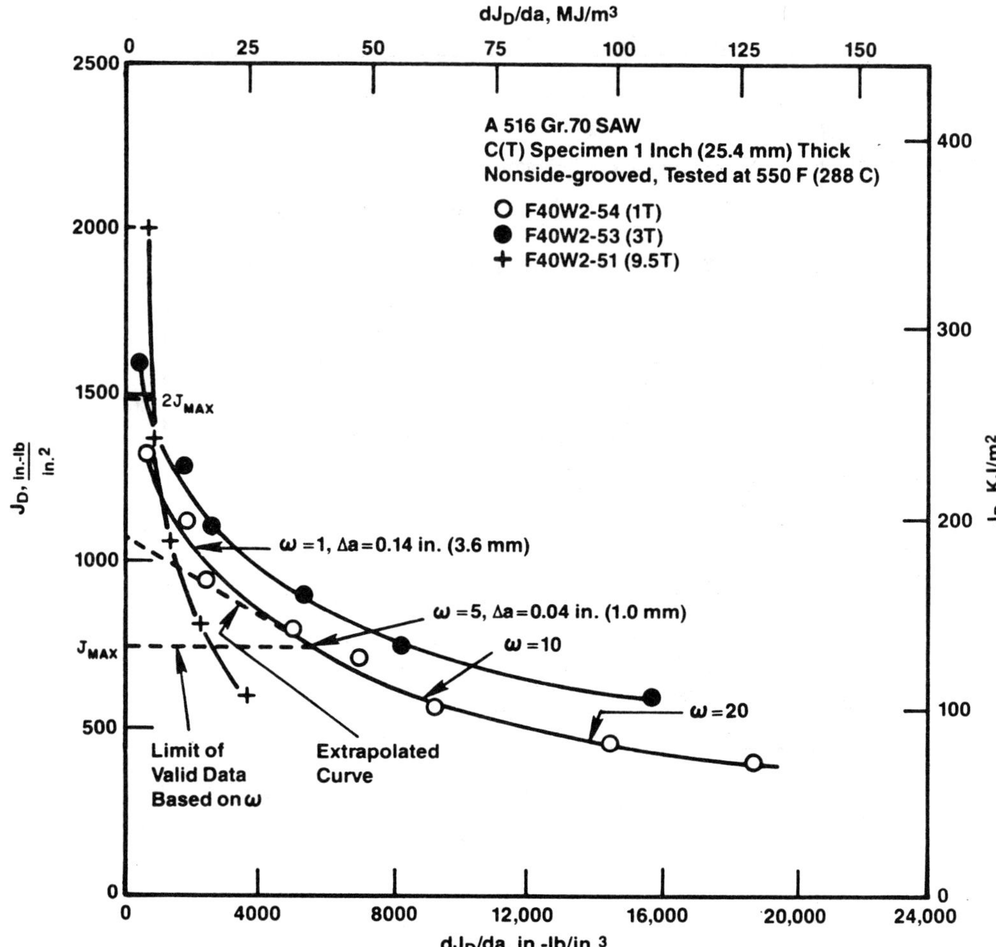

FIG. 20—*Evaluation of NUREG 1061 extrapolation method using A516 Grade 70 SAW nonside-grooved C(T) data.*

are far less than what is needed for pipe flaw evaluations. The extrapolated 1T specimen data being above some of the 10T data also raises some question about this procedure.

Tentative NRC A11 Extrapolation Method

The ASME Section XI Committee is currently developing a criterion to assess low upper-shelf fracture criteria for nuclear reactor pressure vessels. This evaluation procedure addresses an NRC Unresolved Safety Issue referred to as A11. In this procedure it is necessary to use either Charpy data or small C(T) specimen *J-R* curve data to predict the fracture behavior of a nuclear reactor pressure vessel. Typically only 1T C(T) data exists; however, analysis of the vessel suggests that with the safety factors on pressure it is necessary to account for 15.24 mm (0.6 in.) of crack growth. Hence some extrapolation procedure is needed. NRC efforts [8] have recently found that using the J_M-R curve may not be satisfactory. Under

NRC support, the David Taylor Research Center (DTRC) has coordinated efforts to develop a new extrapolation procedure where J_D-R curves are used.

This procedure involves the following steps:

1. The ASTM E 1152 J_D-R curve equations are used.
2. Test specimen results with crack growth of up to 30% of the ligament can be used.
3. Fit a power law to the J_D-R curve data from initiation to Δa of $0.3b$.
4. Use the power law curve to extrapolate the J-R curve to larger crack growth.

To evaluate this procedure, three sets of the planform tests conducted in this effort were used. In this evaluation the A516 Grade 70 base metal, A516 Grade 70 SAW, and the TP304 stainless steel SAW planform test sets were used. The smallest specimens were the 1T C(T) specimens in each set. The procedure was used to develop an extrapolated J-R curve from the 1T data for comparison to the larger specimen data.

The results of the comparisons are illustrated in Figs. 21 to 23. Figure 21 shows that the Tentative NRC A11 Extrapolation Method is conservative relative to the 3T and 10T planform A516 Grade 70 specimen J_D-R curves. It was very conservative for the 10T specimen J_D-R curve data at large amounts of crack growth.

For the A516 Grade 70 SAW specimen data, the Tentative NRC A11 Extrapolation

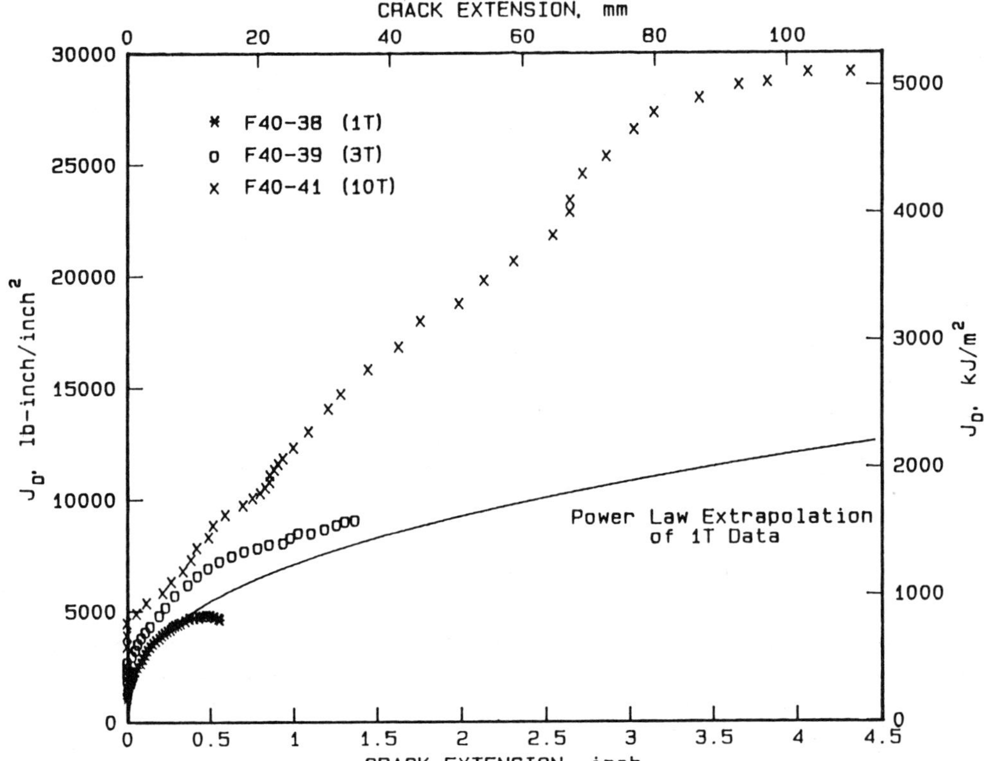

FIG. 21—*Evaluation of tentative NRC A11 extrapolation method using A516 Grade 70 20% side-grooved C(T) data.*

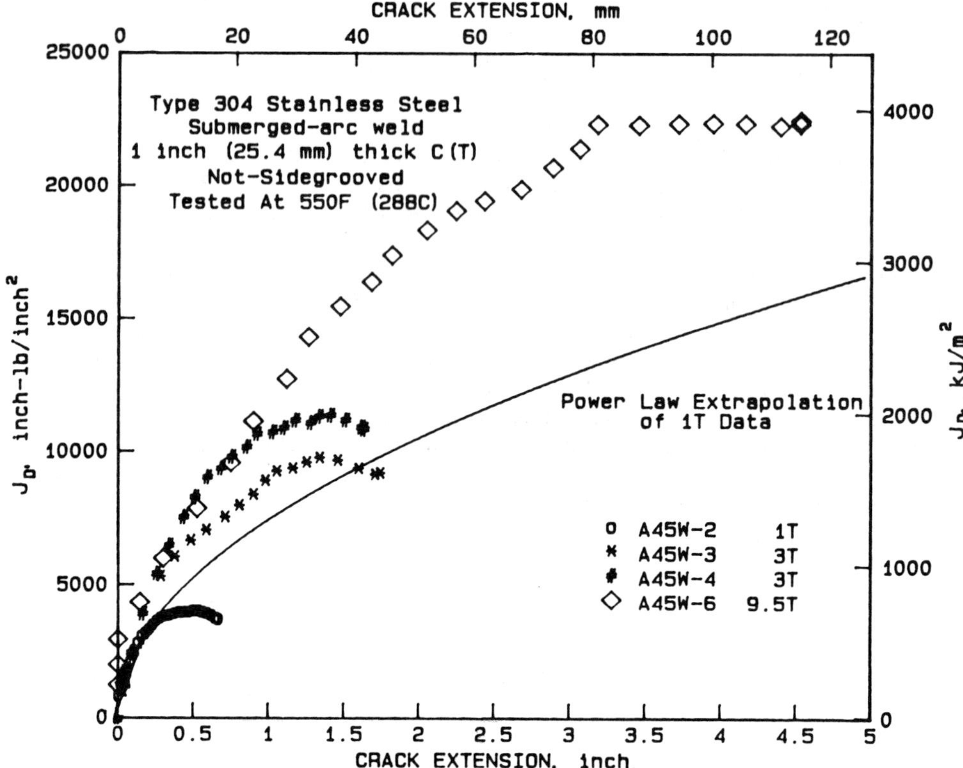

FIG. 22—*Evaluation of tentative NRC A11 extrapolation method using A516 Grade 70 SAW nonside-grooved C(T) data.*

Method is conservative for all the 9.5T data and the 3T data if Δa is less than $0.3b$. A crack growth of $0.3b$ for the 3T specimen is 22.86 mm (0.9 in.) (Fig. 22).

For the TP304 stainless steel SAW specimen data, the Tentative NRC A11 Extrapolation method is conservative for all the 10T data and the 3T data as long as Δa is less than $0.3b$. The extrapolated *J-R* curve is shown in Fig. 23.

These comparisons show that for the planform C(T) specimens the Tentative NRC A11 Extrapolation method is conversative for all the data evaluated. It was found to be more conservative for the largest planform C(T) specimens.

J_M-R Curve Correction for Planform Specimens

The J_M-R curves for each of the planform test sets were given previously. It can generally be seen that the smaller specimens gave a higher J_M-R curve than the larger specimens. This is schematically shown in Fig. 24. Only the stainless steel SAW small specimen J_M-R curve was lower than the other larger planform specimen size J_M-R curves. Hence extrapolating the small specimen J_M-R curve directly in most cases may be erroneous.

The potential geometry effects in the J_M-R curves were assessed by evaluating the constant value of the J_M-R curve slope in a normalized fashion. This slope was evaluated for crack growths of Δa less than $0.3b$. One normalizing parameter was the slope of the J_M-R curve of any specimen relative to the slope from the largest planform specimen tested. For simplicity T_M was used rather than dJ_M/da, where T_M is the tearing modulus using the J_M-R curve. The other normalizing parameter was the ratio of the planform specimen size to the standard specimen size for the planform specimen thickness; for example, a 25.4-mm (1-in.)-thick 10T specimen has a normalized specimen size ratio of 10.

The normalized results of the different series of planform tests are shown in Fig. 25. This shows that as the planform specimen becomes smaller, the T_M relative to that for the large specimen increases. Furthermore, Fig. 25 suggests that, if possible, a planform specimen should be four times larger than the standard specimen to get a lower bound T_M. Note that for the standard specimen size, a ratio of 1.0 in Fig. 25, there is considerable scatter in the ratio of T_M for the standard specimen to the T_M of the large planform specimen. If a large planform specimen test is not possible, one approach would be to use a correction of the small specimen T_M value. The average trend curve in Fig. 25 suggests that a reduction of the small specimen T_M value by a factor of 2.5 would be reasonable.

Since J_i is also somewhat sensitive to the planform size, possibly due to changes from plane strain to plane stress conditions, it is suggested to use the standard specimen J_i value to be

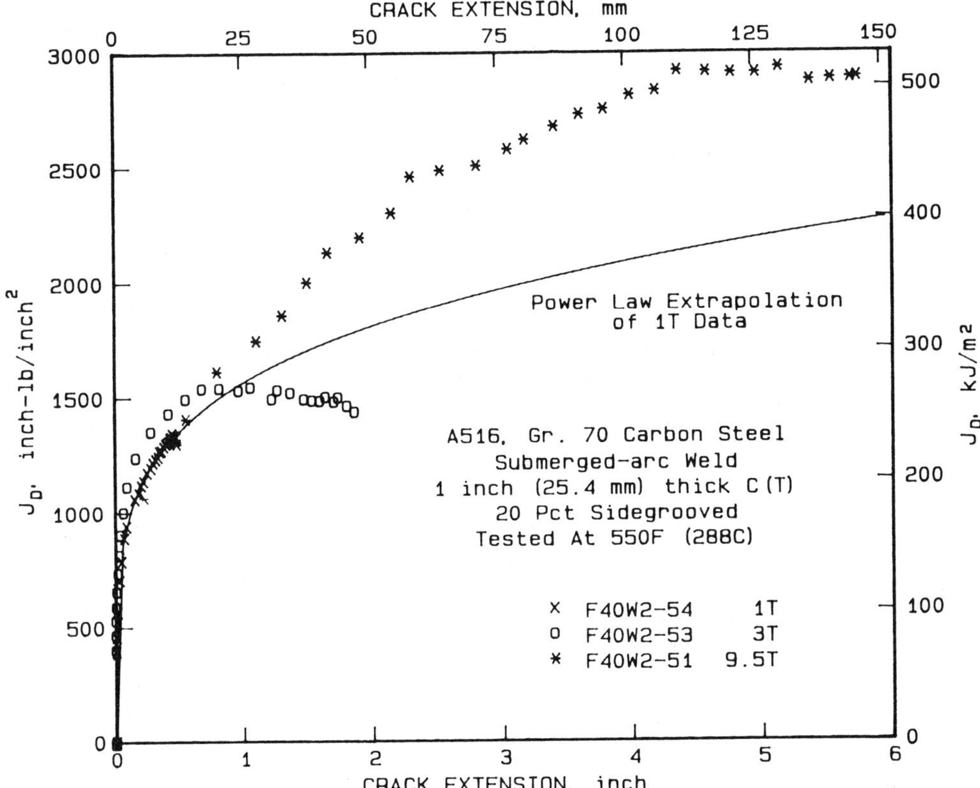

FIG. 23—*Evaluation of tentative NRC A11 extrapolation method using TP304 stainless steel 70 SAW nonside-grooved C(T) data.*

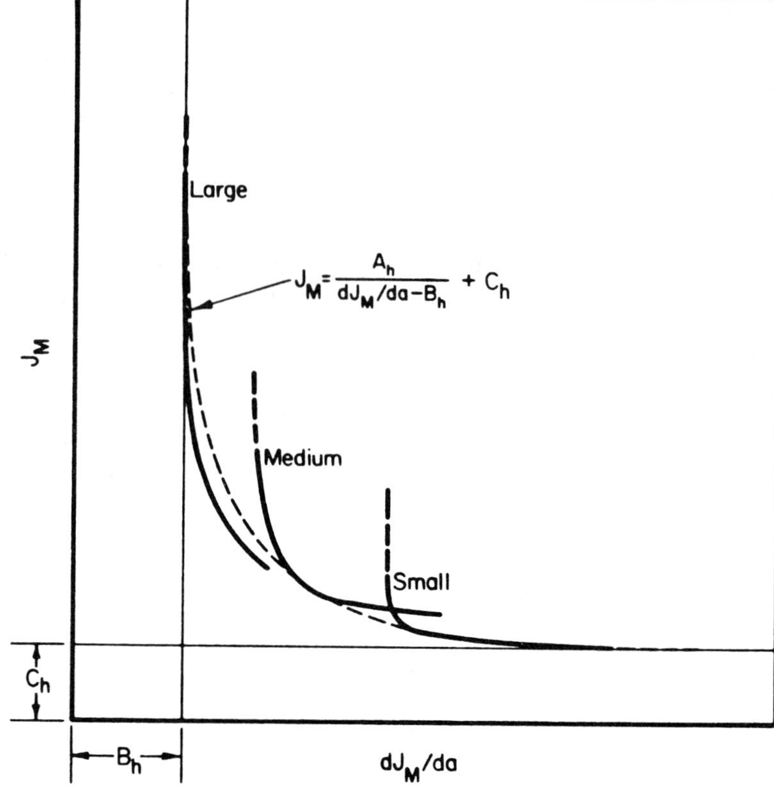

FIG. 24—*Schematic of trends of* J_M-R *curves from planform specimen tests.*

conservative. As Marschall suggested [2], the following relation could then be used to obtain a J_M-R curve to eliminate planform specimen size/constraint effects:

$$J_M = J_i + A_h/(dJ_M/da - B_h) \qquad (5)$$

where

A_h = a curve-fitting parameter (Fig. 24), and
B_h = large planform specimen (dJ_M/da) or the small specimen $(dJ_M/da)/2.5$.

Figures 26 and 27 show this curve-fitted equation as fit to the 25.4-mm (1-in.)-thick stainless steel planform specimen data and the 25.4-mm (1-in.)-thick A516 Grade 70 side-grooved planform C(T) specimen data.

Discussion and Conclusions

The results of this investigation were as follows:

1. From the planform specimen tests, it was found that for many of the materials, the initiation toughness increased with the size of the planform specimen size. This may be the

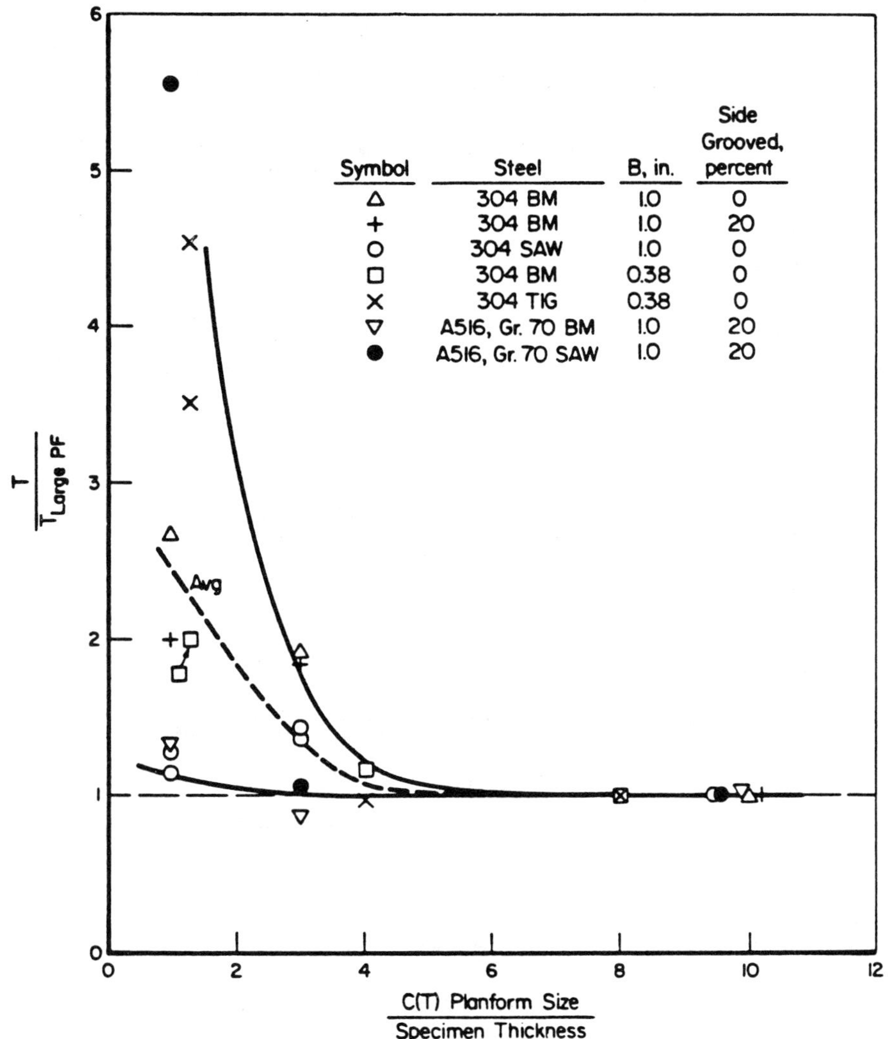

FIG. 25—*Geometry/constraint effects on* T_M *from planform specimen tests.*

result of changes from plane strain to plane stress conditions. It is suggested for conservative evaluations to use a smaller or standard size specimen.

2. As with some experiments from DTRC, it was found that some of the J_M-R curves hooked upward after Δa was greater than 30% of the ligament. Circumferentially cracked pipe test data by Kramer [10] also showed unusual crack growth behavior when Δa was greater than 30% of the ligament.

3. Smaller specimen J_D-R curves in the majority of cases were lower than those from larger specimens.

4. Smaller C(T) specimens generally gave higher dJ_M/da than larger planform specimens. This makes extrapolation of small specimen J_M-R curves nonconservative for large crack growth.

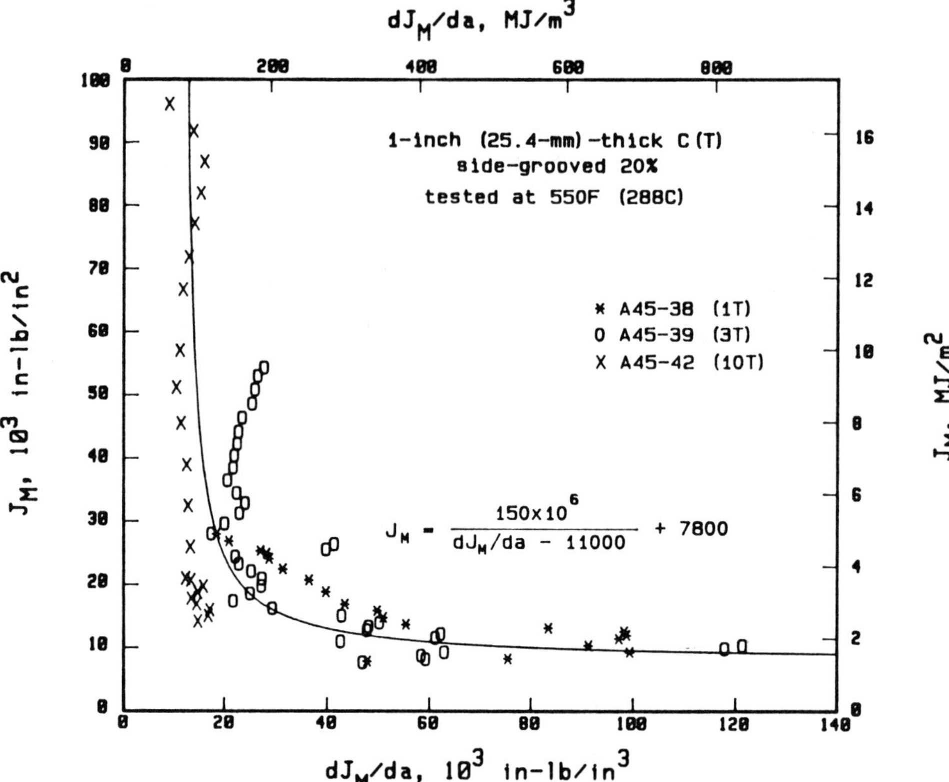

FIG. 26—*Hyperbolic fit of 25.4-mm (1-in.)-thick nonside-grooved TP304 stainless steel* J_M-R *curve data.*

5. The NUREG/CR-1061 *J-R* curve extrapolation was found to be too restrictive on the amount of crack growth allowed.

6. The tentative NRC A11 method was conservative but not as restrictive as the NUREG/CR-1061 method.

7. An extrapolation procedure was developed from the planform test results to account for geometry/constraint effects in J_M-R curves.

Another important point that needs to be considered in using an extrapolated *J-R* curve is the accuracy of the estimation scheme used in the structural analysis. This is to be addressed in a future paper.

Acknowledgments

This research was supported by the U.S. Nuclear Regulatory Office of Nuclear Reactor Research, Materials Branch. The NRC program monitor is Mr Michael Mayfield. We thank the NRC and Mr Mayfield for their support.
Others at Battelle who contributed to these research results are: J. Ahmad, N. Ghadiali,

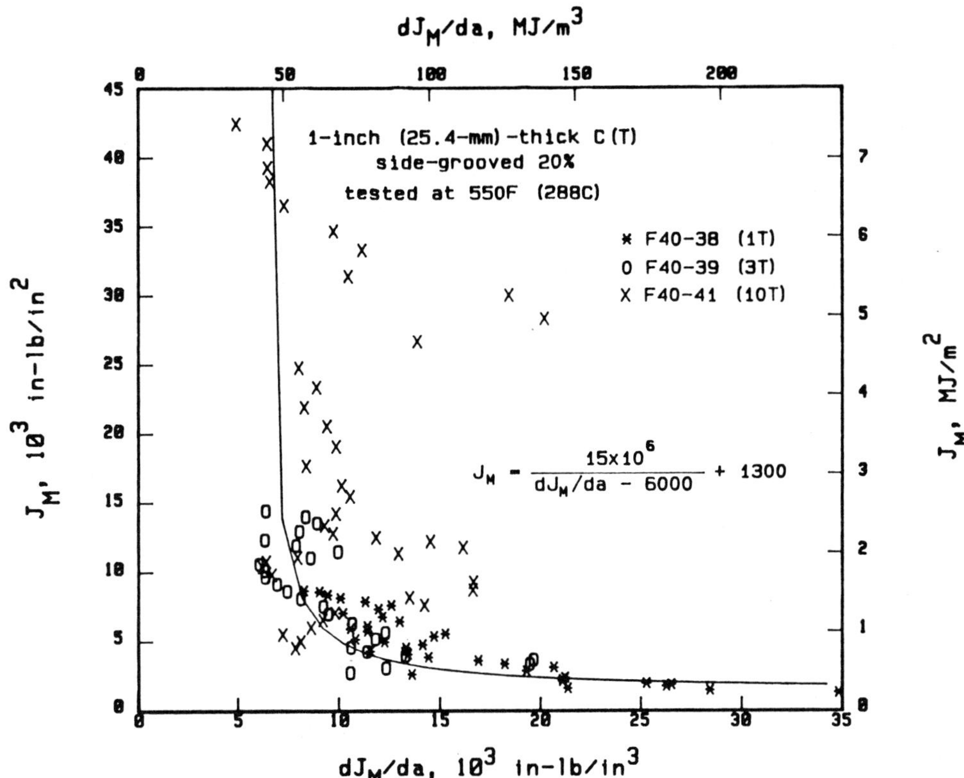

FIG. 27—*Hyperbolic fit of 25.4-mm (1-in.)-thick 20% side-grooved A515 Grade 70* J_M-R *curve data.*

V. Papaspyropoulos, P. Held, P. Mincer, and B. Blanton who helped prepare this manuscript.

References

[1] Wilkowski, G. M. and others, "Degraded Piping Program—Phase II," Program Report, Oct. 1986–Sept. 1987, NUREG/CR-4082, Vol. 6, April 1988.

[2] Papaspyropoulos, V., Marschall, C., and Landow, M., "Predictions of *J-R* Curves with Large Crack Growth from Small Specimen Data," NUREG/CR-4575, Sept. 1986.

[3] Paris, P. C., Brunett, J. V., and Cotter, K. H., "The Effect of Large Crack Extension on the Tearing Resistance of Stainless Steel Piping Materials," in *Proceedings,* CSNI Specialist Meeting on Leak-Before-Break in Nuclear Reactor Piping, Monterey, Calif., 1–2 Sept. 1983.

[4] Johnson, H. H., "Calibrating the Electric Potential Method for Studying Slow Crack Growth," *Materials Research and Standards,* Vol. 5, No. 9, Sept. 1965, pp. 442–445.

[5] Marschall, C. W., Held, P. R., Landow, M. P., and Mincer, P. N., "Use of the Direct-Current Electric Potential Method to Monitor Large Amounts of Crack Growth in Highly Ductile Metals," this publication, pp. 581–593.

[6] Ernst, H. A. and Paris, P. C., "Techniques of Analysis of Load—Displacement Records by *J*-Integral Methods," U.S. Nuclear Regulatory Commission Report NUREG/CR-1222, Jan. 1980.

[7] Ernst, H. A., "Material Resistance and Instability Beyond *J*-Controlled Crack Growth," in *Elastic-*

Plastic Fracture: Second Symposium: Vol. I—Inelastic Analysis, ASTM STP 803, C. F. Shih and J. P. Gudas, Eds., American Society for Testing and Materials, Philadelphia, 1983, pp. I-191–I-213.

[8] Hays, R. and Hackett, E., *Proceedings of a Specialist Meeting on J-R Curves,* to be published as a NUREG report.

[9] "Evaluation of Potential for Pipe Breaks," U.S. Nuclear Regulatory Commission Report NUREG-1061, Vol. 3, Nov. 1984.

[10] Kramer, G. and Papaspyropoulos, V., "An Assessment of Circumferentially Complex-Cracked Pipe Subjected to Bending," NUREG/CR-4687, Oct. 1986.

J. A. Joyce,[1] *D. A. Davis,*[2] *E. M. Hackett,*[2] *and R. A. Hays*[2]

Application of *J*-Integral and Modified *J*-Integral to Cases of Large Crack Extension

REFERENCE: Joyce, J. A., Davis, D. A., Hackett, E. M., and Hays, R. A., "**Application of *J*-Integral and Modified *J*-Integral to Cases of Large Crack Extension,**" *Fracture Mechanics: Twenty-First Symposium, ASTM STP 1074*, J. P. Gudas, J. A. Joyce, and E. M. Hackett, Eds., American Society for Testing and Materials, Philadelphia, 1990, pp. 85–105.

ABSTRACT: The *J*-integral is widely accepted as a measure of elastic-plastic fracture toughness of engineering alloys. Specimen size and geometry dependence were first noted in fracture toughness measurements using the ASTM E 813 calculation of the deformation *J*-integral (J_d) by McCabe and Landes [1] in 1983. The modified *J*-integral (J_m) was introduced by Ernst [2] to attempt to minimize or eliminate size and geometry dependence. Since J_m was introduced, questions have arisen regarding the proper parameter to describe the response of a flawed body to loading. The objective of this research task is to investigate the crack growth and specimen *J* capacity limitations of J_d and to verify the accuracy and specimen independence of the current J_m formulation. This research is expected to affect decisions concerning the use of small laboratory specimens to predict elastic-plastic crack growth resistance in engineering structures.

The *J-R* curve tests have been conducted on ½T, 1T, and 2T compact specimens of materials having critical fracture toughness values ranging from $J_{Ic} = 140$ to 455 kJ/m^2. These materials include HSLA-80 steel, A106 steel, 3-Ni steel, and two A533B steels. These tests were conducted in accordance with ASTM E 1152 except that specimen loading was continued until large crack extensions were present, in many cases exceeding 50% of the initial uncracked ligament (b). All specimen data were then analyzed using the equations of ASTM E 1152 for the standard deformation *J*-integral (J_d) resistance curve. Additional analysis was conducted as well in terms of the modified *J*-integral (J_m) using equations proposed by Ernst and Landes [3]. This latter quantity was possibly useful to larger crack extensions. The test procedure was directed towards finding the useful limits of the two *J* quantities for the materials listed above or for materials of equivalent strength and toughness.

KEY WORDS: fracture, *J*-integral, elastic-plastic, resistance curve, ductile fracture, modified *J*, crack growth, fracture toughness

The *J* resistance curve (*J-R*) standard recently approved by ASTM (E 1152) defines a methodology for developing *J-R* curves. The standard, however, allows only small crack extensions ($\Delta a < 0.1b_0$) and is limited to materials of modest elastic-plastic toughness when standard 1T or smaller specimens are used. The validity region defined by ASTM E 1152 is shown in Fig. 1. This validity region is very restrictive, and much work has been directed towards an enlargement of the box in both the amount of crack extension which is allowed and the maximum *J* that can be developed in a particular size specimen.

In this task a series of compact specimens has been tested using ½T, 1T, and 2T specimens of A710 steel, 3-Ni steel, A106 steel, and two A533B steels. The A106 steel was used exclusively for the blunt notch study described below. These materials have distinctly different

---

[1] United States Naval Academy, Annapolis, MD 21402.
[2] David Taylor Research Center, Bethesda, MD 20084.

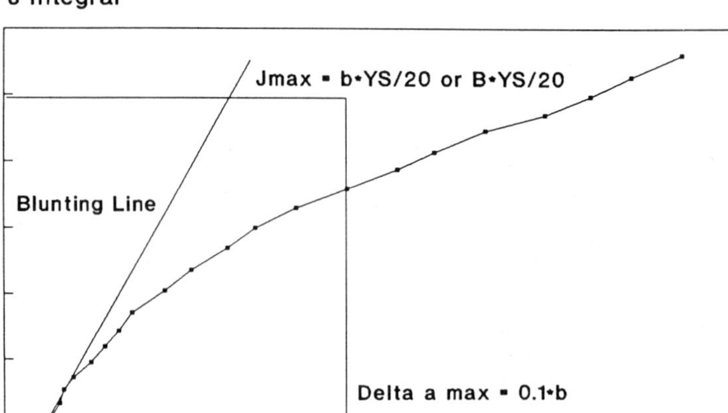

FIG. 1—J-R *curve validity ranges as defined by ASTM E 1152.*

ASTM E 813 J_{Ic} values ranging from 140 to 455 kJ/m². Tests were conducted on these specimens using ASTM E 1152 except that specimen loading was continued until large crack extensions were present, in many cases exceeding 50% of the initial uncracked ligament.

All specimen data were then analyzed using the equations of ASTM E 1152 for the standard deformation *J*-integral resistance curve. Additional analysis was conducted in terms of a modified *J*-integral quantity proposed by Ernst and Landes [3]. The equations are described in subsequent sections.

Experimental Program

All specimens used in this study were compact specimens machined in accordance with the specimen drawings of ASTM E 1152. All specimens had a thickness equal to $W/2$ and were side-grooved with a Charpy notch to a total reduction of 20%. Tests were conducted at quasti-static rates using a computer interactive unloading compliance test procedure.

Material

The five materials used were ASTM A710 steel, A106 steel, 3-Ni steel, and two A533B steels. The chemical compositions for these steels are given in Table 1; the standard tensile mechanical properties are presented in Table 2. The A106 steel was sectioned from a pipe with a 0.77 m inner diameter and a 0.17 m wall thickness. The A710 and 3-Ni steel plate were available in 50 mm thickness, the A533B-02 plate was initially 0.3 m thick, and the A533B-H13 was initially 0.19 m thick. All specimens were oriented as shown in Table 3 where the designations are as defined by ASTM E 399.

Test Matrix

A test matrix of all samples is shown in Table 4. The blunt notch specimens were used to verify the accuracy of deformation plasticity assumptions and were available for the A106,

TABLE 1—*Chemical compositions of alloys (wt%).*

	C	Mn	P	S	Cu	Si	Ni	Cr	Mo	V	Ti	Cb
A710	0.04	0.59	0.005	0.004	0.17	0.25	0.90	0.70	0.19	0.003	0.06	0.03
3-Ni	0.153	0.33	0.012	0.013	0.033	0.18	2.55	1.66	0.37	0.003	<0.001	...
A533B-02	0.22	1.48	0.012	0.18	...	0.25	0.68	...	...	...	..	...
A533B-H13	0.19	1.28	0.012	0.013	...	0.21	0.64	...	0.55	...	...	...
A106	0.26	0.9	0.028	0.017	...	0.022	0.26	0.11	...	...	...	...

TABLE 2—*Tensile mechanical properties of alloys.*

	0.2% Yield Strength		Ultimate Tensile Strength		% Elongation	% R.A.
	psi	MPa	psi	MPa		
A710	74 000.	510.	87 000.	600.	32.	80.
3-Ni	89 000.	613.	106 000.	730.	23.	63.
A533B-02	65 000.	448.	90 000.	620.	19.	60.
A533B-H13	64 150.	442.	87 100.	600.	26.	68.
A106	47 000.	333.	81 000.	558.	32.	67.

TABLE 3—*Orientation of specimens.*

Material	Crack Plane Orientation
A710	T-L
3-Ni	T-L
A533B-02	T-L
A533B-H13	L-T
A106	L-C

TABLE 4—*Test matrix showing number of specimens of each material and specimen type.*

	Unloading Compliance			Blunt Notched
Material	½T	1T	2T	1T
A710	2	5	2	4
3-Ni	2	5	0	3
A533B-02	3	2	0	0
A533B-H13	0	6	0	3
A106	0	6	0	6

3-Ni, A533B-H13, and A710 materials. At the completion of testing, all unloading compliance specimens were heat tinted, broken open after immersion in liquid nitrogen, and the crack lengths measured optically in accordance with ASTM E 1152 to verify the accuracy of the unloading compliance test procedure. All data were stored on magnetic floppy disks for post-test analysis.

Analysis

Experimental J equations have developed continuously since about 1973 when Rice, Paris, and Merkle [4] took the basic definition of the J integral that

$$J = \int \left[\frac{\partial \theta}{\partial a} \right]_M dM \tag{1}$$

and integrated it for the special case of a deeply cracked bend bar to give

$$J = \frac{2A}{Bb} \tag{2}$$

where

    a = crack length,
    θ = total bend angle,
    M = applied bending moment,
    B = specimen thickness at the crack plane,
    b = remaining ligament, and
    A = area under the load versus load-line displacement record obtained experimentally.

This equation was initially applied to compact specimens, but a correction to approximately account for the tensile component present in the compact specimen was developed by Merkle and Corten [5] giving a J expression of the form:

$$J = \frac{\beta A}{Bb} \tag{3}$$

$$\beta = 2(1 + \alpha)/(1 + \alpha^2) \tag{4}$$

$$\alpha = 2\sqrt{(a/b)^2 + (a/b) + 1/2} - 2(a/b + 1/2) \tag{5}$$

The above equations apply rigorously if no crack extension is present but need a further correction if crack growth is present as is the case for J resistance curve calculations. This effect is shown schematically in Fig. 2. The upper load displacement record labelled $ABCD$ corresponds to the real specimen record including a load drop corresponding to a crack extension of $\Delta a = a_f - a_D$. The J value desired at D, since it is a path-independent deformation plasticity quantity, should, however be calculated using the load displacement record

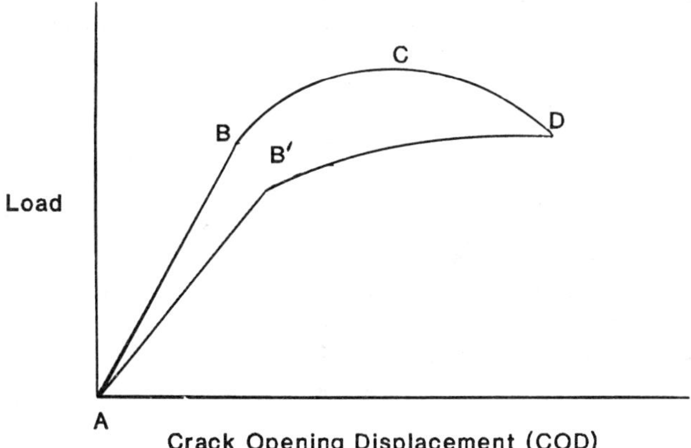

FIG. 2—*Schematic of load displacement records showing the experimental record and the deformation plasticity record.*

$AB'D$ which would have been present if the specimen had started with an initial crack length of a_f and then been loaded to point D without crack extension. Ernst et al. [6] obtained an expression for this "crack growth corrected" J value as

$$J_i = \left[J_{i-1} + \left(\frac{\eta_i}{b_i}\right) \frac{A_{i-1,i}}{B} \right] \left[1 - \frac{\gamma_i}{b_i}(a_i - a_{i-1}) \right] \tag{6}$$

where η_i and γ_i are geometry-dependent parameters. The specimen thickness B is usually replaced here by B_N, the net specimen thickness, if side-grooved specimens are used. The $i - 1$ and i notation usually correspond to the locations of elastic unloadings at which b_i, A_i, η_i, and γ_i are calculated from the specimen compliance. $A_{i-1,i}$ is the area under the experimental load displacement record between the unloading $i - 1$ and the unloading i. The accuracy of this equation is checked in a subsequent section of this report.

Equation 6 was modified slightly in ASTM E 1152 to produce a better agreement with elastic fracture quantities by separating J into two components call J_{EL} and J_{PL} to give

$$J = J_{EL} + J_{PL} \tag{7}$$

$$J_{EL} = \frac{K^2(1 - \nu^2)}{E} \tag{8}$$

$$J_{PLi} = \left[J_{PL(i-1)} + \left(\frac{\eta_i}{b_i}\right) \frac{A_{PL(i-1,i)}}{B_N} \right] \left[1 - \frac{\gamma_i}{b_i}(a_i - a_{i-1}) \right] \tag{9}$$

J-integral calculations using the above equations have generally been limited to small amounts of crack extension. Partly this limit is applied because some areas of the specimen are unloading and this violates a precept of the deformation J theory on which the J-integral exists, and partly it is based on the finite element work of Shih [7]. Rigorous experimental verification of the need for, or applicability of, such limits has not yet been accomplished. ASTM E 1152, for example, limits crack growth to 10% of the initial crack length, based on Booth et al. [8].

Many applications, however, require J resistance data to much larger crack extensions, and recent work has been directed towards increasing the amount of crack extension allowed. One suggestion has been developed by Ernst [9] based on work of Rice and Sorenson [10] which describes a J-type parameter defined by

$$J_m = J_{EL} + J_{PL} + \int_{a_0}^{a} \frac{m}{b} J_{PL} \, da \tag{10}$$

in which m is a geometry factor and other quantities are as defined previously. Ernst [9] and Ernst and Landes [11] proposed that J_m should be less sensitive to large amounts of crack extension and should allow collection of more data from small size specimens, though no estimate is given as to how much might be possible.

In the sections that follow, experimental work is presented in terms of both the deformation theory J calculated using Eqs 7 to 9 and the modified J evaluated using Eq 10.

Effects of Unloadings

All tests conducted to obtain J-R curves were done using an unloading compliance technique. This procedure involves generating small unloadings along the load displacement record and the use of the slope of these partial unloads to estimate the present specimen crack length. These unloadings, if large enough or numerous enough, can affect the resulting *J-R* curve [*10*]. If the unloadings do not exceed 15 to 20%, however, little effect is to be expected. A check of the unload effect was performed as a part of this project, and the result for the 1T compacts of the ASTM A710 alloy is shown in Fig. 3. Specimens J9 and J5 were tested with standard 15% unloadings, while J4 was tested without unloadings. Differences between the load-COD records of these specimens lie in the range expected due to normal material scatter. No significant effect of the unloadings on the load-COD curve could be discerned.

J *Deformation* J-R *Curves*

Unloading compliance tests were conducted on the specimens shown in Table 4 using a computer interactive procedure and a methodology consistent with ASTM E 1152. All compliance equations corresponded to ASTM E 1152 and the deformation *J* calculations were done in accordance with ASTM E 1152.

Results are shown for four materials in terms of deformation *J* in Figs. 4 to 7. The ASTM limits presently in E 1152 are shown in each figure. For most cases all specimens were geometrically similar with $a/W = 0.61$, with the exception of the A533B-H13 specimens, which were all 1T CTs with various a/W ratios. Clearly, the geometry variables produce little difference in the resulting deformation *J* resistance curve at crack extension values and *J* values far in excess of what is presently allowed by ASTM E 1152.

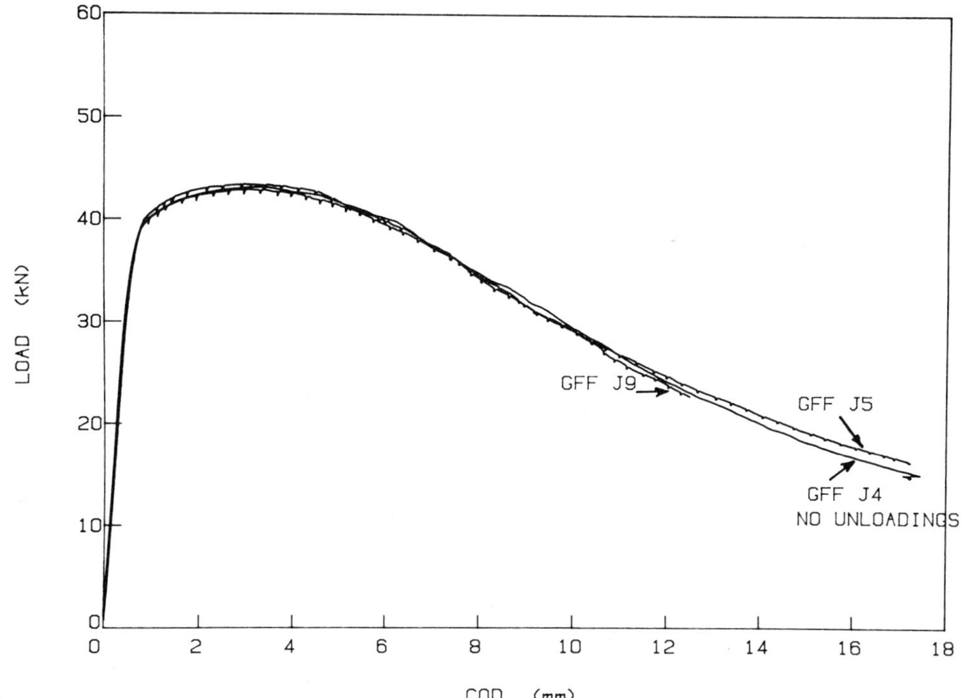

FIG. 3—*Load displacement record showing that unloadings have no effect even when large COD displacements are applied.*

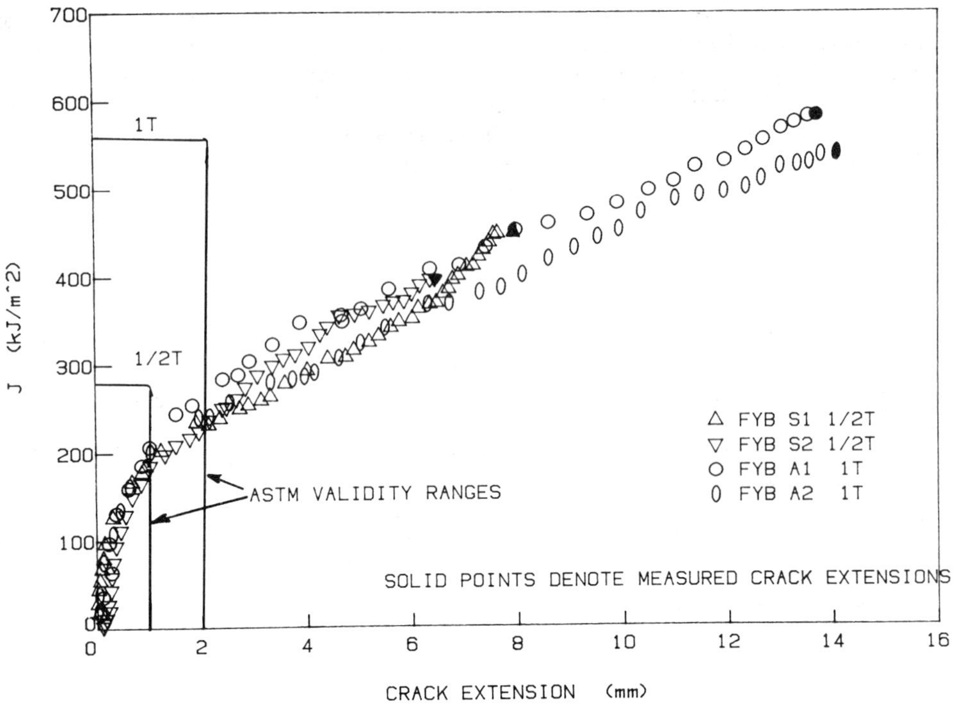

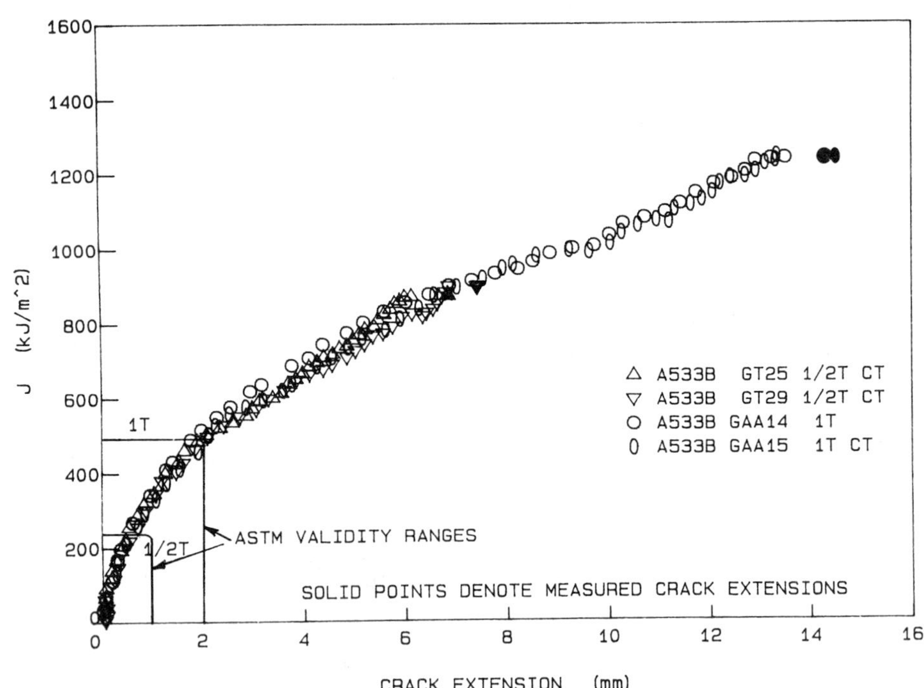

FIG. 4—*Deformation* J-R *curves for 3-Ni steel compact tests.*

FIG. 5—*Deformation* J-R *curves for A533B-02 steel compact tests.*

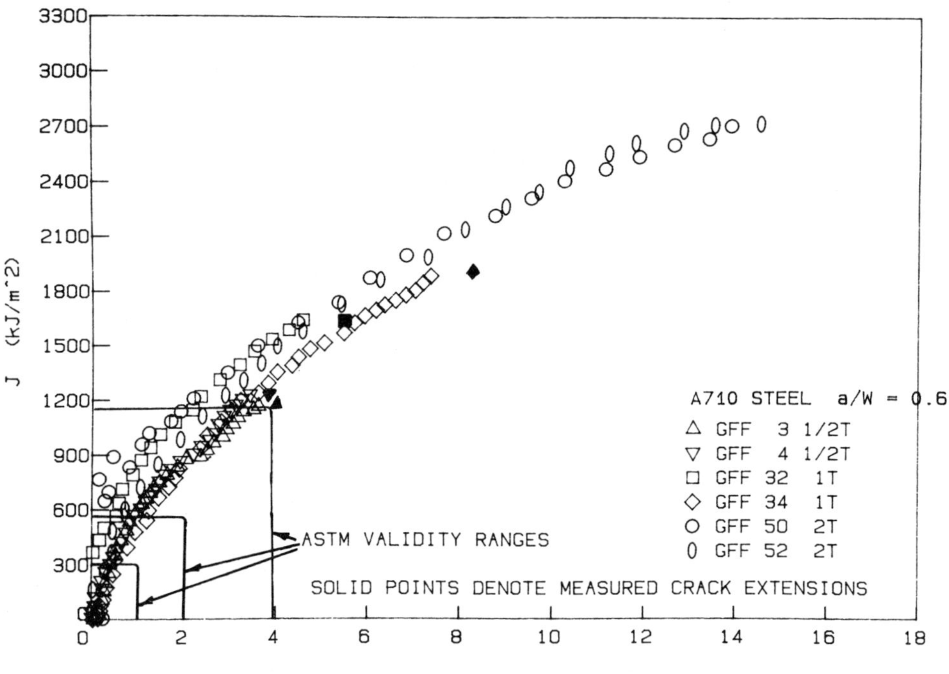

FIG. 6—*Deformation* J-R *curves for A710 steel compact tests.*

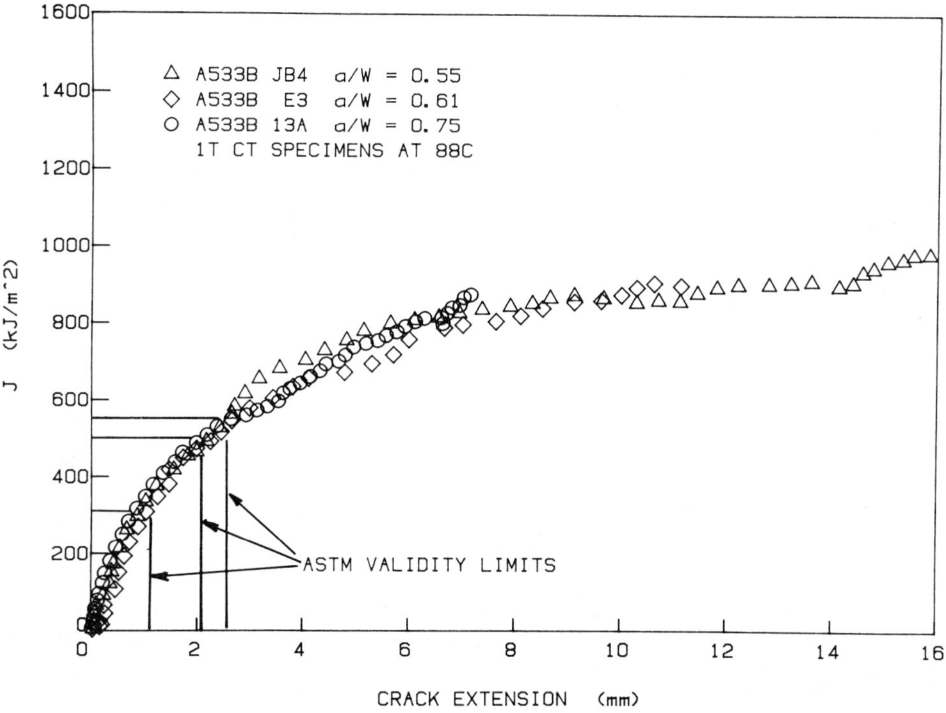

FIG. 7—*Deformation* J-R *curves for A533B-H13 steel compact tests.*

TABLE 5—*Crack length measurement tabulation.*

SPEC ID	W (mm)	B (mm)	Bn (mm)	TYPE	Ai(est) (mm)	Ai(meas) (mm)	Af(est) (mm)	Af(meas) (mm)	bf(est) (mm)	bf(meas) (mm)
A710										
GFF-3	25.4	12.7	10.1	1/2T	16.0	16.5	19.4	20.5	6.0	4.9
GFF-4	25.4	12.7	10.2	1/2T	16.0	16.2	19.3	20.2	6.1	5.2
GFF-30	50.8	25.5	20.3	1T	34.2	34.8	41.3	43.0	9.5	7.7
GFF-31	50.8	25.5	20.4	1T	34.3	34.6	41.6	42.5	9.2	8.3
GFF-32	50.8	25.4	20.3	1T	33.8	35.0	38.8	40.5	12.0	10.3
GFF-33	50.9	25.5	20.3	1T	34.3	34.9	41.6	43.1	9.3	7.7
GFF-34	50.9	25.5	20.3	1T	34.3	35.2	41.6	43.4	9.3	7.5
GFF-L8	101.7	50.8	40.6	2T	60.9	62.0	66.0	67.7	35.7	34.0
GFF-L9	101.7	50.8	40.6	2T	61.6	62.2	71.6	73.6	30.1	28.0
3-Ni										
S-1	25.4	12.7	10.0	1/2T	15.4	15.2	23.0	23.3	2.4	2.1
S-2	25.4	12.7	10.1	1/2T	15.4	15.5	21.7	21.9	3.7	3.5
A-1	50.8	25.4	20.0	1T	30.7	30.7	44.2	44.5	6.6	6.3
A-2	50.8	25.4	20.0	1T	30.9	31.1	44.7	45.3	6.1	5.5
A533B-02										
GT-25	25.4	12.7	10.2	1/2T	15.8	15.7	21.9	22.6	3.5	2.8
GT-26	25.4	12.7	9.9	1/2T	15.3	15.2	16.5	16.4	8.9	9.0
GT-29	25.4	12.7	10.0	1/2T	15.5	15.4	22.4	23.0	3.0	2.4
GAA-14	50.8	25.4	20.3	1T	31.1	30.9	44.6	45.3	6.2	5.5
GAA-15	50.8	25.4	20.1	1T	31.1	31.1	44.4	45.5	6.4	5.3

NOTES: (1) Ai indicates INITIAL crack length
         (2) Af indicates FINAL crack length

Table 5 compares the results of post-test crack length measurements done on these specimens and the unloading compliance estimates used for the J-R curves of Figs. 4 to 7. The errors are generally less than a few percent except for the high toughness A710 specimens which showed up to 20% error after a net ligament reduction of 40 to 60%. These specimens demonstrated the largest amount of specimen deformation including an approximately 10% reduction in specimen thickness (B_N) near the crack front. If the true specimen thickness was monitored and included in the compliance calculation, more accurate crack length estimates might be possible for this material.

Modified J, J_m-R Curves

Re-analyzing these results using the J_m quantity, using Eq 10 with $m = (1 + 0.86\,b/W)$ for the CT specimen [9], gives the resistance curves shown in Figs. 8 to 11. A strong geometry dependence is shown to be a striking feature of these resistance curves except for the high toughness A710 alloy. For this latter material, the small specimens have been shifted from being on the lower side of the large specimen results, which is typical of deformation J results, to being on the upper side of the large specimen curves. Note also that the y-axis scales have been changed in going to J_m, in most cases corresponding to the elevation of J_m produced by Eq 10, though over the first 10% of crack extension both J-integral quantities are essentially identical.

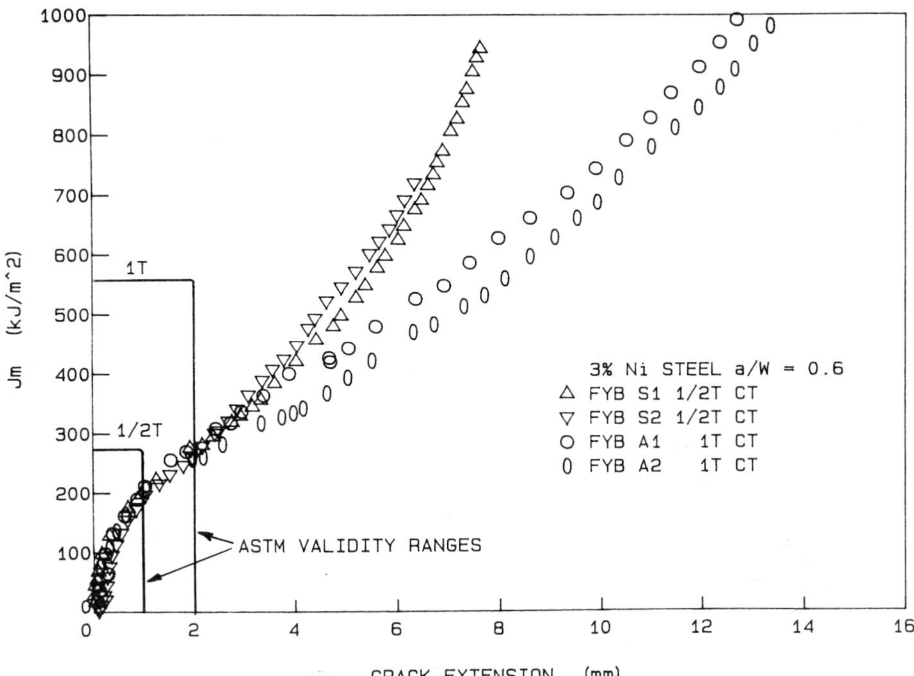

FIG. 8—*Modified* J-R *curves for 3-Ni steel compact tests.*

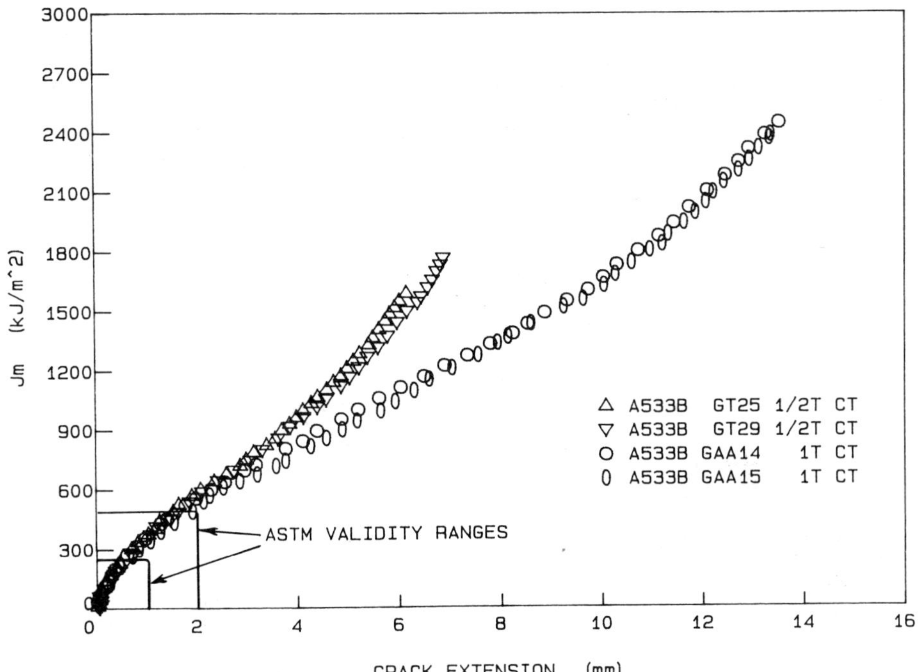

FIG. 9—*Modified* J-R *curves for A533B-02 steel compact tests.*

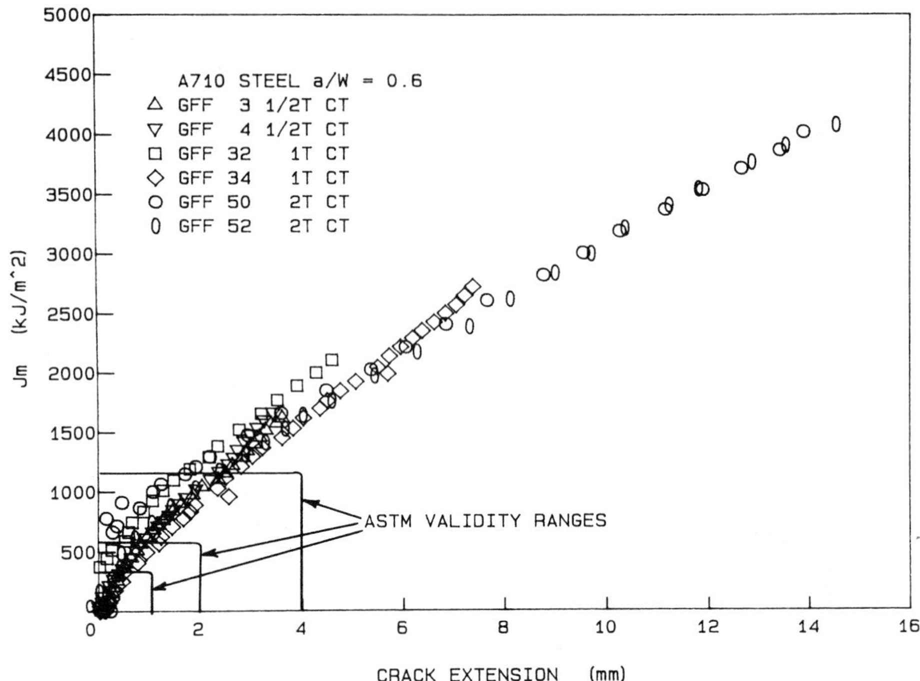

FIG. 10—*Modified* J-R *curves for A710 steel compact tests.*

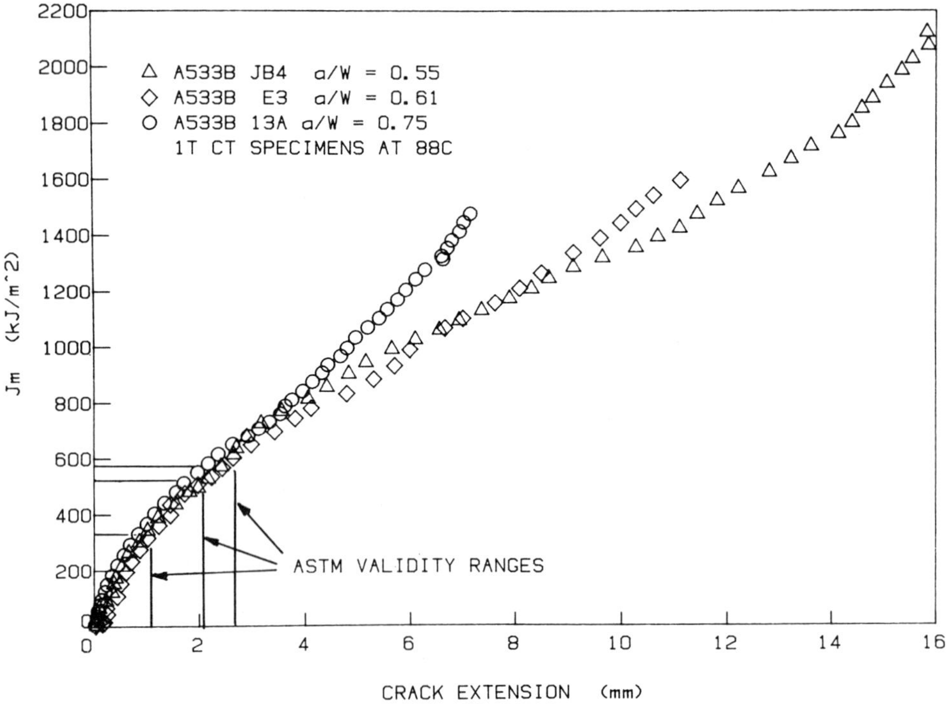

FIG. 11—*Modified* J-R *curves for A533B-H13 steel compact tests.*

Blunt Notched Specimens

Tests of blunt notch specimens were conducted as a part of this test program for two reasons. First, it was desired to verify the accuracy of the crack growth corrected J-integral calculations of Eq 9, especially for the cases of large crack extension and large specimen deformation expected in this program. The second reason was to explore the applicability of a deformation approach to tests exhibiting a large amount of crack extension and consequently a large overall drop in load, large residual stresses, compressive plastic zones away from the crack tip region, gross specimen deformation, and large lateral specimen contraction. Any of these could invalidate the deformation philosophy on its own and in many J-R curve tests they appear in combination.

The test procedure used here was as follows:

1. Standard 1T CT unloading compliance (UC) specimens with initial a/W values of approximately 0.6 were tested in accordance with ASTM E 1152 to crack extension values of 2.5, 5.0, and 7.5 mm.

2. These specimens were heat tinted to mark the crack extent and then broken open at liquid nitrogen temperature.

3. ASTM E 1152 eight-point average optical crack length measurements were made of the final crack lengths present in these specimens.

4. Blunt notched 1T CT specimens were then machined to match either the measured final crack length or 0.75 mm shorter than the measured final crack length. The shorter lengths were found to give a compliance match between the crack length estimated at the end of the unloading compliance test and the compliance estimated at the start of the blunt notch specimen test. All specimens were blunt notched with a 2.4 mm thick slitting saw with a 1.2 mm radius tip.

5. The blunt notch specimens were tested with only initial compliance estimates and final compliance estimates being taken.

The A106 and A533B-H13 materials showed some material variability, as demonstrated by differences in the load displacement curves obtained, while the A710 and 3-Ni show very little material variability.

Figures 12 to 16 show the results obtained by the series of blunt notch tests conducted here. Tabulated results are presented in Tables 6 to 9. Comments about each material are presented in separate sections below.

3-Ni Steel—For this material, 1T CT unloading compliance tests were run to measured final crack extensions of 2.7, 5.3, and 7.7 mm. Specimens of identical geometry were then blunt notched to the final measured crack length of the unloading compliance tests using a slitting saw with a round end radius of 1.2 mm. These specimens were then loaded, without unloadings, giving results as shown in Fig. 12.

The initial compliance of the blunt specimens was found to estimate an initial crack length of approximately 0.75 mm more than the saw cut depth; this is felt to be due to the presence of the blunt notch. The comparison of final loads between the fatigue precracked specimens and the blunt notched specimens is shown in Table 6. In each case the blunt notched specimen load is lower at the end of test than the fatigue cracked specimen with a percentage difference of between 3 and 12%. The corresponding comparison of deformation J-integral quantities is very good even for the specimens that model 7.7 mm of crack extension, which is about 40% ligament reduction in these specimens.

A106 Steel—The A106 steel was tested in a fashion very consistent with the 3-Ni steel material (i.e., the blunt notched specimens were cut to match, as closely as possible, the final measured crack lengths of the unloading compliance tests). These results are shown in Fig.

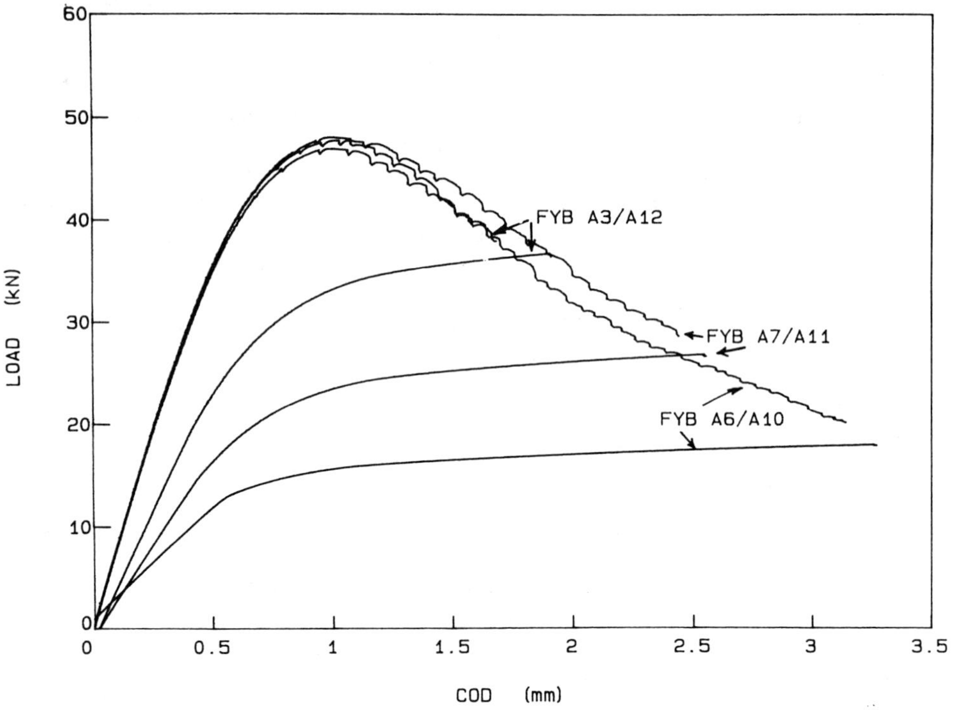

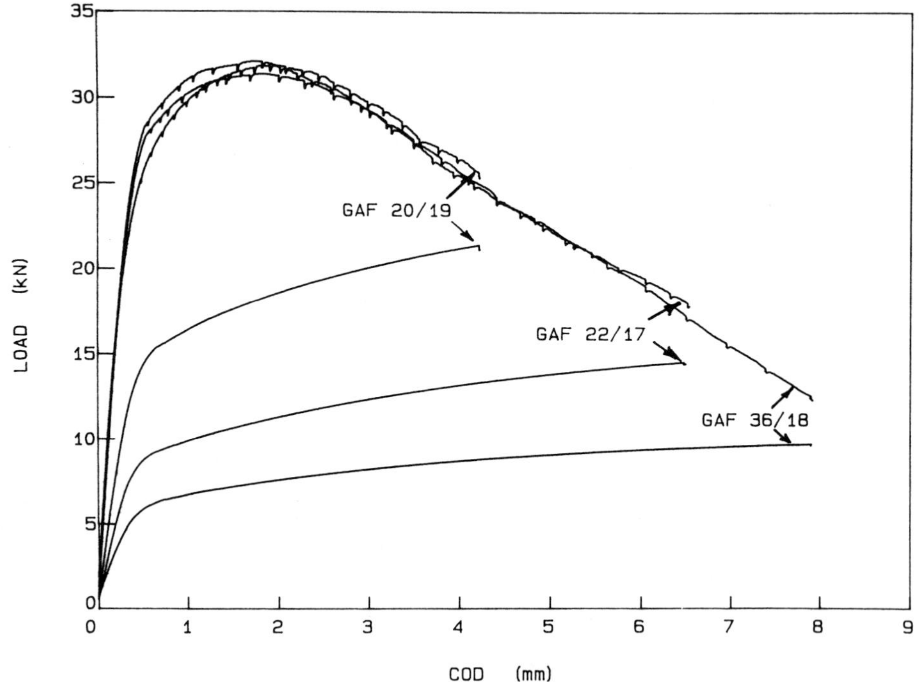

FIG. 12—*Load COD records showing agreement between unloading compliance and blunt notch results for 3-Ni (FYB) steel compact.*

FIG. 13—*Load COD records showing agreement between unloading compliance and blunt notch results for A106 (GAF) steel compact tests.*

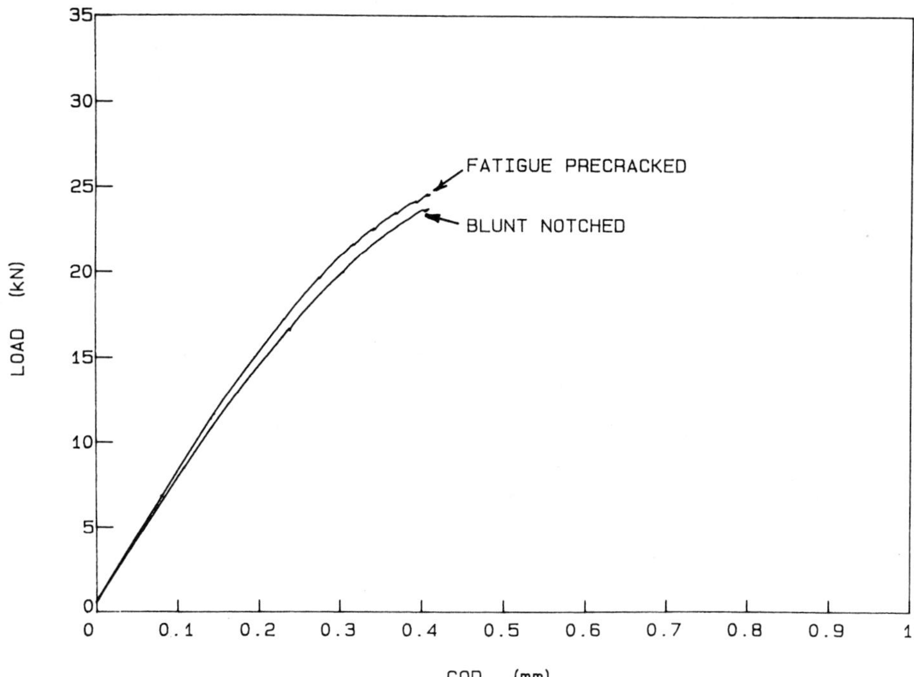

FIG. 14—*Load COD records for a pre-initiation case showing the small effect of blunt notch geometry.*

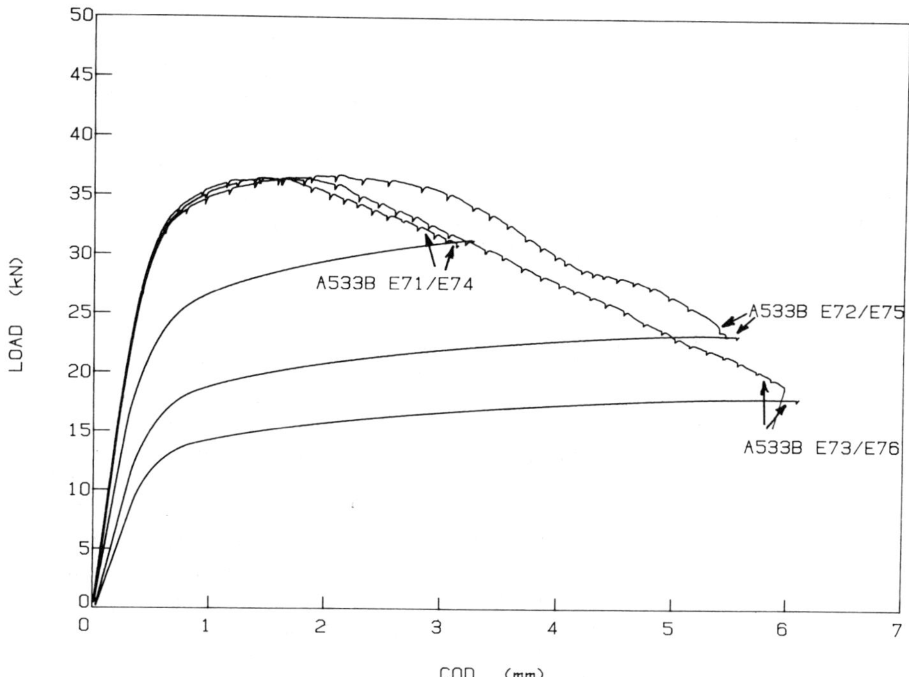

FIG. 15—*Load COD records showing agreement between unloading compliance and blunt notch results for A533B-H13 steel compact tests.*

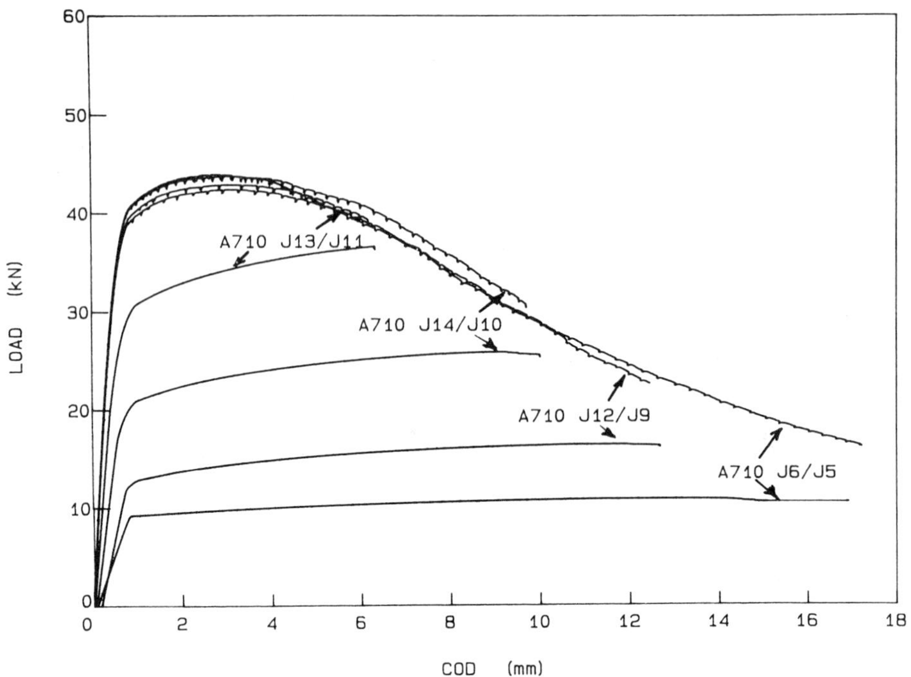

FIG. 16—*Load COD records showing agreement between unloading compliance and blunt notch results for A710 (GFF) steel compact tests.*

13 and Table 7. Larger errors are shown in Table 7 for this material, and some of this error appears attributable to material variability. At the end of the tests, loads from blunt notch tests fell short of the final loads from the unloading compliance tests by approximately 10%. This test series involved an additional data set for near initiation behavior. In this case, the *J-R* curve test was taken to the point on the load-COD where crack initiation was expected and then stopped. The corresponding blunt notch test was taken to the same point in COD. The purpose of this comparison was to assess possible load-crack extension history effects on the overall test matrix. In such a case, final load differences between a *J-R* curve test and a blunt notch test can obviously not be attributed to crack extension effects. This result is presented in Fig. 16 where the final load of the blunt notched test was shown to be about 900 N less than that of the *J-R* curve test. This is less than a 5% error and would be still less if the blunt notched specimen had been machined to match the final compliance of the *J-R* curve test.

A533B-H13 Steel—The shortfall in load in the previous cases could result from two sources: (1) a real effect of material history present in the fracture tests, or (2) an effect simply of the different geometry present in the blunt notch specimens. In order to address this issue, at least in part, this set of tests was done by matching the compliance of the blunt notched specimen to the final compliance of the fracture test rather than matching the notch depths as done in the data sets above. Based on the observations made in testing the 3 Ni and A106 steel specimens, blunt notches were cut 0.75 mm short for these tests to match the initial compliance of the blunt notched specimen to the final compliance of the unloading compliance specimen. The effects of this compliance matching is shown in Fig. 15 and Table 8. Clearly, matching compliancies has produced enhanced agreement between the unloading compliance specimens and the blunt notched specimens at the end of test, resulting, in some

TABLE 6—*3-Ni summary of blunt notch study.*

Specimen I.D.	Specimen Type	a_{0_meas} (mm)	a_{0_est} (mm)	a_{f_meas} (mm)	a_{f_est} (mm)	Δa_meas (mm)	Δa_est (mm)	b_{f_meas} (mm)	b_{f_est} (mm)	P_end (kN)	J_end (kJ/m^2)
FYBA12	UC	30.6	30.6	33.4	33.3	2.71	2.62	17.4	17.5	38.1	292.
FYBA3	Blunt	33.4	33.9	...	...	0.0	0.28	...	...	36.8	298.
FYBA11	UC	30.6	30.5	35.6	35.8	5.26	5.13	15.2	15.0	28.6	388.
FYBA7	Blunt	35.8	36.3	...	...	0.0	0.20	...	...	27.2	386.
FYBA10	UC	30.7	30.5	38.4	38.2	7.72	7.62	12.4	12.6	20.4	421.
FYBA6	Blunt	38.4	39.0	...	...	0.0	0.18	..	...	18.2	426.

TABLE 7—*A106 summary of blunt notch study.*

Specimen I.D.	Specimen Type	$a_{0\text{meas}}$ (mm)	$a_{0\text{est}}$ (mm)	$a_{\text{final meas}}$ (mm)	$a_{\text{final est}}$ (mm)	Δa_{meas} (mm)	Δa_{est} (mm)	P_{end} (kN)	J_{end} (kJ/m^2)
GAF-34	UC	31.2	30.5	33.9	33.2	2.74	2.70	30.9	585.
GAF-34	Blunt	34.0	34.0	34.0	34.5	0.0	0.56	26.4	532.
GAF-19	UC	31.4	30.5	35.3	34.2	3.94	3.76	25.8	578.
GAF-20	Blunt	35.6	35.7	35.7	36.2	0.0	0.53	21.4	518.
GAF-17	UC	31.1	30.4	38.2	37.1	7.1	6.65	18.0	768.
GAF-22	Blunt	38.3	38.0	38.3	38.6	0.0	0.56	14.5	652.
GAF-18	UC	31.2	30.8	40.5	39.8	9.35	9.00	12.6	749.
GAF-36	Blunt	40.9	40.9	40.9	41.4	0.0	0.41	9.76	673.

TABLE 8—*A533B-H13 summary of blunt notch study.*

Specimen I.D.	Specimen Type	$a_{0\text{meas}}$ (mm)	$a_{0\text{est}}$ (mm)	$a_{f\text{meas}}$ (mm)	$a_{f\text{est}}$ (mm)	Δa_{meas} (mm)	Δa_{est} (mm)	P_{end} (kN)	J_{end} (kJ/m^2)
E71	UC	30.9	30.7	33.3	33.3	2.44	2.62	31.2	481.
E74	Blunt	32.6	33.2	. . .	. . .	0.0	0.46	31.3	509.
E75	UC	30.8	30.7	36.1	35.9	5.31	5.16	23.6	771.
E72	Blunt	35.4	35.8	. . .	. . .	0.0	0.61	23.3	781.
E76	UC	30.8	30.6	37.8	37.4[a]	6.96	6.76	19.1	737.
E73	Blunt	37.0	37.6	. . .	. . .	0.0	0.51	17.7	749.

[a] Loaded slightly beyond this unloading.

cases, with the blunt notched specimen load exceeding that of the unloading compliance specimen.

A710 Steel—The A710 specimen tests were conducted in a fashion identical to the A533B-H13 tests described in the previous section (i.e., the blunt notch lengths were cut approximately 0.75 mm short to match the compliance of the blunt notch specimen to the final compliance of the unloading compliance test). The results of these tests are shown in Fig. 16 and Table 9.

These results again show a shortfall of load, as was present in the A106 tests, except that the percentage load shortfall is growing steadily as the specimen deformation increases (i.e., as the blunt notches attempt to model deeper and deeper crack extensions). These specimens were also characterized by larger deformations than any of the above materials. For these tests, the load was as much as 33% low at the end of test, with *J* values low by the end of test by about 19%.

Discussion

This work has demonstrated that *J* resistance curves can be accurately evaluated for many materials to crack extension and *J* values well beyond the present ASTM limits. This is not to say that the *J-R* curve to this extent is still representing a singularity controlled crack growth; this would require more than the experimental testing done here. However, several necessary features have been demonstrated by this work. First, it has been demonstrated that an unloading compliance technique can be used to give accurate crack extension measure-

TABLE 9—*A710 summary of blunt notch study.*

Specimen I.D.	Specimen Type	a_{0meas} (mm)	a_{0est} (mm)	$a_{f_{meas}}$ (mm)	$a_{f_{est}}$ (mm)	Δa_{meas} (mm)	Δa_{est} (mm)	$b_{f_{meas}}$ (mm)	$b_{f_{est}}$ (mm)	P_{end} (k/N)	J_{end} (kJ/m^2)
GFFJ11	UC	31.2	30.9	33.8	33.5	2.54	2.62	17.0	17.3	39.5	1260.
GFFJ13	Blunt	33.0	33.6	...	...	0.0	0.67	...	...	37.0	1250.
GFFJ10	UC	31.2	30.7	36.5	35.9	5.26	5.18	14.3	14.9	30.9	1780.
GFFJ14	Blunt	35.7	36.4	...	...	0.0	0.91	...	...	26.4	1665.
GFFJ9	UC	31.7	31.2	39.3	38.1	7.6	7.0	11.5	12.7	22.7	1960.
GFFJ12	Blunt	38.5	39.3	...	...	0.0	0.79	...	...	16.4	1654.
GFFJ5	UC	31.6	31.1	41.2	40.2	9.8	9.1	9.1	10.6	16.4	2320.
GFFJ6	Blunt	40.7	41.5	...	...	0.0	...	...	...	10.9	1880.

ments to a 50 or 60% reduction of the initial specimen ligament. For most materials errors do not exceed 10% after these substantial crack growths. For highly ductile materials, errors on the order of 20% were found, corresponding to the gross specimen geometry changes observed in these specimens. The major features observed are lateral contractions near the crack tip and side groove closing near the specimen back surface. Inaccuracy of the rotation correction applied to the compliance equation is also a possible contributing factor.

A second point is that blunt notch specimen tests show that deformation theory is applicable for elastic-plastic materials of the type tested here, even when large crack extensions and the corresponding material unloading has occurred. Load history effects generally seem to elevate the load of the specimen undergoing ductile crack growth above the load found for the blunt notched specimen without crack growth, but in most cases the load elevation is less than 10% even when ligament reductions of 40% or more have taken place. When the blunt notch effect is minimized by matching specimen compliance even this 10% elevation seems not to be present. This observation should not be taken as a statement that a singularity field is present to these large crack extensions; however, no clear and direct proof is present from tests of this type to prove that J-type conditions could not still exist.

A third point, related to the previous one, is that deformation J calculated using the crack growth corrected equations of ASTM E 1152 gives accurate results, at least results comparable to what would have been found using specimens of the final crack length loaded without crack growth (i.e., blunt notched as done here).

A fourth point is that, as far as deformation J is concerned, none of the tests completed here seem to distinguish a location where J-controlled crack growth ceases to exist. The development of an experimental measurement which demonstrates the existence of singularity conditions and subsequently the loss of such conditions is badly needed and is still the subject of ongoing research.

The fifth and final point is that the modified J seems to demonstrate a high degree of size dependence for the specimens and materials tested in this program. It works well for high toughness materials like the A710 tested here and the A508 used in the original Westinghouse [1] study but, for less tough materials, the size dependence and the troublesome tendency for the J_m-R curves to tail up need further analysis and explanation.

Conclusions

The following conclusions are based on the work described above:

1. The unloading compliance method can accurately estimate crack extension for compact specimens to crack extensions of 50% even for relatively tough materials.

2. The J equations used in ASTM E 1152 give very accurate J values even when relatively large amounts of crack extension and a corresponding large load drop are present. By "accuracy" it is meant that J values calculated from equations for blunt notched specimens not demonstrating crack growth agree closely with J values calculated by the ASTM E 1152 equations for fatigue precracked specimens which undergo large amounts of crack extension.

3. The deformation J resistance curves appear to apply far beyond the size limitation currently in ASTM E 1152. This conclusion is based on size independence demonstrated for the materials tested here and the observation that the deformation J-R curves continue to rise for crack extensions well beyond the present ASTM E 1152 validity regions. This conclusion appears to be equally true for specimens whose R curves exit out the end of the validity regions and for specimens whose R curves exit out the top of the validity region.

4. The modified J resistance curves are very size dependent beyond 30% crack extension except for the case of the highest toughness materials. The J_m resistance curves for small

specimens, or specimens with deep cracks, deviate in an upward sweeping fashion (i.e., an unconservative fashion) from the J_m-R curves of larger or shorter cracked specimens.

References

[1] McCabe, D. E., Landes, J. D., and Ernst, H. A., "An Evaluation of the J_R-Curve Method for Fracture Toughness Characterization," in *Elastic-Plastic Fracture: Volume II—Fracture Resistance Curves and Engineering Applications, ASTM STP 803*, C. F. Shih and J. P. Gudas, Eds., American Society for Testing and Materials, Philadelphia, 1983, pp. II-562–II-581.

[2] Ernst, H. A., "Material Resistance and Instability Beyond *J*-Controlled Crack Growth," in *Elastic-Plastic Fracture: Volume I—Inelastic Crack Analysis, ASTM STP 803*, C. F. Shih and J. P. Gudas, Eds., American Society for Testing and Materials, Philadelphia, 1983, pp. 1-191–I-213.

[3] Ernst, H. A. and Landes, J. D., "Predictions of Instability Using the Modified *J*, J_M-Resistance Curve Approach," in *Elastic-Plastic Fracture Mechanics Technology, ASTM STP 896*, J. C. Newman, Jr., and F. J. Loss, Eds., American Society for Testing and Materials, Philadelphia, 1985, pp. 128–138.

[4] Rice, J. R., Paris, P. C., and Merkle, J. G., "Some Further Results of *J*-Integral Analysis and Estimates," in *Progress in Flaw Growth and Fracture Toughness Testing, ASTM STP 536*, American Society for Testing and Materials, Philadelphia, 1973, pp. 231–245.

[5] Merkle, J. G. and Corten, H. T., "A *J*-Integral Analysis for the Compact Specimen, Considering Axial Force as well as Bending Effects," *Transactions of ASME*, Nov. 1974, pp 286–292.

[6] Ernst, H. A., Paris, P. C., and Landes, J. D., "Estimations on *J*-Integral and Tearing Modulus *T* from a Single Specimen Test Record," in *Fracture Mechanics: Thirteenth Conference, ASTM STP 743*, Richard Roberts, Ed., American Society for Testing and Materials, Philadelphia, 1981, pp. 476–502.

[7] Shih, C. F., DeLorenzi, H. G., and Andrews, W. R., "Studies on Crack Initiation and Stable Crack Growth," in *Elastic-Plastic Fracture, ASTM STP 668*, J. D. Landes, J. A. Begley, and G. A. Clarke, Eds., American Society for Testing and Materials, Philadelphia, 1979, pp. 65–120.

[8] Newman, J. C., Jr., Booth, B. C., and Shivakurman, K. N., "An Elastic-Plastic Finite-Element Analysis of the *J*-Resistance Curve Using a CTOD Criterion," in *Fracture Mechanics: Eighteenth Symposium, ASTM STP 945*, D. T. Read and R. P. Reed, Eds., American Society for Testing and Materials, Philadelphia, 1988, pp. 665–685.

[9] Rice, J. R. and Sorensen, E. P., *Journal of the Mechanics and Physics of Solids*, Vol. 26, 1978, pp. 163–186.

[10] Joyce, J. A., "Development of a Criterion for the Effect on the *J*-*R* Curve of Elastic Unloadings," in *Fracture Mechanics: Eighteenth Symposium, ASTM STP 945*, D. T. Read and R. P. Reed, Eds., American Society for Testing and Materials, Philadelphia, 1988, pp. 647–662.

Dynamic Fracture

A. S. Douglas[1] and M. S. Suh[2]

Impact Fracture of a Tough Ductile Steel

REFERENCE: Douglas, A. S. and Suh, M. S., **"Impact Fracture of a Tough Ductile Steel,"** *Fracture Mechanics: Twenty-First Symposium, ASTM STP 1074*, J. P. Gudas, J. A. Joyce, and E. M. Hackett, Eds., American Society for Testing and Materials, Philadelphia, 1990, pp. 109–125.

ABSTRACT: The impact fracture of a tough ductile material is examined by comparing the results of static and dynamic viscoplastic finite element computations with experimental observations of the crack tip opening displacement (CTOD) during static and dynamic (impact) fracture. Experiments measured the relative displacement of two small indents 100 μm behind a fatigue crack tip in a three-point bend specimen to determine the CTOD as a function of load (in static tests) or as a function of time (in dynamic tests).

Static computations show that a two-dimensional simulation can capture both the overall structural response, as measured by boundary quantities (such as the applied load and the crack mouth opening displacement), and the near-tip CTOD when crack closure is accounted for. Dynamic finite element simulations also use a two-dimensional model of the three-point bend specimen, but must include the projectile and loading tup to account for the complex stress wave interactions generated during impact loading in order to simulate the CTOD versus time response. Since the finite element model correlates with the observed CTOD history, it can be used to determine relevant field quantities, such as the J-integral, as well as stresses and strains of interest.

The finite element model shows that it is essential to include the effects of viscoplasticity in order to capture the correct CTOD response in the dynamic experiments. These calculations show that the rate of change of the stress intensity factor is $\dot{K} \sim 5 \times 10^6$ MPa $\sqrt{m}$/s in the impact fracture tests. The fracture initiation time has not been unambiguously determined; therefore a unique fracture toughness has not been established.

KEY WORDS: fracture, impact fracture, dynamic fracture initiation, fracture toughness, finite element method, J-integral, crack tip opening displacement, stress waves

One of the primary goals of fracture mechanics is to develop methodologies for predicting crack initiation and growth in structural systems and components under given operating conditions. For brittle materials, the stress intensity factor approach has been used with remarkable success under both static and dynamic loading conditions [1,2]. For ductile materials under quasi-static loading, J-integral resistance (J-R) curves are used to evaluate fracture toughness [3].

However, when impact loading rates are high enough for stress-wave effects to dominate, remote measures of load and displacement (such as instrumented loading tups and crack mouth opening displacement [CMOD] gages) cannot accurately reflect the state of loading (or crack driving force) and deformation at the crack tip. But it is precisely these local crack

---

[1] Associate Professor, Department of Mechanical Engineering, The Johns Hopkins University, Baltimore, MD 21218.

[2] Post-Doctoral Fellow, Department of Mechanical Engineering, The Johns Hopkins University, Baltimore, MD 21218. Current address: Aerospace Engineering Department, University of Maryland, College Park, MD 20742.

tip quantities that govern crack growth. Impact conditions therefore require measurements to be made very close to the crack tip.

Background

Dynamic plane strain fracture initiation toughness (K_{Id}) values for three-point bend steel specimens subjected to the impact of a falling weight were reported by Shoemaker and Rolfe [4]. Their specimens were instrumented with foil strain gages placed adjacent to the crack tip and near the point of impact and loading rates reported corresponded to a rate-of-change of the stress intensity factor, $\dot{K}_I = 2 \times 10^5$ MPa $\sqrt{m}$/s. Strains recorded dynamically were converted to load by assuming that the strain distribution was the same in the dynamic fracture event as that in the equivalent quasi-static test. The strain gage nearest the crack tip was used to determine dynamic crack initiation. Thus stress wave loading effects were ignored and conditions of linear elastic fracture mechanics (LEFM) were assumed to provide an adequate description of the structural response of the specimen.

Extension of the ASTM E 399 standard for static fracture toughness testing to dynamic testing was evaluated in a round-robin program [5] conducted in high-speed closed-loop testing machines using clip gages to record the crack mouth opening displacement (CMOD). Compact and bend specimens of A533 steels were tested at loading rates corresponding to a rate-of-change of the stress intensity factor, $\dot{K}_I \sim 1 \times 10^4$ MPa $\sqrt{m}$/s. In such tests the "ringing" observed as a result of the dynamic response of the clip gage restricts the loading rate for which meaningful results are still obtainable.

The method of caustics can be used to determine stress intensity factors for both transparent and opaque materials. For a review of the method, including dynamic testing applications, see Kalthoff [6,7]. Ravichandar and Knauss [8] have measured the stress intensity factor history in transparent specimens at a rate on the order of $\dot{K}_I = 1 \times 10^6$ MPa $\sqrt{m}$/s by loading their double-cantilever specimens with an exploding foil. Rosakis et al. [9] have used reflected caustics for determining stress intensity factors during dynamic crack propagation in steel specimens. While the authors cited above have used the method of caustics to great effect, there are practical considerations which limit its use for dynamic fracture initiation toughness measurements. The first limitation of the method of caustics is that the "initial curve" (which is the locus of illuminated points on the specimen which, on reflection, form the caustic curve) must be on the order of one half of the specimen thickness away from the crack tip if current models of near-tip deformation are to be used to obtain the stress intensity factor from the caustic curve [10]. If a steel specimen has a thickness of 10 mm, then the time taken for an elastic wave to travel from the crack tip to the initial curve (this is the minimum delay time between an event at the crack tip and its effect on the caustic curve) is on the order of 1 μs. Thus fracture initiation events which take place within a few microseconds cannot accurately be measured by the method of caustics, because stress wave loading effects causing crack initiation cannot be accurately correlated with the deformation seen by the initial curve. The second limitation is that determination of the stress intensity factor K (or the J-integral) from the caustic curve requires a model of the near-tip deformation field. Under conditions of LEFM or for ductile materials subjected to quasi-static loading, these models are well developed [11,12]. Under dynamic loading of blunted notches for which crack initiation took place about 700 μs after drop-weight loading, Rosakis et al. [13] used a three-dimensional static finite element analysis to correlate observed caustics with J-integral measures of toughness. For stress wave dominated loading and for crack initiation on the order of 1 to 20 μs severe difficulties arise in the use of caustics, particularly in ductile materials.

Some of the difficulties associated with dynamic fracture toughness testing of the more standard bend and compact specimens are overcome by the technique used by Costin et al. [14] and Wilson et al. [15]. The specimen is a long rod with a deep annular notch from which a precrack is grown in fatigue in a specially developed rotating bending machine. For dynamic testing, the specimen is loaded by an explosive charge. The rod is instrumented with strain gages on the rod on either side of the annular notch and with a unique optical gage for obtaining the opening displacement of the crack. Loading rates greater than 1×10^6 MPa $\sqrt{m}$/s, corresponding to a loading pulse risetime on the order of 25 μs, can be obtained. For brittle materials, fracture initiation occurs under conditions of small-scale yielding, and remote measurements of load and CMOD are sufficient to determine the initiation fracture toughness using a stress intensity factor approach. For ductile materials, the J-integral can be estimated from remote measurements using the deep crack formula by Rice et al. [16]. A dynamic elastic-plastic finite element analysis of this experiment, conducted by Nakamura et al. [17], shows that, for a relative crack depth greater than 0.7, deep crack formulae based on the load transmitted across the ligament and the notch opening displacement yield J-integral results within 20% of the actual values.

Nakamura et al. [18,20] have analyzed the three-point bend specimen subject to smoothed drop-weight loading using both simple beam models and finite element computations. The motivation for their work is the interpretation of dynamic fracture results obtained from this simple commonly used specimen [21]. They conclude that for fracture initiation occurring at time $t_i > 2t_T$, experimentally measured quantities such as CMOD and load can be used to obtain a measure of the dynamic fracture toughness, J, where the transition time, $t_T \sim 24H/c_o$ (here H is the specimen thickness and c_o the longitudinal elastic bar wave speed for the specimen). For remote measures to give the fracture toughness, $t_i > 600$ μs for a 50 mm wide specimen. For the specimens used in the experimental work cited here [22], with a width of 12.5 mm, $t_i > 150$ μs for remotely measured field quantities to reflect the stress and deformation states at the crack tip. When $t_i < 2t_T$, discrete wave effects need to be taken into account; for this reason the computations described in the Finite Element Procedures section were performed.

Impact fracture toughness of structural materials is of great technological importance, and the studies mentioned above do not cover the full spectrum of investigations. For example, Server [23] and Chevallier [24] report fracture toughnesses under high-rate loading. Novel experimental methods, other than those already cited, often have advantages over more standard techniques, such as those proposed by Klepaczko [25] and Giovanola [26]. For dynamic loading conditions *not* dominated by stress wave loading effects, the simple mass-spring analysis of Williams [27,28] can produce surprisingly useful results.

Basic Experimental Procedure

Measurement of Crack Tip Opening Displacement (CTOD)

The crack tip opening displacement (CTOD) was measured at 100 μm behind the fatigue crack tip by means of the interferometric strain/displacement gage (ISDG) [29,30] technique. This laser-based optical system measures the relative displacement between two reflecting indentations on a specimen. These indentations are impressed into the polished specimen surface with a Vickers microhardness tester and are very small (about 20 μm square and 4 μm deep) and close together (they were placed 70 μm apart for these CTOD measurements). When the indents are illuminated with a laser, interference patterns form in space. The motion of these fringe patterns is related to the change in distance between the indentations. This principle is simply the two-slit interference phenomenon of Young [31] except that it is in reflection. More details of the optical principles are given in Ref *32*.

The resolution of the ISDG system depends on the wavelength of the laser and the geometry of the setup [29,22]. For the He-Ne laser used (with a wavelength of 0.6328 μm) one fringe motion (dark-light-dark) corresponds to a relative indent displacement of 0.94 μm. Since the optical detectors used to monitor fringe motion are at a fixed position, an intensity change from dark to bright and back to dark corresponds to a relative change in displacement between indents of 0.94 μm. The intensity changes recorded by the detector will have a generally sinusoidal shape, although the period will not be constant unless the displacement is changing linearly with load or time. Since it is easy to identify the maxima and minima of the recorded signal, the basic resolution of the measurement is approximately 0.5 μm.

The recording system for the fringe motion was a minicomputer-based data acquisition system. This allowed easy plotting of the CTOD versus load for static tests and CTOD versus time for dynamic tests. In both static and dynamic tests, motion of the specimen perpendicular to the CTOD measurement (i.e., in the direction of loading) does not cause fringe motion. Note also that it is the location of the maximum or minimum that is measured, not its magnitude; thus the illuminating laser beam does not have to be of uniform intensity.

Static Experiments

The static tests on the three-point bend (3PB) specimens were conducted in the electro-hydraulic test machine with the fringe motions measured by photoresistors and recorded by a minicomputer. At the same time, the crack mouth opening displacement was recorded with a clip-gage; this load-displacement plot was used to establish the critical load, P_q, by following the ASTM E 399 procedures.

Clip-gage data for these HY-100 specimens show that the conditions of plane strain fracture toughness are not met [22]. Fractographic analysis of the fracture surfaces showed large amounts of plastic deformation, indicating tough ductile fracture.

Impact Fracture Experiments

The three-point bend specimens were impacted by an aluminum projectile, 25 mm in diameter and 150 mm long, fired by a gas gun at a velocity of about 47 m/s. The aluminum projectile has a blunt nose, and it impacts a hardened steel tup that is initially placed in contact with the specimen. The tup has a wedge-shaped end lined up with the crack in the specimen. The projectile velocity was measured by two wires spaced 25 mm apart and connected to a resistance divider network. The time between the jumps in voltage as a projectile touched each wire was used to determine velocity.

Fractographic analysis of the fracture surfaces revealed a morphology very similar to that exhibited in the static experiments, again indicating tough ductile fracture.

Finite Element Procedures

Basic Modeling Procedures

In order to simulate the experiments [22] outlined in the Basic Experimental Procedure section, elastic-plastic quasistatic and dynamic elastic-viscoplastic finite element calculations were performed. The major assumption used in both static and dynamic computations is that the three-point beam structural response is predominantly that of plane stress [10] and that conditions of plane stress exist on the surface of the specimen very close to the crack tip. Thus plane stress elements were used throughout, rendering the finite element simulation two-dimensional. Other assumptions were that small deformations occur, allowing the decomposition of the small strain tensor ε_{ij} into elastic ε_{ij}^e and plastic ε_{ij}^p components (i.e., $\varepsilon_{ij} = \varepsilon_{ij}^e + \varepsilon_{ij}^p$, and that crack curvature does not play a significant role in the CTOD as measured on the surface.

Since each specimen tested had a fatigue crack grown (per ASTM E 399) under low cyclic stress from a machined notch, a certain amount of crack closure was observed. Since the indents used to observe the CTOD are only 100 μm behind the fatigue crack tip, closure is much more evident than it is when CMOD measurements are made. To account for this closure, the closure load (or the load required to bring the crack surfaces out of contact) was added to the static response.

Plasticity Model

A fairly simple elastic-plastic J_2-flow theory model of plasticity, with isotropic hardening and, for the dynamic computations, power-law viscoplasticity, was adopted here, since the loading caused almost proportional stress increments without significant unloading. For plastic loading, it is useful to introduce the deviatoric stress, s_{ij}, and the hydrostatic pressure, p, which are defined by

$$s_{ij} = \sigma_{ij} + p\,\delta_{ij} \quad \text{and} \quad 3p = -\sigma_{kk} \tag{1}$$

where σ_{ij} is the Cauchy stress, δ_{ij} is the Kronecker delta, and Latin subscripts take the value 1, 2, or 3 with repeated subscripts summed. Similarly, the deviatoric strain, e_{ij}, is defined by

$$e_{ij} = \varepsilon_{ij} - \varepsilon_{kk}\,\delta_{ij}/3 \tag{2}$$

Since the strain is composed of elastic and plastic parts (viz., $\varepsilon_{ij} = \varepsilon_{ij}^e + \varepsilon_{ij}^p$), so is the deviatoric strain (viz., $e_{ij} = e_{ij}^e + e_{ij}^p$).

In the case of infinitesimal plastic deformations, plastic straining results in no volume change ($\varepsilon_{kk}^p = 0$) and elastic response is used to relate the hydrostatic pressure to the volume change; that is, $p = -\kappa\varepsilon_{kk}$, where κ is the bulk modulus ($3\kappa = E/(1 - 2\nu)$). Also, for stresses below the current yield stress, σ_Y, and for unloading from a stress state on the yield surface, changes in the deviatoric components of elastic strain are related to changes in the deviatoric stress through

$$\dot{s}_{ij} = 2G\,\dot{e}_{ij}^e \tag{3}$$

where $2G = E/(1 + \nu)$ is the shear modulus, E is Young's modulus, and ν is Poisson's ratio. Here the superposed dot ($\dot{\ }$) indicates an increment in the static computations and the time rate of change in the dynamic calculations. For active plastic loading only, increments (or the rate of change) in the deviatoric plastic strain are related to the deviatoric stress such that the normality condition is satisfied, giving

$$\dot{e}_{ij}^p = \dot{e}^p\,\frac{3}{2}\frac{s_{ij}}{\sigma_e} \tag{4}$$

where $\dot{e}^p$ is the equivalent plastic strain rate (a scalar) and σ_e is the von Mises effective stress defined by $2\sigma_e^2 = 3s_{ij}\,s_{ij}$.

For rate-independent plasticity (used in the static calculations) the effective stress is given by the yield stress, σ_Y, which is a function of a single isotropic hardening parameter, the effective plastic strain; that is,

$$\sigma_e = \sigma_Y(e^p) \tag{5}$$

where $\sigma_Y(e^p)$ is given by the uniaxial stress-strain response of the relevant material (Fig. 1) [22].

In the dynamic computations, the effects of viscoplasticity (i.e., the increase in yield stress with increasing strain rate), as discussed by Cowper and Symonds [33], Perzyna [34], and Rice [35] must be accounted for. Here the effective strain rate is related to the ratio of difference between the effective stress and the static yield stress for the given plastic strain (i.e., $\sigma_Y(e^p)$) to that static yield stress through the viscosity coefficient D and the exponent q by

$$\dot{e}^p = D\,(\sigma_e/\sigma_Y - 1)^q \tag{6}$$

Since the projectile impacts the loading tup and specimen, all three components undergo high strain-rate deformation. The parameters D and q in Eq 6 were obtained by a careful review of high strain-rate testing [36–38] and are given in Table 1 (where σ_0 is the 0.2% proof stress). The strain-rate sensitivity and yield stresses are plotted in Fig. 2, which shows that the aluminum is not sensitive to strain rate while the tup and specimen are moderately sensitive to strain rate.

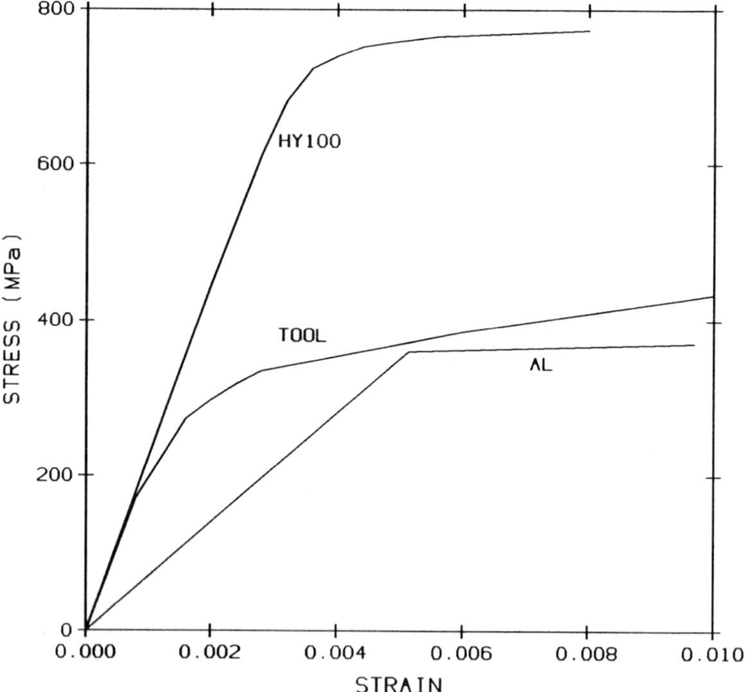

FIG. 1—*Stress versus strain response for materials.*

TABLE 1—*Material properties for dynamic test components.*

Component	Material	E	σ_0	D	q
3PB specimen	HY-100 steel	223 GPa	445 MPa	15.0/s	6.6
Loading tup	SAE01 tool steel	213 GPa	171 MPa	10.0/s	10.0
Projectile	6061T6 aluminum	70 GPa	360 MPa	6500/s	4.0

J-*Integral Calculations*

Since the loading experienced by the crack is essentially monotonic and stresses are nearly proportional, the energy integral J, as discussed by Nakamura et al. [*19*], is a measure of the near-tip deformation and was used to characterize the local crack tip fields. Following Ref *19*, the J-integral was computed using

$$J = \int_{\Gamma} \left(U\, n_1 - \sigma_{ij} n_j \frac{\partial u_i}{\partial x_1} \right) d\Gamma + \int_{A} \rho\, \frac{\partial^2 u_i}{\partial t^2} \frac{\partial u_i}{\partial x_1}\, dA \tag{7}$$

where U is the stress work density, u_i are the displacements, ρ is the material mass density, x_i is the spatial position vector, n_i is the outward normal to the contour Γ which surrounds

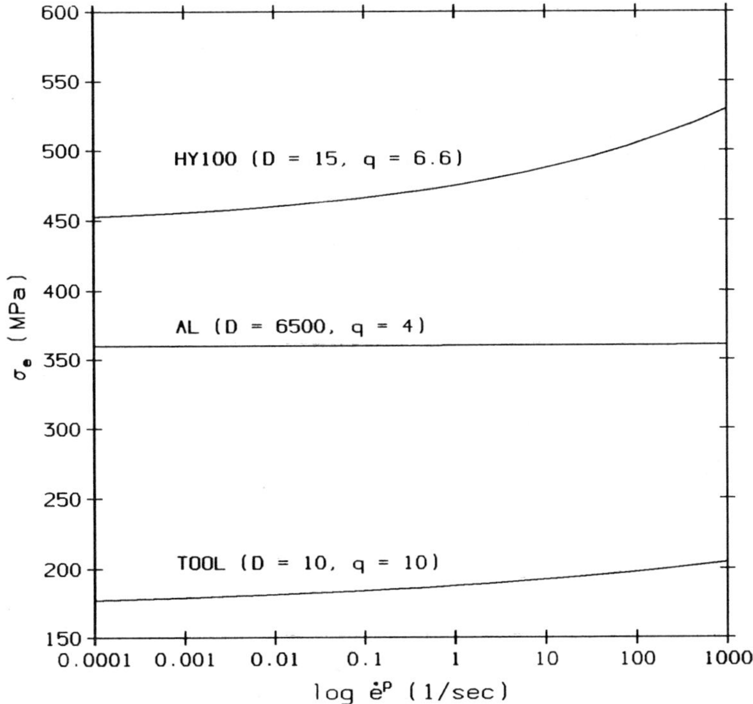

FIG. 2—*Effective stress versus strain rate for materials.*

the crack tip and encloses the area A, and t is time. Here the stress work density is given by

$$U = \int_0^t \sigma_{ij} \frac{\partial^2 u_i}{\partial \tau \partial x_j} \, d\tau \tag{8}$$

Static Model

The simulation of the quasi-static experiments requires only half of the three-point bend specimen to be modeled, which was done using 150 four-noded quadrilateral plane stress elements, with 18 eight-noded quadrilateral plane stress elements adjacent to the crack and 260 nodes (Fig. 3). The two eight-noded elements adjacent to the crack tip were singular elements, with the quarter-point technique used to simulate the $1/\sqrt{r}$ singularity in the stresses and strains for elastic response [*39*].

Dynamic Model

The dynamic finite element simulations of the impact fracture toughness experiments [*22*] described in the Basic Experimental Procedure section must include the projectile and loading tup as well as the three-point beam specimen in order to capture the details of the impact loading on the fracture specimen. Since a two-dimensional model is desirable for reasons of computational economy, plane stress elements are used throughout. The projectile and tup actually have rounded cross sections, and each element modeling these components had a thickness appropriate for its position and consistent with the mass distributions of the projectile and tup. In the dynamic model, the projectile had 55 elements and 77 nodes, the tup had 40 elements and 54 nodes, and the specimen had 178 elements and 273 nodes (Fig. 4).

The details of the mesh in the near-tip region used in the dynamic calculations are shown in Fig. 5. Note the position of the indents used to determine the CTOD relative to the crack tip. The proximity of the indents to the crack tip sets the scale of the near-tip mesh and also the time increment used to integrate the equations of motion.

The Hilber-Hughes-Taylor (HHT) implicit time integration scheme [*40*] was used to integrate the semi-discrete equations of motion. A small amount of structural damping was used to inhibit high-frequency oscillations by setting the HHT parameter $\alpha = -0.05$. A careful check on overall energy balance showed that this caused insignificant energy dissipation. A fixed time increment $\Delta t = 10^{-7}$ s was used for 150 increments to model the first 15 μs of

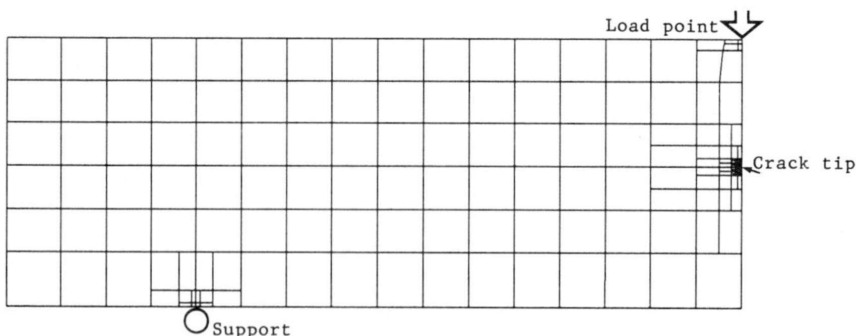

FIG. 3—*Mesh used in static calculations.*

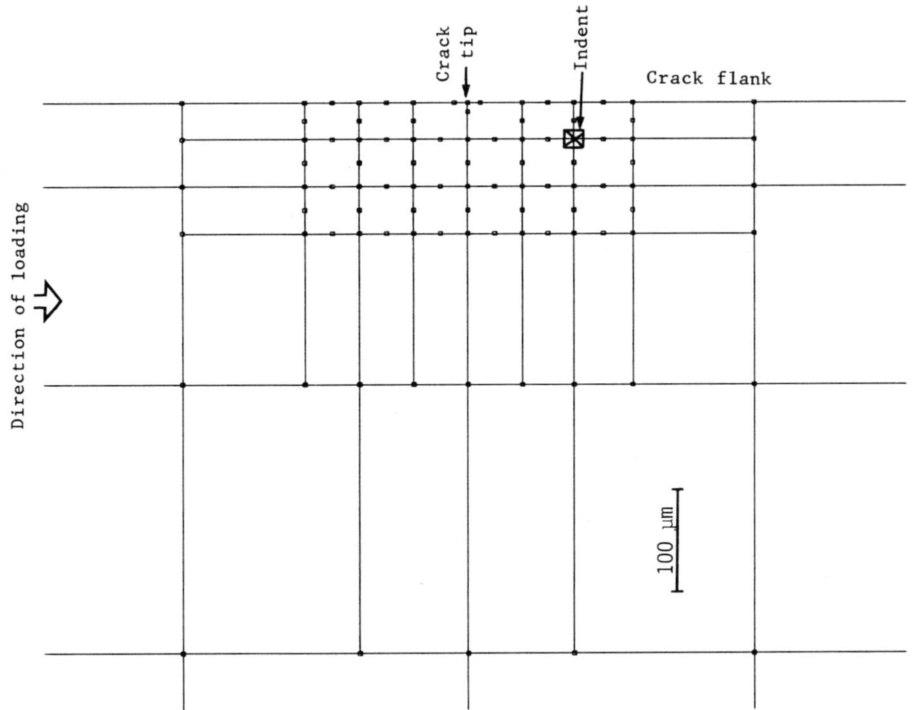

FIG. 5—*Detail of mesh in near-tip region.*

FIG. 4—*Overall mesh of projectile, tup, and specimen used in dynamic computations.*

the impact response. Note that this is significantly larger than the time-step that would be required if an explicit time integration scheme was used. Since the maximum time increment allowed for an explicit scheme is governed by the smallest dimension in any element, time-steps less than 5×10^{-9} s would be required. However, explicit time integration requires significantly reduced computational effort for each time step. Test calculations done with time steps less than 10^{-8} s and with variable time steps gave results identical to those reported.

All computations were done using the ABAQUS finite element program [41]. The computation of the J-integral, defined in Eq 7, was done using a locally written post-processing program. Most computations were done on a MicroVaxII, but a few larger cases with small time increments were run on the NRL Cray. For a discussion on both explicit and implicit time integration see Belytschko [42].

Results

In both the static and dynamic fracture toughness simulations the CTOD computed is compared with the experimentally observed CTOD. Since the CTOD is a key fracture parameter, the agreement between measured and calculated values of CTOD indicates the level of fidelity in the computational model. The finite element results contain information on all field quantities and can therefore be used to extract whatever details are required.

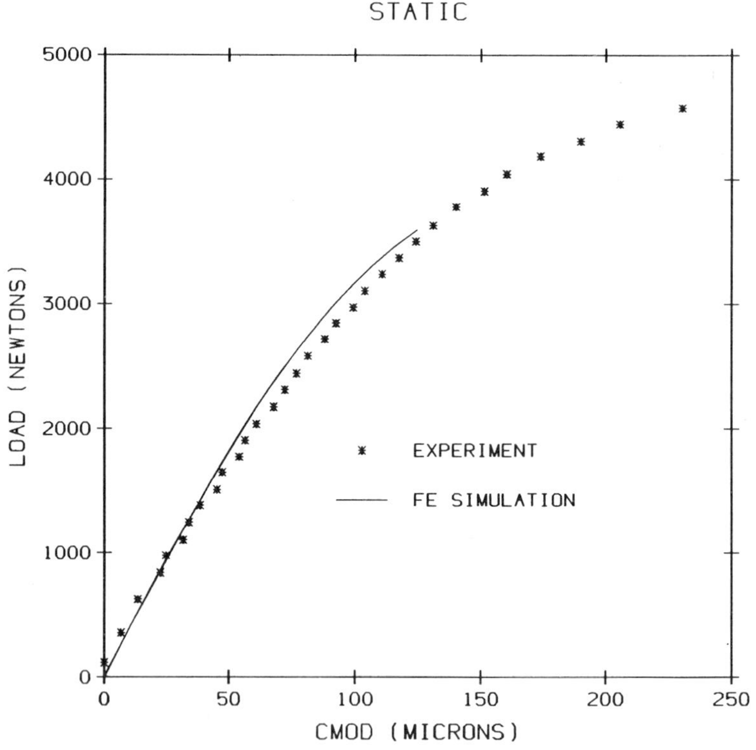

FIG. 6—*CMOD versus load—static case.*

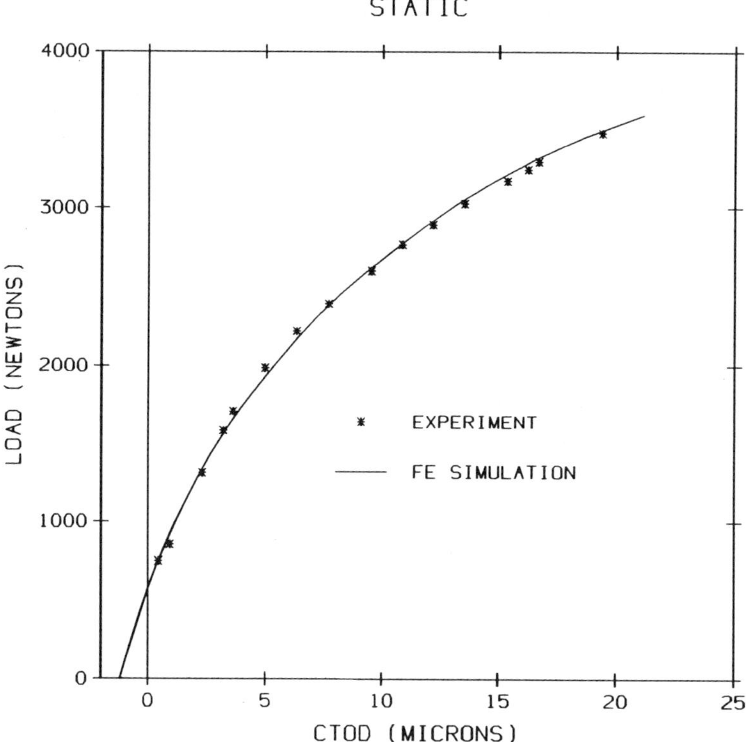

FIG. 7—*CTOD versus load—static case.*

Static Computations

The static experiments were instrumented to measure the CMOD (per ASTM E 399) and the CTOD as a function of load. Comparison of the computed and observed CMOD and CTOD is shown in Figs. 6 and 7 respectively. Note that even though the finite element simulation is carried out beyond the ASTM E 399 point of fracture initiation (P_q = 3.1 kN), there is no divergence of the computed CTOD (for which the crack tip is held stationary) and the observed CTOD (which, after crack initiation, should be larger than the computed CTOD for a given load). To account for the closure load of 570 N, which was determined from the CTOD measurements, the finite element results in Fig. 7 have been shifted parallel to the loading axis.

Figures 8 and 9 show the *J*-integral and the stress intensity factor $K = \sqrt{JE}$ as a function of the applied remote loading. Computation of the *J*-integral was done using two different contours (Γ) surrounding the crack tip and passing through elements not adjacent to the crack tip. The difference between the values given by these two contours is small but should be zero since the *J*-integral is path independent in this case. Since the *K* versus load response in Fig. 9 is linear up until the load level reaches 3 kN, small-scale yielding must dominate in this load range. It should be noted that the CTOD at P_q is approximately 14 μm. If a critical CTOD (Δ_{crit}) criterion is used to establish the onset of fracture, then Δ_{crit} = 14 μm.

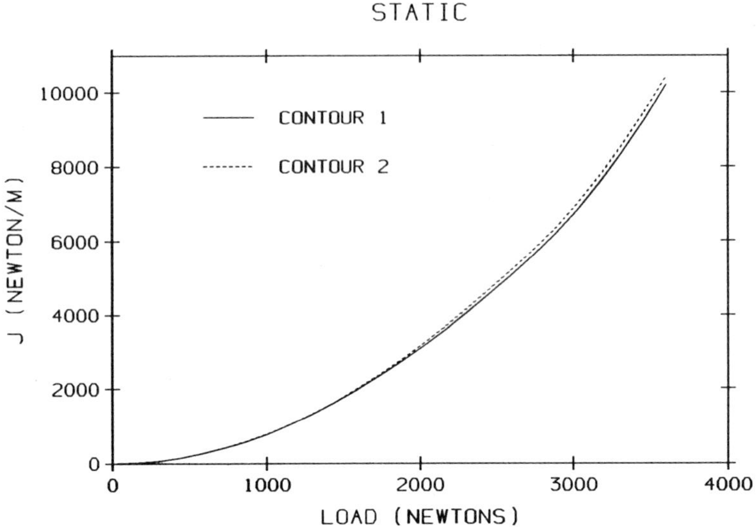

FIG. 8—*J-integral versus load—static case.*

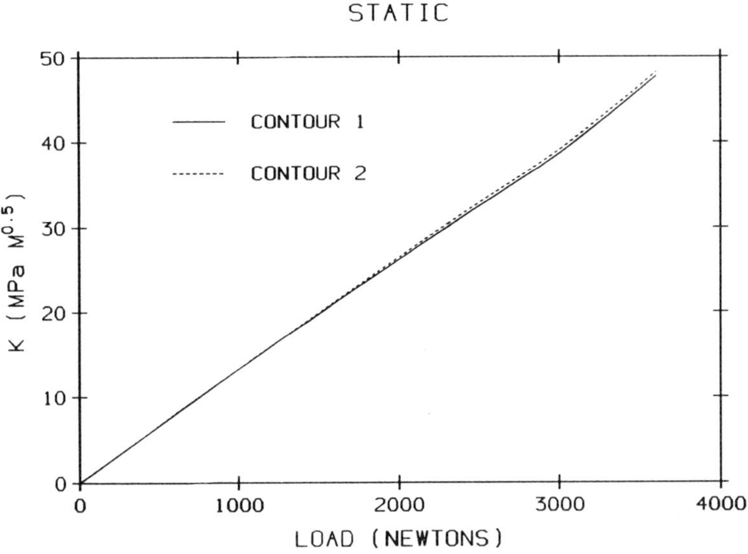

FIG. 9—*Stress intensity factor* K *versus load—static case.*

Dynamic Calculations

The only observed quantities during the dynamic experiments were the projectile impact velocity (which was used as the initial condition for the computational simulation) and the CTOD as a function of time. The exact time of impact is difficult to define experimentally, and the results shown are chosen such that the data correlate best. In the discussion that follows, time, t, refers to the time after the projectile impacts the loading tup and the time rate of change of the CTOD is denoted $\dot{\Delta}$.

As can be seen from Fig. 10, the finite element results show a slower increase in CTOD with time, for the first 2 μm of CTOD, than observed in the experiments. In fact, the experimental CTOD rate is about twice the computed opening rate ($\dot{\Delta}^{exp} \simeq 2 \times \dot{\Delta}^{fe}$). This may be due to the fact that crack closure would cause no CTOD to be observed until a sufficient stress level had developed to open the crack, at which time the local rate of change of field quantities has increased to cause a higher CTOD rate than would occur around a crack without prestrain and some strain hardening. It may also be due to the fact that the effective crack tip is further from the indent than the 100 μm used in the finite element model due to crack curvature. After the crack has opened 2 μm ($\Delta^{exp} = \Delta^{fe} = 2$ μm), the rate of crack tip opening is identical for the next 6 μs ($\dot{\Delta}^{exp} = \dot{\Delta}^{fe} = 1.32$ μm/μs).

Figure 10 also shows the marked difference between the CTOD calculated using the *static* stress versus strain response without viscosity (assuming therefore a purely elastic-plastic material), $\dot{\Delta}^{ep}$, and the CTOD computed when viscoplastic effects were accounted for, $\dot{\Delta}^{vp}$. Since the arrival time, t_A, of the load pulse is governed by the elastic waves, both rate-independent and rate-dependent models, as expected, show the same response, $t_A = 4$ μs. However, the rate of change of CTOD in the rate-insensitive case ($\dot{\Delta}^{ep} = 0.12$ μm/μs for $t_A < t < 8$ μs and $\dot{\Delta}^{ep} = 0.33$ μm/μs for 8 μs $< t < 15$ μs) is much lower than either the observed or numerically computed CTOD rates. This is because material in the tup and specimen is just not stiff enough without viscosity to transmit the correct loading rate.

The J-integral was computed as a function of time using Eq 7 [*19*] for two different con-

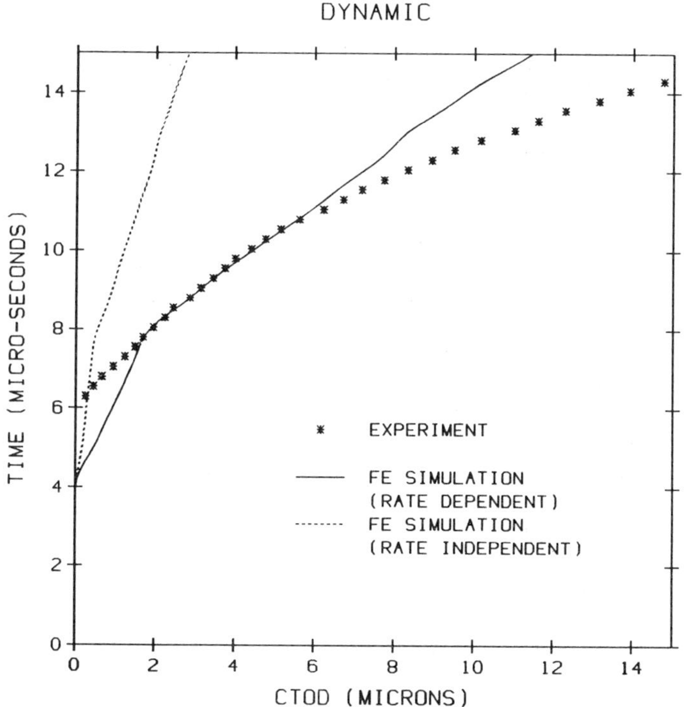

FIG. 10—*CTOD versus time—dynamic case.*

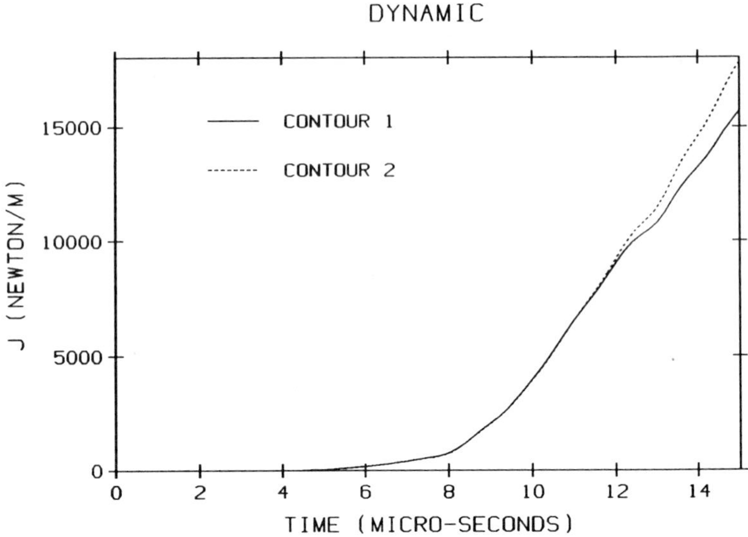

FIG. 11—*J-integral versus time—dynamic case.*

tours, Γ, as shown in Fig. 11. For dynamically loaded stationary cracks the *J*-integral should remain independent of Γ unless stress waves give rise to significant jumps in the stress and velocity fields (see the comparison of computed versus analytical results in Ref *19*). Figure 11 shows that the two contours give the same results for the first 12 μs after impact, which is approximately the time at which fracture initiation occurs. This means that, although the fracture initiation is dominated by stress wave effects, there are no significant discontinuities in stress influencing dynamic fracture initiation.

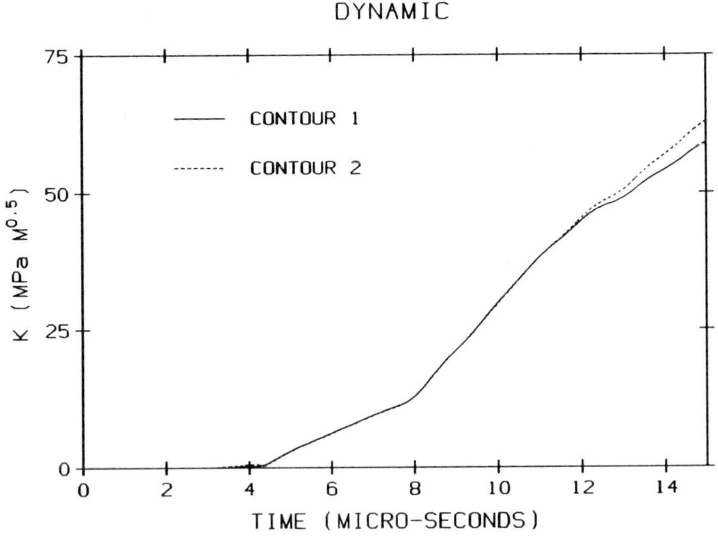

FIG. 12—*Stress intensity factor K versus time—dynamic case.*

If the J-integral is used to compute an equivalent stress intensity factor $K = (\sqrt{JE})$, then, as shown in Fig. 12, the stress intensity factor is seen to rise at two constant rates. For the first 4 μs the crack is unaware of the impact. For the next 2 μs $\dot{K} \simeq 3.2 \times 10^6$ MPa$\sqrt{\text{m}}$/s, after which the loading rate increases to $\dot{K} \simeq 8.4 \times 10^6$ MPa$\sqrt{\text{m}}$/s for the next 4 μs.

The most difficult task in determining the fracture toughness of the tested steel is identifying the onset of crack initiation. A purely experimental method, suggested by Sharpe et al. [22], assumes that the loading rate $\dot{K}$ is constant. The CTOD versus time response is converted to CTOD versus effective load using the initial elastic specimen behavior. The ASTM E 399 offset technique is then used to determine the point of initiation and the fracture toughness. Sharpe reports average dynamic fracture toughness values for the HY-100 steel simulated here of $K_d \simeq 64$ MPa$\sqrt{\text{m}}$. A critical CTOD criterion, which assumes that the crack initiates at a given CTOD regardless of loading rate, would give a dynamic fracture toughness $K_d \simeq 56$ MPa$\sqrt{\text{m}}$.

Comparing the finite element simulation with the observed CTOD response shows that after about 11.5 μs (or 6.5 μm) the observed CTOD diverges significantly from the computed CTOD indicating crack growth because the observed CTOD is larger. If the crack did indeed initiate at the point of divergence of these two curves, then the dynamic fracture toughness is only $K_d \simeq 42$ MPa$\sqrt{\text{m}}$, or only about 80% of the static fracture toughness $K_q \simeq 54$ MPa$\sqrt{\text{m}}$ [22].

Conclusions

It has been demonstrated that simple finite element computations can capture the CTOD history for fracture specimens subjected to impact loading. It is essential to include all the components used in the testing (the projectile, loading tup, and fracture specimen) and to use a viscoplastic material law which describes the stain-rate sensitivity of each of these components. Field quantities provided by the numerical model can be used to compute information of interest, such as the J-integral.

One difficulty that needs to be addressed is that there is no clearly observable point at which the crack initiates under impact loading. This means that there is no independent verification of crack initiation, which is taken to be the time at which computed and observed CTOD responses diverge. However, it is exactly this point of initiation which must be used to quantify dynamic fracture toughness. One possible means of verifying a particular toughness is to use a multispecimen approach where each specimen is impacted with a projectile of different length. If no crack growth was observed for some range of projectile length, then this could be used to place bounds on the fracture toughness.

Using the divergence of computed and observed CTOD as the fracture initiation point, these computations show that the dynamic fracture toughness of the HY-100 steel modeled here, with a static $K_q \simeq 54$ MPa$\sqrt{\text{m}}$, is reduced by about 20% to an equivalent dynamic fracture toughness $K_d \simeq 42$ MPa$\sqrt{\text{m}}$. However, using other criteria for the onset of fracture, the dynamic fracture toughness is somewhat higher than the static toughness (as discussed in the Results section).

Acknowledgments

This research was supported by the Office of Naval Research through Contract N00014-85-K-0711. The experimental work was done under Grant DAAG-29-85-K-0218 from the Army Research Office in conjunction with the David Taylor Research Center (Annapolis) and the Aberdeen Proving Grounds. We would also like to thank Professor Sharpe for all the information and helpful suggestions that he has provided.

References

[1] Kanninen, M. F. and Popelar, C. H., *Advanced Fracture Mechanics*, Oxford, New York, 1985.

[2] Rolfe, S. T. and Barsom, J. M., *Fracture and Fatigue Control in Structures*, Prentice-Hall, Englewood Cliffs, N.J., 1977.

[3] Gudas, J. P. and Davis, D. A., "Evaluation of the Tentative $J(I)$-R Curve Testing Procedure by Round Robin Tests of HY-130 Steel," *Journal of Testing and Evaluation*, Vol. 10, 1982, pp. 252–262.

[4] Shoemaker, A. K. and Rolfe, S. T., "The Static and Dynamic Low-Temperature Crack-Toughness Performance of Seven Structural Steels," *Engineering Fracture Mechanics*, Vol. 8, 1971, pp. 319–339.

[5] Shoemaker, A. K. and Seeley, R. S., "Summary Report of Round Robin Testing by the ASTM Group E24.01.06 on Rapid Loading Plane-Strain Fracture Toughness Testing," *Journal of Testing and Evaluation*, Vol. 11, 1983, pp. 261–272.

[6] Kalthoff, J. F., "On the Mechanics of Dynamic Fracture Phenomena," in *Elastic-Plastic Fracture Mechanics*, L. H. Larsson, Ed., Riedel, Dordrecht, 1985, pp. 479–508.

[7] Kalthoff, J. F., "Shadow Optical Method of Caustics," *Handbook on Experimental Mechanics*, Prentice-Hall, Englewood Cliffs, N.J., 1987, Chapter 9.

[8] Ravi-Chandar, K. and Knauss, W. G., "Dynamic Crack-Tip Stresses under Stress Wave Loading—A Comparison of Theory and Experiment," *International Journal of Fracture*, Vol. 18, 1982, pp. 209–222.

[9] Rosakis, A. J., Duffy, J., and Freund, L. B., "The Determination of Dynamic Fracture Toughness of AISI 4340 Steel by the Shadow Spot Method," *Journal of the Mechanics and Physics of Solids*, Vol. 32, 1984, pp. 443–460.

[10] Rosakis, A. J. and Ravi-Chandar, K., "On Crack-Tip Stress State: An Experimental Evaluation of Three-Dimensional Effects," *International Journal of Solids and Structures*, Vol. 22, 1986, pp. 121–134.

[11] Rosakis, A. J. and Freund, L. B., "The Effect of Crack-Tip Plasticity on the Determination of Dynamic Stress-Intensity Factors by the Optical Method of Caustics," *Journal of Applied Mechanics*, Vol. 48, 1981, pp. 302–308.

[12] Rosakis, A. J. and Freund, L. B., "Optical Measurement of the Plastic Strain Concentration at a Crack Tip in a Ductile Steel Plate," *Journal of Engineering Materials and Technology*, Vol. 104, 1982, pp. 115–120.

[13] Rosakis, A. J., Zehnder, A. T., and Narasimhan, R., "Dynamic Measurement of the J-Integral by the Method of Caustics," GALCIT Report SM-87-27, CalTech, 1987.

[14] Costin, L. S., Duffy, J., and Freund, L. B., "Fracture Initiation in Metals under Stress Wave Loading Conditions," in *Fast Fracture and Crack Arrest, ASTM STP 627*, G. T. Hahn and M. F. Kanninen, Eds., American Society for Testing and Materials, Philadelphia, 1977, pp. 301–318.

[15] Wilson, M. L., Hawley, R. H., and Duffy, J., "The Effect of Loading Rate and Temperature on Fracture Initiation in 1020 Hot-Rolled Steel," *Engineering Fracture Mechanics*, Vol. 13, 1980, pp. 371–385.

[16] Rice, J. R., Paris, P. C., and Merkle, J. G., "Some Further Results of J-Integral Analysis and Estimates," in *Progress in Flaw Growth and Fracture Toughness Testing, ASTM STP 536*, American Society for Testing and Materials, Philadelphia, 1973, pp. 231–245.

[17] Nakamura, T., Shih, C. F., and Freund, L. B., "Elastic-Plastic Analysis of a Dynamically Loaded Circumferentially Notched Round Bar," *Engineering Fracture Mechanics*, Vol. 22, 1985, pp. 437–452.

[18] Nakamura, T., Shih, C. F., and Freund, L. B., "Analysis of a Dynamically Loaded Three-Point-Bend Ductile Fracture Specimen," *Engineering Fracture Mechanics*, Vol. 25, 1986, pp. 323–339.

[19] Nakamura, T., Shih, C. F., and Freund, L. B., "Computational Methods Based on an Energy Integral in Dynamic Fracture," *International Journal of Fracture*, Vol. 27, 1985, pp. 229–243.

[20] Nakamura, T., Shih, C. F., and Freund, L. B., "Three-Dimensional Transient Analysis of a Dynamically Loaded Three-Point-Bend Ductile Fracture Specimen," Brown University Report No. ONR0365/3, Providence, R.I., 1986.

[21] Joyce, J. A. and Hackett, E. M., "An Advanced Procedure for J-R Curve Testing Using a Drop Tower," in *Nonlinear Fracture Mechanics: Vol. I—Time-Dependent Fracture, ASTM STP 995*, American Society for Testing and Materials, Philadelphia, 1989, pp. 298–317.

[22] Sharpe, W. N., Jr., Douglas, A. S., and Shapiro, J. M., "Dynamic Fracture Toughness Evaluation by Measurement of CTOD," Johns Hopkins Mechanical Engineering Report No. WNS/ASD-88-02, Baltimore, Md., Feb. 1988.

[23] Server, W. L., "Impact Three-Point Bend Testing for Notched and Precracked Specimens," *Journal of Testing and Evaluation,* Vol. 6, 1978, pp. 29–34.

[24] Chevallier, J. M., Ansart, J. P., and Dormeval, R., "Fracture Toughness of Some Metals under High Loading Rate Conditions," Institute of Physics, Conference Series Number 70, 1984, pp. 229–236.

[25] Klepaczko, J. R., "Discussion of a New Experimental Method in Measuring Fracture Toughness Initiation at High Loading Rates by Stress Waves," *Journal of Engineering Materials and Technology,* 1982, pp. 29–35.

[26] Giovanola, J. G., "Investigation and Application of the One-Point-Bend Impact Test," in *Fracture Mechanics: Seventeenth Volume, ASTM STP 905,* American Society for Testing and Materials, Philadelphia, 1986, pp. 307–328.

[27] Williams, J. G., "The Analysis of Dynamic Fracture Using Lumped Mass-Spring Models," *International Journal of Fracture,* Vol. 33, 1987, pp. 47–59.

[28] Williams, J. G., "The Analysis of Instrumented Impact Tests Using a Mass-Spring Model," *International Journal of Fracture,* Vol. 33, 1987, pp. 209–222.

[29] Sharpe, W. N., Jr., "Dynamic Strain Measurement with the Interferometric Strain Gage," *Experimental Mechanics,* Vol. 10, 1970, pp. 89–92.

[30] Sharpe, W. N., Jr., "Applications of the Interferometric Strain/Displacement Gage," *Optical Engineering,* Vol. 21, 1982, pp. 483–488.

[31] Jenkins, F. A. and White, H. E., *Fundamentals of Optics,* McGraw-Hill, New York, 1957.

[32] Sharpe, W. N., Jr., "Interferometric Surface Strain Measurements," *International Journal of Nondestructive Testing,* Vol. 3, 1971, pp. 59–76.

[33] Cowper, G. R. and Symonds, P. S., "Strain Hardening and Strain-Rate Effects in the Impact Loading of Cantilever Beams," Technical Report No. 28, Brown University, Providence, R.I., Sept. 1957.

[34] Perzyna, P., "Fundamental Problems in Viscoplasticity," *Advances in Applied Mechanics,* Vol. 9, Academic Press, New York, 1966.

[35] Rice, J. R., "On the Structure of Stress-Strain Relations for Time-Dependent Plastic Deformation in Metals," *Journal of Applied Mechanics,* Vol. 37, 1970, pp. 728–737.

[36] Kendall, D. P., "The Effect of Strain Rate and Temperature on Yielding in Steels," *Journal of Basic Engineering (Transactions of ASME),* Vol. 94, 1972, pp. 207–212.

[37] Lindholm, U. S. and Johnson, G. R., "Strain-Rate Effects in Metals at Large Shears," in *Material Behavior under High Stress and Ultrahigh Loading Rates,* J. Mescall and V. Weiss, Eds., Plenum Press, New York, 1983, pp. 61–79.

[38] Marchand, A. and Duffy, J., "An Experimental Study of the Formation Process of Adiabatic Shear Bands in a Structural Steel," Brown University Technical Report, Providence, R.I., April 1987.

[39] Akin, J. E., *Application and Implementation of Finite Element Methods,* Academic Press, New York, 1982.

[40] Hilber, H. M., Hughes, T. J. R., and Taylor, R. L., "Improved Numerical Dissipation of Time Integration Algorithms in Structural Dynamics," *Earthquake Engineering and Structural Dynamics,* Vol. 5, 1977, pp. 283–292.

[41] *ABAQUS Theory Manual,* Hibbitt, Karlsson and Sorensen, Inc., Providence, R. I., 1982.

[42] Belytschko, T., "A Survey of Numerical Methods and Computer Programs for Dynamic Structural Analysis," *Nuclear Engineering and Design,* Vol. 37, 1976, pp. 23–34.

K. Cho,[1] *J. P. Sklenak,*[1] *and J. Duffy*[1]

Dynamic Fracture Behavior of a Structural Steel

REFERENCE: Cho, K., Sklenak, J. P., and Duffy, J., **"Dynamic Fracture Behavior of a Structural Steel,"** *Fracture Mechanics: Twenty-First Symposium, ASTM STP 1074,* J. P. Gudas, J. A. Joyce, and E. M. Hackett, Eds., American Society for Testing and Materials, Philadelphia, 1990, pp. 126–143.

ABSTRACT: Fracture experiments are described in which stress wave loading is employed to attain a stress intensity rate of about $2 \cdot 10^6$ MPa $\sqrt{m}$/s. The specimens consisted of prefatigued notched round bars of a high strength, low alloy martensitic steel. The tests covered the temperature range $-150°C$ to $200°C$. Comparison is made with corresponding results of quasi-static tests on specimens of the same geometry as well as with results of other investigators who tested similar steels. The present steel contains manganese sulfide inclusions in the form of stringers and globules. Their role in the fracture process is examined by optical and scanning electron microscopy.

KEY WORDS: dynamic fracture, structural steel, microstructural inclusions, mechanical behavior

This paper presents a study of the dynamic fracture toughness of a high strength structural steel with a yield stress of 690 MPa, identified as nominally 100 ksi. Throughout this paper, the steel is referred to as HS Steel. Testing was carried out over a range of temperatures covering the transition from ductile to brittle fracture under both dynamic and quasi-static loading. Other investigators [1–4] have performed experiments with similar steels, although at somewhat lower loading rates. Hasson and Joyce [1] studied the effects of loading rate on the J_{Ic} fracture toughness of HY-80 and HY-130 steels, using *J*-resistance (*J-R*) analysis. Their tests were performed in a high rate servo-hydraulic loading machine. Gudas [2], by using a crack arrest toughness test procedure, obtained lower bound values of dynamic fracture toughness of an HY-80 steel over a range of testing temperatures. In both these investigations a modified compact specimen was employed and the results were supplemented by Charpy tests. More recently, Joyce and Hackett [4] investigated the fracture toughness of an HY-100 steel whose chemical composition and mechanical properties should resemble closely those of the present steel. In their tests, they used a modified three-point bend specimen with a fatigue crack grown in at the root of the notch.

The experimental technique used in the present study was introduced by Costin et al. [5] for the measurement of fracture toughness at high loading rates, and is based on the principle of Kolsky's split-Hopkinson bar. Its advantages, as compared with those of the Charpy test, are detailed in Ref 5. In this test, a sharp-fronted tensile pulse is initiated explosively at one end of a long cylindrical bar of the specimen material. This pulse propagates down the bar to a circumferential precracked fatigue notch, thus fracturing the specimen. A stress intensity

---

[1] Research Associate, Undergraduate Research Assistant, and Professor of Engineering, respectively, Brown University, Providence, R.I. 02912.

rate of approximately 2.0×10^6 MPa $\sqrt{\text{m}}$/s is achieved by this method compared with a rate of about 1 MPa $\sqrt{\text{m}}$/s in quasi-static fracture. It should be noted that, among other requirements, the incident tensile pulse must be made sufficiently large to fracture the specimen on the rising portion of the initial pulse.

The present test method has been analyzed in detail using finite element methods by Nakamura et al. [6]. They employ the same specimen geometry as in the present tests as well as the same form for the incident pulse. Their results indicate that stress wave loading with this test method provides reliable values of the stress intensity factors: K_{Id} for brittle and J_{Id} for ductile fracture. The analysis reveals the importance of the depth of the fatigue notch employed. It shows also that the stress state at fracture initiation is symmetric to either side of the fracture plane, at least locally near the crack tip, and that plane strain conditions prevail near the crack tip.

As expected, the experimental results show that the fracture toughness of the present steel has a lower shelf and upper shelf behavior that depends on test temperature. The transition temperature for quasi-static loading is about $-125°C$ and increases only to about $-110°C$ under dynamic loading. While fracture toughness above the transition temperature is sensitive to loading rate, below the transition temperature it does not appear greatly affected by the rate of loading.

After completion of the fracture tests, fractography was employed to establish correlations between metallurgical factors and the fracture initiation behavior. A characteristic feature of the fracture surface, found in specimens fractured at all temperatures in the testing range and for both loading rates, is a layering structure consisting of alternating sets of troughs and ridges. This layering is attributed to the presence of manganese sulfide (MnS) stringers aligned parallel to one another and to the rolling direction of the original steel plate; the stringers are normal to the axis of the bar and hence lie within the fracture plane.

Description of Experiments

The material used in this study is a low carbon, high strength structural steel, consisting of tempered martensite (Fig. 1). It is received as plate stock about 5 cm (2 in.) thick and tested in the as-received condition. Since the plates in the as-received condition are flame cut from larger plates, all material lying within 7 or 8 cm of any edge is sawn off and rejected. In addition, a layer 1.25 cm thick is removed from each surface of the remaining plate to avoid testing the more heavily work-hardened surface layers (Fig. 2). The hardness of the remaining core material was measured as 25 HRC. Table 1 gives the chemical composition of the steel, which should be compared with that of the steel tested by Joyce and Hackett [4]. The optical micrographs show MnS in the form of globular inclusions and stringers, and the banded microstructure (Fig. 1a) is aligned parallel to the direction of the MnS stringers (Fig. 1b) and parallel to the rolling direction. Figure 3 shows Charpy impact energy as a function of testing temperature. As may be seen, the ductile-brittle transition temperature of the present steel is approximately $-135°C$, which agrees closely with the Charpy tests results in Joyce and Hackett [4], although the values of Charpy energy that they obtain are greater than ours. Though Joyce and Hackett's steel and the present steel are quite similar, the fact that the present specimens are taken from the core of the plate may explain this difference. There are also small differences in chemical composition. The present steel has a higher sulfur content than Hackett's steel, and it has been recognized for some time that an increase in the sulfur content lowers Charpy energy values on the upper shelf [7]. There is also some difference in phosphorous content.

For the dynamic plane strain fracture test, the plate stock is machined into round bars 116 cm long and centerless ground to a diameter of 2.54 cm. To prepare the specimens a circum-

(a)

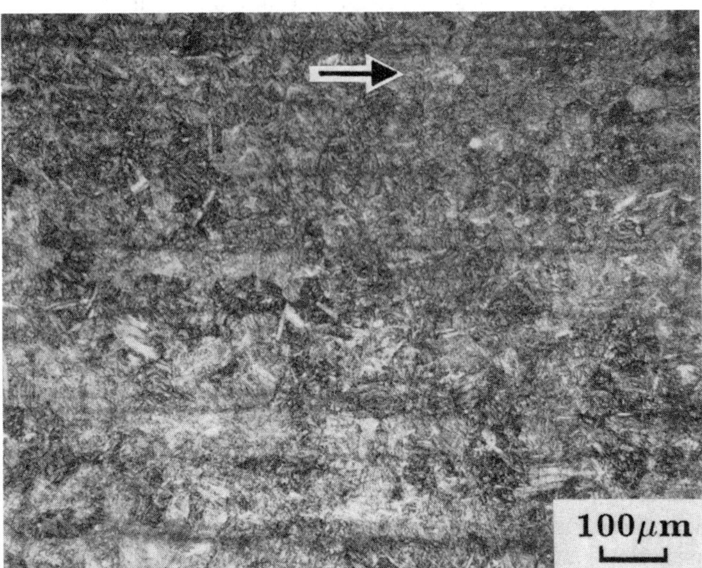

(b)

FIG. 1—*Underformed microstructure of the high strength steel showing the tempered martensite structure. (a) The banded microstructure shown lies parallel to the rolling direction (arrow) of the original plate. (b) MnS inclusion in the form of stringer, as indicated by the arrow.*

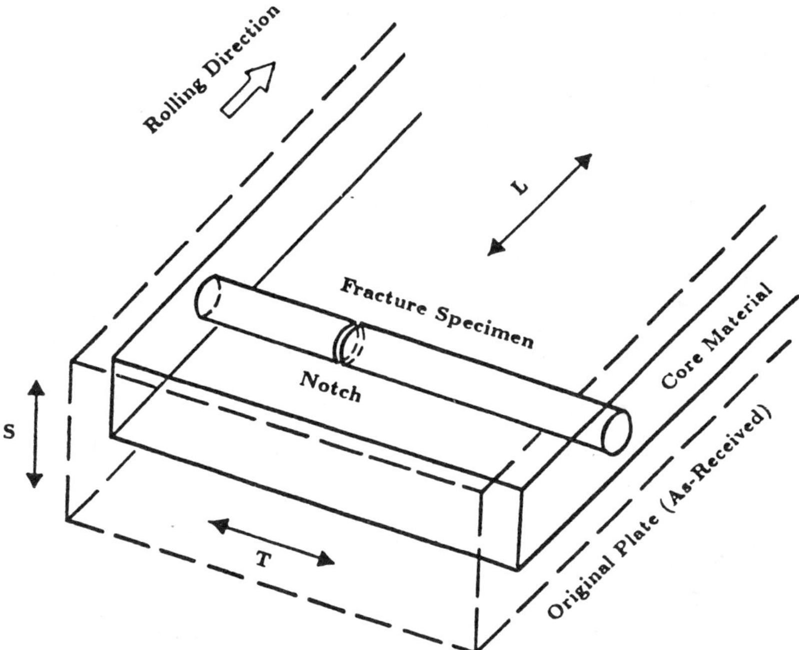

FIG. 2—*Orientation of the dynamic fracture specimen relative to the rolled plate. The MnS stringers lie in the rolling direction. Dashed lines indicate the as-received plate; solid lines the final plate after removing the flame cut ends and work-hardened surface.*

TABLE 1—*Chemical composition (wt %) of high strength steels.*

Alloy	C	Mn	P	S	Si	Ni	Cr	Mo	Pb	V
HS	0.19	0.29	0.006	0.014	0.23	2.56	1.68	0.45	0.013	...
HY100[a]	0.164	0.26	0.003	0.009	0.19	2.78	1.57	0.42	...	0.003

[a] Data from Ref 4.

ferential notch with a sharp root is machined into the bar 77 cm from the loading end, and a fatigue crack is grown in from the root of the notch. The notch has a width of 2.5 mm and is machined with a 60° angle at its tip. The orientation of the fracture plane relative to the rolling direction and to the stringers is shown in Fig. 2. In order to obtain a state of plane strain at fracture, it is essential that the fatigue annulus be of a uniform width and concentric with the outer diameter of the bar. This is achieved by employing a special rotating beam fatigue machine [5] that imposes a set displacement at the bar's center, rather than a set load. In addition, centerless grinding improves concentricity by producing a bar considerably straighter than does turning in a lathe. In the present experiments the fatigue crack is grown to a depth of about 0.10 cm, leaving an unfractured ligament approximately 0.80 cm in diameter.

The apparatus for the dynamic fracture test is shown in Fig. 4. A tensile loading pulse is initiated at one end of the bar by the detonation of an explosive charge placed against the specially shaped loading head. This rapidly rising tensile pulse propagates down the bar from

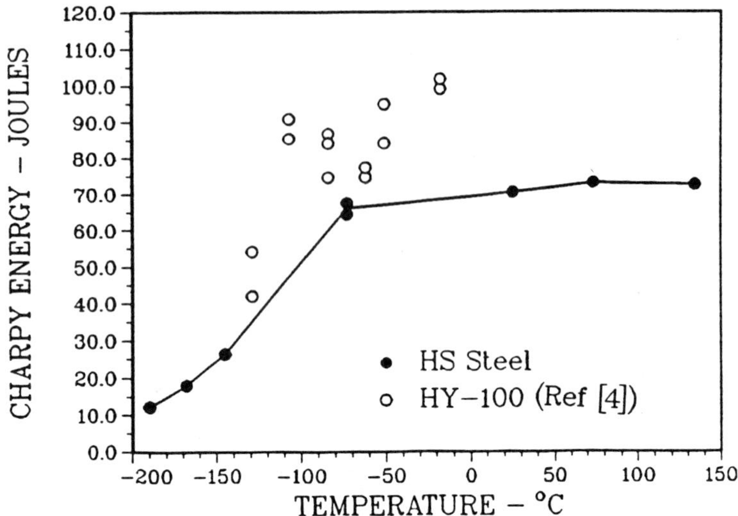

FIG. 3—*Comparison of present Charpy results with those obtained by Joyce and Hackett for an HY-100 steel* [4].

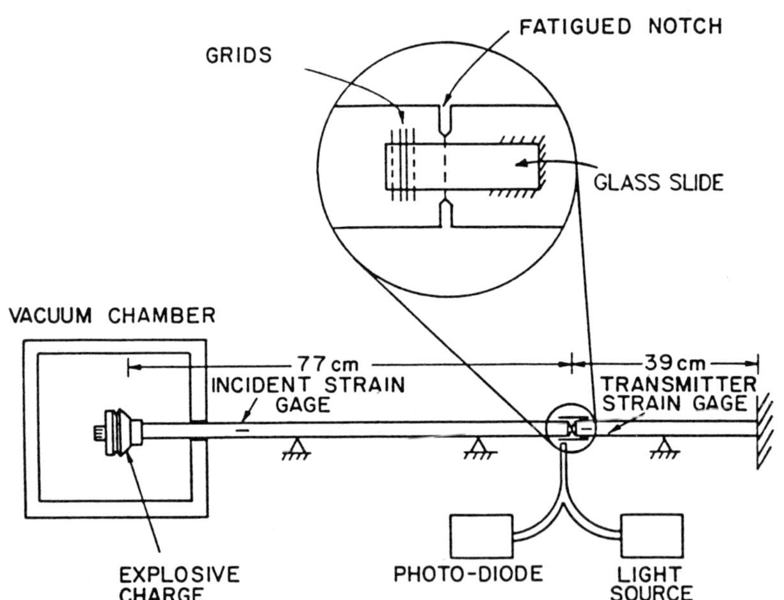

FIG. 4—*Schematic diagram of apparatus employed for dynamic fracture experiments.*

the loading head to the prefatigued section where fracture ensues. In the present tests the tensile pulse has a rise time of 35 to 40 μs, fracturing the specimen 10 to 15 μs after its arrival at the fracture site, thus assuring that fracture occurs on the rising portion of the incident pulse. The transmitted portion of the incident pulse is proportional to the load carried at the fracture site and thus also provides a measure of average stress at the fracture site as a function of time. An optical adaptation of the moiré technique is used to measure crack opening displacement (COD), or more precisely notch opening displacement, as a function of time. A more complete description of the test can be found in Costin et al. [5].

The material for the quasi-static fracture test is taken from the previously tested dynamic fracture specimen. It should be noted that the material outside the fracture region remains elastic in both the static and dynamic fracture tests. The geometry of the specimen employed in the quasi-static fracture initiation test is the same as for the dynamic, except that the overall length is reduced to 71 cm. A circumferential notch is machined near the center of this new bar and a pre-fatigued crack grown at the root of this notch. Thus static and dynamic fracture are performed with like specimens both loaded in tension. In both tests, plane strain conditions prevail at the crack tip. An environmental chamber is employed for tests at other than room temperature, and fracture in the static test is accomplished by using a Riehle universal testing machine. The notch opening displacement in the static test is determined optically by an MTI photonic sensor. Measurement of changes in electrical resistivity across the notch determines the load at the instant of fracture initiation, following the method described by Couque et al. [8].

Metallographic and fractographic investigations were conducted by using optical and scanning electron microscopes (SEM). The optical microscope was used mainly for observations of the undeformed microstructure of the specimen material. For this purpose, samples of undeformed material were polished and etched with a Nital solution. The fracture surfaces themselves were observed by SEM, using an AMR-1000A SEM and an attached EDAX (energy dispersive analysis of X-rays) unit.

Results and Discussion

Determination of Dynamic and Quasi-Static Fracture Toughness in Plane Strain

Fracture Parameters—For a notched round bar the stress intensity factor is given by Tada et al. [9] as

$$K_\mathrm{I} = \frac{P}{\pi R^2} \sqrt{\pi R}\, F\,(2R/D) \tag{1}$$

where R is the radius of the uncracked ligament after fatiguing, P is the applied load, D is the outer diameter of the bar, and $F(2R/D)$ is a dimensionless size function whose value for the present geometry is approximately 0.48. In order to apply linear fracture mechanics concepts, the size of the plastic zone at the crack tip must be small compared with the nominal dimensions of the specimen. The size requirement for a valid K_Ic test is

$$R \geqslant 2.5 \left[\frac{K_\mathrm{Ic}}{\sigma_\mathrm{y}} \right]^2 \tag{2}$$

where σ_y is the initial yield stress of the material obtained at a strain rate comparable to that attained near the root of the notch in the fracture test.

The HS steel tested in the present experiments is quite ductile and does not meet the requirements for a valid K_Ic test, as expressed by Eq 2, even for temperatures as low as $-150°C$ where the fracture mode is one of cleavage. As a result, a J integral approach was adopted. As a criterion for a valid J_Ic test, Paris [10] suggested that the ligament dimension satisfy

$$R \geqslant \alpha \left(\frac{J_\mathrm{Ic}}{\sigma_\mathrm{y}} \right) \tag{3}$$

with $\alpha = 50$. In the present tests this size requirement is met only at $-100°C$ and $-150°C$, statically or dynamically. Over the rest of the temperature range, the requirement is met if $\alpha = 25$, a value suggested by Landes and Begley [11] and by Hutchinson [12]. Rice et al. [13] have shown that the value of J for a notched round bar may be determined from a load-displacement curve by the deep crack formula

$$J_{dc} = \frac{1}{2\pi R^2} \left[3 \int_0^\delta Pd\delta - P\delta \right]$$ (4)

where δ is the critical value of the load-point displacement due to the presence of the crack. This expression for J_{dc}, while approximate, is appropriate for an interpretation of present results, since it is expressed in terms of P, δ, and R, all of which are measured in the experiment.

In using Eq 4, an appropriate value must be used for the load-point displacement, δ. Since the crack notch opening displacement is easily measured in the experiment, this value has been adopted in using Eq 4. A justification for this choice is provided by the finite element analysis of Nakamura et al. [6]. These investigators calculate J by employing a rigorous expression that includes dynamic effects, and compare results with those found using Eq 4 in which they used crack notch opening displacement for δ. Their results show that a slightly conservative value of J is obtained through the use of Eq 4, and that the accuracy of this equation improves with increasing plastic deformation prior to fracture initiation. This latter feature of their analysis implies good accuracy for the present results, since the present HS steel is highly ductile.

Once a reliable value of J_{Ic} has been determined, an equivalent fracture toughness, K_{Ic} is calculated from

$$K_{Ic} = \left[\frac{EJ_{Ic}}{1 - \nu^2} \right]^{1/2}$$ (5)

where E and ν are elastic constants of the material.

Use of the above equations in interpreting fracture data requires knowledge of the stress-strain behavior of the material at strain rates comparable with those expected in the neighborhood of the crack tip and over the range of testing temperatures. Fortunately, these data were available from previous work with the material in this steel plate [14]. As a result, only a few supplementary tests were necessary. These were performed with thin-walled tubular specimens loaded in a torsional Kolsky bar, following the method described in Ref 14. A dynamic strain rate in shear (about $10^3/s$) was imposed to correspond more nearly with the strain rate in the neighborhood of the crack tip in a dynamic fracture test. Quasi-static deformation tests ($10^{-4}/s$) were also performed for reduction of the quasi-static fracture data. Values of initial yield stress in shear are presented in Fig. 5 as a function of test temperature. The values of σ_y in Eq 3 are obtained by multiplying the yield stress in shear by $\sqrt{3}$, under the assumption that the von Mises yield criterion is applicable. Additional details on the stress-strain behavior of the material may be found in Ref 15.

Determination of the Time at Fracture Initiation—In previous experiments using the present fracture test method, fracture initiation in the dynamic test occurred at the instant of peak load, as determined by the transmitted pulse [5]. This was evident, for instance, in testing brittle steels, since a sudden drop occurs in the magnitude of the reflected pulse coinciding with peak load, and a sudden decrease in this pulse indicates a change in mechanical impedance at the fracture site, such as would be produced by a change in the cross-sectional

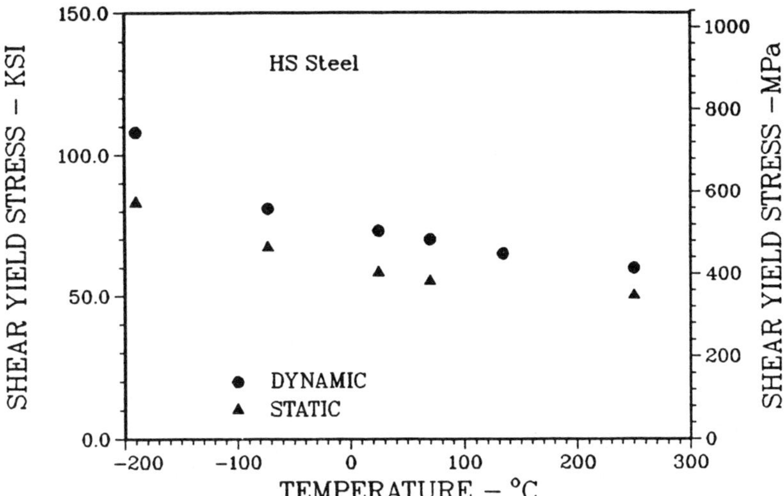

FIG. 5—*Yield stress in shear as a function of test temperature. Dynamic values (defined by the lower yield stress) are obtained at nominal strain rates of $10^3/s$ and quasi-static values (defined by 0.2% offset) at $10^{-4}/s$ [14].*

area due to fracture initiation. For more ductile steels, particularly at elevated temperatures, it was noted in previous tests that this change in impedance might precede the peak load by a short interval and this was taken into account in calculating J. The present steel, however, shows an even greater ductility than previous steels, and no sudden change in impedance is visible in tests conducted at $-20°C$ and above. As a result, for these dynamic tests the values of J_1, and the equivalent K_1, are all calculated on the basis of the maximum load transmitted at the fracture site. They will be referred to as J_{max} and K_{max}, respectively. It should be noted that the load-deformation curve in dynamic fracture tests with the present steel at temperatures of $-20°C$ and above can extend to considerably larger deformations than the static curve. The very large strains observed in shear banding experiments with the present steel [14], and its low hardness, are consistent with high ductility and, particularly, with high ductility in dynamic deformation. Dynamic tests at $-70°C$ and below, as well as all quasi-static tests, however, do provide the measured fracture toughness, K_{Ic}, corresponding to fracture initiation. In the dynamic tests at these low temperatures fracture initiation coincides with peak load, and the dynamic load deformation curve is of about the same length as the static.

The method employed in quasi-static testing is that adopted by Couque et al. [8], who performed separate tests to ensure that a measurement of electrical resistivity does correspond with fracture initiation. In their tests, subsequent microscopic observation of the material behind the crack tip served to distinguish crack blunting from fracture initiation. In summary, while the quasi-static values J_{Ic} and K_{Ic} are intended to represent values calculated at fracture initiation, the dynamic values, J_{max} and K_{max}, are measured at peak load, where some crack growth may have occurred. An exception occurs in dynamic tests at $-70°C$ and below, where the material is brittle enough for K_{Ic} values to be determined.

Results of Fracture Experiments—Calculated values of the stress intensity factors, based on the experimental results, are given in Fig. 6 as a function of test temperature. It is noted that none of the tests satisfies the ASTM requirement for minimum specimen thickness, as specified by Eq 2. As a result it was necessary to calculate a J-integral by means of Eq 4. An equivalent value of the fracture toughness was then determined through use of Eq 5.

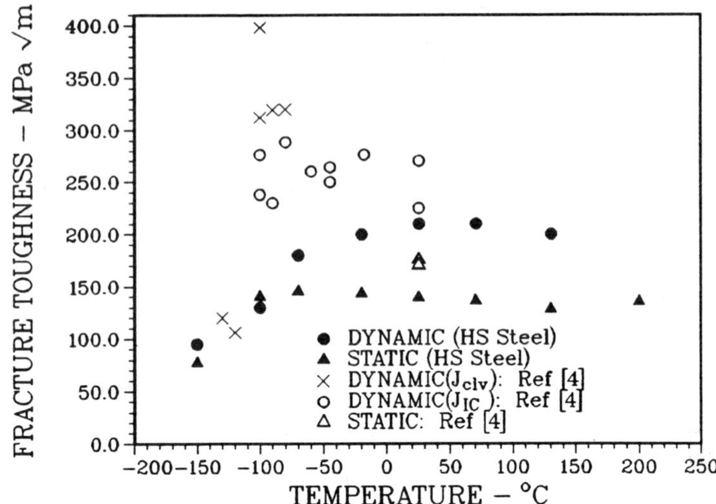

FIG. 6—*Comparison of present results with those obtained by Joyce and Hackett for an HY-100 steel* [4].

The upper shelf dynamic fracture toughness is approximately 50% higher than the quasi-static value. Only to a comparatively minor extent can this be attributed to the higher flow stresses that prevail at the higher strain rate [8]. The principal reason for the higher dynamic stresses probably lies in the time required for the initiation of voids at inclusions, and for their coalescence, during the plastic deformation that precedes fracture. A third factor, as was noted in the previous section, is the difference in definition between dynamic and quasi-static toughness values. Toughness for quasi-static fracture is defined at the point of crack initiation, whereas dynamic toughness is measured at the maximum load, allowing for contributions from both crack initiation and extension processes.

Our measured values of fracture toughness may be compared with these of Joyce and Hackett [4]. Their experiments were performed with a modified three-point bend specimen loaded in a drop tower. However, the comparison with their data is made difficult by the different definitions of toughness in the static and dynamic cases. Joyce and Hackett's results provide two values of fracture toughness for each dynamic test: the lower (J_{Ic}) corresponds to fracture initiation, while the upper (J_{clv}) is the maximum J achieved prior to cleavage. On the upper shelf, our measured values of fracture toughness are lower than either of these measurements (Fig. 6). Aside from microstructural factors, possible sources of the difference between the two sets of dynamic data are the difference in loading rate (at least two orders of magnitude), the specimen geometry, and the fact that the present tests were conducted with core material, while Joyce and Hackett used surface material. This difference between core material and surface material is consistent with the different Charpy test results. The differences in heat treatment or composition alluded to earlier may also be a factor.

According to the present results, the ductile-to-brittle transition temperature is nearly insensitive to the loading rate, increasing perhaps from $-125°C$ in quasi-static fracture to $-110°C$ at the higher rate. (For present purposes the transition temperature was defined as the average stress intensity value, between lower and upper shelf. Thus in dynamic fracture the transition temperature is the average of $-150°C$ and $-70°C$. For quasi-static fracture, the corresponding temperatures are $-150°C$ and $-100°C$.) This transition temperature is considerably lower than in other steels tested using the present experimental method. In

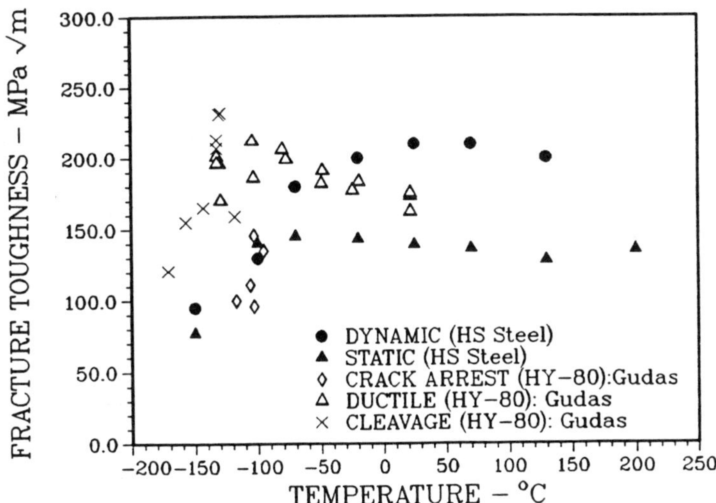

FIG. 7—*Comparison of present results with those obtained by Gudas for an HY-80 steel* [2].

addition, for these other steels, the general trend is for the transition temperature to shift more strongly toward higher temperatures with the higher loading rates. The fracture toughness values of the present steel for static and dynamic loading conditions are compared with those of HY-80 in Fig 7. This steel has a composition similar to that of the present steel and its transition temperature again is not strongly sensitive to loading rate.

It is clear that the high toughness range for this material extends to much lower temperatures than in an AISI 1018 cold-rolled steel [8] or in AISI 4340 VAR steel in the hardness range 45 to 55 HRC [16]. It would appear, therefore, that the present HS steel would perform better in applications requiring high toughness under impact loading where the expected temperatures range from −70°C to room temperature or even to 100°C. It will be seen in the next section that a largely fibrous fracture mode is obtained throughout this ductile range. As expected, at the lowest test temperature, −150°C, the fracture mode is one of cleavage.

Fractography

The fracture mode observed in the experiments is described in Table 2 as it depends on temperature and rate of loading. Fracture on the lower shelf is entirely by cleavage, while on the upper shelf it is a fibrous process. In the transition temperature range, the fracture mode is mixed, changing from fibrous to cleavage. Figures 8 to 10 are SEM fractographs of the fractured notched round bar specimens. The specimens are taken from both dynamic and quasi-static fracture tests and cover the temperature range −150°C to 200°C. At low magnification, as for instance in Fig. 10a, one can observe part of the fatigue annulus, which in this case is about 1.5 mm wide, with the fractured surface on one side and the machined notch on the other. The most characteristic feature of the fracture surface that is evident at low magnification is a layered structure consisting of parallel sets of troughs and ridges. This layered structure also lies parallel with the banding in the microstructure in Fig. 1a. The degree of layering is highly dependent on testing temperature, showing a trend toward a greater density of finer, narrower layers at higher testing temperatures. The observations described below were made primarily within the fracture initiation zone near the prefatigued

TABLE 2—*Mode of fracture in the region of fracture initiation and propagation.*

Test Temp. (°C)	Dynamic Loading		Static Loading	
	Fracture Surface	K_I Level	Fracture Surface	K_I Level
−150	Fully Cleavage	Lower Shelf	Fully Cleavage	Lower Shelf
−100	Fibrous Initiation then Cleavage	Transition	Fibrous Initiation then Cleavage	Upper Shelf
−70	Fibrous Initiation then Cleavage	Transition	Fibrous Initiation then Cleavage	Upper Shelf
−20	Fully Fibrous	Upper Shelf	Fully Fibrous	Upper Shelf
+25	Fully Fibrous	Upper Shelf	Fully Fibrous	Upper Shelf
+70	Fully Fibrous	Upper Shelf	Fully Fibrous	Upper Shelf
+130	Fully Fibrous	Upper Shelf	Fully Fibrous	Upper Shelf
+200	. . .	. . .	Fully Fibrous	Upper Shelf

annulus. The fracture propagation area was also studied for a more detailed examination of the layering structure. Higher magnification was employed to give greater detail (e.g., Figs. 8 and 9). Observations were made of the fracture initiation zone, of ductile dimples surrounding MnS inclusions, etc.

Lower-Shelf Behavior—The fractographs of the dynamic and static specimens tested at −150°C display fracture surfaces characteristic of a brittle fracture mode with a minimum of layering (Fig. 8). Cleavage facets are seen in both the initiation and propagation zones of fracture. Shear walls are evident between these cleavage facets, indicating a more ductile deformation that occurred as part of the overall brittle fracture process. The presence of shear walls indicates multiple cleavage crack paths with eventual linkage between cleavage planes. The ductile deformation within the shear walls probably contributes to the relatively high fracture toughness of high-strength steel at low temperature. Gudas [2] reported observing a similar feature, which he referred to as fluted appearance, in the fracture surface of HY-80 steel.

Transition Temperature Behavior—The fracture surfaces of specimens tested dynamically in the range of the transition temperatures (−100°C and −70°C) show a transition in the fracture mode from ductile to brittle behavior (i.e., the fracture initiates by void growth and nucleation and then propagates by cleavage) (e.g., Fig. 9). Ductile fracture initiation as measured from the tip of the prefatigued notch extends uniformly over a distance of about 200 μm at a test temperature of −100°C and 400 μm at −70°C. In static fracture the transition temperature, as defined by the value of K_I, is somewhat lower. However, even in static fracture at −100°C and −70°C the fracture mechanism initially is one of ductile fracture changing to cleavage. In static tests, the ductile fracture initiation zone is not as uniform as in the dynamic tests. Thus the width of the annulus showing initiation by void growth and coalescence varied from 200 to 500 μm at −100°C and from 400 to 1000 μm at −70°C.

Upper Shelf Behavior—Figure 10a shows the layered structure of the fracture surface for a specimen tested at room temperature. Broken pieces of MnS stringers, identified by EDAX, are visible at the bottom of the troughs (Figs. 10c and 10d). Presumably these stringers were broken during the fracturing of the specimen. Briant and Banerji [17,18] also observed a layered fracture surface in steels with MnS stringers. Based on their observations, they concluded that the layering fracture mode is due to the presence of the MnS stringers. Present results are in agreement and indeed give perhaps deeper troughs or at any rate a more accentuated surface appearance. In the present experiments the layered appearance becomes progressively more pronounced the higher the test temperature. Layering occurs in both

(a)

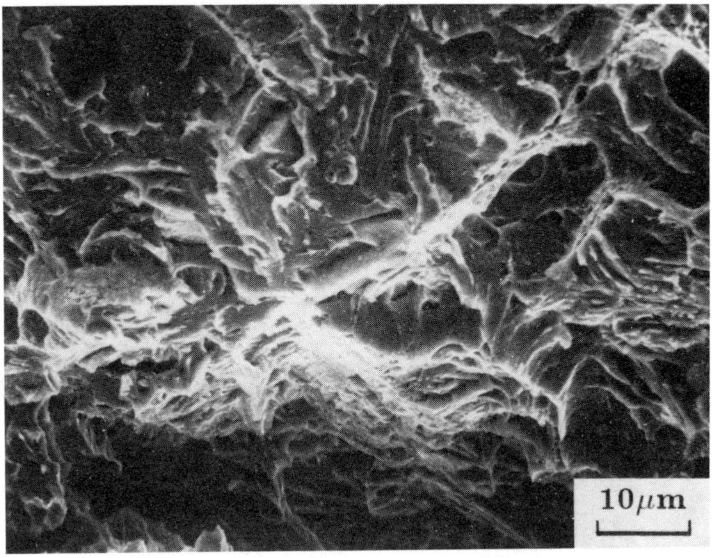

(b)

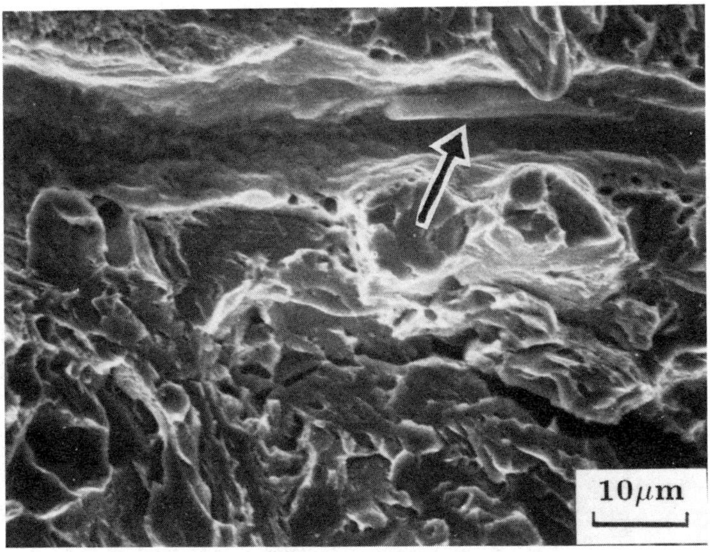

FIG. 8—*SEM fractographs of high strength steel specimens tested at −150°C, showing details of the cleavage facets linked by shear walls. (a) Dynamic and (b) quasi-static fracture tests. In (b) a MnS stringer can be seen producing the layered structure across the top of the photograph.*

(a)

(b)

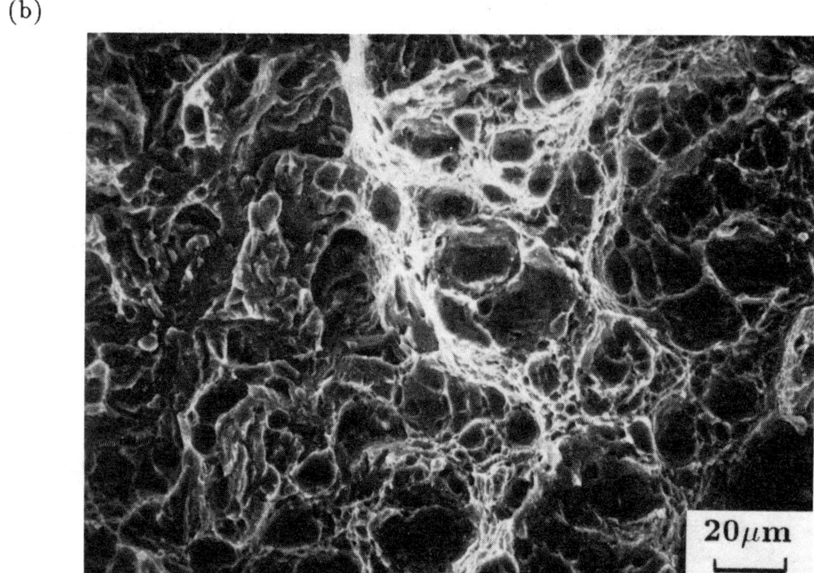

FIG. 9—*SEM fractographs of high strength steel specimens tested at −100°C, showing the ductile-brittle transition near the region of the fracture initiation. (a) Dynamic and (b) quasi-static fracture tests.*

(a)

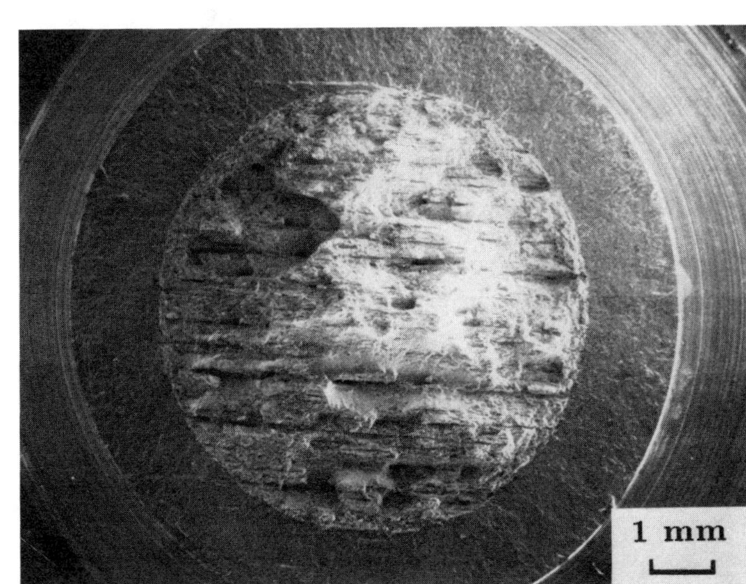

(b)

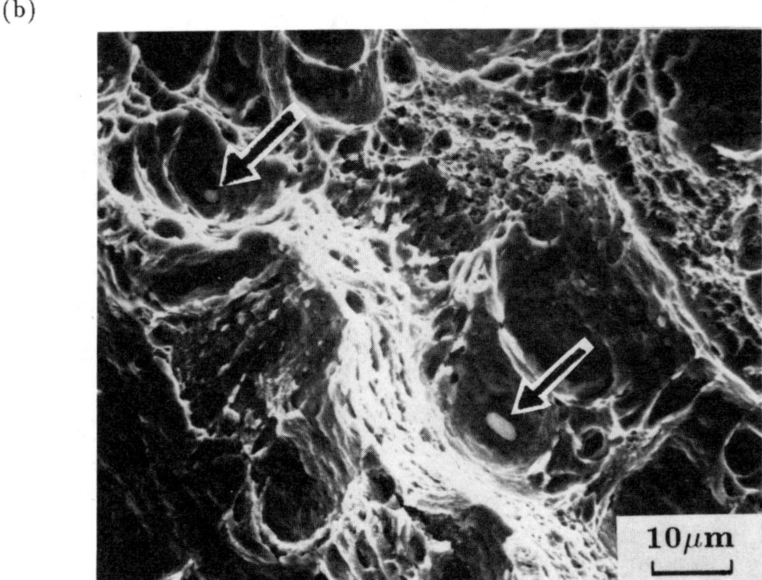

FIG. 10—*SEM fractographs of the high strength steel specimen tested dynamically at room temperature. (a) The layered structure of the fracture surface can be seen, as well as part of the fatigue annulus. (b) Ductile dimples containing MnS globules near their centers. (c) Broken pieces of a MnS stringer in the layered structure. (d) EDAX scan for the MnS stringer in (c).*
[Figs. 10c and 10d are on p. 140.]

(c)

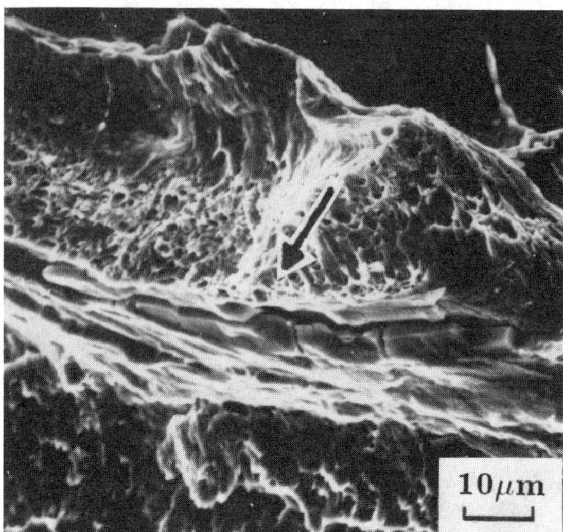

(d)

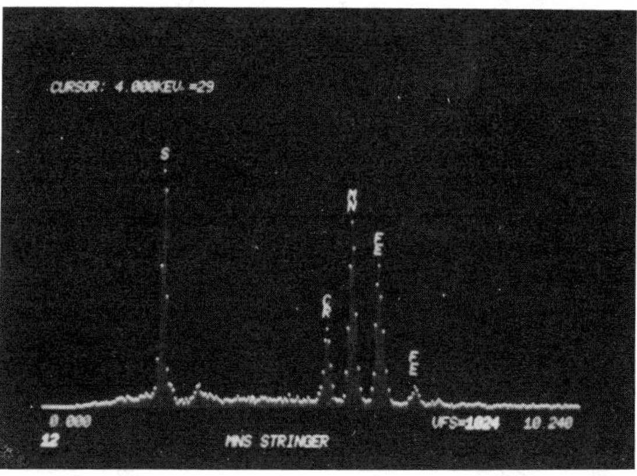

[Caption for Figs. 10c and 10d is on p. 139.]

dynamic and quasi-static fracture, showing little difference between the two. However, in nearly all cases the layering effect seems more pronounced near the center of the specimen than in the fracture initiation zone. It is not clear whether this is due to a difference between fracture initiation and crack propagation or whether it is due to a difference in the void formation process in the two regions, associated perhaps with the difference in the stress states in these two regions.

In addition to the characteristic layering structure, fibrous dimples representative of a ductile fracture mode are also visible. Formed by the initiation and coalescence of microvoids, these dimples were found to be dominant in fractures occurring above the ductile-to-brittle transition temperature. The large dimples contain globular MnS particles at the bottom of the cavity, identified by EDAX, and shown in Fig. 10b. The tiny dimples, considerably less than 1 μm in diameter, can be attributed to fine carbine particles that are dispersed uniformly throughout the microstructure [19,20]. The role of MnS particles in the formation of dimples is probably considerably more important than that of the carbides in the early stages of deformation, since voids generally initiate much more easily at large particles. At a later stage in the deformation process when the strains are greater, voids will begin to initiate around smaller particles as well.

A number of investigators (e.g., Briant and Banerji [17,18] and King and Knott [21]) describe a mechanism for ductile fracture initiation in steels containing MnS inclusions. They show that voids first form around the largest inclusions in a region ahead of the crack tip. Their presence reduces the fracture energy of the steel during ductile fracture. As fracture proceeds, the voids become linked by microshear bands. Many of these bands are oriented parallel to the loading direction, as is seen in the shear walls made visible by SEM (Figs. 8, 10b, and 10c). The macroscopic stress state, as well as the presence of these shear bands combined with an overall requirement of displacement compatibility, results in tensile stresses both in the axial direction and normal to the shear bands, which in turn lead to fracture. At larger strains this process is aided by the presence of microvoids surrounding the carbide particles within the microshear bands. If an impurity element such as phosphorous is segregated to the carbide-matrix interface, these microvoids will form at lower strains. The layering structure may be formed by cracks appearing between neighboring MnS stringers located one above the other. The SEM fractographs of the present HS steel show that linkage of the large voids (due to MnS inclusions) may be associated with the coalescence of carbide nucleated microvoids that are formed in the presence of the high strains within the microshear bands.

Conclusions

Tests have been performed to measure the dynamic fracture toughness over the temperature range $-150°C$ to $200°C$ of a high strength low alloy structural steel with a nominal yield stress of 690 MPa (100 ksi). The test method makes use of stress wave loading initiated explosively, and in this sense is a truly dynamic test. The stress intensity rate attained is about 2 $\times 10^6$ MPa $\sqrt{m}$/s. Plane strain conditions obtain over the entire crack length, and the instrumentation gives the data necessary for a calculation of a stress intensity factor.

Results give a dynamic fracture toughness of about 90 to 100 MPa $\sqrt{m}$ at $-150°C$ (lower shelf) and a transition temperature of about $-110°C$ to a toughness level averaging about 200 MPa $\sqrt{m}$ (upper shelf). The transition temperature is nearly independent of loading rate, rising from $-125°C$ under quasi-static loading to the $-110°C$ dynamic value. Thus loading rate affects principally the level of fracture toughness on the upper shelf where dynamic values exceed the quasi-static by about 50%. It should be noted that in tests on the lower shelf both dynamic and static values of fracture toughness correspond to fracture initiation. Indeed, for static testing the measured values of fracture toughness correspond to fracture initiation over the entire test temperature range to 200°C. However, in the dynamic tests on the upper shelf, the fracture toughness was evaluated at maximum load where some amount of crack growth may have occurred. Nevertheless, results show that present values of the dynamic fracture toughness on the upper shelf are lower than those obtained by Joyce and Hackett with a steel having a similar composition [4].

The fracture mechanism at a test temperature of $-150°C$ appears to be almost entirely one of cleavage, whether loading be dynamic or static, although some shear walls are seen separating cleavage facets. In the transition range, fracture initiates by void growth and nucleation, but beyond the initiation annulus the fracture propagates by cleavage. However, for test temperatures of $-20°C$ and above, the fracture is entirely by void growth and coalescence. A characteristic feature of the fracture surfaces is a layered appearance, consisting of alternating ridges and troughs lying within the plane of rolling of the original plate. While this layered appearance occurs for all test temperatures it is particularly pronounced at the higher test temperatures. Fractography shows that the layering is associated with the presence of MnS stringers, through the initiation of voids at the large MnS inclusions. Fine shear bands then connect the voids with one another aided by the presence of smaller carbides [17–21]. This process leads ultimately to fracture and to the layered appearance of the fracture surface. While MnS inclusions lower the fracture toughness on the upper shelf, their presence does not seem to influence the temperature at which the transition occurs from lower to upper shelf behavior. However, sulfur itself when segregated to the grain boundaries will lead to intergranular embrittlement, quickly raising the transition temperature [22].

In future work, the role of the MnS inclusions during fracture should be looked at in more detail particularly as it relates to orientation. Their role in adiabatic shear banding seems to be considerable, and adiabatic shear banding plays an important role in the failure process and again is related to material orientation. In any case, it is clear that the high toughness range for this material extends to lower temperatures than with many other steels. In this respect, therefore, the present steel should prove particularly useful in applications where the expected temperature range is from about $-70°C$ to room temperature or even to $100°C$.

Acknowledgments

The research support of the Office of Naval Research, Mechanics Division, through Contract N00014-85-K-0597, and the NSF Materials Research Group at Brown University, through Grant NSF/DMR-8714665, is gratefully acknowledged. The authors also wish to thank Professors C. F. Shih and J. Gurland for their advice, and Mr. George LaBonte for his help in the experiments. Thanks are also due Dr. Clyde Briant, General Electric Corporate Research, for trans-Atlantic advice communicated from Churchill College, Cambridge University. One of the authors (JPS) also is grateful to the Westinghouse Corporation for providing fellowship support under the URTA program.

References

[1] Hasson, D. F. and Joyce, J. A., "The Effect of a Higher Loading Rate on the J_{Ic} Fracture Toughness Transition Temperature of HY Steels," *Journal of Engineering Materials and Technology*, Vol. 103, 1981, pp. 133–141.

[2] Gudas, J. P., "Micromechanisms of Fracture and Crack Arrest in Two High Strength Steels," Ship Materials Engineering Department Research and Development Report, SME-87/20, David Taylor Research Center, Annapolis, Md., 1987.

[3] Harding, J., "Effect of Temperature and Strain Rate on Strength and Ductility of Four Alloy Steels," *Metals Technology*, Vol. 4, Jan. 1977, pp. 6–18.

[4] Joyce, J. A. and Hackett, E. M., personal communication.

[5] Costin, L. S., Duffy, J., and Freund, L. B., "Fracture Initiation in Metals under Stress Wave Loading Conditions," in *Fast Fracture and Crack Arrest, ASTM STP 627*, American Society for Testing and Materials, Philadelphia, 1977, pp. 301–318.

[6] Nakamura, T., Shih, C. F., and Freund, L. B., "Elastic-Plastic Analysis of a Dynamically Loaded Circumferentially Notched Round Bar," *Engineering Fracture Mechanics*, Vol. 22, 1985, pp. 437–452.

[7] Franklin, A. G. and Tegart, W. J. McG., "Effect of Desulphurization on the Impact Properties of Some Low-Alloy Steels," *Journal of the Iron and Steel Institute*, Vol. 202, July 1964, pp. 588–592.

[8] Couque, H., Asaro, R. J., Duffy, J., and Lee, S., "Correlations on Microstructure with Dynamic and Quasi-Static Fracture in Plain Carbon Steels," *Metallurgical Transactions A*, Vol. 19A, No. 9, Sept. 1988, pp. 2179–2206.

[9] Tada, H., Paris, P. C., and Irwin, G. R., in *The Stress Analysis of Cracks Handbook,* Del Research Corp., Hellertown, Pa., 1973.

[10] Paris, P. C., in written discussion to J. A. Begley and J. D. Landes, "The *J*-Integral as a Fracture Criterion," in *Fracture Toughness, ASTM STP 514,* American Society for Testing and Materials, Philadelphia, 1972, pp. 21–23.

[11] Landes, J. D. and Begley, J. A., "The Effect of Specimen Geometry on J_{Ic}," in *Fracture Toughness, ASTM STP 514,* American Society for Testing and Materials, Philadelphia, 1972, pp. 24–39.

[12] Hutchinson, J. W., "Fundamentals of the Phenomenological Theory of Nonlinear Fracture Mechanics," *Journal of Applied Mechanics,* Vol. 50, Dec. 1983, pp. 1042–1051.

[13] Rice, J. R., Paris, P. C., and Merkle, J. G., "Some Further Results of *J*-Integral Analysis and Estimates," in *Progress in Flaw Growth and Fracture Toughness Testing, ASTM STP 536,* American Society for Testing and Materials, Philadelphia, 1973, pp. 231–245.

[14] Marchand, A. and Duffy, J., "An Experimental Study of the Formation Process of Adiabatic Shear Bands in a Structural Steel," *Journal of the Mechanics and Physics of Solids,* Vol. 36, No. 3, 1988, pp. 251–283.

[15] Cho, K., Sklenak, J. P., and Duffy, J., "The Dynamic Fracture Behavior of a Structural Steel," Brown University Technical Report, Providence, R.I., Sept. 1988.

[16] Chi, Y. C., Lee, S., Cho, K., and Duffy, J., "The Effects of Tempering and Test Temperatures on the Dynamic Fracture Initiation Behavior of an AISI 4340 VAR Steel," *Materials Science and Engineering,* Vol. A114, 1989, pp. 105–126.

[17] Briant, C. L., and Banerji, S. K., "Tempered Martensite Embrittlement and Intergranular Fracture in an Ultra-High Strength Sulfur Doped Steel," *Metallurgical Transactions A,* Vol. 12A, 1981, pp. 309–319.

[18] Briant, C. L. and Banerji, S. K., "The Fracture Behavior of Quenched and Tempered Manganese Steels," *Metallurgical Transactions A,* Vol. 13A, 1982, pp. 827–836.

[19] Cox, T. B. and Low, J. R., Jr., "An Investigation of the Plastic Fracture of AISI 4340 and 18 Nickel-200 Grade Maraging Steels," *Metallurgical Transactions A,* Vol. 5, 1974, pp. 1457–1470.

[20] Clayton, J. Q. and Knott, J. F., "Observations of Fibrous Fracture Modes in a Prestrained Low-Allow Steel," *Metal Science,* Feb. 1976, pp. 63–71.

[21] King, J. E. and Knott, J. F., "Effects of Temper-Embrittling Heat Treatment on Ductile Failure in High-Strength Low-Alloy Steel," *Metal Science,* Jan. 1981, pp. 1–6.

[22] Briant, C. L., "Intergranular Fracture in Ferrous Alloys in Nonaggressive Environments," in *Treatise on Materials Science and Technology,* C. L. Briant and S. K. Banerji, Eds., Academic Press, New York, 1983, pp. 21–58.

DISCUSSION

J. J. Krupowicz[1] *(written discussion)*—The authors emphasized the role of manganese sulfide inclusions in the fracture process of HY-100 structural steel. The manganese and sulfur contents of the alloy were significantly lower than some of the commonly used structural steels such as ASTM A516. The marked influence of inclusions in this study probably resulted from the relatively small net diameter of the notched and precracked round bar specimens. A randomly occurring inclusion within the small fracture cross section can drastically affect results.

The study illustrates how influential are inclusions in the fracture of engineering alloys. Fractographic examinations of service failures often reveal inclusions associated with the crack propagation. The role of inclusions in ordinary structural steels with relatively poor microcleanliness merits further study.

K. Cho et al. (authors' closure)—The authors appreciate the comments made by Dr. Krupowicz and certainly agree with his second paragraph. The variation in the values of the measured stress intensity factor was small. Thus the inclusions do not seem to play a significant role there. One should remember that the largest stringers are only 50 or 60 μm long (and 2 or 3 μm in diameter) compared with a fractured ligament of about 8 mm.

---

[1] Mobil Research and Development Corporation, Paulsboro Research Laboratory, Paulsboro, NJ 08066.

W. Böhme[1]

Dynamic Key-Curves for Brittle Fracture Impact Tests and Establishment of a Transition Time

REFERENCE: Böhme, W., **"Dynamic Key-Curves for Brittle Fracture Impact Tests and Establishment of a Transition Time,"** *Fracture Mechanics: Twenty-First Symposium, ASTM STP 1074*, J. P. Gudas, J. A. Joyce, and E. M. Hackett, Eds., American Society for Testing and Materials, Philadelphia, 1990, pp. 144–156.

ABSTRACT: Impact tests with precracked bend specimens were performed, and the crack tip loading was measured utilizing the shadow optical method of caustics and high-speed photography. Some results are presented in comparison to a quasi-static solution for the crack loading history, which is based on a simple mass-spring model. The deviations obtained are a quantitative measure of dynamic effects acting at the crack tip and they define dynamic correction functions k^{dyn}. In a normalized form these time-dependent functions are very general curves for impact tests; therefore they can be considered as dynamic key-curves. They allow the prediction of the dynamic crack tip loading history for arbitrary impact test conditons so far as the usual restrictions of the linear elastic fracture mechanics are fulfilled.

The dynamic key-curves show clearly the decrease of dynamic effects with increasing time, allowing us to define transition times t_{dyn}. For times larger than t_{dyn} the crack loading history can be described sufficiently accurately by the quasi-static solution. The established transition time t_{dyn} is compared with transition times defined by Ireland and by Nakamura, Shih, and Freund.

KEY WORDS: impact test, fracture toughness, impact fracture toughness, dynamic key-curves, transition times, stress intensity factors, caustics, high-speed photography

Impact tests are often used to characterize material behavior under dynamic loading conditions [1,2]. These tests are usually performed with drop weight towers or pendulums. Prenotched bend specimens are impacted at velocities of several meters per second. The absorbed impact energy or the obtained damage can be used for a qualitative comparison of different materials.

During the last two decades such tests have been modified and extended (e.g., in order to determine impact fracture toughness values K_{Id}), specimens with fatigued precracks instead of notches have been tested, and the "instrumented" impact test has been developed [2–4], where the impacting tup is instrumented by strain gages to measure the externally applied, time-dependent load, $P(t)$. It was then proposed to determine K_{Id} from the critical load at the moment of crack initiation following the static evaluation procedure of ASTM E 399.

This evaluation procedure can only be applied if—as in statics—an overall equilibrium of forces is given. However, at the very beginning of an impact test dynamic effects such as propagating elastic waves occur and, as a consequence, vibrations, which cause an oscillating load signal. Therefore Ireland [3] restricted the range of applicability of this procedure to

---

[1] Fraunhofer-Institut für Werkstoffmechanik, Wöhlerstr. 11, D-7800 Freiburg, West Germany.

times to fracture, t_f, which are larger than the duration 3τ of three oscillations of the measured load $P(t)$. This time 3τ can be considered as a transition time for impact tests that separates the initial, dynamically influenced time range from the following time range, where approximately an overall quasi-static state of loading can be assumed, as indicated by the decreased amplitude of the oscillations of the load signal.

For the determination of K_{Id} values even for times to fracture t_f less than 3τ, the concept of impact-response curves was developed, especially for application to steels [5,6]. The impact-response curve is the history of the dynamic crack tip stress intensity factor, $K_I^{dyn}(t)$. For given impact test conditions this curve has to be established only once (e.g., by the shadow optical method of caustics). This curve depends on the elastic reaction of the specimen-striker system, but it applies to steels of different toughnesses. For experiments performed under the same test conditions with other steels of interest, only the time to fracture, t_f, has to be measured. Then the dynamic fracture toughness K_{Id} can be easily determined from the pre-established impact-response curve following the equation $K_{Id} = K_I^{dyn} (t = t_f)$.

A method of predicting $K_I^{dyn}(t)$ from only the knowledge of the impact test conditions has been presented and discussed in detail [7,8]. It is the basic assumption of this method that $K_I^{dyn}(t)$ can be described as a product of a quasi-static part $K_I^{qs}(t)$ and a dynamic correction function $k^{dyn}(t)$, which in a normalized form is called the dynamic key-curve:

$$K_I^{dyn}(t) = K_I^{qs}(t) * k^{dyn}(t) \tag{1}$$

The first right-hand term can be calculated by an analytic formula, which results from a simple mass-spring model. The second right-hand term was determined once in model experiments and resulted in a set of dynamic key-curves. The transferability of these results to arbitrary impact test conditions and the application to the determination of dynamic fracture toughnesses (e.g., of steels) have been demonstrated elsewhere [7,8].

In the following sections the basic ideas of this concept of dynamic key-curves will be summarized so far as seems to be necessary for the understanding of the next step, the definition of transition times. Based on these dynamic key-curves transition times t_{dyn} have been defined and determined previously by the author [7]. These results will then be compared to a transition time defined by Ireland [3] and another one defined by Nakamura et al. [9].

Mass-Spring Model

In general the crack loading history in impact tests is determined by many system parameters. For a first estimate of the influence of these parameters on the crack loading behavior a simple mass-spring model has been introduced for the impact arrangement sketched in Fig. 1. The hammer is replaced by a point mass, M, and the specimen by a massless spring with a compliance C_s. The compliance of the loading device, C_m, can be taken into account by a second spring if C_m can not be neglected in comparison to the specimen compliance C_s. The machine compliance C_m is considered here as an integral compliance resulting from the compliances of the striking system, the supports, and the three "points" of contact. A procedure to experimentally determine C_m is given in Ref *10*.

It may be mentioned, however, that in contrast to other mass-spring models [*11–14*] it is not the objective of this model to describe the complete impact event as accurately as possible. It is rather intended to find a solution as a function of the test parameters for the quasi-static part in Eq 1. Afterwards this solution can be used as a basis to study dynamic effects like wave propagation within the specimen which have been neglected so far and which are the deviations from this quasi-static solution.

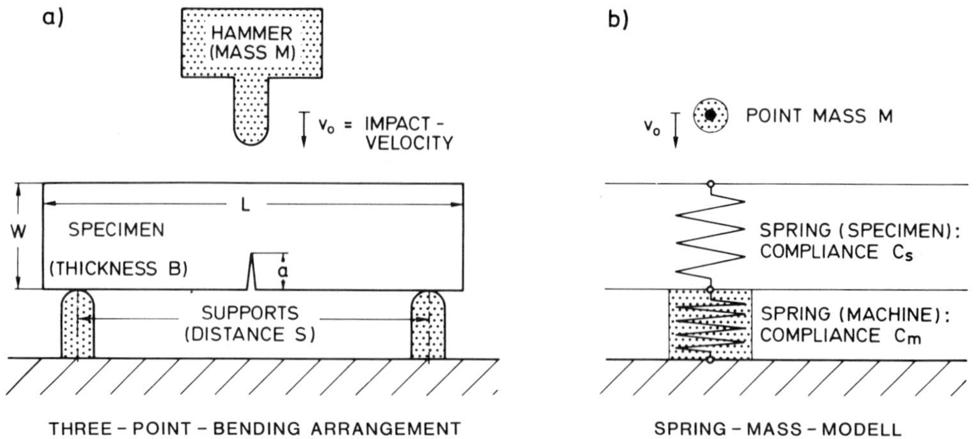

FIG. 1—*Mass-spring model for the impact test.*

The result of the mass-spring model is given for the commonly used relative support distance $S/W = 4$ and for $C_m \ll C_s$ by

$$K_I^{qs}(t) = v_o\, Y\, \sqrt{\frac{E\,M}{W\,B\,C_s^*}}\, \sin\left(\sqrt{\frac{E\,B}{M\,C_s^*}}\, t\right) \tag{2}$$

where

$$
\begin{aligned}
v_o &= \text{impact velocity,} \\
M &= \text{hammer mass,} \\
W &= \text{specimen width,} \\
B &= \text{specimen thickness,} \\
E &= \text{specimen modulus,} \\
a/W &= \text{relative initial crack length,} \\
Y &= Y(a/W), \text{ see below, and} \\
C_s^* &= C_s^*(a/W), \text{ see below.}
\end{aligned}
$$

$Y = Y(a/W)$ is the well known static relationship for K_I determination of three-point bend specimen after Srawley [15]:

$$Y(a/W) = \frac{6(a/W)^{1/2}\,[1.99 - a/W(1 - a/W)\,(2.15 - 3.93a/W + 2.7(a/W)^2)]}{(1 + 2a/W)\,(1 - a/W)^{3/2}}$$

$C_s^* = C_s^*(a/W) = EBC_s$ is the dimensionless specimen compliance for plane stress and the Poisson ratio $v = 0.33$ after Bucci et al. [16]:

$$
\begin{aligned}
C_s^*(a/W) = 20.1 + 135(a/W)^2\,[1 &- 2.11(a/W) + 8.76(a/W)^2 - 19.9(a/W)^3 \\
&+ 41.4(a/W)^4 - 67.7(a/W)^5 + 92.1(a/W)^6 - 76.7(a/W)^7 + 35.6(a/W)^8]
\end{aligned}
$$

Equation 2 explicitly describes the quasi-static crack loading behavior as a function of time and the test parameters. The machine compliance can be taken into account in Eq 2 by replacing C_s^* by $C_s^*(1 + C_m/C_s)$. Dynamic effects acting at the crack under impact loading can now be quantified by a comparison of experimental data with the result calculated by Eq 2.

Experimental Procedure and Results

The experimental setup of the performed model experiments is schematically shown in Fig. 2. The specimens were machined from the epoxy-resin Araldite B. In some experiments crack-similar notches instead of natural cracks were used in order to avoid early fracture of the specimen. The specimens were impacted in a drop weight tower with a hammer mass of 4.9 kg and impact velocities ranging from 0.5 to 5 m/s.

The hammer was instrumented by strain gages and calibrated statically in order to measure the applied, time-dependent load ($P(t)$). These signals are discussed in detail in Refs 7 and 17 in comparison to the actual dynamic crack tip loading $K_I^{dyn}(t)$, which is of main interest in this context. $K_I^{dyn}(t)$ was determined by the shadow optical method of caustics (see e.g., [18]). During the impact tests the caustics are registered by a 24-spark high-speed camera with a maximum time resolution of $1\mu s$. One series of caustics is shown in Fig. 3. The actual stress intensity factor can be determined quantitatively from the diameter of the caustic [18]. From this evaluation results the time-dependent dynamic crack tip loading $K_I^{dyn}(t)$.

Some measured data are given in Fig. 4 for one special set of test conditions (see insert in Fig. 4). The load-signal $P(t)$ has been transferred to stress intensity factors $K_I^{stat}(P(t))$ by applying the static formula given in ASTM E 399 or in Ref 15. The pronounced oscillations of the $K_I^{stat}(P(t))$ signal, the less pronounced oscillations of $K_I^{dyn}(t)$, and the missing proportionality between the two signals are caused by wave propagation effects and in this special case also by loss of contact at the supports as has been discussed in detail in previous publications [7,17,19].

Dynamic Key Curves

The quasi-static result $K_I^{qs}(t)$ as calculated by Eq 2 is also plotted in Fig. 4 as a dashed line, which obviously represents a good approximation of the mean dynamic crack loading history $K_I^{dyn}(t)$. Details like the obtained oscillations can not be described by this approximation since dynamic effects are neglected in $K_I^{qs}(t)$. However, the ratio between the measured $K_I^{dyn}(t)$ sig-

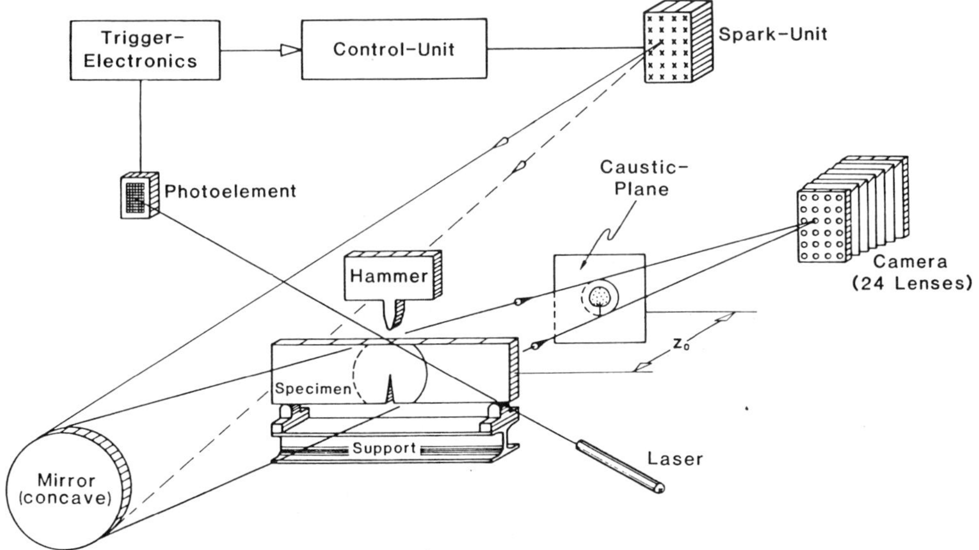

FIG. 2—*Experimental setup.*

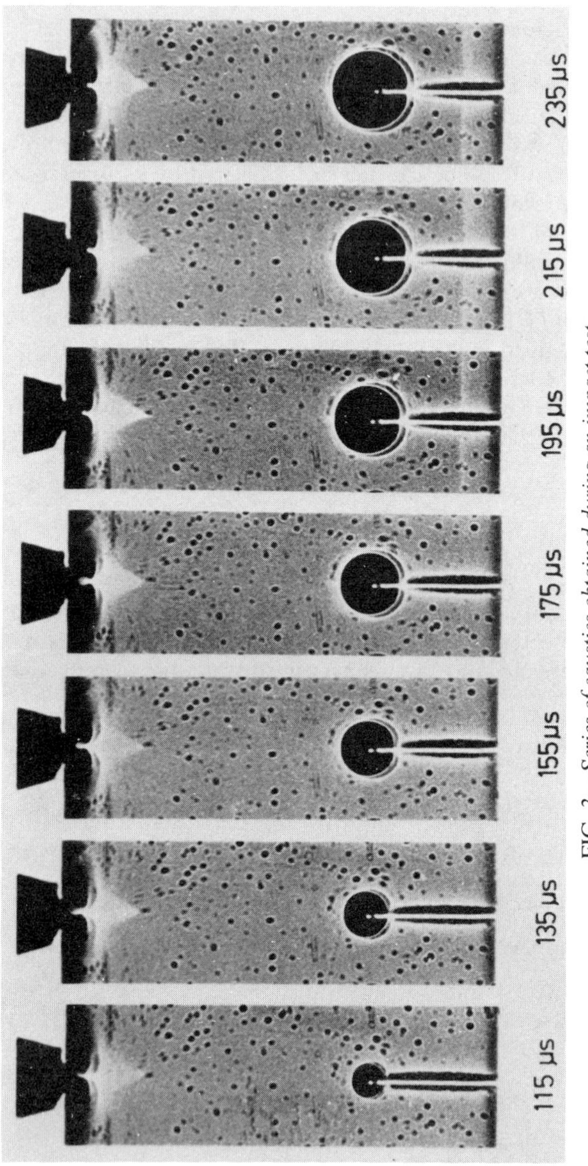

FIG. 3—*Series of caustics obtained during an impact test.*

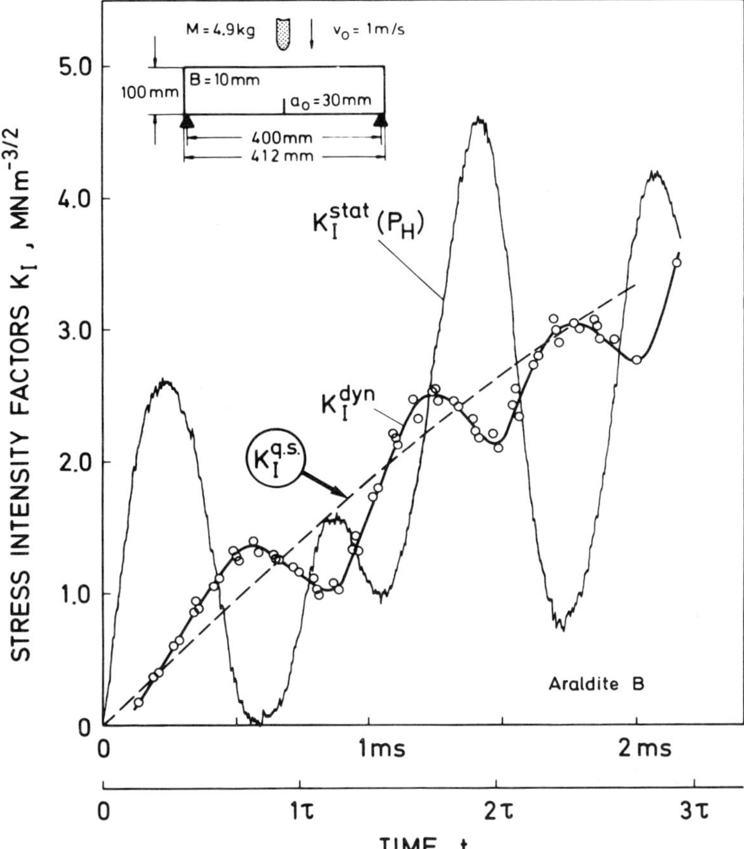

FIG. 4—*Quasi-statically calculated and measured loading histories.*

nal and the calculated $K_I^{qs}(t)$ values defines the time-dependent dynamic correction function $k^{dyn}(t)$, which is used to quantify the dynamic effects:

$$k^{dyn}(t) = \frac{K_I^{dyn}(t)}{K_I^{qs}(t)} \tag{3}$$

The dynamic correction function $k^{dyn}(t)$ as determined from the results presented in Fig. 4 is shown in Fig. 5 together with additional results especially for the short time range. The value $k^{dyn} = 1$ represents the quasi-static behavior. The deviations from this value quantitatively describe the influence of dynamic effects, which as expected are large in the beginning of the impact event. They decrease with increasing time and approach the quasi-static state via an oscillation with damped amplitude.

In Fig. 5 the time axis is normalized by the time needed by longitudinal waves to propagate across the specimen width W, where c_1 is the longitudinal wave propagation velocity for plane stress (c_1 = 1766 m/s for Araldit B and 5390 m/s for steels). In this normalized form $k^{dyn}(c_1 t/W)$ is independent of most of the test parameters [7,8]. Especially $k^{dyn}(c_1 t/W)$ is independent of the special features of usual impact loading systems such as, for example, the

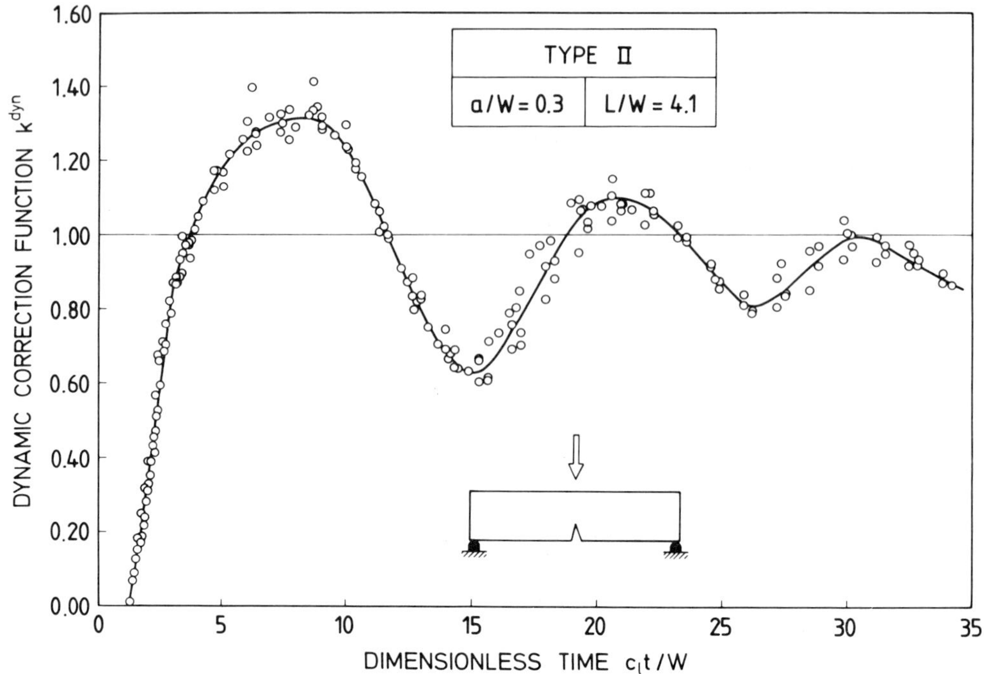

FIG. 5—*Dynamic correction function determined from data given in Fig. 4.*

mass, M, of the striker (insofar as it is large compared to the mass of the specimen), the impact velocity, v_o, or the machine compliance, C_m. This seems to be meaningful since these parameters will not change the wave propagation behavior within the specimen. The impact velocity will determine the intensity of the whole process in a linear manner, but this is already involved in the quasi-static part of the complete solution (see Eq 2).

In a first approach, dynamic key-curves are also independent of the thickness B of the specimen, since the same arguments hold for this parameter. On the other hand, for specimens of different sizes or different elastic properties the wave propagation behavior will be different. But this effect will be linear in time and is included in the dynamic key-curve by the normalization of the time axis.

The fact that the normalized k^{dyn} curve is not dependent on the aforementioned parameters was verified experimentally [7,8]. Therefore $k^{dyn}(c_1t/W)$ can be considered as a dynamic key-curve for brittle fracture impact tests, and following Eq 1 it can be used for sufficiently accurate predictions of the dynamic crack tip loading under arbitrary test conditions. This is demonstrated for steel and relatively large specimens in Fig. 6 and for the Charpy test in Refs 7 and 8.

The shape of such a dynamic key-curve is determined by wave propagation and vibration effects. By detailed experimental investigations different characteristic wave fronts were detected which at specific times significantly contribute to the increase of the dynamic crack tip loading [7,8,20]. These are, for example, shear waves propagating in crack direction, Rayleigh waves propagating along crack surfaces, and longitudinal waves propagating perpendicular to the crack plane. The normalized arrival times of these wave fronts at the crack tip are correlated to three geometrical parameters: the relative crack length (a/W), the relative specimen length (L/W), and the relative support distance (S/W) (since additional impact events occur at the anvils as a consequence of initial loss of contact effects [*19*]). Since S/W = 4 holds for most test devices, only the two parameters a/W and L/W are to be considered as dynamically relevant parameters which can be used to classify specimens as dynamically different types.

For four different types of specimen (Types I to IV), characterized by certain combinations of a/W and L/W, different dynamic key-curves were experimentally determined; the data are given in Fig. 7. The crack tip loading of a specimen of Type IV is obviously strongly affected by dynamic effects as indicated by strong deviations of the k^{dyn} curve from the value $k^{dyn} = 1$. Only reduced dynamic effects are obtained for Type III specimens which differ from Type IV only by the larger specimen length L/W (i.e., by the overhang over the supports). It is worth noting that the Type III specimen corresponds to the Charpy specimen.

In general an interpolation between these different curves is not allowed since these are non-linear curves. However, previous experience [*7,8*] indicates that the usual experimental scatter of the initial fatigue crack length for a series of specimens of about ±5% around a mean value (e.g., $a/W = 0.5$) will affect the curves only moderately within their own accuracy of about ±10%.

Transition Times

As can be seen from the set of dynamic key-curves given in Fig. 7 a quasi-static crack tip loading with $k^{dyn} = 1$ is reached at different times for different types of specimens. Therefore

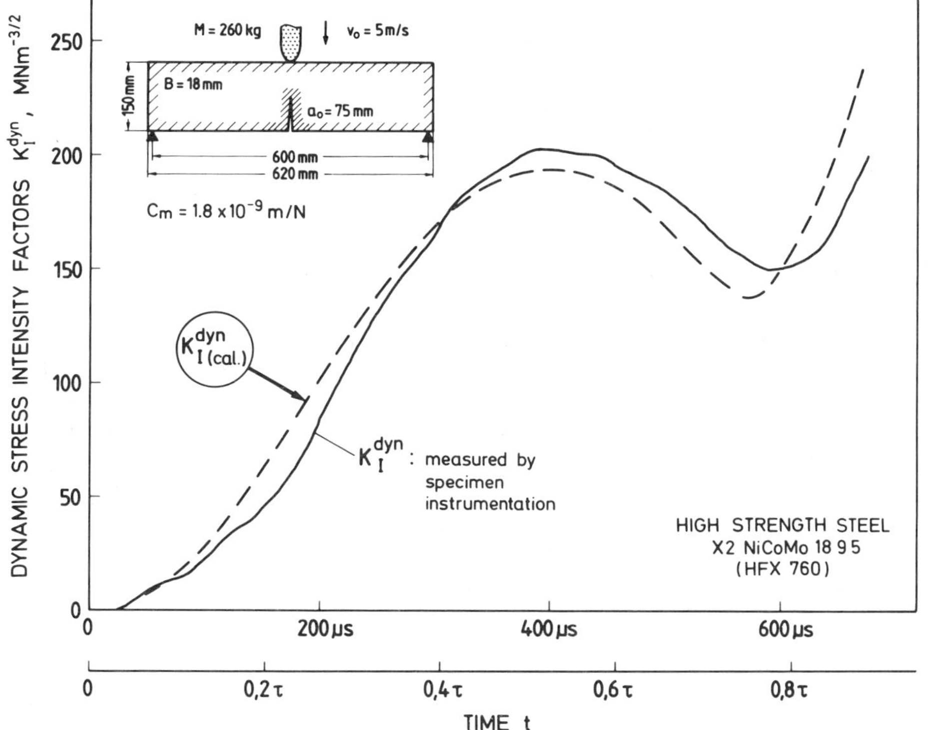

FIG. 6—*Comparison of the measured dynamic crack tip loading* (———) *and the prediction* (– – –) *based on dynamic key-curves.*

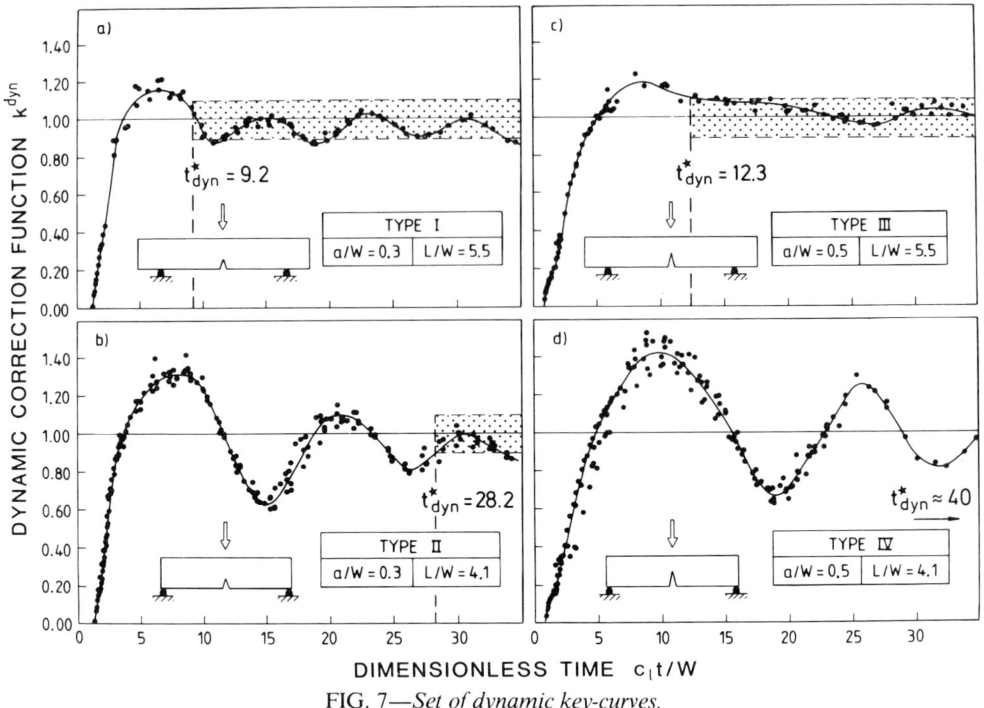

FIG. 7—*Set of dynamic key-curves.*

a characteristic transition time can be defined on the basis of the dynamic key-curves. This transition time (denoted as t_{dyn}) is defined [7] as that time beyond which $k^{dyn}(t)$ differs from the value $k^{dyn} = 1$ by less than 10%:

$$0.9 \leq k^{dyn}(t) \leq 1.1 \qquad \text{for} \qquad t \geq t_{dyn} \qquad (4)$$

Based on this definition, transition times t_{dyn} or dimensionless transition times $t^*_{dyn} = t_{dyn}c_1/W$ have been determined from the dynamic key-curves (Fig. 7) and are listed in Table 1 for the four types of specimen. The transition times increase with increasing crack length a/W and with decreasing specimen length L/W (for fixed support span $S/W = 4$). These transition times are compared in the next section with transition times defined independently by other authors.

TABLE 1—*Transition times* t^*_{dyn}.

Specimen	a/W	L/W	t^*_{dyn}
Type I	0.3	5.5	9.2
Type II	0.3	4.1	38.2
Type III	0.5	5.5	12.3
Type IV	0.5	4.1	≈40

Discussion

Based on impact tests on Charpy specimens an empirical formula was introduced by Ireland [3] in order to estimate the transition time 3τ:

$$3\tau = \frac{1.68}{c_\mathrm{o}} \sqrt{S\,W\,C_\mathrm{s}^*} \tag{5}$$

where c_o = longitudinal bar wave speed (for steels it is convenient to use $c_\mathrm{o} = 5000$ m/s).

For the most common support span ($S = 4W$) and for plane stress with $c_1/c_\mathrm{o} = 1.06$ for $\nu = 0.33$, the dimensionless transition time $3\tau^* = 3\tau c_1/W$ can be derived from Eq 5:

$$3\tau^* = 10.7 \sqrt{C_\mathrm{s}^*} \tag{6}$$

This equation holds for the special case of the Charpy test and can be used to estimate the transition time beyond which the amplitude of the hammerload oscillations has been sufficiently diminished. The transition time $3\tau^*$ depends on the specimen compliance, C_s^*, and therefore on the relative crack length a/W. The result of Eq 6 is plotted as a full line in Fig. 8.

For comparison the experimentally determined transition times t_dyn^* (Table 1) must be considered. These results are plotted as data points in Fig. 8. In order to get a solution which can be directly compared with the result of Eq 6, the two data points for $L/W = 4.1$ and those for $L/W = 5.5$ are fitted in a linear manner. The resulting relations are (see dashed lines in Fig. 8):

For $L/W = 4.1$:

$$t_\mathrm{dyn}^* = 5.21 \sqrt{C_\mathrm{s}^*} \tag{7a}$$

For $L/W = 5.5$:

$$t_\mathrm{dyn}^* = 1.64 \sqrt{C_\mathrm{s}^*} \tag{7b}$$

These relations can be used to estimate the transition time t_dyn^* beyond which the amplitude of the crack tip loading oscillations has been sufficiently diminished. The transition time t_dyn^* for short specimens ($L/W = 4.1$) is about three times larger than for elongated specimens ($L/W = 5.5$). This indicates enlarged dynamic effects for short specimens, which is in agreement with earlier observations [19].

The influence of L/W was not considered by Ireland [3] during the establishment of the transition time 3τ. Since 3τ has been derived from Charpy test results, the transition time 3τ corresponds to $L/W = 5.5$. Therefore Eq 6 must be compared to Eq 7b; it can then be concluded that $3\tau^*$ is about six times larger than t_dyn^*. This means that dynamic effects acting at the crack tip diminish six times earlier than dynamic effects acting at the point of impact.

Another transition time was defined by Nakamura, Shih, and Freund [9]. This definition is based on results of simple models established to calculate the ratio of kinetic energy and strain energy as a function of time for impact tests. This ratio was found to be large in the very beginning of an impact event and decreases with increasing time down to values less than 1. The time where this ratio equals 1 is denoted as t_T. The authors stated that $2t_\mathrm{T}$ is a transition time beyond which interpretable fracture toughness data can be determined experimentally. The equation to calculate t_T is given in Ref 9 in a similar form to Eq 5. Under

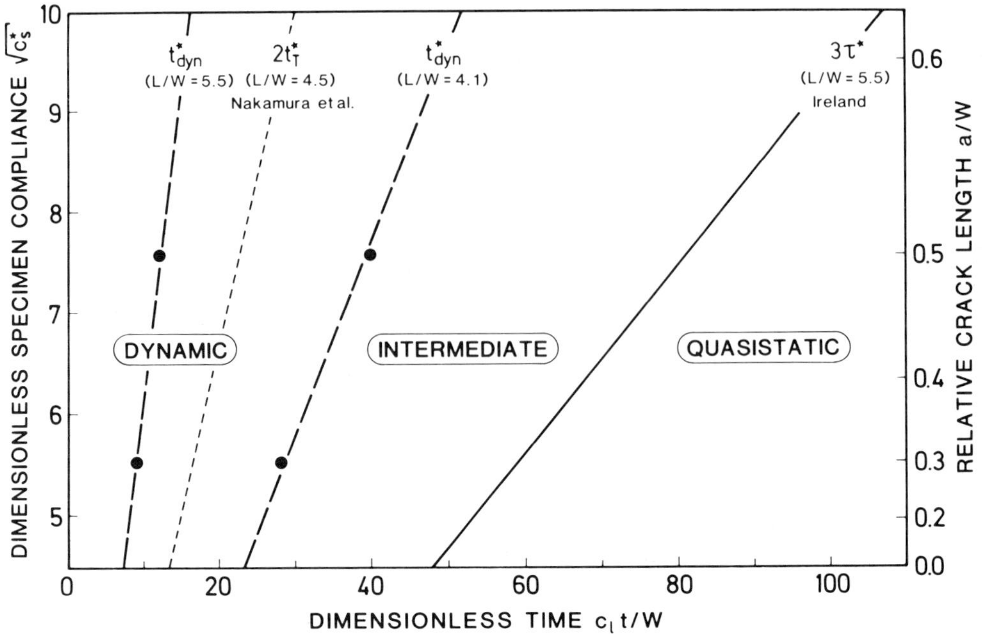

FIG. 8—*Different time ranges in impact tests separated by characteristic transition times.*

the same assumptions as used above the following equation for the dimensionless transition time $2t_T^*$ can be derived:

$$2t_T^* = 3.00 \sqrt{C_s^*} \tag{8}$$

This result is shown as a dotted line in Fig. 8. The transition time $2t_T^*$ is very close to and just in between the experimentally determined transition times t_{dyn}^*. This seems to be meaningful since Nakamura et al. [9] considered a specimen with $L/W = 4.5$. Despite the different definitions there is astonishing agreement between the transition times $2t_T^*$ and t_{dyn}^*, which indicates a certain relationship between the two different definitions. However, the significant influence of the relative length of the specimen, L/W, and loss of contact effects at the supports are not taken into account by Nakamura et al. Therefore the transition time t_{dyn}^* gives more detailed information.

The aforementioned time t_{dyn}^* (or $2t_T^*$) is related to the transition from a fully dynamic to a quasi-static crack tip loading, whereas $3\tau^*$ is related to the transition to an overall quasi-static loading behavior. Therefore an impact test of precracked specimens divides into three time ranges:

- Fully *dynamic* time range, limited by

$$0 \leq c_1 t/W < t_{dyn}^* \tag{9}$$

where t_{dyn}^* depends strongly on L/W. In this short time range the actual crack tip loading is significantly influenced by dynamic effects. (A subdivision of this time range is given elsewhere [7,8,20].) Fully dynamic measuring and evaluation procedures are necessary to obtain reliable results. For engineering purposes an evaluation based on the dynamic key-curves can be used with sufficient accuracy [7,8].

- *Intermediate* time range, limited by

$$t^*_{\mathrm{dyn}} \leq c_1 t / W < 3\tau^*$$ (10)

In this time range the dynamic effects acting at the crack tip already have decreased significantly; therefore the actual crack tip loading can be described sufficiently accurately by a formula (Eq 2) based on quasi-static assumptions. However, externally applied and measured loads still are influenced significantly by dynamic effects.

- Overall *quasi-static* time range, limited by

$$c_1 t / W \geq 3\tau^*$$ (11)

In this time range an overall quasi-static behavior can be assumed; therefore even static evaluation procedures may be applied with sufficient accuracy.

Which kind of measuring and evaluation procedure must be applied in an impact test depends on the time range where failure will occur. If a certain value of the expected impact fracture toughness K_{Id} can be assumed, the time to fracture, t_f, can be estimated roughly from a simplified version of Eq 2 using $t = t_\mathrm{f}$ and $K_\mathrm{I}^{\mathrm{dyn}} = K_{\mathrm{Id}}$:

$$t^*_\mathrm{f} = \frac{c_1 t_\mathrm{f}}{W} = \frac{c_1 K_{\mathrm{Id}} C^*_\mathrm{s}}{E \, Y \, v_\mathrm{o} \, \sqrt{W}}$$ (12)

Depending on the estimated value of t^*_f in comparison to the transition times, it can be easily decided in the planning phase of an impact experiment whether special dynamic evaluation and measuring procedures will be necessary.

Conclusions

The developed and experimentally determined dynamic key-curves for brittle fracture impact tests are a quantitative measure of dynamic effects acting at the crack tip. They allow the prediction of the dynamic crack tip loading history for arbitrary impact test conditions with sufficient accuracy for engineering purposes and can be applied to the determination of impact fracture toughnesses. Based on these dynamic key-curves, transition times t_{dyn} are established beyond which the actual dynamic crack tip loading can be described approximately in a quasi-static manner. The results have also been compared to transition times defined by other authors. It can be concluded:

- The transition time t_{dyn} established by the author [7] is comparable to the independently and differently defined transition time $2t_\mathrm{T}$ of Nakamura, Shih, and Freund [9].
- The transition time t_{dyn} is strongly dependent on the specimen length L/W of the fixed support span (i.e., on the overhang of the specimen), which indicates the significant influence of dynamic effects like loss of contact at the anvils which also varies with L/W.
- The transition time 3τ defined by Ireland [3], which is related to reduced hammerload oscillations, is about six times larger than t_{dyn}, which indicates that dynamic effects diminish six times earlier at the crack tip than at the impacting hammer.
- The transition time t_{dyn} (or t_T) and the transition time 3τ are limits defining three time

ranges (fully dynamic, intermediate, and quasi-static) with different amounts of dynamic effects.

• A simple equation used to estimate the expected time to fracture, t_f, only from the knowledge of the impact test conditions enables one by comparison with the related transition times to decide before an experiment whether fully dynamic evaluation and measuring procedures are necessary or if quasi-static procedures can be applied with sufficient accuracy.

Acknowledgments

The author is grateful to the Deutsche Forschungsgemeinschaft and the Stiftung Volkswagenwerk, who supported parts of these investigations.

References

[1] *Impact Testing of Metals, ASTM STP 466,* American Society for Testing and Materials, Philadelphia, 1969.

[2] *Instrumented Impact Testing, ASTM STP 563,* American Society for Testing and Materials, Philadelphia, 1973.

[3] Ireland, D. R., "Critical Review of Instrumental Impact Testing," presented to Int. Conf. on Dynamic Fracture Toughness, London, 1976.

[4] Proposed Standard Method of Test for Instrumented Impact Testing of Precracked Charpy Specimens of Metallic Materials, ASTM E24.03.03, Draft 2, 1981.

[5] Kalthoff, J. F., Winkler, S., Böhme, W., and Klemm, W., "Determination of the Dynamic Fracture Toughness K_{Id} in Impact Tests by Means of Response Curves," in *Advances in Fracture Research,* D. Francois et al., Eds., Pergamon Press, Oxford, 1980, pp. 363–373.

[6] Kalthoff, J. F., Winkler, S., and Böhme, W., "A Novel Procedure for Measuring the Impact Fracture Toughness K_{Id} with Precracked Charpy Specimens," *Journal de Physique,* Coll. C5, No. 8, Tome 46, 1985, pp. 179–186.

[7] Böhme, W., "Experimentelle Untersuchungen dynamischer Effekte beim Kerbschlagbiegeversuch," Dissertation, Darmstadt, W. Germany, 1985, or Scientific Report W1/85, Fraunhofer-Institut für Werkstoffmechanik, Freiburg, W. Germany, 1985.

[8] Böhme, W. and Kalthoff, J. F., "On the Quantification of Dynamic Effects in Impact Loading and the Practical Application for K_{Id}-Determination," *Journal de Physique,* Coll. C5, No. 8, Tome 46, 1985, pp. 213–218.

[9] Nakamura, T., Shih, C. F., and Freund, L. B., *Engineering Fracture Mechanics,* Vol. 25, No. 3., 1986, pp. 323–339.

[10] Ireland, D. R. in Ref 2, pp. 3–29.

[11] Turner, C. E., "Dynamic Fracture Toughness Measurements by Instrumented Impact Testing," presented to Adv. Seminar on Fracture Mechanics, Joint Research Center, Ispra, Italy, 1975.

[12] Peuser, T. in *Proceedings, AFMMS Int. Conf.,* Freiburg, W. Germany, G. C. Sih, et al., Eds., Martinus Nijhoff, 1983, pp. 455–465.

[13] Wich, G., "Ein Beitrag zur Ermittlung von Risszaehigkeiten bei schlagartiger Belastung," Dissertation, München, W. Germany, 1986.

[14] Williams, J. G. and Adams, G. C., *International Journal of Fracture,* Vol. 33, 1987, pp. 209–222.

[15] Srawley, J. E., *International Journal of Fracture,* Vol. 12, 1976, pp. 475–476.

[16] Bucci, R. J., Paris, P. C., Landes, J. D., and Rice, J. R. in *Fracture Toughness, ASTM STP 514,* American Society for Testing and Materials, Philadelphia, 1972, pp. 40–69.

[17] Kalthoff, J. F., Böhme, W., Winkler, S., and Klemm, W. in *Proceedings,* CSNI Specialists Meeting on Instrumented Precracked Charpy Testing, Palo Alto, Calif., 1980, also EPRI Report NP-2102-LD, 1981, pp. 1–17.

[18] Kalthoff, J. F. in *Handbook on Experimental Mechanics,* A. S. Kobayashi, Ed., Prentice-Hall, Englewood Cliffs, N.J., 1986, pp. 430–500.

[19] Böhme, W. and Kalthoff, J. F., *International Journal of Fracture,* Vol. 20, 1982, pp. R139–R143.

[20] Böhme, W., "The Influence of Stress Waves on the Dynamic Crack Tip Loading in Three-Point Bend Impact Testing," in *Proceedings,* IMPACT '87, Bremen, W. Germany, 1987, C. Y. Chiem et al., Eds., DGM Informationsgesellschaft, Oberursel, Vol. 1, 1988, pp. 305–311.

L. Nash Gifford,[1] *Joan R. Carlberg,*[1] *Archie J. Wiggs,*[1] *and J. Bart Sickles*[1]

Explosive Testing of Full Thickness Precracked Weldments

REFERENCE: Gifford, L. N., Carlberg, J. R., Wiggs, A. J., and Sickles, J. B., **"Explosive Testing of Full Thickness Precracked Weldments,"** *Fracture Mechanics: Twenty-First Symposium, ASTM STP 1074,* J. P. Gudas, J. A. Joyce, and E. M. Hackett, Eds., American Society for Testing and Materials, Philadelphia, 1990, pp. 157–177.

ABSTRACT: We describe an explosive test which uses small charges to induce plastic strain up to a few percent in a cracked structural element with thickness on the order of 50 mm (2 in.). The boundary conditions of the test section, its nominal strain versus time history, and peak strain as a function of charge weight are all determined experimentally. Finite element models capable of reproducing this behavior are also described; their purpose is to allow computation of possible fracture-governing parameters and to correlate response of the element to full-scale structural behavior. Some early fracture test results for shallow cracks in HY-80 GMA welds are presented and discussed. We find that these tests, which reflect severe but realistic service conditions, are well beyond the limits of current ASTM elastic-plastic fracture tests such as E 813-87, Test Method for J_{Ic}.

KEY WORDS: fracture mechanics, ductile fracture, dynamic fracture, explosive tests, shallow cracks, full-scale simulation, J-integral, finite element analysis

The hulls of Naval ships frequently employ welded construction of high yield (HY) and high strength-low alloy (HSLA) steels. Such steels and their weldments, with yield strengths in the neighborhood of 700 MPa (100 ksi), can be remarkably tough at service temperatures. High toughness is needed, because extreme service application can involve static and dynamic loadings which produce gross section yielding accompanied by peak strain on the order of several percent.

Standards for acceptance of these materials in hull applications have been set on the basis of what can be *achieved* in toughness rather than on toughness actually *required* for structural integrity. This philosophy [1–4] is reflected in the stringent requirements set for the ASTM Dynamic Tear Test (E 604-83) and for the "Explosion Bulge" test as prescribed in military standards [5]. Such requirements assure that the best materials are used, but they are loosely qualitative about the service performance of a material when fabrication defects or fatigue-induced cracks are present. Nor do such tests allow us to determine if less tough materials will perform just as well, in fact exceeding service requirements.

We have much to learn about the deleterious effects of cracks on the fracture response of tough hull structures subjected to large dynamic strains such as those associated with explosive attack [6,7]. This is due in large part to the fact that state-of-the-art principles of linear-elastic and elastic-plastic fracture mechanics are rarely applicable. The J-integral (ASTM E

---

[1] Ship Structures and Protection Department, David Taylor Research Center, Bethesda, MD 20084-5000.

813-87) and tearing modulus concepts [8,9] can be used to characterize the static toughness of ductile HY and HSLA steels [10–12]. Under certain limitations [13,14], the dynamic fracture toughness can be characterized as well [15,16]. In our work, however, we are dealing with shallow cracks (crack depth/thickness $\leq$ 0.25) typical of service conditions and with gross section yielding (GSY) in the sense defined by Dodds et al. [17] (yielding through the thickness even without a flaw). Under conditions of GSY, low strain-hardening modulus typical of HY and HSLA steels, and nominal peak strain beyond yield, the "HRR-singularity" J-fields do not dominate [18]. Calculated values of the J-integral are well beyond upper-shelf values measured in valid J_{Ic} and tearing tests. The same applies to the crack tip opening displacement fracture criterion [19]. These facts, coupled with the need to evaluate the fracture resistance of full-scale *structures* as opposed to full-thickness *specimens,* provided reason to develop an experimental method by which comparative dynamic hull toughness could be inexpensively demonstrated.

In this paper, we discuss the configuration, testing procedure, calibration, and fracture test results for shallow cracks in a full-thickness, explosively loaded "structural element" designed to simulate prototype structural behavior. We also briefly discuss inelastic dynamic finite element analyses of the element, which are useful for quantifying the element response and serve as a tool to compare conditions in the element to those calculated for the prototype. Our purpose is to outline our approach to dealing with structural fracture problems for which cracks are usually shallow, materials are usually tough, and dynamic strains are usually large.

Mode I Hull Toughness Structural Element

Our goal in developing the hull toughness structural element was to simulate dynamic fracture in an actual hull by means of a relatively simple, full-thickness structural test in the laboratory. The motivation for this derives from the fact that realistic crack tip constraint is not readily achieved in scale models (even if we could build small models whose weld properties matched the prototype point for point) and from the impracticability of testing actual hulls (or hull sectors) to high inelastic strain levels. Our approach is thus not to model the ship hull itself, but rather to simulate the local conditions likely to exist in the hull near a crack.

Design of the Mode I structural element was guided by the following constraints:

1. Materials must duplicate those which are found in the prototype. To test actual welds and to develop realistic constraint at the crack site, material thickness must be at full scale.

2. Dynamic stress and strain fields, and strain rise time, must approximate the severity of those found at the flaw site in the prototype.

3. Peak dynamic strain (at the crack site on the element) varying from elastic levels up to several percent plastic strain should readily be accommodated.

4. The element should be relatively inexpensive to fabricate and test at service temperature.

5. The choice of crack should be conservative. We chose a Mode I crack because the loading necessary for its extension is considered to be less than for other likely crack orientations.

6. The response of the element should be analytically reproducible. This done, we can compare fracture-producing conditions existing in the structural element to those existing in the prototype.

The simple explosively loaded structural element and test procedure resulting from this thinking are shown schematically in Fig. 1 for a 51 mm (2 in.) thickness. This particular

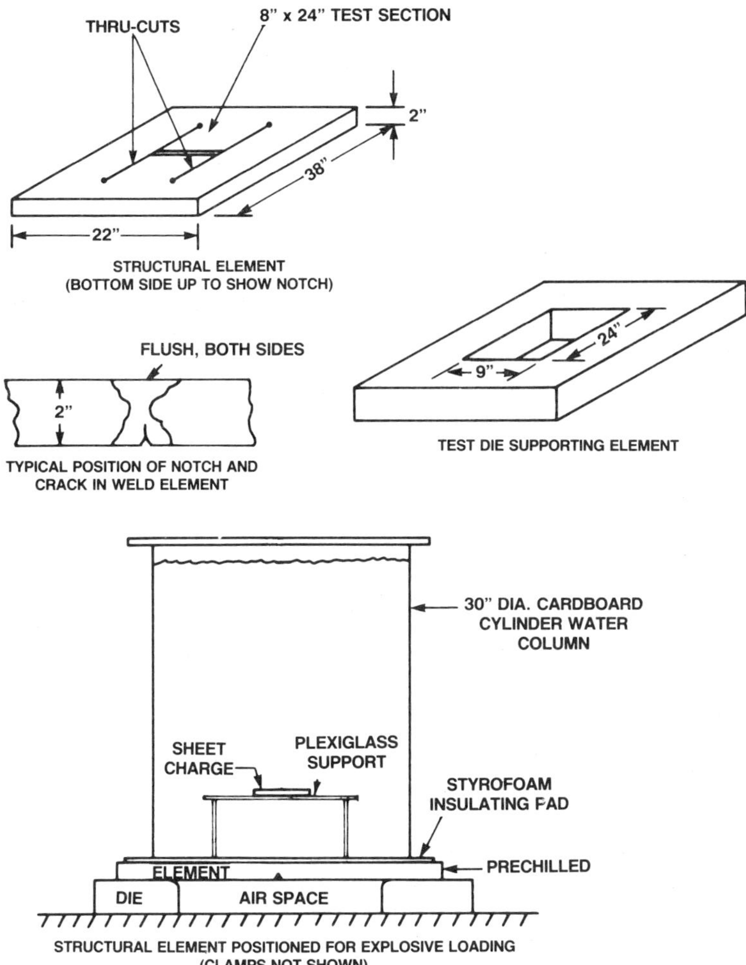

FIG. 1—*Schematic of structural element, test die, and configuration for explosive hull toughness testing.*

element weighs about 213 kg (470 lb), but the size of the element can be adjusted to test at thicknesses in the 19 to 76 mm (¾ to 3 in.) range. For tests of hard-to-get weld metal or other scarce materials, only the test section (central portion) need be fabricated. The ends of the test section can subsequently be welded into a separate carrier plate using matching yield strength welds.

The element is basically a derivative of the old Naval Research Laboratory "Explosion Tear" element [5], having a narrower but longer test section which gives a slower strain rise time. The structural element also differs from that test by the introduction of controlled cracks, which, of course, are subjected to only one shot. These changes, and the use of a water column which surrounds a sheet charge at close range, represent the major advances. The water column provides an efficient coupling path between the charge and the element, eliminating the need for large charges and remote test sites.

Clamping System

A convenient clamping arrangement to secure the structural element to the test die is shown in Fig. 2. It consists of an over-sized steel cover plate which has a cutout to expose the test section to the blast. Eight bolts are used to hold the cover plate securely to the test die. The entire assembly weighs on the order of 900 kg (1 ton).

Water Column

The water column used in our tests is 762 mm (30 in.) in diameter and is cut from cardboard cylinders used in forming concrete columns. Since it is open at both ends, it is lined with thin plastic sheet to make it watertight. We have been unable to discern any effect of the depth of the water column on peak strain response of the element in our tests, all of which used at least 203 mm (8 in.) of water above the charge standoff level.

Explosive Charge

In the development of a structural element for use with small charges, we experimented with different explosives. We found that we could induce a peak strain of a few percent at the center of the element of Fig. 1 with charges on the order of 500 g (1 lb) at a 200 mm (8 in.) standoff. We also found that peak strain is not only sensitive to charge weight and standoff but also to charge configuration. For example, a sheet charge spread out over a 180 mm (7 in.) square does not produce as great a peak strain as a concentrated cylindrical charge of the same weight and standoff.

The outcome of this experimentation was a decided preference for sheet charges. Adaptability and availability alone justify this choice, but sheet charges are cheaper also. We use Du Pont Deta Sheet C, which comes in differing weights designated in the unique units of grams per square inch (1 g/in.^{2} = 1550 g/m^2 or 0.32 lb/ft^2). The sheet is readily cut to any desired size and stacked to any desired weight. To distribute the blast over the center portion of the test section evenly, we use a square charge with sides varying from 152 mm (6 in.) to 190 mm (7.5 in.). This dimensional latitude allows us flexibility to quickly construct charges of any desired weight, from which peak strain is reliably predictable. Accurate charge position and standoff are achieved by supporting the charge on a thin Plexiglas table above the element.

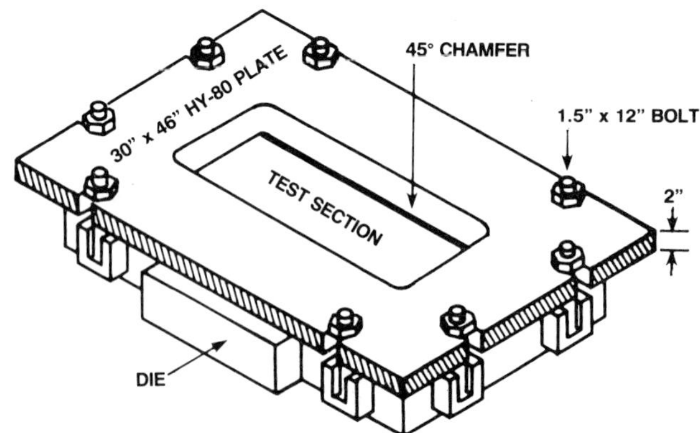

FIG. 2—*A clamping arrangement for restraint of structural elements during explosive test.*

Instrumentation

Establishment of the relations between charge shape, charge weight, charge standoff, element thickness, element yield stress, and the peak nominal strain occurring at the flaw site (center) of the structural element is of paramount importance in our experimental program. We originally used several approximations to relate charge weight and standoff to either load versus time history or to initial velocity conditions in the test section of the element. This information provided input for use in inelastic dynamic finite element models to predict strain versus time in the test section [20]. Such analyses guided our early tests, but their verification required the measurement of inelastic dynamic strain histories. Because of the violent explosive environment, this is a difficult proposition.

In our experience using electrical strain gages to measure strain versus time histories, we have learned three important things:

1. It is impossible to measure strain on the explosive side of the element (even at low strain levels) without some form of exotic gage protection. Both the gage wires and the gages themselves are instant victims of the explosive blast.

2. Gages attached to the underside of the element will remain bonded under the shock of a blast which will produce more than 3% peak strain. But even this requires close attention to special procedures. We have had success using Micro-Measurements Type EP-08-125AD strain gages installed with Type AE-10/15 adhesive cured at 66°C (150°F) for 2 h.

3. The gage wires are the Achilles' heel of any strain measurement system. Special wiring procedures (use of ultralight wiring, low residue soldering, blast-resistant heavy cables, etc.) were developed to withstand the blast environment and motion of the test assembly.

Temperature Control

Testing at or just below the least anticipated service temperature is important to assure that results are representative. We achieve this by pre-chilling the entire structural element in a freezer to a temperature about 8°C (15°F) below the desired test temperature. The water column is insulated from the element test section by means of a thin Styrofoam sheet, slowing the rapid heating that would otherwise occur. As a result, the total time required to prepare for test is comfortably less than that required for the structural element to warm to test temperature.

Calibration of Structural Element

In order to understand the results from our explosive fracture tests and to relate them to fracture response in an actual ship hull, a complete understanding of the deformation mechanisms and strain fields developed in the element is necessary. Tests conducted for this purpose are described here.

Static Tests to Establish Boundary Conditions

We have confined our finite element analyses (discussed later) to the central test section of the element. In so doing, we must exercise great care in specifying the end boundary conditions if accurate results are to be obtained. Unknown elastic support of the carrier plate by the die, and the continuous connection of the test section to the carrier plate, lead to complex boundary conditions. The ends of the test section are neither simply supported nor fully fixed, but rather lie between these classical descriptions.

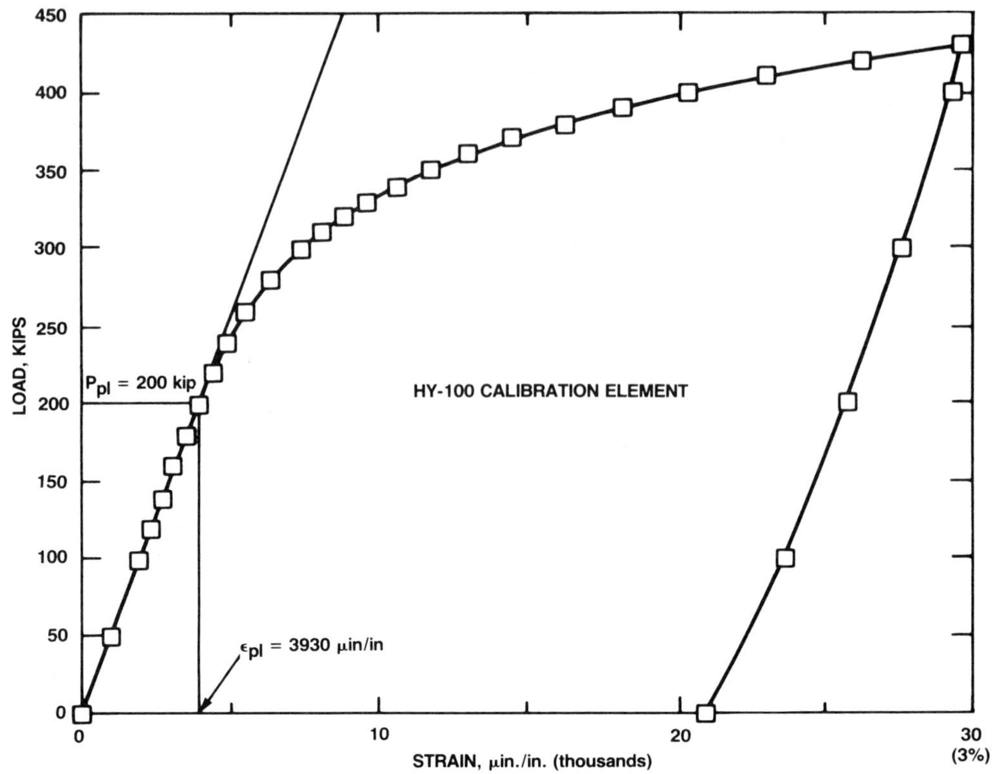

FIG. 3—*Typical strain response at center of a statically loaded 51 mm (2 in.) thick HY-100 calibration structural element.*

To establish the boundary conditions and deformation response in a nominal (uncracked) structural element, three 51 mm (2 in.) thick HY-100 baseplate elements were instrumented along their lengths with strain gages and tested statically. The elements were supported by the usual test die and were loaded in four-point bending using a 272 000 kg (600 000 lb) universal testing machine. The spacing between the two loading points—89 mm (3.5 in.)— was chosen to approximate the distributed loading condition anticipated for explosive loading. Eleven strain gages were positioned along the length of the element at distances of 0, 38, 64, 254, 279, and 305 mm (0, 1.5, 2.5, 10, 11, and 12 in.) from the element midspan. For each element, several elastic tests were performed, followed by a fully plastic test to a peak center strain of 3%. A typical experimental strain versus load history at the center of an element is shown in Fig. 3, while the experimental axial distribution of strain at increasing load levels is shown in Fig. 4.

Analyses and data from these tests were documented in an internal report.[2] Among the conclusions were:

1. The deformation mechanisms observed in the static tests duplicate those calculated by finite element methods for explosive loading. (These calculations are discussed subse-

[2] Dally, J. W., Gifford, L. N., and Sickles, J. B., "The Nonlinear Boundary Conditions of the Hull Toughness Structural Element," DTRC Ship Structures and Protection Dept. Rept. SD-87-172-64, July 1987.

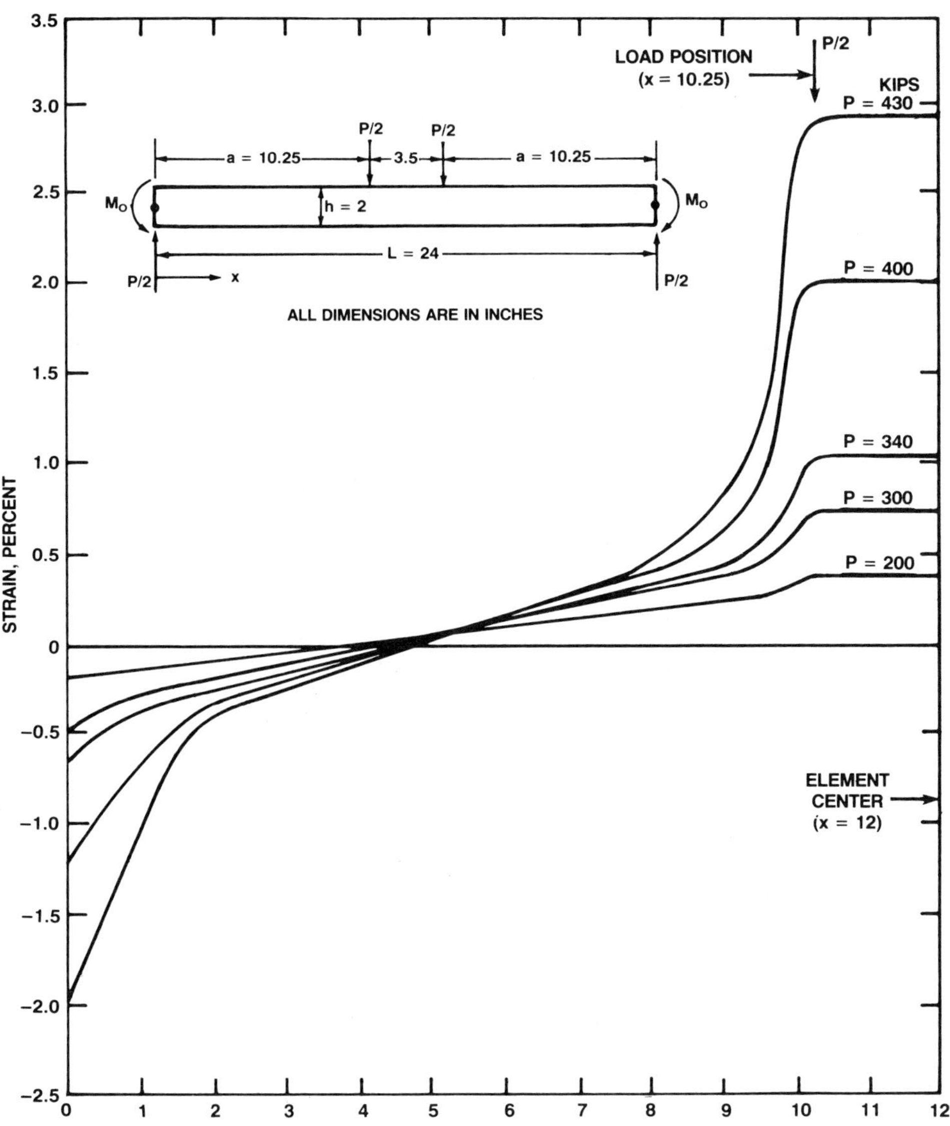

FIG. 4—*Strain distributions at increasing load severity typical of uncracked HY-100 structural element response.*

quently.) Both static and dynamic deformations follow a sequence of: (1) elastic response everywhere, (2) yield zones initiate on the surface at the element center, (3) yield zones initiate at the ends of the test section, (4) elastic-plastic bending continues, (5) a plastic hinge forms at the element center, (6) more elastic-plastic bending, (7) additional plastic hinges form at the ends of the test section, and (8) further deformation is resisted by membrane forces.

2. The behavior of the structural element is distinctly different from that of common fracture mechanics test specimens which are statically determinate. More like a structure, the

element is capable of supporting a reduced load even when severed at the center of the test section. (Load is then supported by the resulting twin cantilever beams.)

3. Bending strain predominates membrane strain. Membrane strain does not begin to develop until a peak bending strain (measured at the center surface of the element) of about 1.5% is reached. At 3% peak surface strain (membrane plus bending), the ratio of bending strain to membrane strain still has a value greater than 4.

4. An analysis based on beam theory, coupled with the experimental data, revealed that the degree of constraint at the ends of the test section is a nonlinear function of the applied load. At elastic load levels, we obtained an end moment which was 67% of the fixed-end value. At a load corresponding to 3% peak center strain, the end moment had increased to 87% of the fixed-end condition.

5. The load dependence of the end constraint is due entirely to the gradual formation of plastic hinges and does not result from a change in support conditions. Thus it is possible to simulate the boundary condition in finite element models by the introduction of linear elastic springs at the ends of the test section. The resulting models (which are not unique) reproduce static experimental response in the linear range and, because yielding is accommodated in the finite element mesh, in the nonlinear range as well. Thus the load dependence of the end condition is automatically accommodated.

Explosive Calibration Tests

We have alluded to the difficulty encountered in measuring large dynamic strains in an explosive environment. As a consequence, our dynamic calibration data are somewhat limited. Our results are sufficient, however, to provide engineering estimates for peak strain at the center of 51 mm (2 in.) thick elements as a function of charge weight for a 203 mm (8

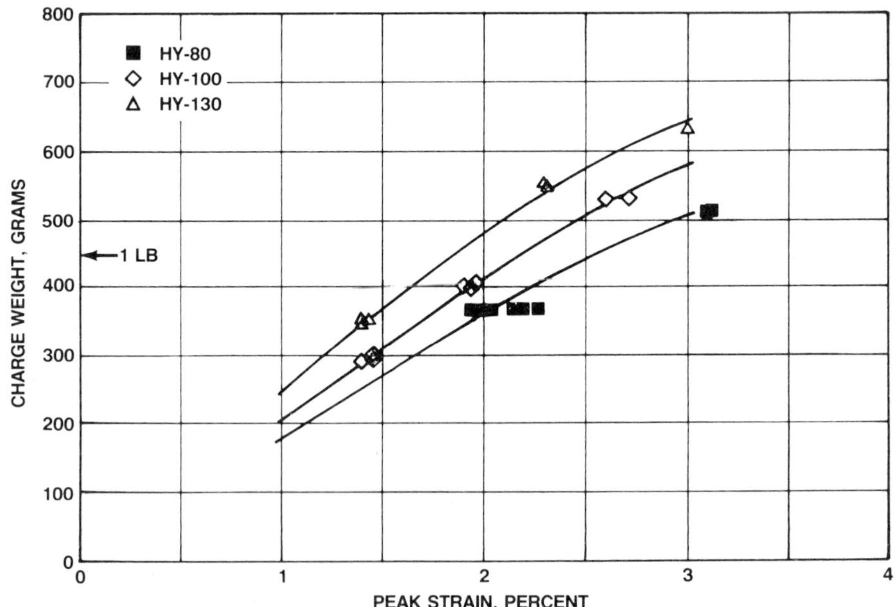

FIG. 5—*Peak strains measured in explosive calibration tests of 51 mm (2 in.) thick HY structural elements shot at 203 mm (8 in.) standoff.*

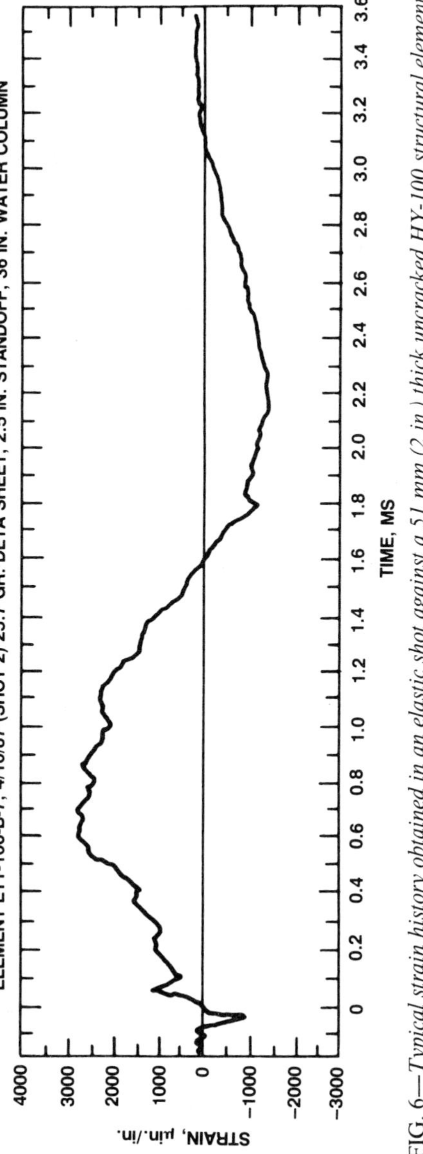

FIG. 6—*Typical strain history obtained in an elastic shot against a 51 mm (2 in.) thick uncracked HY-100 structural element.*

in.) standoff. Our measurements of peak strains achieved in HY-80, HY-100, and HY-130 steel structural elements are shown in Fig. 5. The results of these tests can be adjusted to calibrate moderately different thicknesses and yield strengths by use of nonlinear static finite element analyses. This can be done approximately by solving for a static "load" which produces the same strain at the center of a calibration element at a given charge weight, and then applying this "load" to an element of differing thickness and/or yield stress and calculating peak strain.

Examples of raw strain histories obtained from some of the calibration tests are shown in Fig. 6 for an elastic shot and in Fig. 7 for three shots into the nonlinear strain range. (All strain histories were measured at the center of the element on the side opposite the explosive charge.) Duplicate elastic tests indicated that strain-time history is highly repeatable. The peak strain developed at the center of the element results from basic bending (cosine displacement shape) and occurs only after several longitudinal transit times of the stress waves induced by the explosive loading. Thus we believe that peak strain is not significantly affected by stress waves which, incidentally, remain elastic for loading severities of interest to us.

Elastic finite element natural frequency calculations indicated a first mode rise time of about 0.54 ms. The actual experimental rise time to peak strain is about 0.6 ms for elastic shots and about 1 ms for plastic shots. The average strain rate (peak strain/rise time), however, is much higher for plastic shots. At 3% peak strain, the average strain rate is on the order of 30/s as compared to about 5/s for elastic shots. This is not very fast in the world of dynamic fracture, however, and this fact may simplify future hull toughness research.

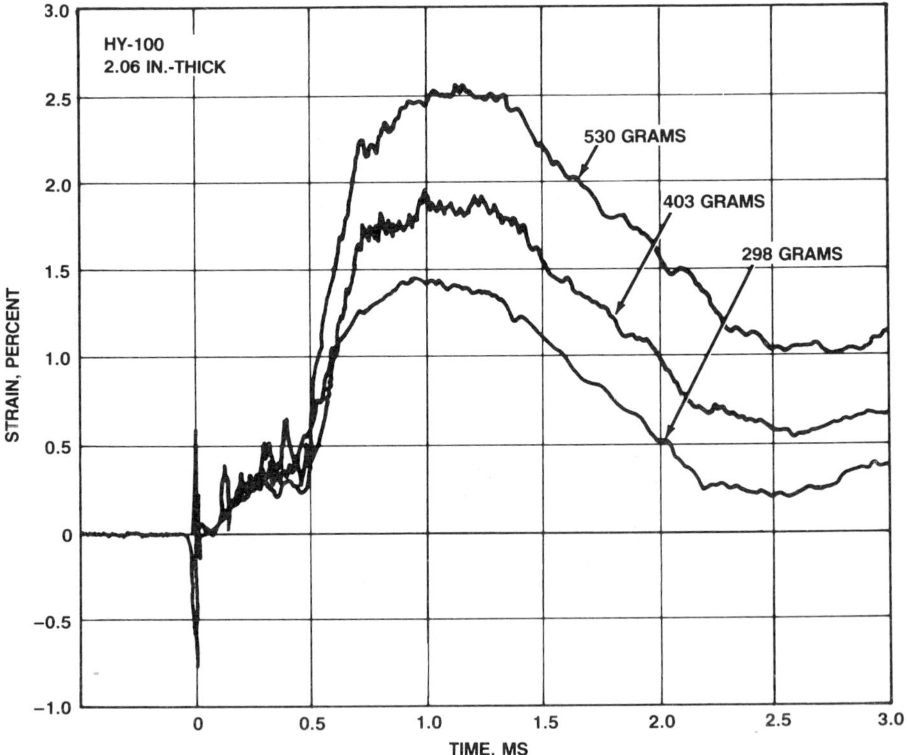

FIG. 7—*Inelastic strain histories measured in three tests of HY-100 uncracked elements shot at 203 mm (8 in.) standoff.*

Analysis of Structural Element

The value of a structural fracture test can be greatly enhanced if we can calculate the associated dynamic stress and strain fields and, perhaps more importantly, calculate some measure of a crack driving "force". The three leading fracture criteria in nonlinear fracture mechanics are presently the J-integral, crack tip opening displacement (CTOD), and strain energy density (SED). As part of our computational program, we have thus far investigated the first two of these parameters. The third, SED [21], is under research at the Naval Research Laboratory [22] and elsewhere.

The purposes of our analyses were three-fold. First, we wished to establish proper finite element boundary conditions necessary to reproduce the experimental behavior of the static tests. This done, we wished to obtain approximate estimates of the J-integral and CTOD as a function of nominal (uncracked) peak strain for crack depths and materials of interest. These objectives were reached by static computations. Finally, we wished to establish a means of estimating the dynamic J-integral and CTOD under explosive load at large strain and shallow crack depths of interest for ship hulls. Such computations are not trivial, so this work is briefly described here.

Static Analysis

We used two elastic-plastic finite element codes—PAPST [23,24] and version 4-5-171 of ABAQUS [25]—to bolster confidence in our analyses. Plane stress conditions were assumed for all computations, although the actual behavior of the structural element falls between the conditions of plane stress and plane strain. We adjusted for this difficulty by adjusting the boundary conditions for the plane stress models to reproduce measured nonlinear strain response as discussed below.

PAPST and ABAQUS analyses employed 12-noded (bi-cubic) and 8-noded (bi-quadratic) isoparametric quadrilateral elements respectively. Because of symmetry, only one half of the structural element was modeled (Fig. 8). Full numerical integration of the element stiffness was used in both codes: 4×4 for PAPST and 3×3 for ABAQUS.

To match our static calibration tests, material properties for HY-100 were used. The 0.2% offset yield stress was 772 MPa (112 ksi), with the nonlinear stress-strain curve being modeled with a four-piece multilinear approximation as tabulated in the inset in Fig. 8. Young's modulus was taken as 207 GPa (30×10^6 psi), Poisson's ratio as 0.3. Incremental plasticity (J_2 flow theory), the von Mises yield criterion and associated flow rule, and isotropic hardening were used in both computer codes.

Because the ends of the test section are neither fixed nor pinned, it was not obvious what boundary conditions for the finite element model would reproduce the experimental response. We had to determine this iteratively. We varied the half-length of the model and used various combinations of clamped, roller, and spring boundary conditions in this process. Good agreement with experimental strains was demonstrated for a half-length of 483 mm (19 in.) when distributed elastic springs with a spring constant of $k = 136$ GN/m/m^2 (0.5×10^6 lb/in./in.^{2}) were placed at the end of the test section (Fig. 8). This boundary condition is not unique; it is just one way of modeling the partially clamped end of the structural element, and it allows a plane stress assumption to reproduce experimental strains.

The introduction of a crack significantly complicated the ABAQUS finite element model. In order for a fracture criterion such as CTOD to be calculable and meaningful, the size of the elements near the crack tip must be on the order of the crack tip opening displacement at fracture initiation [26]. An extremely fine finite element mesh was thus needed as shown in Fig. 8. Elements right at the crack tip were 0.159 mm (0.00625 in.) square. The crack tip opening displacement, δ, was approximated by the Rice 90-deg intercept method [27,28].

Nonlinear static computations were performed for full-width cracks in a 51 mm (2 in.) thick element. A uniform load distributed over the central 254 mm (10 in.) of the test section was chosen as a static approximation of the explosive load. J versus applied load was determined from one analysis, while the corresponding strain versus load for an uncracked element was determined in a baseline analysis. Elimination of the load parameter allowed us to construct plots of J for cracked elements versus the nominal applied tensile strain at the center of an uncracked element. The PAPST code computed J directly (by path integration), while the ABAQUS model first calculated the crack tip opening displacement. We then related J to δ by

$$J = m\sigma_y\delta \tag{1}$$

where σ_y is the yield stress and m was approximated as 1.0, the value for an elastic-perfectly plastic material in plane stress [29].

Results for J versus applied nominal strain for a 6.35 mm (0.25 in.) deep crack are shown

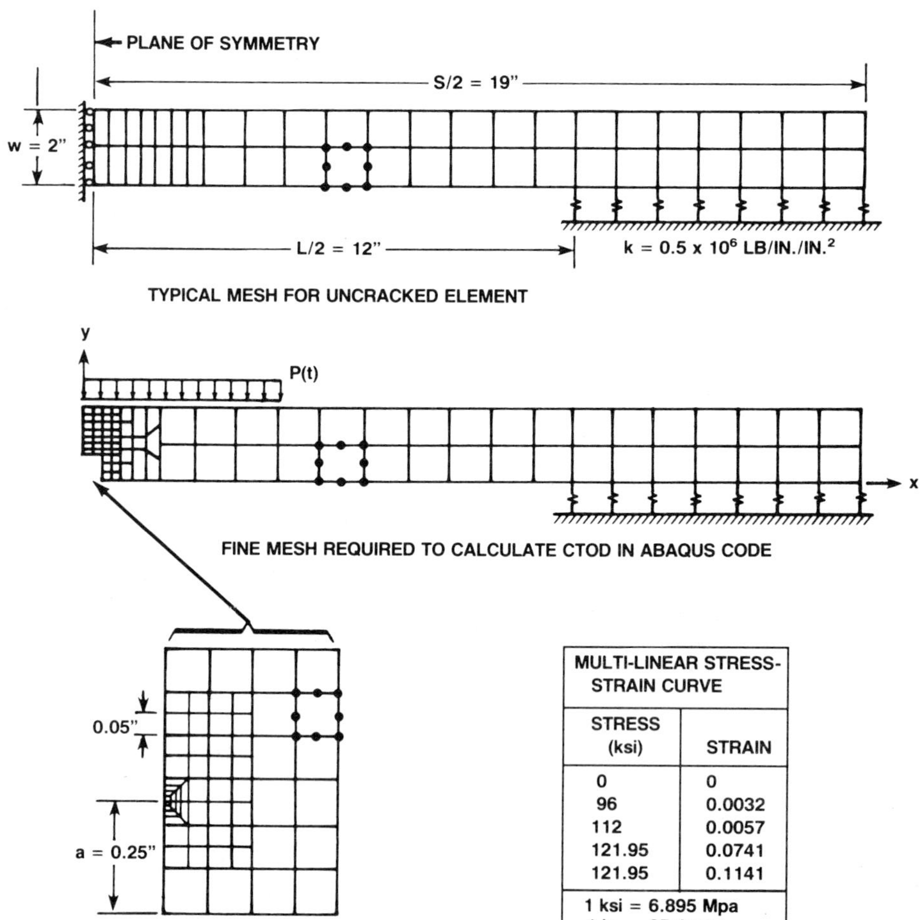

FIG. 8—*Typical finite element meshes used in static and dynamic analyses.*

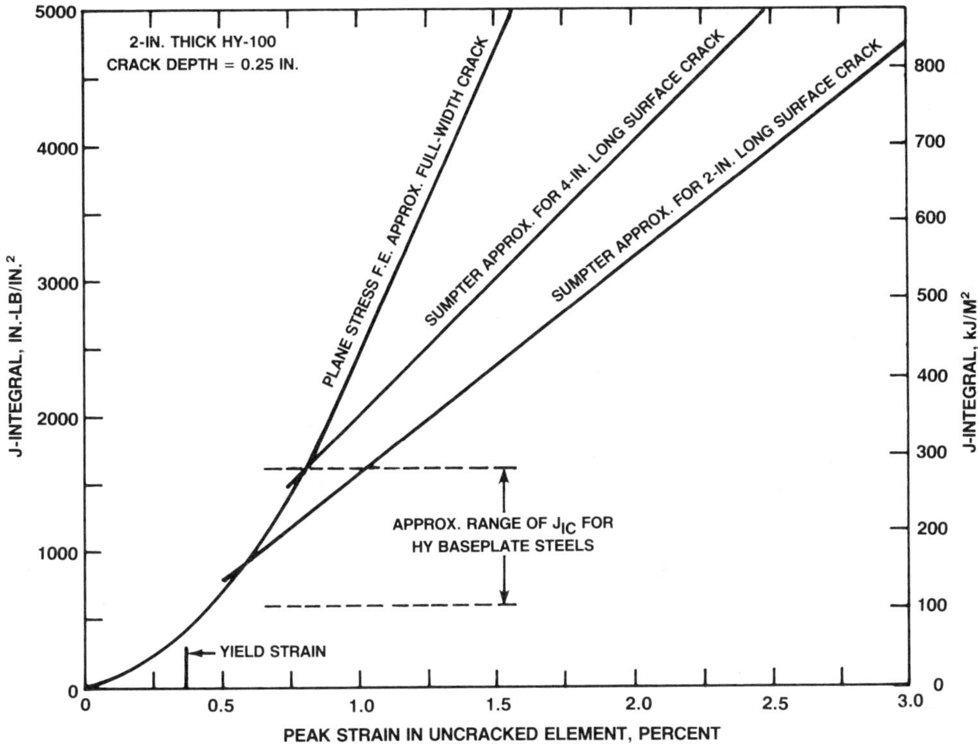

FIG. 9—*Estimates of the J-integral as a function of nominal applied strain for HY-100 structural elements containing 6.35 mm (0.25 in.) deep cracks.*

in Fig. 9. We have also plotted estimates of J from Turner [30] and Sumpter [31] for 51 and 104 mm (2 and 4 in.) long surface cracks of the same depth. These estimates of J are significantly different, the full-width crack apparently being much more severe. Nonetheless, the results serve to roughly quantify the conditions at the crack tip in the structural element, which is the only intent here. Figure 9 shows that as the nominal center strain approaches 1%, the value of the J-integral exceeds the upper-shelf range of toughness values measured by ASTM E 813 for HY steels. For deeper (but still shallow) cracks, this will occur at even smaller nominal strain levels.

Dynamic Analysis

Transient response was first calculated for a 51 mm (2 in.) thick uncracked element using a two-layer finite element model as shown at the top of Fig. 8. We used this model to establish computational strategy for the ABAQUS code. The cracked finite element model contained 1028 degrees of freedom and required significantly more computer time. A consistent mass, implicit time integration scheme was employed in the numerical calculations. We used the Hilber-Hughes-Taylor operator [32] (an extension of the Newmark β operator) to introduce some artificial damping to eliminate high frequency noise. Time increment choices typically ranged from 1 to 3 μs during analysis of the uncracked element and from 0.5 to 3 μs for the cracked element. More than 400 time steps were typically required to trace nonlinear response through the time of peak strain.

Because spatial and temporal pressure variations on the structural element have not been experimentally quantified, we approximated the dynamic loading history by Taylor theory for a plane exponential underwater shock wave. The pressure-time history has the form

$$P(t) = P_0 e^{(-t/\theta)} \qquad (t \geq 0) \tag{2}$$

where the maximum pressure, P_0, is a function of charge weight, charge standoff, and experimental constants, and θ, the "time constant", governs the pressure decay and is also an experimental constant.

The Taylor theory loading was modified for our analyses by changing the rise time from 0 to 5 μs. The decay curve was also approximated by a straight line falling to zero pressure at 55 μs. These adjustments were based on our early pressure-time measurements which were made for compact pentolite charges in a water column. The resultant loading consisted of a triangular pressure pulse of 55 μs duration, which we applied uniformly over the 254 mm (10 in.) central span of the top surface of the element as shown in Fig. 10.

Typical finite element transient histories for the bottom and top surface strains at the center of an uncracked element are also shown in Fig. 10. The peak pressure for this analysis was 517 MPa (75 ksi), which corresponds to a charge weight of about 500 g (1.1 lb). Peak strain of about 2.5% is reached at about 0.8 ms, a little earlier than the experimentally determined time. Also shown in the figure is the top surface (backface) compressive strain for the same element containing a full-width crack 6.35 mm (0.25 in.) deep. Peak strain is about

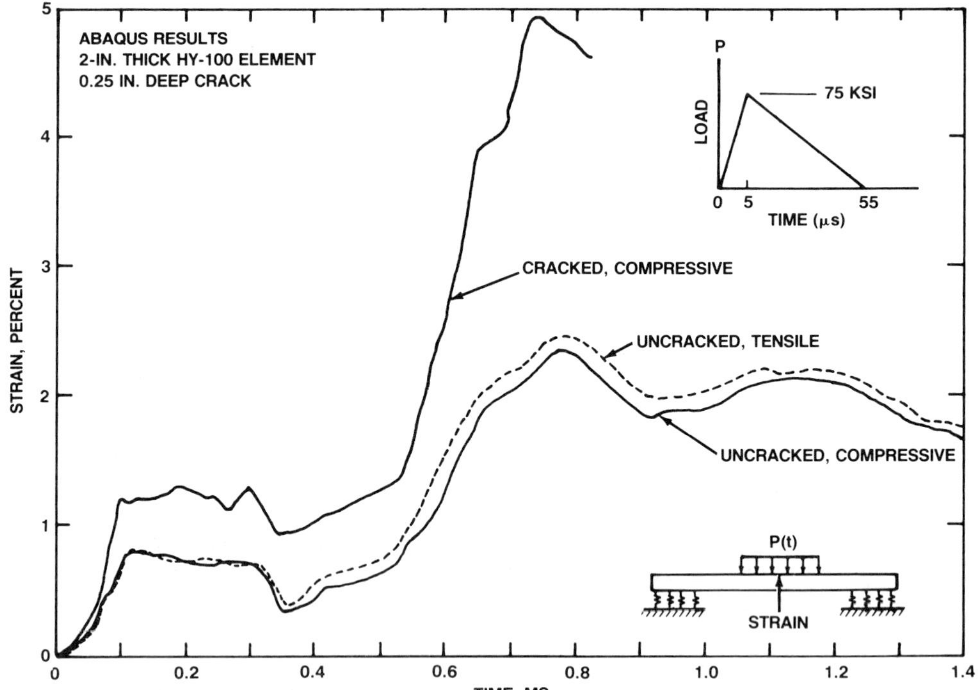

FIG. 10—*Calculated dynamic strain response for cracked and uncracked HY-100 structural elements at identical load severity.*

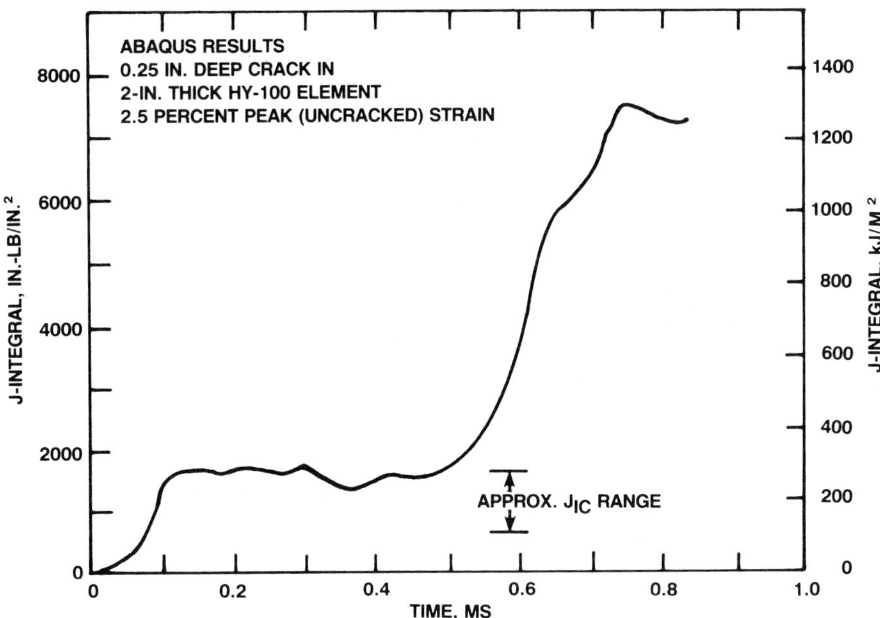

FIG. 11—*Dynamic J-integral history calculated for a cracked HY-100 structural element with* a/W = 0.125.

doubled. Figure 11 shows what we are really after—the *J*-integral time history, through the time of peak strain, for the cracked element. Again, *J* significantly exceeds maximum toughness values measured in valid ASTM tests.

Results of Shots Against Cracked Elements

Exploratory Tests

Following calibration tests and analyses, we were ready to begin actual fracture testing. We were faced with two important questions:

1. What is the proper relationship between peak nominal strain (at the crack site) and crack depth which will allow us to compare the relative fracture resistance of different materials? Blowing everything into two pieces or, at the other extreme, never extending cracks, were situations we obviously wanted to avoid since they do not assist in quantifying conditions of interest.
2. What types of cracks should we use? We have options ranging from surface breaking cracks of various lengths (the most realistic choices) to full-width cracks (the most conservative choice).

To gain an engineering "feel" for these issues, we tested 14 cracked structural elements, each 51 mm (2 in.) thick. Materials ranged from HY-80 to HY-140 baseplate, cracks ranged in length from full-width down to 51 mm (2 in.) long surface cracks, crack depths ranged from 5.1 to 25.4 mm (0.2 to 1.0 in.), and nominal peak strain ranged from 1.8 to 6.5%. The

TABLE 1—*Material properties for HY-80 GMA weldments.*

PROPERTY	AIRCO AX-90	LINDE 095046	U.S. NAVY REQ'MT. (BASEPLATE)
YIELD STRENGTH, KSI (Mpa) (AVERAGE OF TWO TESTS)	97.8 (674)	90.2 (623)	80 TO 99.5 (552 TO 686)
TENSILE STRENGTH, KSI (Mpa) (AVERAGE OF TWO TESTS)	114.8 (792)	106.8 (736)	INFO ONLY
ELONGATION IN 2 IN. (51mm), PERCENT	18	23	20
CHARPY VEE AT 0 F (−18C), FT-LBS (J) (AVERAGE OF THREE TESTS)	112 (152)	116 (157)	60 (81)
CHARPY VEE AT −60 F (−51C), FT-LBS (J) (AVERAGE OF THREE TESTS)	72 (98)	89 (121)	35 AT −120F (47 AT −84 C
DYNAMIC TEAR AT 30 F (−1C), FT-LBS (J) (AVERAGE OF THREE TESTS)	1020 (1383)	1121 (1520)	−
DYNAMIC TEAR AT −20 F (−29C), FT-LBS. (J) (AVERAGE OF THREE TESTS)	673 (913)	975 (1322)	450 AT −40F (610 AT −40C)
WELDS PASSED CRACK STARTER AND EXPLOSION BULGE TESTS.			

results of these tests were documented in an internal report,[3] the conclusions of which can be summarized briefly as follows:

1. Complete fracture of the element is possible, but only at very severe loading levels, crack depths, or low temperatures.

2. For crack depths greater than 5 mm (0.2 in.), the useful upper limit on nominal crack site strain is on the order of 3%. This loading severity is sufficient to cause crack extension in all but the toughest of materials.

3. We observed no striking differences between the behavior of elements containing full-width cracks and those which contained shorter surface cracks. The results indicated that the structural element is equally well suited to accommodate both edge (full width) cracks and surface cracks.

Tests of HY-80 GMA Weldments

Double-V HY-80 GMA butt welds 51 mm (2 in.) thick were produced by Mare Island Naval Shipyard in plates of sufficient size to provide 20 structural element test sections. The GMA spray transfer process was used to form the welds in 19 passes using 135 to 149°C (275 to 300°F) preheat and interpass temperature, and 3.94 kJ/mm (100 kJ/in.) heat input. Consumables were Airco AX-90 and Linde 095046. The resulting material properties are given in Table 1. The Charpy and dynamic tear test results indicate that we had high quality welds, far exceeding minimum Navy toughness requirements.

A total of 17 of these elements were ground flush, notched, prefatigued within limits of ASTM E 399, and explosively tested to a nominal crack site strain of 2.5%. The elements

---

[3] Gifford, L. N., "Tests and Results for Dynamic Mode I Submarine Hull Toughness," DTRC Ship Structures and Protection Dept. Rept. DTRC-SSPD-88-172-6, Sept. 1987.

were prechilled to permit tests near the freezing point of seawater. Following testing, the elements were heat-tinted, soaked in liquid nitrogen, and then broken open to reveal the extent of crack growth.

Eleven of the elements contained full-width cracks ranging in depth from 4.3 to 12.7 mm (0.17 to 0.5 in.). The remainder contained circular-section surface cracks having a length of 63.5 or 102 mm (2.5 or 4.0 in.) and a depth from 5.6 to 10.7 mm (0.22 to 0.42 in). The ratio of crack depth to element thickness (a/W) for these tests thus ranged from 0.085 to 0.25. Results of these tests are given in Table 2 and plotted in Fig. 12.

Figure 12 indicates that the peak strain level for these tests is about right, in the sense that some of the cracks extended and some did not. However, the figure does not suggest a well-

TABLE 2—*Explosive test results for 51-mm (2-in.) thick HY-80 GMA cracked welds shot at 2.5% nominal peak strength.*

ELEMENT ID ETT-80	WELDING CONSUMABLE	CRACK LENGTH, IN. (mm)	AT CENTER OF NOTCH			SET DEFECTION, IN. (mm)	TEST TEMP., °F (°C)
			NOTCH + FATIGUE CRACK DEPTH, IN. (mm)	CRACK GROWTH, IN. (mm)	FINAL CRACK DEPTH, IN. (mm)		
W-4	AIRCO	(FULL WIDTH) 8.0 (203)	0.17 (4.3)	0	NO GROWTH	0.22 (5.6)	32 (0)
W-5	AIRCO	(FULL WIDTH) 8.0 (203)	0.19 (4.8)	0	NO GROWTH	0.27 (6.9)	32 (0)
W-11	LINDE	(FULL WIDTH) 8.0 (203)	0.25 (6.4)	0.05 (1.3)	0.30 (7.7)	0.27 (6.9)	30 (−1)
W-6	AIRCO	(FULL WIDTH) 8.0 (203)	0.31 (7.9)	0	NO GROWTH	0.25 (6.4)	25 (−4)
W-7	AIRCO	(FULL WIDTH) 8.0 (203)	0.31 (7.9)	0	NO GROWTH	0.28 (7.1)	29 (−2)
W-9	LINDE	(FULL WIDTH) 8.0 (203)	0.32 (8.1)	0.08 (2.0)	0.40 (10.1)	0.37 (9.4)	30 (−1)
W-10	LINDE	(FULL WIDTH) 8.0 (203)	0.40 (10.2)	1.30 (33.0)	1.70 (43.2)	0.37 (9.4)	31 (−1)
W-15	LINDE	(FULL WIDTH) 8.0 (203)	0.40 (10.2)	1.60 (40.6)	COMPLETE SEPARATION	0.62 (15.7)	27 (−3)
W-8	AIRCO	(FULL WIDTH) 8.0 (203)	0.44 (11.2)	1.34 (34.0)	1.78 (45.2)	0.61 (15.5)	40 (4)
W-13	LINDE	(FULL WIDTH) 8.0 (203)	0.49 (12.4)	1.28 (32.5)	1.77 (44.9)	0.84 (21.3)	44 (7)
W-14	LINDE	(FULL WIDTH) 8.0 (203)	0.50 (12.7)	1.32 (33.5)	1.82 (46.2)	0.69 (17.5)	53 (12)
W-18	LINDE	4.0 (102)	0.22 (5.6)	0.13 (3.3)	0.35 (8.9)	0.25 (6.4)	31 (−1)
W-17	LINDE	4.0 (102)	0.25 (6.4)	0.06 (1.5)	0.31 (7.9)	0.31 (7.9)	30 (−1)
W-20	LINDE	4.0 (102)	0.40 (10.2)	0.10 (2.5)	0.50 (12.7)	0.29 (7.4)	26 (−3)
W-19	LINDE	4.0 (102)	0.42 (10.7)	1.18 (30.0)	1.60 (40.7)	0.25 (6.4)	47 (8)
W-22	LINDE	2.5 (63.5)	0.25 (6.4)	1.10 (27.9)	1.35 (34.3)	0.54 (13.7)	25 (−4)
W-23	LINDE	2.5 (63.5)	0.32 (8.1)	1.13 (28.7)	1.45 (36.8)	0.34 (8.6)	32 (0)

Charge: 6.15 IN. (156 mm) SQUARE DETA SHEET, 1 lb. (454g).
8 IN. (203 mm) STANDOFF IN 36 IN. (914 mm) DEEP WATER COLUMN.

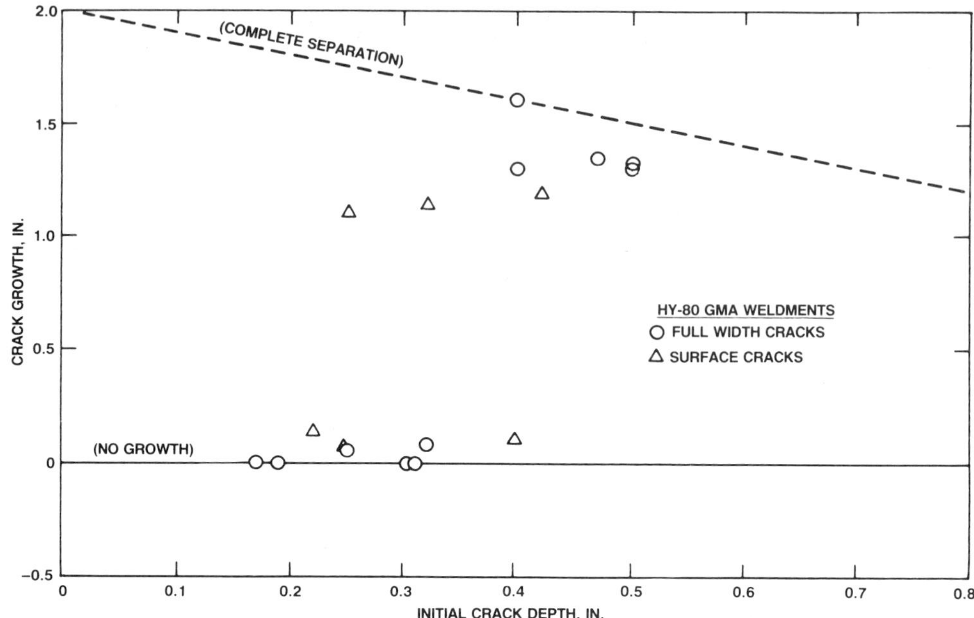

FIG. 12—*Crack growth results for high-quality HY-80 GMA weldments explosively loaded to 2.5% peak nominal strain.*

defined curve for predicting the amount of crack extension as a function of initial crack depth at a 2.5% peak strain level. In fact, Fig. 12 does not even clearly indicate the crack depth beyond which fracture is sure to initiate. We would be surprised if it did, since we are dealing with multi-pass weld metal in which variability is to be expected. We can nonetheless draw three conclusions:

1. The data do not show any clear difference between the behavior of full-width cracks and the shorter surface cracks of comparable depth. Crack growth in the surface cracked elements was every bit as severe as in the full-width cracked elements. This unexpected behavior can be attributed to local variability in toughness exaggerated by limited data, but that seems unlikely to be the sole explanation. We suspect that shallow surface cracks, under dynamic, high-strain loading, may be more prone to crack extension than intuition or extrapolation from conventional experience would suggest. We need to compare tests of baseplate elements containing both full-width and surface cracks to investigate this interesting possibility.

2. Very importantly, the data show that crack extension is either insignificant or large (more than an inch), with nothing falling between. This "all-or-none" behavior is entirely consistent with data obtained by Sumpter [*31*] using an explosively loaded flawed bulge element. This implies, at least for this material, that the energy to propagate the crack is small compared to that absorbed in reaching the state where propagation can initiate. This, in turn, suggests that it may be sufficient to design against fracture initiation alone, ignoring the more complicated crack extension problem. Again, systematic tests of weld and baseplate materials can address this possibility.

3. The data are insufficient to set a definite initial depth below which cracks in this material will not extend at the 2.5% peak strain level. Additional tests will be required to set this limit. The data are nonetheless quite useful for design purposes and serve as an excellent reference for comparative tests of other materials.

Discussion

The analyses of the structural element, coupled with early test results, have made it apparent that a fracture criterion (beyond peak applied strain) descriptive of our tests of shallow cracks is not to be found among those customarily used in elastic-plastic fracture mechanics. For the combination of shallow (but realistic) cracks, high toughness, and high strain level employed in our tests, Figs. 9, 10, and 11 collectively show that the J-integral and the corresponding CTOD are far beyond their critical values determined in standard ASTM tests. Yet crack extension is very limited or does not occur at all in many cases (Fig. 12).

The requirements for J-controlled crack growth have been investigated by Shih et al. [18,33], and our experiments are well beyond those bounds. The cracks we considered are sufficiently shallow, and the extent of plasticity is sufficiently large (breaking to the free surface at relatively low load) that crack tip constraint is lost within the early stage of loading. Thus a one-parameter fracture criterion like the J-integral could not be expected to govern fracture.

This is not a consequence of the test itself, and we would not encounter this difficulty for deeper cracks in less tough materials. For the tough materials we tested, however, the predominant effect of a shallow crack is apparently limited to dictating the site for initial tearing while reducing the load-resisting cross-sectional area of the element. This implies that limit analysis, using the reduced cross-sectional area, coupled with an empirical factor to relate limit load to peak strain, might be about as good as anything else for predicting shallow crack extension in structures made of the materials we tested. If this is so, then we might also use sharp notches instead of fatigue precracks without significantly affecting test results. These are open and interesting questions which we intend to pursue in the future. As for right now, however, we are taking a purely empirical approach, constructing failure envelopes in terms of peak strain, crack depth, and thickness for materials of differing toughness. Such data have immediate engineering applications.

This interpretation of the fracture performance of shallow cracks results from tests of structure-like elements of the kind described here. Application of J-integral testing of HY steels to predict fracture in our tests (or in a real structure) would lead to seriously low estimates for the loading (peak nominal strain) the structural element (or structure) can resist. This statement is *not* as true for less-tough, deeply cracked materials, of course, and a structural element test of this type, coupled with analysis, can indeed serve as an excellent vehicle to couple laboratory toughness tests to actual structural performance. In fact, structural tests of this type appear to be *necessary* to show when and how ASTM-type toughness data can be used as a basis for predicting full-scale fracture performance.

The fact remains that we cannot at present use valid ASTM fracture toughness data to estimate fracture performance in the tests we described here. Nor can we use results of our tests (for one material) to project full-scale fracture performance in another material for which we have only baseline ASTM fracture toughness properties. This fact motivates our ongoing tests of differing weld and baseplate structural elements, and justifies our perception that full-scale, structure-like fracture testing is necessary to be assured of actual fracture performance. There is awareness of this in ASTM, reflected by a recently proposed international research project to develop fracture mechanics tests for shallow cracks.

Conclusion

We have discussed the configuration, calibration, loading, analysis, and some fracture test results for an explosively loaded cracked structural element capable of developing strains of several percent in thicknesses on the order of 51 mm (2 in.). The element is structure-like (statically indeterminate) and represents a radical departure from conventional ASTM fracture toughness tests in the sense that a large sample of material is tested at full thickness and in the presence of cracks that represent service conditions instead of maximum crack tip constraint. The test can be performed at laboratory sites capable of detonating charges on the order of 450 g (1 lb).

The element provides us with a "right-now" means to directly evaluate the in-service dynamic fracture performance of full thickness cracked weldments (or other product forms) when subjected to strain rates on the order of 30/s. Although sample results included here are for shallow cracks in tough welds, the test is not restricted to these. The test serves as a structural-type "shoot-and-see" proof experiment.

Beyond tests described here, we have also used the structural element to evaluate the fracture resistance of low-toughness castings [34] and of uncracked but pre-fatigued weldments containing various levels of weld porosity. There are a number of other near-term applications. For example, the effect of mismatched weld yield strength on flaw tolerance at inelastic strain levels can be evaluated. The influence of weld joint configuration (straight-through walls, narrow gaps, weld joint reinforcement, etc.) on resistance to dynamic fracture can be investigated. The fracture performance of different weld filler metals, used in conjunction with different weld processes, can also be demonstrated at full scale.

References

[1] Vanderveldt, H. H. and Gudas, J. P., "Application of Fracture Control Technology in Navy Ships and Submarines," in *Fracture Mechanics,* N. Perrone et al., Eds., University of Virginia Press, Charlottesville, 1978.

[2] Palermo, P. M., "Structural Integrity Procedures for Naval Ships," *Transactions of American Society of Mechanical Engineers, Journal of Engineering Materials and Technology,* Vol. 102, No. 1, Jan. 1980, pp. 7–14.

[3] Judy, R. W. et al., "Review of Fracture Control Technology in the Navy," Naval Research Laboratory Report 4170, Feb. 1980.

[4] Pellini, W. S., *Principles of Structural Integrity Technology,* Office of Naval Research, Arlington, Va., 1976.

[5] "Military Standard: Standard Procedures for Explosion Testing Ferrous and Non-Ferrous Metallic Materials and Weldments," MIL-STD-2149(SH), 14 Nov. 1983.

[6] "Response of Metals and Metallic Structures to Dynamic Loading," National Materials Advisory Board Report NMAB 341, National Academy of Sciences, Washington, D.C., May 1980.

[7] "Establishing Toughness Requirements for Low and Intermediate Strength Steels in Weapon Systems," National Materials Advisory Board Report NMAB 360, National Academy of Sciences, Washington, D.C., 1980.

[8] Hutchinson, J. W. and Paris, P. C., "The Theory of Stability Analysis of *J*-Controlled Crack Growth," in *Elastic-Plastic Fracture, ASTM STP 668,* J. D. Landes, J. A. Begley, and G. A. Clarke, Eds., American Society for Testing and Materials, Philadelphia, 1979, pp. 37–64.

[9] Ernst, H. A., "Material Resistance and Instability Beyond *J*-Controlled Crack Growth," in *Elastic-Plastic Fracture: Second Symposium, Vol. I—Inelastic Crack Analysis, ASTM STP 803,* C. F. Shih and J. P. Gudas, Eds., American Society for Testing and Materials, Philadelphia, 1983, pp. I-191—I-213.

[10] Joyce, J. A. and Gudas, J. P., "Computer Interactive J_{Ic} Testing of Naval Alloys," in *Elastic-Plastic Fracture,* J. D. Landes, J. A. Begley, and G. A. Clarke, Eds., American Society for Testing and Materials, *ASTM STP 668,* Philadelphia, 1979, pp. 451–468.

[11] Davis, D. A., Vassilaros, M. G., and Gudas, J. P., "Specimen Geometry and Extended Crack Growth Effects on J_I-R Curve Characteristics for HY-130 and ASTM A533B Steels," in *Elastic-Plastic Fracture: Second Symposium, Vol. II—Fracture Resistance Curves and Engineering Applications, ASTM STP 803,* C. F. Shih and J. P. Gudas, Eds., American Society for Testing and Materials, Philadelphia, 1983, pp. II-582—II-610.

[12] Hasson, D. F., Zanis, C. A., and Anderson, D. R., "Fracture Toughness of HY-130 Steel Weld Metals," *The Welding Journal,* Vol. 63, No. 6, June 1984, pp. 197s–202s.

[13] Nakamura, T., Shih, C. F., and Freund, L. B., "Elastic-Plastic Analysis of a Dynamically Loaded Circumferentially Notched Round Bar," *Engineering Fracture Mechanics,* Vol. 22, No. 1, 1985, pp. 437–452.

[14] Nakamura, T., Shih, C. F., and Freund, L. B., "Three-Dimensional Transient Analysis of a

Dynamically Loaded Three-Point-Bend Ductile Fracture Specimen," Division of Engineering Report ONR0365/3, Brown University, Providence, R.I., Sept. 1986.

[15] Joyce, J. A. and Hackett, E. M., "Dynamic *J-R* Curve Testing of a High Strength Steel Using the Key Curve and Multispecimen Techniques," in *Fracture Mechanics: Seventeenth Volume, ASTM STP 905*, J. H. Underwood et al., Eds., American Society for Testing and Materials, Philadelphia, 1986, pp. 741–774.

[16] Kishimoto, K. et al., "On Path-Independent Integral $\hat{J}$," *Engineering Fracture Mechanics,* Vol. 13, No. 4, 1980, pp. 841–850.

[17] Dodds, R. H., Read, D. T., and Wellman, G. W., "Finite-Element and Experimental Evaluation of the *J*-Integral for Short Cracks," in *Fracture Mechanics: Fourteenth Symposium, Vol. I—Theory and Analysis, ASTM STP 791*, J. C. Lewis and G. Sines, Eds., American Society for Testing and Materials, Philadelphia, 1983, pp. I-520—I-542.

[18] Shih, C. F. and German, M. D., "Requirements for a One-Parameter Characterization of Crack-Tip Fields by the HRR Singularity," *International Journal of Fracture Mechanics,* Vol. 17, 1981, pp. 27–43.

[19] Harrison, J. D., Dawes, M. G., Archer, G. L., and Kamath, M. S., "The COD Approach and Its Application to Welded Structures," in *Elastic-Plastic Fracture, ASTM STP 668,* J. D. Landes, J. A. Begley, and G. A. Clarke, Eds., American Society for Testing and Materials, Philadelphia, 1979, pp. 606–631.

[20] Carlberg, J. R., "Dynamic Response of Explosively Loaded Beam Specimens," in *Proceedings, SESA/Japan Society of Mechanical Engineers, 1982 Joint Conference on Experimental Mechanics,* Part I, May 1982, pp. 488–497.

[21] Sih, G. C., "Classical Concepts and Strain Energy Density Theory," in *Handbook of Stress Intensity Factors,* Lehigh University, Bethlehem, Pa., 1973.

[22] Jolles, M. I., Matic, P., Kirby, G. C., Gensheimer, V. M., and Harvey, D. P., "Critical Issues and Directions for Fracture Mechanics and Structural Integrity," Naval Research Laboratory Memorandum Report 6035, 1988.

[23] Gifford, L. N. and Hilton, P. D., "Preliminary Documentation of PAPST—Nonlinear Fracture and Stress Analysis by Finite Elements," David Taylor Research Center Structures Department Report m-43, Jan. 1981.

[24] Hilton, P. D. and Gifford, L. N., "Elastic-Plastic Finite Element Analysis for Two-Dimensional Crack Problems," in *Elastic-Plastic Fracture: Second Symposium, Vol. I—Inelastic Crack Analysis, ASTM STP 803,* C. F. Shih and J. P. Gudas, Eds., American Society for Testing and Materials, Philadelphia, 1983, pp. I-256—I-273.

[25] "ABAQUS User's Manual," Version 4, Revised, Hibbitt, Karlsson, and Sorenson Inc., Providence, R.I., Aug. 1982.

[26] Hoff, R., Rubin, C. A., and Hahn, G. T., "A New Finite-Element Technique for Modelling Stable Crack Growth," *Engineering Fracture Mechanics,* Vol. 23, No. 1, 1986, pp. 105–118.

[27] Rice, J. R., "A Path Independent Integral and the Approximate Analysis of Strain Concentration by Notches and Cracks," *Journal of Applied Mechanics, Transactions of the American Society of Mechanical Engineers,* Vol. 34, June 1968, pp. 379–386.

[28] Tracey, D. M., "Finite Element Solutions for Crack-Tip Behavior in Small Scale Yielding," *Journal of Engineering Materials and Technology,* Vol. 98, 1976, pp. 146–151.

[29] Shih, C. F., "Relationship Between the *J*-Integral and the Crack Opening Displacement for Stationary and Extending Cracks," *Journal of the Mechanics and Physics of Solids,* Vol. 29, 1981, pp. 305–326.

[30] Turner, C. E., "Further Developments of a *J*-Based Design Curve and Its Relationship to Other Procedures," in *Elastic-Plastic Fracture: Second Symposium, Vol. II—Fracture Resistance Curves and Engineering Applications, ASTM STP 803,* C. F. Shih and J. P. Gudas, Eds., American Society for Testing and Materials, Philadelphia, 1983, pp. II-80—II-102.

[31] Sumpter, J. D. G., "Prediction of Critical Crack Size in Plastically Strained Welded Panels," Third ASTM International Conference on Nonlinear Fracture Mechanics, Knoxville, Oct. 1986.

[32] Hilber, H. M., Hughes, T. J. R., and Taylor, R. L., "Collocation, Dissipation, and 'Overshoot' for Time Integration Schemes in Structural Dynamics," *Earthquake Engineering and Structural Dynamics,* Vol. 6, 1978, pp. 99–117.

[33] Shih, C. F., Dean, R. H., and German, M. D., "On *J*-Controlled Crack Growth: Evidence, Requirements, and Applications," General Electric Company Report T13, Schenectady, N.Y., 1981.

[34] Kirk, M. T. and Hackett, E. M., "Fracture Behavior Prediction for Rapidly Loaded Surface Cracked Specimens," presented at the American Society for Testing and Materials Symposium on Surface Crack Growth: Models, Experiments, and Structures, Sparks, Nev., Apr. 1988.

Siegfried R. Winkler[1]

Magnetic Emission Detection of Crack Initiation

REFERENCE: Winkler, S. R., **"Magnetic Emission Detection of Crack Initiation,"** *Fracture Mechanics: Twenty-First Symposium, ASTM STP 1074*, J. P. Gudas, J. A. Joyce, and E. M. Hackett, Eds., American Society for Testing and Materials, Philadelphia, 1990, pp. 178–192.

ABSTRACT: A new technique has been developed to measure the onset of fracture events. Steel structures or fracture mechanics steel specimens emit magnetic signals during rapid mechanical loading and when material failure occurs. These signals can be utilized for the detection of the originating events. In particular, the initiation of crack propagation, which is of basic interest for the determination of critical material parameters in fracture mechanics, can be determined. Reported are first results and efforts to understand the origin of the signals by a comparison with fracture surface appearances from fracture mechanics specimens.

KEY WORDS: magnetic emission, ferromagnetic materials, magnetic field, magnetic detector, logarithmic signal processor, initiation of stable crack extension, ductile fracture, brittle fracture

Fracture mechanics is based substantially on the knowledge of the instant of failure occurrence or the onset of crack propagation for the determination of characteristic material parameters. Sophisticated techniques for the detection of these events have therefore been developed. Potential drop methods, acoustic emission, and strain gage techniques are examples, each covering an application field which is determined by the special properties of the method.

This paper describes a novel technique, the magnetic emission (ME) method, which is in some respects analogous to the acoustic emission method. Mechanical events cause alterations of the magnetic structure of the material which, in turn, give rise to the emission of magnetic signals. These can be used for the detection of the originating events. The method is applicable with all ferromagnetic materials, of which iron, nickel, and cobalt are best known. Of commercial importance is the wide class of ferromagnetic steels. Information regarding internal events is transported by the magnetic field of the structure from the inside to outside. This is the reason that information can be picked up by contactless detectors outside a specimen, or a structure, with almost no delay. The application is simple since specimen preparations are not required.

Reported are the principles of the detection technique and attempts for an understanding of the origin of the signals in fracture mechanics material tests. For this purpose the magnetic emission method has been applied with the Charpy impact test as well as with fracture experiments under quasi-static loading conditions. Large ME signals were obtained with rapid crack propagation; however, ME could also be correlated to microfracturing during stable crack extension in the Charpy tests and in common fracture mechanics experiments.

---

[1] Fraunhofer-Institut fuer Werkstoffmechanik, Freiburg, Federal Republic of Germany.

Origin of Magnetic Signals

Magnetic signals are generated when the magnetic structure of a ferromagnetic material is rearranged. This can be caused by different influences. Best known is the magnetization of a material by an external field. Barkhausen [1] discovered in 1919 that even with a continuously increasing magnetizing field the magnetization curve is discontinuous. It is assembled by small steps, called, *Barkhausen jumps,* and the generation of each step is accompanied by a rapid change of the external magnetic field.

The elements of the magnetic structure in ferromagnetic materials are domains which are spontaneously magnetized to saturation. Size and orientation follow energy minimizing principles (see, for instance, [2]). If the internal energy is changed—for instance, by an external magnetic field, by a static mechanical pressure, by a stroke with a hammer (stress waves), by yielding processes, or by material separation (grain cleavage, void nucleation, micro- and macro-fracture)—a new state of minimum energy will be sought by the domains. This is achieved by a sudden rearrangement. Those domains which are in a suitable orientation grow, others shrink, and some may have to change their magnetization orientation. Movements in the energy balance are instantaneously succeeded by rapid changes of the external magnetic field and are therefore observable.

Events accompanied by material separation (e.g., the formation of voids and cracks) may exhibit an additional effect caused by a drastic change in the permeability. The magnetic resistance increases if a piece of ferromagnetic material is replaced by air or vacuum. This is because of the permeability being several orders of magnitude larger in the ferromagnetic material. A magnetic field present in the sample will therefore experience an observable alteration. Also, in order to make use of this second effect, it is helpful to have a magnetization of the sample.

Thus far, magnetic signals in dynamic experiments have been registered as disturbances if electronic sensors, such as strain gages or thermocouples, were used. The frequently small profitable signals can be completely concealed in huge magnetic disturbances with the consequence that electronic methods have to be replaced by, for instance, optical methods. The new technique presented in the following sections attempts to take advantage of the disturbances which have previously been considered as useless.

Magnetic Emission Method

The magnetic emission method was detected in a drop weight test with the specimen being bent and broken rapidly into two parts by a brittle fracture. Load was measured by strain gages glued on the tup of the hammer. The histories of this signal and of the output of a coil which was placed near the notch tip of the specimen were registered by an oscilloscope (Fig. 1). The load curve is typical for a specimen showing brittle fracture. This is indicated by a sudden drop of the load. The magnetic trace shows essentially two signals, a small one correlated to the beginning of the load signal and a large one in coincidence with the load drop due to the brittle fracture event. Similar ME signals have been observed in many experiments. It is concluded therefore that the small signal at the beginning is generated by the pressure wave traversing the specimen immediately after the impact. Experience has indicated that a large signal coinciding with the load drop varies in its magnitude by the amount, and also by the fall time, of the load drop. As magnitude increases and fall time decreases, a corresponding increase in the magnetic signal results.

With other specimens which do not show brittle fracture and therefore do not exhibit fracture markings in the load curve, magnetic signals could only be detected if the amplification of the magnetic channel was drastically increased. An example is given in Fig. 2.

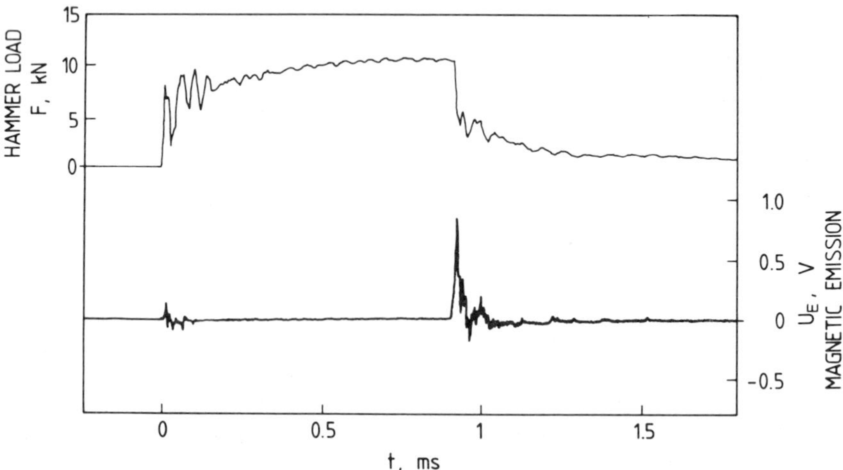

FIG. 1—*Load history and magnetic signal indicating a rapid crack event.*

Although the upper load curve is rather smooth and gives only minute hints for fracture events, the lower magnetic trace shows many different signals. It is initiated by a narrow spike, about 10 μs wide, beginning with the onset of load. Comparison of a number of experiments showed that under constant conditions the magnitude of this first spike is a linear function of the impact velocity. From this it is deduced that the spike may be generated by the impact pressure wave.

In the following a great variety of magnetic signals accompany the fracture process which is almost completed at the end of the diagram. Most of the details are still not understood and present an open field for research. In all experiments with similar load signals, however,

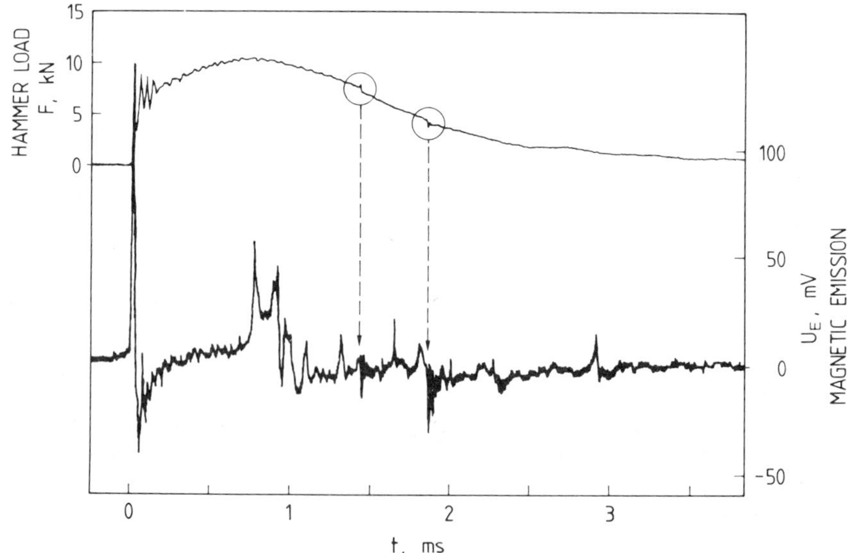

FIG. 2—*Load history and magnetic signal indicating a stable crack event.*

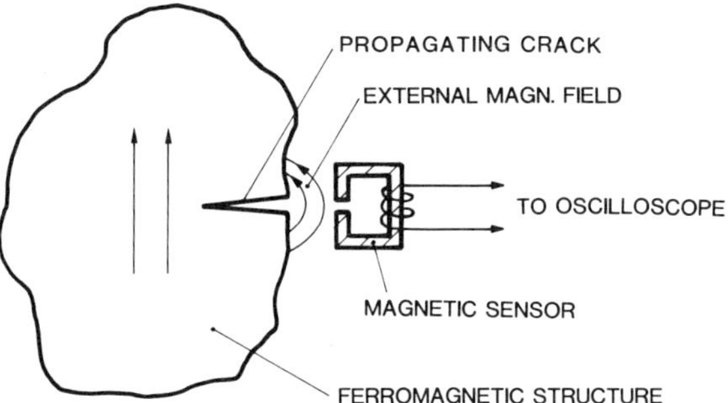

FIG. 3—*Receiving magnetic emission signals.*

it was observed that, after the initial signal has settled, at about the maximum point of the load a larger signal or burst of magnetic signals appears. These results lead to the conclusion that this indicates the onset of stable crack propagation. Furthermore, two tiny dents in the load trace (Fig. 2) which mark two fracture events are accompanied by high frequency oscillations in the magnetic trace. At the end of the experiment when the specimen is broken into two parts, no further magnetic signals are present. The entire fracturing process consisting of plastic flow and stable crack extension must have been represented by the observed magnetic signals.

To receive these signals a coil, the pickup of a tape recorder, was used (Fig. 3). The two ends of the coil were connected to an oscilloscope. However, the two examples in Figs. 1 and 2 also show the problems associated with this simple method. The spectrum of observed signal amplitudes can range over several orders of magnitude and signals may be missed because of a wrong oscilloscope setting. Amplitudes from the microvolt range up to the volt range have been observed. They depend on the speed and extent of the events. In most cases, the magnitude of the expected signal will be unknown before the test.

In order to overcome this potential problem, a logarithmic amplifier was used in our experiments. This prototype was designed for four decades of the input signal, from 100 μV to 1 V, covering an output range of 20 V. Each decade spans 5 V. The 100 μV level of the input voltage corresponds to -10 V of the output range, the 1 mV level to -5 V, and so on, up to the 1 V level, which corresponds to the output $+10$ V level. The bandwidth of this logarithmic signal processing is about 1 MHz. Details of this technique are given in the Appendix.

The intent of the initial step in the utilization of ME signals is the measurement of the instant of their occurrence (i.e., their utilization as a trigger). With these data critical fracture parameters (e.g., the critical values in load and COD histories) can be determined for evaluation (for example, the fracture toughness K_{Ic} of a material). For an automatic evaluation the signals seem to be best suited for trigger purposes after a logarithmic treatment. This ensures that small, as well as large, signals are of comparable magnitude and therefore handled equally well by an electronic device, since each decade spans over the same voltage range determined by the constant U_D (Eqs 1 and 2 of the Appendix). One more reason for their qualification for trigger purposes is that the ME signals occur very fast, with the speed of the magnetic field.

Magnetic Emission in the Charpy Test

The Charpy test is well suited for the application of the ME method, since the same class of signals can be expected during a test series. Differences are due to variations in the material and can easily be detected. The method is useful to determine the instant of events which characterize a material. This, in particular, can be the onset of stable crack propagation. The Charpy hammer is instrumented by a magnetic sensor (Fig. 4). This is fixed to the pendulum and moves, therefore, together with the hammer. During the loading phase of the specimen the sensor is located at a position near the fracture event (distance about 3 mm). To protect the sensor against impacting specimen fragments it is housed in a steel box. A window of nylon allows the magnetic field to enter the housing (Fig. 4b).

Experimental results of Burget and Blauel [3] obtained on Charpy specimens taken from submerged arc welded double V joint are reported. The base material was a modified T StE 355 steel with an original plate thickness of 60 mm. The welds (Fig. 5) were fabricated with a tandem wire system at a heat input of 3 kJ/mm using a wire/flux combination of Type S3/OP121TT. They were tested in the post-weld heat-treated (p.w.h.t., Fig. 7) condition (570°C/2.5h/air). The specimens were located so that the V-notch sampled either reheated or as-deposited microstructure only (Fig. 6). Charpy energy versus temperature curves (Fig. 7) have been determined for the two different weld metal microstructures, the reheated (upper curve) and the as-deposited (lower curve).

The magnetic emission fracture detection method was applied with all Charpy experiments (about 60) of this program. The results of two typical experiments of the two different materials are presented here. Both specimens were tested at the same temperature ($-30°C$) and with the same striking velocity (5 m/s). The fracture surface appearances (Figs. 8 and

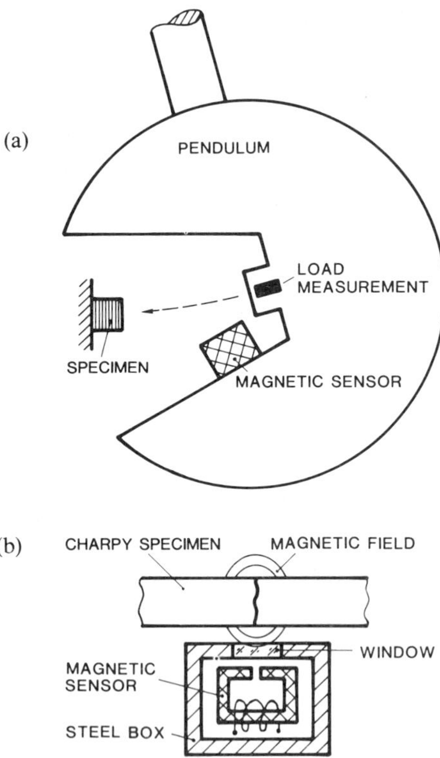

FIG. 4—(a) *Charpy pendulum instrumented for load and magnetic emission measurement.* (b) *Magnetic sensor in pick-up position.*

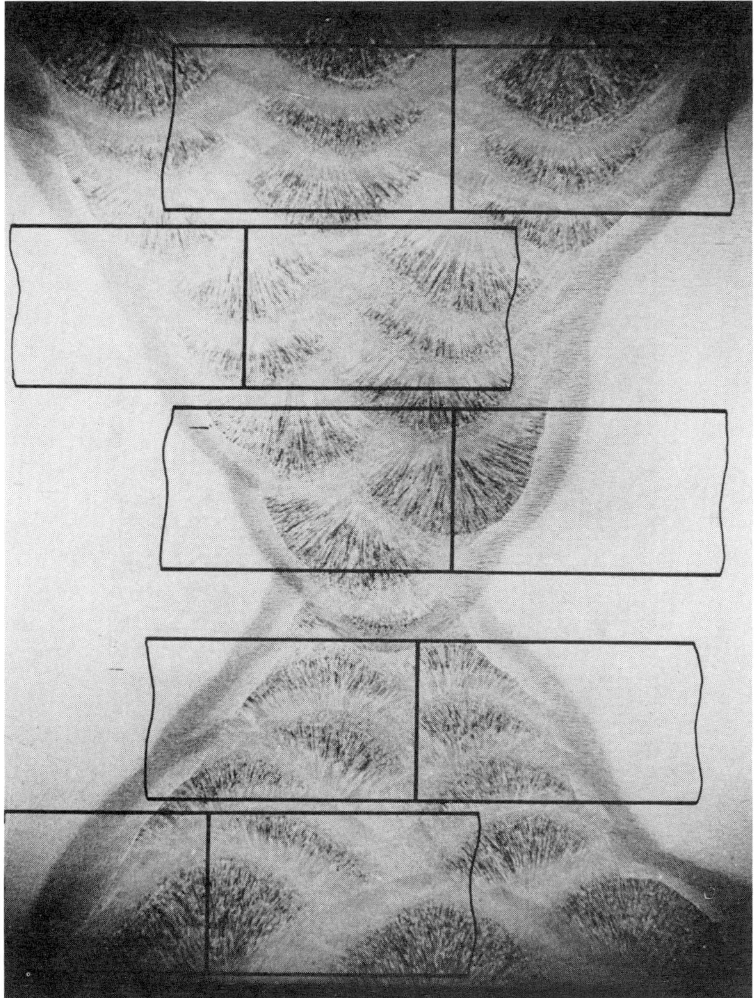

FIG. 5—*Weld joint and Charpy specimen location.*

10, see also the drawings to elucidate the different areas) are compared with the corresponding load and ME signals (Figs. 9 and 11).

Three main areas can be identified on the fracture surface of the specimen from the reheated material (Fig. 8). Area (I) is the notch tip region exhibiting large plastic deformations and a curved crack front due to multiaxial stresses. Area (II) is the area of ductile stable crack growth with a fibrous appearance. Area (III) is the area of the shiny, crystalline brittle fracture.

The corresponding load signal (Fig. 9) does not show all these details. Only the brittle fracture is represented by a load drop. The magnetic emission trace, however, shows many details, beginning with a strong pulse due to the impact pressure wave. The signals produced by the plastic notch tip and surface deformation follow immediately with almost no visible transition. Close to the maximum load a new burst of magnetic signals indicates stable crack extension with an expected speed of somewhat less than hammer impact velocity (5 m/s).

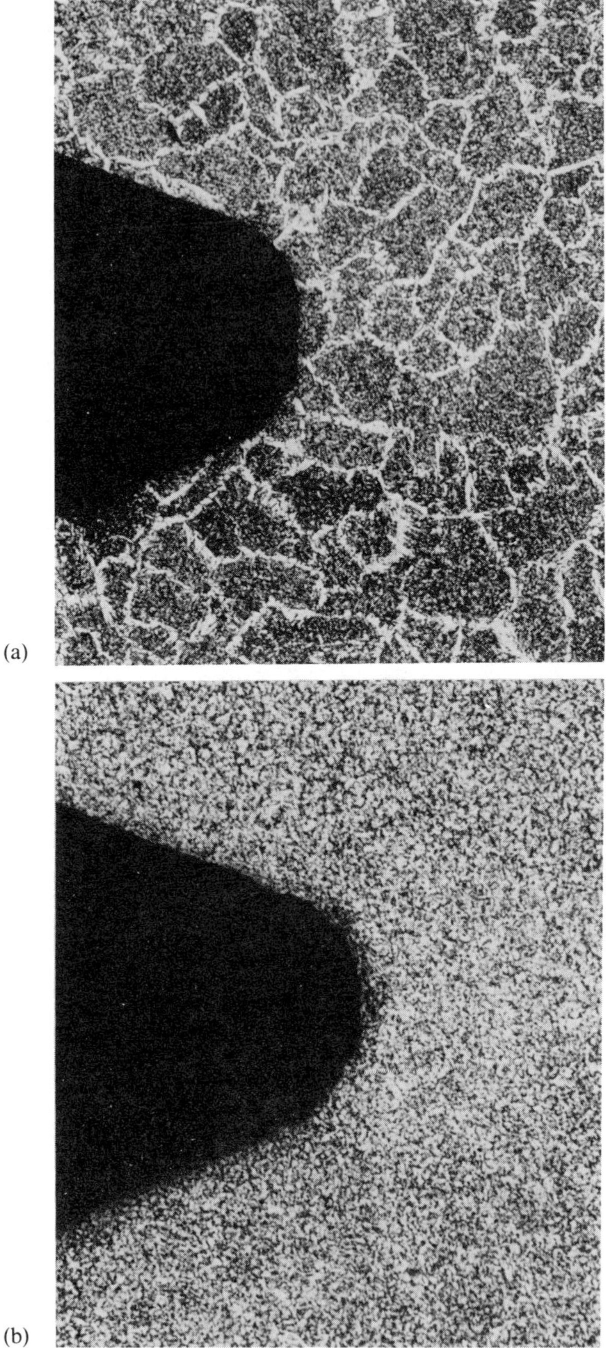

FIG. 6—*Weld metal microstructures at the notch tips.* (a) *Reheated material.* (b) *As-deposited material.*

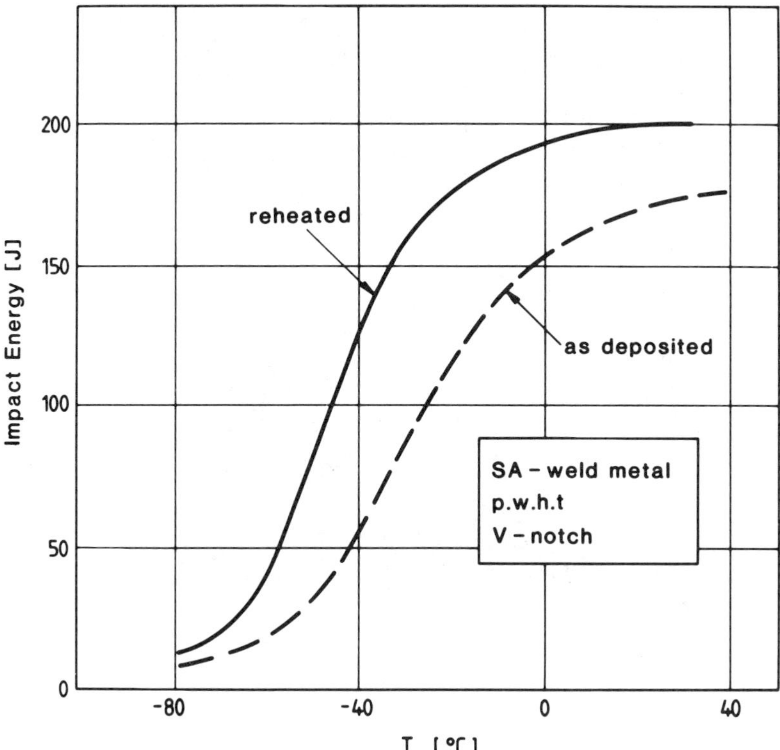

FIG. 7—*Charpy energy versus test temperature.*

With a burst duration, slightly exceeding half a millisecond, and a crack length of about 2 mm (Fig. 8), an average crack speed of 3 to 4 m/s is calculated. A third burst appears concurrently with the load drop (i.e, the brittle fracture) by a large signal and is followed by a block of signals due to the succeeding ductile rupture of the remaining material. This is demonstrated by the non-zero load signal. Figure 9 shows on the left vertical axis the scale for the upper load trace and on the right vertical axis the logarithmic scale of the magnetic input signal, U_E, ranging from 100 μV to 1 V and corresponding to the displayed oscilloscope scale, U_A, of ± 10 V. The magnetic impact signal, as well as the signal from the brittle fracture, exceed the 10 mV level, whereas the signals resulting from low speed fracture events remain in the 1 mV range.

The specimen with the notch location in the as-deposited weld metal microstructure has a similar fracture appearance (Fig. 10) with the exception of the missing area (II). Instead, the brittle area is much larger. This is truly demonstrated by the ME signal (Fig. 11). Again there is the impact pulse at the very beginning followed directly by the plastification period. However, as expected from the surface appearance, the middle part is missing. Instead, a large burst of signals exceeding the 10 mV level indicates the brittle fracture with the succeeding completion of specimen fracturing.

Magnetic Emission in Quasi-Static Experiments

With three-point-bend tests used to determine static fracture toughness values the magnetic emission method has been applied in addition to other measurement techniques. Local

(a)

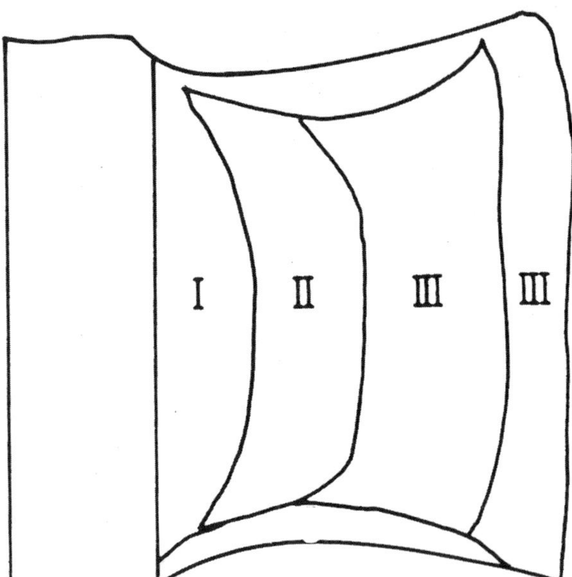

(b)

FIG. 8—(a) *Fracture surface of a Charpy specimen (reheated material).* (b) *Drawing for identification of fracture areas.*

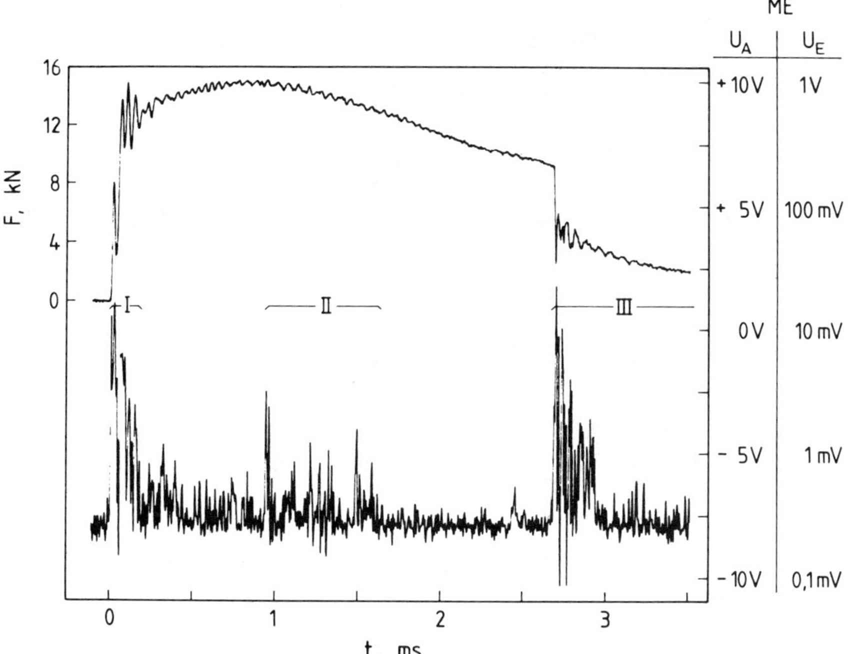

FIG. 9—*Load history and magnetic emission of the reheated material (specimen of Fig. 8).*

brittle fracture (pop-in) events were expected, and magnetic emission signals were provided for triggering the transient recorders with their occurrence. Additionally, it was anticipated that information on the event would be obtained from the ME signals.

For one of these experiments the load-time curve, a photograph of the fracture surface (including a drawing to make clear where the pop-in events are located), and the magnetic emission signals are shown in Figs. 12, 13, and 14, respectively. Three fracture events have been observed with the load curve (load drops, Fig. 12), and, accordingly, three isolated fracture areas can be seen on the photograph (Fig. 13) with areas of 12, 17, and 23 mm². Also three ME signals have been registered. The signals differ somewhat in size from each other as do the fracture areas. In this case a direct correlation between signal and fracture area could not be found, since the succession of the cracks remained unknown.

For an analysis of the ME signals it must be remembered that the detector has a differentiating characteristic and can detect only transitions of the field. A registered rectangular pulse, therefore, would be the derivation of a ramp function (i.e., an increasing or decreasing field). The slope of the ramp would determine the amplitude of the signal, and its duration would be equal to the duration of the transition. The pop-in signals of Fig. 14, therefore, indicate that a steadily increasing magnetic field must have generated them. In this case the detector was placed across the crack mouth similar to Fig. 3. It is assumed that in all three cases the growing and opening crack is responsible for an increasing field (the specimen was magnetized before the test). Each signal is expected to tell the entire history of the crack development. Each one starts very fast (large amplitude), slows down, and has a total duration of about 100 to 200 μs. From the fracture surface it follows that the length of each crack is between 1.5 and 15 mm, depending on the unknown propagation direction. The corresponding crack speeds are calculated to be in the range of 15 and 150 m/s.

(a)

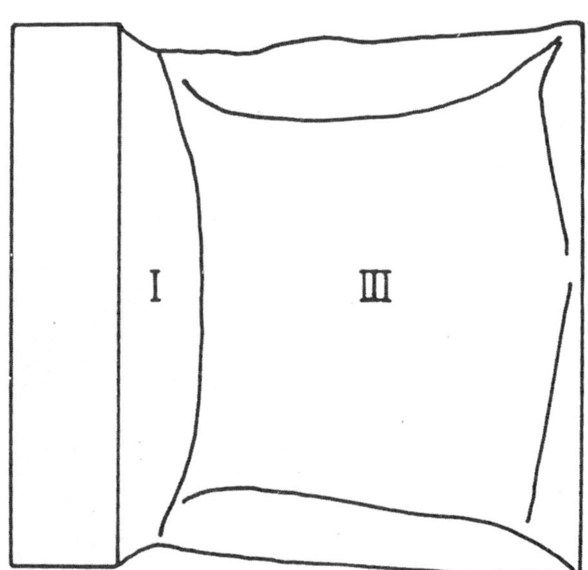

(b)

FIG. 10—(a) *Fracture surface of a Charpy specimen (as-deposited material).* (b) *Drawing for identification of fracture areas.*

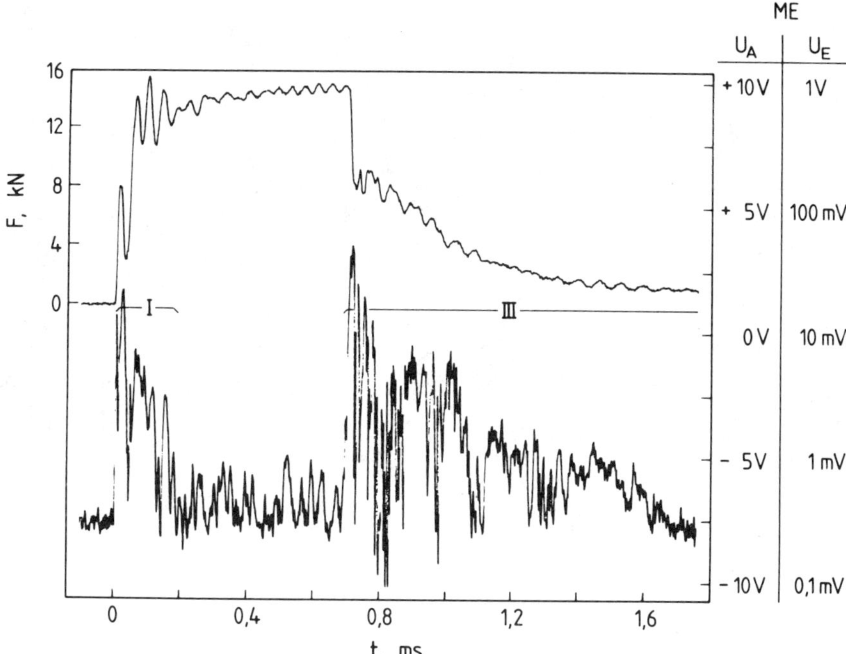

FIG. 11—*Load history and magnetic emission of the as-deposited material (specimen of Fig. 10).*

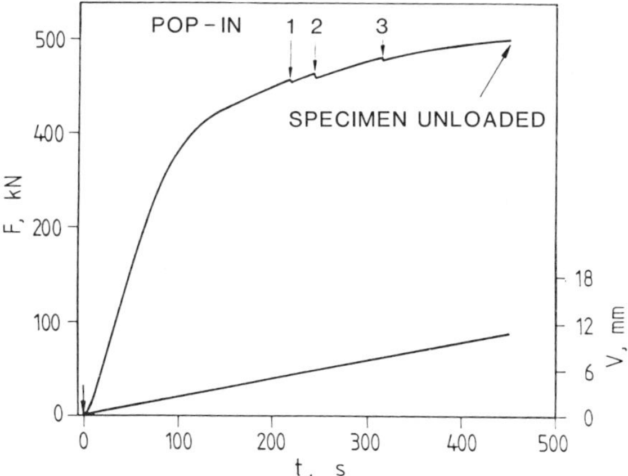

FIG. 12—*Static three-point-bend test: load and deflection versus time, exhibiting three pop-in events.*

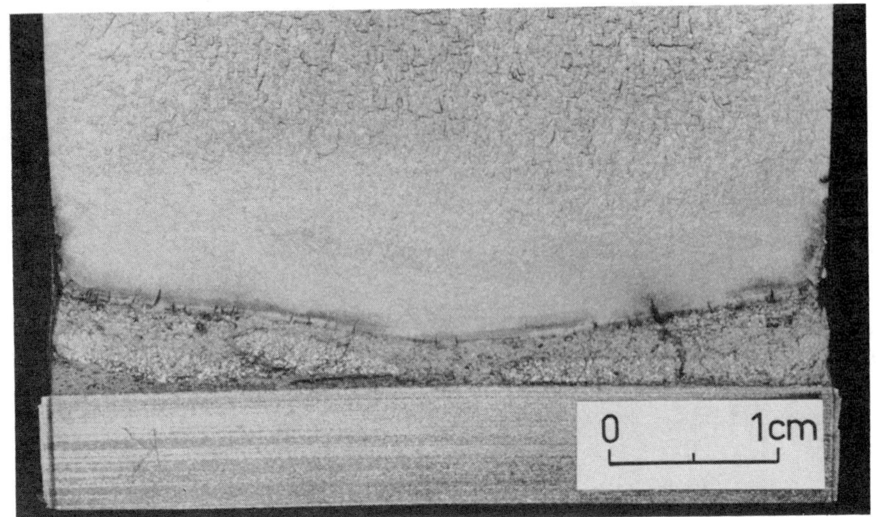

(a)

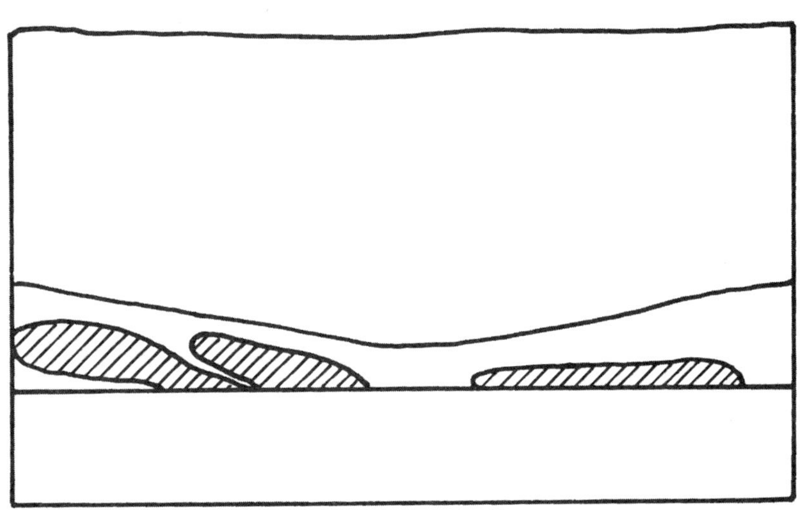

(b)

FIG. 13—(a) *Fracture surface of the specimen of Fig. 12 and location of the pop-in events.* (b) *Drawing for identification of fracture areas.*

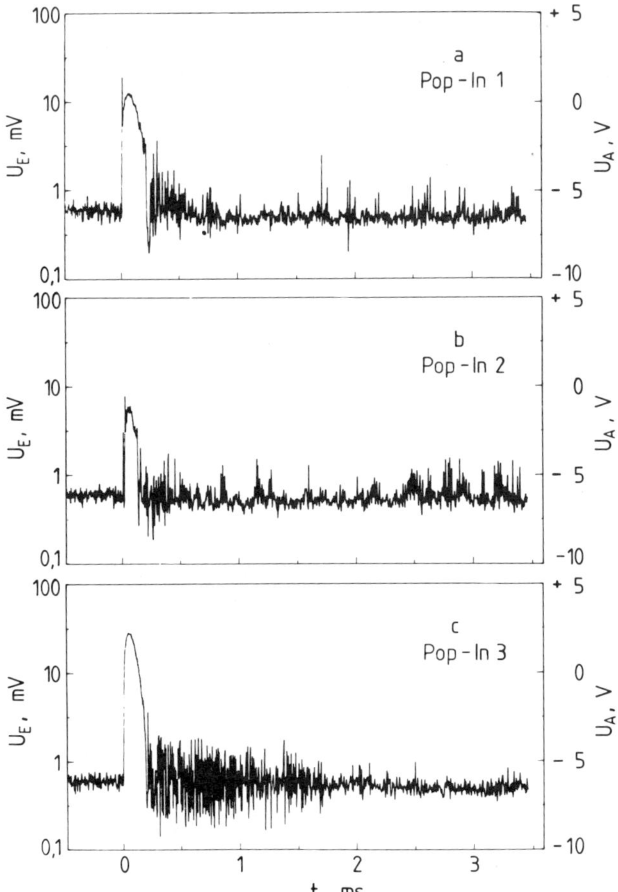

FIG. 14—*Magnetic emission signals corresponding to the three pop-in events of Figs. 12 and 13.*

Conclusions

Although, up to now, only a limited number of experiments has been carried out, it may be concluded that the magnetic emission method could become a useful technique for trigger purposes in fracture mechanics. The experiments reported have already shown that ME is a convenient and inexpensive method to indicate fracture events. A possible application of this method with constructions under load is to use magnetic signals to trigger electronic warning systems at the onset of critical situations in order to avoid catastrophic accidents. It is also expected that, with increasing experience and an improved measuring technique, many more details of complex fracture processes can be read from the magnetic signal. This method is especially easy to apply with routine tests (like the Charpy) and a good opportunity exists to understand the signals. Working contactless, avoiding specimen preparation, and yielding fast and large signals are the outstanding and convincing features of this new method.

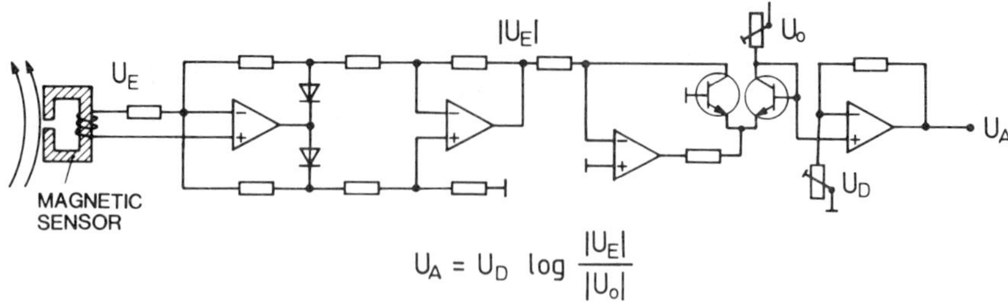

FIG. 15—*Logarithmic signal processing.*

APPENDIX

Logarithmic Signal Processing

The transfer characteristic of the logarithmic amplifier used in this work is given by Eq 1, where care was taken for a positive argument of the log function by forming the absolute value of the input voltage U_E by a special rectifier technique (Fig. 15):

$$U_A = U_D \log \frac{|U_E|}{U_0} \tag{1}$$

where U_D and U_0 are constants for the determination of the voltage span for one decade and for the zero crossing point, respectively. The achieved bandwidth was nearly 1 MHz. The disadvantages with the formation of the absolute value in Eq 1 are the loss of the sign in the input signal U_U (negative parts become positive) and the increased density of signals which makes reading more complicated. In future applications this will be avoided by a more complex treatment of the input signal:

$$U_A = U_D \left[\log \frac{+U_E}{U_0} - \log \frac{|-U_E|}{U_0} \right] \tag{2}$$

This method separates the positive $(+U_E)$ and the negative $(-U_E)$ signal parts and treats them independently until they are combined again by subtraction.

References

[1] Barkhausen, H., "Zwei mit Hilfe der neuen Verstaerker entdeckte Erscheinungen," *Physikalische Zeitschrift*, Vol. 20, 1919, pp. 401–403.
[2] Kittel, C., *Introduction to Solid State Physics*, Wiley, New York, 1956.
[3] Burget, W. and Blauel, J. G., "Fracture Toughness of MMA- and SA-Welded Joints in Normalized C-Mn Steels," in *Fatigue and Fracture Testing of Weldments, ASTM STP 1058*, American Society for Testing and Materials, Philadelphia, forthcoming.

Transition Fracture

S. J. Garwood,[1] *T. G. Davey,*[1] *and Y. C. Wong*[2]

Effect of Biaxial Loading on A533B in Ductile-Brittle Transition

REFERENCE: Garwood, S. J., Davey, T. G., and Wong, Y. C., **"Effect of Biaxial Loading on A533B in Ductile-Brittle Transition,"** *Fracture Mechanics: Twenty-First Symposium, ASTM STP 1074,* J. P. Gudas, J. A. Joyce, and E. M. Hackett, Eds., American Society for Testing and Materials, Philadelphia, 1990, pp. 195–214.

ABSTRACT: A series of large-scale wide plate tests have been carried out on the nuclear reactor pressure vessel steel A533B to investigate the effect of biaxial loading on fracture performance at temperatures towards the lower shelf of the ductile-brittle transition. These tests have been conducted using a purpose built test rig in which a cruciform specimen is loaded so that a central region (approximately 500 mm^2) experienced uniform (equibiaxial) stressing. These tests are compared to the behavior of wide plates under uniaxial loading. Small-scale three-point-bend tests have also been performed to aid the interpretation of the large-scale test results. This paper describes the experimental method and draws the following conclusions from the test results:

1. Failure conditions toward the lower shelf of toughness for equibiaxial and uniaxial loadings may be safely predicted from full thickness three-point-bend specimens.
2. The stress intensity factor (K_c) at failure would appear to be lower under equibiaxial than under uniaxial conditions. In addition, much greater plasticity (as measured by CTOD and stretch zone width) occurs in the uniaxial case.
3. Under equibiaxial conditions, preloading at $+70°C$ to loads at K levels which could cause final fracture at $-130°C$ would appear to have no significant effect on the fracture performance at the low temperature.

KEY WORDS: fracture mechanics, cracks, toughness, biaxial loading, A533B steel

Laboratory tests are conventionally carried out under uniaxial conditions, whereas most pressurized applications involve biaxial loading. To investigate the relevance of using uniaxial wide plate data and fracture toughness derived from the conventional laboratory specimens to predict the fracture performance of biaxially loaded components, The Welding Institute has carried out a series of large-scale wide plate tests on the nuclear reactor pressure vessel steel A533B, under uniaxial and biaxial conditions. These tests have been conducted using a purpose built test rig in which a cruciform specimen is loaded so that a central region (approximately 500 mm^2) experienced uniform biaxial (equibiaxial) stressing. Tests have been carried out to compare behavior under biaxial loading with that under uniaxial loading. In addition, the tests have given information on the effect of a warm prestress on subsequent low temperature performance under biaxial loading. Small-scale three-point-bend tests have also been performed to aid interpretation of the large-scale test results. This paper describes the experimental method and the tests carried out to date and interprets the data.

---

[1] The Welding Institute, Abington Hall, Abington, Cambridge, CB1 6AL, United Kingdom.
[2] Central Electricity Generating Board, Generation Design and Construction Division, Barnwood, Barnet Way, Gloucester, United Kingdom.

TABLE 1a—*Summary of plate characterization data: chemical composition.*

Sample Reference	Element, wt %														
	C	S	P	Si	Mn	Ni	Cr	Mo	V	Cu	Nb	Ti	Al	Sn	Co
1A248	0.19	0.004	0.010	0.23	1.42	0.57	0.20	0.48	0.002	0.14	<0.002	0.003	0.027	0.01	0.02

TABLE 1b—*Summary of plate characterization data: tensile properties at +70°C.*

Specimen Orientation	Yield Strength (σ_Y), N/mm^2	Tensile Strength (σ_u), N/mm^2	Elongation, %	Reduction of Area, %
Longitudinal	471	591	25	70
Transverse	475	597	22	66

FIG. 1—*Central region of biaxial specimen, showing gages and notch location.*

Test Material

All specimens were extracted from a single plate of 50-mm-thick A533B Class 1 material specifically purchased for this project. The material was subjected to an extensive program of characterization tests, the main results from which are summarized in Tables 1a and 1b.

Test Equipment

The configuration of the biaxial test specimens is shown in Fig. 1. The part-through surface notch was arranged to lie at right angles to the rolling direction, at the center of a 500 mm^2 section of the 50-mm-thick plate. Side arms extended from this square portion, raising the effective area around the notch to approximately 1 by 1 m. The design of the specimen was developed using Perspex models to give the most uniform biaxial loading, whilst promoting the development of reasonably high strains within the specimen. These preliminary design tests resulted in a cruciform-type specimen with slits in the loading arms. The final arrangement is shown in its supporting frame in Fig. 2. The specimens were surface notched with a final defect size approximately 150 mm long by 25 mm deep, fatigue sharpened in bending from a 135 mm long by 25 mm deep machined starter notch. From the use of "time of flight" ultrasonic inspection, fatigue crack growth was ensured along the entire crack front.

FIG. 2—*Biaxial specimen with transverse loading beams in place.*

This notch geometry was chosen to be consistent with 25-mm-thick specimens in A533B from the same heat which were tested at $+70°C$ [1].

A transverse loading beam was welded to each of the four side arms (Fig. 2). This consisted of a deep stiffened I beam on which the loading rams could bear, and it was designed to take a maximum total load of 15 MN for the outermost capsule arrangement. The load was applied to the beams using four 1000 tonne capacity load capsules from one of The Welding Institute's wide plate rigs. One pair of capsules was mounted horizontally, the other pair vertically (Fig. 3). All biaxial specimens were loaded with equal pressures in the horizontal and vertical capsules (i.e., under equibiaxial conditions).

Test Specimens

Ten three-point-bend specimens had been tested before the commencement of the large-scale testing to produce a fracture toughness transition curve. These specimens were 50-mm-square cross-section specimens in the L-S orientation tested in accordance with BS

FIG. 3—*Biaxial specimen with vertical and horizontal load capsules in place. A corresponding pair is mounted behind the specimen.*

5762:1979 ("Methods for Crack Opening Displacement (COD) Testing") and BS 5447:1977 ("Methods of Test for Plane Strain Fracture Toughness (K_{Ic}) of Metallic Materials"). A total of four large-scale tension tests have now been performed. Since much of the large-scale work involved δ tests at around $-135°C$ and $-155°C$, two further sets of three-point-bend specimens were carried out, one set at each of these temperatures.

The first large-scale wide plate specimen (M01-10) had been loaded previously at $+70°C$ as part of a project for the Nuclear Installations Inspectorate [1]. This test plate had been subjected to various loading sequences under both uniaxial and biaxial loading, but no crack growth had been induced. This specimen was then progressively cooled to lower temperatures and reloaded equibiaxially to marginally below the maximum equibiaxial load achieved at $+70°C$. The plate eventually fractured at $-129°C$ at 10.2 MN.

The second large-scale specimen (M01-20) had not been subjected to any prior loadings, and was cooled directly to $-129°C$ and loaded equibiaxially until it fractured at 10.5 MN.

Specimen M01-30 was of the same configuration as used in the previous two tests, but was loaded uniaxially (Fig. 4) using one pair of hydraulic capsules at right angles to the notch plane. This plate survived loads of 10.7 MN and 10.8 MN at $-116°C$ and $-136°C$ respectively. The specimen was then cooled to $-157°C$ and reloaded. Failure occurred at 10.7 MN.

The fourth large-scale specimen (M01-40) was designed as an exact repeat of M01-20 in that it was loaded equibiaxially at $-132°C$. However, this plate survived a load of 11.5 MN at this temperature. The plate was then unloaded, cooled to $-163°C$, and reloaded, failure occurring at 11.2 MN.

On the first specimen (M01-10), deformation at $+70°C$ was monitored using 63 central post-yield strain gages, 72 far-field strain gages, 3 clip gages, 10 displacement transducers, and 6 thermocouples, all of which were logged by computer together with load and time readings. On subsequent specimens slightly fewer gages were used. The initial large-scale test had a single central mouth opening clip gage, but in all subsequent tests a double-clip gage arrangement was employed (Fig. 5) to enable the interpretation of crack tip opening displacement (CTOD).

Results

Small-Scale Bend Tests

The results of the 16 three-point-bend specimens are given in Table 2.

The CTOD values given in Table 2 were calculated using load, mouth opening clip gage, and crack length measurements following the procedures of BS 5762:1979. The type of behavior of the load displacement record is also noted in the table. Yield strength (σ_Y) values used in these calculations were based on the data obtained at $+70°C$ corrected for temperature effects using the equation

$$\sigma_Y = \sigma_{Y(RT)} + \frac{10^5}{(491 + 1.8T)} - 189 \ (N/mm^2) \tag{1}$$

where $\sigma_{Y(RT)}$ is the yield strength at room temperature (no effect of temperature is assumed between room temperature and $+70°C$), and T is the test temperature in $°C$.

K values were determined using BS 5447 procedures. All six additional tests failed by cleavage fracture (δ_c) with entirely elastic load-clip gage displacement traces. Two specimens achieved the validity requirements of BS 5447 and thus are K_{Ic}'s.

The δ and K results from these tests can be compared in Table 2 with those from the specimens tested in the transition range. As these latter specimens exhibited increasing plas-

FIG. 4—*Test geometry of uniaxial specimen M01-30 (shown after fracture at* −157°C).

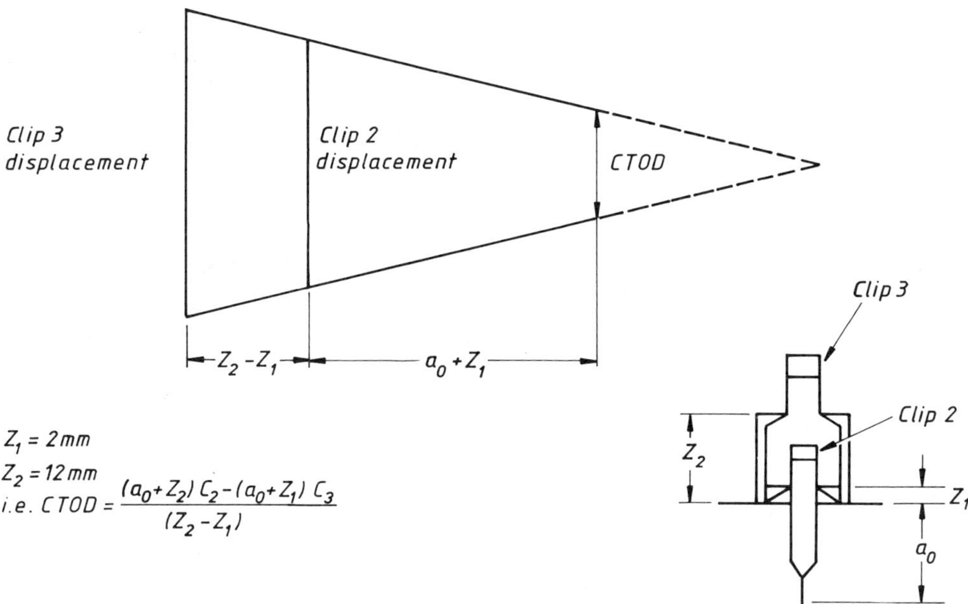

$Z_1 = 2\,mm$

$Z_2 = 12\,mm$

$i.e.\ CTOD = \dfrac{(a_0 + Z_2)\,C_2 - (a_0 + Z_1)\,C_3}{(Z_2 - Z_1)}$

FIG. 5—*Crack tip opening displacement inferred from double-clip gage arrangement.*

TABLE 2—*Fracture toughness data from tests on 50-mm-square three-point-bend specimens (L-S orientation).*

Test Temp., °C	CTOD, mm	Type of Behavior to BS 5762	K, N mm$^{-3/2}$ (MPa$\sqrt{m}$)	$K_{(J)}$, N mm$^{-3/2}$ (MPa$\sqrt{m}$)
0	1.30	δ_m	4348 (139)	21 720 (695)
−25	1.66	δ_m	4479 (143)	20 353 (651)
−40	1.48	δ_u	4560 (146)	18 373 (588)
−55	1.20	δ_u	4630 (148)	16 747 (536)
−60	0.60	δ_u	4409 (141)	11 506 (368)
−70	1.06	δ_u	4703 (150)	15 997 (512)
−80	0.16	δ_u	3889 (124)	5 919 (189)
−90	0.27	δ_u	4039 (129)	7 646 (245)
−100	0.055	δ_c	3460 (111)	3 898 (125)
−110	0.036	δ_c	3004 (96)	2 967 (95)
−135	0.024	δ_c	2468 (79)	2 468 (79)
−135	0.028	δ_c	2488 (80)	2 488 (80)
−135	0.005	δ_c	1530[a] (49)	1 530[a] (49)
−155	0.021	δ_c	2548 (82)	2 548 (82)
−155	0.015	δ_c	2025[a] (65)	2 025[a] (65)
−155	0.008	δ_c	1624 (52)	1 624 (52)

[a] Valid K_{Ic} to BS 5447.

ticity with increasing temperature (specimens tested at $\geqq$ $-90°C$ also showed evidence of ductile crack extension, denoted by the δ_u symbol in the table), K estimates from J were obtained using the equation

$$K_{(J)} = \sqrt{\left(\frac{JE}{(1 - \nu^2)}\right)} \tag{2}$$

where E is the elastic modulus (taken as 207 000 N/mm^2) and ν is Poisson's ratio (0.3).

J measurements were based on the area (U) under the load-cross head displacement records (corrected for extraneous displacement) using the formula

$$J = \frac{2U}{B(W - a_0)} \tag{3}$$

where

B = specimen thickness,
W = specimen width, and
a_0 = initial fatigue crack length (i.e., any ductile extension is not included).

The additional six low temperature data points are consistent with the earlier set as can be seen in Figs. 6 to 8, where the complete transition curves are shown in terms of K, $K_{(J)}$ and δ.

Large-Scale Wide Plate Tests

Test Data

Stresses at various positions on the four wide plate tests were calculated from strain gage readings for the following loading conditions:

1. Equibiaxial loading of 10.2 MN at $+70°C$ for plate M01-10.
2. Uniaxial loading of 10.9 MN at $+70°C$ for plate M01-10.
3. Equibiaxial loading of 10.5 MN at $-128°C$ (failure condition) for plate M01-20.
4. Uniaxial loading of 10.7 MN at $-116°C$ for plate M01-30.
5. Uniaxial loading of 10.8 MN at $-136°C$ for plate M01-30.
6. Uniaxial loading of 10.7 MN at $-157°C$ (failure condition) for plate M01-30.
7. Equibiaxial loading of 11.5 MN at $-131°C$ for plate M01-40.
8. Equibiaxial loading of 11.2 MN at $-163°C$ (failure condition) for plate M01-40.

For the first large-scale test (M01-10) post-yield gages were employed for the measurements at $+70°C$, but no strain readings were taken at lower temperatures (since these gages do not operate below 0°C). For subsequent tests KFL-5 gages were employed and appropriate temperature corrections made to the strain measurements, in accordance with the makers' calibration, to enable accurate stress estimates to be made at the appropriate test temperature. For uniaxial loadings, stress/strain conversion was via Young's modulus, but under biaxial conditions account had to be taken of Poisson's contraction, giving the following relationships:

$$\sigma_{xx} = \frac{E}{(1 - \nu^2)}(e_{xx} + \nu e_{yy}) \tag{4}$$

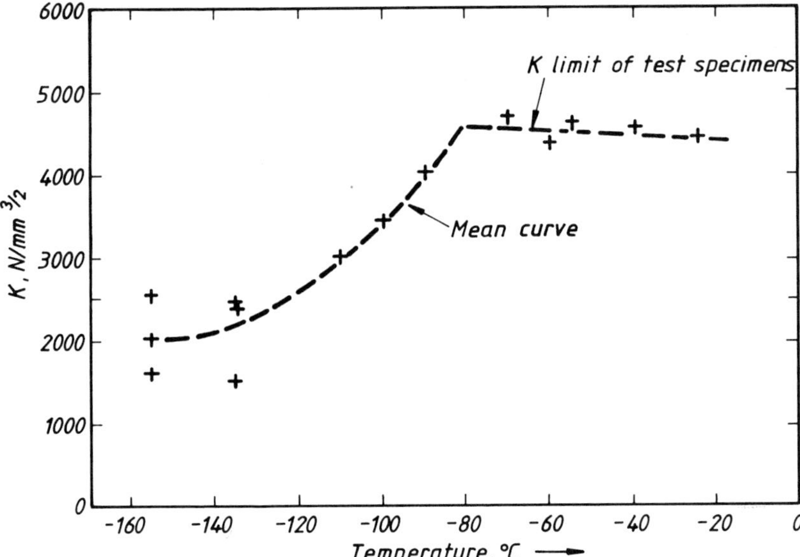

FIG. 6—K *transition curve of three-point-bend data.*

and

$$\sigma_{yy} = \frac{E}{(1 - \nu^2)} (e_{yy} + \nu e_{xx}) \tag{5}$$

where e_{xx} and e_{yy} are the strains, and σ_{xx} and σ_{yy} are the corresponding stresses in the two perpendicular directions on the plate surface. None of the specimens showed any ductile crack growth before final failure. Typical fracture faces are shown in Fig. 9. Several specimens showed the branching fracture path expected under biaxial loading, as seen for example in Fig. 10. The depth and width of the initiating cracks on the plates were measured using a travelling microscope with digital read-out, and the same system was used to measure the stretch zone width (SZW).

Calculation of Stress Intensity Factors

To enable the stress intensity (K) induced at the maximum crack depth to be calculated, the average gross section stress in the vicinity of the crack had to be determined. Two methods were used for this.

First, the stress was found directly from the strain gage readings as described above. Unfortunately there were no readings at low temperatures for plate M01-10. For plate M01-20 it was only possible to record values for one side of the plate, since the other gages did not give reliable readings. The average values of stress determined from the gage readings of the loading conditions (1 to 8 above) are compared to the stress in the specimen loading arm (defined as load/plate thickness (51 mm) and nominal arm width (500 mm) in Table 3).

Secondly, the comprehensively strain gaged plate M01-10, which was subjected to uniaxial and equibiaxial loadings at +70°C, was used to determine a calibration factor relating the load applied to the loading arms to the average stress generated in the central region of the

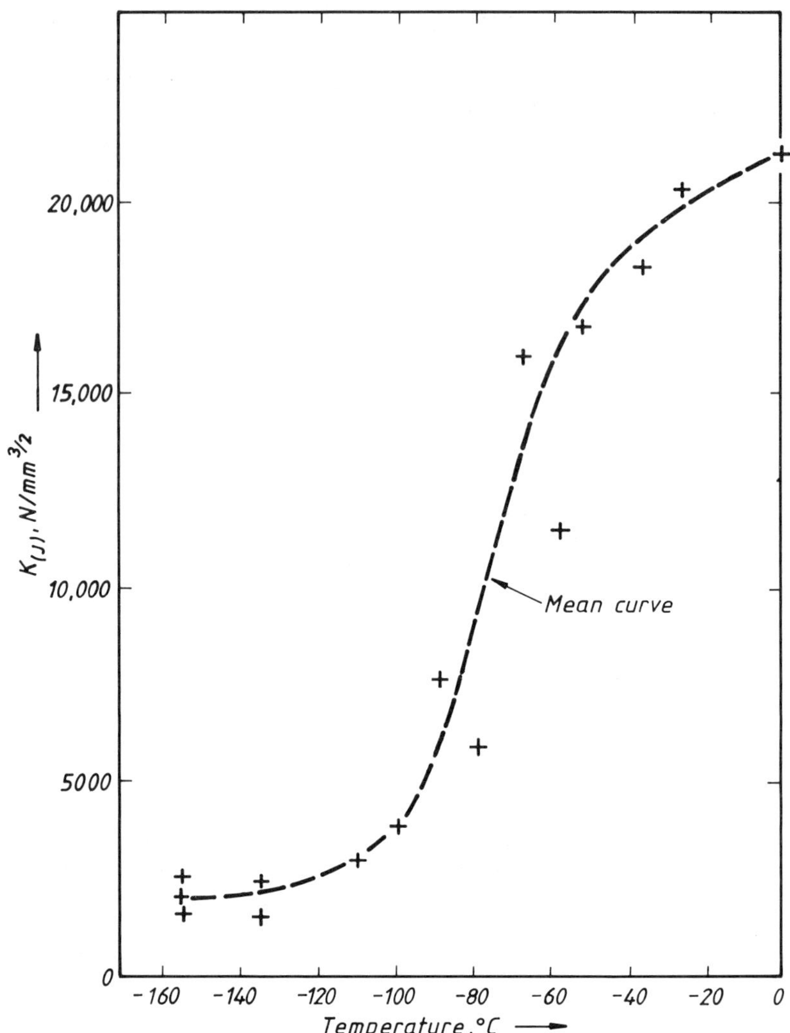

FIG. 7—*Transition curve of* $K_{(J)}$ *determined using* J *integral measurements from three-point-bend data.*

plate. Using data from the foregoing loading conditions 1 and 2 respectively the following factors were determined:

For Equibiaxial Loading:

$$\sigma_{\text{plate}} \ (\text{N/mm}^2) = 27.3 \times \text{load in arm (MN)}$$

For Uniaxial Loading:

$$\sigma_{\text{plate}} \ (\text{N/mm}^2) = 31.0 \times \text{load in arm (MN)}$$

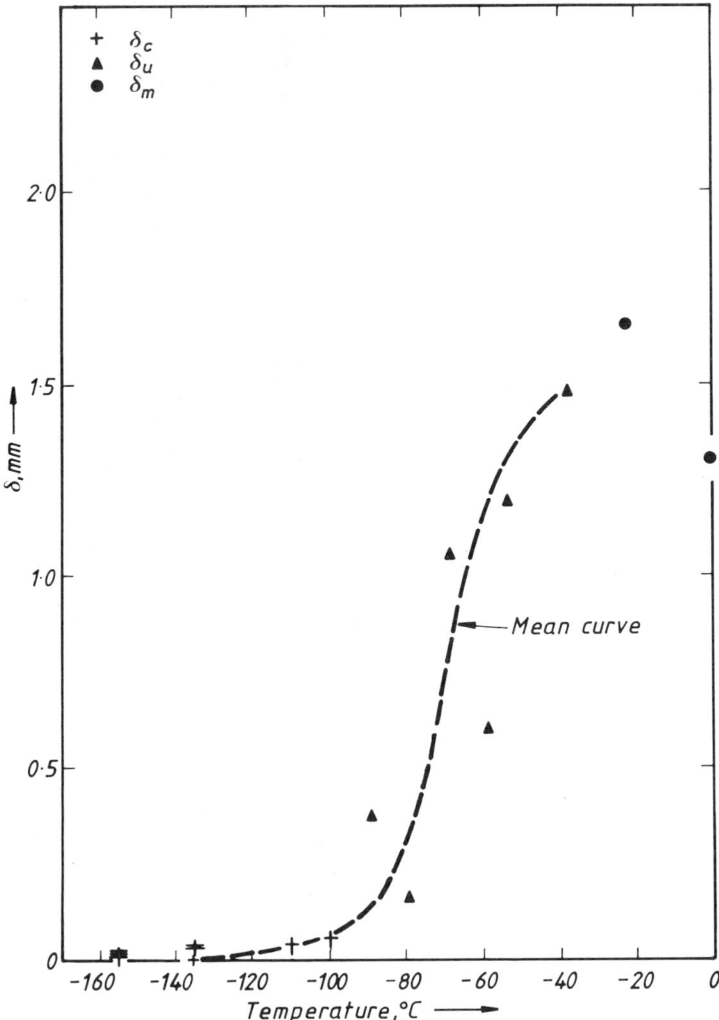

FIG. 8—*CTOD transition curve determined from three-point-bend specimens.*

The plate stresses determined for the maximum load conditions at the various temperatures for the four plates are also given in Table 3. Inspection of this table indicates that the two procedures for stress estimates are in close agreement. For consistency, the second method was adopted to determine the stress intensity factors relating to the various loadings for the four plates.

The average plate stress from Table 3 was used together with $\bar{a}$ (the equivalent half-through thickness crack size for the surface defects actually employed) in the expression

$$K = \sigma \sqrt{(\pi \bar{a})} \tag{6}$$

The parameter $\bar{a}$ was determined from the crack depth and width data from Fig. 12 of BS PD6493:1980 [2] with the crack measurement data for each of the wide plates.

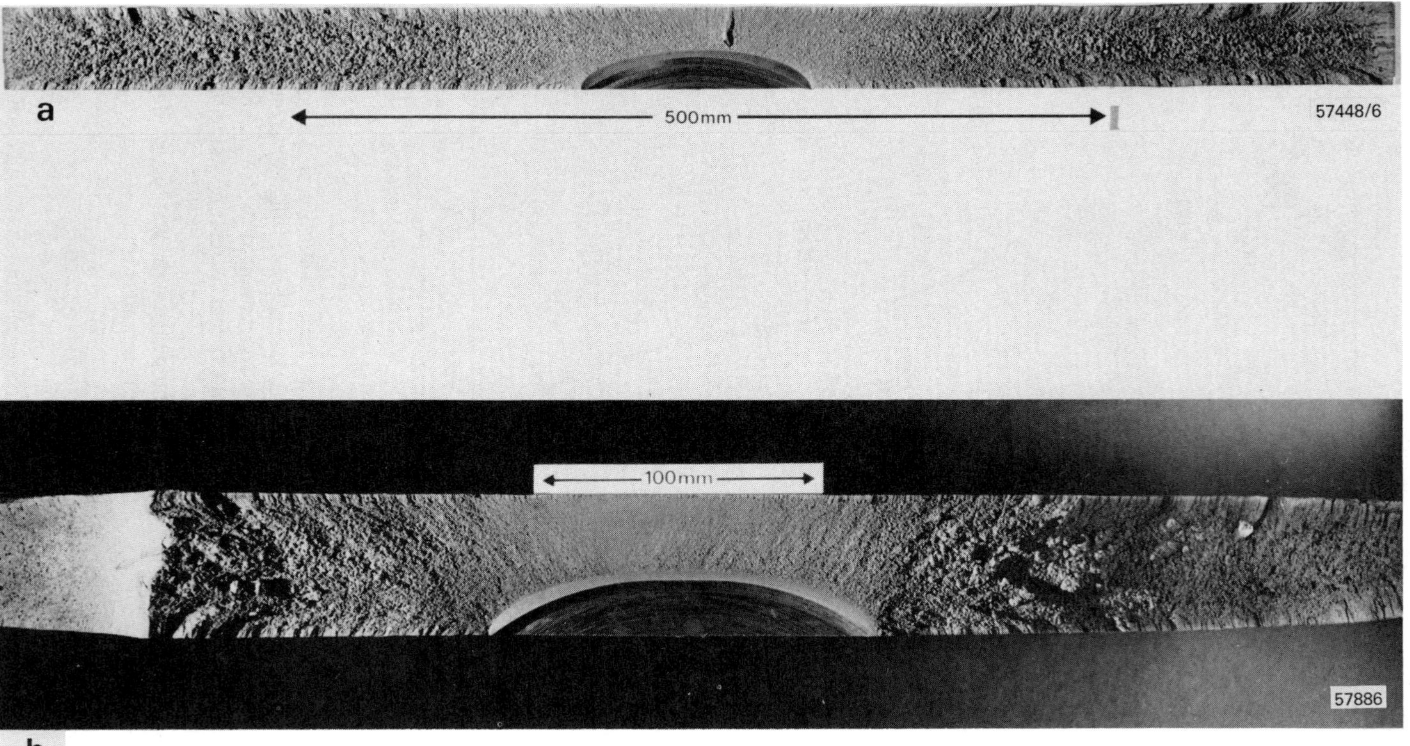

FIG. 9—(a) Fracture face of uniaxial wide plate M01-30 after failure at −157°C. (b) Fracture face of biaxial wide plate M01-40 after failure at −163°C.

FIG. 10—*Biaxial specimen M01-20 after fracture at −128°C.*

TABLE 3—*Calculation of stresses in notch region (σ_{plate}) for use in stress intensity factor calculations.*

Specimen No.	Loading Type	Temperature, °C	Load, MN	Mean Stress from Strain Gages,[a] N/mm²	Nominal Stress in Arm, N/mm²	σ_{plate} Stress from Calibration, N/mm²
M01-10	Biaxial	+70	8.031	220	312	[b]
M01-10	Biaxial	+70	10.226	278	398	[b]
M01-10	Biaxial	−129	10.239	...	398	279
M01-20	Biaxial	−128	10.536	308	410	287
M01-40	Biaxial	−132	11.477	316	446	313
M01-40	Biaxial	−163	11.189	323	435	305
M01-10	Uniaxial	+70	10.908	338	424	[b]
M01-30	Uniaxial	−136	10.830	336	421	336
M01-30	Uniaxial	−157	10.747	345	418	333

[a] Corrected for temperature effects.
[b] These specimens were used to set the calibration factors of 27.3 for biaxial and 31.0 for uniaxial tests.

This procedure was also compared with the K determinations for biaxial loading using the CEGB program BERSAFE on a defect 25.88 mm deep by 150 mm long in a 50-mm-thick plate subjected to 1000 N/mm². Figure 12 of PD6493 gives $\bar{a}$ = 36.55 mm, which converts to a K value at the deepest part of the crack of 10.686 N mm$^{-3/2}$ using Eq 6. The numerically determined K at this point using BERSAFE was 12.188 N mm$^{-3/2}$. The difference in estimation can be attributed to the boundary conditions used. For the numerical estimates a constant stress at faces 500 mm away from the center of the defect was assumed, whereas the Newman and Raju [3] solutions adopted for PD6493 are for an infinite plate. In general, it can be assumed that the plates analyzed here would see an equivalent increase in K of ~12% if calculated using the numerical procedures adopted.

On plates which had been fitted with a double-clip gage arrangement, the crack tip opening displacement (CTOD) was calculated using the formula given in Fig. 7 and the central crack depth measurements. The results are given in Table 4. These values of CTOD were found to give reasonable agreement with those calculated from the stretch zone width (SZW) measured at the maximum depth position of the cracks using

$$\delta = 2 \times SZW \tag{7}$$

The values of $\bar{a}$, K, SZW, and CTOD found by the above methods are given in Table 4.

A double-clip gage arrangement was not used for the first specimen (M01-10); thus no estimate of CTOD is given for that specimen.

Discussion

Figures 11 to 13 show the transition curves of the bend specimens in terms of K, $K_{(J)}$ and δ respectively with the results of the wide plate tests superimposed. These figures indicate that all four large-scale specimens failed at temperatures towards the lower shelf of toughness as determined by three-point-bend specimens, in the regime where at least some small-scale specimens gave valid K_{Ic} results. Of particular interest in Figs. 11 and 12 is the locus of the points relating to specimen M01-10 as it was loaded at progressively lower temperatures, failure occurring once the bend transition curve had been crossed. This failure point was confirmed by specimen M01-20, which had no prior preloading, the failure points being almost coincident on Figs. 11 and 12.

Thus these figures and the results of Tables 2 and 4 show no beneficial effect due to the warm prestressing of plate M01-10 that had been preloaded to a K of 3000 N mm$^{-3/2}$ biaxially and 3462 N mm$^{-3/2}$ uniaxially at +70°C. This specimen actually failed at a K of 2863 N mm$^{-3/2}$ at −129°C, which was consistent with the failure conditions of the non-prestressed plates M01-20 and M01-40.

Based on the results contained in a recent review of prior overload effects by Smith [4], the preload K level under biaxial loading incurred by specimen M01-10 would not be expected to have any significant effect as it is of the same level as the failure conditions at low temperatures achieved in specimens M01-20 and M01-40 without preloading. The stretch zone width measured on the preloaded specimen was significantly higher than the other two tests, however, but this obviously occurred at +70°C under uniaxial loading. The plasticity achieved under uniaxial conditions at +70°C appears to have had no effect on the subsequent low temperature performance under biaxial conditions.

Inspection of Tables 2 and 4 and Figs. 11 to 13 allows the direct comparison of uniaxial and biaxial loading failure conditions with those of the three-point-bend tests. At approximately −135°C an applied K of ~3000 N mm$^{-3/2}$ ($\delta \simeq 0.06$ mm) is sufficient to cause failure under equibiaxial loading conditions. This compares with a mean K of ~2500 N mm$^{-3/2}$ (δ

TABLE 4—*Loading conditions for wide plate tests.*[a]

Test Specimen	Loading Conditions	Temp., °C	Maximum Load, MN	Central Crack Mouth Opening Displacement (Upper/Lower Clip Gage Positions), mm	$\bar{a}$, mm	K, N mm$^{-3/2}$	Stretch Zone Width at Max. Depth of Crack SZW, mm	CTOD from Double-Clip Gage at Center of Crack, mm
M01-10	Equibiaxial	+70	10.6	.../0.21 +	33.4	3001	...	...
	Uniaxial	+70	10.9	.../0.34 +	33.4	3462	0.07	...
	Equibiaxial	+60	10.6	.../0.20 +	33.4	2950	...	...
	Equibiaxial	+7	10.0	.../0.19 +	33.4	2796	...	...
	Equibiaxial	−19	10.0	.../0.20 +	33.4	2796	...	...
	Equibiaxial	−39	9.9	.../0.18 +	33.4	2755	...	...
	Equibiaxial	−60	10.0	.../0.20 +	33.4	2796	...	...
	Equibiaxial	−80	10.0	.../0.19 +	33.4	2796	...	...
	Equibiaxial	−104	10.4	.../0.20 +	33.4	2889	...	...
	Equibiaxial	−129	10.2	.../0.20 +	33.4	2848	...	...
M01-20	Equibiaxial	−128	10.5	0.25/0.20 x	33.0	2863	0.03	0.07
M01-30	Uniaxial	−116	10.7	0.35/0.31 x	31.9	3314	...	0.21
	Uniaxial	−136	10.8	0.38/0.33 x	31.9	3361	0.06	0.20
	Uniaxial	−157	10.7	0.37/0.32 x	31.9	3335	...	0.19
M01-40	Equibiaxial	−132	11.5	0.29/0.22 x	30.2	3051	0.03	0.05
	Equibiaxial	−163	11.2	0.25/0.19 x	30.2	2975	...	0.04

[a] Knife edge heights: + = 3.14 mm.
x = 12.00 mm and 2.00 mm respectively.

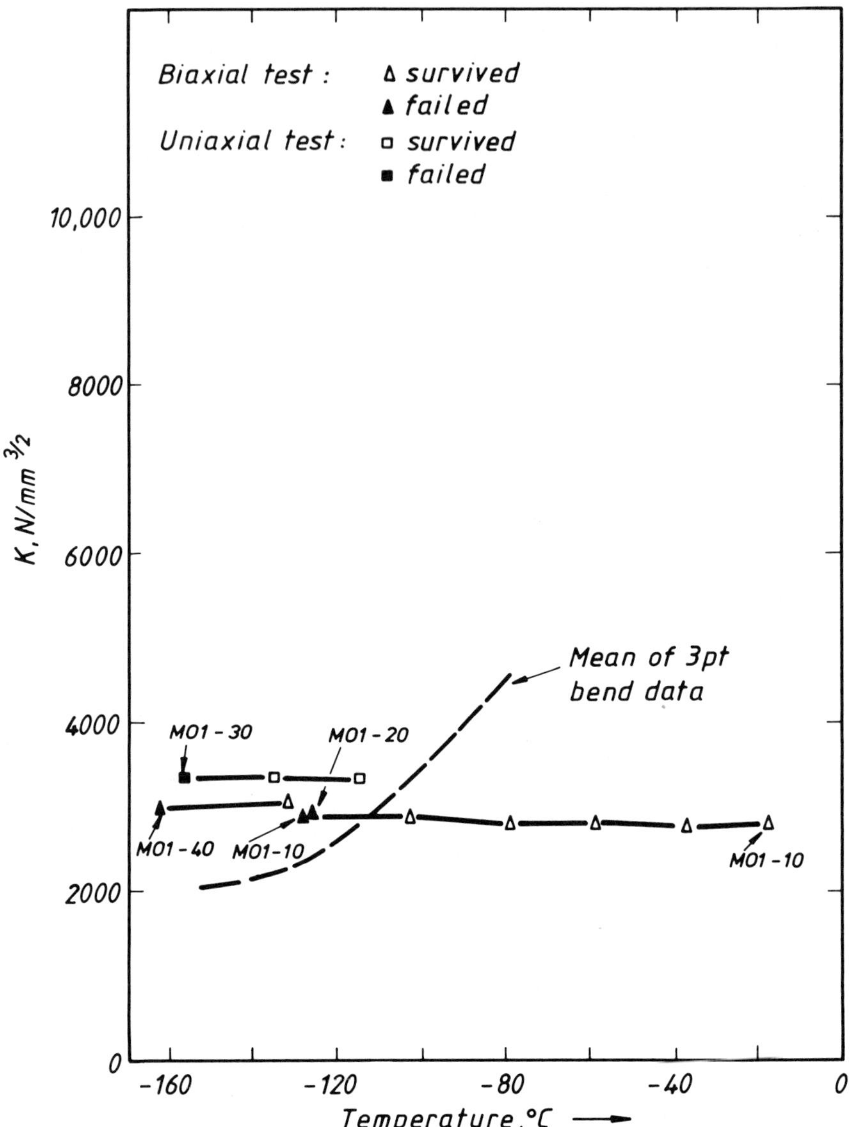

FIG. 11—*Comparison of* K *determinations for wide plate and three-point-bend tests.*

$\simeq 0.02$ mm) from the three-point-bend tests. The uniaxial specimen survived a K of ~ 3400 N mm$^{-3/2}$ ($\delta \simeq 0.20$ mm) at this temperature.

At approximately $-155°$C, the three-point-bend tests give an average K of ~ 2000 N mm$^{-3/2}$ ($\delta \simeq 0.015$ mm) compared to ~ 3000 N mm$^{-3/2}$ ($\delta \simeq 0.05$ mm) for biaxial specimen M01-40 and ~ 3300 N mm$^{-3/2}$ ($\delta \simeq 0.19$ mm) for the uniaxial test.

It would therefore appear that biaxial loading, as well as causing a significant decrease in CTOD for a given K level [1], causes a reduction in the critical K at a given temperature when compared to uniaxial behavior.

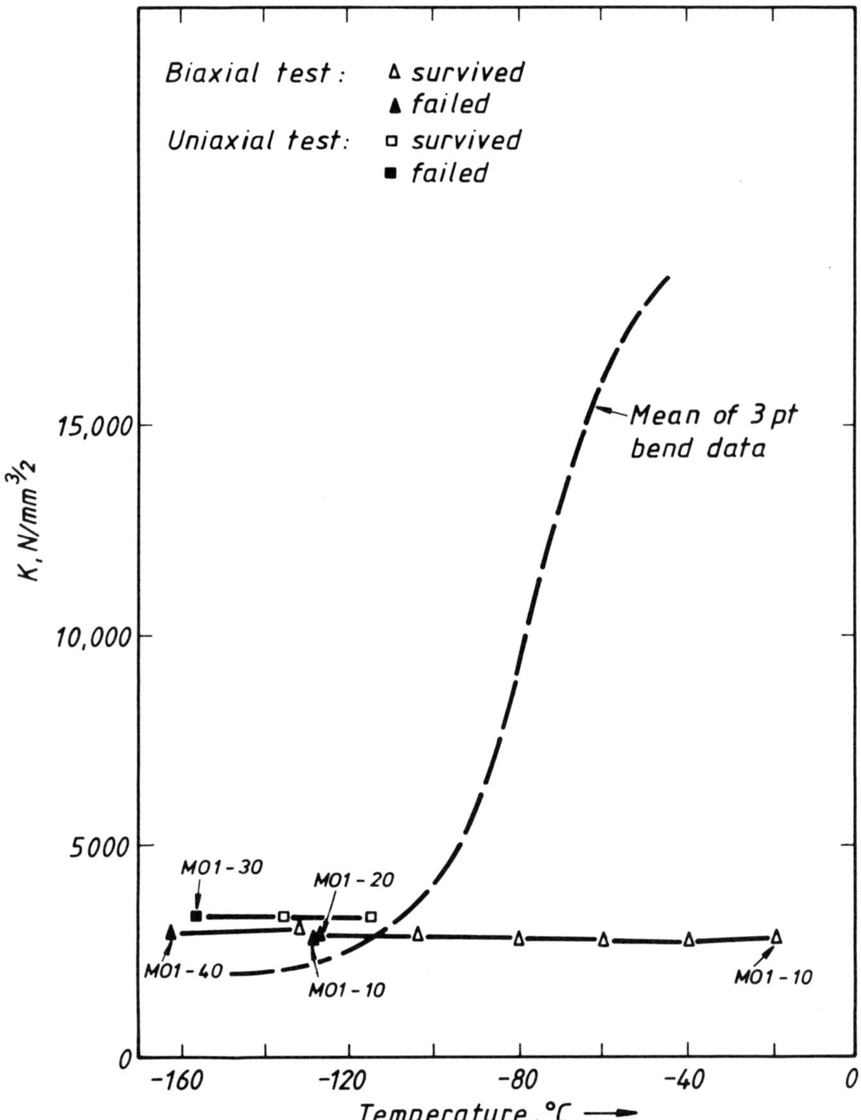

FIG. 12—*Comparison of* K *from wide plates with* $K_{(J)}$ *from three-point-bend tests.*

Both uniaxial and biaxial tension failure conditions occur at higher K and δ values than the critical conditions obtained from three-point-bend specimens at the same temperature.

Using R6 Rev. 2 procedures [5], data points from the four wide plate specimens are plotted on the failure assessment (FAD) diagram in Fig. 14. K_r (i.e., $K_{applied}/K_{MAT}$) values were estimated using the $K_{applied}$ values from the wide plates given in Table 4 and K_{MAT} from the $K_{(J)}$ measurements from the bend tests at the appropriate test temperature from Table 2. (An average K value was used from the three data points at $-135°C$ and $-155°C$.) S_r was taken as the ratio of the stress in the region of the crack (σ_{plate}) for the appropriate load in the arms

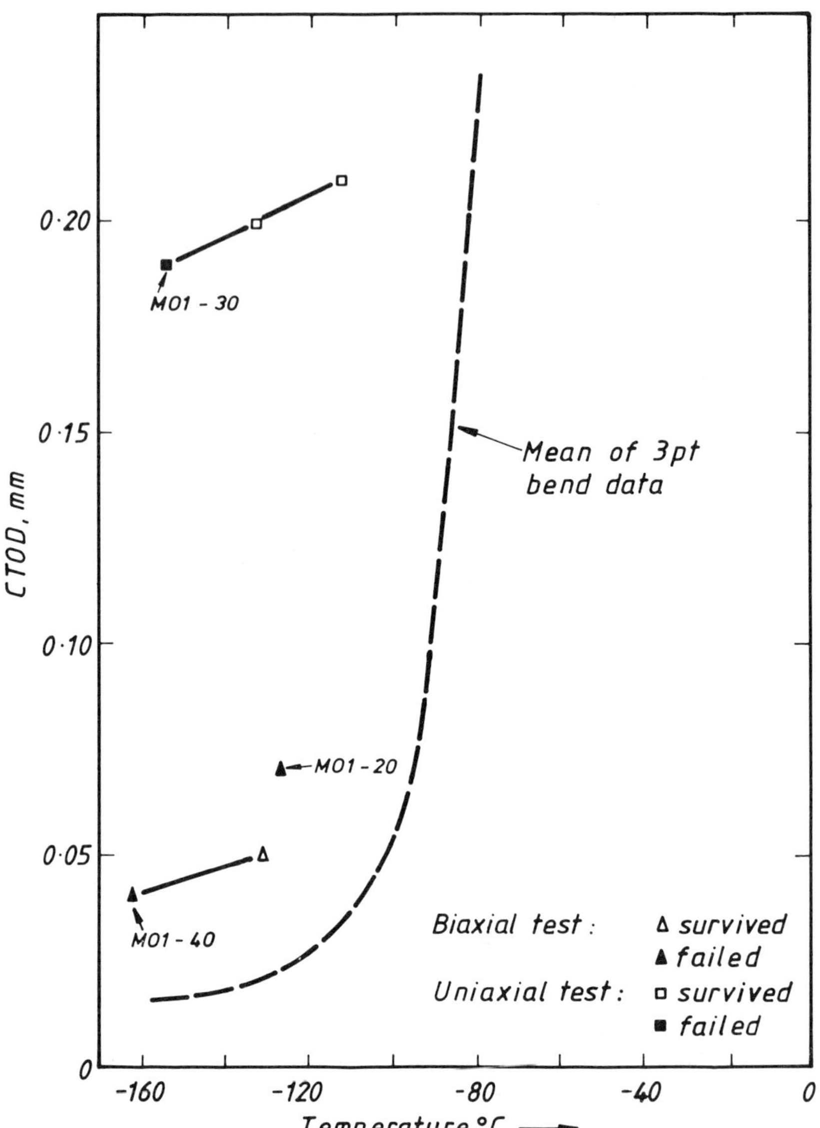

FIG. 13—*Comparison of CTOD from wide plate and three-point-bend tests.*

and the flow stress ($\bar{\sigma}$) taken as the average of yield and ultimate strengths corrected for temperature using Eq 1. Inspection of Fig. 14 shows the locus of K_r and S_r points for specimen M01-10 as it was cooled. The FAD illustrates that failure would not have been expected until the loading at $-130°C$. The biaxial plates fail at a K_r value of ~ 1.3. The uniaxial plate shows a larger factor of safety, surviving a K_r value of over 1.5 before failure at a $K_r \simeq 1.6$. Even larger factors of safety are derived using $\sqrt{\delta_r}$ (defined as the square root of the ratio of CTOD measured from the plates and the average CTOD from the three-point-bend data at the wide plate test temperature) in place of K_r. Values of $\sqrt{\delta_r}$ of 1.6 and 1.9 are obtained for

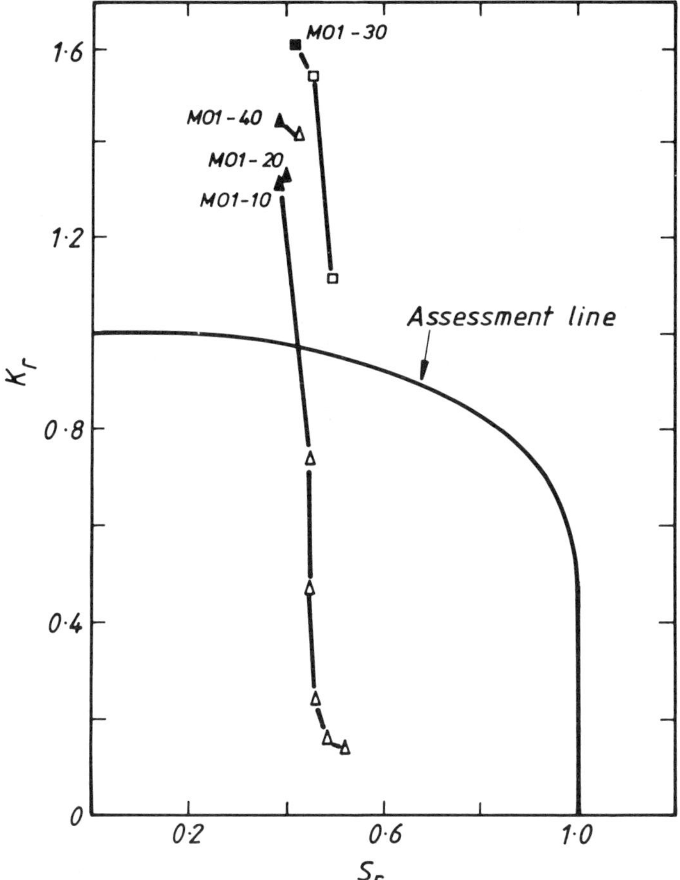

FIG. 14—*Comparison of wide plate results with the failure assessment diagram from Ref 5.*

equibiaxial plates M01-20 and M01-40 respectively, whilst a $\sqrt{\delta_r}$ of 3.6 is obtained for the uniaxial plate M01-30 at failure.

Conclusions

From the results of large-scale biaxial and uniaxial tests and smaller scale three-point-bend specimens on A533B Class 1 plate, the following conclusions can be drawn:

1. Failure conditions on the lower shelf of toughness for equibiaxial and uniaxial loadings may be safely predicted from full thickness three-point-bend specimens.

2. The stress intensity factor (K) to cause brittle fracture would appear to be lower under equibiaxial than under uniaxial conditions. In addition, much greater plasticity (as measured by CTOD and stretch zone width) occurs in the uniaxial case.

3. Under equibiaxial conditions, preloading at $+70°C$ to loads at K levels similar to those which cause final fracture at $-130°C$ would appear to have no significant effect on the fracture performance.

Acknowledgments

The authors would like to thank the Central Electricity Generating Board who sponsored this program of tests carried out at The Welding Institute. The opinions expressed are those of the authors and are not necessarily those of the sponsoring organizations.

References

[*1*] Garwood, S. J. and Davey, T. G., "A Study of Biaxial Loading in A533B," in *Proceedings,* 9th Int. Conf. on Structural Mechanics in Reactor Technology, Lausanne, 17–21 Aug. 1987, Vol. G, pp. 15–20.

[*2*] PD6493:1980, "Guidance on Some Methods for the Derivation of Acceptance Levels for Defects in Fusion Welded Joints," British Standard Institution.

[*3*] Newman, J. C., "A Review and Assessment of the Stress Intensity Factors for Surface Cracks," in *Part-Through Crack Fatigue Life Prediction, ASTM STP 687,* American Society for Testing and Materials, Philadelphia, 1979, pp. 16–42.

[*4*] Smith, D. J., "A Critical Review of the Use of Prior Overload Techniques to Reduce the Risk of Subsequent Fracture in Pressure Vessels," Welding Institute Confidential Report 5561/4/87, April 1987, to be published in *International Journal of Pressure Vessels and Piping.*

[*5*] Harrison, R. P., Loosemore, K., Milne, I., and Dowling, A. R., "Assessment of the Integrity of Structures Containing Defects," CEGB Report R/H/R6—Rev. 2, April 1980.

E. Morland[1]

Fracture Toughness in the Transition Regime for A533B-1 Steel: The Effect of Specimen Sidegrooving

REFERENCE: Morland, E., **"Fracture Toughness in the Transition Regime for A533B-1 Steel: The Effect of Specimen Sidegrooving,"** *Fracture Mechanics: Twenty-First Symposium, ASTM STP 1074,* J. P. Gudas, J. A. Joyce, and E. M. Hackett, Eds., American Society for Testing and Materials, Philadelphia, 1990, pp. 215–237.

ABSTRACT: This report describes the results of an experimental test program to investigate the influence of specimen sidegrooving on the cleavage fracture toughness of A533B-1 steel. The work represents a companion study to a previous investigation into the effects of specimen size on the cleavage fracture toughness of the same A533B-1 plate material.

Compact specimens were prepared in two sizes (12.5 and 25 mm thick) in the 0, 20 or 50% sidegrooved conditions. A total of 160 tests were performed in the transition regime with a minimum of 29 specimens being tested for each specimen size and depth of sidegrooving. Variations in cleavage fracture toughness were characterized using exponential curve fits to data across a temperature range and arithmetic mean values calculated at discrete test temperatures.

The results show that sidegrooving has the effect of decreasing the mean level of cleavage toughness in the upper transition regime (high mean K levels) and increasing it in the lower transition regime (low mean K levels). These results are explained in terms of the competing effects of increased specimen constraint and reduced net thickness in sidegrooved specimens. Reduced resistance to ductile tearing in sidegrooved specimens is also shown to be a contributing factor to decreased toughness levels in the upper transition regime.

KEY WORDS: brittle fracture, cleavage, sidegrooving, K_{Ic}, A533B, brittle-ductile transition, fracture toughness

The probability of obtaining cleavage fracture in specimens, or components, is influenced by two phenomena: (1) the material's inherent resistance to cleavage (i.e., its toughness) and (2) the way in which applied stresses are distributed throughout the body. If the material has low toughness, or if high levels of stress intensification occur, then the probability of cleavage is high. Both of these phenomena also contribute to the frequently observed effect that, at a given temperature within the brittle-ductile transition regime, increasing specimen size decreases (on average) fracture toughness at failure. Material properties are important in the sense that real (heterogeneous) materials have a distribution of toughness properties, ranging from "tough" to "weak". Consequently, larger specimens have an increased statistical probability of sampling localized regions of "weak" material, which could trigger global fracture of the specimen under applied load. Similarly, levels of constraint are generally higher in

---

[1] Senior Research Officer, Structural Performance Group, Northern Research Laboratories–AEA Technology, Risley, Warrington, WA3 6AT, United Kingdom.

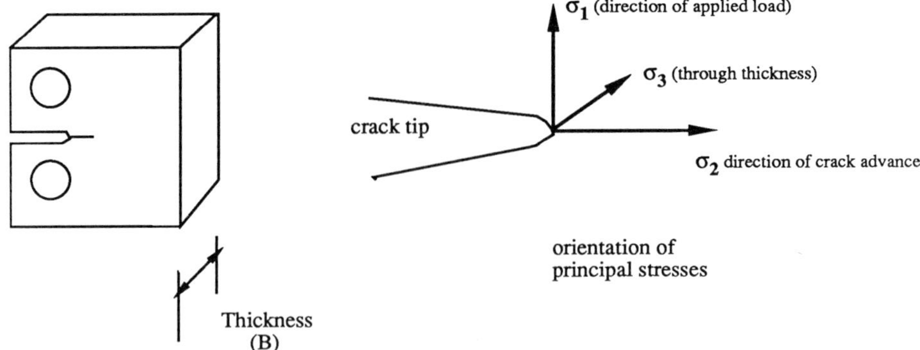

Any elevation in the principal tensile stress (σ_1) causes associated increases in the values of both stress intensity factor (K) and the J integral. The value of σ_1 increases as the crack tip stress state changes from plane stress ($\sigma_3=0$; $\sigma_1=\sigma_y^*$) to plane strain ($\sigma_3=\upsilon(\sigma_1+\sigma_2)$; $\sigma_1=\sigma_y^*+\sigma_2$). So that a convenient index of stress intensification is the ratio

$$I = \sigma_3/\upsilon(\sigma_1+\sigma_2)$$

which equals zero for plane stress and unity for plane strain.

[σ_y^* = uniaxial yield stress defined by the Tresca criterion.]

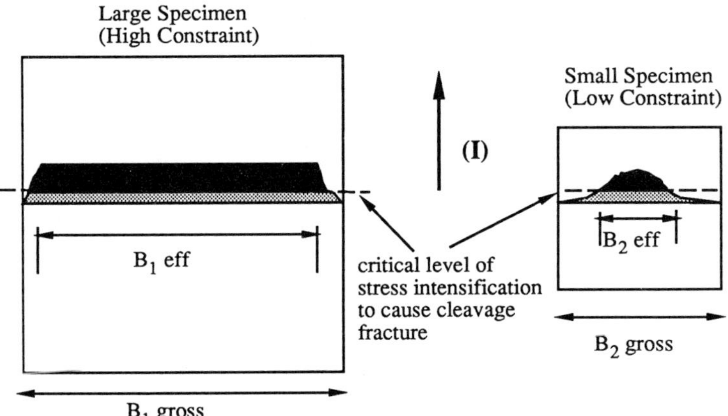

Specimen constraint increasing with specimen size ie.
(B_1 eff / B_2 eff > B_1 gross / B_2 gross)
But <u>maximum</u> stress intensification (I) the same in both sizes.

FIG. 1—*Schematic representation of the concepts of stress intensification and constraint.*

TABLE 1—*Test matrix.*

B (mm)	% SG	No. of Specimens
12.5	0	35
	20	29
25	0	30
	20	30
	50	36

large specimens (Fig. 1), leading to suppression of plastic yielding and higher stress intensifications.[2]

The effect of specimen size on transition toughness was clearly demonstrated (and quantified) in a recent comprehensive study on A533B-1 pressure vessel steel [*1*]. Unfortunately, despite the large number of specimens tested in that program, the relative influences of statistical or constraint-based contributions to the size effect were not clear. Nonetheless, if size effects in the transition regime are to be fully understood, and possibly mitigated, an important objective remains, namely the isolation of the contributions of statistical sampling and changing constraint on transition toughness.

One long-established [*2,3*] method of increasing specimen constraint without increasing specimen size is to place grooves along the specimen sides, in line with the crack plane. Sidegrooving (SG) therefore provides a practical opportunity for differentiating between statistical and constraint-based contributions to shifts in transition temperature. This report presents details of an experimental investigation of the influence of sidegrooving on transition toughness. The material studied was the same 257-mm-thick plate of A533B-1 material used in Ref *1*. Tests were performed on compact specimens (CS) of two sizes (12.5 and 25 mm gross thickness) and sidegrooved to different values of net thickness. This allowed comparisons to be made between specimens of either the same gross or the same net thickness, but with different levels of constraint. The test matrix is presented in Table 1.

Experimental Details

Material

All specimens were manufactured from a single 257-mm-thick A533 Grade B Class 1 steel plate produced by the basic electric process. The chemical composition of the product and details of the heat treatment are presented in Tables 2*a* and 2*b*. Further details of material condition, tensile, Charpy V-notch impact, and drop weight NDT properties are presented in Ref *1*. For the sake of convenience, relevant tensile properties are reproduced here in Table 3.

Specimens

Compact specimens were manufactured in accordance with either ASTM Test for Plane-Strain Fracture Toughness of Metallic Materials (E 399-83) or ASTM Test for J_{Ic}, A Measure

---

[2] The terms **stress intensification** and **constraint** are often used interchangeably. Within the context of the present work **stress intensification** refers to an elevation in the principal tensile stress σ_1 at a *single position* on a crack front (Fig. 1). The term **constraint** is used in the general sense suggested by Knott (in *Fundamentals of Fracture Mechanics,* Butterworths Press, 1973); i.e. ". . . the *integrated* effect of stress intensification *across the* [specimen] *cross-section.*"

TABLE 2a—*Chemical composition (wt%) by product analysis.*

C	Mn	Mo	Ni	Si	Cr	S	P	Cu	Al
0.21	1.44	0.48	0.67	0.28	0.18	0.005	0.006	0.05	0.021

TABLE 2b—*Heat treatment.*

Operation	Temp. (°C)	Time/Cooling Method
Normalize	880/940	9h
Temper	640/685	8.5h/air cool
Austenitize	844/928	8.3h/water quench
Temper	635/670	6.75h/air cool

TABLE 3—*Tensile properties.*

Test Temp. (°C)	Yield Stress (MPa)	Tensile Strength (MPa)	Elongation (%)	Reduction of Area (%)
20	470	620	25	66
−10	491	653	24	65
−50	522	693	24	64
−90	560	740	ND[a]	ND[a]

[a] Not Determined.

of Fracture Toughness (E 813-81). The spatial distribution of specimens within the plate is shown in Figure 2.

For compatibility with the original program on size effects, all specimens were manufactured in the T-S orientation and were taken from the mid-depth of the plate. Final crack length (a) to width (W) ratios of between 0.52 and 0.54 were obtained for all specimens. Specimens were produced in the 0, 20, and 50% sidegrooved conditions (Table 1). Percentage sidegrooving is defined as

$$\%SG = (1 - B_n/B_g) \times 100 \tag{1}$$

where

B_g = gross specimen thickness before sidegrooving, and
B_n = net specimen thickness after sidegrooving.

All sidegrooving operations were performed after fatigue pre-cracking using a 45° groove profile and a 0.25 mm root radius.

Test Procedure and Analyses

All testing was performed within the temperature range −90°C to +10°C using an environmental chamber with electronically controlled liquid/gaseous nitrogen feed. Before testing, specimens were held at the target temperature (±2°C) for approximately 15 to 20 min.

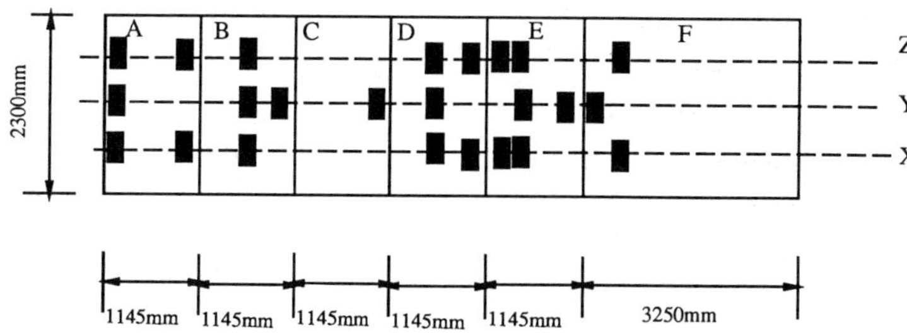

FIG. 2—*Location of compact specimens within the overall plate.*

Tests were carried out in displacement control at elastic loading rates of between 2.5 and 5.0 MPa $\sqrt{m}$ s^{-1}. All but three specimens failed in a brittle manner. Those three specimens were unloaded at displacements well in excess of the maximum load condition, heat tinted, and subsequently broken open at liquid nitrogen temperature.

Consistent with the procedures used in Ref *1*, fracture toughness was quantified in terms of the stress intensity factor *K*, derived from values of the *J* integral using the expression

$$K_{Jc}, K_J = \sqrt{[EJ/(1 - \nu^2)]} \tag{2}$$

where

K_{Jc} = stress intensity factor at the point of brittle fracture,
K_J = stress intensity factor at the point of final unloading (no failure),
E = Young's modulus (for consistency with Ref *1* this was taken as $E = 210\text{-}0.054T$, where E is in GPa and T is in °C), and
ν = Poisson's ratio (taken as 0.3).

J was evaluated using the expression

$$J = J_o[1 - (0.75 f(a/W) - 1)\Delta a/(W - a)] \tag{3}$$

where

J_o = value of *J* uncorrected for crack growth = $U_{LL} \cdot f(a/W)/[B_n(W - a)]$,
U_{LL} = area under the specimen load/load-line displacement trace,
Δa = amount of stable crack extension present at the point of brittle fracture or the point of unloading, and
$f(a/W)$ = a dimensionless coefficient correcting for the tensile component of loading in a compact specimen [5].

Values of J_o derived using Eq 3 are very similar to values derived using the *J* formulation given in ASTM E 813-87, Annex A2. Comparative calculations, between Eq 3 and the ASTM E 813-87 formulation, have been performed for specimens exhibiting widely varying levels of plasticity. For example, for the predominantly linear elastic test on Specimen

HE103 (see Appendix) the ASTM E 813-87 formulation gives a J value 3% lower than the J_o value provided by Eq 3, whilst for Specimen HE90 (gross plasticity) the ASTM J value is 4% lower than the J_o value from Eq 3.[3] In all cases, values of specimen crack length and stable crack extension were calculated using a nine-point average across the net thickness per ASTM E 813-37.

For specimens of the K_{1c} design (ASTM E 399-83), where displacements were monitored on or above the specimen front face, load-line displacements were inferred by correcting measured displacements using the geometry of similar triangles and assuming that the compact specimen rotates around a hinge point located at $0.4(W - a)$ below the original crack length.

Results

The 160 compact specimens were separated into five separate sets according to specimen thickness and depth of sidegrooving. Each set contained a minimum of 29 specimens. A transition curve was developed for each set such that at the lowest temperature the mean K_{Jc} level was ≈ 100 MPa$\sqrt{m}$ and at the highest temperature specimens on average achieved final displacements (Δ_f) well beyond the maximum load condition (typically $\Delta_f \approx 2\Delta_{max\ load}$). The number of temperatures sampled in each transition curve was deliberately kept small, at between three and five, thereby maximizing the accuracy of any data derived at individual test temperatures (e.g., mean K_{Jc} values).

All values of Δa, J, K_{Jc} and K_J are tabulated in the Appendix. The data have been used to examine the effects of sidegrooving on values of fracture toughness measured in the transition regime. Some comment is also made upon the effect of sidegrooving on resistance to ductile crack extension before specimen failure.

Effects of Sidegrooving on Transition Toughness (K_{Jc}, K_J)

There are many ways of representing fracture toughness values in the brittle-ductile transition regime [1]. Two approaches have been examined in the present study: (1) fitting an exponential curve to fracture toughness data across a range of test temperatures and (2) calculating means and variances of fracture toughness values at individual test temperatures. In both cases analyses were restricted to specimens of a given size and sidegrooved condition.

Transition Curves—Transition curves were defined using two-parameter exponential curve fits of the form

$$K_{Jc} = A \exp(\phi T) \tag{4}$$

where

T = temperature, °C, and

A and ϕ = regression coefficients obtained from least squares regression analysis.

For compatibility with data presented in Ref *1*, only those data which failed by cleavage fracture were included in the analyses. This does, however, account for 98% of all tests performed. The resulting exponential curves are shown for each specimen condition, together

---

[3] For the case of Specimen HE90, the crack growth correction of Eq 3 means that the *final J* (i.e., not J_o) value is 17% lower than that provided by the ASTM E 813-87 formulation, which does not correct for crack growth. There was no ductile crack growth on Specimen HE103.

(a)

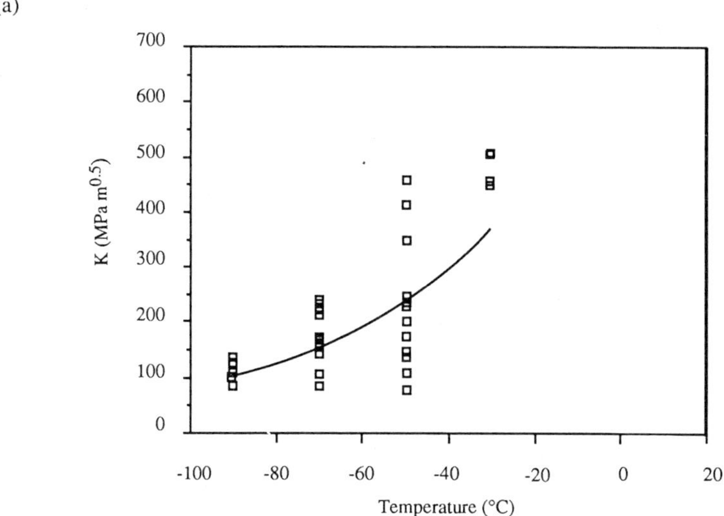

(b)

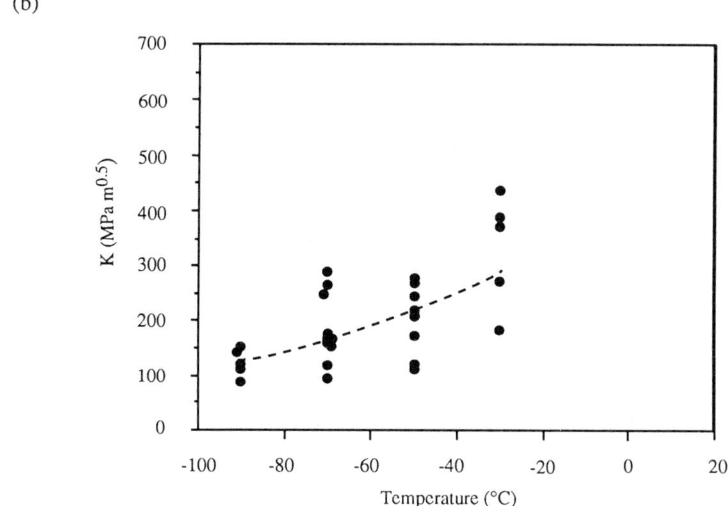

FIG. 3—*Fracture toughness data for 12.5 mm-thick compact specimens: (a) 0% and (b) 20% sidegrooved data.*

with all test data, in Figs. 3 and 4. Values of the coefficients A and ϕ are given in Table 4 together with values of the correlation coefficient (r). Exponential transition curves obtained for specimens of the same gross thickness (B_g) but sidegrooved to different net thicknesses (B_n) are compared in Figs. 5a and 5b.

The results indicate that sidegrooving decreases transition toughness at high K levels/temperatures and increases it at low K levels/temperatures. For example, at the *highest* common temperatures at which plain and sidegrooved specimens were tested, exponential curve fits to 0% and 20% SG data (either 12.5 or 25 mm CS) predict K_{Jc} levels to be ≈ 80 MPa $\sqrt{m}$ lower for the sidegrooved specimens. A larger decrease of ≈ 150 MPa $\sqrt{m}$ is obtained in an

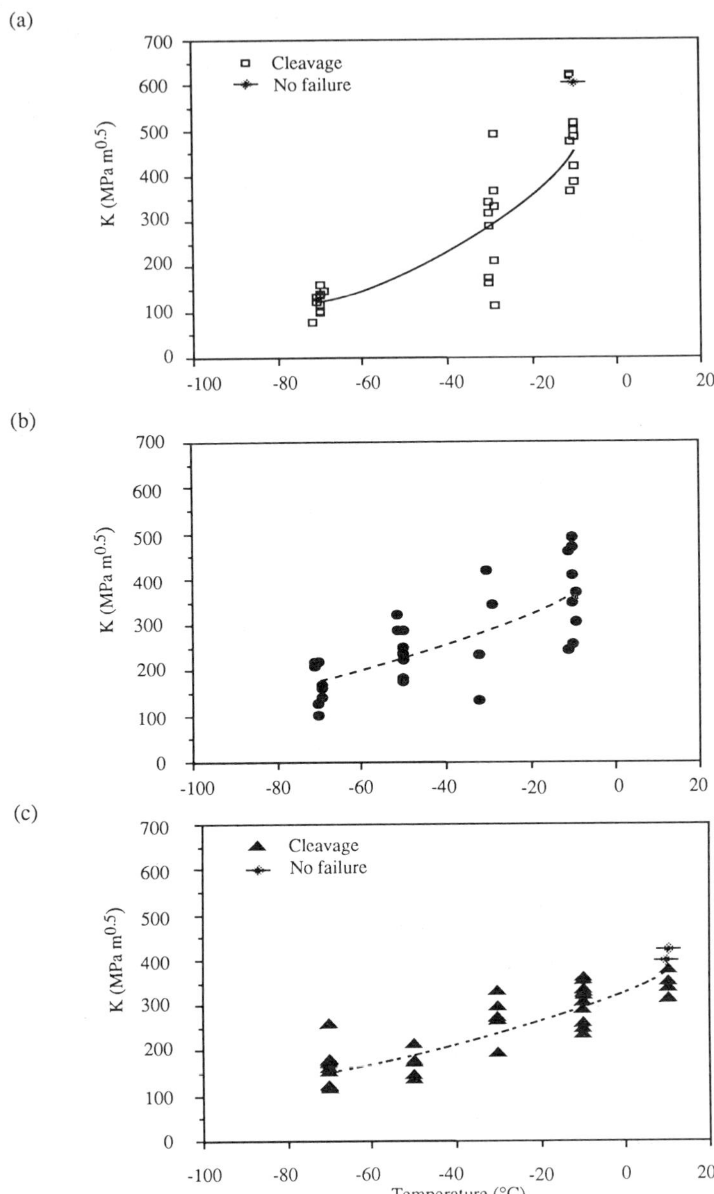

FIG. 4—*Fracture toughness data for 25-mm-thick compact specimens:* (a) *0%,* (b) *20% and* (c) *50% sidegrooved data.*

equivalent comparison between plain and 50%SG 25mmCS. In contrast, at the *lowest* common test temperatures exponential curve fits to 0% and 20% SG data predict K_{Jc} *increases* of ≈ 60 MPa$\sqrt{\text{m}}$ and 25 MPa$\sqrt{\text{m}}$ for 25 mm CS and 12.5 mm CS respectively, whilst for the case of 50%SG 25mmCS the predicted K_{jc} increase is ≈ 34 MPa$\sqrt{\text{m}}$.

A comparison of exponential curves for specimens of the same *net* thickness ($B_n = 12.5$

TABLE 4—*Regression coefficients in the expression* $K_{Jc} = A \exp(\phi T)$.[a]

Specimen Thickness (mm)	% Sidegrooving	A	ϕ	Correlation Coefficient
12.5	0	714.9	0.022	0.73
12.5	20	441.1	0.014	0.66
25	0	554.5	0.022	0.87
25	20	413.0	0.012	0.74
25	50	330.1	0.011	0.86

[a] K_{Jc} in MPa$\sqrt{m}$ and T in °C.

mm) but with different levels of sidegrooving is presented in Fig. 5c. These results show that constraint alone can cause a substantial change in transitional behavior. For example, at a K_{Jc} level of 300 MPa$\sqrt{m}$ the mean exponential curve for 50%SG 25mmCS is located some 30°C higher than that for 0%SG 12.5mmCS of the same net thickness.

Means and Standard Deviations of Toughness Data at Discrete Temperatures—Arithmetic means (K_{mean}) and standard deviations (s) of all fracture toughness data of a given specimen size, sidegrooved condition, and test temperature are presented in Table 5. Figures 6 and 7 show variations in K_{mean} with increasing levels of sidegrooving for 12.5 mm and 25 mm compact specimens respectively.

Comparison between K_{mean} values for plain and sidegrooved specimens of the same gross thickness reveals the same trend as that observed in the previous section; that is, sidegrooving decreases toughness at high K levels (temperatures) and increases it at low K levels (temperatures). However, the decreases are larger and the increases smaller than those predicted on the basis of exponential curve fits (Table 5).

The statistical significance of differences between K_{mean} values of plain and sidegrooved specimens has been examined using tests based upon Student's t distribution. Full details of the tests, their results, and the underlying assumptions on which they are based are presented in Ref 6. However, a brief summary of the statistical test results is presented below.

1. At the highest common test temperatures K_{mean} values of smooth-sided specimens are significantly higher (in a statistical sense) than those of sidegrooved specimens of the same gross thickness.

2. At the lowest common test temperatures K_{mean} values of smooth-sided specimens are significantly lower than those of sidegrooved specimens of the same gross thickness.

3. At intermediate common test temperatures there are no statistically significant differences between the K_{mean} values of smooth or sidegrooved (20% or 50%) specimens of the same gross thickness.

Resistance to Ductile Crack Extension

Previous work [7] has shown that J-Δa data acquired from tests in the transition regime can be used to construct a J-resistance curve. Within the present study resistance curves have been constructed for each specimen size and sidegrooved condition and represented by power laws of the form

$$J_R = D (\Delta a)^\theta \tag{5}$$

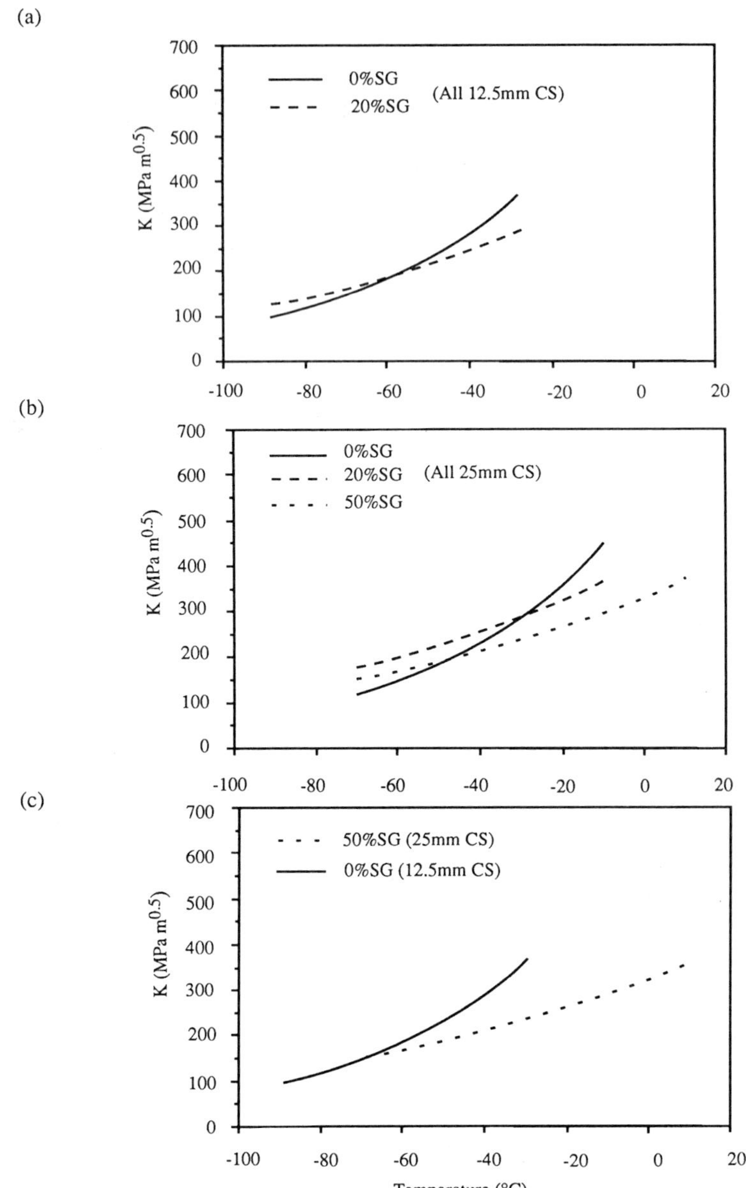

FIG. 5—*Effect of sidegrooving on transition toughness:* (a) B_{gross} = *12.5 mm,* (b) B_{gross} = *25 mm,* (c) B_{net} = *12.5 mm.*

The curves were obtained by linear regression analysis of $\log(J)$ on $\log(\Delta a)$ data using all data with $\Delta a \geq 0.2$ mm (i.e., even where cleavage fracture did not occur). Figure 8a shows the effect of sidegrooving on J_R for 12.5 mm-thick specimens, and Fig. 8b presents equivalent data for 25-mm-thick specimens. Calculated values of regression coefficients D and θ are given for each data set in Table 6.

TABLE 5—*Means and standard deviations of fracture toughness at different test temperatures.*

Specimen Thickness (mm)	Percentage Sidegrooving	Test Temp. (°C)	Sample Size (N)	K_{mean} (MPa $\sqrt{m}$)	Standard Deviation (MPa $\sqrt{m}$)
12.5	0	-30	5	472.7	30.2
	0	-50	12	231.2	119.6
	0	-70	12	171.7	48.0
	0	-90	6	108.4	19.8
	20	-30	5	330.4	101.0
	20	-50	9	205.0	58.8
	20	-70	10	185.1	63.5
	20	-90	5	123.9	25.7
25	0	-10	10	499.0	93.5
	0	-30	10	279.9	114.2
	0	-70	10	124.0	24.7
	20	-10	10	372.0	86.2
	20	-30	4	283.8	124.0
	20	-50	8	248.3	52.0
	20	-70	8	170.2	45.0
	50	$+10$	6	367.3	41.0
	50	-10	10	302.4	44.7
	50	-30	5	271.9	50.4
	50	-50	5	170.3	31.1
	50	-70	10	156.4	42.3

Table 6 shows that the influence of sidegrooving on fracture toughness values at the initiation of ductile tearing, approximated by $J_{0.2}$ (the value of J at $\Delta a = 0.2$ mm), is small. In contrast, Figs. 8*a* and 8*b* show that resistance to tearing (dJ/da) is definitely reduced by sidegrooving at the 20% level. At the 50% level, sidegrooving causes a dramatic decrease in resistance to ductile crack extension (Fig. 8*b*). However, it is worth restating that the *J-R* curves presented in Fig. 8 represent curve fits to *all* data with $\Delta a > 0.2$ mm. If the regression analyses had been restricted, for example, by using the $J_{max}, \Delta a_{max}$ limits suggested in the ASTM Test for Determining *J-R* Curves (E 1152-87), then the above effects would have been less pronounced.

Discussion

A large experimental test matrix (160 specimens) has allowed a systematic examination of the influence of specimen sidegrooving upon fracture toughness in the transition regime. The results have been assessed using two separate forms of representation: (1) exponential curve fits to data across a temperature range and (2) calculation of mean and standard deviation toughness values at individual test temperatures. Neither of these methods is without its shortcomings, not the least of which is the fact that both methods imply particular distributions of fracture toughness values. Use of the exponential representation implies a lognormal distribution of K_{Jc} values throughout the transition regime, whilst statistical comparison of sample K_{Jc}, K_J means and standard deviations requires that population means are normally distributed. Nevertheless, both of these representations are widely used, their deficiencies well understood, and as such they facilitate comparison with other studies. Attempts at defining actual fracture toughness distributions, for example, by estimating Weibull Probability Density Functions, were considered unsuitable for this particular study where sample sizes at individual test temperatures were small (<15). The dangers of using small sample

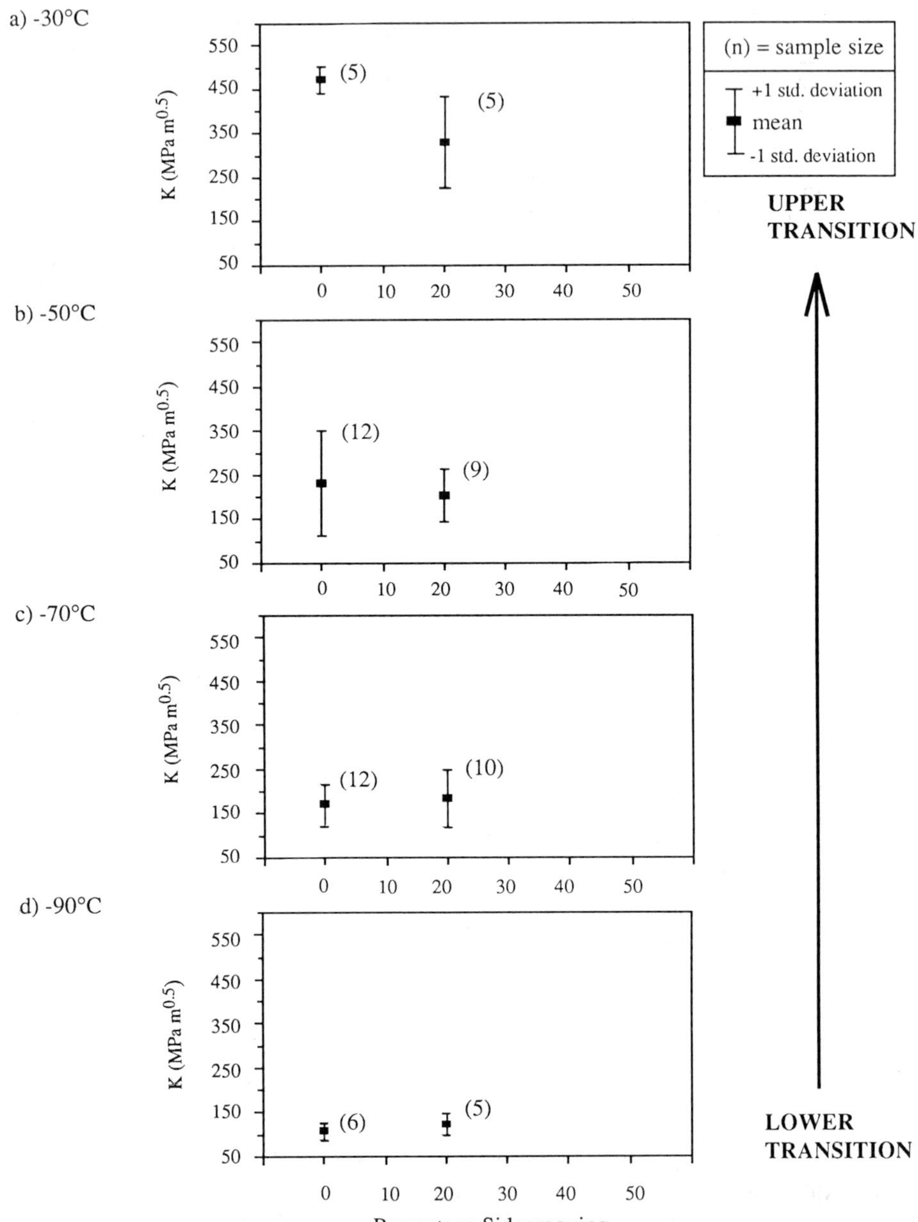

FIG. 6—*Influence of sidegrooving on the mean fracture toughness of 12.5-mm-thick compact specimens tested at different temperatures.*

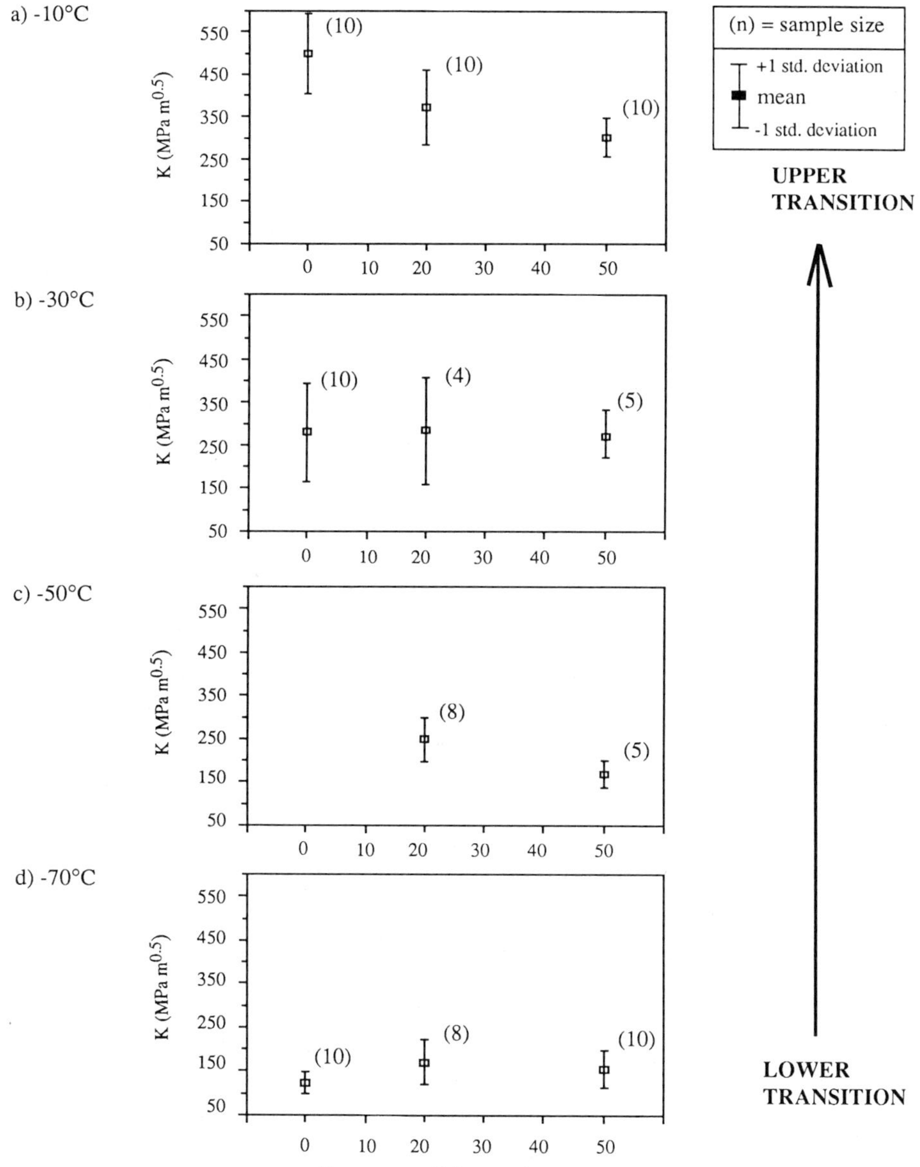

FIG. 7—*Influence of sidegrooving on the mean fracture toughness of 25-mm-thick compact specimens tested at different temperatures.*

data to predict population distributions having been previously described by Slatcher [8] and also by Wallin [9].

The main finding of the experimental program is that, for specimens of identical gross thickness, sidegrooving decreases mean fracture toughness in the upper transition regime and increases it in the lower transition. Although the magnitude of the effect seems to depend

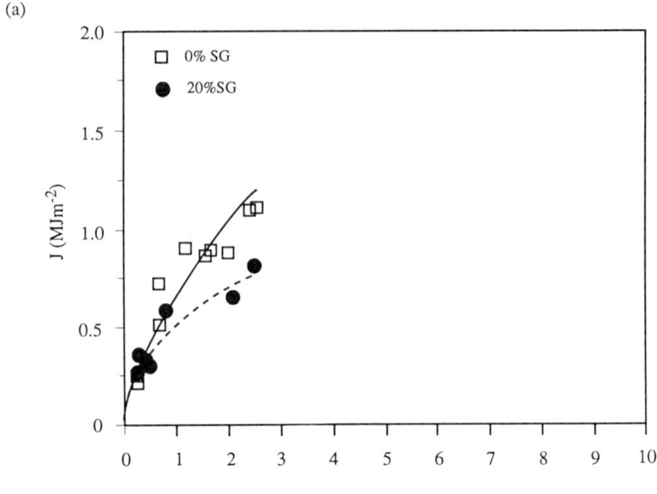

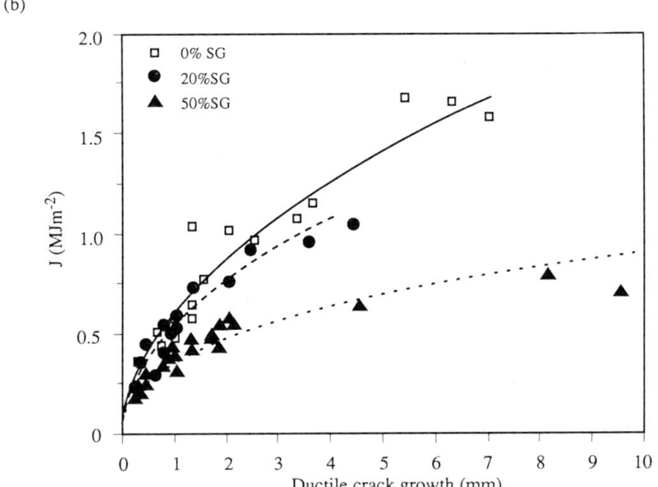

FIG. 8—*Effect of sidegrooving on the J resistance curve;* (a) *12.5 mm CS and* (b) *25 mm CS.*

TABLE 6—*Regression coefficients in the expression* $J_R = D (\Delta a)^{\theta}$.[a]

Specimen Range Thickness (mm)	Percentage Sidegrooving	D	θ	$J_{0.2}$ (MJ m^{-2})	Temp. (°C)
12.5	0	0.645	0.668	0.220	$-30/-90$
12.5	20	0.504	0.463	0.239	$-30/-90$
25	0	0.600	0.530	0.256	$-10/-70$
25	20	0.534	0.501	0.238	$-10/-70$
25	50	0.362	0.400	0.190	$+10/-70$

[a] J in MJ m^{-2} and Δa in mm.

upon which way the data are analyzed (exponential curve or discrete mean values), the *overall* trend is the same regardless of the form of representation. The effects are shown particularly clearly in Fig. 5, where it can be seen that exponential curve fits to 0% and 20% SG data "cross-over" between high and low K levels.

Cleavage fracture occurs when a material is subjected to stresses in excess of its critical fracture stress over a microstructurally significant volume [10]. Therefore the *effective* specimen thickness for cleavage fracture is that proportion of available (net) thickness which is subject to high stress levels (i.e., a combination of actual net thickness and constraint) (Fig. 1). Relative levels of constraint in plain and sidegrooved specimens have been examined in a variety of finite element studies [11–16]. The experimental results obtained in the present work are now considered in light of these numerical studies. Distinction has been drawn between sidegrooving at the 20% and 50% levels.

Comparison of 0% and 20% Sidegrooved Transition Data

At low levels of stress intensity factor (K), finite element studies suggest only small differences in constraint between smooth and moderately (up to 25%) sidegrooved specimens [11]. Consequently, greater net thickness in smooth specimens should lead to a greater effective thickness and lower mean fracture toughness at failure. Conversely, at higher levels of K, numerical studies predict increasing constraint in sidegrooved specimens relative to plain-sided specimens [12]. At some level of K this increase in constraint will more than compensate for reduced net thickness, leading to lower mean fracture toughness in *sidegrooved* specimens, as observed in the present program. The competing influences of decreased thickness and increased constraint on fracture probability, and thus fracture toughness at failure, are illustrated schematically in Fig. 9. The figure is not intended as an exact representation of the rate of change in constraint nor of the relationship between constraint and fracture probability (mean fracture toughness). Rather, it serves as a model to explain the observed experimental results. There is, however, one additional consideration which further helps to explain the results, namely the presence of stable ductile tearing prior to brittle fracture.

It has been shown in this and other studies that tearing resistance is definitely reduced by moderate sidegrooving. Therefore, at a given J level *beyond* crack initiation, more potentially weak microstructures will be sampled by sidegrooved specimens (due to increased crack growth). The mean fracture toughness at failure in sidegrooved specimens should be increasingly reduced relative to smooth specimens. It has already been noted above that, for a given specimen size, K_{Jc}, K_J population means of 0% SG data were *significantly* higher than corresponding 20% SG values *only* at the highest test temperature. At these temperatures the ratio $K_{mean}/K_{0.2}$ ($K_{0.2}$ = value of $J_{0.2}$ converted using Eq 2) falls within the range 1.40 to 2.09, depending upon specimen size and percentage sidegrooving. On the other hand, at the next highest temperatures, where differences between plain and sidegrooved K_{mean} values were *not* statistically significant, the ratio $K_{mean}/K_{0.2}$ is reduced to the range 0.87 to 1.20. The suggestion that differences in brittle fracture probabilities between 0% and 20% SG specimens in the upper transition regime are linked to respective differences in resistance to stable ductile tearing therefore appears to be substantiated by the experimental results.

Comparison of 0% and 50% Sidegrooved Transition Data

Relationships between the transition toughnesses of plain and 50% sidegrooved specimens are similar in trend to those observed in the previous section. Within the upper transition however, the reduction in K_{mean} values due to 50% sidegrooving is greater than that due to 20% sidegrooving. To the author's knowledge no three-dimensional elastic-plastic finite ele-

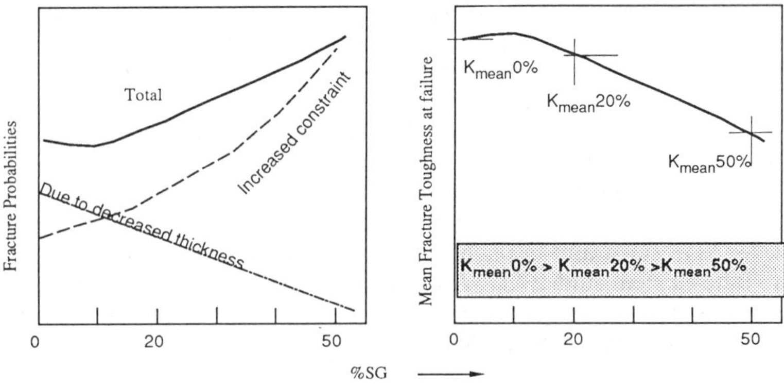

Large increases in constraint (Upper Transition)

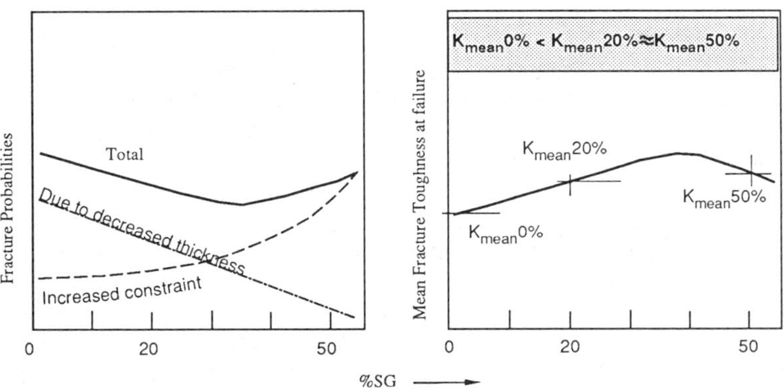

Moderate increases in constraint (Lower Transition)

FIG. 9—*Schematic diagram showing the competing influences on fracture probability and mean toughness at failure, of the increased constraint and reduced net thickness caused by sidegrooving.*

ment studies of 50% SG specimens have been published. Consequently, levels of constraint in 50% SG specimens at high levels of K can only be surmised. However, on the basis of comparisons at lower K levels [*11*] it is probably safe to assume that constraint in 50% SG specimens is much higher than in plain-sided specimens. This, allied to an observed reduction in tearing resistance in 50% SG specimens, should dramatically increase the volume of material sampled by high local stresses, thereby reducing K_{mean} values.

At the lowest common test temperature ($-70°C$) the population mean of 0% SG data was found to be significantly lower than that for 50% SG data. Within this lower transition regime ductile tearing does not occur ($K_{mean} < K_{0.2}$). It would therefore appear that even the large increase in constraint noted for 50% SG specimens [*12*] is insufficient (on its own) to compensate for the reduced net thickness in that specimen type.

Values of fracture toughness for 50% SG specimens were calculated using Eqs 2 and 3. In fact, experimental work by Gibson et al. [*17*] suggests that, due to extensive deformation of the sidegrooved flanks in 50% SG specimens, the plastic work of which does not contribute to the true value of J, direct use of Eq 3 is inappropriate and calculated J values should be reduced by $\approx 24\%$. If the 50% SG results are corrected as suggested, then differences between K_{mean} values of 0% and 50% SG data in the upper transition are accentuated, whilst differences at the lowest common test temperature cease to be statistically significant. Clearly the contrasting effects of increased constraint and reduced thickness are closely matched in 50% SG compact specimens tested in the lower transition regime.

One additional effect of sidegrooving at the 50% level seems to be a reduction in data scatter in the upper transition regime. Statistical tests [6] showed that at the highest mutual test temperature ($-10°C$) the variance of 50% SG data was significantly lower than that for smooth specimens. This reduction in data scatter is probably a combination of two phenomena: (1) increased constraint and (2) reduced physical capacity of 50% SG specimens to support high J values. This latter point will itself be partly due to constraint-induced reductions in tearing resistance.

Conclusions

1. Within the upper transition regime (for 0% SG CS: $K_{mean} \approx 2K_{0.2}$), sidegrooving at the 20% or 50% level causes a small, though statistically significant, decrease in mean fracture toughness at brittle fracture.

2. Within the lower transition regime (for 0% SG CS: $K_{mean} \approx 0.5K_{0.2}$), the same sidegrooving causes a corresponding increase in mean fracture toughness at brittle fracture.

3. Within the upper transition regime, reductions in mean fracture toughness at brittle fracture are more pronounced in 50% sidegrooved specimens than in 20% sidegrooved specimens. These additional reductions are linked to decreased resistance to stable ductile tearing in 50% SG specimens.

4. The experimental results are consistent with the relative increases in constraint (predicted by several numerical studies) and decreases in net thickness in sidegrooved specimens.

Acknowledgments

The author would like to acknowledge Dr D. P. G. Lidbury for useful discussions. This work was partly funded by the United Kingdom Central Electricity Generating Board.

References

[1] Ingham, T., Knee, N., Milne, I., and Morland, E., "Fracture Toughness in the Transition Regime for A533B Steel: Prediction of Large Specimen Results from Small Specimen Tests," in *Fracture Mechanics: Perspectives and Directions (Twentieth Symposium), ASTM STP 1020*, American Society for Testing and Materials, Philadelphia, 1989, pp. 369–389.

[2] Freed, C. N. and Krafft, J. M., "Effect of Side Grooving on Measurements of Plane Strain Fracture Toughness," report to ASTM Committee E-24, 12 May 1965.

[3] Brown, W. F. and Srawley, J. E., *Plane Strain Crack Toughness Testing of High Strength Metallic Materials, ASTM STP 410*, American Society for Testing and Materials, Philadelphia, 1969, pp. 51–53.

[4] Merkle, J. G. and Corten, H. T., "A J Integral Analysis for the Compact Specimen, Considering Axial Force as Well as Bending Effects," *Transactions of the American Society of Mechanical Engineers, Journal of Pressure Vessels and Testing*, Vol. 96, 1976, p. 49.

[5] Schwalbe, K-H., Ingham, T., and Neale, B. K., "The Development of an European Group on Fracture Procedure to Measure Elastic-Plastic Fracture Parameters," in *Proceedings*, 6th European Conference on Fracture—Fracture Control of Engineering Structures, H. C. van Elst and A. Bakker, Eds., Vol. 1, Engineering Materials Advisory Services, 1986, pp. 277–299.

[6] Morland, E., "The Effect of Sidegrooving on the Fracture Toughness of A533B-1 Steel in the Transition Regime," to be published by United Kingdom Atomic Energy Authority as Report No. NRL-R-1006(R).

[7] McCabe, D. E., Landes, J. D., and Ernst, H. A., "Prediction of Heavy Section Performance of Nuclear Vessel Steels from Surveillance Sized Specimens," *Structural Mechanics in Reactor Technology,* Vol. 7, No. G214, 1983, pp. 47–54.

[8] Slatcher, S., "The Probabilistic Modelling of Fracture Toughness," in *Proceedings,* 6th European Conference on Fracture—Fracture Control of Engineering Structures, H. C. van Elst and A. Bakker, Eds., Vol. 1, Engineering Materials Advisory Services, 1986, pp. 551–561.

[9] Wallin, K., *Engineering Fracture Mechanics,* Vol. 22, No. 1, 1985, pp. 149–163.

[10] Ritchie, R. O., Knott, J. F., and Rice, J. R., *Journal of the Mechanics and Physics of Solids,* Vol. 21, 1973, p. 395.

[11] Shih, C. F. and deLorenzi, H. G., *International Journal of Fracture,* Vol. 13, No. 4, Aug. 1977, pp. 544–548.

[12] deLorenzi, H. G. and Shih, C. F., *International Journal of Fracture,* Vol. 21, No. 3, March 1983, pp. 195–220.

[13] Mendelson, A. and Alam, J., *International Journal of Fracture,* Vol. 22, No. 2, June 1983, pp. 105–116.

[14] Sakata, M., Aoki, S., Kishimoto, K., and Takagi, R., *International Journal of Fracture,* Vol. 23, No. 3, Nov. 1983, pp. 187–200.

[15] Kikuchi, M. and Miyamoto, H., *International Journal of Pressure Vessels and Piping,* Vol. 16, No. 1, 1984, pp. 1–16.

[16] Neale, B. K., *International Journal of Pressure Vessels and Piping,* Vol. 10, 1982, pp. 375–398.

[17] Gibson, G., Wilmshurst, M., and Druce, S. G., "An Experimental Evaluation of the *J* Integral in 50% Sidegrooved Compact Geometry Specimens," United Kingdom Atomic Energy Authority Report No. AERE-R 11202, May 1984.

APPENDIX

Fracture toughness data for 12.5 mm and 25 mm compact specimens containing different percentages of sidegrooving and tested in the transition region are presented in Tables A.1 to A.5 (pp. 233–237).

Table A.1

Transition data for 0% sidegrooved 12.5mm Compact Specimens

Specimen Ident.	Temp (°C)	J (MJm^{-2})	K_{Jc} (MPa$\sqrt{m}$)	Ductile Growth (mm)	Cleavage Failure
FX90	-30	1.095	504.5	2.40	√
FY94	-30	1.106	507.0	2.54	√
FY102	-30	0.887	454.1	1.69	√
FY103	-30	0.874	450.8	1.99	√
FZ98	-30	0.860	447.3	1.55	√
FX98	-50	0.092	147.0	0	√
FX100	-50	0.078	135.0	0.04	√
FX104	-50	0.514	346.6	0.68	√
FX105	-50	0.128	172.6	0.08	√
FY95	-50	0.172	200.2	0.15	√
FY90	-50	0.222	227.6	0.28	√
FY100	-50	0.026	77.9	0	√
FY101	-50	0.052	110.8	0	√
FZ101	-50	0.731	413.4	0.70	√
FZ102	-50	0.241	237.3	0.18	√
FZ104	-50	0.899	458.5	1.19	√
FZ105	-50	0.261	247.0	0.23	√
FX91	-70	0.243	238.7	0.15	√
FX99	-70	0.126	172.0	0.09	√
FX103	-70	0.116	164.8	0.10	√
FY96	-70	0.031	85.3	0	√
FY98	-70	0.121	168.5	0.06	√
FY99	-70	0.190	211.0	0.17	√
FY107	-70	0.111	161.7	0.05	√
FY108	-70	0.102	154.6	0.07	√
FY109	-70	0.213	223.6	0.15	√
FZ92	-70	0.087	143.3	0.05	√
FZ93	-70	0.047	105.4	0	√
FZ99	-70	0.227	231.1	0.26	√
FX101	-90	0.035	91.1	0	√
FX102	-90	0.068	126.5	0	√
FY104	-90	0.078	135.4	0.04	√
FY105	-90	0.031	84.9	0	√
FY106	-90	0.044	102.1	0	√
FZ103	-90	0.051	110.1	0	√

Table A.2

Transition data for 20% sidegrooved 12.5mm Compact Specimens

Specimen Ident.	Temp (°C)	J (MJm^{-2})	K_{Jc} (MPa√m)	Ductile Growth (mm)	Cleavage Failure
HE3	-30	0.816	435.7	2.51	√
HE10	-30	0.318	271.9	0.38	√
HE15	-30	0.650	388.8	2.10	√
HE20	-30	0.147	184.7	0.14	√
HE26	-30	0.591	370.9	0.84	√
HE1	-50	0.131	175.0	0.10	√
HE4	-50	0.211	222.0	0.13	√
HE6	-50	0.312	270.2	0.48	√
HE7	-50	0.256	244.4	0.33	√
HE11	-50	0.186	208.6	0.11	√
HE17	-50	0.190	210.5	0.14	√
HE21	-50	0.064	122.1	0.06	√
HE22	-50	0.055	113.7	0.03	√
HE30	-50	0.333	278.9	0.45	√
HE8	-69	0.117	165.4	0.07	√
HE2	-70	0.124	170.6	0.06	√
HE12	-70	0.360	290.8	0.29	√
HE13	-70	0.300	265.4	0.45	√
HE16	-70	0.061	119.4	0	√
HE18	-69	0.099	152.3	0.06	√
HE24	-70	0.136	178.9	0.08	√
HE25	-70	0.040	97.2	0.02	√
HE27	-70	0.109	160.4	0.05	√
HE29	-71	0.267	250.5	0.22	√
HE5	-90	0.097	152.7	0.05	√
HE9	-91	0.086	142.4	0.06	√
HE14	-90	0.054	112.9	0.01	√
HE23	-90	0.032	87.2	0.02	√
HE28	-90	0.066	124.4	0.04	√

Table A.3

Transition data for 0% sidegrooved 25mm Compact Specimens

Specimen Ident.	Temp (°C)	J (MJm^{-2})	K_{Jc} (MPa√m)	Ductile Growth (mm)	Cleavage Failure
HE113	-10	0.767	421.3	1.58	√
HE71	-10	1.012	483.9	2.07	√
HE76	-10	0.644	386.0	1.35	√
HE90	-10	1.579	604.4	7.02	X
HE108	-11	1.676	622.8	5.40	√
HE44	-11	1.658	619.4	6.30	√
HE104	-11	0.968	473.3	2.55	√
HE85	-11	0.578	365.7	1.33	√
HE112	-10	1.071	497.8	3.34	√
HE41	-10	1.147	515.1	3.66	√
HE137	-29	1.034	490.3	1.34	√
HE36	-30	0.503	342.0	0.69	√
HE135	-29	0.192	211.3	0.17	√
HE74	-30	0.130	173.9	0.10	√
HE65	-30	0.115	163.5	0.05	√
HE82	-29	0.054	112.0	0	√
HE58	-30	0.435	318.1	0.73	√
HE119	-30	0.362	290.1	0.28	√
HE99	-29	0.476	332.7	1.02	√
HE63	-29	0.573	365.0	1.01	√
HE131	-70	0.109	160.0	0.07	√
HE103	-72	0.027	79.7	0	√
HE94	-71	0.063	121.7	0	√
HE32	-69	0.093	147.8	0.05	√
HE102	-71	0.076	133.6	0.04	√
HE80	-70	0.041	98.1	0.03	√
HE97	-70	0.059	117.7	0.03	√
HE68	-70	0.085	141.3	0.06	√
HE83	-70	0.079	136.2	0.06	√
HE67	-70	0.046	104.0	0.03	√

Table A.4

Transition data for 20% sidegrooved 25mm Compact Specimens

Specimen Ident.	Temp (°C)	J (MJm^{-2})	K_{Jc} (MPa√m)	Ductile Growth (mm)	Cleavage Failure
HE130	-11	0.921	461.7	2.47	√
HE106	-9	0.405	306.1	0.78	√
HE62	-10	1.040	490.5	4.44	√
HE116	-9	0.599	372.2	1.03	√
HE43	-10	0.288	258.1	0.63	√
HE72	-10	0.727	410.1	1.33	√
HE138	-10	0.955	470.1	3.60	√
HE34	-10	0.525	348.5	1.04	√
HE132	-10	0.546	355.4	0.79	√
HE61	-11	0.264	247.2	0.31	√
HE56	-30	0.760	420.4	2.05	√
HE73	-32	0.236	234.3	0.24	√
HE87	-29	0.507	343.3	0.92	√
HE107	-32	0.081	137.3	0.06	√
HE125	-51	0.362	290.9	0.36	√
HE55	-51	0.449	324.0	0.43	√
HE42	-50	0.145	184.1	0.08	√
HE79	-50	0.246	239.8	0.23	√
HE118	-50	0.219	226.2	0.15	√
HE49	-50	0.272	252.1	0.28	√
HE93	-50	0.136	178.3	0.11	√
HE81	-50	0.362	290.9	0.34	√
HE115	-70	0.072	130.1	0.07	√
HE136	-70	0.208	221.1	0.14	√
HE98	-71	0.210	222.1	0.17	√
HE69	-70	0.044	101.7	0.03	√
HE89	-69	0.122	169.3	0.10	√
HE50	-69	0.114	163.6	0.08	√
HE54	-71	0.191	211.9	0.10	√
HE35	-69	0.086	142.1	0.03	√

Table A.5

Transition data for 50% sidegrooved 25mm Compact Specimens

Specimen Ident.	Temp (°C)	J (MJm^{-2})	K_{Jc} (MPa√m)	Ductile Growth (mm)	Cleavage Failure
HE111	+10	0.693	399.3	9.55	X
HE51	+10	0.776	422.5	8.17	X
HE57	+10	0.628	380.2	4.55	√
HE129	+10	0.494	337.0	1.68	√
HE127	+10	0.424	312.5	1.88	√
HE75	+10	0.538	352.0	1.87	√
HE64	-10	0.291	259.7	0.43	√
HE133	-10	0.541	353.7	2.15	√
HE48	-10	0.251	240.9	0.46	√
HE77	-10	0.561	360.4	2.05	√
HE59	-10	0.239	235.3	0.43	√
HE95	-10	0.457	325.0	1.32	√
HE134	-10	0.430	315.3	0.97	√
HE84	-10	0.367	291.4	0.91	√
HE123	-10	0.486	335.4	1.70	√
HE40	-10	0.408	307.2	1.35	√
HE121	-30	0.304	265.9	1.05	√
HE92	-30	0.468	329.8	1.68	√
HE53	-30	0.161	193.5	0.23	√
HE100	-30	0.379	297.0	1.03	√
HE101	-30	0.321	273.4	0.77	√
HE86	-50	0.199	215.8	0.32	√
HE128	-50	0.132	175.6	0.14	√
HE39	-50	0.079	136.0	0.06	√
HE117	-50	0.092	147.0	0.14	√
HE60	-50	0.134	177.3	0.14	√
HE78	-70	0.137	179.5	0.09	√
HE110	-70	0.059	117.4	0.03	√
HE91	-70	0.101	153.7	0.03	√
HE120	-70	0.114	164.0	0.10	√
HE47	-70	0.285	258.7	0.49	√
HE114	-70	0.066	124.5	0.01	√
HE52	-70	0.114	164.0	0.04	√
HE109	-70	0.064	122.2	0.00	√
HE122	-70	0.109	159.5	0.09	√
HE37	-70	0.062	120.8	0.04	√

M. T. Miglin,[1] C. S. Wade,[1] and W. A. Van Der Sluys[1]

Analysis of Fracture Toughness Data for Modified SA508 Cl2 in the Ductile-to-Brittle Transition Region

REFERENCE: Miglin, M. T., Wade, C. S., and Van Der Sluys, W. A., **"Analysis of Fracture Toughness Data for Modified SA508 Cl2 in the Ductile-to-Brittle Transition Region,"** *Fracture Mechanics: Twenty-First Symposium, ASTM STP 1074,* J. P. Gudas, J. A. Joyce, and E. M. Hackett, Eds., American Society for Testing and Materials, Philadelphia, 1990, pp. 238–263.

ABSTRACT: Fracture toughness testing of steels in the ductile-to-brittle transition region is complicated by a large amount of data scatter. There are many available methods of analyzing transition region toughness data with the objective of reducing the data scatter and arriving at a representative value for the lower bound toughness. Ten 1T and five 4T compact specimens of SA508 Cl2 pressure vessel steel given a special heat treatment were tested in the transition region, and the data were analyzed according to six of the available methods. Of three purely statistical methods, one based on a large collection of pressure vessel and rotor steel data is preferred. Of three analytical methods, one which subtracts the contributions of plasticity and crack growth from measured toughness is preferred.

KEY WORDS: K_{Ic}, J_{Ic}, fracture toughness, ductile, brittle, cleavage, steel, A508, SA508, pressure vessel steel, low alloy steel

Fracture toughness testing of ferrous materials in the ductile-to-brittle transition region is complicated by an extreme amount of data scatter. Data scatter in transition region testing is governed predominantly by material response. It cannot be completely alleviated by improving control over experimental variables. Failure of statically loaded compact fracture specimens in the transition region often occurs by ductile tearing followed by cleavage failure. The data scatter results, in part, from large specimen-to-specimen differences in the amount of ductile tearing which precedes the initiation of cleavage fracture. Landes and Shaffer [1] hypothesized that cleavage initiates at a weak point in the material lying ahead of the fatigue precrack; if the weak point is distant from the fatigue precrack, some ductile tearing may occur prior to cleavage. Specimen-to-specimen variability in the distance between the weak point and the fatigue precrack causes scatter in the toughness measurements.

There are several methods of dealing with transition region data scatter. Some are purely statistical manipulations of conventional measures of toughness, while others are new methods of calculating toughness in the transition region. Among the statistical methods are Weibull analysis, log-normal analysis, and a method developed by Iwadate et al. [2]. New techniques for calculating toughness in the transition region include the β_{Ic} constraint correction [3], available energy analysis [4], and a procedure developed by Watanabe et al. [5].

The objective of this study is to evaluate the suitability of the above methods for determining lower bound fracture toughness in the ductile-to-brittle transition region. The test

---

[1] Babcock and Wilcox Company, Alliance Research Center, Alliance, OH 44601.

material is SA508 C12, specially tempered to produce a high transition temperature for use by Oak Ridge National Laboratories (ORNL) in the Pressurized Thermal Shock (PTS) project of the Heavy-Section Steel Technology Program [6].

Experimental Procedures

The test material was taken from a forged hollow cylinder of SA508 C12 (Bethlehem Steel heat no. 121S163) normalized at 893°C (1640°F) for 13 h and air cooled, austenitized at 857°C (1575°F) for 8¾ h and water quenched. Tempering was done at 523°C (973°F) for 9.5 h, followed by a post-weld heat treatment at 561°C (1041°F) for 12 h. The yield strength at room temperature is 620 MPa (90 ksi) and the ultimate strength is 779 MPa (113 ksi). The fracture toughness as a function of temperature for this material is shown in Fig. 1. The data in Fig. 1 were generated for the PTS project by the same laboratory in which this work was done [7]. Fifty-six 1T compact specimens were tested in accordance with ASTM E 399-81,

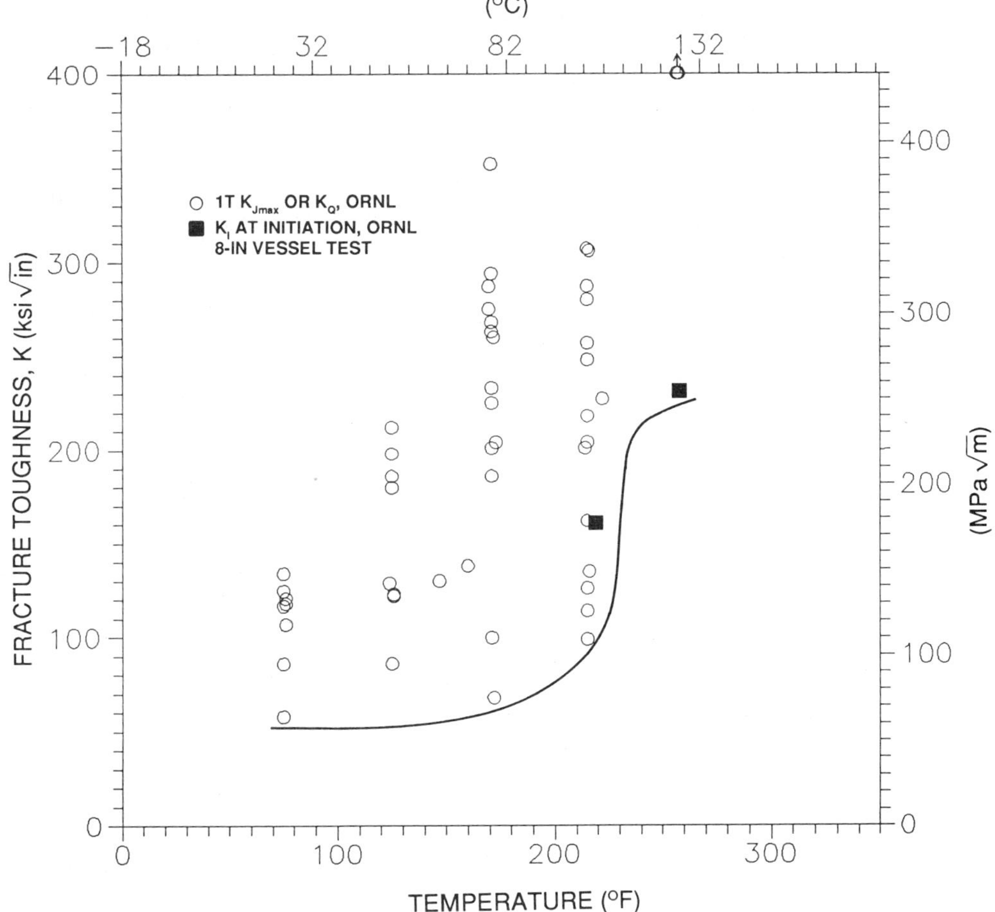

FIG. 1—*Fracture toughness as a function of temperature for the SA508 C12 steel used in this program. Data are from a program by ORNL. Included are two data points from tests of a 203-mm (8-in.)-thick vessel [6].*

and J_{max} was calculated in accordance with ASTM E 813-81 by measuring the area under the load-displacement curve. J_{max} was converted to $K_{J_{max}}$ by

$$K_{J_{max}} = \sqrt{\frac{J_{max}E}{(1 - \nu^2)}}$$ (1)

where the elastic modulus E = 202 000 MPa and Poisson's ratio ν = 0.3.

For this program, fifteen additional tests were conducted, including ten one-inch-thick (1T) and five four-inch-thick (4T) compact specimens. All tests were conducted at 80°C (175°F), the temperature at which data scatter is greatest. Specimen orientation is the same as for the PTS program (Fig. 2). The 1T specimens were fatigue precracked such that the a/W ratio, where a is crack length and W is specimen width, was between 0.52 and 0.54. The 4T specimens were precracked to an a/W ratio between 0.54 and 0.57, except for Specimen 4 which was precracked to a/W = 0.61. All specimens were side grooved to a depth of 10% of total thickness on each side.

Tests were conducted by loading each specimen slightly beyond the 5% secant point required by ASTM E 399 prior to performing any partial elastic unloads. Subsequent loading was interrupted regularly for partial elastic unloads to measure specimen compliance for use in crack length calculations. For specimens which passed the 5% secant point, an R-curve was established and J_{Ic} was determined if possible. The loading rate was 0.005 mm/s (0.0002 in./s) for all 1T specimens and 0.010 mm/s (0.0004 in./s) for the 4T specimens, except for one 4T specimen which was loaded rapidly due to an equipment malfunction. For all specimens, J_{max} was calculated at instability in accordance with ASTM E 813 and converted to a K value per Eq 1.

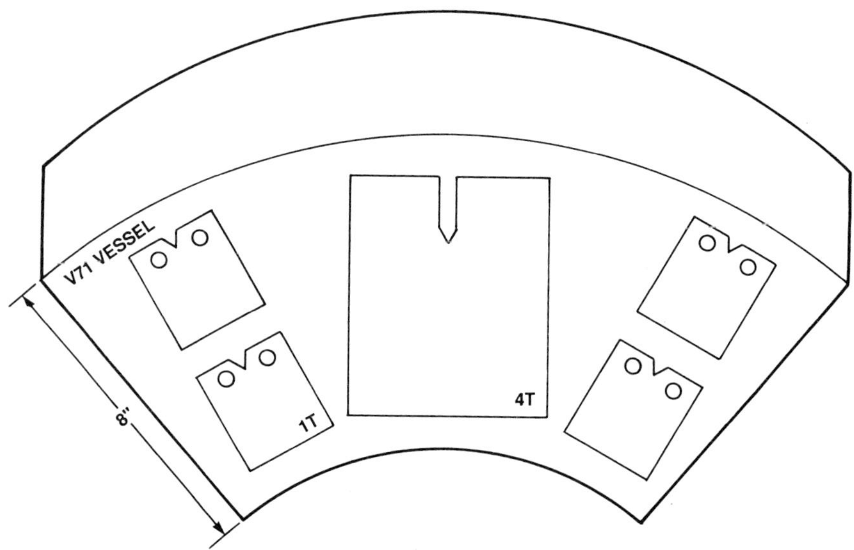

FIG. 2—*Schematic diagram showing specimen orientation relative to the SA508 C12 vessel forging.*

TABLE 1—*Results of fracture toughness tests at 80°C.*

1T Specimens

Specimen Number	J_{IC} (KJ/m^2)	$K_{J_{\text{max}}}$ (MPa $\sqrt{\text{m}}$)
1-1	...	102
1-2	140	290
2-1	139	297
2-2	126	246
3-1	157	244
3-2	...	191
4-1	140	343
4-2	...	138
5-1	116	205
5-2	136	308

4T Specimens

Specimen Number	K_Q (MPa $\sqrt{\text{m}}$)	Remarks
1	37	Rapid loading rate
2	82	Irregular precrack
3	160	
4	149	Irregular precrack
5	162	

Results of Fracture Toughness Tests

The test results are given in Table 1 and shown graphically in Fig. 3. All of the measured $K_{J_{\text{max}}}$ values violate the ASTM E 399 thickness requirement. For the 1T specimens, the specimen-to-specimen variation in material behavior is large. Figure 4 shows several of the load-displacement curves for the 1T specimens. The specimen with the lowest toughness failed in the elastic portion of the load-displacement curve, several specimens failed near maximum load, and some specimens were far beyond maximum load when ductile crack extension was interrupted by cleavage instability.

In Fig. 4 the amount of ductile crack growth measured after testing is indicated. Ductile crack lengths vary from negligible (0.03 mm) to extensive (4.70 mm). In all cases, ductile crack extension was terminated by conversion to cleavage failure.

For the 4T specimens, four out of five failures occurred before the 5% secant line, and the fifth occurred just beyond the 5% secant. Specimen 1 was loaded very rapidly due to an equipment malfunction, and the measured toughness was extremely low at 37 MPa $\sqrt{\text{m}}$ (34 ksi $\sqrt{\text{in}}$.). Specimens 2 and 4 had irregularities in the fatigue precrack and the measured toughness values, 82 and 149 MPa $\sqrt{\text{m}}$ (75 and 135 ksi $\sqrt{\text{in}}$.) respectively, were lower than those obtained with the remaining two specimens. (Specimen 2 was in violation of ASTM E 399 requirements for precrack straightness.) The two specimens with relatively straight fatigue precracks and no testing irregularities had measured toughness values of 160 and 162 MPa $\sqrt{\text{m}}$ (146 and 147 ksi $\sqrt{\text{in}}$.). These are reported as K_Q values because the thickness requirement of ASTM E 399 was violated. Only the data for the latter two specimens are shown in Fig. 3.

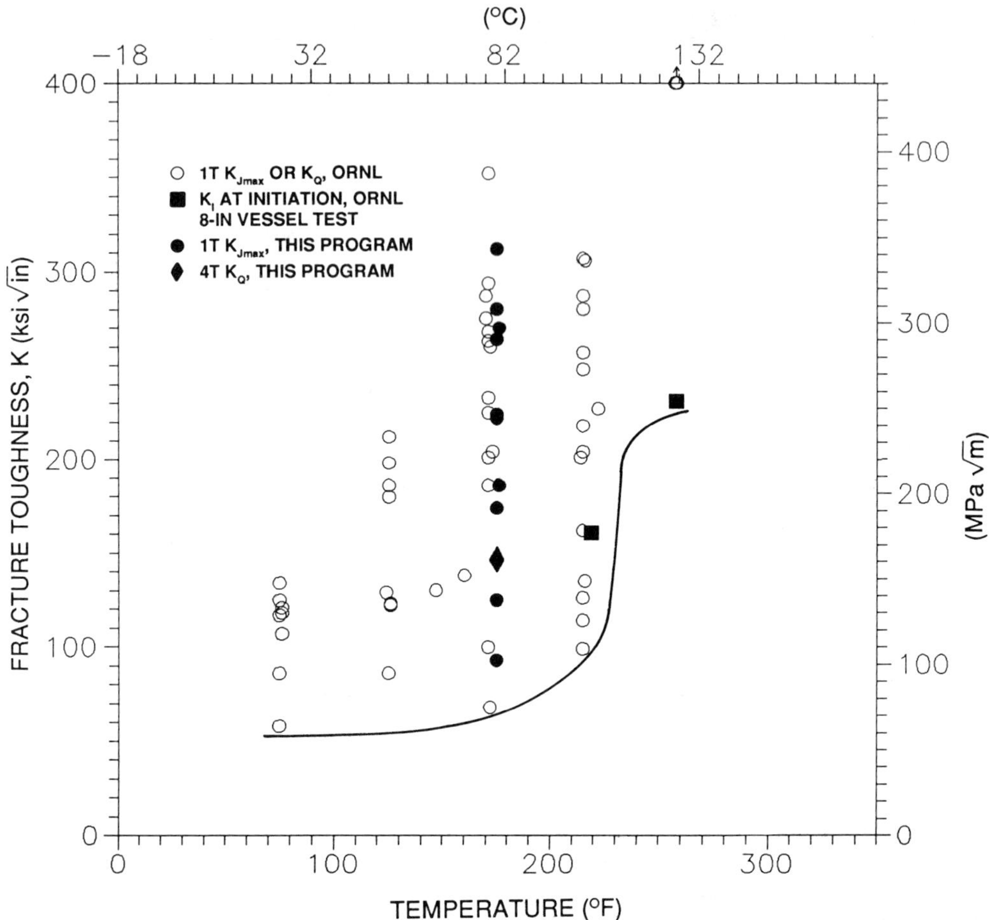

FIG. 3—*Results of fracture toughness tests conducted under this program combined with the data in Fig. 1 from the ORNL program.*

Fractography

The failed specimens were examined in a scanning electron microscope (SEM) to measure the extent of ductile crack advance and identify the origins of the cleavage instability. The ductile crack extension, Δa, and the distance from the ductile crack tip to the cleavage origin, X, as defined in Fig. 5, were measured. The results are reported in Table 2. Figure 5 also shows a plot of X versus Δa similar to one reported by Rosenfield and Shetty for another heat of specially prepared SA508 from the PTS program [4]. Rosenfield and Shetty found that the amount of stable crack growth, Δa, was slightly less than half of the precrack/origin separation, $\Delta a + X$. This appears true for the data in Fig. 5 as well, but the data scatter in Fig. 5 is greater.

Figure 6 shows the variation in $K_{J_{max}}$ with precrack/origin separation, $\Delta a + X$. It is apparent that higher toughness is correlated with greater distance of the cleavage origin from the precrack, in agreement with Rosenfield and Shetty's results [4].

For some specimens, multiple cleavage origins were observed and close observation of the

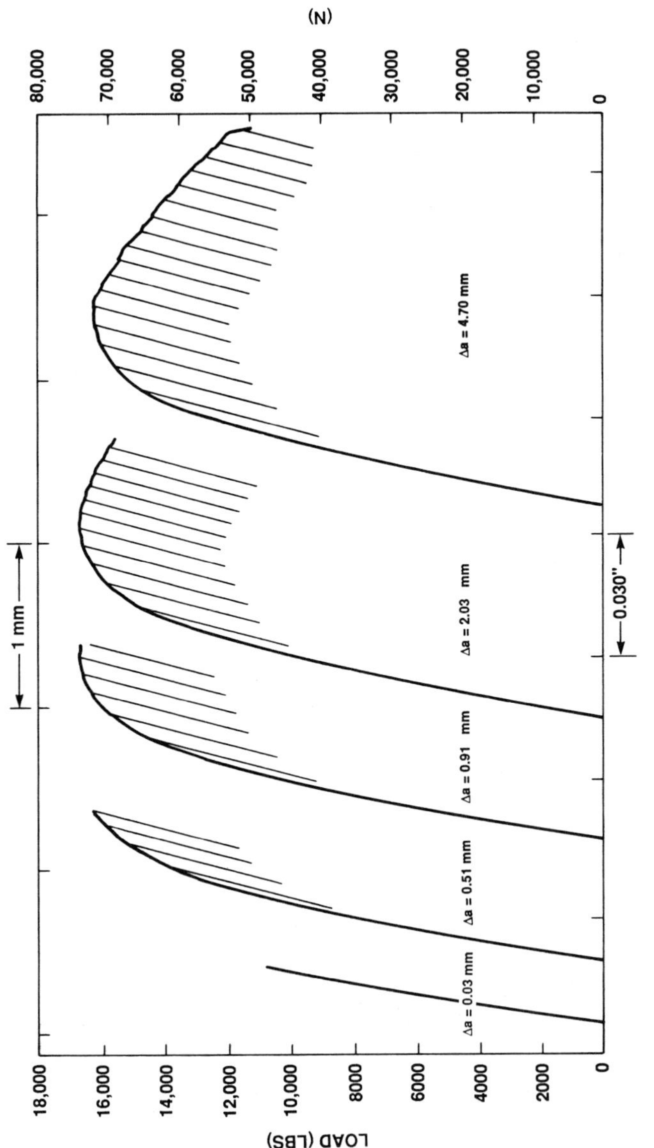

FIG. 4—*Several load-displacement curves from 1T tests at 80°C (175°F) showing the wide range of material behavior observed at this temperature (Δa = extent of ductile crack growth).*

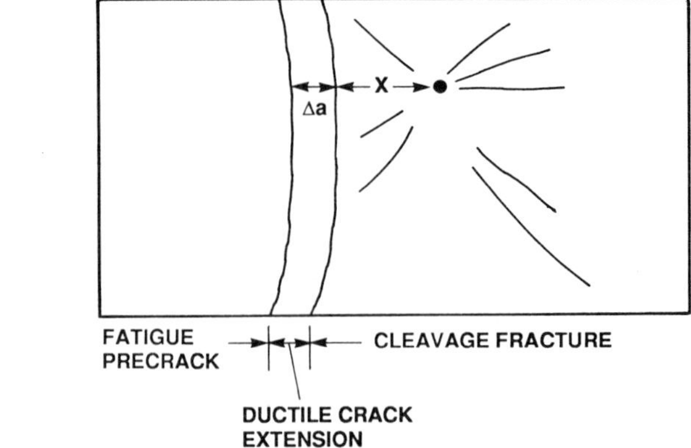

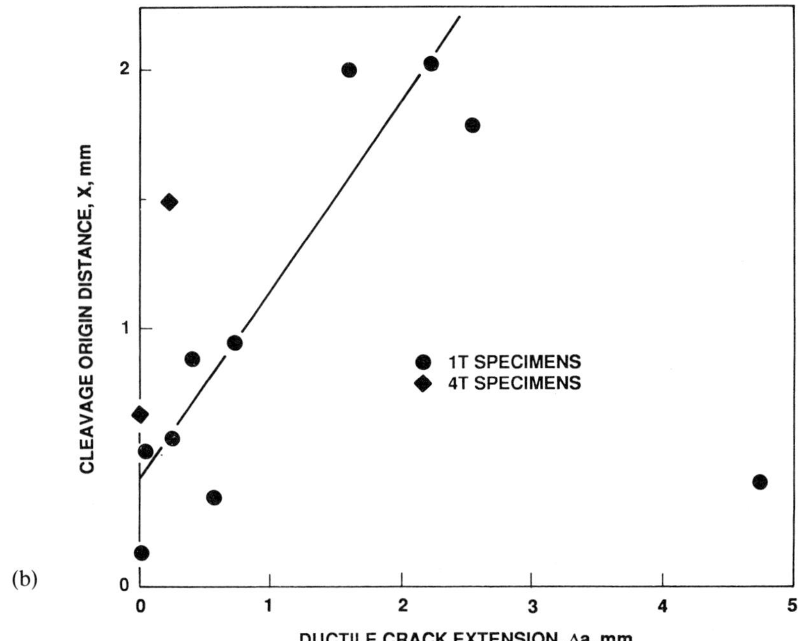

FIG. 5—(a) *Schematic drawing of fractured specimen showing ductile crack extension, Δa, and cleavage origin distance, X. (b) Cleavage origin distance versus ductile crack extension.*

river patterns was required to determine which cleavage event occurred first. Figure 7 is an example of multiple cleavage origins. This is quite common in the 1T specimens, less common in the 4Ts. Cleavage initiated at carbides, sulfides, and, for one specimen, cleavage initiated at a grain boundary. Figures 8, 9, and 10 show examples of cleavage initiation at a carbide, sulfide, and grain boundary, respectively. For several specimens, no distinct microstructural feature is identifiable at the cleavage initiation site. In particular, for Specimen 2-2, cleavage initiated within a grain with no indication of any inhomogeneity in the matrix

TABLE 2—*Fractographic measurements.*

1T Specimens

Specimen Number	Ductile Crack Extension (mm)	Cleavage Origin Distance (mm)	Cleavage Origin
1-1	0.03	0.13	?
1-2	1.59	2.00	inclusion
2-1	2.22	2.03	inclusion
2-2	0.71	0.96	?
3-1	0.59	0.35	grain boundary
3-2	0.39	0.88	inclusion
4-1	4.77	0.40	?
4-2	0.06	0.52	carbide
5-1	0.24	0.58	inclusion
5-2	2.55	1.78	?

4T Specimens

Specimen Number	Ductile Crack Extension (mm)	Cleavage Origin Distance (mm)	Cleavage Origin
1	0.03	0.09	inclusion
2	0	0.78	inclusion
3	0.22	1.48	?
4	0.38	0.31	carbide
5	0	0.66	inclusion

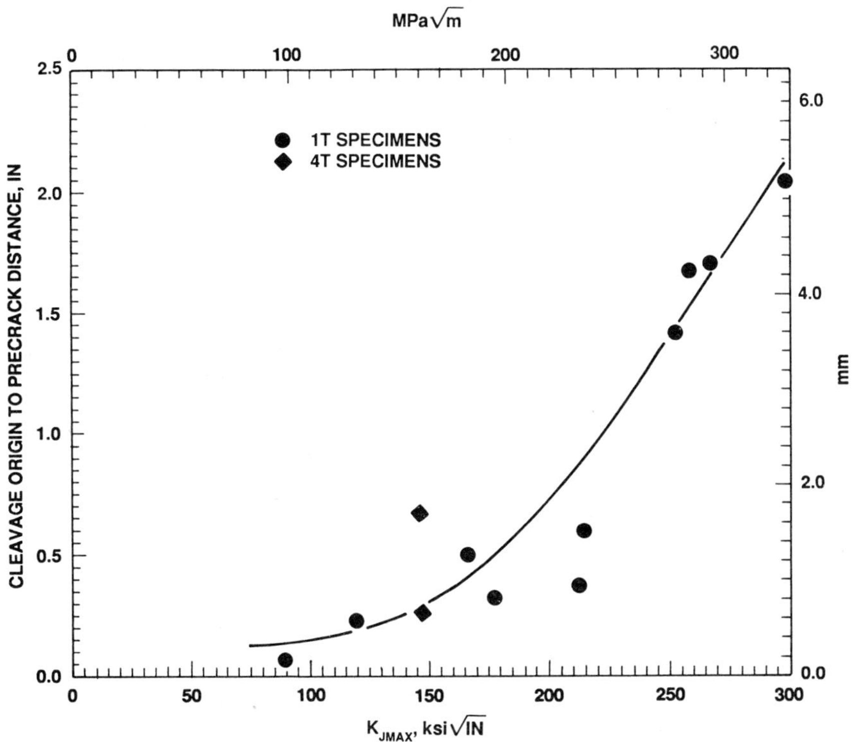

FIG. 6—*Distance from the cleavage origin to the fatigue precrack tip versus* K$_{J_{max}}$.

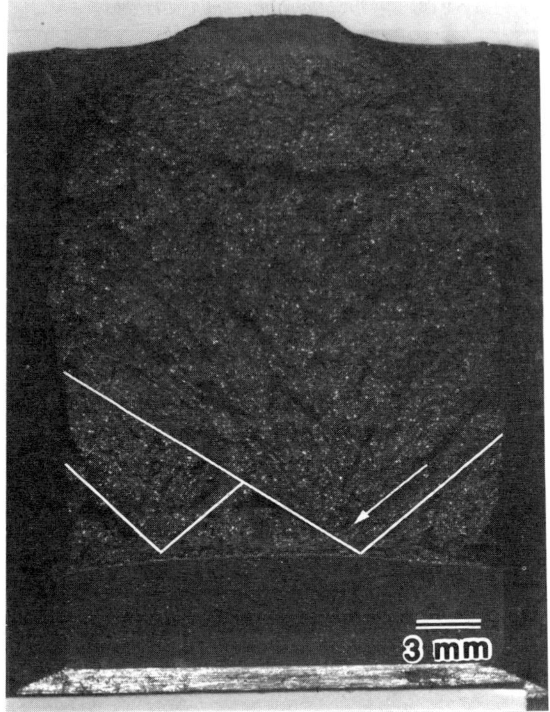

FIG. 7—*Example of a specimen with multiple cleavage origins. The river patterns are traced to show that cleavage initiated first at the arrow.*

at the initiation site (Fig. 11). Rosenfield et al. report that MnS is the most common cleavage origin in the upper transition region for their heat of specially-treated SA508 [7], and the results reported in Table 2 are in agreement.

Two of the 1T specimens were heat tinted after testing and prior to separation of the specimen halves. For Specimen 5-2, cleavage initiated ahead of the ductile crack tip and propagated both away from and back toward the crack tip, as observed in the other specimens, but did not connect with the ductile crack tip. Instead, a ligament of uncracked material separated the cleavage crack from the ductile crack tip (Fig. 12). This may also have occurred in other specimens, but because they were not heat tinted it is not discernible.

Two of the 4T specimens show an influence of precrack irregularities on the location of the cleavage initiation site. Figures 13 and 14 are fractographs from Specimens 2 and 4 showing cleavage initiation near curved portions of the fatigue precracks. For both specimens the measured toughness values are lower than the values measured with Specimens 3 and 5. At

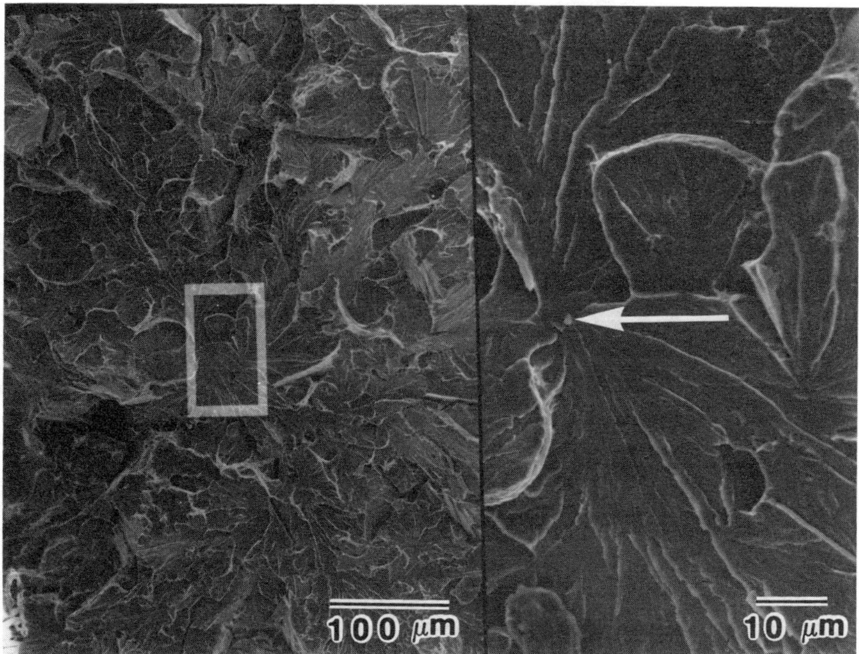

FIG. 8—*Cleavage initiation at a carbide particle in Specimen 4–2.*

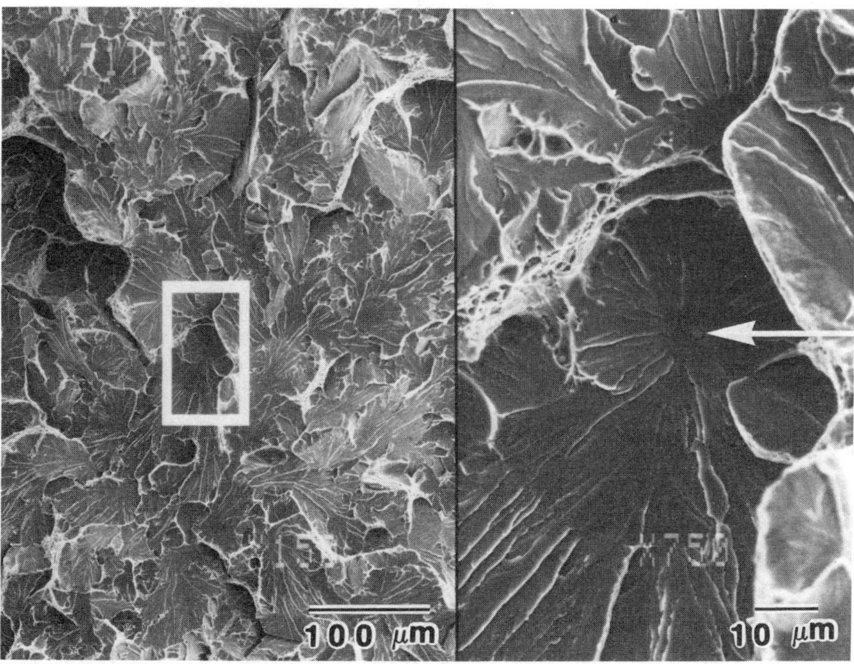

FIG. 9—*Cleavage initiation at a manganese sulfide inclusion in Specimen 2–1.*

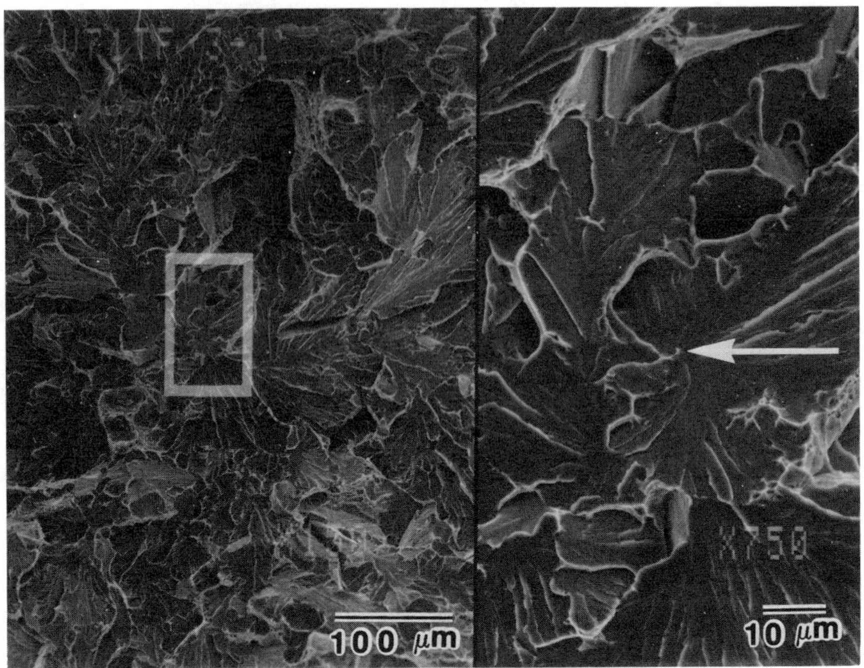

FIG. 10—*Cleavage initiation at a grain boundary in Specimen 3-1.*

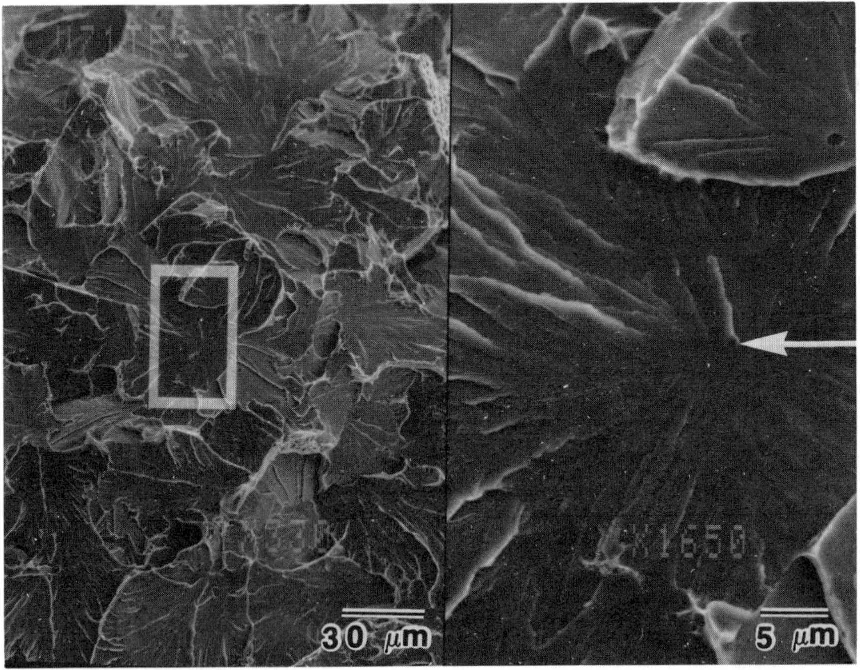

FIG. 11—*Cleavage initiation at the center of a grain in Specimen 2-2.*

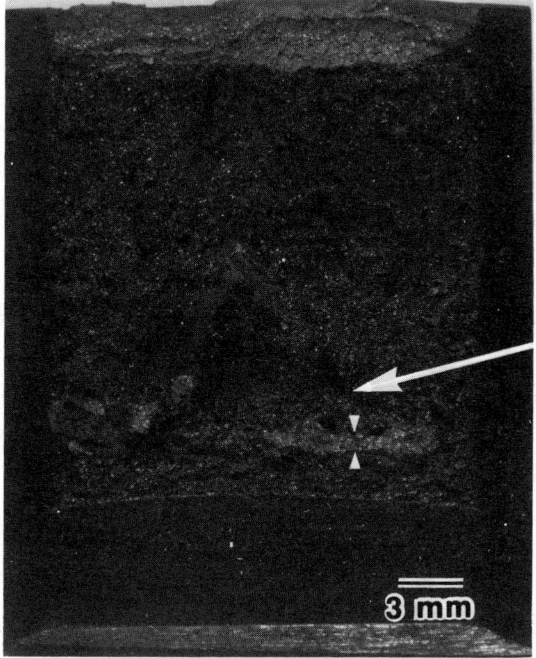

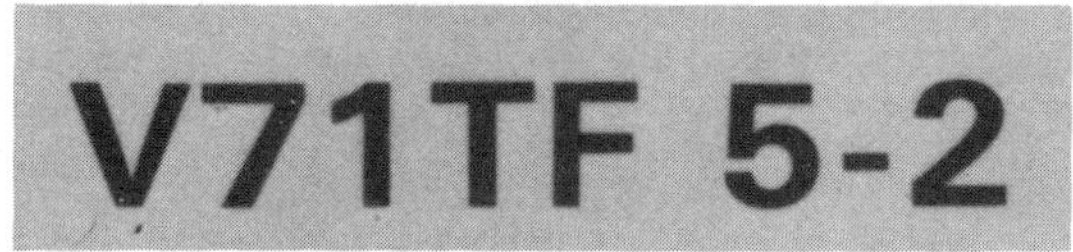

FIG. 12—*Cleavage initiation which did not propagate all the way back to the ductile crack front. Small arrows denote uncracked ligament.*

the precrack irregularity the local stress intensity is higher than the measured stress intensity for the specimen. Therefore the local stress intensity at the cleavage initiation site is higher than the measured stress intensity at failure.

Data Analysis

Several techniques are used for evaluating toughness in the transition region. Discussed below are methods for determining the lower bound toughness at temperatures for which the data scatter is large.

Statistical Methods

Several statistical methods have been suggested for calculating lower bound toughness in the transition region. These methods include Weibull and log-normal statistics, and a statistical technique developed by Iwadate et al. [2] especially for the transition region. The data

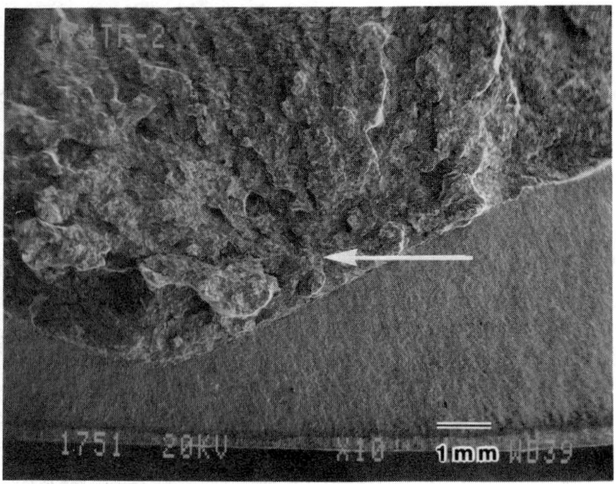
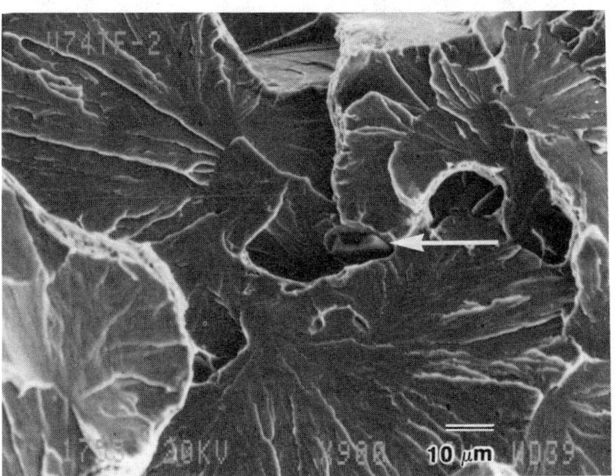

FIG. 13—*Four-inch-thick compact specimen (No. 2) showing location of cleavage origin influenced by irregularity of precrack.*

are presented as $K_{J_{max}}$ values calculated in accordance with ASTM E 813. No changes are made in the $K_{J_{max}}$ calculation; only the distribution in $K_{J_{max}}$ values is considered.

Weibull Analysis

Weibull statistics are used for modeling processes governed by a weakest link phenomenon. Weibull statistics can be used not only to model the distribution of a given data set, but also to explain the difference between data sets obtained with specimens of different sizes. Larger specimens generally yield lower toughness values than smaller specimens when tested in the transition region [1–3]. Landes and Shaffer proposed that the fracture toughness of any specimen is governed by the point or region of lowest toughness along the crack front. Larger specimens contain a larger sampling of low toughness points and therefore exhibit lower measured toughness values [1].

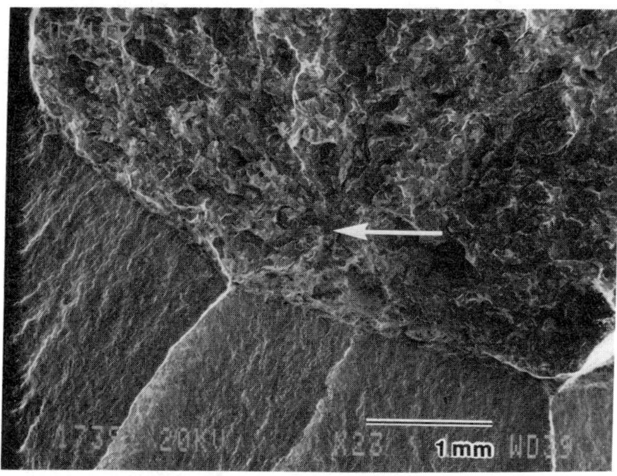
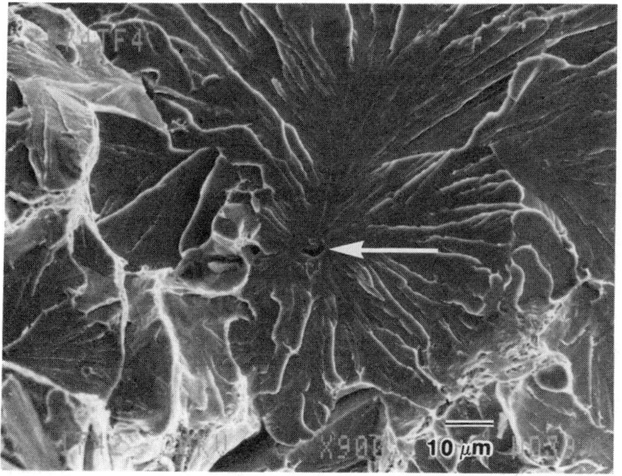

FIG. 14—*Four-inch-thick compact specimen (No. 4) showing location of cleavage origin influenced by irregularity of precrack.*

If a large specimen behaves like a set of equivalent small specimens, as suggested by Landes and Shaffer in Fig. 15, then the probability of survival of a 4T specimen should be the fourth power of the survival probability of a 1T specimen. The Weibull function for a set of 1T data can be represented as:

$$F_1(x) = 1 - \exp -(x/b)^c \qquad (2)$$

where c is the slope of the Weibull distribution, b is the location parameter, and x is the measured toughness. Figure 16 shows the 1T data collected under this program, plus the data from Fig. 1 for 80°C plotted on a Weibull graph. The $K_{J_{max}}$ values for the 80°C in Fig. 1 were calculated using ASTM E 399. In order to make them better comparable to the ten 1T $K_{J_{max}}$ values from this program, which were calculated using ASTM E 813, the Fig. 1 data were recalculated using the ASTM E 399 $K_{J_{max}}$ equation and the crack length at onset of

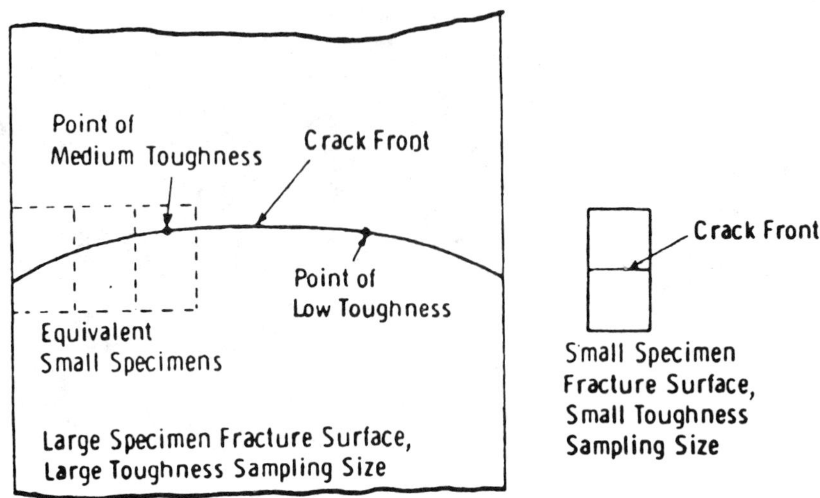

FIG. 15—*Schematic diagram illustrating Landes and Shaffer hypothesis that larger speci-mens yield lower toughness values because of a larger sampling size* [1].

cleavage (crack length at J_{max}), rather than the fatigue precrack length as called for in E 399. The values for b and c were calculated for the 1T data using the maximum likelihood method and used to predict the toughness for a 4T specimen:

$$F_4(x) = 1 - \exp - \left[\frac{4^{1/c}x}{b}\right]^c \tag{3}$$

where the second term in Eq 3 equals $[1 - F_1(x)]^4$ [1]. This method predicts a mean of 171 MPa$\sqrt{m}$ (155 ksi$\sqrt{in.}$) for a set of 4T specimens, which is close to the mean of 161 MPa$\sqrt{m}$ (145 ksi$\sqrt{in.}$) obtained from Specimens 3 and 5 tested in this program.

While the Weibull parameters for the 1T data in Fig. 16 provide a good prediction for the 4T measured toughness, the data do not fit the Weibull distribution as well as might be expected. The straight line in Fig. 15 is a best fit line for the entire data set. The four lowest toughness values fall to the left of the line, and only those four specimens had no macro-scopic ductile crack growth. The remaining specimens underwent visible ductile crack growth prior to cleavage. It is apparent from Fig. 16 that, if only the data from the four specimens without visible ductile crack growth were fitted to the Weibull distribution, a more conservative lower bound would result. Anderson and Stienstra [8] discourage use of Wei-bull analysis for specimens with ductile crack extension prior to cleavage. This appears to be a theoretically sound recommendation in agreement with the results of this program.

Doig [9] suggested that a three-parameter rather than a two-parameter Weibull function be fitted to the data [9]. However, for this data set the two-parameter function provided a better fit to the data. Anderson [10] advises against the use of a three-parameter Weibull analysis, claiming that it yields nonconservative values of lower bound toughness which are artifacts of the curve-fitting procedure.

Wallin [11] also supports the use of a two-parameter Weibull function for modeling tran-sition region data, and predicts a Weibull slope of 4 for K data based on theoretical argu-ments and an extensive literature review. Anderson and Stienstra [8] surveyed the data in

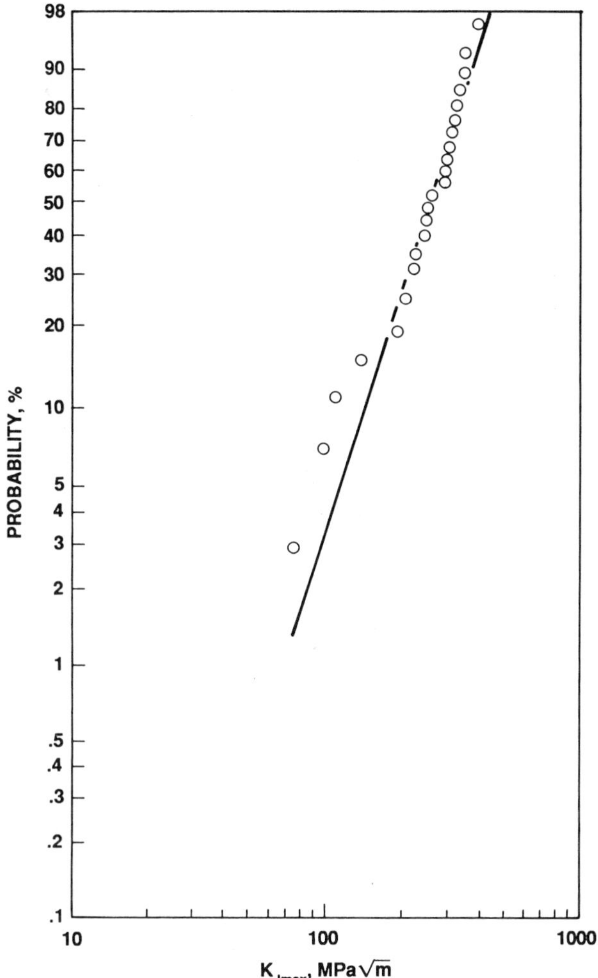

FIG. 16—$K_{J_{max}}$ *data from this program plus 80°C data from Fig. 1 plotted on a Weibull graph.*

the literature and found support for a Weibull slope of 4. A Weibull slope of 3.6 was calculated for the data shown in Fig. 16.

Log-Normal Analysis

Anderson [10] suggests a log-normal fit to transition region data, as shown in Fig. 17 using the same data set as for Fig. 16. For this data set the log-normal fit is not as good as the Weibull fit.

Iwadate Method

The method of Iwadate et al. [2] is based on the results of approximately 300 fracture toughness tests of A508 C13 and an A470 Ni-Cr-Mo-V rotor steel. Iwadate et al. determined

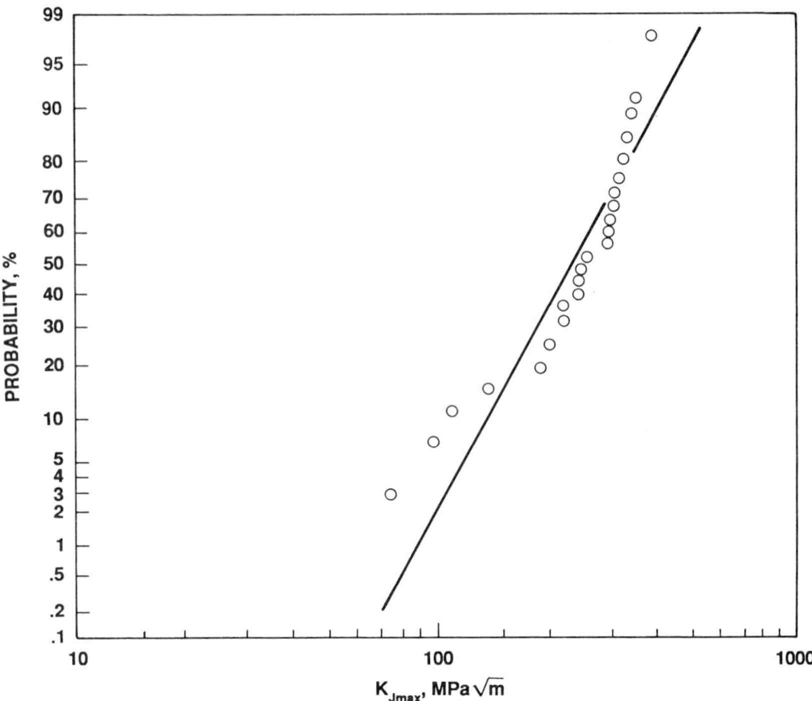

FIG. 17—$K_{J_{max}}$ data from this program plus 80°C data from Fig. 1 plotted on a log-normal graph.

the lower bound toughness value by taking the lowest $K_{J_{max}}$ value from a prescribed number of tests. Iwadate et al. assumed that the toughness of one specimen of thickness N can be modeled by N 1T specimens or, equivalently, $2N$ 0.5T specimens. This is based on the model of Landes and Shaffer [1]. The lower bound toughness value is simply the lowest $K_{J_{max}}$ value from a prescribed number of tests, N, of specimens of thickness B (i.e., a prescribed value of NB). Iwadate et al. determined that the effect of NB on lower bound toughness disappears when

$$NB > 3000\, J_c/\sigma_y \tag{4}$$

where J_c is the lowest measured J_{max} value and σ_y is the yield strength (Fig. 18). For the ten 1T specimens tested in this program, the lowest $K_{J_{max}}$ value is 102 MPa $\sqrt{m}$ (93 ksi $\sqrt{in.}$). The yield strength at 80°C is 603 MPa (88 ksi). According to Eq 4, a minimum of nine 1T specimens is required for 102 MPa $\sqrt{m}$ to be considered a lower bound toughness value at 80°C. The Iwadate lower bound for all the 80°C data shown in Fig. 3 would simply be 75 MPa $\sqrt{m}$ (68 ksi $\sqrt{in.}$), the lowest measured value at that temperature.

Analytical Methods

The following methods of dealing with transition region data scatter involve recalculation of the individual data points to correct for the excessive plasticity occurring in transition

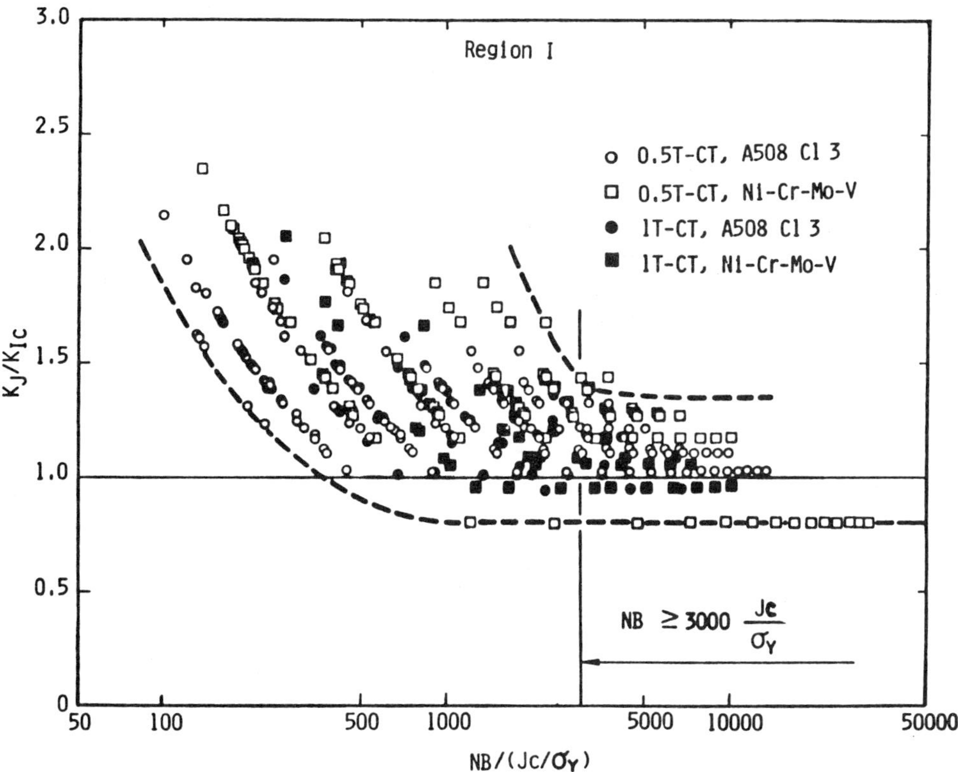

FIG. 18—*Data for A508 C13 and a Ni-Cr-Mo-V rotor steel showing independence of the minimum measured* $K_{J_{max}}$ *from specimen number, N, and size, B, when NB exceeds 3000 J_{max}/σ_y [2].*

region tests. Included are the β_{Ic} constraint correction, which corrects for loss of plastic constraint in small specimens, and the available energy and Watanabe methods, which correct for ductile crack growth. Because the available energy and Watanabe methods require fractographic information, only the ten 1T specimens tested for this program were analyzed using these methods.

β_{Ic} Constraint Correction

The β_{Ic} correction [3] adjusts the K values obtained using small specimens for insufficient through-thickness plastic constraint. The β_{Ic} constraint correction is purely empirical; Merkle [12] used it to predict large specimen results from small specimen tests for several materials in the low temperature portion of the transition region. At higher temperatures where extensive ductile crack growth occurs prior to cleavage, a strain-rate correction is applied in addition to the β_{Ic} constraint correction.

Applied to the ten 1T specimens tested in this program, the β_{Ic} constraint correction reduces the measured values as shown in Fig. 19. The β_{Ic}-corrected 1T data do not predict the 4T results; however, the 4T results are well above the lower bound for the data set. The

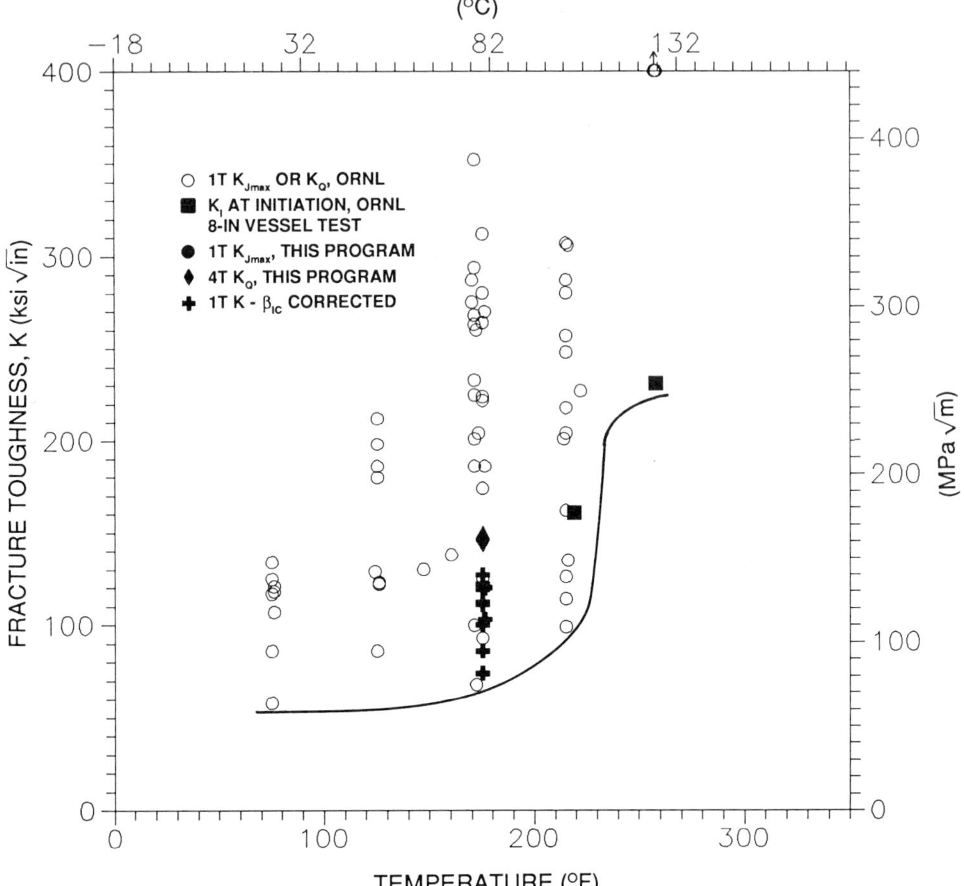

FIG. 19—*Data set from Fig. 3 plus β_{Ic}-corrected data at 80°C.*

β_{Ic}-corrected data show less scatter and lie closer to the lower bound curve than the uncorrected data.

Available Energy

The available energy method developed by Rosenfield and Shetty [4] is based on Seidl's [13] assumption that the toughness available to drive a cleavage crack is the elastic energy stored in the specimen. The available energy method uses the load and crack length at the onset of cleavage fracture to calculate a K_{Ic} value, rather than the initial crack length and the load, P_Q, as defined by ASTM E 399. In this manner, the energy expended in crack tip blunting and ductile tearing is not included in the fracture toughness calculation. Rather, only the elastic energy expended in cleavage fracture contributes to K_{Ic}, as illustrated schematically in Fig. 20.

Available energy estimates of K_{Ic} were made for the ten 1T specimens tested under this program and are shown graphically in Fig. 21. The available energy calculations greatly reduce the data scatter.

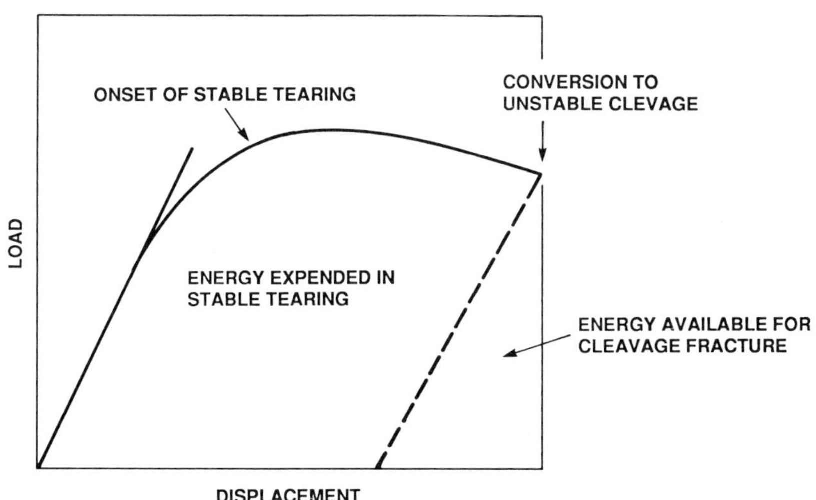

FIG. 20—*Schematic load-displacement curve illustrating available energy calculation* [4] .

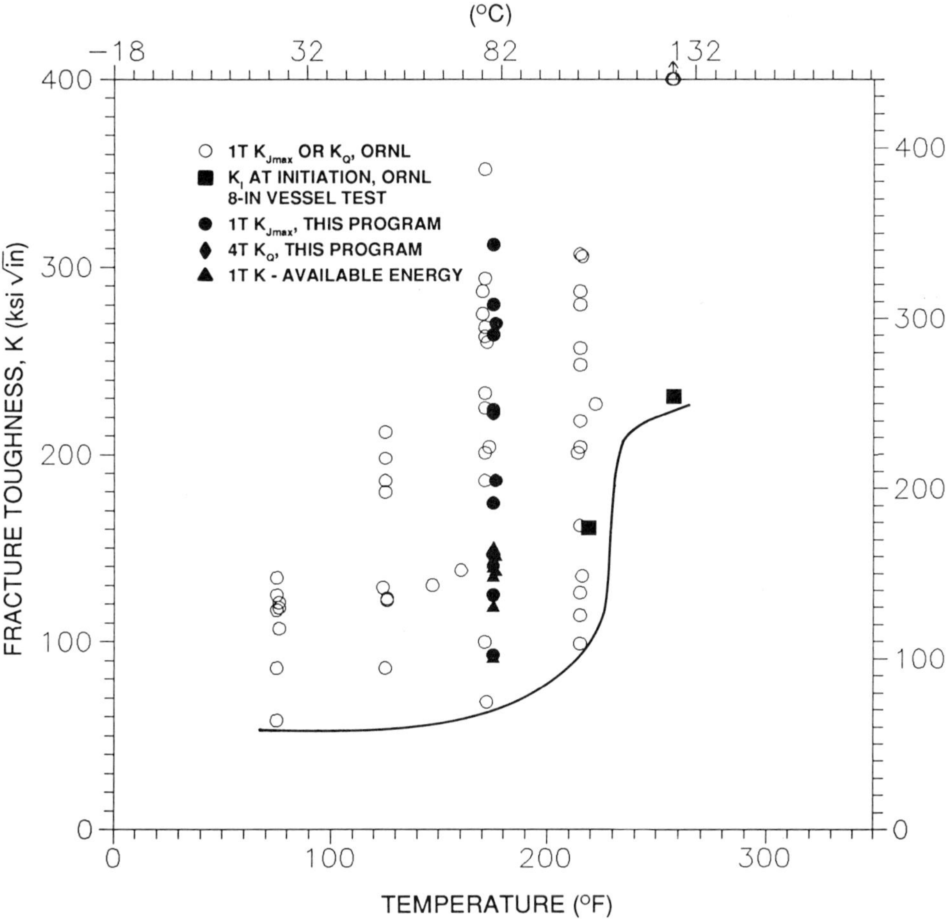

FIG. 21—*Data set from Fig. 3 plus available energy corrections for the ten 1T specimens tested in this program.*

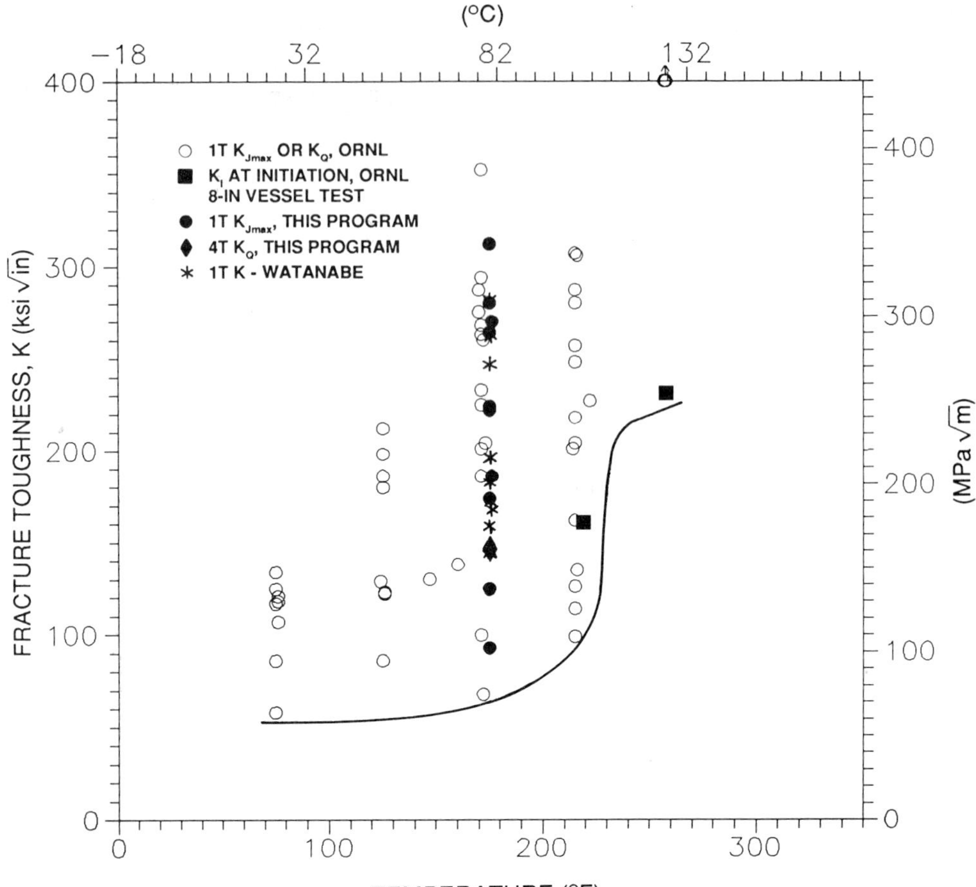

FIG. 22—*Data set from Fig. 3 plus Watanabe corrections for the ten 1T specimens tested in this program*

Watanabe Method

The Watanabe method [5] is based on Weibull analysis of $K_{J_{max}}$ data, after the contribution from ductile tearing to J_{max} is removed. The J_{max} data are fitted to the equation

$$J_{max} = A + B\Delta a + CX \tag{5}$$

where Δa is the ductile crack length and X is the distance from the ductile crack tip to the cleavage initiation site, as shown in Fig. 5a, and A, B, and C are constants. The J_{max} data are recalculated to J_1 values, which is J_{max} minus the ductile tearing energy, according to

$$J_1 = A + C(\Delta a + X) \tag{6}$$

The J_1 data are then fitted to a Weibull distribution. Figure 22 shows the J_1 values calculated according to the Watanabe method for the ten specimens in this program.

Discussion

The following is a discussion of the six methods for dealing with transition region data scatter. They are evaluated based upon their effectiveness in reducing the data scatter and in providing an estimate of lower bound toughness which is conservative and consistent with the remainder of the data set.

Statistical Methods

Statistical methods, including Weibull, log-normal, Iwadate, and any other statistical method which might be used, are not the method of choice for determining lower bound toughness in the transition region. The cleavage resistance in the absence of plastic blunting or tearing should be determined. At temperatures where some test specimens cleave without ductile tearing, it is possible that the structure will cleave without tearing. Therefore, including data from specimens which tear prior to cleavage without correcting for plasticity in the determination of the lower bound toughness is potentially nonconservative. However, in the absence of a proven analytical method of determining the lower bound toughness, statistical analysis is the only option.

Fitting a statistical distribution to a data set, such as a Weibull distribution or a log-normal distribution, is only as good as the size of the data set. A data set that is too small will yield a poor fit, no matter how appropriate the choice of distribution. Fitting the data to a statistical distribution does nothing to reduce the scatter; it simply provides a means for estimating the lower bound. The quality of the lower bound estimate is also only as good as the size of the data set.

The Iwadate method is an improvement over straight statistical methods in that it relies not only on the current data set but on a collection of past data for prediction of the lower bound toughness. The Iwadate method can be improved by expanding the data set with which the fundamental equation was derived, and re-evaluating the constant term in the equation. Because the amount of data scatter varies with test temperature in the transition region, a coefficient which varies with temperature may improve the predictive capability. The Iwadate lower bound for the 10 1T specimens in this program, 102 MPa $\sqrt{m}$, is 36% higher than 75 MPa $\sqrt{m}$, the lowest measured value at 80°C. Differences of this magnitude may render the Iwadate approach unacceptable for some applications, unless the coefficient is re-evaluated.

Analytical Methods

An analytical method for correcting transition region data scatter is preferred, perhaps followed by a statistical method to arrive at a lower bound estimate. Data should be corrected to remove contributions to measured fracture toughness from crack tip blunting and ductile crack growth. Even after these corrections, there will likely be more scatter in the corrected data than the scatter observed on the lower shelf. On the lower shelf, it appears that there are several low-energy sites which initiate cleavage in the specimen nearly simultaneously. In the transition region, cleavage initiation at only one low-energy site can cause the entire specimen to cleave. Thus there is more specimen-to-specimen scatter in the cleavage energy because the initiation sites will vary in type and energy from one specimen to the next.

The analytical methods for dealing with transition region data scatter are the β_{Ic} constraint correction developed by Merkle [3], available energy analysis [4], and the Watanabe method [5]. The β_{Ic} constraint correction reduces data scatter, but by purely empirical means. The β_{Ic} correction is based on an equation developed by Irwin [14] to correct small specimen data for insufficient through-thickness plastic constraint. As such, one would expect the β_{Ic} correction factor to have the same value for specimens of the same thickness, but this is not the case. Wallin [15] analyzed data for specimens of thickness ranging from 2 mm to 100 mm, comparing the size effect predicted by the Landes and Shaffer weakest-link theory [1] with that predicted by the Irwin constraint equation [14]. He found that the weakest-link approach predicted the observed size effect more accurately than did the constraint equation.

The available energy method appears promising because it reduces data scatter by removing the contribution from plastic blunting and tearing from each toughness calculation. However, there is some question whether the elastic stress singularity at the tip of a crack in a small specimen which has undergone macroscopic ductile tearing is representative of the stress singularity at the tip of a crack in a thicker structure which undergoes less plastic deformation in achieving the same K level [16]. The available energy method does not include a procedure for predicting lower bound toughness. A Weibull fit to the available energy corrected data is shown in Fig. 23. As for the Weibull fit to uncorrected data shown in Fig. 16, the data from specimens with no macroscopic ductile crack growth (the two data points at lower left in Fig. 23) fall to the left of the best fit line and, if only data from specimens with no ductile crack growth were used to predict a lower bound, a more conservative prediction would result. Available energy corrected data may better fit a statistical function other than the Weibull function.

The Watanabe method subtracts plastic blunting and tearing from measured toughness, but not on a per specimen basis. Instead, the data from all specimens are combined to determine an equation for plasticity correction of all the data in like manner. The Watanabe method of plasticity correction does not reduce data scatter (Fig. 22), while the available energy method does (Fig. 21). The amount of plasticity appears to depend upon the distance of the cleavage origin from the fatigue precrack. This distance varies widely from specimen to specimen (Fig. 5). Therefore, it is preferable to correct each datum separately for plasticity, as by the available energy method.

Of the three analytical methods for reducing data scatter and predicting lower bound toughness, the available energy method, combined with statistical analysis to identify a lower bound, appears to be the most logical. It is not perfect, however, but analytical correction for ductile crack growth has its limits. Further improvements in lower bound toughness prediction in the transition region may require experimental suppression of ductile crack growth, perhaps by dynamic loading.

Of these six methods for reducing data scatter in the transition region and determining a conservative value of lower bound toughness, the best statistical method appears to be the Iwadate method, while the best analytical method appears to be the available energy method combined with statistical analysis to identify the lower bound. The advantages of each are that the Iwadate method is based on a broad data base, and the available energy method corrects each datum separately for energy contributions from crack tip blunting and tearing. The shortcomings of each are that the Iwadate method would benefit from re-evaluation of the constant term in the equation to avoid potentially nonconservative lower bound predictions, and that the available energy method may not be completely correct in making an elastic calculation of fracture toughness for specimens which have undergone ductile tearing.

Conclusions

Based on the above reported tests and analysis of SA508 C12 in the ductile-to-brittle transition region, the following conclusions can be drawn:

1. Fracture of statically loaded compact specimens of SA508 C12 tested in the transition region generally occurs by ductile tearing followed by cleavage.

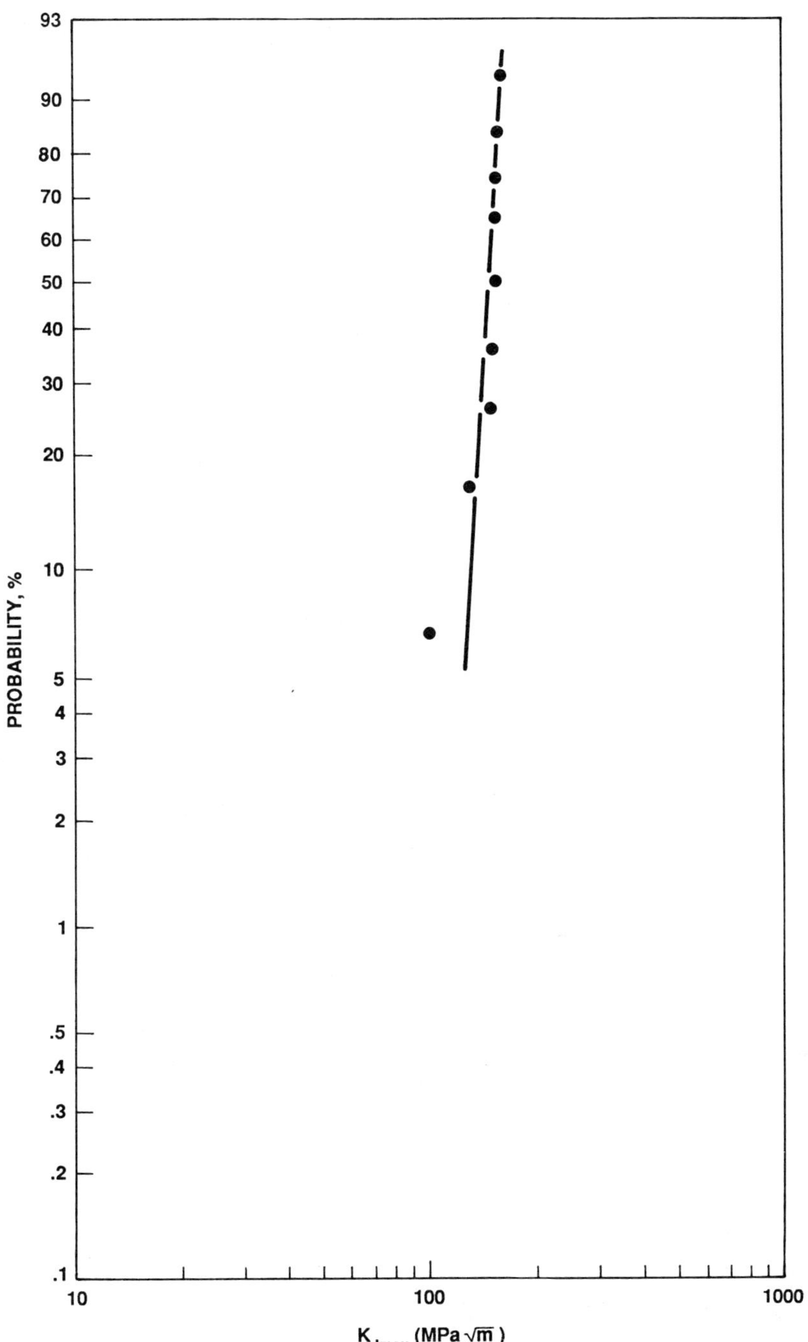

FIG. 23—*Weibull graph of available energy-corrected data.*

2. Cleavage may initiate at only one or at several sites ahead of the crack tip.

3. Cleavage can initiate at various metallurgical features and second phases.

4. Weibull and log-normal fits to uncorrected transition region toughness values are poor because of plastic blunting and tearing prior to cleavage.

5. Of the three purely statistical methods of determining lower bound toughness in the transition region—Weibull analysis, log-normal analysis, and the Iwadate method—the Iwadate method is preferred because it is based upon a large data set. It can be improved by re-evaluation of the constant term in the equation to avoid potentially nonconservative lower bound predictions.

6. Of the three analytical methods for reducing data scatter and determining lower bound toughness in the transition region—the β_{Ic} constraint correction, available energy analysis, and the Watanabe method—the available energy method is preferred because it corrects each result individually for plasticity and ductile crack growth. It can be improved by including a statistical means for determining lower bound toughness values.

Acknowledgments

The authors would like to acknowledge the support of L. A. Oberjohn of Babcock & Wilcox for assistance with statistical analysis, and of S. K. Schillig for preparation of the manuscript. Discussions with and comments by Dr. A. R. Rosenfield of the Battelle Memorial Institute were most helpful and are greatly appreciated.

References

[1] Landes, J. D. and Shaffer, G. H., "Statistical Characterization of Fracture in the Transition Region," in *Fracture Mechanics: Twelfth Conference, ASTM STP 700,* American Society for Testing and Materials, Philadelphia, 1980, pp. 368–382.

[2] Iwadate, T., Tanaka, Y., Ono, S., and Watanabe, J., "An Analysis of Elastic-Plastic Fracture Toughness Behavior for J_{Ic} Measurement in the Transition Region," in *Elastic-Plastic Fracture: Second Symposium, Volume II—Fracture Resistance Curves and Engineering Applications, ASTM STP 803,* C. F. Shih and J. P. Gudas, Eds., American Society for Testing and Materials, Philadelphia, 1983, pp. II-531-561.

[3] Merkle, J. G., "An Examination of the Size Effects and Scatter Observed in Small-Specimen Cleavage Fracture Toughness Testing," ORNL Report TM-9088, NUREG CR-3672, 1984.

[4] Rosenfield, A. R. and Shetty, D. K., "Cleavage Fracture of Steel in the Upper Ductile-Brittle Transition Region," *Engineering Fracture Mechanics,* Vol. 17, No. 5, 1983, pp. 461–470.

[5] Watanabe, J., Iwadate, T., Tanaka, Y., Yokobori, T., and Ando, K., "Fracture Toughness in the Transition Region," *Journal of Engineering Fracture Mechanics,* (in press).

[6] Bryan, R. H. et al., "Pressurized-Thermal-Shock Test of 6-in.-Thick Pressure Vessels. PTSE-1: Investigation of Warm Prestressing and Upper-Shelf Arrest," ORNL Report 6135, NUREG CR-4106, 1985.

[7] Domian, H. A., "Vessel V-7 and V-8 Repair and Characterization of Insert Material," ORNL Report 82-52845-1, NUREG CR-3771, 1984.

[8] Anderson, T. L. and Stienstra, D., "A Model to Predict the Sources and Magnitude of Scatter in Toughness Data in the Transition Region," *Journal of Testing and Evaluation,* (in press).

[9] Doig, P., "Evaluation of Lower-Bound Fracture Toughness Values using Weibull Analysis of Single Specimen Data," *Engineering Fracture Mechanics,* Vol. 21, No. 5, 1985, pp. 963–987.

[10] Anderson, T. L., "A Combined Statistical/Constraint Model for the Ductile-Brittle Transition Region," *Third National Symposium on Nonlinear Fracture Mechanics,* 6–8 Oct. 1986, Knoxville, TN.

[11] Wallin, K., "The Scatter in K_{Ic} Results," *Engineering Fracture Mechanics,* Vol. 21, No. 5, 1985, pp. 963–987.

[12] Merkle, J. G., "Evidence Concerning Crack Tip Constraint and Strain Rate Effects in Fracture Toughness Testing," *Third National Symposium on Nonlinear Fracture Mechanics,* 6–8 Oct. 1986, Knoxville, TN.

[*13*] Seidl, W., "Specimen Size Effects on the Determination of K_{Ic}-Values in the Range of Elastic-Plastic Material Behavior," *Engineering Fracture Mechanics,* Vol. 12, 1979, pp. 581–597.
[*14*] Irwin, G. R., "Fracture Mode Transition for a Crack Traversing a Plate," *Journal of Basic Engineering,* Vol. 82, 1968, pp. 417–425.
[*15*] Wallin, K., "The Size Effect in K_{Ic} Results," *Engineering Fracture Mechanics,* Vol. 22, No. 1, 1985, pp. 149–163.
[*16*] Macdonald, B. D., Knolls Atomic Power Laboratory, private communication, Jan. 1989.

DISCUSSION

Joseph M. Krafft[1] (written discussion)—This excellent paper, as well as those in this session by Anderson and Stienstra and by Moreland, deals with the excessive $K_{Ic}/J_{Ic}(T)$ data scatter, generally regarded as a statistically random effect. Some years ago, this discusser and NRL associates were surprised by the large scatter in K_{Ic} values of an X-52 line pipe steel loaded at varied speeds in search of K_{Ic} minima (*ASME J. Engr. Ind.,* B92;330, May 1970). It appeared eventually that the scatter could be a consequence of an inherent sensitivity of K_{Ic} to crack loading rate dK/dt: a strong, complex, yet deterministic pattern of $K_{Ic}(\dot{K})$. Since the X-52 tests (1965) preceded establishment of the ASTM E 399 K_{Ic} test method, the evidence of a patternal behavior was regarded cautiously.

Subsequently, however, two major tasks of the HSST (Heavy Section Steel Technology) Program were carried out seeking dynamic K_{Ic} minima. Large specimens of A553 Class B steel at near-ambient temperatures generally met ASTM E 399 size requirements. Both studies showed large variation of K_{Ic} as a function of $\dot{K}$. The data of Crosley and Ripling (*ASME J. Basic Engr.,* 91;525, Sept. 1969) were found to be in remarkable correspondence with our X-52 data, compared both graphically and by the autocorrelation statistical analysis of Hettche (*Met. Trans.,* 2; 1905, July 1971). The even more monumental effort of Shabbits, using specimens as large as 8T, was never published except as an internal-to-HSST Technical Report (#6, Dec. 1969). Among 43 total tests, Shabbits produced six duplicate as well as one triplicate $K_{Ic}(\dot{K})$ test points of nil scatter. I feel that the evidence now at hand indicates $\dot{K}$ sensitivity as the most likely source of K_{Ic} transitional scatter. Its further study could prove invaluable.

M. T. Miglin et al. (authors' closure)—It is the authors' conclusion that the data scatter observed in these static tests is primarily a result of variations in the amount of ductile tearing prior to cleavage initiation. The authors are currently engaged in a program investigating the effects of strain rate on data scatter in the ductile-brittle transition region. Preliminary results show that increasing the loading rate decreases data scatter from that observed in static tests, but that large increases in loading rate are required before scatter is significantly reduced. Increasing loading rate promotes cleavage directly from the fatigue precrack, without the prior ductile tearing which produces high measured fracture energy. The references provided by Dr. Krafft will assist in this program and are greatly appreciated. The sensitivity of K_{Ic} to $\dot{K}$ will also be investigated.

---

[1] Consultant (formerly of NRL, Washington, DC), Alexandria, VA 22302.

David Lidbury[1] *and Peter Birkett*[1]

Effects of Warm Pre-Stressing on the Transition Toughness Behavior of an A533 Grade B Class 1 Pressure Vessel Steel

REFERENCE: Lidbury, D. and Birkett, P., **"Effects of Warm Pre-Stressing on the Transition Toughness Behavior of an A533 Grade B Class 1 Pressure Vessel Steel,"** *Fracture Mechanics: Twenty-First Symposium, ASTM STP 1074,* J. P. Gudas, J. A. Joyce, and E. M. Hackett, Eds., American Society for Testing and Materials, Philadelphia, 1990, pp. 264–285.

ABSTRACT: It is now well recognized that pre-loading a ferritic steel structure at a temperature above its ductile-brittle transition range may result in an effective increase in toughness on loading at a lower temperature where fracture occurs by cleavage. Of particular significance is the fact that cleavage failure has never been reported in a cracked structure at a low temperature, following pre-loading at a higher temperature, for cases where the applied crack driving force was constant or decreasing with time. This is the basis of the so-called "conservative" warm pre-stressing (WPS) principle. Although this principle is of considerable practical importance, much work is still needed to further understand all aspects of WPS phenomena in order that analysts be better able to assess their effects quantitatively in safety cases.

In the present work, WPS experiments have been carried out using 20-mm-thick, plain-sided compact specimens sectioned from 150-mm-thick A533 Grade B Class 1 pressure vessel steel. The effects of load-unload-cool-fracture (LUCF) and load-cool-fracture (LCF) pre-loading cycles ($K_{WPS} \approx 100$ to 350 MPa $\sqrt{m}$; $T_{WPS} = 25$ or $290°C$) on cleavage fracture resistance have been investigated over a range of temperatures spanning the ductile-brittle transition range.

A particular objective of the work was to investigate the combined effects of WPS and strain ageing on transition toughness behavior; therefore, in addition to WPS experiments, tests were carried out on "virgin" specimens to obtain base-line data and on specimens sectioned from material which had been uniformly pre-strained by 2% at ambient temperature and then aged for 1.25 h at 290°C to obtain control data.

The main results were as follows. The validity of the conservative WPS principle was confirmed: In every test performed, an increase in loading was required to cause cleavage fracture at low temperature following a severe pre-load at a higher temperature. Moreover, in the case of pre-loading carried out at 25°C, conferred WPS margins were observed even for those tests where pre-loading had resulted in ductile tearing ($0 \leq \Delta a \leq 1.0$ mm) under conditions of ligament yielding; and the Case 1 and Case 2 solutions of Chell's Q_1-integral theory of WPS yielded reasonably accurate predictions of final fracture. A deleterious effect of pre-loading at 290°C on effective transition toughness below a level of ≈ 200 MPa $\sqrt{m}$ was demonstrated by a combination of control and LUCF and LCF WPS data, and found to be consistent with the expected embrittling effect of strain ageing. However, all toughness data generated by the study were bounded by the ASME K_{Ic} versus temperature curve indexed to the value of RT_{NDT} ($= -30°C$) appropriate to the material in the non-strain-aged condition.

KEY WORDS: fracture mechanics, *J*-integral, transition fracture, ductile-brittle transition, warm pre-stressing, strain ageing, pressure vessel, A533 Grade B Class 1 steel

---

[1] Section Leader and Research Officer, respectively, Engineering Metallurgy Group, Northern Research Laboratories—Risley, United Kingdom Atomic Energy Authority, Risley, Warrington, Cheshire, WA3 6AT, U.K.

Nomenclature

A	Total work evaluated from the area under a specimen load versus load-point displacement curve
A_{el}	Elastic component of work evaluated from the area under a specimen load versus load-point displacement curve
A_{pl}	Plastic component of work evaluated from the area under a specimen load versus load-point displacement curve
A_{WPS}	Work associated with warm pre-stressing, evaluated from the prescribed area under a specimen load versus load-point displacement curve
A_f	Work associated with final fracture following warm pre-stressing, evaluated from the prescribed area under a specimen load versus load-point displacement curve
A'	Pre-exponential coefficient in two-parameter curve fit to base-line cleavage toughness versus temperature data
a	Crack size
a_0	Initial crack size prior to growth
Δa	Crack size change
B	Specimen thickness
B'	Exponential coefficient in two-parameter curve fit to base-line cleavage toughness versus temperature data
b	Uncracked ligament size
b_0	Initial uncracked ligament size
c	Uncracked ligament half size
E	Modulus of elasticity
J	J-integral
J_R	Value of the J-integral corrected for ductile crack growth
J_c	Critical value of J-integral associated with cleavage failure in base-line fracture toughness data
J_{WPS}	J-based parameter associated with A_{WPS}
J_f	J-based effective toughness parameter associated with A_f
K	Elastic stress intensity factor
K_{max}	Maximum value of elastic stress intensity factor associated with specimen fatigue pre-cracking
ΔK	Elastic stress intensity factor range associated with specimen fatigue pre-cracking
K_Q	Non-valid linear-elastic toughness associated with cleavage failure in base-line fracture toughness data
K_J	Non-valid elastic-plastic toughness calculated from J
K_{Jc}	Non-valid elastic-plastic toughness calculated from J_c
K_{crit}	Non-valid toughness based on a two-parameter curve fit to the variation of K_Q, K_{Jc} with temperature
$K_{crit}(\text{pred})$	Predicted toughness corresponding to K_{crit}
K_{WPS}	Elastic-plastic stress intensity factor calculated from J_{WPS}
$K_{WPS, el}$	Elastic stress intensity factor corresponding to K_{WPS}
$K_f (\equiv K_f(\text{obs}))$	Elastic-plastic stress intensity factor calculated from J_f
$K_f(\text{pred})$	Predicted effective toughness corresponding to $K_f(\text{obs})$
P_{app}	Applied load
P_{WPS}	Load associated with initial step in WPS cycle

P_L	Limit load
P_f	Load associated with final step in WPS cycle
RT_{NDT}	Reference nil-ductility transition temperature
S_1	Plastic zone size associated with initial loading step in LUCF and LCF WPS cycles
S_2	Compressive plastic zone size associated with unloading step in LUCF WPS cycles
S_3	Plastic zone size associated with final loading step in LUCF and LCF WPS cycles
T	Temperature
ΔT	Temperature shift relative to a base-line toughness versus temperature curve
T_{WPS}	Temperature associated with the initial loading step in a WPS cycle
T_f	Temperature associated with the final loading step in a WPS cycle
W	Specimen width
X_0	Microstructural scale parameter in the Ritchie, Knott and Rice local criterion for cleavage fracture
α	Derived parameter used in J- estimation procedure $(=f(2a_0/b_0))$
σ_f^*	Critical stress in the Ritchie, Knott and Rice local criterion for cleavage fracture
η	Derived parameter used in J-estimation procedure $(= f(\alpha))$
σ_{ys}	Yield strength, or 0.2% offset yield strength
σ_u	Ultimate strength
σ_y	Effective yield strength $(= (\sigma_{ys} + \sigma_u)/2)$
$\sigma_y(T_{WPS})$	Effective yield strength at $T = T_{WPS}$
$\sigma_y(T_f)$	Effective yield strength at $T = T_f$
ν	Poisson's ratio

Introduction

The potential benefits of warm pre-stressing (WPS) in relation to the structural integrity assessment, testing, and operation of large ferritic steel pressure vessels are now well recognized [1]. These benefits have been deduced mostly from simple tests on pre-cracked specimens [2] and large-scale thermal shock and pressurized thermal shock experiments and associated analyses [3–5].

A typical laboratory demonstration would involve pre-loading a specimen at a temperature where toughness behavior was on the ductile upper-shelf, unloading and cooling to a temperature below the ductile-brittle transition range, and then re-loading until brittle fracture occurred. An effective increase in toughness would be found, provided the level of WPS was to a higher value of crack driving force than that corresponding to the toughness at the lower temperature. An increase would also be found if only partial or zero unloading took place, or indeed if unloading and cooling took place simultaneously. The bounds of possible behavior described above are delineated experimentally by the LUCF (load-unload-cool-fracture) and LCF (load-cool-fracture) WPS cycles shown schematically in Fig. 1.

The observation of benefits associated with LCF cycles in particular has led to the "conservative" WPS principle being propounded as follows [6]. Following an initial pre-load at an elevated temperature, fracture of a cracked specimen or structure will not occur at a lower temperature, provided the applied crack driving force remains constant or decreases with time. It is pertinent to note the mechanistic justification for this. Immediately following WPS, a state of stress equilibrium exists due to non-linear elastic-plastic deformation processes. The material yield strength increases with falling temperature, and so the equilibrium is maintained. Since localized crack-tip plastic flow is a necessary precursor to brittle (cleav-

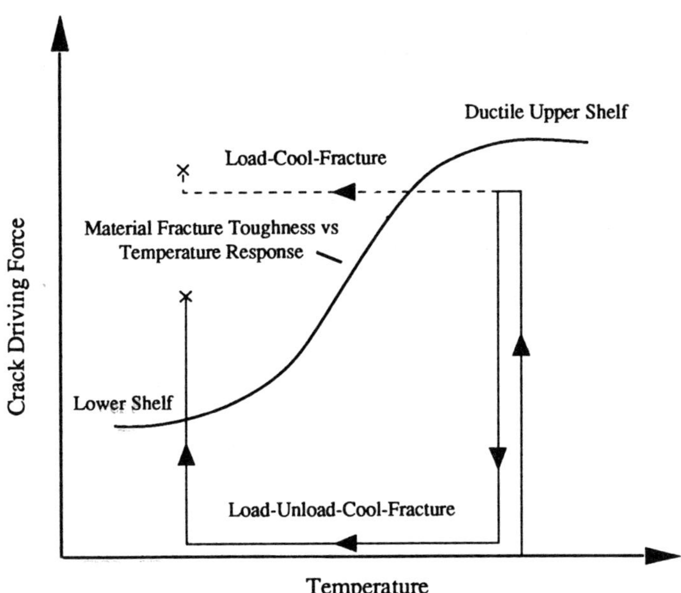

FIG. 1—*Schematic illustration of warm pre-stressing benefits for a ferritic material: LUCF (load-unload-cool-fracture) and LCF (load-cool-fracture) WPS cycles.*

age) fracture, this separation process cannot occur at the lower temperature so long as the level of loading is not increased. Implicit within this brief description is the key physical concept upon which current predictive theories of WPS phenomena are based [7–13], namely, the creation of a zone (or zones) of residual crack-tip plasticity following WPS due to a reduction in temperature, reversed-yielding with unloading, or a combination of these factors.

Warm pre-stressing effects are of particular interest to the nuclear industry and, within this context, have been discussed in detail in relation to light water reactor (LWR) pressure vessels subject to the postulated large loss-of-coolant-accident (LOCA) [1,14]. Specifically, the conservative WPS principle may be invoked with advantage for this particular design basis transient—especially in relation to the initiation behavior of shallow, surface-breaking defects located in the vessel belt-line region. Possible benefits with respect to pressurized thermal shock (PTS) transients (where some reloading occurs following WPS) have also been intimated [13], although at present there is considerable uncertainty in assessing these accurately because of possible complicating factors such as the inherent susceptibility of certain materials with low upper-shelf Charpy energies to pre-cleavage ductile tearing [5].

As noted in Refs 12 and 13, a better understanding of WPS effects and a codified predictive capability are clearly needed before benefits can be argued within the context of failure prediction rather than that of failure avoidance. As a step in this direction, the present paper describes the results of an experimental program which was carried out with the specific aim of advancing understanding of the combined effects of WPS and strain ageing on transition fracture toughness behavior.

Material Details and General Experimental Approach

The material used in this investigation was from a heat of A533 Grade B Class 1 pressure vessel steel, the product form being 150-mm-thick plate. Details of material chemistry and

TABLE 1—*Material chemistry and heat treatment.*[a]

C, wt%	Mn, wt%	Si, wt%	Ni, wt%	Cr, wt%	Mo, wt%	Cu, wt%	S, wt%	P, wt%	Al, wt%	V, wt%	O, wt%	N (mobile), ppm
0.19	1.40	0.20	0.66	0.13	0.49	0.095	0.006	0.010	0.020	<0.010	...	<3.0

[a] Compositions, other than N, obtained from heat analysis.

Heat treatment: (*i*) austenitized 4.8 h at 895°C, then water quenched; (*ii*) tempered 5.3 h at 650°C; (*iii*) stress relieved 25 h at 615°C.

TABLE 2—*Variation of effective yield strength with temperature.*[a]

Temperature (°C)	−150	−125	−100	−75	−50	−25	0	25	290
σ_y (MPa)	827	747	683	631	591	560	535	523	500

[a] Effective yield strength is here defined as the average of the yield strength, or the 0.2% offset yield strength, and the ultimate strength—i.e., $\sigma_y = (\sigma_{ys} + \sigma_u)/2$.

Two tests were performed at each temperature. Specimen gage length = 50.5 mm; cross-sectional area = 50 mm^2.

heat treatment are presented in Table 1. Values of effective yield strength (equal to the average of the yield strength, or the 0.2% offset yield strength, and the ultimate strength—i.e., $\sigma_y = (\sigma_{ys} + \sigma_u)/2$) appropriate to the longitudinal (L) orientation[2] and plate midsection are presented in Table 2 for the temperature range of interest ($-150 \leq T \leq 290°C$).

Plain-sided compact fracture toughness specimens (thickness (B) = 20 mm and width (W) = 40 mm) were extracted in the L-T orientation from midsection regions of the plate. Specimens were fatigue pre-cracked at ambient temperature. During the final stages of pre-cracking, values of maximum stress intensity factor (K_{max}) and stress intensity factor range (ΔK) were ≈ 19 to 21 MPa$\sqrt{m}$ and ≈ 17 to 19 MPa$\sqrt{m}$ respectively. The resultant crack length to width ratios (a_0/W) were in the range 0.51 to 0.53. Specimens were tested in batches using either a screw-driven, servo-electric machine or a servo-hydraulic machine. Displacement control was used throughout all test sequences, except where load-holding was required in respect of LCF WPS cycles. In these latter cases, a transfer to load control was made for the duration of the load-hold period. Testing was mostly carried out at a load-point displacement rate of 1.0 mm min^{-1}; however, as explained fully in the next section, pre-loading in certain tests was carried out at a load-point displacement rate of 0.01 mm min^{-1}.

The crack driving force, and values of toughness equated with critical values of the crack driving force, were established for base-line tests and control tests as follows.[3] Where specimen load versus load-point displacement traces were sensibly linear, values of stress intensity factor (K) and toughness (K_Q) were calculated in accordance with ASTM Test for Plane-Strain Fracture Toughness of Metallic Materials (E 399-83). Where specimen load versus load-point displacement traces were non-linear, values of the elastic-plastic stress intensity factor ($K_J = K(J)$) and cleavage toughness ($K_{Jc} = K(J_c)$) were obtained from values of the *J*-

---

[2] L = longitudinal or rolling direction; T = width or transverse direction.

[3] See Table 3 for distinction between "base-line" and "control" tests.

integral calculated using the expressions [*15*]

$$J = \eta A / b_0 B \tag{1}$$

$$\eta = 2(1 + \alpha)/(1 + \alpha^2) \tag{2}$$

$$\alpha = [(2a_0/b_0)^2 + 2(2a_0/b_0) + 2]^{1/2} - [(2a_0/b_0) + 1] \tag{3}$$

$$K_J = [EJ/(1 - \nu^2)]^{1/2} \tag{4}$$

where $A = A_{el} + A_{pl}$ is the sum of elastic and plastic work evaluated from the area under the load versus load-point displacement curve; E = elastic modulus, ν = Poisson's ratio = 0.3; and $b_0 = W - a_0$ = initial uncracked ligament size.

For tests involving other than monotonic loading, values of the crack driving force associated with pre-loading (K_{WPS}) and final loading (K_f) were calculated in a way similar to that described above for non-linear behavior, but with the values of J_{WPS} and J_f respectively related to the shaded areas A_{WPS} and A_f in Fig. 2a (LUCF WPS cycles) or Fig. 2b (LCF WPS cycles).

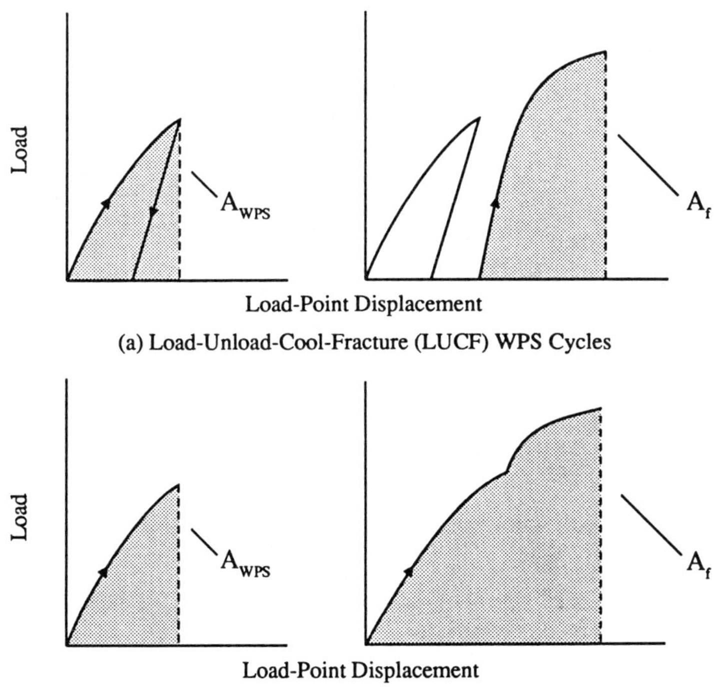

FIG. 2—*Schematic illustration of the areas under load versus load-point displacement curves used in the calculation of* J_{WPS} *and* J_f *for* (a) *LUCF WPS cycles and* (b) *LCF WPS cycles.*

Results

A summary of the various compact specimen tests carried out is presented in Table 3.

The results of base-line tests carried out on "virgin" specimens in the temperature range -150 to $0°C$ ($RT_{NDT} = T_{NDT} = -30°C$) are shown in Fig. 3. In order to compare the results of the WPS experiments with the predictions of Chell's Q_1-integral theory of warm pre-stressing [7–11], it was necessary for the parameters K_{WPS} and K_f (here understood to be associated with cleavage failure) to be normalized with respect to the value of toughness ($K_{crit} = K_Q$, K_{Jc}) appropriate to the temperature of final loading. To this end, a least squares two-parameter exponential curve fit of the form $K = A' \exp(B'T)$, with A' in MPa$\sqrt{m}$ and B' in $°C^{-1}$, was performed to describe the variation with temperature of mean toughness for which K_Q, $K_{Jc} < 300$ MPa$\sqrt{m}$. However, the resulting equation

$$K_{crit} = 309 \exp(0.0123T) \tag{5}$$

was not meant to describe the rapid upswing in toughness that occurred with temperatures greater than $-25°C$. It should be noted that the above treatment of base-line data has the virtue of being broadly consistent with the J-measurement capacity of the 20-mm-thick compact specimens (Fig. 3). Also shown in Fig. 3 is a calibration based on the amount of ductile crack extension associated with a given test result (K_{Jc} or K_J). This calibration justified the omission of a crack growth correction in estimating the J-integral for cases where $\Delta a \leq 1.0$ mm ($1.0 \geq J_R/J \geq 0.96$ and K_{Jc}, $K_J < 400$ MPa$\sqrt{m}$). Strictly speaking, a crack growth correction was necessary for cases where $\Delta a \geq 2.0$ mm ($J_R/J \leq 0.93$ and K_{Jc}, $K_J > 500$

TABLE 3—*Summary of compact specimen tests.*[a]

Test Code	Test Type	Nominal Pre-Load Level (K_{WPS}), MPa$\sqrt{m}$	Load-Point Displacement Rate During Pre-Load, mm min^{-1}	Temperature During Pre-Load, °C	Number of Tests
0	Base-line	...	...	...	29
0/SA	Control	...	...	...	18
1A	WPS, LUCF	100 (0.73)	1.0	25	2
1B	WPS, LUCF	200 (0.90)	1.0	25	19
1C	WPS, LUCF	250 (0.92)	1.0	25	2
1D	WPS, LUCF	300 (0.95)	1.0	25	2
1E	WPS, LUCF	350 (0.97)	1.0	25	1
2A	WPS, LCF	100 (0.75)	1.0	25	2
2B	WPS, LCF	200 (0.91)	1.0	25	7
2C	WPS, LCF	250 (0.91)	1.0	25	2
2D	WPS, LCF	300 (0.94)	1.0	25	2
2E	WPS, LCF	350 (0.97)	1.0	25	1
1/B/HT	WPS, LUCF	200 (0.95)	1.0	290	21
1/B/HT/LR	WPS, LUCF	200 (0.95)	0.01	290	18
2/B/HT/LR	WPS, LCF	200 (0.95)	0.01	290	6

[a] Control tests were carried out on specimens sectioned from material which had been uniformly pre-strained by 2% at ambient temperature and then aged for 1.25 h at 290°C.

A figure in parentheses is the average value of the ratio, P_{WPS}/P_L, for a given nominal pre-load level. The limit load is calculated from the relationship $P_L = 1.455\alpha b_0 B \sigma_y$ where $\alpha = [(2a_0/b_0)^2 + 2(2a_0/b_0) + 2]^{1/2} - [(2a_0/b_0) + 1]$ and $a_0/W = 0.52$.

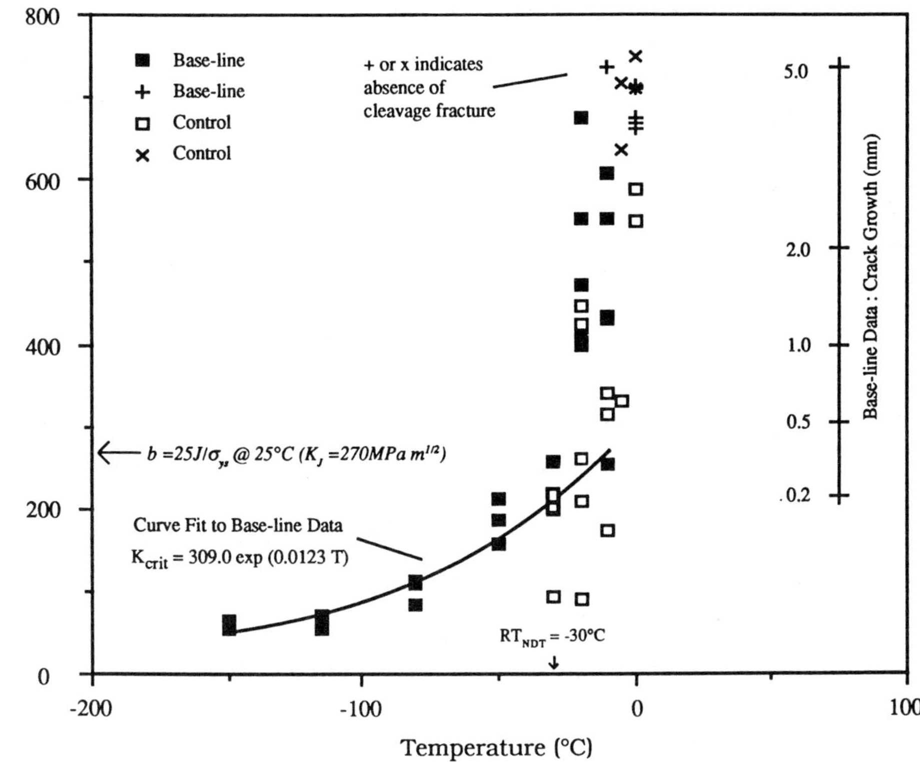

FIG. 3—*Fracture toughness versus temperature response: base-line and control data.*

MPa $\sqrt{m}$); however, for the purposes of comparison, there was little to be gained from this, since the most significant variations in WPS and control data occurred at toughness levels ≤ 300 MPa $\sqrt{m}$.

Also plotted in Fig. 3 are control data for the temperature range -30 to $0°C$. As noted in Table 3, control tests were conducted on specimens sectioned from material which had been uniformly pre-strained by 2% at ambient temperature and then aged for 1.25 h at 290°C. Figure 3 clearly shows that the distribution of control data points extends below that of base-line data points within the same temperature range. Moreover, two of the control data points lie well below the fitted curve to base-line data. Overall, this is consistent with the expected embrittling effect of strain ageing on transition toughness. Above a toughness level of ≈ 200 MPa $\sqrt{m}$, however, control and base-line data (K_{Jc} and K_J) are distributed similarly.

Figures 4 and 5 summarize the results of warm pre-stressing experiments—LUCF and LCF WPS cycles respectively—for which the temperature of pre-loading (T_{WPS}) was 25°C, the specimen load-point displacement rate was 1.0 mm min^{-1} and nominal pre-load levels (K_{WPS}) were in the range of 100 to 350 MPa $\sqrt{m}$. In each figure the fitted curve to base-line data has been drawn for comparison with the observed failure levels (K_f(obs)), and those data where final cleavage fracture did not occur have been identified. There are a number of points to note concerning these figures.

In Fig. 4, WPS benefits are evidence, in the sense that pre-loading is seen to have resulted in an effective increase in cleavage toughness, for those data where the temperature of final loading (T_f) was $\leq -80°C$. However, the indicated margins (judged relative to the curve fit

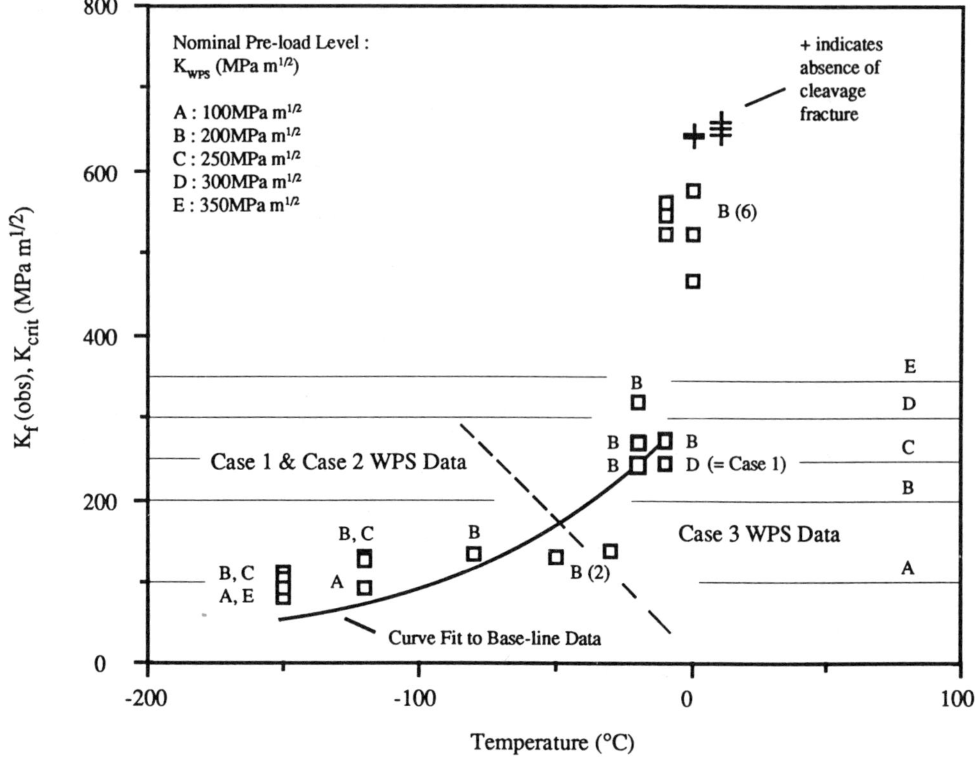

FIG. 4—*Observed failure levels, K_f(obs), in warm pre-stressing experiments: LUCF cycles; $T_{WPS} = 25°C$; load-point displacement rate = 1.0 mm min^{-1}.*

to base-line data) do not increase systematically with pre-loading level. This is almost certainly due to inherent scatter and the limited number of data points involved. Nevertheless, benefits are in evidence for pre-loading levels which were below (Type A), approximately equal to (Type B), and in excess of (Types C, D, and E) the level corresponding to ductile crack initiation (≈ 200 MPa $\sqrt{m}$). This estimation of initiation toughness was derived from the ductile crack growth calibration in Fig. 3, which further shows that a nominal pre-load level of 350 MPa $\sqrt{m}$ (Type E) at 25°C is associated with ≈ 1.0 mm ductile tearing. No WPS benefits are evident for those data where the temperature of final loading was $\geq -20°C$. Here, data are distributed similarly to those base-line data for which K_{Jc}, $K_J \geq 250$ MPa $\sqrt{m}$, and, generally, are associated with significant amounts of pre-cleavage ductile tearing ($1.0 < \Delta a < 5.0$ mm) during final loading.[4] The data relating to the intermediate temperature range ($-80 < T_f < -20°C$) lie below the fitted curve to base-line data; however, in the absence of further results, it is not possible to say whether this is an effect due to pre-loading at 25°C.

In Fig. 5, WPS benefits are shown for those data where the temperature of final loading

---

[4] In the most general case, ductile tearing occurred during WPS and also during final loading, although experimentally it was not always possible to distinguish between the two. Because of this, the amount of tearing associated with a given nominal pre-load level was estimated with reference to the crack growth calibration shown in respect of base-line data in Fig. 3.

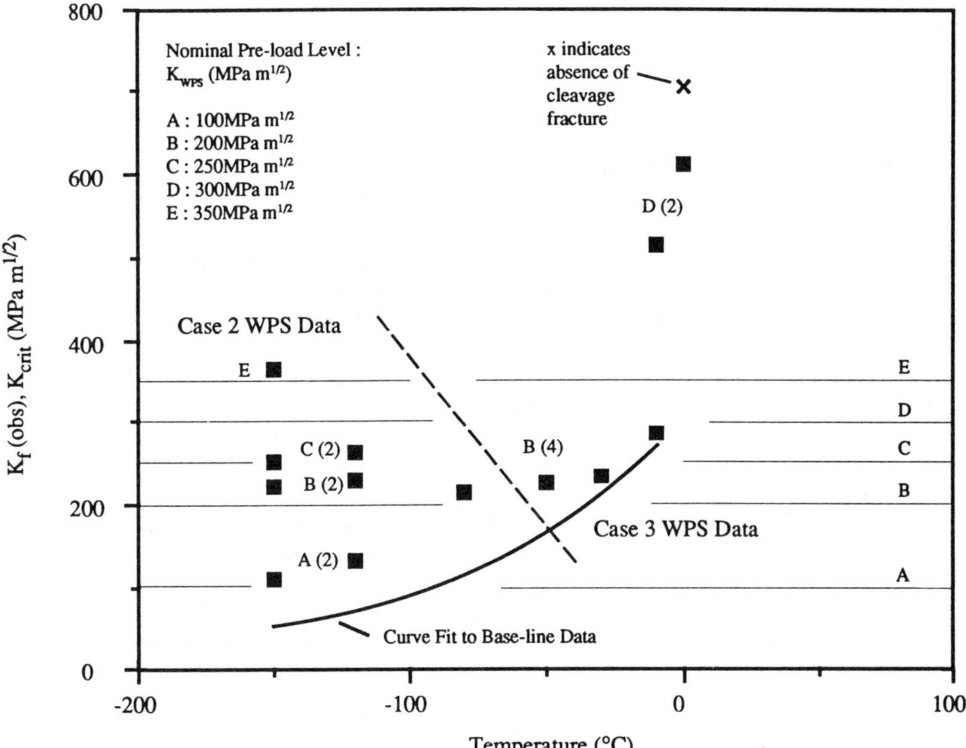

FIG. 5—*Observed failure levels,* K_f(obs), *in warm pre-stressing experiments: LCF cycles;* T_{WPS} = *25°C; load-point displacement rate* = *1.0 mm min^{-1}.*

was $\leq -50°C$. As would be expected in the case of LCF WPS cycles, where the added loading increment required to cause final fracture is small, the increases in effective toughness are directly related to pre-loading level. No WPS benefits are indicated for those data where the temperature of final loading was $\geq -30°C$.

Consistent with the conservative WPS principle, the results presented in Fig. 4 and, in particular, Fig. 5 were such that, in each case, an increase in load was required to cause cleavage fracture following pre-loading and subsequent cooling.

The reference to Case 1, Case 2, and Case 3 data in Fig. 4, and Case 2 and Case 3 data in Fig. 5, follows the classification scheme of Chell's WPS theory [7–11]. Within this context, let S_1 be the plastic zone size associated with pre-loading, S_2 that following unloading, and S_3 that associated with final loading. The zones S_1 and S_2 are zones of residual plasticity, S_1 because of the reduction in temperature following pre-loading, and S_2 because of the reversed-yielding caused by, and the reduction in temperature following, unloading. The three cases are defined as follows.

LUCF WPS Cycles:

$$\text{Case 1:} \quad S_1 > S_2 > S_3 \tag{6a}$$

$$\text{Case 2:} \quad S_1 > S_3 > S_2 \tag{6b}$$

$$\text{Case 3:} \quad S_3 > S_1 > S_2 \tag{6c}$$

LCF WPS Cycles:

$$\text{Case 2:} \quad S_1 > S_3 \, (S_2 = 0) \qquad\qquad (7a)$$
$$\text{Case 3:} \quad S_3 > S_1 \, (S_2 = 0) \qquad\qquad (7b)$$

In Eqs 6 and 7 above, Case 3 corresponds to the final loading step wiping out the residual plasticity associated with previous loading steps. Case 3 data points should therefore be distributed similarly to base-line data points. The truth of this prediction is confirmed by comparing the WPS data in Figs. 4 and 5 with the base-line data in Fig. 3. In particular, it should be noted that those data points associated with specimens which exhibited significant amounts of pre-cleavage tearing during the final loading are all Case 3 data points.

Chell's WPS theory is based upon the principle of plastic superposition [16] and the small-scale yielding solutions of the BCS crack-tip model [17]. As such, it is not strictly applicable to the analysis of WPS test data where significant plasticity was generated during pre-loading (Table 3). Accepting this limitation, it is nevertheless worthwhile comparing the Case 1 and Case 2 WPS test data presented in Figs. 4 and 5 with the theory's predictions. These comparisons are made in Figs. 6 to 8.

The above comparisons were made in two ways. First, a given observed failure level (K_f(obs)) and the relevant values of effective yield strength ($\sigma_y(T_f)$ and $\sigma_y(T_{WPS})$) were used to

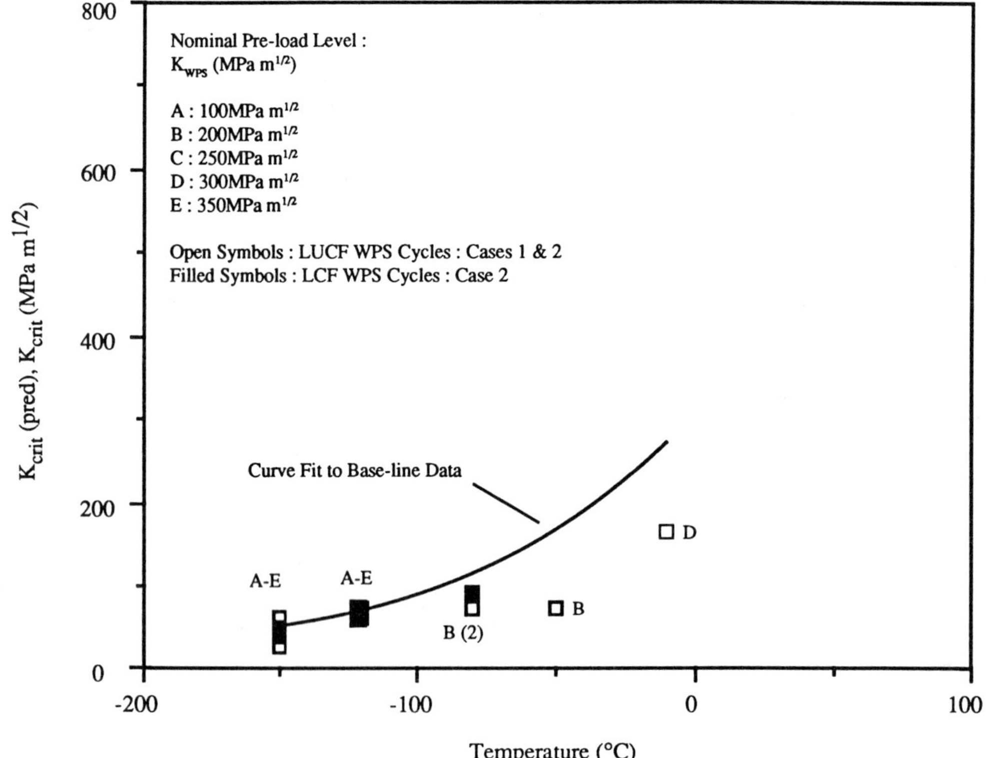

FIG. 6—*Predicted fracture toughness levels for Case 1 and Case 2 WPS data:* T_{WPS} = 25°C; *load-point displacement rate* = 1.0 mm min^{-1}.

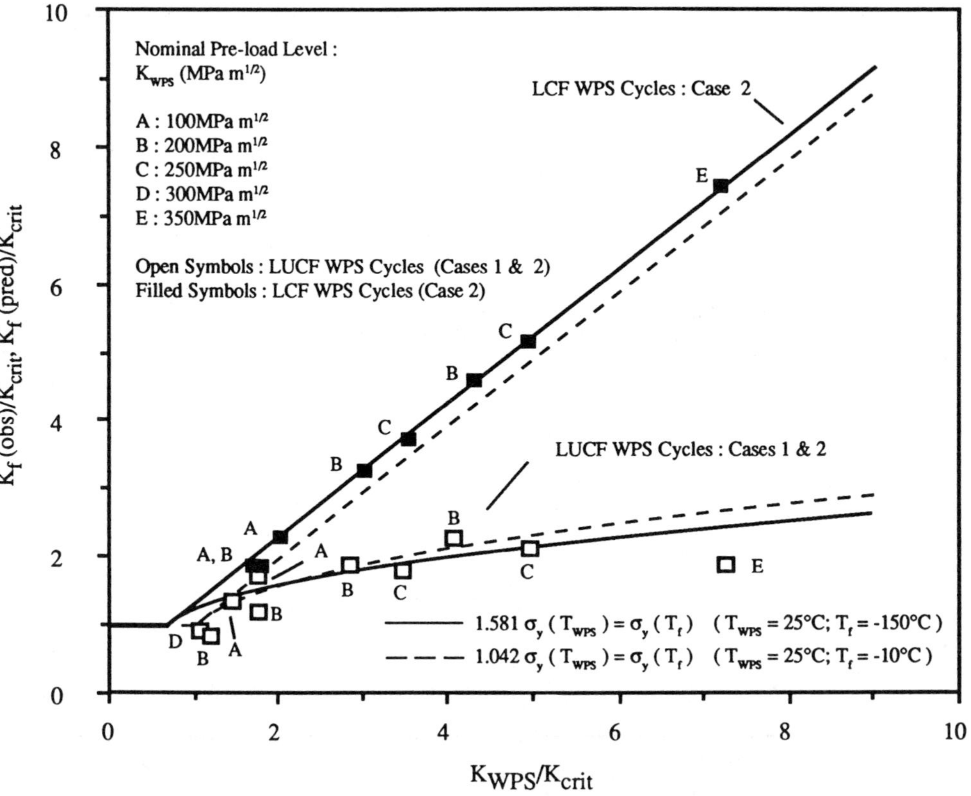

FIG. 7—*Normalized failure levels, K_f(obs)/K_{crit}, versus normalized WPS levels, K_{WPS}/K_{crit}, compared with predictions of Q_I-integral theory [7–11]: LUCF and LCF WPS cycles; T_{WPS} = 25°C; load-point displacement rate = 1.0 mm min^{-1}.*

calculate the toughness at the temperature of final loading (K_{crit} (pred)). Values of K_{crit}(pred) were calculated from an explicit functional relationship of the form

$$K_{crit}(\text{pred}) = F_1 \{K_{WPS}, K_f(\text{obs}), \sigma_y(T_{WPS}), \sigma_y(T_f)\} \tag{8}$$

Predicted toughnesses were then compared with the corresponding toughnesses indicated by the curve fit to base-line data (Fig. 6). Secondly, a given base-line toughness value (assessed in terms of K_{crit}) and the relevant values of effective yield strength were used to calculate the failure level at the temperature of final loading (K_f(pred)). Values of K_f(pred) were calculated by iterative solution of an implicit functional relationship of the form

$$F_2 \{K_f(\text{pred}), K_{WPS}, K_{crit}, \sigma_y(T_{WPS}), \sigma_y(T_f)\} = 0 \tag{9}$$

and then compared with the corresponding observed failure levels (K_f(obs)) (Figs. 7 and 8).

In Fig. 7, both predicted and observed failure levels are normalized with respect to K_{crit} and shown plotted as a function of WPS level, also normalized with respect to K_{crit}. The predicted failure levels, however, are not presented as discrete points, rather they are shown as a series of bounding curves. The solid curves show predictions of the Chell theory for the

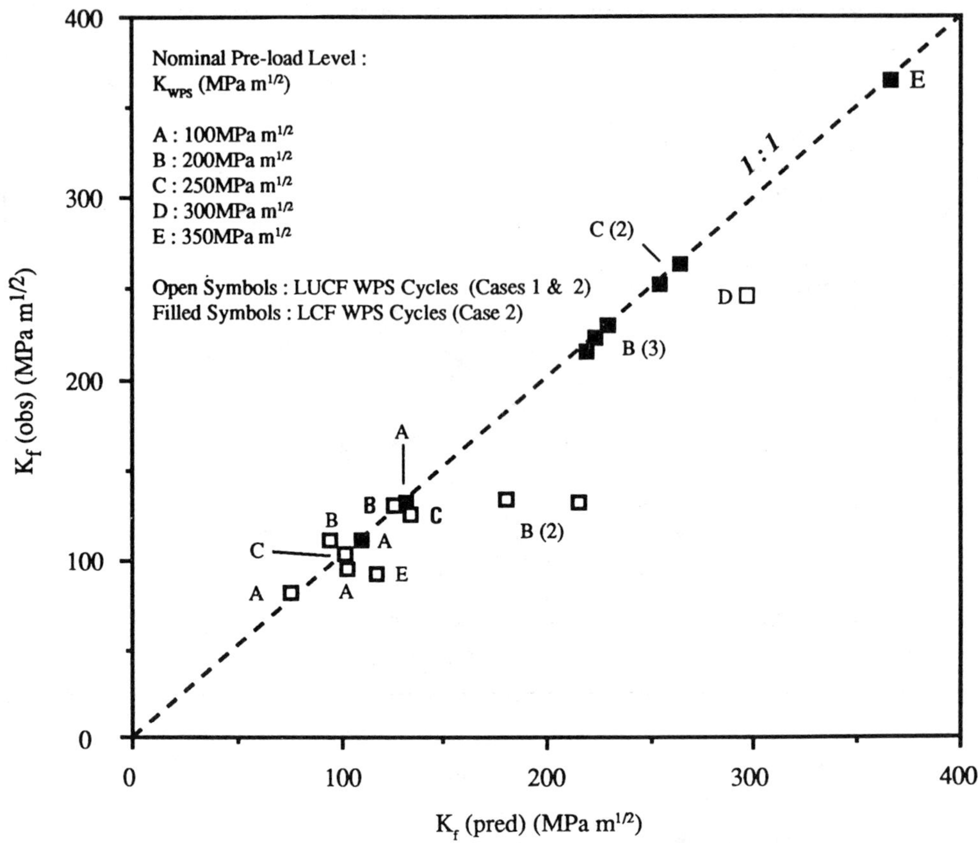

FIG. 8—*Observed failure levels, K_f(obs), compared with predicted failure levels, K_f(pred), calculated from Q_I-integral theory [7–11]: LUCF and LCF WPS cycles; T_{WPS} = 25°C; load-point displacement rate = 1.0 mm min^{-1}.*

lower temperature bounding case where T_{WPS} = 25°C and T_f = −150°C, which corresponds to $1.581\sigma_y(T_{WPS})$ = $\sigma_y(T_f)$. The dashed curves show predictions for the higher temperature bounding case where T_{WPS} = 25°C and T_f = −10°C, corresponding to $1.042\sigma_y(T_{WPS})$ = $\sigma_y(T_f)$. The predictions cover both LUCF WPS cycles (lower solid and dashed curves) and LCF WPS cycles (upper solid and dashed curves). It should be noted that all four curves merge for the Case 3 solution K_f(pred) = K_{crit}. Implicit within the presentation of predicted failure levels in Fig. 7 is acknowledgment of the fact that Case 1 solutions are physically inadmissible in respect of LCF WPS cycles (see Eqs 7a and 7b above).

Figure 8 portrays a more straightforward comparison of predicted and observed failure levels, in which values of K_f(obs) are plotted as a function of the corresponding values of K_f(pred). These latter values of K_f(pred) represent explicit solutions of the Chell theory, as distinct from the bounding solutions discussed in relation to Fig. 7.

Figure 6 shows that Chell's WPS theory produced accurate predictions of toughness for temperatures in the lower transition range—i.e., T_f < 100°C. At higher temperatures, however, toughnesses were underpredicted. Figure 7 and, in particular, Fig. 8 show that the theory produced reasonable predictions of failure level, with the exception of two results (T_f =

-80 and $-50°C$ respectively) relating to a Type B LUCF WPS cycle (nominal pre-load level $= 200$ MPa$\sqrt{m}$), where the observed failure levels were less than the predicted failure levels by ≈ 25 to 40%.

It is important to note in relation to the above comparisons between experimental results and theoretical predictions that by far the most discriminating were those made in respect of LUCF WPS cycles and Case 1 solutions.

Although not presented in Figs. 7 and 8, predictions of failure level were also made with reference to a parametric representation of the lower bound to base-line toughness data. The exercise showed that failure levels predicted in this way were generally less accurate than those predicted with reference to a parametric representation of the mean to base-line toughness data—i.e., Eq 5. A similar conclusion was reached by Stonesifer and Rybicki in their analysis of plate HSST 02 WPS data [13].

Lastly, Fig. 9 summarizes the results of warm pre-stressing experiments in which the temperature of pre-loading was 290°C, specimen load-point displacement rates were either 1.0 or 0.01 mm min^{-1}, and the nominal pre-load level was 200 MPa$\sqrt{m}$. Also shown plotted are control data for which $K_{Jc} < 300$ MPa$\sqrt{m}$. Below a toughness level of ≈ 200 MPa$\sqrt{m}$, WPS and control data points lie consistently below the curve fit to base-line data. Phenomenologically, this observation is consistent with the expected embrittling effect of strain ageing on transition toughness. Above a toughness level of ≈ 200 MPa$\sqrt{m}$, however,

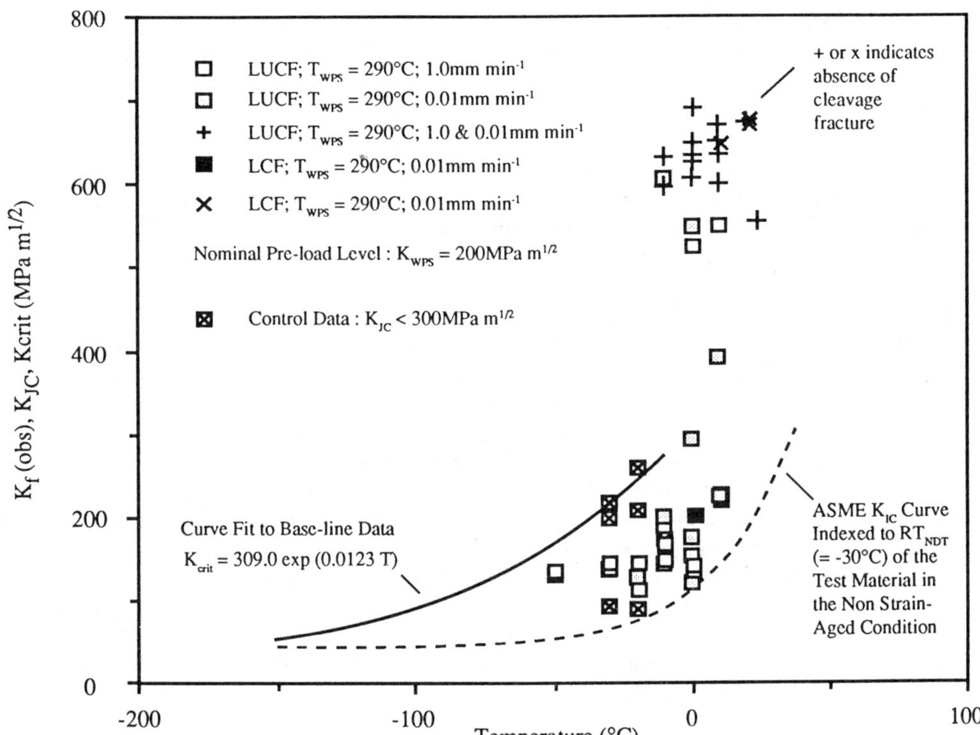

FIG. 9—*Observed failure levels, K$_f$(obs), in warm pre-stressing experiments: LUCF and LCF cycles; K$_{WPS}$ = 200 MPa$\sqrt{m}$; load-point displacement rate = 1.0 or 0.01 mm min⁻¹: comparisons with control data (K$_{Jc}$ < 300 MPa$\sqrt{m}$) and ASME K$_{Ic}$ curve appropriate to the test material in the non-strain-aged condition.*

the WPS data are distributed similarly to base-line data, as a comparison between Fig. 9 and Fig. 3 shows. Lastly, it is noted that all of the data presented in Fig. 9 are bounded by the ASME K_{Ic} versus temperature curve indexed to the value of RT_{NDT} ($= -30°C$) appropriate to the material in the non-strain-aged condition.

Discussion

There are a number of points arising from the above results which will be discussed under the following general headings: (i) the J-estimation procedure adopted for evaluating the parameters J_{WPS} and J_f; (ii) the effects of WPS in the elastic-plastic regime; and (iii) the interaction between WPS and strain ageing.

J-Estimation Procedure

The ability in fracture mechanics to make meaningful measurements of toughness depends upon the macroscopic loading parameter chosen adequately characterizing crack-tip stresses, strains, and displacements. In small-scale yielding, the crack-tip field may be calculated in terms of the linear-elastic stress intensity factor (K) even when previous load or yield stress changes, or both, have occurred. Under these circumstances, however, the stress intensity factor will no longer be directly proportional to the currently applied load. Instead, it must be calculated by applying the principle of plastic superposition [16] to the crack-tip fields associated with the component (monotonic) loading states [7–12]. Beyond small-scale yielding, the crack-tip field is no longer characterized by the linear-elastic stress intensity factor. However, it is characterized by the J-integral. This remains true even under general yielding, provided, of course, that certain specimen size criteria are met [18].

Equation 1 shows that the J-integral may be estimated from the work performed on a cracked specimen during loading. But the estimation is strictly valid, in terms of accuracy and crack tip characterization, only if the applied loading is monotonic and isothermal. If previous load or yield stress changes, or both, have occurred, then there are two plausible schemes by which the J-integral may be estimated, although neither can be justified rigorously.

In the first scheme, the J-integral is estimated from the total work performed under the component load versus load-point displacement curves, with the proviso that elastic work components are only counted once. With reference to the LUCF WPS cycle shown in Fig. 2a, J_{WPS} is defined by the area A_{WPS}, but J_f is defined by the area A_f plus the unshaded area drawn immediately to the left of A_f. With respect to the LCF WPS cycle depicted in Fig. 2b, J_{WPS} and J_f are defined by the areas A_{WPS} and A_f respectively. If a number of loading-unloading steps take place isothermally so that no changes in yield stress occur, then any residual plasticity resulting from a given unloading step will be wiped out by the next loading step, provided it exceeds the previous maximum level. Under these restricted circumstances, and assuming the temperature is sufficiently high for upper-shelf conditions to apply, it may be argued that this J-estimation scheme has some relevance to the characterization of ductile fracture.

In the second scheme, which has been adopted for the purposes of this study, the J-integral is estimated from the total work performed under the current load versus load-point displacement curve. This is the scheme shown explicitly in Fig. 2. Scheme 1 and Scheme 2 estimations only differ if unloading occurs, as in the case of LUCF WPS cycles.

Some intuitive justification for the use of Scheme 2 in presenting WPS data may be argued from consideration of Curry's WPS theory [12], which is based on the Ritchie, Knott and Rice (RKR) local criterion for cleavage fracture [19] and the principle of plastic superposition. Under conditions of small-scale yielding, Chell's theory and Curry's theory yield closely similar predictions. However, Curry's theory is, in principle, capable of extension to large-scale yielding—assuming, of course, that the loading characterizing the crack-tip field remained proportional, or nearly proportional, so that the principle of plastic superposition continued to hold [16]. The prediction of failure under these more general circumstances would still depend upon a detailed description of the crack-tip stress distribution associated with the component loading states. Therefore, by analogy with the way in which predictions of K_f are made and compared with experimental data for small-scale yielding, it seems plausible that predictions of the generalized theory would relate more closely with experimental values of $K_f(J_f)$ estimated by Scheme 2 than by Scheme 1.

In summary, it is only fair to point out that the issues raised in the above discussion warrant a far more rigorous and detailed theoretical examination than has been possible within the overall scope of the present work.

Effects of WPS in the Elastic-Plastic Regime

It has already been intimated that current predictive theories of WPS are based on small-scale yielding crack-tip models [7–13]. In general, however, possible elastic-plastic effects need to be taken into account in interpreting data. This is true not only for data obtained using sub-size fracture toughness test specimens (as discussed here) but for service conditions should WPS produce appreciable yielding.

The first, and most obvious, point to note is that crack-tip plasticity under conditions of small-scale yielding is completely contained by elastically deformed material. With sub-sized specimens (0.8T-CT in the present case), small-scale yielding can only be anticipated up to relatively low loading levels. With a medium strength, high toughness pressure vessel steel, ligament yielding can be expected well before the specimen loading has reached the level corresponding to ductile crack initiation (≈ 200 MPa $\sqrt{m}$ in the A533 Grade B Class 1 steel under discussion).

Figure 10a shows the distribution of plasticity in a compact specimen at ligament collapse, as deduced from simple limit load considerations. The important point to note is the reversed-yielding predicted to occur in the back-face region to the left of the rotation center, distance $(1 + \alpha)c$ from the crack tip (with $c = b/2$ and α as defined in Eq 3). Figure 10b shows the distribution of ligament plasticity in a compact specimen, $a/W = 0.5$, and work hardening, $\sigma_u/\sigma_{ys} = 1.45$, for various values of the ratio P_{app}/P_L ($P_{app} = $ applied load, $P_L = $ limit load), based on a plane strain finite element analysis by Neale [20]. In the present experiments, the value of P_{WPS}/P_L ranged from 0.73 to 0.97 at 25°C, where $\sigma_u/\sigma_{ys} = 1.29$, and equalled 0.95 at 290°C, where $\sigma_u/\sigma_{ys} = 1.51$. To the extent that Neale's analysis can be considered as applicable to the present discussion, Fig. 10b indicates that, despite the occurrence of back-face yielding, the crack-tip plasticity generated at all pre-loading levels was essentially "contained". It should also be noted, in the case of LUCF WPS cycles, that the residual plastic zone generated by reversed-yielding during unloading would have been approximately one quarter of the size of the corresponding crack-tip plastic zone generated during the initial loading step.

Elastic-plastic effects of WPS have been estimated by Chell using the BCS crack-tip model solutions for cracks in bodies of finite width [7]. Figure 11 shows the predicted failure levels as the degree of plasticity resulting from LUCF WPS load cycles increases. Also shown in Fig. 11 are the relevant WPS data from the present work.

Figure 11 indicates that, for a given value of K_{WPS}/K_{crit}, the predicted effective toughnesses decrease as the ratio $K_{WPS}/K_{WPS,el}$ increases ($K_{WPS,el} = $ elastically calculated stress intensity factor corresponding to K_{WPS}), until eventually a deleterious effect of WPS on K_f occurs. This

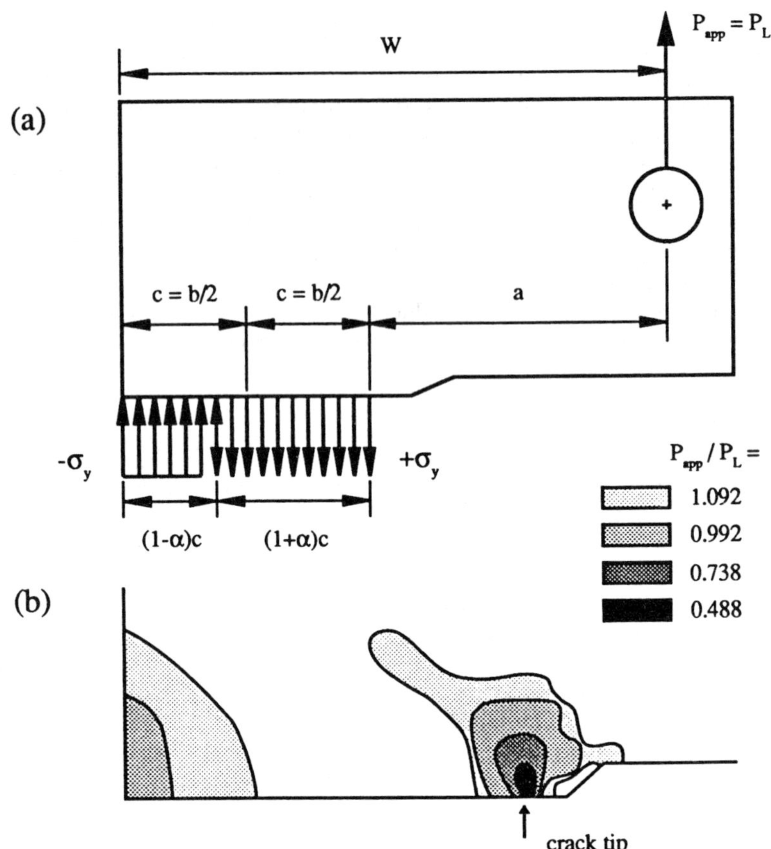

FIG. 10—*Plasticity distribution in a compact specimen:* (a) *limit load analysis;* (b) *plane strain finite element analysis with* a/W = 0.5 *and work hardening,* σ_u/σ_{ys} = 1.45 *(after Neale* [20]).

predicted fracture behavior is, strictly speaking, only applicable to the regime where failure is covered by Case 1 (i.e., the zone of reversed-yielding arising from the second loading step is not wiped out by the final loading step) and J_{WPS} is less than, or approximately equal to, $X_0\sigma_y$ (X_0 is the microstructural scale parameter in the RKR local criterion for cleavage fracture [19] over which the crack-tip opening stress must exceed the critical fracture stress (σ_f^*)).For quenched and tempered bainitic steels similar to the A533 Grade B Class 1 steel under consideration $X_0 \approx 50~\mu m$ [21]. Therefore the corresponding upper limit on K_{WPS} is in the range $\approx$ 70 to 90 MPa $\sqrt{m}$.

In view of the above restrictions, the correlation shown in Fig. 11 between experimental results and theoretical predictions for conditions of small-scale yielding ($K_{WPS}/K_{WPS,el}$ = 1.0) is remarkably good. Whilst two of the data points are such that K_f(obs) < K_{crit}, these both correspond to the regime where $K_{WPS}/K_{crit} \approx$ 1.0. Here, values of final, reloading temperature, T_f, are relatively high, and therefore the effects of material scatter are likely to be most pronounced. Some rationalization of the experimental data within the overall context of the theoretical predictions is possible, as suggested by the "dashed" curves drawn respectively through the Type A, Type B, and Type C data points. The overall conclusion to be drawn from Fig. 11, however, is that the present investigation has produced no firm evidence to suggest the occurrence of deleterious elastic-plastic effects of WPS on effective toughness in sub-size compact specimens.

Interaction between WPS and Strain Ageing

It is well known that embrittlement may result if strain ageing takes place in a crack-tip plastic zone. In ferritic materials ductile tearing and cleavage fracture may be considered as competing modes of separation. The strengthening associated with strain ageing promotes cleavage fracture by inhibiting yielding and so allows crack-tip stresses to build up to a critical level. The RKR local criterion for cleavage fracture may be used to quantify the resultant reductions in toughness [19,7], provided (*i*) they are primarily due to an associated increase in flow properties, and (*ii*) the micro-mechanisms of fracture remain unchanged—i.e., σ_f^* and X_0 remain unchanged. Curry's warm pre-stressing theory [12] has been extended on this basis to make predictions of fracture behaviour resulting from combined WPS and strain ageing [7]. In particular, the extended theory has shown that if WPS is followed by ageing, then the overall effect on effective toughness may be deleterious.

Basically, there are two ways in which the interaction between WPS and strain-ageing may be investigated experimentally. In the first, WPS is carried out at, say, ambient temperature, followed by ageing at some higher temperature. Strain ageing therefore takes place in a resid-

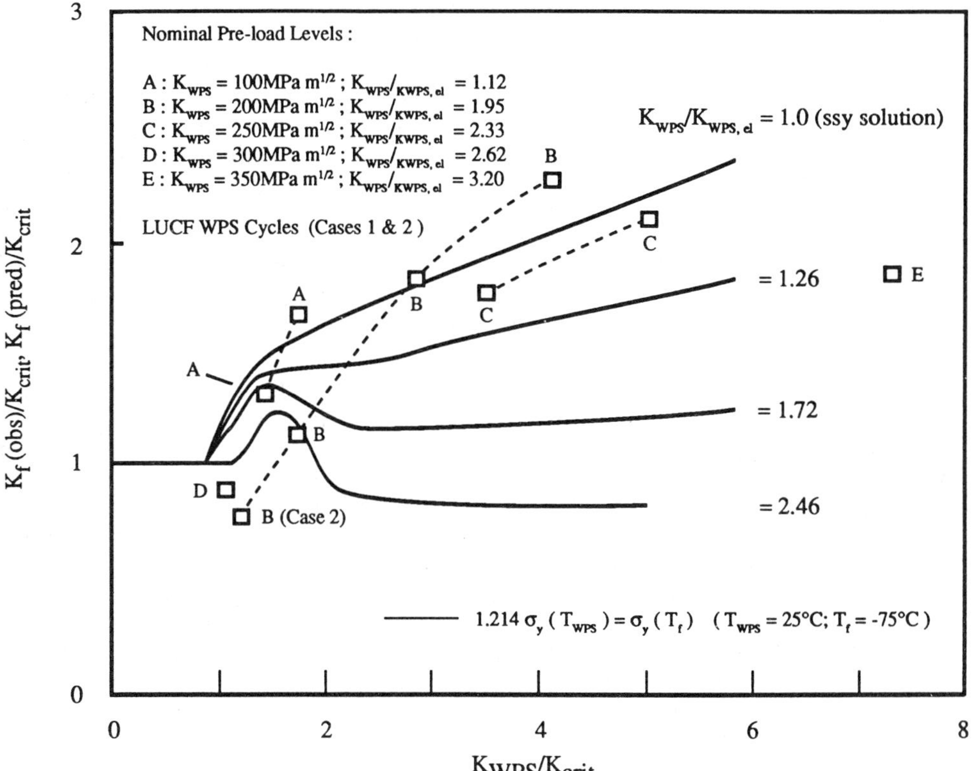

FIG. 11—*Normalized failure levels, K_f(obs)/K_{crit}, versus normalized WPS levels, K_{WPS}/ K_{crit}, compared with predictions of Q_I-integral theory as degree of plasticity increases [7]: T_{WPS} = 25°C; load-point displacement rate = 1.0 mm min^{-1}.*

ual crack-tip plastic zone where reversed-yielding has taken place. In the second, WPS and ageing are carried out simultaneously, and dynamic strain ageing occurs in a crack-tip zone of developing plasticity. The first approach is relevant, for example, to the simulation of effective toughness in a defected pressure vessel during a periodic low temperature in-service proof test; the second relates to a safety assessment following survival of a transient in a temperature range where strain ageing occurs. With either approach, the pre-strained and aged specimens must then be tested over a range of temperatures where cleavage occurs and the results compared with base-line toughness versus temperature data. The task of discriminating between beneficial effects due to WPS and detrimental effects due to strain-ageing is not straightforward. In principle, this can be eased somewhat if, in addition, control experiments are carried out on specimens sectioned from material which has previously been uniformly pre-strained and aged. Again, two basic approaches are possible in sequencing the processes of pre-straining and ageing.

The first of the above two approaches is illustrated by the work of Succop et al. [2]. These authors carried out tests on two steels, A516 Grade 70 and A533 Grade B. Single edge-notched bend (SENB) specimens were pre-strained by various amounts in the range 44 to 100 MPa $\sqrt{m}$ (40 to 90 ksi$\sqrt{in}$.) at 93°C (200°F). Some of the specimens were then aged for 1000 h at 343°C (650°F). All of the specimens were subsequently tested at -100°C (-150°F). Effective toughnesses were enhanced by WPS in all cases, but the conferred margins were slightly reduced in the case of those specimens which had been both pre-strained and aged.

The second approach is illustrated by the work of Haigh [22], who investigated the combined effects of WPS, ductile crack growth, and strain ageing on the effective toughness and load-bearing capacity of 50-mm-thick compact specimens sectioned from C-Mn steel plate. The key findings of this work were as follows. (i) WPS enhanced effective lower shelf toughnesses even when ductile tearing ($\Delta a \leq 4.0$ mm) or strain ageing or both had occurred during pre-loading; (ii) strain ageing shifted effective transition toughnesses relative to the base-line transition curve by amounts in the range $55 < \Delta T \leq 110$°C; and (iii) combined pre-straining and ageing was more severe in shifting the transition curve than pre-straining followed by ageing.

As noted in an earlier section, the second approach was followed in the present investigation in WPS tests where pre-loading was carried out at 290°C. However, in the case of control tests, the first approach was used because practicalities precluded specimens being fabricated from thick-section material which had been simultaneously pre-strained and aged.

In addition to observations of fracture behavior, evidence for the occurrence of strain ageing was obtained by analyzing specimen load versus load-point displacement traces. For example, in the case of LUCF tests where $T_{WPS} = 290$°C and $T_f \approx -10$°C, estimates of the elevation in yield stress (σ_{ys}) at the lower temperature produced by pre-straining and ageing at the higher temperature were in the range 15 to 20%. These particular observations show that appreciable strain-ageing effects can occur in a steel where the mobile nitrogen content is apparently low (< 3.0 ppm in the A533 Grade B Class 1 tested). Despite this response, corresponding test traces and effective toughnesses did not display any significant strain-rate sensitivity: Serrated flow behavior was not observed in respect of tests where the load-point displacement rate was 1.0 mm min^{-1} and the total pre-loading time was ≈ 300 s; it was only *just* discernible when the load-point displacement rate was 0.01 mm min^{-1} and the total pre-loading time was ≈ 8.3 h. The elevation in yield stress produced by a uniform 2% pre-strain at ambient temperature followed by ageing for 1.25 h at 290°C was assessed by comparing traces relating to control and base-line tests carried out in the temperature range 0 to -10°C. The estimated yield stress appropriate to control tests was found to be $\approx 10\%$ higher than that for base-line tests. The consistency between the WPS and control data in Fig. 9, together with associated observations in respect of specimen load versus load-point displacement behavior, further support the conclusion that the observed deleterious effects on toughness behaviour are due to strain-ageing embrittlement.

Lastly, the question of crack growth requires brief mention. There are two aspects of this: ductile tearing associated with WPS and pre-cleavage tearing during final loading. With respect to the first, it has already been noted that effective toughnesses were increased despite WPS having involved pre-load levels sufficient to have caused ductile crack growth (≈ 1.0 mm in the case of nominal pre-load levels of 350 MPa $\sqrt{m}$). The significance of this observation is that WPS benefits do not appear to be strongly dependent upon the crack-tip geometry changes associated with pre-loading, the extremes being a well-blunted region *vis-à-vis* a sharp tear. With respect to the second, Chell has generalized his original theory and demonstrated theoretically [*11*] that WPS benefits are maintained, provided crack growth does not extend beyond the residual crack-tip zone formed during unloading.[5] Within the context of this paper, it seems likely that this conclusion would remain true, even had strain ageing occurred. The argument is clearly only applicable to Case 1 LUCF WPS cycles and therefore relevant, in the strictest sense, to cases where crack extension had occurred by a sub-critical mechanism (e.g., by fatigue)—an aspect of WPS behavior not covered by this investigation.

Concluding Remarks

The present investigation has confirmed the validity of the conservative WPS principle. In every test performed an increase in loading was required to cause cleavage failure at low temperature following a severe pre-load at a higher temperature; see, in particular, Fig. 12 which clearly demonstrates this with respect to LCF WPS cycles.

In the case of pre-loading carried out at 25°C, conferred WPS margins were observed even for those tests where pre-loading had resulted in ductile tearing ($0 \leq \Delta a \leq 1.0$ mm) under conditions of ligament yielding. The Case 1 and Case 2 solutions of Chell's Q_1-integral theory of WPS yielded reasonably accurate predictions of final fracture. This occurred despite (*i*) the theory being strictly applicable only in the regime of small-scale yielding, and (*ii*) values of the experimentally determined parameter K_f(obs) being derived from a *J*-based estimation procedure for which no rigorous theoretical justification can be claimed.

The results show that whilst WPS phenomena may occur in isolation to strain-ageing effects, the converse is not true. Below a toughness level of ≈ 200 MPa $\sqrt{m}$, both control and WPS data (LUCF and LCF cycles) show a deleterious effect of pre-loading at 290°C on effective transition toughness which is consistent with the expected embrittling effect of strain ageing. However, all data generated by the study are bounded by the ASME K_{Ic} versus temperature curve indexed to the value of RT_{NDT} ($= -30$°C) appropriate to the material in the non-strain-aged condition.

Acknowledgments

This work was partly funded by the U.K. Central Electricity Generating Board under the terms of the Thermal Reactor Agreement. The authors wish to thank Dr G. G. Chell (CEGB—Leatherhead) and Dr L. G. Taylor (CEGB-PMT—Knutsford) for their helpful comments on the manuscript.

---

[5] Large-scale experimental confirmation of this was provided by the results of pressurized thermal shock experiment PTSE-2A [*5*], where 2.9 mm of precleavage ductile tearing occurred during the reloading phase. Nevertheless, K_f/K_{crit} was observed to be ≈ 1.3, which is in reasonable agreement with Chell's extended theory [*11*].

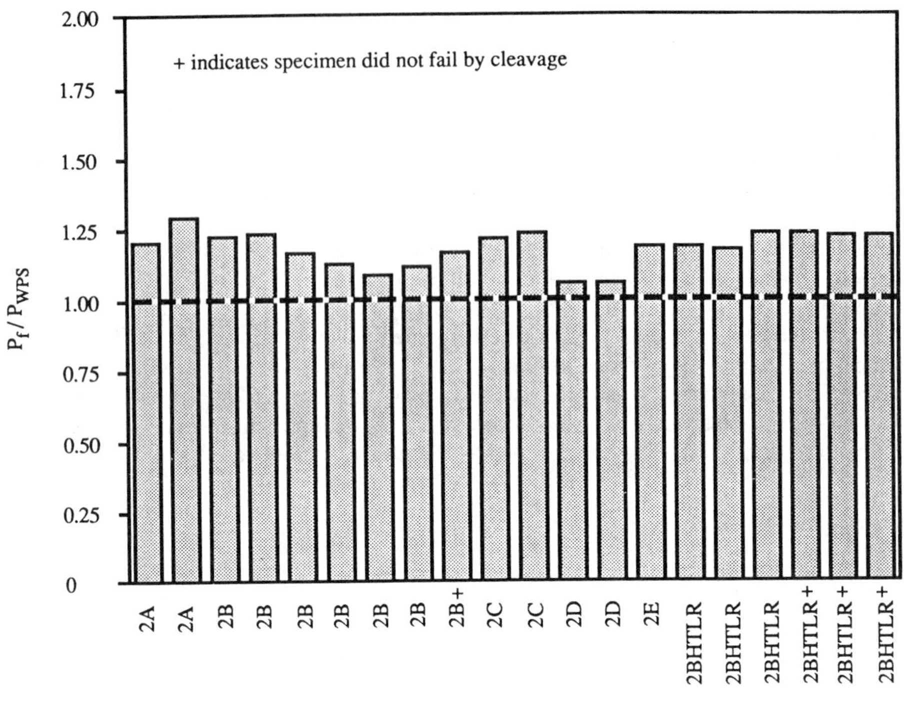

FIG. 12—*Confirmation of conservative warm pre-stressing principle in respect of LCF cycles.*

References

[1] Pickles, B. W. and Cowan, A., *International Journal of Pressure Vessels and Piping,* Vol. 14, 1983, pp. 95–131.

[2] Succop, L. N., Pense, A. W., and Stout, R. D., "The Effects of Warm Overstressing on Pressure Vessel Steel Properties," HSST 5th Annual Information Meeting, Paper 29, March 1971. (See also *Welding Journal, Research Supplement,* Vol. 51, 1970, p. 354-S.)

[3] Loss, F. J., Gray, R. A., and Hawthorne, J. R., "Significance of Warm Prestress to Crack Initiation during Thermal Shock," NRL/NUREG Report 8165, Sept. 1977.

[4] Cheverton, R. D. et al., *Journal of Pressure Vessel Technology,* Vol. 105, 1983, pp. 102–110.

[5] Bryan, R. H. et al., "Pressurized-Thermal-Shock Test of a 6-in.-Thick Pressure Vessel; PTSE-2—Investigation of Low Tearing Resistance and Warm Prestressing," NUREG/CR-4888 (ORNL-6377), Oak Ridge National Laboratory, Dec. 1987.

[6] McGowan, J. J., "An Assessment of the Beneficial Effects of Warm Prestressing on the Fracture Properties of Nuclear Reactor Vessels under Severe Thermal Shock," Westinghouse Report WCAP-9178, March 1978.

[7] Chell, G. G. and Curry, D. A. in *Developments in Fracture Mechanics—2,* G. G. Chell, Ed., Applied Science Publishers, New York, 1981, pp. 101–192.

[8] Chell, G. G., Haigh, J. R. and Vitek, V., "A Theory of Warm Prestresing: Experimental Validation and the Implications for Elastic-Plastic Failure Criteria," CEGB CERL Lab Note RD/L/ND63/79, April 1979. (See also *International Journal of Fracture,* Vol. 17, 1981, p. 61.)

[9] Chell, G. G., "A Theory for Predicting Failure Loads of Cracked Structures Subjected to Warm Prestressing: 1—Load Changes at Constant Temperature," CEGB CERL Lab Note RD/L/N78/79, Sept. 1979.

[10] Chell, G. G., "Some Fracture Mechanics Applications of Warm Prestressing to Pressure Vessels," 4th International Conference on Pressure Vessel Technology, Paper C22/80, London, May 1980.

[*11*] Chell, G. G., "The Effects of Sub-Critical Crack Growth on the Fracture Behaviour of Cracked Ferritic Steels after Warm Prestressing," *Fatigue and Fracture of Engineering Materials,* Vol. 9, No. 4, 1986, pp. 259–274.

[*12*] Curry, D. A., *International Journal of Fracture,* Vol. 17, No. 3, 1981, pp. 335–343.

[*13*] Stonesifer, R. B. and Rybicki, E. F., "Development of Models for Warm Prestressing," NUREG/CR-4491, MEA-2122, Jan. 1987.

[*14*] United Kingdom Atomic Energy Authority, "An Assessment of the Integrity of PWR Pressure Vessels: Addendum to the Second Report of the Study Group," since 1982 under the Chairmanship of Professor Sir P. B. Hirsch, FRS, Her Majesty's Stationery Office, April 1987.

[*15*] Clarke, G. A. and Landes, J. D., *Journal of Testing and Evaluation,* Vol. 7, No. 5, 1979, pp. 264–269.

[*16*] Rice, J. R. in *Fatigue Crack Propagation, ASTM STP 415,* J. C. Grosskreutz, Ed., American Society for Testing and Materials, Philadelphia, 1967, pp. 247–311.

[*17*] Bilby, B. A., Cottrell, A. H., and Swinden, K. H., *Proceedings of the Royal Society A,* Vol. 272, 1963, pp. 304–314.

[*18*] McMeeking, R. M. and Parks, D. M. in *Elastic-Plastic Fracture, ASTM STP 668,* J. D. Landes and J. A. Begley, Eds., American Society for Testing and Materials, Philadelphia, 1979, pp. 175–194.

[*19*] Ritchie, R. O., Knott, J. F., and Rice, J. R., *Journal of the Mechanics and Physics of Solids,* Vol. 21, 1973, pp. 395–410.

[*20*] Neale, B. K., "Elastic-Plastic Analysis of Cracked Bodies Using the *J*-Integral Method," CEGB Report RD/B/N/3253, Jan. 1975.

[*21*] Bowen, P., "Effects of Micro-Structure on Toughness in Pressure Vessel Steel," Ph.D. thesis, Churchill College, University of Cambridge, Oct. 1984.

[*22*] Haigh, J. R., "Effects of Combined Warm Prestressing, Strain Ageing and Ductile Crack Growth on the Toughness and Load-Bearing Capacity of a C-Mn Steel," CEGB Report NWR/SSD/84/0012/N, Jan. 1984.

Elastic-Plastic Fracture Mechanics (II)

M. R. Etemad[1] and C. E. Turner[1]

Unique Elastic-Plastic *R*-Curves: Fact or Fiction?

REFERENCE: Etemad, M. R. and Turner, C. E., **"Unique Elastic-Plastic *R*-Curves: Fact or Fiction?"** *Fracture Mechanics: Twenty-First Symposium, ASTM STP 1074,* J. P. Gudas, J. A. Joyce, and E. M. Hackett, Eds., American Society for Testing and Materials, Philadelphia, 1990, pp. 289–306.

ABSTRACT: A simple model is presented for the effect of size and geometry on fracture toughness when represented as *J-R* curves in the fully plastic region. The model, which is little more than dimensional analysis, refers to growth beyond any *J*-controlled regime. It suggests that the abscissa of a *J*-based *R*-curve should be $\Delta a/c$, where c may be thickness, ligament, or material toughness (expressed as a plastic zone size) according to circumstance. The analysis is consistent with the well known picture in the LEFM regime where thickness is the only geometric term that affects toughness.

Comparison is made with many published results, in most cases satisfactorily. The agreement is best for tests of various ligament widths with constant thickness. The reason for this is the limited understanding of how the shear lips develop before a steady-state pattern of growth is reached and the likelihood that the controlling factor may change when thickness is a variable. Since for the fully plastic state the increment of work from which toughness is derived involves only the limit load and dq/da, the relation between an increment of displacement and the corresponding crack growth, the latter seems to be the factor that contributes most to the differing trends of fully plastic *J-R* curves as initial geometry is varied. The near total lack of published data for dq/da needs urgent remedy if these geometric effects are to be understood.

KEY WORDS: fracture toughness, elastic-plastic fracture mechanics (EPFM) test methods, *R*-curves, size effects

Fracture resistance *R*-curves measured in the fully plastic range have become well established as a concept where the failure mechanism is ductile in the micromode. Nevertheless, many uncertainties remain over the circumstances in which an *R*-curve for a given material is independent of geometry. However, an ASTM method of measurement has emerged recently for a restricted range of test conditions [1].

The uncertainty encompasses both the choice of the most suitable parameter and the necessary limitations of geometry. Within the *J*-based methods, one of the main proposals is to keep absolute size of thickness and ligament greater than $25J/\sigma_\text{fl}$, where σ_fl is a flow stress usually taken as the mean of yield and tensile strength. Limiting the amount of growth to 6% of the ligament was proposed [2] as a way of ensuring so-called *J*-controlled growth that would be geometry independent. Figure 1 uses data from [3] to show that the 6% restriction is not always necessary, since in some cases similar curves can be obtained well beyond that limit. In Fig. 2*a*, however, which uses data from [4], the restriction would have to apply to 6% of the smallest ligament but to only 1.5% of the largest.

Much data in the literature show marked effects of size and proportion. One proposal to minimize geometric effects was to use the term J_m [5]. Apart from the original data, results

---

[1] Mechanical Engineering Department, Imperial College, London SW7 2BX, England.

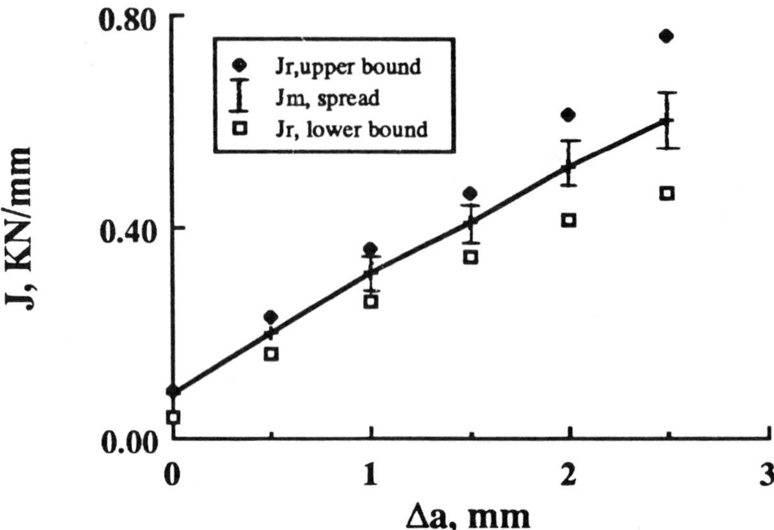

FIG. 1—*Date from [3] for bend and tension specimens of BS 4360 steel, 25 mm thick, non-side-grooved. For* J$_r$ *the upper bound is CCT and DEN data (b = 12.5 mm); the lower bound is SENB-3, SENB-4, and CT (b = 25 mm). For* J$_m$ *the data are barely separable, with CCT tending to be highest and DEN lowest.*

such as seen in Fig. 1 might give credence to the statement that use of J_m was sufficient to give geometry independence. As seen in Fig. 2a, however, that is not the case, even within a group of deep notch three-point bend configurations of fixed thickness with various ligaments, unless other restrictions are introduced. The smallest specimen is larger than required to meet the $25J/\sigma_{fl}$ restriction based on initiation toughness, although if 25 were increased to 50 or the current rather than the initiation value were used for J then the smallest specimen would not be regarded as valid.

Since none of the definitions of J used to analyze conventional experimental data are the original contour integral definition, it would no doubt be better if the terminology J were dropped in favor of R (for resistance). The current experimentally derived R (or J) value is based on the summation of functions of the increments of work done, dU, normalized by a shape factor, η, and net section area, Bb, where B is thickness and b is the remaining ligament (i.e., W-a, where W is width and a is crack length so that $da = -db$). This term is "corrected" by a variety of secondary terms in da/b, originally intended to accommodate the change in ligament size but in [5] intended to discount it. In all cases the dominant term is

$$dR = \eta dU/Bb \qquad (1a)$$

The various secondary terms omitted from Eq 1a do not affect the trends of the following arguments. However, some definitions of J are based on a summation of internal energy rates, dw, rather than of work increments, dU, where essentially

$$dw = dU - BRda \qquad (1b)$$

and that distinction does alter the trends.

Before initiation the distinction between the work done and the internal energy affects the physical meaning of J as to whether the term is recoverable, but its value is not altered for

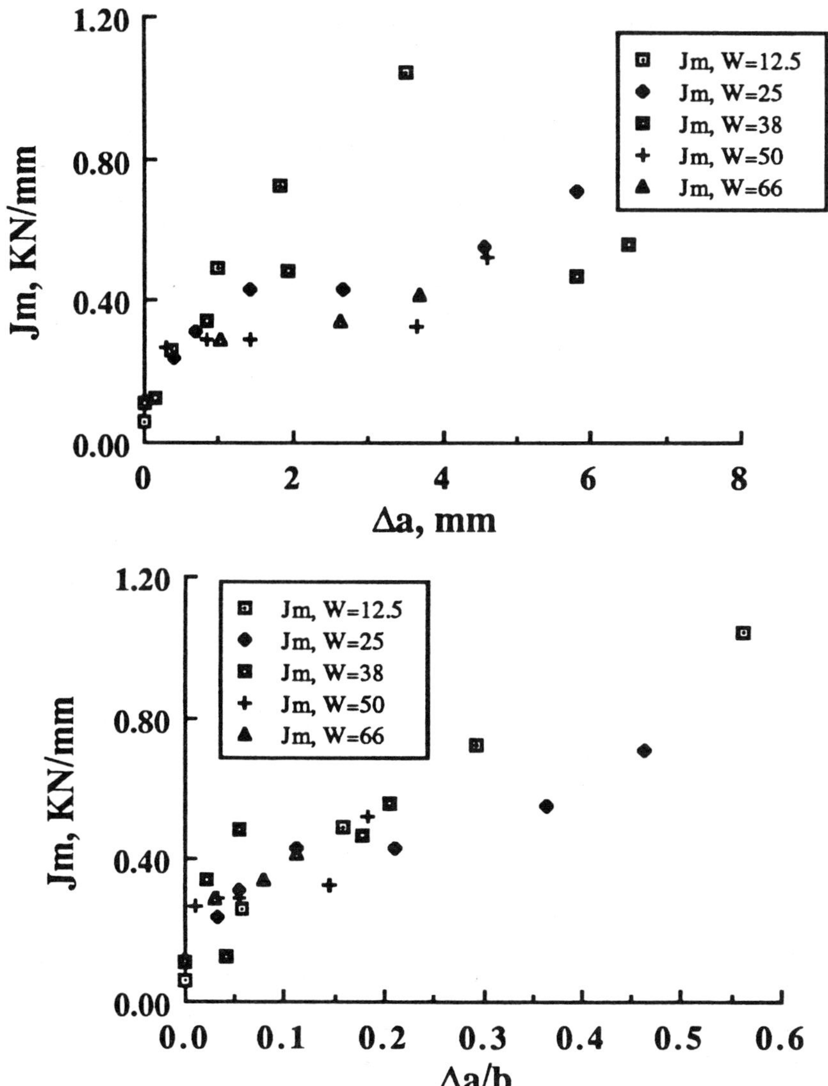

FIG. 2—*Data from* [4] *for HY130 steel, 50 mm thick, three-point bend tests,* a/W = 0.5, *with various widths,* W, *in millimetres.* (a) *Conventional* R-*curves where* J_m *does not give a unique result.* (b) *Same data with abscissa scaled to* Δa/b.

any given plasticity model. After initiation that is no longer so and for large amounts of growth the use of internal energy, dw, may give a reducing *R*-curve, whereas the work increment, dU, must always give a rising curve [6]. Use of dw also begs the issue of the relevance of a term derived from non-linear elastic (NLE) theory to real plasticity materials, but that larger question is only entered indirectly in that the U-based (work) terms are used in much of the following. It is also noted that although dJ_m was expressed in [5] in terms of the deformation theory term J_d, which is a measure of the NLE change in internal energy, it was derived to satisfy the condition in [7] that as plasticity dominated so dJ should become a measure of dU rather than dw. Thus it can be expressed as

$$dJ_m = \eta dU_{tot}/Bb - Gda/b \qquad (2)$$

where U_{tot} is the combined elastic plus plastic increment of work.

If a state is envisaged where there is no change of internal energy, dw, then from Eq 1b the R-value implied is dU/Bda. That term, multiplied by η, was called J_{ss} in [6] and was shown to fall towards a steady state as dw indeed reduced towards zero at large growths. The same data are shown in Fig. 3 (Curves 1, 2, and 3) but expressed simply as work, internal energy, or work increment per unit crack area (i.e., U, w, or dU/Bda respectively). If total work done is recorded in the test for each growth, then Curve 1 is a direct plot of the actual data. If increments of work and growth are recorded in the test, then Curve 3 is a direct record of the test data. Curve 2 is obtained via Eq 1b, so the one set of data can be displayed in three quite different styles. The reducing curve of dU/Bda will be referred to again later. Additional normalization by η/Bb would convert Curve 1 to the conventional J value, J_0, and Curve 2 to J_d from which several other well known versions, J_{i+1} or J_m, can then be found.

This paper sets out several cases of geometric effects in R-curves for bending (including compact tension [CT]) and offers some reasons for the behaviors found. It must be emphasized, however, that the remarks made are very "broad brush." Even where the general trends are correctly identified the transition from one case to another must include much detail not yet perceived.

Conventional View of Size Effects

In any discussion of size effects it is convenient to characterize toughness in terms of a length dimension, possibly COD or here a size of plastic zone. The ratio between such a

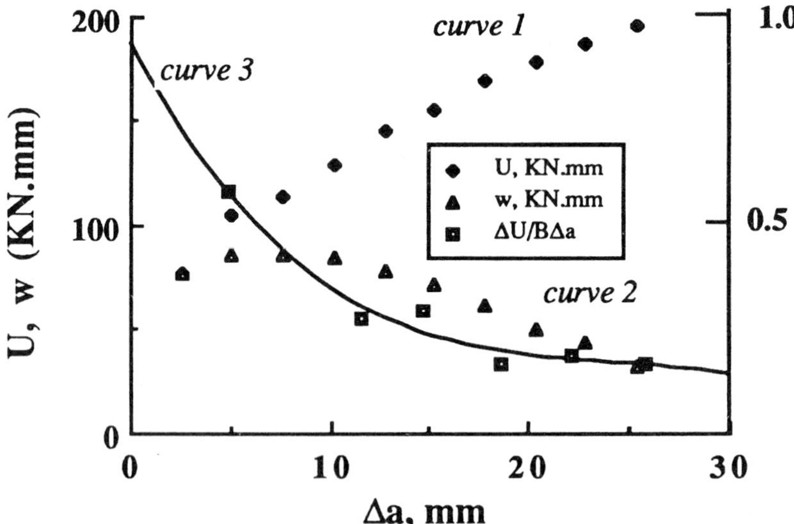

FIG. 3—*Data from* [14] *reanalyzed in* [6] *here presented as work,* U, *which always rises, and internal energy,* w, *which first rises then falls. Also shown is the normalized increment of work done,* dU/Bda, *that in plasticity always falls.* dw/Bda (not shown) *would change from a positive to a negative value and then rise to zero if a steady state is reached.*

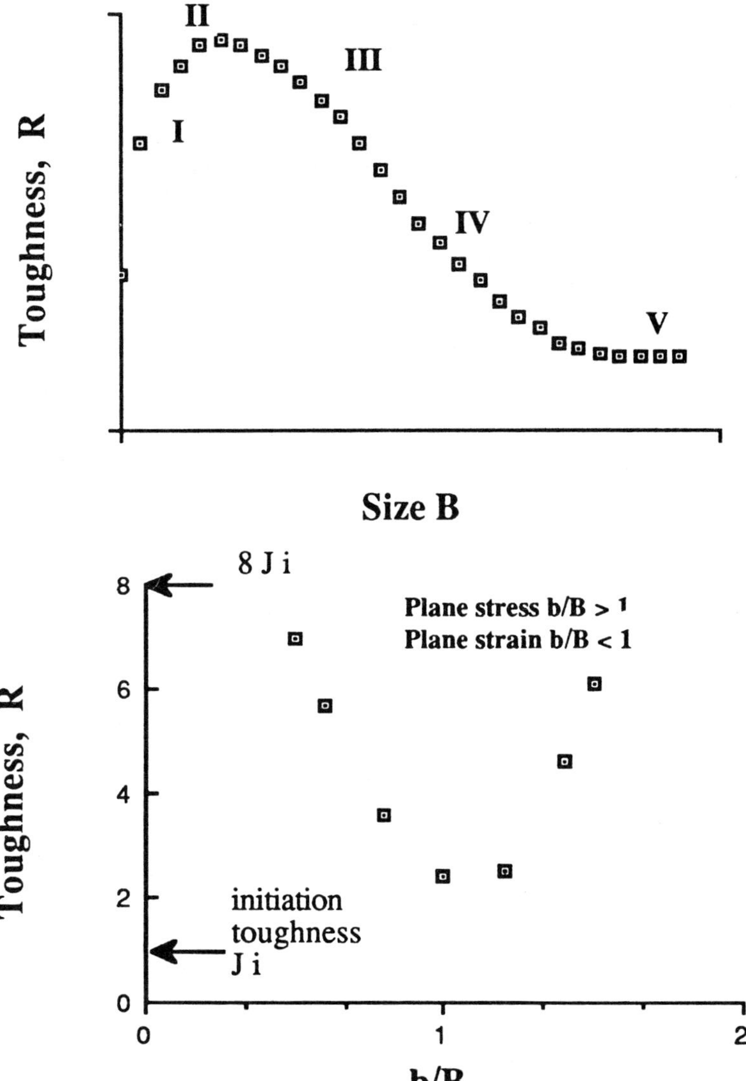

FIG. 4—*Schematic of toughness versus thickness. (a) Conventional LEFM picture for increase of thickness at large widths. (b) "Bath-tub" picture from [4] for fully plastic behavior of HY130 after 2 mm growth (falling limb from B = constant, variable b; rising limb from b = constant, variable B).*

measure and the several dimensions of a specimen or component then becomes a convenient parameter in the discussion. The effect of thickness on toughness is usually presented schematically as in Fig. 4*a*, which is based on the well known diagram for LEFM behavior. The first rising limb (I) describes where toughness (i.e., a zone size) is inhibited by thickness, the peak of the curve (II) occurring notionally where the maximum possible zone radius, here denoted r_{mm} (*m*aximum for the *m*aterial in question) equals half the thickness, $B/2$. For thicker sections the toughness reduces (III and IV) (the distinction between these two relates only to later fully plastic arguments) as the plane stress surface shear lips each of nominal extent $r_{mm}/2$ occupy an ever smaller proportion of the increasing thickness until LEFM valid plane strain test conditions are reached (V) when the plane stress shear lip zone size is no more than $B/50$ in extent.

Conceptually, for the fully plane strain LEFM condition fracture would occur as soon as initiation of crack growth occurred, when the zone size could be denoted r_{ie} (*initiation in plain strain, e*). In reality, for material that is ductile on the microscale it is now generally accepted that there is both a tearing and zone growth before the formal K_{Ic} value, as defined in the 5% offset procedure [8], is reached.

This view relates plane stress or plane strain to a material toughness property in which the dimensions of the specimen enter only through thickness. Provided that the crack length, a, is sufficient for the LEFM singularity analysis to hold and that the ligament, b, is appreciably greater than r_p so that the plastic zone does not interact with the "far" width boundary, the ratios of the geometric sizes are not directly relevant.

In contrast to this view, Fig. 4b shows a "bath-tub" effect for toughness at a given growth level with general yield [4], where the behavior for a range of proportions are unified on an abscissa of b/B. This arises because in the fully plastic case it is the b/B ratio rather than a material toughness term that controls the plane stress or plane strain mode of deformation.

EPFM View of Tearing Toughness Based on Contained Plasticity

For contained plasticity, ranging from r_p negligible in relation to width up to an unspecified significant proportion, an explanation of the interaction between size and R-curve toughness has been set out [7], for the case of plane strain, on a persuasive, though not completely proven, basis. In relation to the present arguments the important feature is that the abscissa of the R-curve is not just Δa but $\Delta a/c$, where c is the maximum size of plastic zone that would have been reached, c_c, in contained yield at a particular J level. That has been estimated as [7]

$$c_c = 0.2 JE/\sigma_y^2 \tag{3}$$

Since the analysis in [7] is for plane strain in contained yield, then the zone size, c, must be less than any dimension, crack length a, thickness B, or ligament b. More strictly, the zone size must be limited to some fraction of these dimensions, as in LEFM, but the actual limitations are not defined.

The present discussion is of data presented in [9] and [10], or other data found since, which are all in uncontained yield, not necessarily in plane strain. The rationale pointed out [11] and pursued here is that, with uncontained yield, the size of the zone of severe deformation, c (which will not be a single term unless the zone is spherical), may be controlled either by inherent material toughness or by geometry. In the former case, where plane-sided specimens large in relation to toughness (i.e., to zone size) are implied, then $c = r_{mm}$. If adequate side grooves were present, then presumably $r_{mm} > c > r_{ie}$. Where specimens are not large in relation to toughness, the zone size is inhibited by one or other of the dimensions, B, b, or, even for shallow cracks, a. An example of this is shown in Fig. 2b, where use of the abscissa $\Delta a/b_o$ (where b_o is the original ligament size) normalizes the sets of data shown in Fig. 2a. Cases where crack length, a, is the limiting factor are not discussed because of the generally accepted belief that constraint is lost with shallow notches and toughness thereby increased.

Clearly it can be argued that where geometry restricts the size of the deformation zone the "full toughness" is not developed. That may well be so but, as already noted, the present *J*-based *R*-curve procedures use increments in work done, *dU* (itself coming from load, *Q*, and increments in displacement, *dq*), to give *dJ*. In short, it is a valid criticism of the use of small specimens for assessment of toughness beyond initiation, but *faute de mieux* it is a method widely used for which understanding of the results is lacking.

Another important point is crack growth measurement, which is usually monitored by one of several possible methods not entered into here. This may influence the perceived geometric variation of the *R*-curve, since the averaging techniques to find growth levels are presently the same for specimens with and without shear lips (side-grooved) regardless of the so-called crack tunneling.

Geometric Effects with Uncontained Plasticity

As just noted, the dominant term in the analysis of experimental data is the increment of work, *dU*, from the load-displacement curve. Using the definition dJ_u [12], where *b* is understood to be the current ligament, we obtain

$$dJ_u/da = \eta(dU/da)/Bb \tag{4a}$$

When the elastic component is negligible, the Gda/b term in Eq 2 can be omitted and J_u is similar in value to J_m. In that case, as is often so for large growth, J_m, J_u and *dU* might then strictly be written with the suffix "pl." For small growth J_u is the same as J_o based on the original ligament and that is usually adequate for trend analysis. For fully plastic three-point bending, the load $Q = \lambda\sigma_y Bb^2/S$ and

$$dJ_u/da = \eta(Q\, dq/da)/Bb = \eta(\lambda\sigma_y Bb^2/BbS)(dq/da) \tag{4b}$$

where η is a proportionality factor (2 for the conventional bend test), λ is a load factor identified with the plastic constraint factor for the fully plastic nonhardening case, and *S* is the span. For the compact tension specimen, η is about 2.2 and, neglecting the direct stress term, the moment, *M*, is related to the force by the load offset from the center-line, $M = Q(a + [b/2])$, rather than by the three-point bend relation, $M = QS/4$. Variation in the value of λ for three-point bending, according to the degree of deformation reached and the effect of hardening on yield or flow stress, will affect the numerical value of *dJ* and *J* but not the geometric dependencies.

Thus in bending there is a trend relationship, with a multiplier $\eta\lambda\sigma_y$ that will increase with load until full plasticity:

$$dJ_u/da = f\{(b/S)(dq/da)\} \tag{5}$$

The precise function found in any analysis would depend upon the particular definition used for *J* after initiation, being linear for dJ_o. For tests with given *S/W* and initial *a/W* ratios but of various values of *W*, this reduces to *dq/da*, or strictly dq_{pl}/da if only the plastic component is considered, the term about which least is perceived. It implies that the normalization of Fig. 2*b* is either fortuitous or a consequence of the variation of $dq/da = f(\Delta a)$ with *B/b* ratio, which indeed changes from 8 to 2.0 for the cases shown; that is, $dq/da = f(\Delta a, B/b)$ of the form $f(\Delta a/b)$ since *B* is constant. Whilst fortuitous scaling can never be excluded, the number of cases reported [10], some of which have been shown diagrammatically [9–13], seems to suggest that there is a reality.

Published data on the behavior of dq/da seems well overdue since such values are buried in all the original test records from which J-R curves have been derived, but in their absence an alternative statement is followed here, developed from the ideas of [9–13].

The term dJ/da has dimensions of stress covered in Eq 4 by σ_y. Since all other terms can be expressed as nondimensional geometric ratios, it seems unreasonable to suppose dJ/da can be other than a function of a normalized crack growth, $\Delta a/c$, as argued above. Thus a J-based R-curve summed from linear functions of the work increment will reflect various functions of Δa according to the physical term that controls c. By dimensional analysis Eq 4 can be regarded as an increment of plastic work dissipation, ρ per unit volume, averaged through a severe deformation zone. The notional extent of the severe deformation from the crack tip for a symmetric quadrant of a specimen is c_1 in the thickness sense, c_2 in the crack growth direction, and c_3 normal to the crack plane, and it is averaged over the extent thickness B, the ligament b, and an axial length (normal to the crack plane) represented in Eq 4 by $S/4$ so that

$$dJ_u/da = 4\rho_b c_1 c_2 c_3/BbS \qquad (6)$$

where the value of ρ_b (the bending case) contains all other constants and the uncertain function dq/da so that ρ_b may be $f(\Delta a/c)$.

The notional extent c_1 can be limited either by r_{mm} or by slip to $B/2$ or about $b/2$. The notional extent c_2 can be limited either by r_{mm} or by b (or perhaps nearer $b/2$ since the tip zone is restricted to the tensile side of the specimen). From a simple slip line argument the extent c_3 might be taken as $b/2$, or rather less, for a hinge-type deformation in plane strain but as $B/2$ for plane stress.

Effect of Ligament Size at Constant Thickness

Some Scale . . .

One case of scaling with $\Delta a/b$ has already been shown (Fig. 2b) for deep notch bend specimens in plane strain (i.e., $B > b$). Another case was shown [13] using side-grooved data in plane strain from [14]; another in plane stress ($B < b$) was given in [15] from [16]. In a rather different case where scaling to $\Delta a/b$ was reported [17], an R-curve for stainless steel was measured on a deeply notched welded compact specimen ($B = 8.6$ mm, $b = 16$ mm) and applied to a circumferentially cracked welded pipe ($B = 8.6$ mm, b [say half the remaining circumference] $= 110$ mm). The data were presented in [17] in terms of pipe deflection versus crack extension, with computed predictions based on the "raw" R-curve giving no agreement with experiment, whereas based on a scaled R-curve good agreement was found (Fig. 5). In [17] this result was presented by analogy to HY130 data from [4] that had been reported in [10] as scaling with $\Delta a/b$, for flat fracture, essentially as in Fig. 2. However, as the authors of [17] comment, the circumstances are quite different, since $B > b$ (Fig. 2) gives plane strain in the fully plastic sense, whereas in [17] $B < b$, implying plane stress proportions. Although the weld metal has a much greater yield strength than the pipe material used for the test plate, it has a lower toughness. Therefore the present writers infer that there will not be sufficient load to spread general yield away from the ligament into the parent material despite its lower yield stress. Thus, provided either the thickness, B, or the extent of the weld metal normal to the crack plane, D (shown [16] as roughly comparable to or less than B), controls the severe deformation zone, then, from Eq 5, $dJ/da \propto$ constant$/b$. With b differing between the specimen and pipe, scaling with ligament is to be expected.

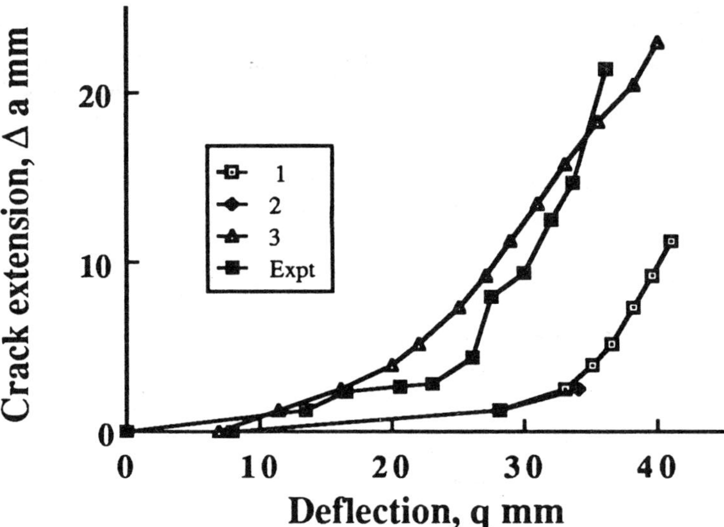

FIG. 5—*Comparison of experimental and computed deflection versus crack growth for a circumferentially cracked pipe as given in* [17]. *Curves 1 and 2 were computed from* J *versus* Δa *data, whereas Curve 3 used* J *versus* Δa/b.

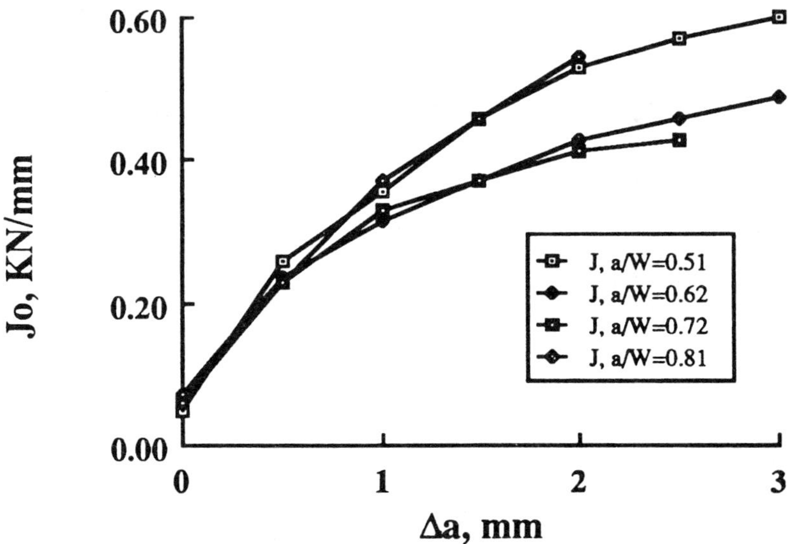

FIG. 6—*Selected 20% side-grooved data for A533B steel with various* a/W *ratios, originally shown in* [18] *as part of families of curves for different amounts of side grooving for four separate* a/W *values. (Data for duplicate tests fall within the scatter shown here.)*

. . . But Some Do Not

Results were given in [*18*] for a series of plane and side grooved 1T CT tests on A533B, 25 mm thick, with various a/W values ranging from 0.5 to 0.8, with two sets of data at each configuration. The *R*-curves were shown as families for given a/W, with various side grooving (0, 10, and 20%); from the case of 20% side grooving, all a/W values are shown in Fig. 6. In this family the only variable is ligament size. For the plane sided and 10% grooved families, results for various a/W values (here thought of as b/B since $W = 2B$) scatter within the repeatability, and in Fig. 6 there is also no trend with a/W (i.e., B/b = 2.63, 1.00, 1.35, and 1.72, based on full thickness B values, or 20% less, based on B_n). All this was rationalized

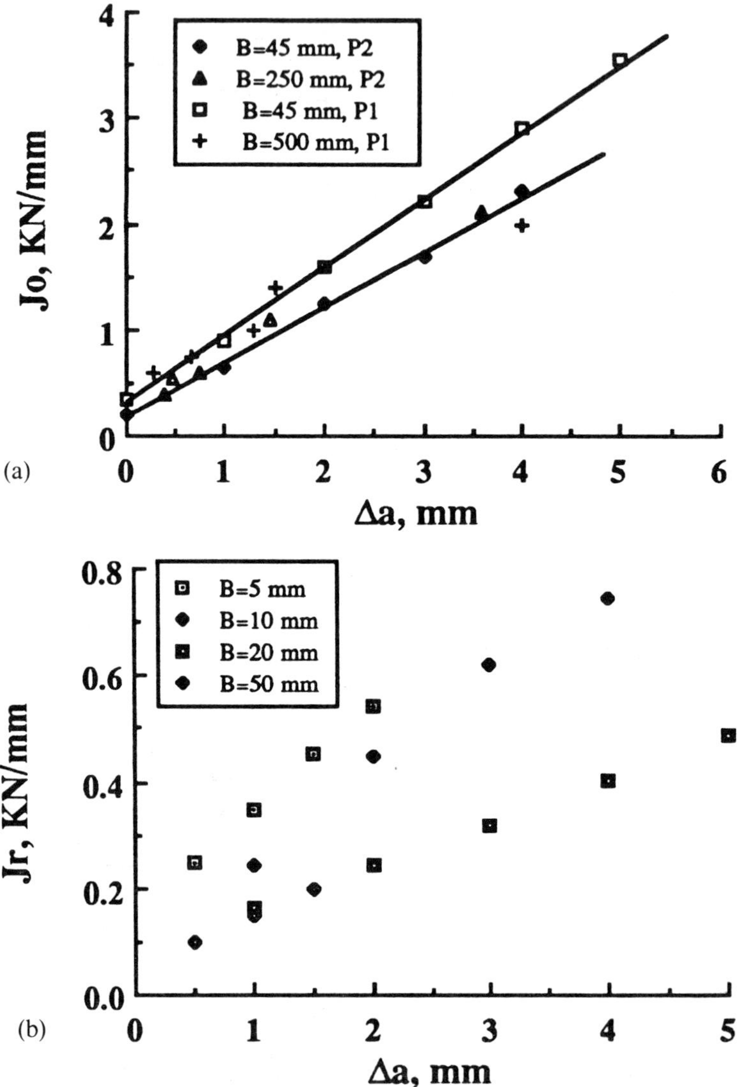

FIG. 7—*Effects of thickness.* (a) *A533B data from* [22] *for exceptionally thick specimens, but small ligaments in bending seem to give similar curves without the need to scale.* (b) *CT*

[*19*] by considering either an effective B larger than the actual value or a flat fracture zone size, r_{ff}, of value nearer r_{ie} rather than r_{mm}, to give b smaller than either r_{ff} or B_{eff}, both of which cases suggest that scaling is not required in order to obtain a common curve if the argument here expressed as $c_3 = b/2$ is used. The 20% side grooved data were re-examined [*20*] in great detail, and those results are commented on later.

One further set of results is mentioned. In [*21*] data are given for bend specimens of A533B, $S/W = 4$, $a/W = 0.4$, $B = 53$ mm, and $W = 65$ mm ($b = 39$ mm) or 125 mm ($b = 75$ mm). The R-curve for the wider specimen rises to double the value of the narrower at about 15 mm growth. None of the arguments adduced here rationalize that behavior. It is noted that a simplified formula is used, specially developed for the fully plastic case, giving J directly in terms of the area under the load displacement curve. The essential terms seem to reduce to $J = f(bq/S)$ where the function is mainly multiplicative. If that is a correct assessment, then, since q will be roughly proportional to S, J indeed seems to be proportional to b in accordance with the experimental results. This seems to throw doubt on the generality of the simplification used, since it does not reflect the dominant, but not clearly understood, trends of dq/da as brought out by Eq 5. To resolve the matter, it would probably be necessary to examine original records and all subsequent steps of analysis in order to see where the two analyses depart from each other or whether the raw experimental data of load, Q, and deflection per crack increment, dq/da, contain terms not accounted for in Eqs 4 to 6.

Effect of Thickness at Constant Ligament Size

In [*22*] bending R-curves are given for A533B for $W = 90$ mm, $a/W = 0.5$, $B = 45$, and 250 mm on one plate and 45 and 500 mm on another (Fig. 7*a*). For the first plate the two thicknesses give the same R-curve, consistent with the arguments here, if $c_1 = B/2$, $c_2 = b/2$, and $c_3 = b/2$. Those factors would also imply the same (near linear) curve for the 45 and 500 mm wide specimens, but the curve for 500 mm then falls below its companion 45 mm specimen after 2 mm growth. To suit that behavior $c_1 \alpha r_{mm}$ or $b_1/2$ must be implied. In fact,

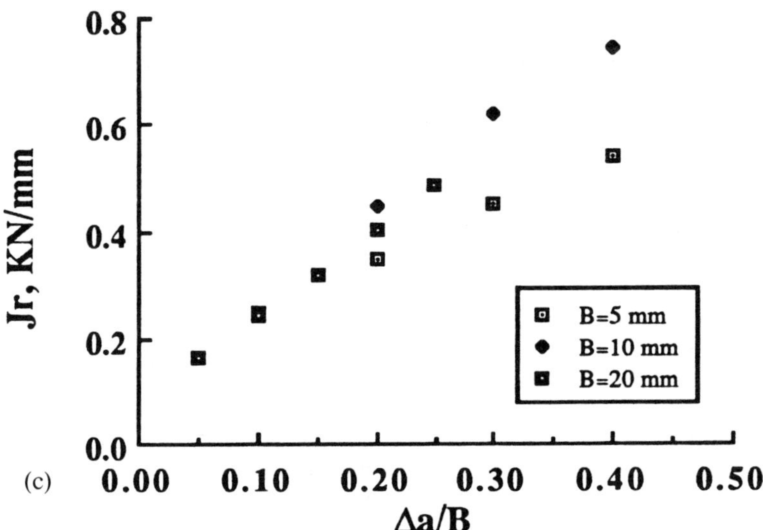

specimens of Cr-Ni-Mo steel from [*23*] show a marked effect of thickness for the thinner specimens. (c) These thinner specimens give near similar curves when scaled by thickness.

all curves are similar to start with, perhaps following J-controlled growth, but there is only one point for each of the thick pieces ($B = 250$ and 500 mm) that defines the two different R-curves beyond $\Delta a/b = 3.5\%$. Indeed, viewed collectively as in Fig. 7a, rather than separately as in [22], perhaps all the data are independent of B within the scatter.

In [4] R-curves for three thickness of HY130 are shown. In [23] four curves for a Cr-Ni-Mo steel of rather moderate toughness are shown. In each case the curves for the two thicker sizes are the same whilst the thinnest pieces with $B < b$ give substantially higher curves, as in Fig. 7b for the data of [23]. In both cases the higher curves for the thinnest pieces scale reasonably with $\Delta a/B$, but that effect is not predicted here unless $c_1 \alpha r_{mm}$ or $b/2$ is used. In both cases the amount of growth reported is less than $B/2$, so it is speculated that the trend is due to the growth of shear lip plausibly over a region of $\Delta a < B/2$ so that a steady-state condition as modelled here has not become established. The two similar curves in [4], for thicker specimens with $B >$ or $= b$, are consistent with present arguments within the uncertainty of estimating r_{mm}. In [23], however, only for the thickest specimen does $B = b$, so that the next thinner piece might be expected to have a higher curve for the same reason as do the two yet thinner. In fact, the curve for the thick specimen extends only to $\Delta a/b = 2\%$ whereas that for the next thinner extends to 10%, so that the statement that the two are the same must be treated with reservation; if the curve for the thickest piece had been continued it might or might not have remained the same.

Effect of Absolute Size (Geometric Similarity)

One of the early sets of J-R experiments [24] showed a case of identical curves for two sizes of En32 low strength ductile steel: $B = W = 10$ mm and $B = W = 24$ mm, $a/W = 0.5$ in three-point bending. More recently, data for A533B have been shown [25] for a range of CT specimens from 25 to 100 mm thick and for a rejected plate of A508 steel [26] in CT specimens 25, 100, and 250 mm thick, $a/W = 0.5$, side grooved. Again both sets of results are independent of size. On the other hand, three-point bend data, $a/W = 0.5$, $S/W = 4$, for three sizes of HY130 in [4] gave the same curve for $B = W = 10$ and 25 mm but a lower curve for $B = W = 50$ mm (Fig. 8).

The results for cases where the toughness (zone size) is large in relation to the dimensions are understood here with $c_1 = B/2$, $c_2 = b/2$, and $c_3 = b/2$, with $b/S = $ constant. For the largest specimen in [4], it is suggested that one of the c terms is controlled by a material factor such as r_{mm} so that dJ/da depends on $\Delta a/$size.

Fact or Fiction?

It will be apparent from the remarks already made that there seem exceptions to every case cited. To briefly repeat a few:

- Are the curves from [22] (Fig. 7a) the same for several thicknesses as one point at $B = 250$ mm implies or lower for yet thicker pieces as one point at 500 mm suggests?
- Why do the J-R curves from [23] (Fig. 7b) increase with a decrease of thickness in some cases despite the only *evident* thickess terms in Eq 4 cancelling?
- Is the scaling with ligament pointed out [17] (Fig. 5) for the cracked pipe accounted for rationally here, probably because of the different properties of weld metal and parent plate affecting the original R-curve, or is it fortuitous?

The authors incline to the opinion that there are so many cases that *do* scale that the simple arguments given here are basically correct within a number of unclear limitations.

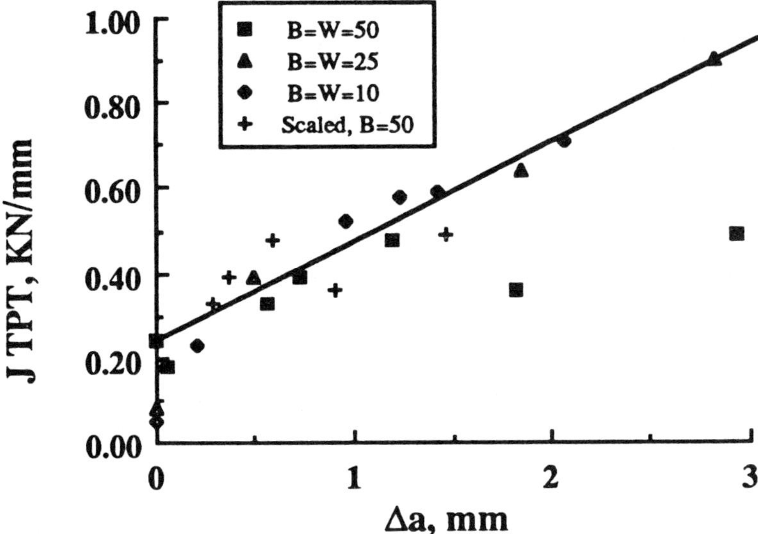

FIG. 8—*Data for HY130 from [4] for which the largest size scales to the medium size. There is no need to scale the medium and small sized pieces.*

The term least understood is dq/da (more strictly dq_{pl} but the differences are small for the fully plastic cases considered here), particularly in the transient state where shear lips are developing. Some light is thrown on this, at least indirectly, by the recent reanalysis of [18] in [20]. The main feature of relevance here is shown in Fig. 9a, where dU_{pl}/Bda is plotted against Δa for four different a/W ratios. In [20] there are many data points at small Δa, indicated by the hatched zone on Fig. 9a. It is believed that these rising parts of the curves are essentially in the elastic or well-contained yield part of the load displacement diagrams and are either in the stretch zone region or correspond to regime (I) of Fig. 4 where the size of plastic zone is still growing with Δa.

The immediate present interest is that the main decreasing trend in Fig. 9a, where in the plastic regime the work rate reduces with crack growth, is similar in all respects to the decreasing curve of work increment for A533B in Fig. 3 here and of J_{ss} in Figs. 3 and 4 of [13] for HY130 and a titanium alloy respectively. However, the data for the different a/W cases show a trend not noticed in the data just cited where in each case only a single geometry was analyzed. It is now observed that the several curves of Fig. 9a reduce to one curve (Fig. 9b) when the abscissa is scaled to $(\Delta a/b)(S/b)$. The rationale for this is not fully clear. From the arguments that lead to Eqs 4 and 5 it must reflect the behavior of dq/da, but the data for the regime where dq/da is changing rather than for the steady state on which the present argument is based. That steady condition is being approached by the data of Fig. 9 and, when reached, must reflect a constant pattern of shear lip development (or crack front curvature since the data in [20] are for side-grooved specimens) whereby dq/da is constant for a given test, implying that a steady state is nearly reached. The value of the constant for steady-state dq/da is dependent, however, on geometry, a/W, in Fig. 9. Using Eq 5 where $Q\alpha(b/S)$ then, if dq/da were the same, dJ/da might be expected to increase with $b^2\Delta a/S$, although the effective S increases according to $(a + [b/2])$ for the CT geometry as already noted.

Contrary to that argument, reconstructing the J_o-R curves (using plastic terms only, from Fig. 9a) gives a higher curve for the smaller ligament (Fig. 10). Moreover, the same data have already been shown in Fig. 6, where the curves just showed scatter with the smallest ligament,

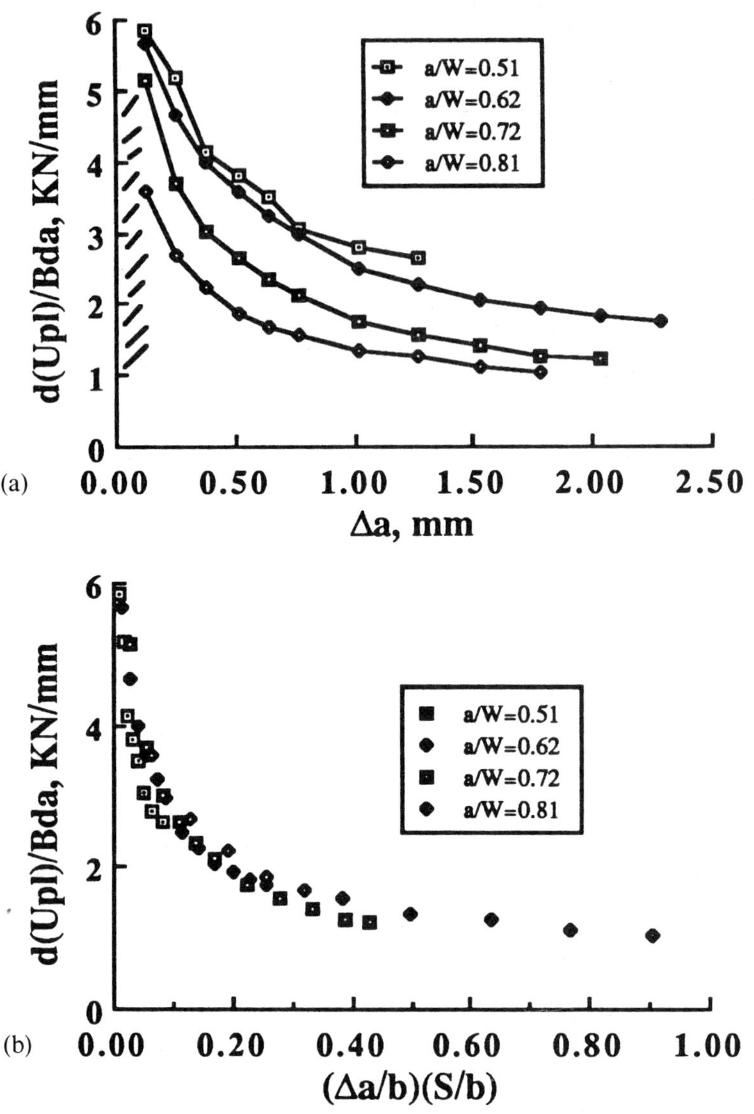

FIG. 9—*A533B data from* [18] *as shown in Fig. 6 for 20% side-grooved CT specimens of various a/W ratios but reanalyzed in* [20] *to show* dU_{pl}/Bda *versus* Δa. (a) *As presented in* [20]; *the cross-hatched region covers points before maximum load, many apparently in the blunting region.* (b) *With abscissa scaled to* $(\Delta a/b)(S/b)$ *here for the plastic regime only.*

$a/W = 0.8$, and the largest, $a/W = 0.5$, at the top of the scatter. The regular trend with a/W in Fig. 10, as distinct from the scatter in Fig. 6 cross-plotted from [18], may well be due to the choice of the initiation value, J_i, here taken as common to all four ligaments from other data given in [20]. Strictly, the different curves in Fig. 9a imply that there is no common region of J-controlled growth with dJ/da independent of ligament as could have been accepted from Fig. 6. The trend in Fig. 10 is the same as in Fig. 2a. Although precise scaling to $\Delta a/b$ is not seen for the J_o curves of Fig. 10, if Fig. 9b is indeed a single curve, then $J_{u,pl}$

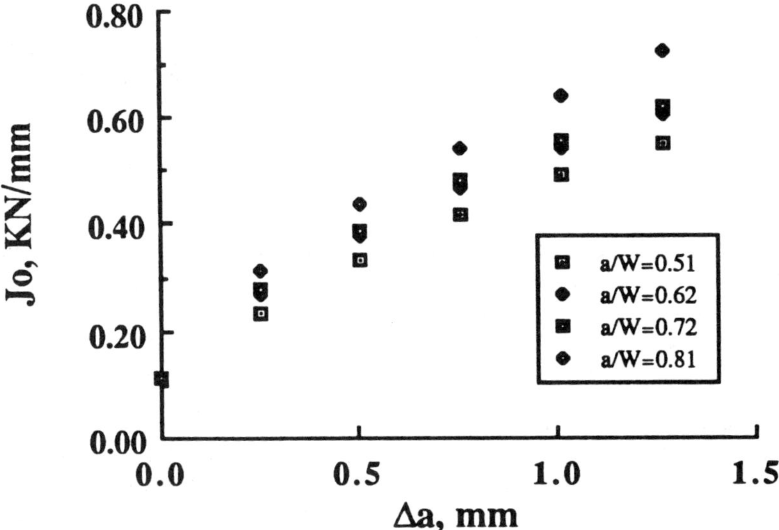

FIG. 10—*J*$_o$–R *curves reconstructed from Fig. 9a by using only the plastic component of the work increment and adding a common value of initiation toughness, now showing a small trend with* a/W *(compare Fig. 6), the smaller ligament giving the higher* R-*curve.*

would scale with $\Delta a/b_o$. On the other hand, Figs. 9 and 10 reflect non-steady-state behavior, and the controlling sizes in the non-steady and steady states may be not the same, perhaps because in Fig. 2, $B > r_{mm}$, whereas in Fig. 10, $r_{mm} \gg B$ or b subject to the uncertain effect of the side grooves, which itself might make use of r_{ie} a more indicative factor than $r_{mm} = 0.2EJ/\sigma_y^2$. The former is smaller than B or b but the latter is larger and both suggest that the 20% side grooves, whilst deep enough to ensure flat fracture, do not ensure full plane strain or steady-state behavior.

It therefore seems to follow that the rapid reduction in dU_{pl}/Bda with growth must, from Eqs 4 and 5, reflect the changing behavior of dq/da both with Δa and a/W. According to Fig. 9b, the change of dq/da (in fact dq_{pl}) must be $f(\Delta a/b)(S/b)$ unless the constraint factor λ is varying unexpectedly and systematically with both growth and ligament so that a fully plastic state does not exist.

Discussion

As already noted, the choice of the most appropriate formula to evaluate dJ and J after initiation is a matter of opinion. None truly reflect the original contour integral meaning of J nor is it at all clear [7] that that term in any form is the correct one for real elastic-plastic materials. These important points have been discussed in many papers (e.g., [27]), but they are not entered into here since the trends under discussion, namely the way in which much existing but inconsistent *R*-curve data can be rationalized, are not much affected by which formulation of J is used. The region of so-called *J*-control for small growth is not discussed beyond noting that it is supported by only some of the data. Where it exists, scaling would apply only beyond the *J*-controlled limit.

The present discussion emphasizes the obvious, namely that whatever was intended, the value of a *J*-based *R*-curve reflects the dissipation of work associated with the whole plastic region of deformation. With increasing plasticity this spreads from local to the tip when characterized by LEFM, to the final limit of full plasticity where slip patterns and component geometry govern the behavior. The commonly used presentation in which an R-curve rises reflects total work dissipation. If an incremental, d/da, work or energy term, is used, consistent with the LEFM concept of work and energy *rates,* it will *decrease* once the maximum size of plastic zone has been reached. The increment remains positive and tends to a limit when based on work but may become negative if based on internal energy. It must then be asked whether engineers intend that the term "toughness" relate to "work (or energy) *rate associated with fracture*" or do they prefer to relate toughness to "*total quantity* of work done in causing fracture" or indeed are both needed as separate factors in fracture behavior? (In the present context the choice of work or internal energy is not important. The question of whether the total [i.e., summation] includes elastic plus plastic or only plastic terms is also not relevant to the present discussion of fully plastic behavior, although it would be crucial to questions of final instability.) Clearly the Griffith concept characterizes the increment or d/da *rate* of either potential energy or work done. On the other hand, the Charpy-type test uses total work *quantity* and J-R-curves as conventionally measured use either total work or an inferred non-linear elastic energy as normalized *quantities* (which increase) rather than as d/da *rates* (which decrease). The distinction is not crucial in LEFM where they are directly related [*28*], but it is argued here that, despite it not apparently being widely recognized, it is crucial in EPFM where the one term increases whilst the other decreases with growth.

The dependence of dJ/da on geometry can be outlined according to whether the limit on plastic work is for reasons of material "toughness" or of size limitation. Three stages can be defined, subject to some of them not being distinct in some cases: (1) J-control for small growth, (2) shear lip control for a transition region (perhaps up to $\Delta a < B/2$ or maybe $b/4$), and (3) steady-state growth with dq/da near constant. The model of Eq 6 is restricted to the steady-state picture, since the trends of dq/da with growth of shear lip are not represented. It also has some application to cases in Stage 2 where even though dq/da is not constant, the variation is controlled by the same geometric factor that controls Stage 3.

Since in the data of [*20*] there are no shear lips, any change in the steady-state value of dq/da (which is approached but not strictly reached in Fig. 9*a*) must reflect changes in the crack front curvature with a/W ratio. If that is so, the differences could be eliminated by an appropriate choice of definition of mean Δa value. Where multiple tests are made and crack growth measured by, say, a nine-point average across the thickness, it is clearly seen that for the early stages of growth there could be large differences in the evaluation of an appropriate mean growth. It appears that by adjusting the definition of the effective Δa two disparate curves could be brought together. That might imply that there is no unique interpretation of the unloading compliance as a measure of crack growth common to various configurations and further suggests that the discrepancy often seen between the final unloading compliance point and a physically measured length is in part genuine. The alternative is that the general interpretation of J-R-curves as a function of Δa is inadequate and that as proposed in [*9, 10*] the abscissa should be scaled. The coming together of the data of Fig. 9*a* into the single curve of Fig. 9*b* is strong support for that interpretation for which Eq 6 is only an incomplete picture.

Finally, if the value of all three c terms is limited by B, then Eq 6 is in accord with the rising limb of the LEFM curve (Fig. 4*a*). If the axial extent, c_3, is taken as arbitrarily small, then ρc_3 reduces to the LEFM plane strain concept of work per unit area, 2γ. If the limit is r_{mm}, then the reducing limb is given as B (and b) increase. In fact, the four thicknesses in Fig. 7 follow a pattern not inconsistent with that limb although an EPFM analysis is relevant to [*23*]. With plasticity, a distinction exists between Stage III of Fig. 4*a* where r_{mm} would be smaller than the next geometric size, b or B, and Stage IV where r_{mm} is less than both; but, as already noted, the plasticity picture of which Fig. 4*b* is one case is better represented against b/B, since that determines the overall plane stress or plane strain nature of the behavior.

Conclusions

1. A J-based R-curve derived from specimens at full plasticity will reflect size factors that limit the extent of dissipation of plastic work.

2. In a steady-state condition of (large) growth, these factors will be some combination of thickness, ligament, and material toughness as represented by the maximum possible size of plastic zone for that material. A number of examples are set out in the paper. The model as presented here seems to explain satisfactorily many of the size effects found in the literature.

3. The present model can be applied more readily to variation of ligament at fixed thickness rather than variation of thickness at fixed ligament. This seems to arise because for ductile materials the ligament may control the extent of shear lip development throughout the test even though dq/da has not reached a steady state. Many cases of geometrically similar increase of size, where there is no change in the governing size factor, are also accounted for.

4. The main perceived limitation of the model where the trend with thickness is not predicted is in the region after J-control, if such is found, of fairly small growth, perhaps $\Delta a < B/2$ when B is the controlling factor, or $b/4$ when b controls, before a steady state of shear lip has developed. Behavior at a certain ligament size will therefore depend on whether it is an initial size with shear lips not yet developed or a current size with shear lips fully developed.

5. The physical response least well understood is the relation between applied displacement and crack growth, dq/da, which is no doubt influenced by the shear lip pattern and crack front curvature. It is in the absence of such data that the present simple steady-state model (or at least growth with control by the same factor, even if not steady state) is seen to be useful.

References

[1] ASTM Test Method for Determining J-R Curves (E 1152).
[2] Shih, C. F., Delorenzi, H. G., and Andrews, W. R., in *Elastic-Plastic Fracture, ASTM STP 668,* American Society for Testing and Materials, Philadelphia, 1979, pp. 65–120.
[3] Gibson, G. P., Druce, S. G., and Turner, C. E., *International Journal of Fracture,* Vol. 32, 1987, pp. 219–240.
[4] Etemad, M. R. and Turner, C. E., *Journal of Strain Analysis,* Vol. 20, No. 4, 1985, pp. 201–208.
[5] Ernst, H. in *Elastic-Plastic Fracture: Second Symposium, Vol. I—Inelastic Crack Analysis, ASTM STP 803,* American Society for Testing and Materials, Philadelphia, 1983, pp. 191–213.
[6] Etemad, M. R. and Turner, C. E., *International Journal of Pressure Vessels and Piping,* Vol. 21, 1985, pp. 81–88.
[7] Rice, J. R., Drugan, W. J., and Sham, T-L. in *Fracture Mechanics (Twelfth Conference), ASTM STP 700,* American Society for Testing and Materials, Philadelphia, 1980, pp. 189–219.
[8] ASTM Test Method for Plane Strain Fracture Toughness of Metallic Materials (E 399).
[9] Etemad, M. R. and Turner, C. E., *International Journal of Pressure Vessels and Piping,* Vol. 20, 1985, pp. 1–5.
[10] Etemad, M. R. and Turner, C. E., *International Journal of Pressure Vessels and Piping,* Vol. 26, 1986, pp. 79–86.
[11] Wei, Y. L., Etemad, M. R., John, S., and Turner, C. E. in *Proceedings,* Int. Conf. on Fracture and Fracture Mech., Fudan Univ., Shanghai, 1987, C. Ouyang, Ed., pp. 60–72.
[12] Turner, C. E., *Size Effects in Fracture,* Institution of Mechanical Engineers, London, 1986, pp. 25–32.
[13] Etemad, M. R., John, S., and Turner, C. E., in *Fracture Mechanics: Eighteenth Symposium, ASTM STP 945,* American Society for Testing and Materials, Philadelphia, 1988, pp. 986–1004.

[*14*] Davis, D. A., Vassilaros, M. G., and Gudas, J. P. in *Elastic-Plastic Fracture: Second Symposium, Vol. II—Fracture Curves and Engineering Applications, ASTM STP 803,* American Society for Testing and Materials, Philadelphia, 1983, pp. II:582–II:610.

[*15*] Etemad, M. R. and Turner, C. E., "Further Evidence on the Normalisation of *J-R* Curves," EGF7 (Sept. 1988), EMAS, Vol. 2, pp. 684–687.

[*16*] Newman, J. C. Jr, in *Elastic-Plastic Fracture Mechanics Methodology, ASTM STP 896,* American Society for Testing and Materials, Philadelphia, 1986, pp. 5–96.

[*17*] Cardinal, J. W. and Kanninen, M. F. in *Nonlinear Fracture Mechanics: Vol. II—Elastic-Plastic Fracture, ASTM STP 995,* American Society for Testing and Materials, Philadelphia, 1989, pp. 320–329.

[*18*] Vassilaros, M. G., Joyce, J. A., and Gudas, J. P. in *Fracture Mechanics (Twelfth Conference), ASTM STP 700,* American Society for Testing and Materials, Philadelphia, 1980, pp. 251–270.

[*19*] Etemad, M. R. and Turner, C. E., "Scaling of *J-R* Curves for Side Grooved Pieces," *International Journal of Pressure Vessels and Piping,* to be published.

[*20*] Mecklenburg, M. F., Joyce, J. A., and Albrecht, P. in *Nonlinear Fracture Mechanics: Volume II—Elastic-Plastic Fracture, ASTM STP 995,* American Society for Testing and Materials, Philadelphia, 1989, pp. 594–612.

[*21*] Garwood, S. J. in *Fracture Mechanics (Twelfth Conference), ASTM STP 700,* American Society for Testing and Materials, Philadelphia, 1980, pp. 271–295.

[*22*] Garwood, S. J. in *Proceedings,* AFMS Int. Conf., Sih et al., Eds., 1983, pp. 935–950.

[*23*] Kaiser, H. J. and Hagedorm, K. E. in *Proceedings,* ICF5, Francois, Ed., Pergamon Press, 1981, pp. 855–862.

[*24*] Garwood, S. J. and Turner, C. E. in *Fracture 1977,* Univ. of Waterloo, Ontario, D. M. R. Taplin, Ed., 1977, pp. 279–284.

[*25*] Ingham, T., Bland, J. T., and Wardle, G. in *Proceedings,* SMIRT 7, 1983, Elsevier Applied Science, 1984, Paper G2/3.

[*26*] McCabe, D. E., Landes, J. D., and Ernst, H. A. in *Elastic-Plastic Fracture: Second Symposium, Vol. II—Fracture Curves and Engineering Applications, ASTM STP 803,* American Society for Testing and Materials, Philadelphia, 1983, pp. II:562–II:581.

[*27*] *Post-Yield Fracture Mechanics,* 2nd ed., D. G. H. Latzko et al., Eds., Elsevier Applied Science, 1984.

[*28*] Turner, C. E. in *Fracture Mechanics (Twelfth Conference), ASTM STP 700,* American Society for Testing and Materials, Philadelphia, 1980, pp. 314–332.

M. F. Mecklenburg,[1] C. O. Arah,[2] D. McNamara,[2] H. Hand,[2] and J. A. Joyce[3]

Adhesive Fracture Testing

REFERENCE: Mecklenburg, M. F., Arah, C. O., McNamara, D., Hand, H., and Joyce, J. A., **"Adhesive Fracture Testing,"** *Fracture Mechanics: Twenty-First Symposium, ASTM STP 1074,* J. P. Gudas, J. A. Joyce, and E. M. Hackett, Eds., American Society for Testing and Materials, Philadelphia, 1990, pp. 307–321.

ABSTRACT: The need for lightweight materials for a variety of applications has resulted in the use of structural adhesives to bond prototype structures. Adhesives developed to accommodate the stringent requirements of these high-technology applications are usually deficient in one or two of three very crucial properties: strength, moisture resistance, and toughness. So far, advances in adhesive formulation that have ameliorated one of these deficiencies have generally adversely affected the others. Hence a considerable amount of effort is being expended in the search for strong, moisture resistant, and tough adhesives.

As adhesives become tougher and less brittle, evaluating their performance in terms of fracture parameters becomes more complicated. Linear elastic fracture mechanics (LEFM), which is widely used to characterize these materials, does not fully describe adhesive performance as more and more ductility or plastic deformation is introduced. In the study reported here, we introduce the energy separation method for characterizing the fracture resistance of adhesives and compare it with currently used elastic and plastic fracture parameters such as G and the J integral. Both neat and bonded ½ CT plan specimens were tested and compared in this work.

KEY WORDS: elastic-plastic fracture, structural adhesive, neat bonded fracture, fracture toughness, energy separation

During a research program to develop a strong, moisture-resistant, ambient-temperature-curing, structural adhesive, we found a consistent reduction in ductility (as measured by neat material tensile tests) as strength and moisture resistance improved. This effect was primarily due to the higher crosslink densities in the modified stock epoxy adhesives we were using. Our immediate concern, however, was the fracture resistance of the modified epoxies and how to reliably characterize it so that it could be related to joint performance (e.g., correlate modifications to adhesive chemistry with adhesive fracture parameters).

A review of the fracture characterization options for our anticipated tougher adhesives indicated that the most useful parameters would be the elastic energy released, the plastic energy dissipated, and the potential energy of the test specimen at any crack extension. Linear elastic fracture mechanics (LEFM), which is widely used to characterize these materials, would not be applicable as ductility or plastic deformation increased.

The J integral [1] was explored because of its application to nonlinear elastic material behavior. For the adhesive geometry of interest the HRR singularity field could not be expected, but the J integral parameter still generally provides a useful experimental parameter to evaluate the relative material toughness of materials falling outside the LEFM regime.

---

[1] Conservation Analytical Laboratory, The Smithsonian Institution, Washington, DC 20560.
[2] Martin Marietta Laboratories, Baltimore, MD 21227.
[3] Department of Mechanical Engineering, U.S. Naval Academy, Annapolis, MD 21402.

The J integral parameter was not fully satisfactory for this application, since it was found to be relatively insensitive to the overall failure mode of the adhesive bond. The J fracture parameter is, in effect, a single-value parameter combining elastic and plastic energy components of crack growth resistance. We needed a clearer distinction of these energies, however, because some types of polymer modification would have a large effect on elastic component while others would influence the plastic component. Ting and Cottington [2] gave clear indications of experimental procedures that might be useful, even though they primarily concentrated on the elastic energy component of adhesive fracture testing. The method we chose for characterization was an energy separation technique [3] that clearly distinguishes crack growth energies and employs a standard, single-specimen, unload compliance test procedure commonly used in J integral testing [4].

Computation Methods

Energy Separation Method

The area under the plot of a load versus load-line displacement record is the sum of all the energies—elastic, plastic, and potential—applied to a fracture test specimen during crack initiation and extension. The work, W, done by the external load, P, can be related to the internal energy, U, by

$$dW = dU_s + dU_e + dU_p \tag{1}$$

where

W = work done by external load,
U_s = stored elastic strain (potential) energy,
U_e = elastic energy released during crack extension, and
U_p = plastic energy dissipated during crack extension.

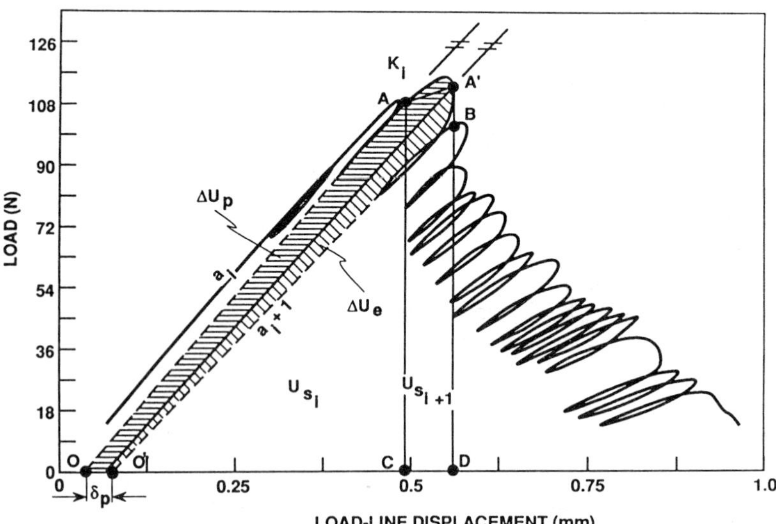

FIG. 1—*Partitioned load versus load-line displacement record for specimen M2A.*

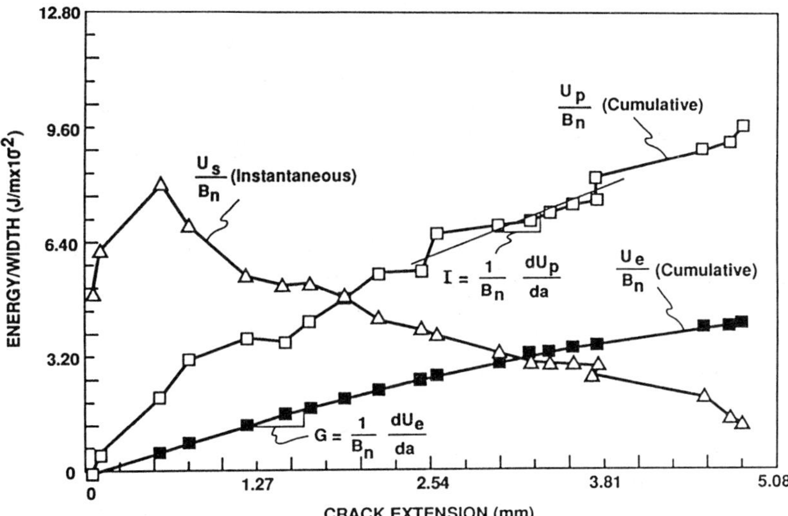

FIG. 2—*Cumulative energies released and stored versus crack extension for adhesive specimen M2A.*

These energies are depicted in Fig. 1, which illustrates an unload compliance, load versus load-line displacement record of a neat adhesive ½ CT plan specimen. Here the specimen is side-grooved 20% giving $B = 6.35$ mm (0.25 in.) and $B_n = 5.08$ mm (0.20 in.). In Fig. 1:

- Area $OAA'O' = U_p$
- Area $O'A'B' = U_e$
- Area $O'BD = U_s$ after crack extension (U_{si+1})
- Area $OAC = U_s$ prior to crack extension (U_{si})

The total resistance to a crack growth increment, da, is the sum of the plastic energy dissipated, dU_p, and the elastic energy released, dU_e. The total rate of energy release and dissipation with respect to crack growth can be expressed mathematically as

$$J_{ES} = G_{ES} + I = 1/B_n \, dU_e/da + 1/B_n \, dU_p/da \qquad (2)$$

where

$$\begin{aligned} J_{ES} &= \text{total energy release rate,} \\ G_{ES} &= \text{elastic energy release rate,} \\ I &= \text{plastic energy dissipation rate, and} \\ B_n &= \text{net specimen width between side grooves.} \end{aligned}$$

Figure 2 shows the width normalized, cumulative sums of the plastic energies dissipated, and the elastic energies released and plotted with respect to the crack extension. As can be seen, G_{ES} is the slope of the cumulative released elastic energy normalized by the net width, and I is the slope of the cumulative dissipated plastic energy normalized by the net width. Also shown in Fig. 2 is the instantaneous value of the stored potential energy, U_s/B_n, at any crack extension. Neither of the curves is linear with respect to crack extension (i.e., their respective slopes are not constant and the values of G and I are not independent of crack length). Nevertheless, by holding specimen size and initial crack length constant, the approach outlined above provides a clear method for comparing the specific effects of modifying the adhesive chemistry to the different aspects of fracture resistance.

The quantities G_{ES} and I are obtained from the test load, load-line displacement records and hence include the full specimen energy release and energy dissipation, respectively. Both quantities come in part from crack growth and in part from elastic energy released and plastic energy dissipated in the specimen well away from the crack tip region. The inability to separate the near and far components is a major drawback to any energy method. In this study, where all inelastic deformation is confined to the adhesive and in which the specimen size and geometry is not varied, it is felt that the energy method provides a useful measure of material toughness. Difficulties encountered with the application of J integral methods are discussed in the following section.

Figure 3 illustrates the crack length dependency with plots of G_{ES} and I against crack extension. After an initial jump both G_{ES} and I decrease with crack extension. To ensure reliable correlations during the adhesive study in the presence of this crack length dependency, all specimens were made identical in size and precracked to the same initial crack length. For comparison, a crack growth of 0.762 mm (0.030 in.) beyond the initial crack length was selected, since that interval generally coincides with the onset stable crack growth. This point is indicated by the vertical dashed line in Fig. 3.

Comparisons with J *Integral*

The usefulness of the energy separation method can be assessed by comparing it with the J integral method. In his original derivation of the J integral, Rice [1] was fairly explicit in relating J to body potential energy, whether the material behavior was linear or nonlinear. By assuming that elastic-plastic material behavior could be treated as nonlinear elastic material behavior in the absence of unloading, Rice, Paris, and Merkle [5] concluded that J for

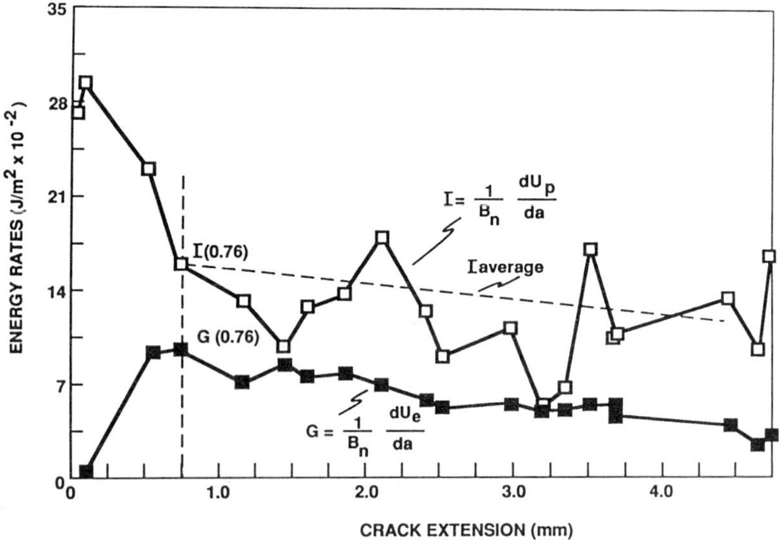

FIG. 3—*Elastic energy release rate and absolute plastic energy dissipation rate versus crack extension for specimen M2A.*

the deeply cracked bend bar could be expressed as

$$J = 2A/B_n b \tag{3}$$

where

A = total area under the load versus load-line displacement record,
b = remaining ligament of test specimen, and
B_n = net width of test specimen.

The coefficient 2 in Eq 3 was modified to account for the tensile component of stress present in the compact tension specimen and designated by η. In general, the current calculation of the J integral takes the form

$$J = \eta U/B_n b = J_{el} + J_{pl} = G_{ES} + J_{pl} = \eta U_s B_n d + \eta U_p/B_n b \tag{4}$$

where

J_{el} = elastic component of J,
J_{pl} = plastic component of J,
η = a coefficient related to the type of specimen,
U = total area under the load versus load-line displacement record, and
U_s, U_p, B_n, b, and G_{ES} are as defined previously.

In the specific case of calculating the elastic energy release rate, G is now taken to be

$$G_q = K_q^2(1 - \nu^2)/E \tag{5}$$

where K_q is the stress intensity factor as calculated from ASTM E 399.

Experimental work, described more fully below, showed that the elastic modulus of the aluminum could be used for bonded specimens in this conversion, while for neat specimens the adhesive elastic modulus was used as obtained from tensile tests of the adhesive material.

The applicability of the adherend elastic modulus for this calculation for the bonded specimens should be similarly verified if either the adherend or the adhesive is changed markedly.

We can now make some observations relating energy separation to the J integral by calculating G in several different ways and by comparing those values to the values obtained by the energy separation method. It may be seen from Eqs 2, 4, and 5 that

$$G = K_q^2/(1 - \nu^2)/E = 1/B_n \, dU_e/da = \eta U_s/B_n b \tag{6}$$

which apparently relates the stress intensity factor K_q, as calculated by ASTM E 399, to the instantaneous potential energy of the test specimen, U_s, and also to the elastic energy released, U_e. Figure 4 illustrates the results of evaluating G by the three expressions: from K_q, as determined from ASTM E 399; from energy separation using the increment of elastic energy release, $G_{ES} = 1/B_n \, dU_e/da$; and from energy separation, but using the instantaneous value of the potential energy, $G_{PE} = \eta \, U_s/B_n$, where $\eta = 2.3$. As can be seen, results of all three calculations compare favorably in spite of plastic deformation, suggesting that the method of partitioning the different energies is consistent with current theory. The last form of calculation (i.e., G_{PE}) permits G to be determined at the initiation of unstable crack growth.

The major drawback found in this work to the application of J to characterize the fracture toughness of the adhesive material and adhesively bonded specimens was the relative insen-

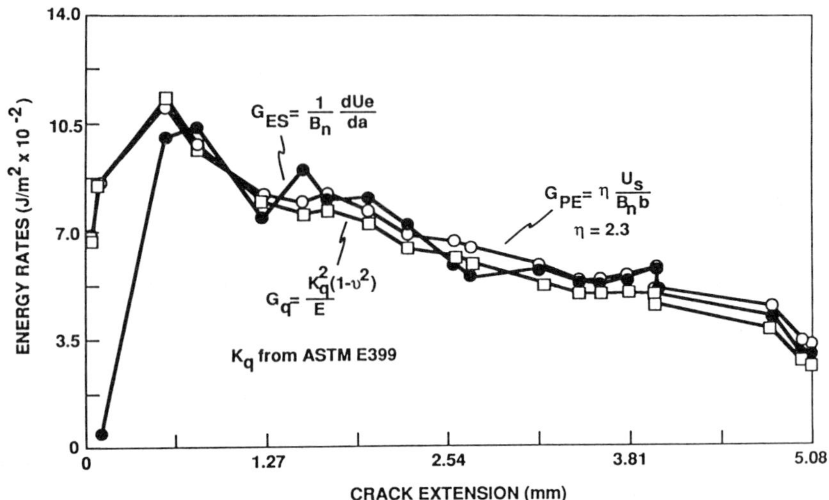

FIG. 4—*Comparison of three methods of evaluating G using neat adhesive M2A.*

sitivity of J to the demonstrated fracture behavior. Generally, for these tests, a tradeoff occurred between the elastic J component and the plastic J component, giving relatively small changes in J at crack initiation in spite of dramatically different specimen behaviors. Utilization of only the plastic J component was considered, but it appeared that the use of I, as given by Eq 2, gave a better measure of material toughness and was more sensitive to changes in adhesive chemistry and moisture content.

The J integral treats U_p (non-recoverable) as a component of the potential energy U_s (recoverable), whereas the energy separation method treats U_p as a post-crack extension quantity similar to U_e. As a consequence, I becomes a large value when U_p is divided by a small quantity, da, whereas J_{pl} is a relatively small quantity since U_p is divided by the total remaining ligament b. Hence the significance of U_p might be lost when it is combined to form the total quantity J. As we will show in our experimental results, G for our tests adhesives does not vary dramatically, while I can show substantial difference from one adhesive type to another.

Experimental Procedures

All specimens, either neat adhesive or bonded, were ½ CT plan with $B = 6.36$ mm (0.25 in.). Neat specimens were molded with a side groove to give $B_n = 5.08$ mm (0.20 in.), while bonded specimens were ungrooved. The dimensions of the specimens are presented in Fig. 5. In both cases blunt notches were fabricated to a depth of about 10.5 mm and subsequently extended by pressing in a razor blade to produce a sharp starter crack of about 12.9 mm total depth.

The clip gage load-line attachment points were made integral to the specimen. The bulk specimens were cast in aluminum molds coated with release agents. Adherends for the bonded specimens were machined from aluminum and bonded in alignment jigs to provide a bondline thickness of about 0.254 mm (10 mils).

Surface preparation for all adhesive bonding in this program was done using Boeing Corporation's phosphoric-acid-anodization (PAA) process. This process has gained wide acceptance as the treatment which provides the most durable and reliable bonding surface for aluminum. By selecting this process, we ensured that we were testing the properties of the adhesive materials, unclouded by uncertainties in adhesive-metal bonding. In addition, the PAA process is not very difficult to implement and is routinely used for large-scale airframe components.

The PAA process consists of the following steps:

1. Degreasing/surface cleaning in an alkaline soap bath.
2. Immersion rinse in distilled water.
3. Deoxidizing in an acid solution (we used the Forest Products Laboratory aluminum etch: 17 vol% concentrated sulfuric acid and 60 g/L sodium dichromate at 66°C (150°F) for 15 min).
4. Immersion rinse in distilled water.
5. Anodizing in 10 wt% phosphoric acid solution at 10 V (specimen attached to the positive terminal) for 20 min at room temperature.
6. Immersion rinse in distilled water.
7. Air dry.

The resultant surface exhibits water-break-free behavior during rinsing and a blue-green iridescence when viewed at grazing incidence.

The oxide surface produced by the PAA process on aluminum consists of a "forest" of extremely fine whiskers (roughly 100 Å across) that can mechanically interlock with the epoxy polymers, producing a very rugged bond. The phosphorous ions in the solution, which are incorporated into the oxide during the anodizing, prevent moisture from attacking the oxide film. Thus bonds to PAA surfaces are also more durable than those surfaces treated by other means.

The crack length, a, was determined using the elastic compliance method and the empirical equations as provided in ASTM E 1152. In general, we found that the crack length could be more accurately obtained if the unloading was at least 20% of the current maximum load. Measured crack lengths agreed well with those calculated by the empirical equation for both the neat and bonded compact specimens. The results of this comparison are presented in Fig. 6.

The modulus of elasticity, E, was determined in a tensile test on neat specimens, and was used for the crack length determination and the calculation of K_q. For the bonded specimens, we used the modulus of the aluminum.

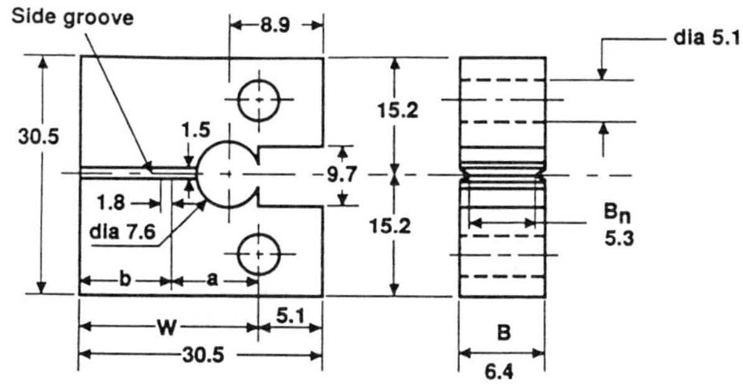

FIG. 5—*Compact tension ½ specimen dimensions.*

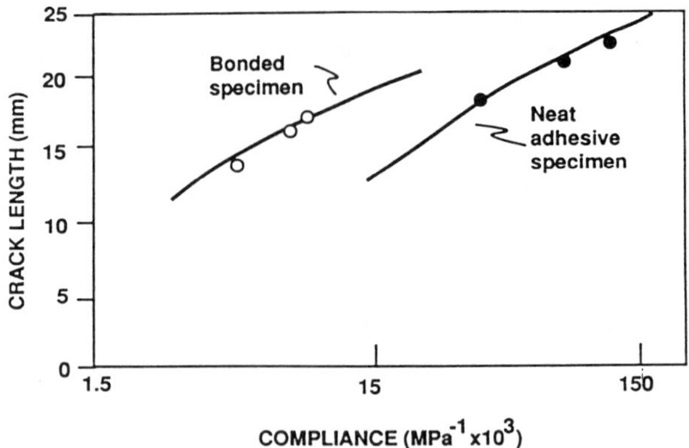

FIG. 6—*Comparison of calculated* (solid line) *and experimental* (circles) *plot of crack length as a function of compliance.*

Results and Discussion

Neat Adhesives

Test results of the different neat adhesives varied widely, displaying considerable differences in load magnitudes at crack initiation as well as plastic deformation. Three examples of typical neat adhesive, unloading compliance, load versus load-line displacement records are shown in Figs. 7a–c, which show respectively: (*a*) material that experienced a brittle,

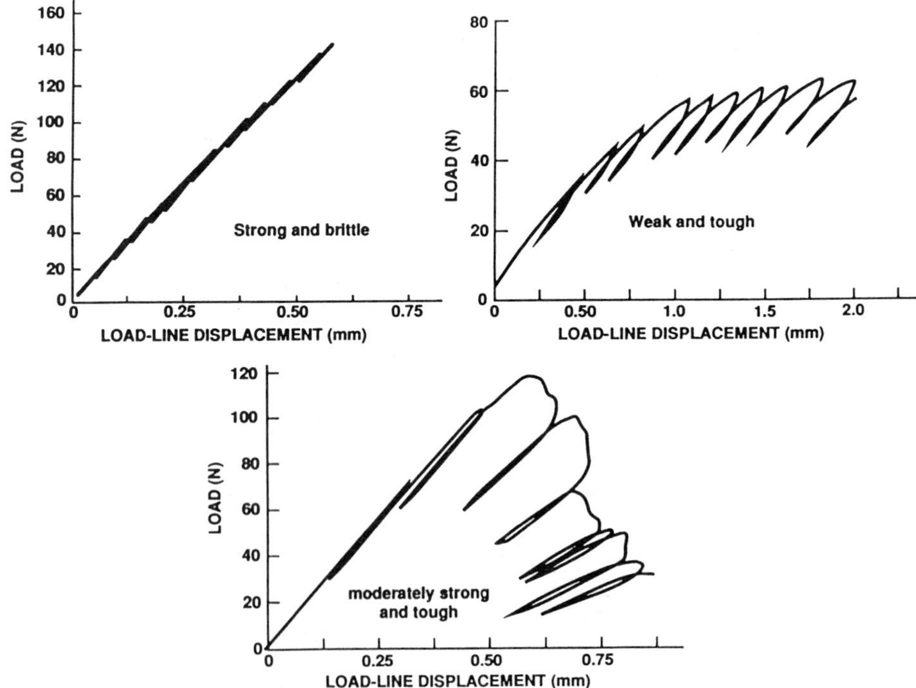

FIG. 7—*Load versus load-line displacement records of different adhesives.*

unstable failure at about 156 N (35 lb) but was quite moisture resistant; (*b*) very tough but weakly adhesive behavior with little resistance to moisture; and (*c*) a moderately tough and strong adhesive with some moisture resistance. Moisture resistance was determined separately in adverse environments using stress relaxation tensile tests on bulk specimens [6].

Some typical results for G_{ES} and I are shown in Table 1 for fracture tests on neat adhesive compact tension specimens; J is included for comparison. All values listed (average of three tests) were taken at a crack extension of 0.762 mm (0.03 in.). G_{ES} and I are calculated by the energy separation method except in the case of brittle fracture, where G was calculated from Eq 5. J was calculated as defined by ASTM E 1152. Within a single specimen, G_{ES} and G_q maintained moderately constant decreasing values over the crack extension examined and had a standard deviation of less than or equal to 20% from specimen to specimen of the same type of adhesive. I values were more variable both within the same specimen and from specimen to specimen. Of all variables influencing the test results, crack extension was the most difficult to measure reliably. Crack length measurements improved with deeper unloadings, better determination of the values of E, and more accurate load-line displacement measurements.

The wide variations in I from adhesive to adhesive shown in Table 1 had nearly no correlation to the elongation to failure data as measured in a tensile test. Table 2 shows typical results of the neat specimen tensile testing. Of particular interest are Adhesives 14 and 96, which show moderate tensile elongations but completely brittle fracture.

TABLE 1—*Fracture toughness results on neat adhesive specimen (average of at least three tests).*

Adhesive	G_{ES}		I		J		Peak Load	
	J/m²	(in.-lb/in.²)	J/m²	(in.-lb/in.²)	J/m²	(in.-lb/in.²)	N	(lb)
1	650	(3.7)	735	(4.2)	825	(4.7)	142	(32)
3	1450	(8.3)	1020	(5.8)	1910	(10.9)	151	(34)
5	615	(3.5)	3270	(18.7)	1540	(8.8)	62	(14)
100	1820	(10.4)	4850	(27.7)	3150	(18.0)	258	(58)
14	1930[a]	(11.0)	0	0	1930[a]	(11.0)	160	(36)
17	630[a]	(3.6)	0	0	630[a]	(3.6)	124	(28)
96	1260[a]	(7.2)	0	0	1260[a]	(7.2)	138	(31)
98	1380[a]	(7.9)	0	0	1380[a]	(7.9)	156	(35)

[a] Specimen showed totally brittle behavior; $G_{ES} = G_{PE} = J$.

TABLE 2—*Tensile test values for neat adhesives.*

Adhesive	Ultimate Tensile Strength		Modulus		Elongation, %
	MPa	(psi)	MPa	(ksi)	
1	61.8	(8960)	2830	(410)	2.8
3	52.8	(7650)	2280	(330)	5.4
5	14.9	(2160)	760	(110)	10.6
100	57.5	(8340)	2830	(410)	2.8
14[a]	60.9	(8830)	2280	(330)	5.4
96[a]	73.9	(10710)	2690	(390)	6.3
98[a]	61.8	(9520)	2280	(330)	4.9

[a] Denotes model adhesives developed for this program. The other adhesives are commercially available room-temperature-cure systems.

Bonded Specimens

Fracture characterization data indicated that, in general, bonded specimens were more stable and demonstrated greater load capacities than neat specimens. Figure 8 shows load versus load-line displacement plots of ½T CT neat and bonded test specimens identical in size and type for the same adhesive. The stiffer bonded specimen exhibits considerably less load-line displacement than the neat specimen and withstands over three times the load at crack initiation.

Typically, bonded specimens remained stable throughout the test procedure, while neat specimens were unstable beyond the initiation of crack extension. Comparing the adhesive toughness in terms of J and G generated the appearance of similar values, an appearance felt to be very misleading in terms of the expected performance in the intended application where strength and stable behavior were desired. Comparison of the adhesive toughnesses based only on I at initiation shows that, on this basis, the bonded specimens are tougher and should be more stable than the neat specimens. Typical results for the bonded specimens are presented in Table 3. For the case where brittle behavior was observed G was taken as G_{PE} and was identical to J. For Adhesives 5 and 14 (Fig. 9) G_{ES} was found to be greater than G_{PE} or J at initiation, while for Adhesives 11, 3, 100, and 96 the reverse was true. In all cases it was I that related directly to the desired application properties of high bond strength and stable crack growth.

The values of G_{ES} (or G_{PE}) for the bonded specimens are similar to, but slightly lower than, those of the neat specimens, and bondline failure was always cohesive in these cases. When the bondline failed at the adhesive-adherend interface, the values of G_{ES} dropped substantially; I for the bonded specimens was routinely higher, consistent with their more stable crack growth.

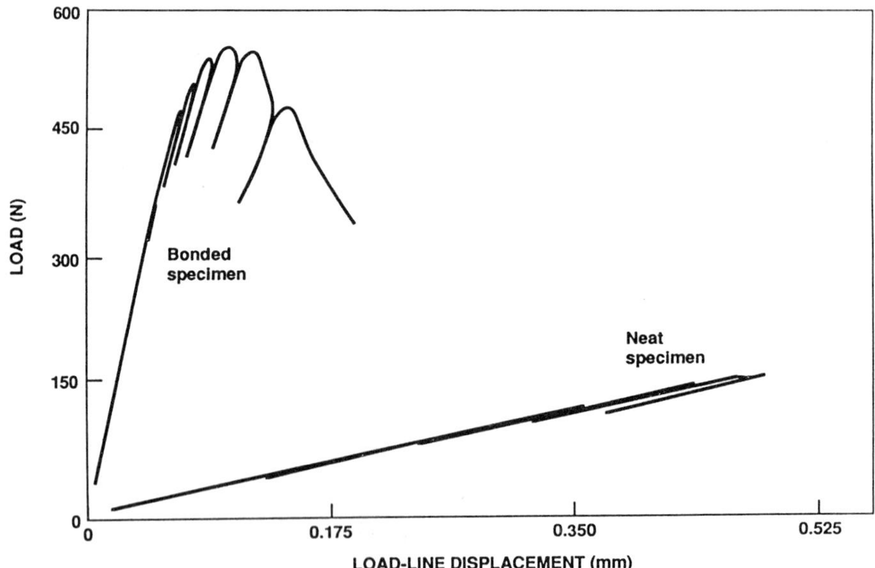

FIG. 8—*Comparison of the load versus load-line displacement records for bonded and neat adhesive specimens.*

TABLE 3—*Fracture toughness results for bonded specimens (average of at least three tests).*

Adhesive	G_{ES}		I		J		Peak Load	
	J/m^2	(in.-lb/in.^{2})	J/m^2	(in.-lb/in.^{2})	J/m^2	(in.-lb/in.^{2})	N	(lb)
1	735	(4.2)	3470	(19.8)	790	(4.5)	663	(149)
3	860	(4.9)	2170	(12.4)	1240	(7.1)	614	(138)
5	610	(3.5)	2080	(11.9)	525	(3.0)	334	(75)
100	1450	(8.2)	7000	(40.0)	1930	(11.0)	952	(214)
14	665	(3.8)	210	(1.2)	370	(2.1)	583	(131)
96	700	(4.0)	470	(2.7)	719	(4.1)	494	(111)
98	1380^a	(7.9)	0	(0)	1380^a	(7.9)	494	(111)

^{a} Specimen showed totally brittle behavior; $G_{ES} = G_{PE} = J$.

Adherend Stiffness Effects

The most probable explanation for the differences in behavior is the substantial difference in the stiffness of the two types of specimens. The aluminum adherends, which are quite stiff relative to the bondline adhesive, can provide considerable mechanical constraints on the adhesive as well as allow for more uniform rotation of the specimen during loading. More uniform rotation of the adherend produces more uniform distribution of the entire bondline normal stresses, which reduces the concentration of stress at the crack tip and allows the specimen to sustain a higher load. On the other hand, the neat specimen is more flexible as demonstrated by the greater load-line displacement and tends to "peel" apart, which concentrates stresses nearer the crack tip and allows a smaller load to initiate crack growth. These effects are illustrated in Figs. 10 and 11. Figure 10 presents the elastic plane-strain finite element bondline stress analysis for the aluminum-bonded specimen. The specimen used as the model was bonded with Adhesive 100, which has a maximum load capacity of 952 N (214 lb) and a modulus of 2830 MPa (410 ksi); the modulus used for the aluminum adherend was 73.1 GPa (10.6 $\times$ 10^3 ksi). The bondline used in the model was 0.381 mm (0.015 in.) thick. The stress distribution along the bondline over the remaining ligament is quite uniform. For the stress distribution to be this uniform, the aluminum adherends must be uniformly rotating during the loading of the test specimen. On the other hand, the neat specimen stress analysis (Fig. 11) presents a stress distribution that is quite non-linear, with a

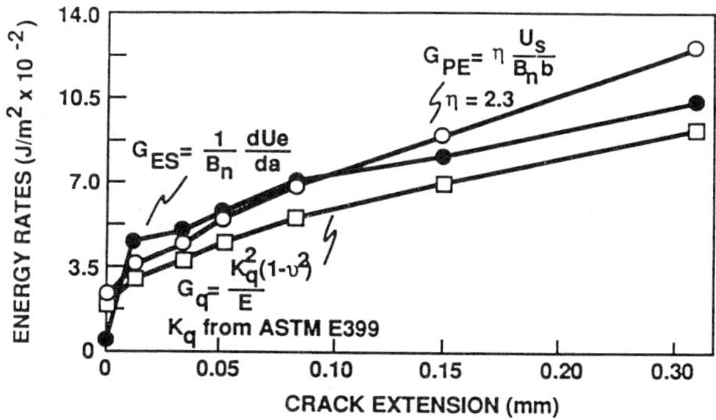

FIG. 9—*Comparison of three methods of evaluating G using bonded adhesive #14.*

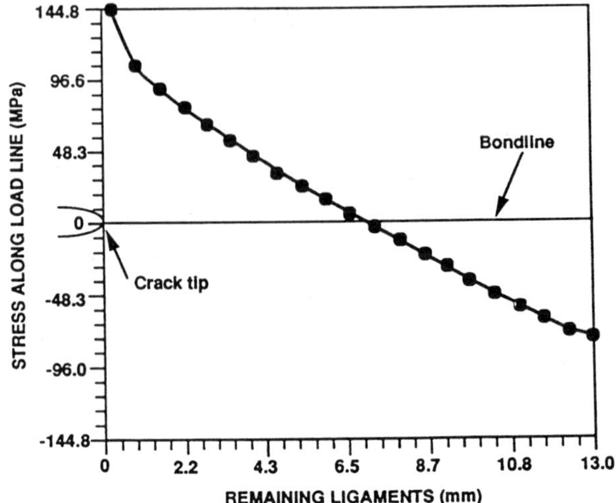

FIG. 10—*Finite element analysis of a bonded ½ CT specimen showing elastic stress distribution.*

severe concentration near the crack tip. The same model geometry was used for both finite element analyses (Figs. 10 and 11), but a value of 2830 MPa (410 ksi) was used as the modulus of the entire neat specimen and the maximum load applied was 258 N (58 lb). The more flexible neat specimen is seen to "peel" apart, leading to progressive crack growth and development of a load-carrying capacity below the full potential of the bondline.

Conclusions

The fracture performance of both neat and bonded adhesives can be effectively compared on a very small compact tension specimen using the unload compliance procedure. Using this approach we can make a better evaluation of changes in adhesive chemistry.

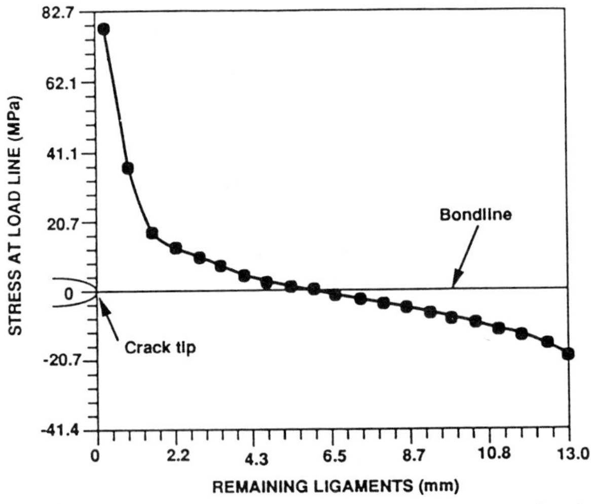

FIG. 11—*Finite element analysis of a neat ½ CT specimen showing elastic stress distribution.*

The energy separation method described here allows G_{ES} to be measured directly from the test record using either the increment of released energy, dU_e, or the instantaneous value of the potential energy, U_s. Either measurement correlates well to current theory.

The parameter I, as derived with the energy separation method, provides a distinct and sensitive measurement of the plastic component of fracture resistance.

Since all the parameters discussed in this study are dependent on the crack length, care must be used to maintain consistent specimen size and geometry including the initial crack length.

Compared with neat adhesive specimens, bonded specimens sustained more than three times the load at crack initiation and consistently exhibited greater crack growth stability, even when the neat specimens showed brittle behavior.

The calculated values of J suggest that the neat adhesive specimens resist crack growth better than the bonded specimens, contrary to experimental observations. This discrepancy is primarily due to the under-representation of the plastic energy component in the computation of J. Results of the energy separation technique, however, clearly show the superiority of the bonded specimens in resisting crack growth, as reflected in the values of I.

Acknowledgments

We gratefully acknowledge financial support for Martin Marietta from the U.S. Army Troop Support Command's Belvoir Research, Development and Engineering Center under Contract DAAK-70-86-0084.

References

[1] Rice, J. R., "A Path Independent Integral and the Approximate Analysis of Strain Concentration by Notches and Cracks," *Journal of Applied Mechanics,* Vol. 35, 1968, p. 379.

[2] Ting, R. Y. and Cottington, R. L., "Comparison of Laboratory Techniques for Evaluating the Fracture Toughness of Glassy Polymers," *Journal of Applied Polymeric Science,* Vol. 25, 1980, p. 1815.

[3] Mecklenburg, M. F., Joyce, J., and Albrecht, P., "Separation of Energies in Elastic-Plastic Fracture," in *Non-Linear Fracture Mechanics: Volume II—Elastic-Plastic Fracture, ASTM STP 995,* American Society for Testing and Materials, Philadelphia, 1989, p. 594.

[4] Joyce, J. A. and Gudas, J. P. in *Elastic-Plastic Fracture, ASTM STP 668,* American Society for Testing and Materials, Philadelphia, 1979, p. 451.

[5] Rice, J. R., Paris, P. C., and Merkle, J. G., in *Progress in Flaw Growth and Fracture Toughness Testing, ASTM STP 536,* American Society for Testing and Materials, Philadelphia, 1974.

[6] Albrecht, P., Mecklenburg, M. F., Wang, J., and Hong, W., "Effect of Environment on Mechanical Properties of Adhesives," Final Report, Office of Research, Development, and Technology, Federal Highway Administration, Contract DTFH-61-84C-00027, Feb. 1987.

DISCUSSION

C. E. Turner[1] *(written discussion)*—This writer entirely supports the authors' use of dU_{pl}/Bda as a meaningful measure of toughness, implying that toughness is a plastic work dissipation rate associated with the separation process. The writer will not, however, use the symbol I for that term, since he has previously used it (e.g., [1] and at more length [2]) for "the other side of" the instability equation, namely, the "elastic energy release rate in the presence

---

[1] Mechanical Engineering Department, Imperial College, London, England.

of plasticity". Thus, by mischance, ductile instability expressed as "energy release rate (including compliance)" = "work absorption rate" would become (Turner)/(geometric factor) = (Mecklenburg et al.)! Moreover, the authors' Fig. 1, showing elastic and plastic energy rates, is misleadingly similar to that used by the writer to define I (Turner). The differences are, however, most important.

It is not clear whether in Fig. 1 the line $O'A'$ is parallel to OA. If it is, there would seem to be no crack growth during the essentially constant load behavior, AA'. If the lines are not parallel, implying that there is growth, then surely there is an elastic component within $OAA'O'$? The analysis assigns the plastic work $OAA'O'$ to growth during the whole period $AA'B$. Strictly, since point B is marked on the next reloading loop, these remarks refer to a point, say B', just above B as the load falls from A'. The combination into one analysis of what appears to be a period of either no growth or stable growth (during AA', according to whether or not the lines are parallel) with a period of apparently unstable growth (during $A'B'$ where the event is at fixed displacement) seems to merit further explanation.

The comparison made between I (Mecklenburg et al.) and J is moreover rather misleading in that dJ/da is not mentioned. Yet I (Mecklenburg) relates to dJ/da rather than J. The precise relationship depends on the definition of J after growth, but the leading term in any definition follows from the first term in the differentiation of $J = \eta U/Bb$ as $dJ_0/da = \eta(dU/da)/Bb$ (here written as dJ_0 to distinguish it from the many other definitions in use). Thus I (Mecklenburg) = $(b/\eta)(dJ_0/da)$, and its decreasing value (shown, for example, in Figs. 3 and 4) would be reflected in the decreasing slope of the conventional J-R curve.

The point at issue appears to be does "toughness" imply to the user a normalized work, as in the evaluation of the conventional J and J-R curve formulae quoted above, or a work rate with respect to crack growth, as in the original LEFM usage and as dJ/da or I (Mecklenburg) would measure it? The emphasis in that question, as discussed in the writer's own (jointly authored) paper presented to this symposium, is not whether work or internal energy should be used, relevant though that is [3], but rather whether the quantity or its crack rate derivative, d/da, is the factor of interest. The question arises because prior to initiation both terms change in proportion so that one encompasses the other. That is also true for growth in LEFM plane stress, whereas in EPFM *after crack growth* work increases whilst work rate decreases. This was illustrated most clearly in a paper written by some of the same authors [4], which case was discussed at some length in the writer's presentation to the symposium. An answer can hardly be drawn here if 20 years of research have not yet produced it, although the writer believes that most would agree that prior to initiation any of the terms will serve as a measure of crack tip severity because work, work rate, and the HRR (including K for LEFM) crack intensity field can all be related, whereas after initiation it is the work rate that controls instability. In that case both terms are required separately since they have different numerical trends and different physical significance. What is lacking is a clear definition and agreed upon usage of two separate words, one of which will be continuous in meaning with LEFM. Since "toughness" there undeniably means "work rate", the implication is that that designation should be retained for dJ_0/da or I (Mecklenburg), subject to whether the geometric term η/b is included or not. A separate word or expression other than "toughness", such as "tearing resistance", should come then into use for J and J-R curve properties based on normalized work. Of course, the definition of η ensures that *up to initiation* the numerical value of the "tearing resistance" from $J = \eta A/Bb$ is the same as the "toughness" from $J = -dP/Bda$ (where A is area under the load-displacement diagram and P is potential energy with the usual reservations on plasticity with no unloading), but after initiation the two terms would differ. Both rise in contained yield but diverge rapidly as full plasticity is reached, "tearing toughness" (from work) continuing to rise but "toughness" (from work rate) falling with growth.

In the writer's opinion this lack of an agreed upon terminology has led to the loose usage of the word "toughness" outside a strict LEFM connotation and has caused much confusion in the study of fracture in the EPFM regime.

Discussion References

[1] Turner, C. E. in *Fracture Mechanics (Eleventh Conference), ASTM STP 677*, 1979, pp. 614–628.
[2] *Post-Yield Fracture Mechanics*, D. G. H. Latzko et al., Elsevier Applied Science, 2nd ed., 1984, Chap. 2.
[3] Etemad, M. R. and Turner, C. E., *International Journal of Pressure Vessels and Piping*, 1985, pp. 81–88.
[4] Mecklenburg, M. F., Joyce, J. A., and Albrecht, P., "Separation of Energies in Elastic-Plastic Fracture," *Non-Linear Fracture Mechanics: Volume II—Elastic-Plastic Fracture, ASTM STP 995*, American Society for Testing and Materials, Philadelphia, 1989, pp. 594–612.

M. F. Mecklenburg et al. (authors' closure)—The authors thank Professor Turner for his comments. Lines OA and $O'A'$ are intended to be parallel, as is clearly stated in Ref *3*. Our use of the geometric construction in Fig. 1 is in no way an attempt to describe the sequence of events leading to elastic and plastic energy releases. This construction was simply a way to illustrate the total difference between the released elastic energy, U_e, and the dissipated plastic energy, U_p.

We use the letter I to mean the plastic energy dissipation rate, as shown in Eq 2; it is more fully described in Ref *3*.

W. R. Lloyd[1] and W. G. Reuter[1]

Evaluation of Elastic-Plastic Surface Flaw Behavior and Related Parameters Using Surface Displacement Measurements

REFERENCE: Lloyd, W. R. and Reuter, W. G., "**Evaluation of Elastic-Plastic Surface Flaw Behavior and Related Parameters Using Surface Displacement Measurements,**" *Fracture Mechanics: Twenty-First Symposium, ASTM STP 1074*, J. P. Gudas, J. A. Joyce, and E. M. Hackett, Eds., American Society for Testing and Materials, Philadelphia, 1990, pp. 322–336.

ABSTRACT: Surface displacement measurements on surface cracked plate specimens loaded in tension were recorded using geometric moiré. An algorithm based on the displacements predicted by HRR fields was developed by the authors and used to evaluate the value of the applied *J*-integral at the crack border/free surface intersection point at various levels of applied remote stress. The values of *J* calculated in this manner were found to be in good agreement with predictions by Parks obtained from a nonlinear three-dimensional finite element model (ABAQUS).

KEY WORDS: surface cracks, *J*-integral, geometric moiré, CTOD, displacement measurements

Nomenclature

a	Crack depth of surface flaw
a/c	Semielliptical surface flaw aspect ratio
$2c$	Crack length of surface flaw
C	Parameter defined in Eq 8
C(T)	Compact tension (specimen geometry)
CTOD	Crack tip opening displacement
D_n	Parameter in Eq 9
E	Young's modulus
EDM	Electric discharge machining
I_n	Parameter in Eq 2
J	Path-independent integral postulated to characterize kinematic behavior of material in the vicinity of a crack tip
n	Strain hardening term in Eq 1
PTC	Partial-through-thickness crack (surface flaw)
r	Distance from crack tip to where displacement was measured
S	σ_∞/σ_0
SEN(B)	Single edge notch bend (specimen geometry)
$\hat{u}_r(\theta,n)$	Parameter in Eq 2, nondimensional displacement magnification
$U_r(r,\theta)$	Radial component of displacement in a polar coordinate system
W	Specimen width

[1] Idaho National Engineering Laboratory, EG&G Idaho, Inc., P.O. Box 1625, Idaho Falls, ID 83415-2218.

$X(J)$	Parameter in Eq 3
α	Coefficient for plastic component of Ramberg-Osgood relationship
δ_t	Total crack tip opening displacement
ϵ	Applied strain
ϵ_0	$\epsilon_0 = \sigma_0/E$
ξ	Parameter in Eq 3 and defined in Eq 5
σ_∞	Remote applied stress
σ_0	Yield strength
σ_Y	Effective yield stress $= (\sigma_0 + \sigma_{uts})/2$
σ_{uts}	Ultimate tensile strength

Introduction

The mechanical integrity of structures containing surface flaws has been receiving particular attention in recent years. While considerable progress has been made in predicting the integrity of high-strength, low-toughness materials, where linear elastic fracture mechanics (LEFM) applies, there is still a great deal to be understood concerning more ductile materials which fail in the elastic-plastic (E-P) or fully plastic (FP) regimes. Since 1968 [1], the J-integral concept has been used with varying degrees of success to measure J_{Ic} and to predict crack initiation in two-dimensional (2D) (i.e. plane stress or plane strain) cracked bodies. The J concept was a tremendous accomplishment in the realm of theoretical mechanics; however, the validity of using 2D J or related concepts, such as the tearing modulus, to predict crack initiation or instability in flawed three-dimensional (3D) structures has yet to be rigorously proven.

Parks and Wang [2], along with others using numerical methods, have begun to develop a database of solutions for local J values as a function of position around semielliptical surface flaws for a few selected materials and flaw geometries. This database provides a good start in understanding and developing predictive capabilities for flawed 3D structures using J-based concepts.

As attractive as the above numerical solutions are, there are still very limited experimental data by which the numerical modeling may be verified. Read [3] and King [4] et al. have provided experimental evaluation of J at maximum crack depth in a surface flawed plate by direct integration of surface strain distributions obtained from bonded strain gages. The results of King et al. [4] at maximum crack depth seem reasonable for flaw aspect ratios (a/c) ranging from 0.3 to 0.7 (the range evaluated in Ref 4); however, assumptions concerning out-of-plane shear stresses and strains occurring normal to the plane on which J is evaluated are required. Additionally, King's results only provide a J value at a single point (actually an average value over a small segment of crack border). The agreement between J_{init} (experimental measurement of J at crack initiation) and J_{Ic} suggests that this approach provides a reliable assessment of J at maximum crack depth.

The objectives of the present work are (1) to introduce a new method of determining the value of J from surface displacement measurements which will (2) provide additional data concerning surface crack behavior that can be used, in addition to the results of King and Read, to establish the validity of the numerical results of Parks and others.

Procedures

Equipment

All tensile tests of the surface flawed specimens were performed on an Instron 1325 servohydraulic load frame (1000 kN capacity). The specimens were loaded through 38.1 mm

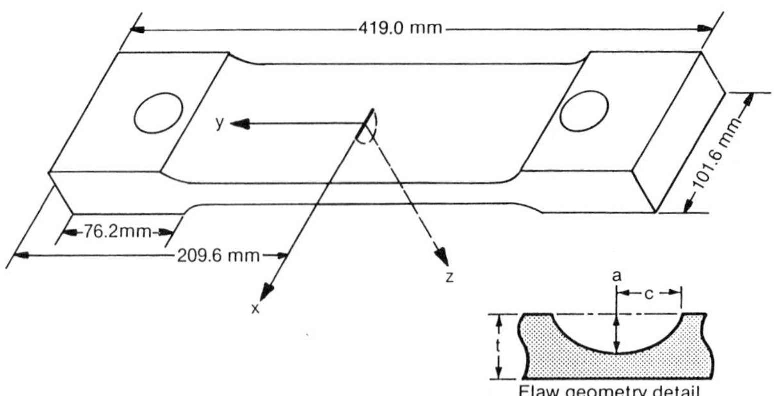

FIG. 1—*Schematic diagram of a typical surface cracked specimen.*

diameter pins (Fig. 1). A ± 2.5 mm extensometer was used to measure the remote uniform strain of the tensile plate specimen during the test.

High-resolution (512 by 480 pixel resolution) black-and-white CCD video cameras were used to make real-time recordings of the optical data generated by moiré methods. The video signals were recorded on Umatic format video recorders; a separate camera and recorder system were employed to record front and back surface displacement fringe patterns simultaneously. In addition, two 35 mm cameras with 105 mm macrolenses and power winders were used to collect data at discrete times during the test. The lens axes of the 35 mm cameras were located normal to the specimen surface and intersected the specimen at its geometric center. The video cameras were placed adjacent to the still cameras, with the lens axes of the video cameras 5 to 6 deg from the still camera axes. Figure 2 shows the physical positions of the cameras with respect to the specimen.

Specimen Description and Preparation

Surface flaw tensile specimens (Fig. 1) were fabricated from ASTM A710 Grade A steel. A crack starter notch was cut into the sample using EDM. The starter notch was then extended by fatigue cycling with a bending load. The resulting crack profiles had a nominal aspect ratio (a/c) of 0.23. For Specimen E-10 (the only specimen tested in this study), the crack depth was 3.90 mm and the plate thickness was 6.35 mm.

The constitutive behavior of the material can be approximated by the power relation

$$\epsilon/\epsilon_0 = \alpha(\sigma/\sigma_0)^n \tag{1}$$

where ϵ, ϵ_0, α, σ, σ_0, and n are as defined in the Nomenclature section. A uniaxial stress-strain curve for A710 steel at room temperature is provided in Fig. 3.

Analysis of stress-strain data for this material shows it to have a room temperature yield stress of 470 MPa and an elastic modulus of 208 GPa. Curve fitting experimental stress-strain data to Eq 1 yields values of α ranging from 0.417 to 0.740 corresponding to values of n ranging from 3.5 to 12 as the maximum value of strain over which the analysis is performed is increased from 1.5 to 8.0%. Using ASTM Standard E 813 to evaluate J_{Ic} yielded values of 240 kJ/m² for 15.9 mm thick SE(B) specimens (all criteria satisfied, $B > 11$ mm, no cleavage fracture) and 40 kJ/m² for 15.9 mm thick SE(B) specimens and 25.4 mm thick C(T) specimens where cleavage fracture was observed. This behavior is in part attributable to the tests

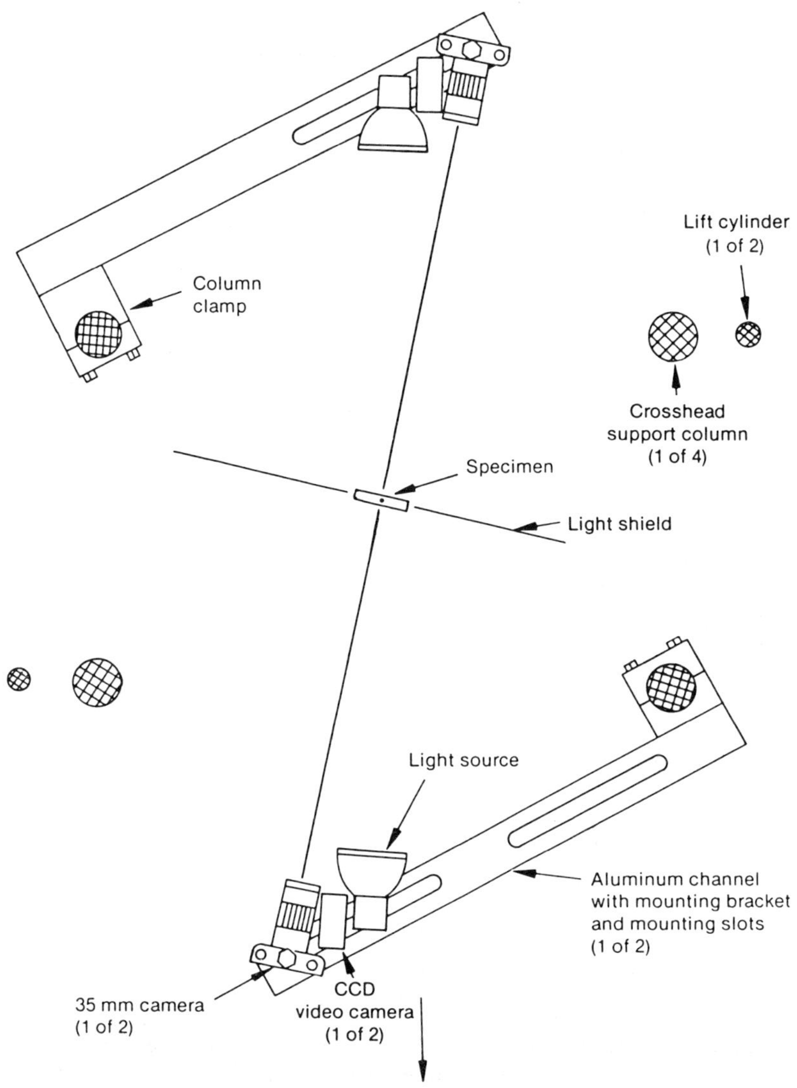

FIG. 2—*Schematic diagram of camera arrangement to record data.*

being run at room temperature, which is in the middle of the ductile-to-brittle transition for this lot of A710 material. A separate project at the Idaho National Engineering Laboratory (INEL) is investigating this.

Amplitude gratings used to obtain in-plane displacement data were replicated onto the front and back specimen surfaces. These gratings, approximately 100 mm square with a pitch of 50.8 μm, were prepared using the procedure outlined in the Appendix. The areas of interest on the specimens were coated with white polyurethane paint (thinned 1:4 with polyurethane thinner) using an air brush. The higher background reflectivity provided by the white paint, versus bare metal, improved the contrast of the resultant experimental fringes. A 20

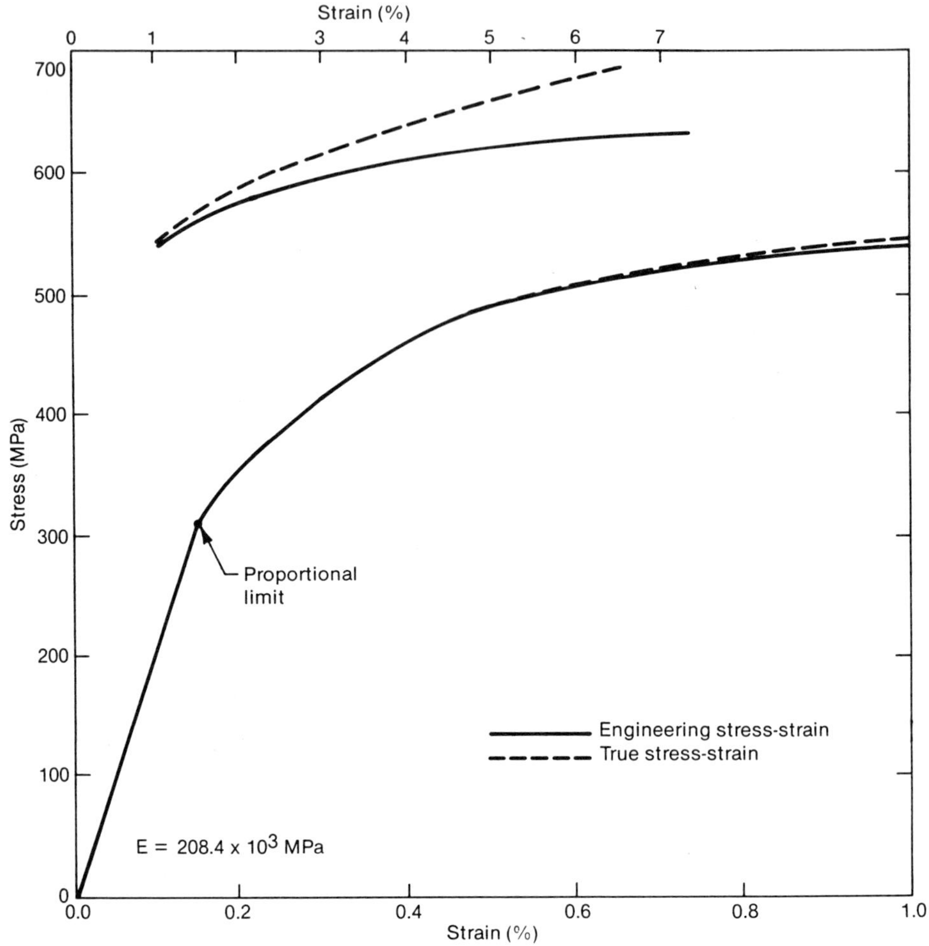

FIG. 3—*Uniaxial stress-strain curve for A710 steel at room temperature.*

power microscope was used to align the specimen grating, transfer side down, with either the crack plane or a reference line on the specimen surface. Cellulose tape was placed along one edge of the transfer grating to maintain alignment. The grating was then folded back and a liberal amount of cyanoacrylate adhesive was placed at the joint between grating and specimen where they were taped together. A rubber-coated print roller was then used to press the grating into the adhesive, making sure not to trap air bubbles under the grating. After 2 or 3 min, the acetate backing on the transfer was peeled away, leaving the grating bonded to the painted surface. The procedure is very similar to the one used for attaching foil strain gages, only on a larger scale.

The master amplitude gratings, on glass photographic plates, were used to make flexible reference gratings on sheet film by contact printing with collimated light. Correct exposure of the print to obtain a 50/50 grating was obtained by making multiple exposures of different durations, developing, and inspecting the resulting film under a microscope.

The reference gratings were placed over the specimen gratings with a thin film of glycerin between them. After aligning the reference grating to obtain a null field, cellulose tape was applied at two points along one edge to maintain alignment during the test.

An extensometer was attached to the edge of the specimen about 50 mm from the crack plane to monitor remote strain as a function of applied load. Inspection of the experimental fringes, recorded during the test, confirmed a uniform strain field at the extensometer location. The specimen was then placed into the load frame, and the extensometer cable was connected to the extensometer signal conditioning unit in the Instron 2150 control console.

Test Procedure

Mounting brackets that clamp to the 101 mm columns on the Instron 1325 load frame were fabricated. Support arms of 76 mm aluminum channel were bolted to the brackets. Slots were milled in the channels to allow for adjustment of the angle and position of the 35 mm and video cameras.

The monotonic loading was done in manual control mode using displacement control. The actuator displacement rate was set at 0.5 mm/min. After starting the test machine, the video recorders ran continuously until the end of the test. Still images were photographed at 15 to 20 s intervals for the duration of the test. Applied load, actuator displacement, and remote strain were recorded autographically for the duration of the test. A remote display voltmeter was used to provide applied load values in the video and still images; a small liquid crystal display (LCD) load read-out was suspended adjacent to the specimen in view of both the still and video cameras.

Experimental Results

Amplitude Moiré (In-Plane) Displacement Data

Full field (spatially continuous) moiré patterns representing surface displacements normal to the crack plane were collected on the front (cracked) and back surfaces of the plate specimens. The fringe patterns covered the entire width of the specimen and extended approximately one half the plate width above and below the crack plane (where uniform remote strain was observed). For this series of tests, the displacement sensitivity is 50.8 μm/fringe (one fringe spacing is the distance from one point of minimum intensity to the next adjacent point of minimum intensity). This sensitivity proved ideal for measuring moderate plastic deformations at remote stress levels approaching yield; however, it did become a limitation at stress levels below $S = 0.75$.

A fringe pattern representing in-plane displacements normal to the crack plane at $S = 0.949$ is shown in Fig. 4. The straight line passing through the crack tip normal to the crack plane in the figure is where the displacement data are collected for use in the new method of analysis to be presented. Figure 5 is the radial displacement versus position graph for three applied stress levels ($S = 0.734, 0.859,$ and 0.949) taken along this line.

In-plane displacement data were also collected on the specimen's back surface; a typical fringe pattern is presented in Fig. 6. Displacements normal to the plane of the crack were measured. The back surface strain profiles obtained from the displacement distributions did compare favorably with nonlinear finite element predictions [2, 5] of plastic zone boundaries on the back surface. The correlation of the experimental and numerical back surface displacements and their significance to the fracture process are currently undergoing further investigation.

Out-of-Plane Displacement Data

Out-of-plane displacement data have also been collected on the back surface (opposite the crack) of replicate specimens. These data were generated using the shadow moiré technique [5]; sensitivity ranged from 15 to 50 μm/fringe. The higher sensitivity provided good data in

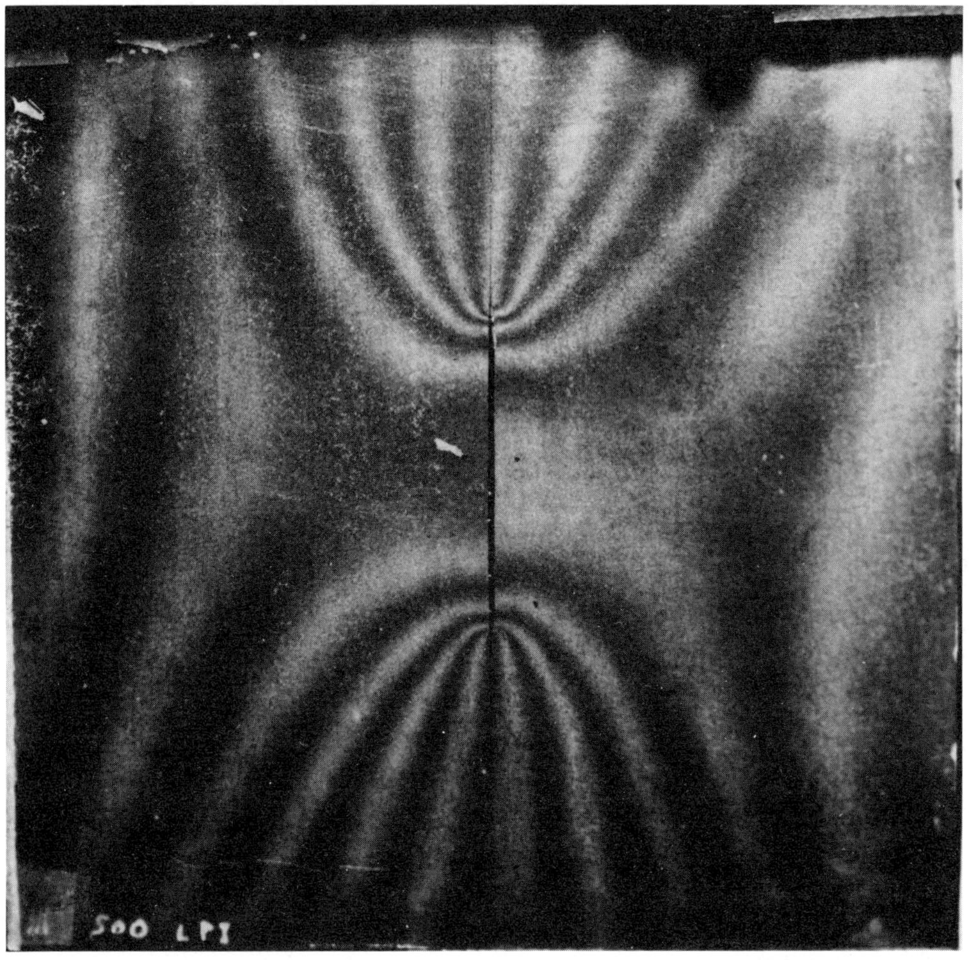

FIG. 4—*Moiré fringe pattern of displacements normal to the crack plane on the cracked surface at S = 0.949.*

the elastic to elastic-plastic regimes, while the 50 μm/fringe sensitivity allowed data collection to be extended well beyond net section yielding.

Epstein et al. [5] have investigated the significance of the out-of-plane displacement data with respect to elastic-plastic fracture and have correlated the data to numerical results. Although these data have not yet been used to directly evaluate J, the results presented by Epstein lend credence to the assumption of a quasi-2D stress state existing in the vicinity of maximum crack depth for surface flaws with $a/c \simeq 0.23$ or less and $S \leq 1.05$.

An algorithm which uses the out-of-plane displacements on the back surface to evaluate J is currently being developed. Results of this technique will be presented in a future publication.

Analyses of Data

For this preliminary investigation, displacement data were analyzed at three discrete levels of applied stress corresponding to $S = 0.734, 0.859,$ and 0.949. These stresses were selected

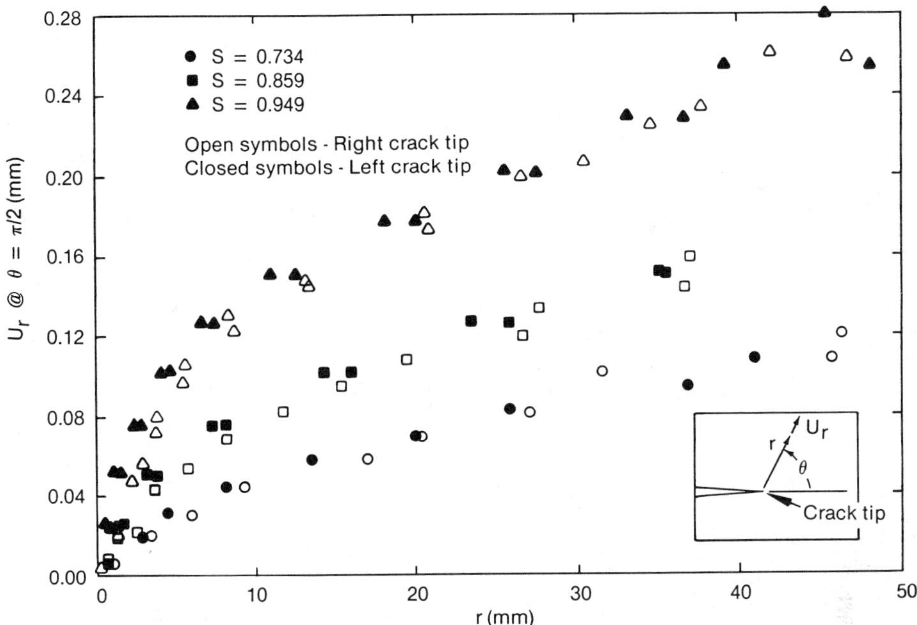

FIG. 5—*Displacements normal to the crack plane at the crack tip versus distance from the crack plane at S = 0.949.*

because Parks [2] used the same stresses to show that, for $S < 0.94$, J is thought to dominate the kinematic behavior of the specimen in the region surrounding the crack tip. The displacement data were used to (1) calculate J at maximum depth using the concept of Read [3] and King [4], and (2) calculate J at the free surface using a new concept.

Evaluation of J at Maximum Crack Depth Using the Integral Definition [3,4]

Surface displacement data were collected on front and back surfaces of the cracked plate specimen, in the plane normal to the crack border which passed through the point of maximum crack depth. Strain distributions were evaluated from the displacements by differentiating a cubic spline interpolation of discrete displacements measured directly from the fringe patterns. Strain energy density profiles were then developed by correlating the derived strains with the strain energy density (SED) function for the material; SED was obtained from uniaxial tensile test data from the same heat of material.

The strain energy density and displacement data were then used in a numerical integration procedure outlined by King [4] and Read [3] to evaluate J around different paths on the specimen symmetry plane, mutually normal to the crack plane and the cracked specimen surface (up the front surface to an arbitrary Y position value from the crack plane, through the plate thickness, and back down the back surface to the crack plane). Assuming the integral defined by King et al. [4] to be path independent, the integrals evaluated using their procedure should, at each load level, all have essentially the same value, regardless of the actual contour of integration selected. Analyses of the data using this method showed a wide variation in the calculated values of J, ranging from about 0 up to values predicted by Parks and Wang [2]. The values of J, however, seemed to be converging as increasingly longer integration paths were used. After analyzing the data at several stress levels using this technique, it became apparent that the sensitivity of the strain measurement in regions of low

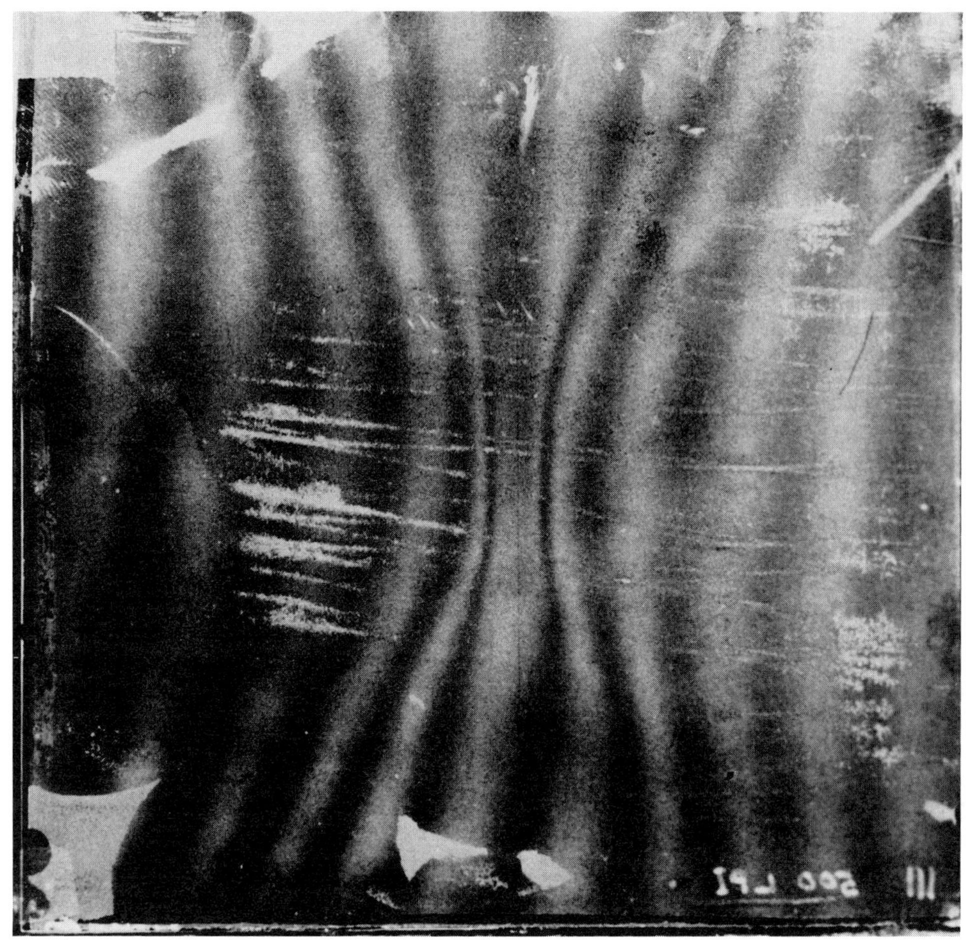

FIG. 6—*Moiré fringe pattern of displacements normal to the crack plane on the back surface at* S = 0.949.

strain ($< 1000 \mu\epsilon$) was not sufficient; therefore further evaluation of J using this method has been deferred until more sensitive measurements can be made. A measurement technique developed by Dadkhah and associates [6], using moiré interferometry, offers higher sensitivity and has been proven to be an effective and reliable way to evaluate J on the surface of C(T) specimens. This suggests that the approach of King [4] and Read [3] will provide useful estimates of J for some surface crack configurations.

Evaluation of J *at the Free Surface/Crack Intersection*

The experimentally determined displacement fields in the region surrounding the crack tip are expected to be closely predicted by the Hutchinson, Rice, and Rosengren (HRR) analytical solution for displacements in materials obeying J_2 deformation theory. Additionally, the 3D numerical solutions for surface flaws developed by Parks [2] confirm general convergence of 3D field quantities with the 2D HRR solutions as the crack tip is approached at maximum crack depth, with only moderate divergence of the two solutions at moderate distances from the crack tip. Additional information on the ABAQUS finite element code used by Parks and Wang may be found in *Users Manual, ABAQUS 4.6-162* (Hibbitt, Karlsson, and Sorensen, Inc., 1987). Additional data from Parks also imply that this convergence of solutions occurs around the remainder of the flaw border. Using this as a basis, and assuming the existence of an HRR field in some zone surrounding the crack tip, an algorithm was developed to evaluate J at the free surface/crack intersection by fitting experimental displacement data to an analytically based function, as follows.

The functional form of the HRR displacement fields may be found in a report by Shih [7] along with a tabulation of related parameters which are dependent on the constitutive behavior of the material and the general state of stress around the crack tip. The general form of the equation for the radial displacement component can be written

$$U_r = X(J, n, \ldots)r^\xi = X(J)r^\xi$$

where

$$X(J) = \alpha\epsilon_0\hat{u}_r(\theta,n)(J/\alpha\epsilon_0\sigma_0 I_n)^{n/(n+1)} \tag{2}$$

Note that $X(J, n, \ldots)$ has been abbreviated to $X(J)$ with the remaining variables assumed implicitly.

I_n and $\hat{u}_r(\theta,n)$ are tabulated in Shih's work for a range of values of the strain hardening exponent, n. The values of these parameters are also dependent upon the state of stress (plane stress or plane strain). Since the experimental data were collected on a traction-free surface, a local state of plane stress is assumed when selecting values from Shih's tables for the analysis.

Preliminary examination of the experimental data showed $U_r \cdot r^{-\xi}$ not to be a constant over any range of raw data. The addition of a linear displacement term (equivalent to the superposition of a constant strain field onto the HRR field) to the right-hand side of Eq 2 was required to accurately represent the real displacements occurring on the specimen's surface. While this correction assists in modeling displacements more accurately, $X(J) \cdot r^\xi$ remains the dominant term on the right-hand side of the new model equation

$$U_r = X(J) \cdot r^\xi + \epsilon \cdot r \tag{3}$$

where ϵ is the remote normal strain, which can be evaluated either directly from the far field displacement data or by mechanical measurements such as with an extensometer. By moving the linear term in Eq 3 to the left-hand side and taking the log of both sides, a linear relationship with a known function of the strain hardening exponent, n, is produced. This relation has the form

$$\ln (U_r - \epsilon \cdot r) = \xi \cdot \ln (r) + \ln \{X(J)\} \tag{4}$$

where $X(J)$ and ϵ are as described in Eqs 2 and 3 and

$$\xi = \frac{1}{n + 1} \tag{5}$$

Equations 4 and 5 can then be used with standard linear regression of the experimental data, $(U_r, r)_i$, $i = 1, 2, 3, \ldots$ and ϵ to evaluate the strain hardening exponent, n, and the

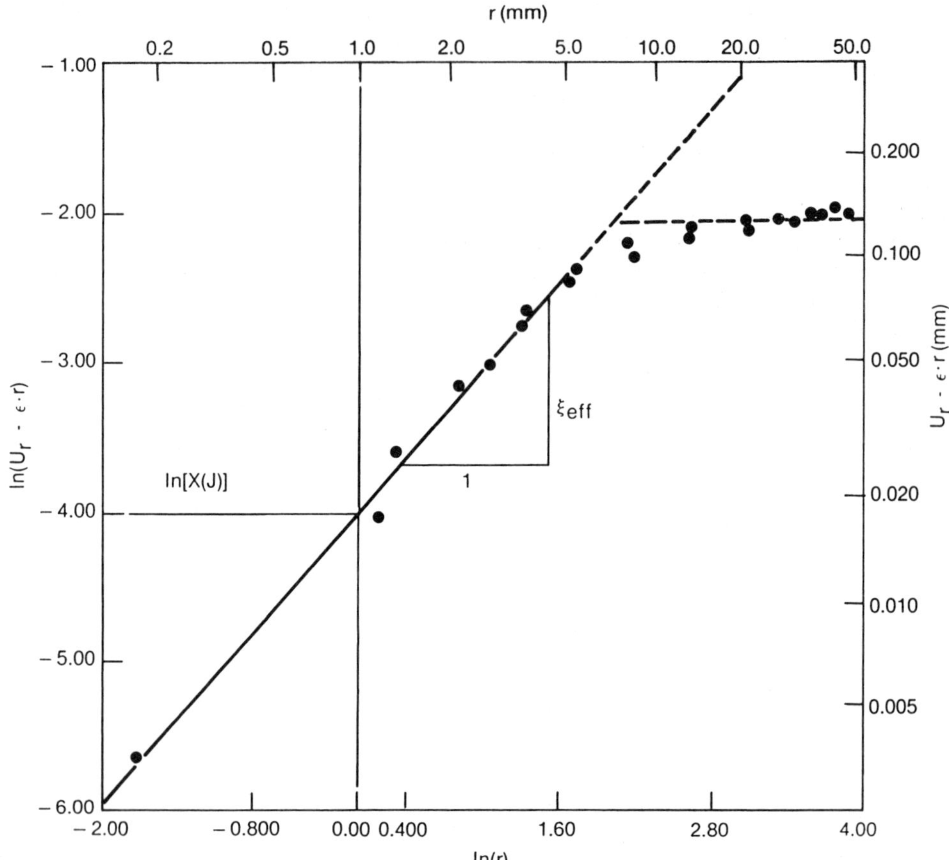

FIG. 7—*Log-log plot of displacement data to determine* X(n) *using linear regression.*

quantity $J = f\{\ln[X(n)]\}$. A representative plot of this procedure is shown in Fig. 7. When Eqs 4 and 5 are used to evaluate the experimental data over the range $10 < r < 40$ mm, values of n consistent with uniaxial stress-strain data are obtained. At this distance range, however, the calculated values of J were 5 to 7 times the magnitude of J values predicted by Parks and Wang [2]. In Fig. 7 a second linear zone is found to exist over a nominal range of $0.2 < r < 5$ mm. Evaluating the data in this zone yielded values of J which were much more consistent with Parks' results, as can be seen by comparing the circle symbols in Fig. 8 with the dotted line (FEM, $n = 10$). It is noted here that Parks and Wang [2] have presented an analysis that indicates that stresses, ahead of the crack tip, are not accurately represented by HRR predictions once S exceeds $\simeq 0.94$ for the flaw geometry currently under investigation. Their work also indicates a zone, which is quite small (a few CTODs), where "real" stresses (by FEM) correspond to the stresses predicted by HRR. Data presented here, though, indicate that good correlation exists between J values calculated using energy principles (Parks' FEM) and with J values calculated by fitting local displacement values to modified HRR displacement field equations, for values of S approaching unity. It is hypothesized that the HRR displacements may be valid over a zone considerably larger than the corresponding HRR stresses. Work by Smith [8], who studied LEFM stresses and displacements, found the displacements to be predicted accurately farther from the crack tip than were the stresses in 2D and 3D finite elastic bodies. These results were obtained from photoelastic (stresses) and moiré (displacement) data for SEN(B) and PTC specimens where the displacement model had a larger range of validity than did the stress model.

To provide a "check" on the values of J calculated using Eq 4, to investigate possible data "smoothing" which occurs by taking the log of the raw data, and to look at possible experimental bias in the raw data, Eq 3 was cast in the linearized form

$$U_r \cdot r^{-\xi} = X(J) + \epsilon \cdot r^{(1-\xi)} \tag{6}$$

Again, standard linear regression is used to obtain $X(J)$ as the intercept of the best fit line through the linear portion of the data. A graphic example of the procedure is given in Fig. 9. As with Eq 4, two fairly distinct and separate linear zones exist in the data when fitted to Eq 6. Having obtained values for $X(J)$ using Eq 6, the value of applied J may be calculated as

$$J = \left[\frac{X(J)}{\alpha \cdot \epsilon_0 \cdot \hat{u}_r(\theta,n) \cdot C^{[n/(n+1)]}} \right]^{(n+1)/n}$$

where

$$C = (\alpha \cdot \epsilon_0 \cdot \sigma_0 \cdot I_n)^{-1} \tag{7}$$

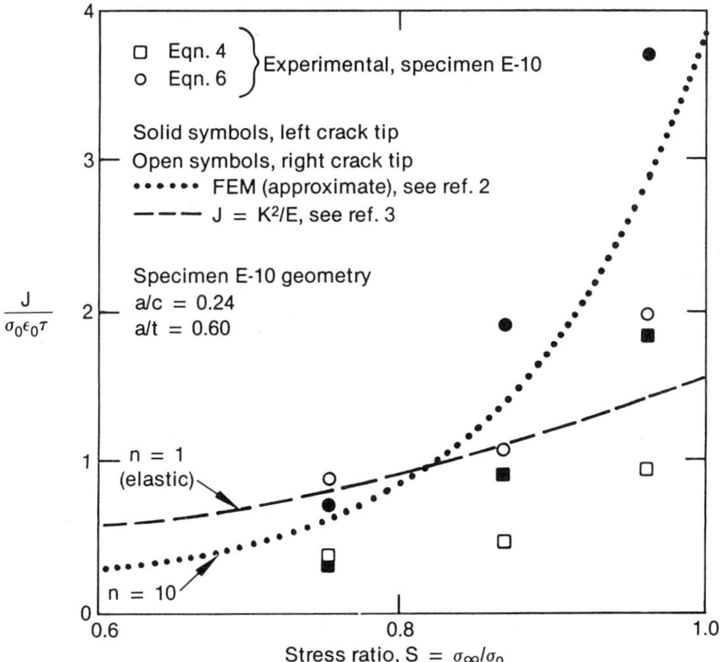

FIG. 8—*Normalized J versus nondimensional stress ratio,* S, *showing various results from different methods.*

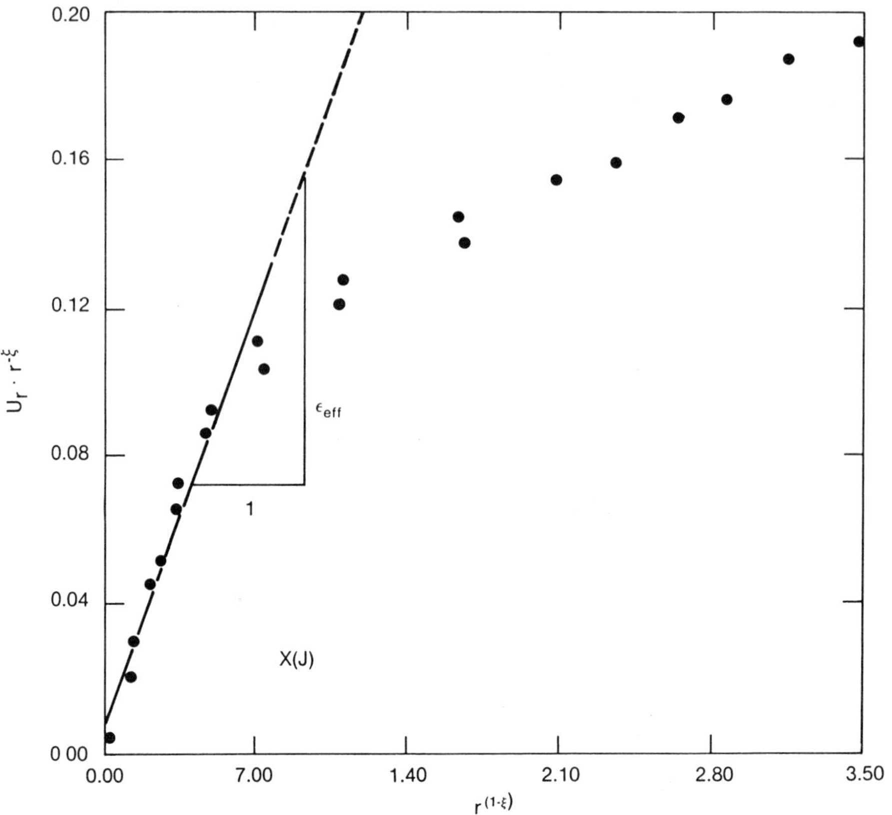

FIG. 9—*Linearized plot of displacement data used to determine* X(n) *by linear regression.*

Values of J obtained using Eq 4 (circles) are plotted in Fig. 8. The values are based on the linear portion in Fig. 9 corresponding to data in the range of $0.2 < r < 5$ mm. These results provide values of J similar to those predicted by Parks [2]. Values of J obtained in this way were consistent for each of the two data sets collected at each discrete load level (a set of data at each of two crack tips) and were also relatively insensitive to moderate variations in the value of n selected for use in Eq 6 (n varying from 8 to 12 only changed J 3% to 4%).

In all cases examined so far where Eqs 4 and 6 are used as model equations, two distinct linear zones are found to exist in the data, one associated with the far field and one associated with the near tip region (Figs. 7 and 9). The J values obtained using Eqs 4 and 6 and the procedures used are listed in Table 1; the same data along with comparisons to the results of Parks and linear theory appear graphically in Fig. 8.

TABLE 1—*Normalized J values,* J/($\sigma_0 \cdot \epsilon_0 \cdot$ t), *for a surface flaw with* a/c = 0.24 *and* a/t = 0.60.

σ_∞/σ_0	Right Tip (Eq 4)	Right Tip (Eq 6)	Left Tip (Eq 4)	Left Tip (Eq 6)
0.734	0.826	0.368	0.675	0.300
0.859	1.087	0.463	1.876	0.877
0.949	1.943	0.919	3.586	1.827

Discussion

Comparison of Experimental J Values to FEM Results

The trend of the experimental values of J compares favorably with the finite element results of Parks [2], especially those values obtained using Eq 4 (Fig. 8). The deviation of the values of J obtained using the different methods of Eqs 4 and 6 may be attributable to the limited spatial resolution (about 0.2 mm) of the moiré method employed to collect the data. Difficulty in measuring the precise location of maximum or minimum intensity of the fringes by direct visual inspection may be introducing spatial shifting (experimental bias) of the whole data set. This shift may affect the linearized data (per Eq 6) more than the log-log data (Eq 4), causing the general shift in the calculated J values shown in Table 1. An automated fringe measuring system is being developed at the INEL which should eliminate this bias if it exists. It is encouraging, however, that the two methods produced preliminary results which bracket the three-dimensional FEM results of Parks.

That the method using Eq 4 always produces results larger in magnitude than the method of Eq 6 is not a matter of great concern at the present time. The new image analysis system mentioned above will significantly improve spatial resolution and displacement sensitivity. The results at the free surface/crack intersection using the two methods of analysis are expected to converge once more accurate data are obtained.

Additionally, the method of King [4] and Read [3], which uses a direct integration described in a previous section to evaluate J at maximum crack depth, is expected to succeed once more accurate data become available. This is based on the work of Dadkhah et al. [6] previously cited.

It is also important to note that the method presented here uses local field quantities to directly evaluate the applied J-integral, while the work of Parks and Wang [2] uses a more global volume integral based on energy balance to predict J. That the two distinctly different approaches produce the same J values, within the accuracy limits of the preliminary analysis, is very encouraging. If future analysis shows the method of King and Read to produce similarly reliable results, a combination of the methods may provide an extremely powerful tool for analysis of elastic-plastic fracture and associated predictive capability.

Summary and Conclusions

The foundation for a new method of experimentally evaluating J has been presented. The method was successfully applied to preliminary sets of data generated using geometric moiré and has been shown to produce reasonable results when compared with three-dimensional FEM results produced by Parks and his associates [2].

The distribution of the preliminary results indicates that the reliability of results obtained using this method may be increased by improving the method of data collection; work is under way to develop a displacement measuring system to increase the sensitivity and spatial resolution of measurements by at least an order of magnitude.

In addition to the evaluation of J at the free surface, accurate measurements of continuous surface displacements on the front and back surfaces of surface flawed specimens, combined with the data analysis techniques currently being developed by the authors, will provide improved understanding of elastic-plastic fracture processes in surface cracked components.

Acknowledgments

This work was supported by the U.S. Department of Energy, Office of Energy Research, Office of Basic Energy Sciences, under DOE Contract DE-AC07-76ID01570.

APPENDIX

Fabrication of 50 μm Amplitude Specimen Gratings

Commercially available graphic arts transfer film for black transfers, e.g. black I.N.T. (a product of the 3M Co.), is used as the basis for the specimen gratings. The I.N.T. consists of a UV sensitive emulsion on an acetate or polyester film backing. UV light exposure causes the emulsion to harden and become relatively impervious to the developing solution. Where the film is exposed to the UV, the black, adhesive emulsion remains during developing; the unexposed areas of emulsion are dissolved away. The result is analogous to contact printing photographic film using white light. Master amplitude gratings of the desired pitch printed on glass plates may be obtained from most any optical supplier.

It is the authors' experience that the most satisfactory gratings are obtained by exposing the I.N.T. contact print in a printed circuit board exposing cabinet containing a uniform UV light source and a vacuum printing frame. As per instructions in the I.N.T. kit, place the emulsion side of the master grating plate against the black side of the I.N.T. film. Place the film on opaque black paper. Place this "sandwich" in the vacuum printing frame of the PC board light cabinet. Determination of the correct exposure time is by trial and error. A special UV density wedge available from 3M may help speed the process.

Following exposure, the film is developed per instructions included with the I.N.T. with the following exceptions. First, use *copious* amounts of *clean* developer, keeping the film fully wetted at all times during developing. Secondly, the developing process must be continued for approximately three to five times longer than recommended in the instructions. The film should be soaked in the developer, without rubbing, for 30 to 45 s. Using a clean developing pad or lint-free graphic arts application pad, such as Webril Appli-pads, finish developing using circular, then linear, wiping motions. This typically takes 4 to 6 min. Finish by rinsing in water and drying per the packaged instructions. The properly developed film should have a uniform, translucent gray appearance. Examination under $20\times$ magnification should reveal evenly spaced dark and clear bands. The exposure time must be varied to yield the correct 50/50 ratio of dark to clear spacing of the grating lines.

The specimen gratings are then applied as stated in the text of this paper.

References

[1] Rice, J. R., "A Path Independent Integral and the Approximate Analysis of Strain Concentration by Notches and Cracks,"*Journal of Applied Mechanics,* June 1968, p. 379.

[2] Parks, D. M. and Wang, Y.-Y., "Elastic-Plastic Analysis of Part-Through Surface Cracks," *Analytical, Numerical, and Experimental Aspects of Three Dimensional Fracture Processes, AMD,* Vol. 91, Joint ASME/SES Applied Mechanics and Engineering Science Conference, Berkeley, Calif., June 1988, pp. 19–32.

[3] Read, D. T., "Experimental Method for Direct Evaluation of the *J*-Contour Integral," in *Fracture Mechanics: Fourteenth Symposium, Volume II—Testing and Applications, ASTM STP 791,* J. C. Lewis and G. Sines, Eds., American Society for Testing and Materials, Philadelphia, 1983, pp. II-199—II-213.

[4] King, R. B. et al., "*J*-Integral Analysis of Surface Cracks in Pipeline Steel Plates," in *Elastic-Plastic Fracture: Second Symposium, Volume I—Inelastic Crack Analysis, ASTM STP 803,* C. F. Shih and J. P. Gudas, Eds., American Society for Testing and Materials, Philadelphia, 1983, pp. I-444—I-457.

[5] Epstein, J. S. et al., "Measurement tof the Elastic-Plastic Surface Flaw Back Surface Topology and Its Relation to Constraint," *SPIE,* Vol. 814, International Conf. on Photomechanics and Speckle Metrology, 1988.

[6] Dadkhah, M. S. et al., "*J*-Integral Measurements Using Moiré Interferometry," in *Proceedings,* 6th International Conference on Experimental Mechanics, Society for Experimental Mechanics, Portland, Ore., 5–10 June 1988.

[7] Shih, C. F., "Tables of Hutchinson-Rice-Rosengren Singular Field Quantities," Report MRL E-147, Materials Research Laboratory, Brown University, Providence, R.I., June 1983.

[8] Smith, C. W. and Olaosebikan, O., "On the Extraction of Stress Intensity Factors from Near Tip Photoelastic Data," presented to SESA Spring Conference, 1983.

Micromechanics of Fracture

C. W. Marschall,[1] *M. P. Landow,*[1] *and G. M. Wilkowski*[1]

Effect of Dynamic Strain Aging on Fracture Resistance of Carbon Steels Operating at Light-Water Reactor Temperatures

REFERENCE: Marschall, C. W., Landow, M. P., and Wilkowski, G. M., **"Effect of Dynamic Strain Aging on Fracture Resistance of Carbon Steels Operating at Light-Water Reactor Temperatures,"** *Fracture Mechanics: Twenty-First Symposium, ASTM STP 1074,* J. P. Gudas, J. A. Joyce, and E. M. Hackett, Eds., American Society for Testing and Materials, Philadelphia, 1990, pp. 339–360.

ABSTRACT: This paper reviews the phenomenon of dynamic strain aging in carbon steels and considers its effects on the fracture behavior of carbon-steel pipes and pressure vessels in light-water reactors operating at elevated temperatures near 290°C (550°F).

Dynamic strain aging is a phenomenon in which aging occurs simultaneously with plastic straining. It occurs over a range of temperatures that depends on strain rate. In tensile tests, it is manifested by increased tensile strength, increased strain-hardening rate, serrated stress-strain curves, and decreased ductility.

Evidence is presented to show that the occurrence of dynamic strain aging can significantly lower the fracture resistance of carbon steels. This lowering of fracture resistance may be manifested in several ways: (1) J_{Ic} is lower at light-water reactor (LWR) temperatures than at room temperature, (2) the tearing modulus is lower at LWR temperatures than at room temperature, and (3) stable ductile crack growth may be interrupted by unstable ductile fracture at LWR temperatures but not at room temperature.

The paper examines probable causes of dynamic strain aging and describes methods for identifying which steels are susceptible to it.

KEY WORDS: dynamic strain aging, embrittlement, fracture toughness, *J-R* curves, unstable crack growth, pressure vessel steels, pipe steels, carbon steels, serrated stress-strain curves

Many of the carbon steels used in coolant pipes and pressure vessels in light-water reactors are susceptible to a phenomenon known as dynamic strain aging (DSA). The most commonly observed indicator of DSA is tensile strength; if the tensile strength at the reactor operating temperature of 290°C (550°F) is greater than or approximately equal to that at room temperature, this behavior is almost certainly a manifestation of DSA.

From strength considerations alone, susceptibility to DSA would appear to be a positive attribute of a steel. For example, a steel that exhibits DSA might be as much as twice as strong at 200 to 300°C (390 to 570°F) as a similar steel that is immune to DSA [1]. In addition, fatigue strength and creep strength in the same temperature range are reported to be improved by DSA [2]. However, as is often the case, the increased strength is obtained at the cost of decreased dutility and toughness. In fact, it is these latter properties that are commonly emphasized in discussing DSA, as evidenced by the use of the terms *blue brittleness* or *dynamic-strain-aging embrittlement* to describe the phenomenon. In the former term, the

---

[1] Battelle, Columbus, OH 43201.

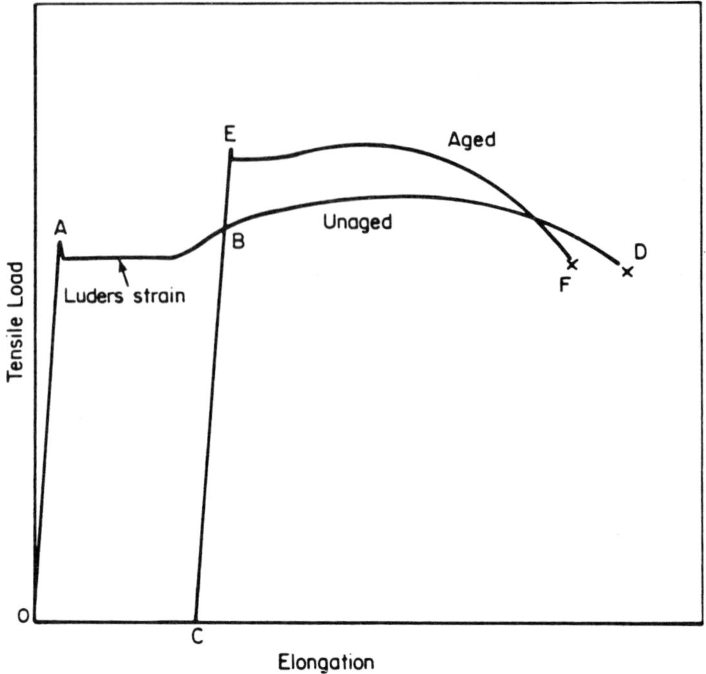

FIG. 1—*Schematic illustration of static strain-aging in a tensile specimen.*

word *blue* refers to the color of the oxide formed in the temperature range of DSA; the word *brittle* does not necessarily imply that cleavage fracture occurs but rather that the steel shows a decrease of toughness or ductility.

The discovery of blue brittleness in steels is not a recent event. According to Ho and Lepik [3], it was observed as early as the late 1800s, but that fact does not allow complacency for today's structures. When dealing with nuclear reactor piping and pressure vessels that operate in the blue-brittle range of temperatures, susceptibility of the steel to DSA and to possible loss of toughness must be considered [3–6].

This paper describes some of the manifestations of DSA, as well as the closely related static strain aging (SSA), and examines probable causes of these phenomena. It also describes a number of observations, apparently associated with DSA, that are not completely understood. The most striking of these is the occurrence of unstable ductile fracture at light-water reactor temperatures, both in tests on small fracture mechanics specimens and in full-scale pipe tests [7].

Both SSA and DSA have been the subject of several excellent reviews [1,2,8] to which the interested reader is referred for mechanistic details, alloying effects, and thermal treatment effects.

Static versus Dynamic Strain Aging

Strain aging in carbon steels is manifested as a change in properties arising from both plastic straining and aging. In static strain aging (SSA), the property changes occur after plastic deformation and depend on the temperature and time (aging) after the deformation occurred. In dynamic strain aging (DSA), on the other hand, the property changes occur during plastic deformation (i.e., aging occurs simultaneously with the straining).

Figure 1 illustrates SSA in a tensile specimen. Assume that the specimen is loaded to produce some plastic strain and then unloaded such that curve *OABC* is produced. If the specimen is reloaded immediately, it will follow curve *CBD;* the *BD* portion of the curve will be the same as if the specimen had not been unloaded. If, however, the specimen is allowed to age for some time, particularly at a slightly elevated temperature, after unloading to Point *C,* reloading will produce a curve similar to *CEF.* Thus SSA increases the yield strength, causes the Luders strain to reappear, increases the ultimate strength, and decreases the fracture elongation. Although not shown in Fig. 1, SSA can also cause a significant decrease of fracture toughness. For example, Baird [2] reported a remarkable increase of about 175°C (315°F) in the 41 J (30 ft-lb) ductile-brittle fracture transition temperature for V-notched impact bend specimens of a semikilled mild steel after straining 10% and aging at 250°C (482°F) for 1 h. Of that shift, about 105°C (190°F) was due to the strain alone and 70°C (125°F) to the aging at 250°C (482°F). It is of interest to note, however, that the same steel showed no significant embrittlement, beyond that from the 10% straining, when it was aged at room temperature for 20 000 h rather than at 250°C (482°F) for 1 h.

DSA differs from SSA in that the aging occurs simultaneously with the straining; that is, in quasi-static tensile tests (strain rates of about 10^{-3} to 10^{-5} s^{-1}) at elevated temperatures, the stress-strain curves are much different from those at room temperature, as is illustrated in Fig. 2 for a low-carbon steel [8]. Notice that, instead of the expected loss of strength and increase of ductility that typically accompany a temperature increase, this material became stronger and less ductile as the temperature was raised between room temperature and approximately 300°C (570°F). In addition, load fluctuations commonly referred to as *serrations* developed on the stress-strain curve and the rate of strain hardening (the slope of the flow curve) was increased.

Figure 3 shows the effect of test temperature on the flow strength at 10% strain of the low-carbon steel whose stress-strain curves were shown in Fig. 2. Note that the serrations do not encompass the entire temperature range of DSA, thus refuting the commonly held supposition that DSA is always evidenced by serrated stress-strain curves.

As will be discussed in the next section, DSA is caused by the diffusion of nitrogen and

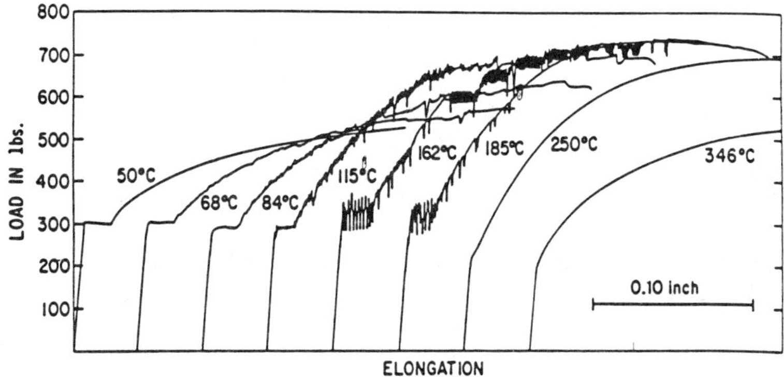

FIG. 2—*Load-elongation curves of 0.035C steel strained in tension at a crosshead speed of 0.0025 in./min. (Cross-section area ∼ 0.014 in.², gage length = 1.00 in.) (Ref 8, Keh and others; from* Dislocation Dynamics, *edited by Rosenfield et al., McGraw-Hill, New York, 1968; used with permission of McGraw-Hill Book Company.)*

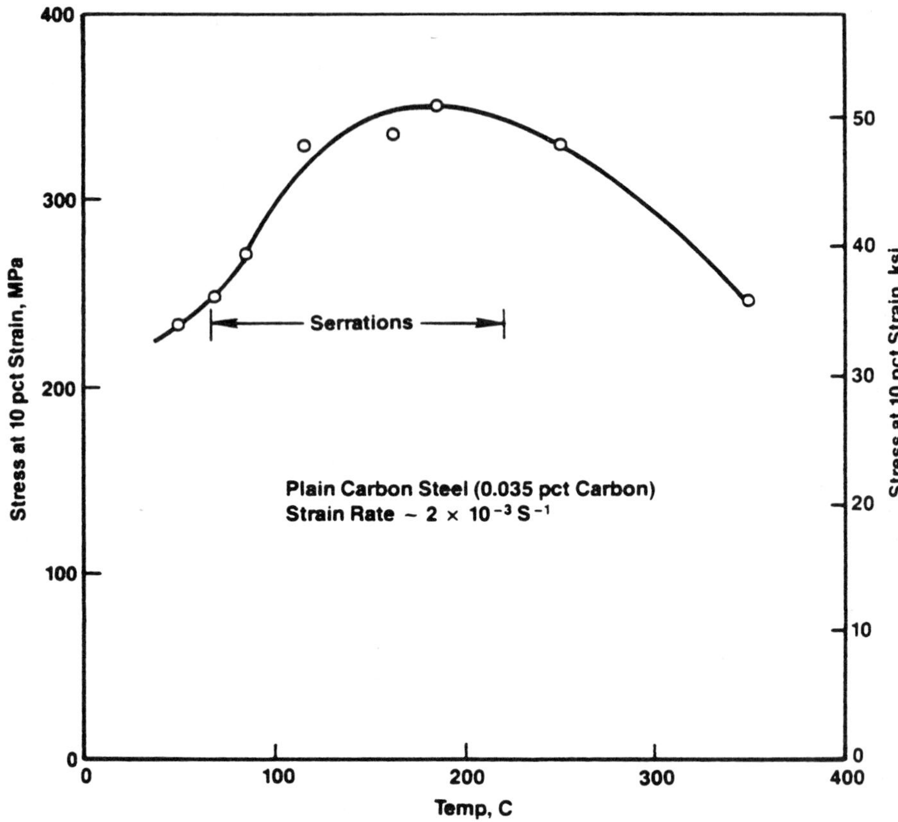

FIG. 3—*Flow strength at 10% strain versus test temperature for the carbon steel in Fig. 2.*

carbon atoms to moving dislocations. Thus, since diffusion requires time, it is to be expected that the stress-strain behavior of a susceptible steel in the DSA range will be dependent on strain rate as well as on temperature. Manjoine [9] demonstrated this over 40 years ago. He showed that the tensile strength peak, such as that shown in Fig. 3, as well as the temperature range in which serrated stress-strain curves are observed, move to higher temperatures as strain rate is increased and to lower temperatures as strain rate is decreased. Some of Manjoine's results are presented in Fig. 4. Note that increasing strain rate, which, in most ductile metals, causes increased tensile strength, can cause a variety of responses in steels that show DSA. At 250°C (480°F), for example, strength decreases with increasing strain rate, whereas strength increases with increasing strain rate below about 100°C (210°F) and above about 500°C (930°F). Other intermediate temperatures show mixed effects.

Keh et al. [8] have studied the occurrence of serrated stress-strain curves as a function of temperature and strain rate in a low-carbon steel susceptible to DSA. Their results, presented in Fig. 5, illustrate the fact that, at a given temperature, the presence or absence of serrations is governed by the strain rate. The form of the graph—log strain rate versus $1/T$—is one often used in trying to establish a mechanism for a phenomenon being studied. In this case, the slope of the upper line strongly suggests that diffusion of nitrogen and/or carbon is responsible for the serrations.

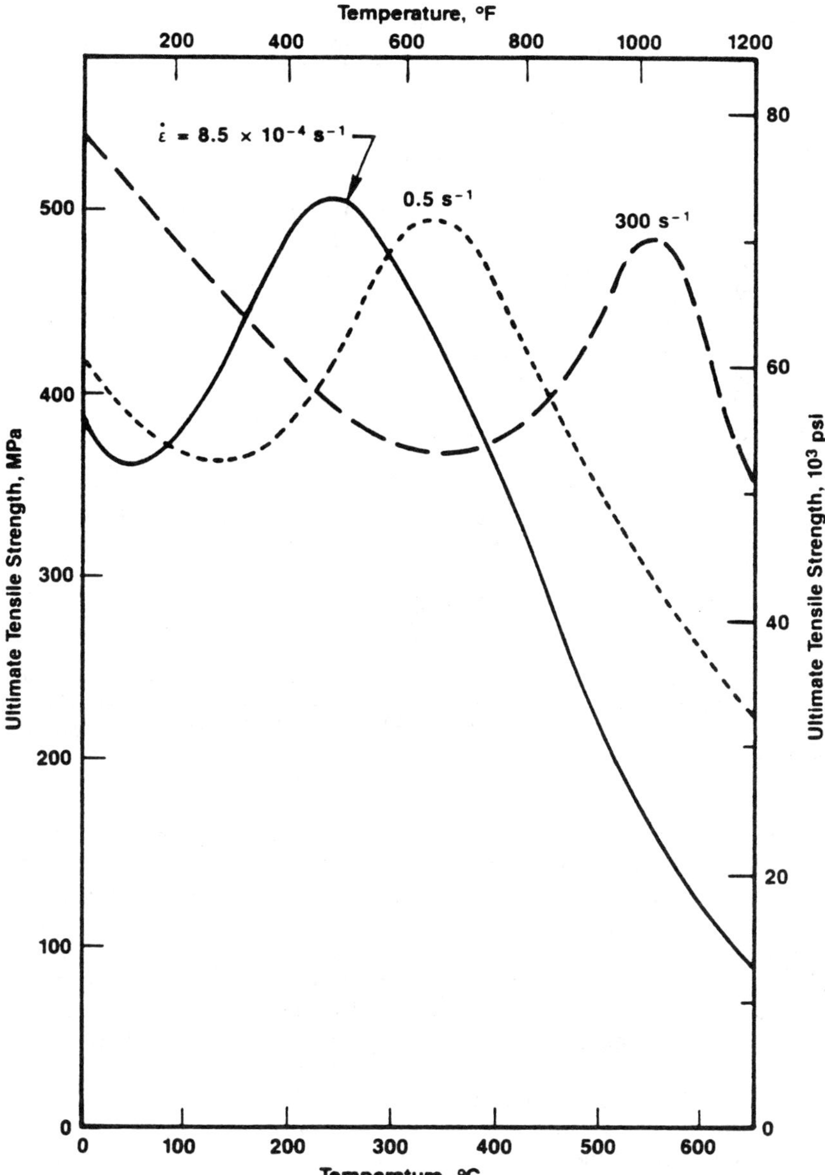

FIG. 4—*Ultimate tensile strength of a mild steel at various temperatures and strain rates* [9].

Causes of Strain Aging

It is well established that both static and dynamic strain aging are the result of interactions between dislocations and dissolved interstitial solute atoms, principally nitrogen and carbon, in the steel [1–3]. Explanations of the various features of the stress-strain curve, particularly in the DSA region, have provided fertile ground for dislocation theoreticians [8,10,11].

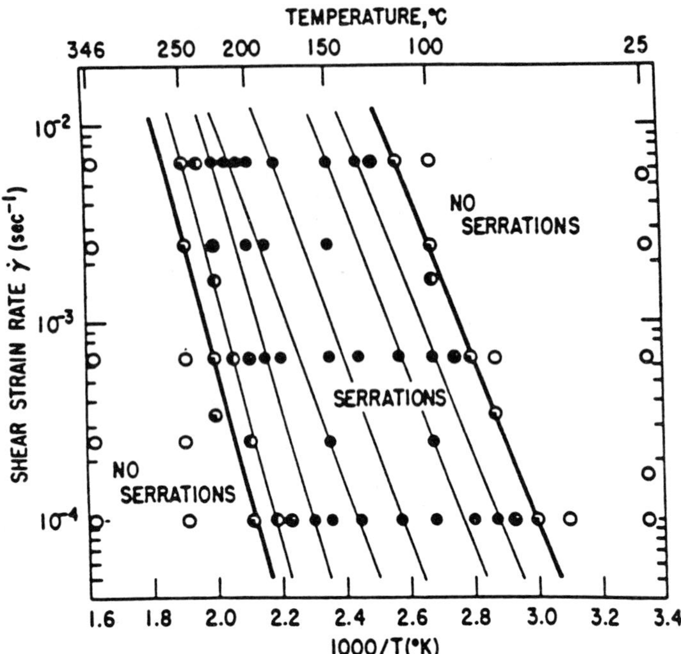

FIG. 5—*Dependence of stress-strain curve serrations on temperature and strain rate for 0.03% carbon steel, quenched from 250°C (480°F). (Ref 8, Keh and others; from* Dislocation Dynamics, *edited by Rosenfield et al., McGraw-Hill, New York, 1968; used with permission of McGraw-Hill Book Company.)*

The key to understanding the interactions between dislocations and interstitial atoms is based on two simple facts: (1) nitrogen and carbon atoms are slightly too large for the interstitial sites that they occupy in the body-centered-cubic iron crystal lattice [12] and thus create a local misfit stress when they are present, and (2) edge dislocations include a region near their core in which the normal atom spacing is expanded slightly (Fig. 6), thus more readily accommodating the slightly oversize nitrogen and carbon atoms in interstitial positions. If sufficient time is available for diffusion (dependent on temperature), the nitrogen and carbon atoms will move to the dislocation core region, thereby lowering the misfit energy.

The accumulation of nitrogen and carbon atoms at edge dislocations, sometimes referred to as *atmospheres* or *clouds,* influences the plastic flow behavior of the steel inasmuch as plastic flow is the result of the movement of dislocations. Because the nitrogen and carbon atoms segregated to dislocations represent a more stable condition (lower misfit energy), they tend to "immobilize" or "lock" or "pin" the dislocations, thus making plastic strain more difficult (i.e., they raise the stress required to cause dislocations to move).

With respect to SSA, the sharp yield point at Point *A* in Fig. 1 and the subsequent stress plateau (Luders strain) are indicators that the dislocations had previously been pinned by dissolved nitrogen and/or carbon atmospheres; application of a sufficiently high stress (Point *A*) has moved the dislocations away from their pinning atmospheres, beyond which they can continue to move at a lower stress. If the specimen is unloaded at Point *B* and allowed to age for time sufficient to allow the interstitial atoms to again diffuse (dependent on temperature) to the dislocation cores, reloading will again reveal a sharp yield point, Point *C,* and a region of Luders strain.

DSA differs from SSA in that the aging, or diffusion of the interstitial atoms to dislocations, occurs simultaneously with the straining. Accordingly, the temperature range of DSA is above that of SSA because rapid diffusion of nitrogen and carbon is required and this is aided by raising the temperature. As is shown in Fig. 2, the result of the dynamic interactions of interstitial atoms and dislocations includes increased tensile strength, decreased ductility, increased rate of strain hardening, and the appearance of serrations on the stress-strain curve. The serrations are believed to result from alternate pinning and unpinning of dislocations, similar to the yield point behavior shown in Fig. 1 for SSA. The increased strain-hardening rate and increased tensile strength are believed to arise from greater-than-normal dislocation densities in steels that exhibit DSA. These high dislocation densities are believed to occur because of the pinning, which requires that fresh dislocations be formed continually to maintain the applied strain rate. Other explanations for increased strength in the DSA range include locking of dislocation sources, precipitation of nitrides and carbides on dislocations, and increased frictional drag on moving dislocations due to the presence of the interstitial atmospheres [2].

Because diffusion of interstitial atoms is responsible for DSA, it is to be expected that DSA will depend on strain rate as well as on temperature. That this is the case has already been illustrated in Figs. 4 and 5.

Up to this point, nothing has been said about the relative importance of nitrogen and carbon in producing strain aging or about the amount of nitrogen or carbon required. According to Baird [2], nitrogen and carbon have similar diffusion coefficients in iron and distort the ferrite lattice nearly identically. Hence it would be expected that the two elements will produce similar strain aging affects in steel. However, near room temperature, the solubility of carbon in ferrite is so small—a factor of perhaps 100 less than the solubility of nitrogen [13]—that it is commonly assumed that nitrogen, rather than carbon, is chiefly responsible for strain aging. While this may be the case for temperatures below approximately 100°C (212°F), at higher temperatures the increased solubility of carbon can cause

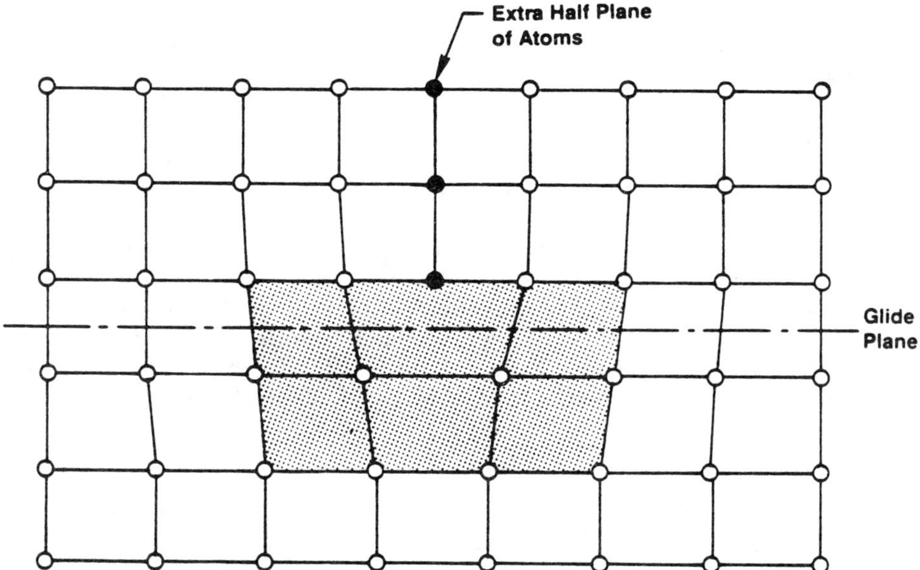

FIG. 6—*Schematic illustration of an edge dislocation in a simple cubic crystal lattice showing the abnormally large amount of interstitial space* (shaded) *near the dislocation.*

strain aging even in the absence of nitrogen [*13*]. Thus, when both elements are present, both can contribute to strain aging at temperatures above about 100°C (212°F), although the nitrogen may play a larger role because of its greater solubility at all temperatures.

The amount of dissolved nitrogen or carbon needed to cause some evidence of strain aging is extremely small. Leslie and Rickett [*13*] state that a nitrogen content of only 0.0004% by weight produces an easily measured amount of strain aging and suggest that the level of dissolved interstitial atoms must be reduced to 0.0001 wt% or less to eliminate strain aging. Wilson and Russell [*14*] suggested that only 0.0005 to 0.002 wt% of dissolved carbon or nitrogen is sufficient to completely lock the dislocations in moderately deformed steels.

Minimization of Strain Aging

Complete elimination of strain aging in carbon steels is difficult because of the remarkably small quantities of nitrogen and/or carbon in solution that can cause a discernible effect. Nonetheless, it is possible to minimize the degree of strain aging by adding elements to the steel that combine with the nitrogen and carbon to form nitrides, carbides, and carbonitrides, thus removing the interstitial elements from solution in the ferrite. For steels exposed only to temperatures near room temperature, it may be possible to essentially eliminate strain aging by removing only the nitrogen, because the solubility of carbon is extremely low at those temperatures.

Killed steels (i.e., steels that have been deoxidized with silicon and/or aluminum) are less susceptible to aging (if properly treated) than are unkilled steels because the deoxidizing agents form nitrides [*13*], thereby removing dissolved nitrogen from the ferrite lattice. However, as noted earlier, because strain aging at elevated temperatures can also result from dissolved carbon, silicon- or aluminum-killed steels will not be free of strain aging at all temperatures. To remove carbon from solution would require the addition of strong carbide formers (e.g., chromium, vanadium, tungsten, molybdenum, or titanium).

Li and Leslie demonstrated the susceptibility of killed steels to dynamic strain aging [*15*]. They conducted quasi-static tensile tests over a range of temperatures on a semikilled AISI 1008 steel, a silicon-killed AISI 1020 steel, and a silicon-aluminum killed AISI 1035 steel. Each steel was in the hot-rolled condition, and total nitrogen contents ranged from 0.004 to 0.007% by weight. At 200°C (392°F), the three steels exhibited flow stress values at 5% strain that were greater than those at room temperature by about 95 MPa (14 ksi), 130 MPa (19 ksi), and 95 MPa (14 ksi), respectively.

Even in the case of a killed steel, its strain aging response will be a function of the heat treatment [*13*]. The highest susceptibility to strain aging will be associated with rapid cooling from an elevated temperature which will cause relatively high levels of nitrogen and carbon to remain in solution in the ferrite. The lowest susceptibility will be associated with treatments that provide precipitation of nitrides and carbides, namely, slow cooling from the hot rolling temperature, or extended treatment near 600°C (1110°F) followed by slow cooling.

From the foregoing, it is evident that it is not the total nitrogen and/or carbon content in the steel that governs strain aging. Rather, it is the amount dissolved in the ferrite, which depends on the steel grade, the deoxidation practice, and the heat treatment.

Identifying Susceptibility to Dynamic Strain Aging

There is no agreed upon method for identifying a steel's susceptibility to DSA nor is there an agreed upon parameter for quantifying a steel's degree of susceptibility. Nonetheless, a relatively simple method that provides information in both areas consists of conducting two quasi-static tensile tests (strain rate $\sim 10^{-4}$ s^{-1}), one at room temperature and one at about

200°C (392°F) [*16*]. If the strength at 200°C (called S_{200}) is equal to or greater than that at room temperature (called S_{22}), it is a certainty that the steel is susceptible to DSA. The value of the ratio S_{200}/S_{22} will provide a crude measure of the degree of susceptibility.

The principal shortcoming of this approach is that the meaning of ratios less than 1.0 is unclear, mainly because there is relatively little information on what the ratio is for nonsusceptible steels. A ratio of 0.9, for example, would not permit a conclusion to be drawn regarding the steel's susceptibility, or lack thereof, to DSA. In such cases, it might be necessary to conduct additional tensile tests at several temperatures in the range from 100 to 350°C (210 to 660°F) to obtain a graph of tensile strength versus temperature. The shape of that curve, as well as the individual stress-strain curves, would permit the steel's susceptibility to be assessed.

Deleterious Effects of Dynamic Strain Aging

As was noted earlier, DSA is accompanied by several favorable features, including increased tensile strength, increased fatigue strength, and increased creep-strength at elevated temperatures. Unfortunately, the story does not end there. Several deleterious effects are known to accompany DSA and these can be separated into two categories:

1. An increase in the ductile-to-brittle fracture transition temperature following plastic deformation in the DSA temperature range.
2. A lowering of the ductile-fracture resistance at temperatures within the DSA temperature range.

Each category is discussed in the following sections.

Increased Transition Temperature

When a steel susceptible to DSA is plastically deformed in the DSA temperature range, the microstructure of the steel is altered: (1) an unusually high dislocation density is produced, and (2) many of the dislocations are immobilized by nitrogen and/or carbon atmospheres. Such microstructural changes might be expected to produce a shift in the ductile-to-brittle fracture transition temperature.

Li and Leslie [*15*] have shown that a tensile prestrain of only 3% at 250°C (482°F) in a steel susceptible to DSA can cause the ductile-to-brittle fracture transition temperature, as determined from subsize Charpy V-notch impact specimens, to increase by as much as 45°C (81°F). Much larger effects were reported by Formby and Charnock [*5*], who strain cycled precracked bend bars of C-Mn pressure vessel steel at a temperature in the range of 275 to 300°C (525 to 570°F). After only 1 cycle of forward and reverse bending [crack-tip opening displacement of 0.12 mm (4.8×10^{-3} in.) per half cycle], the ductile-to-brittle transition temperature was raised by 76°C (137°F). After 5 cycles, it was raised by 118°C (212°F), which is a remarkable effect. Formby and Charnock likened this situation to a pressure vessel containing a sharp defect that experiences plastic strain near the crack tip when subjected to service pressurizations in the DSA range. Repeated cycles of pressurization at elevated temperatures could lead to severe embrittlement of the notch tip region and possibly lead to brittle fracture if pressurization were to occur at ambient temperatures.

Another equally insidious manifestation of embrittlement accompanying prestraining in the DSA range pertains to sharp defects or discontinuities located in close proximity to a weld. If the defect is present prior to welding, the material near the tip of the defect will experience plastic strains, both tensile and compressive, at an elevated temperature as each welding pass is made, due to the thermal expansion and contraction [*17–24*]. The amount of strain developed in, and the temperatures experienced by, the material will depend on several variables, including the distance from the notch tip to the fusion zone, the welding current and speed, the number of passes, and the amount of constraint. If the steel is susceptible to DSA, it is almost certain that the notch tip region following welding will have a ductile-to-brittle transition temperature that is *much* higher than that of the bulk of the structure. Dawes [*23*] has demonstrated this embrittlement for a semikilled C-Mn steel in the as-rolled condition.

In Dawes' experiment, welds were prepared in wide plate specimens after first introducing a series of sharp notches in the edges of the plates to be welded. These notches had a tip radius of about 0.08 mm (0.003 in.) and depths that varied from 2.5 to 8.9 mm (0.1 to 0.35 in.). After the plates were welded, Charpy-size bend specimens were machined from the plates such that the tip of the original notch was also the tip of the standard 2 mm (0.079 in.) deep notches in the bend specimens. The specimens were tested quasi-statically at $-20°C$ ($-4°F$) to determine the critical crack-opening displacement (COD) to initiate cracking. The results are shown in Fig. 7, where other results obtained from the as-rolled plate are included for comparison. Note the striking effect of welding in close proximity to the notch tip, particularly for the two intermediate-depth notches. Those two notch depths had critical COD values of only 0.02 to 0.12 mm (0.0008 to 0.0047 in.) at $-20°C$ ($-4°F$), which are less than the COD value for the as-rolled plate at a temperature of $-100°C$ ($-148°F$). That result indicates that the transition temperature of the two intermediate-depth-notch-tip regions near the butt weld was as least 80°C (144°F) higher than that of the as-rolled plate.

The occurrence of highly localized embrittlement at flaws located near welds, along with the tensile residual stresses that exist near welds, are thought to be responsible for many of the "low-stress fractures" that have occurred in steel storage tanks and steel ships at temperatures believed to be safe.

Decreased Resistance to Ductile Fracture at DSA Temperatures

It is a widely held belief that carbon steels tested above their ductile-to-brittle fracture transition temperature will show constant or slightly increasing fracture resistance as the temperature is raised. This assumption definitely is not true of steels that are susceptible to DSA, as was demonstrated over 60 years ago by Maurer and Mailander [*25*]. Figure 8 shows that the energy required to break a keyhole-notched, mild-steel bend specimen in quasi-static tests at 200 to 250°C (390 to 480°F) was about half that required at room temperature. Maurer and Mailander also reported that the minimum in the curve was shifted upward to approximately 550°C (1020°F) if the specimens were loaded in impact, as would be anticipated from arguments presented earlier regarding combined strain-rate/temperature effects.

In view of Maurer and Mailander's results, it is perhaps not surprising that more recent studies using fracture mechanics specimens have found reduced fracture resistance in carbon steels tested in the DSA temperature range. Miglin et al. [*26*], in a study of strain aging effects in unloading-compliance J tests, reported that several steels susceptible to DSA displayed lowered J_{Ic} values in the DSA temperature range. An example of their findings is shown in Fig. 9; both heats of SA 106 Grade C steel showed a J_{Ic} minimum value near 200°C (390°F), while J_{Ic} values for the SA 516 Grade 70 steel were still declining at 290°C (550°F). Miglin et al. reported also that the tearing modulus, proportional to the slope of the J-resistance curve, diminished significantly with increasing temperature in the DSA temperature range for a steel that was susceptible to DSA. Low values of tearing modulus favor the occurrence of ductile crack instabilities in compliant structures. Mukherjee [*27*] also has reported that

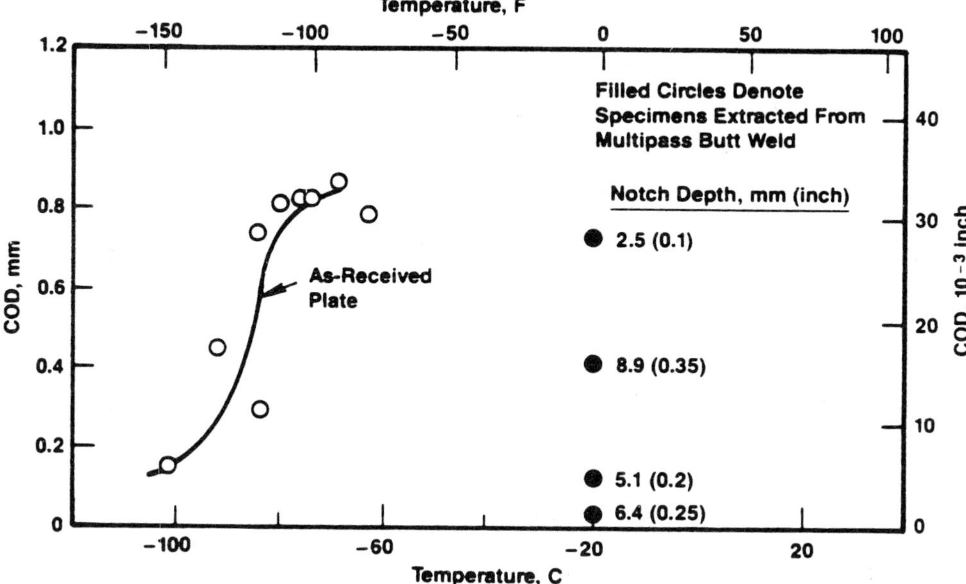

FIG. 7—*Effect of welding near a notch on COD in slow-bend, Charpy-size specimens* [23].

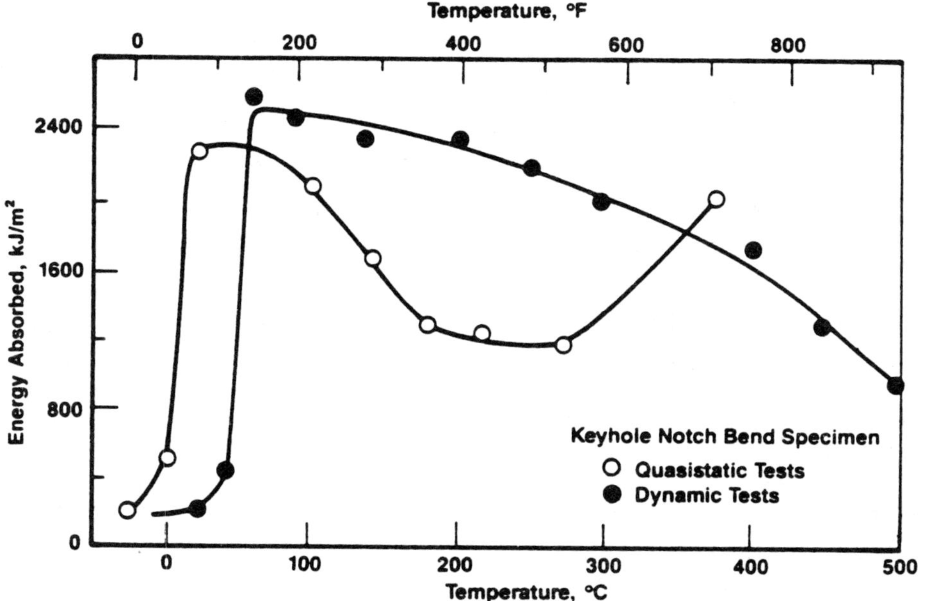

FIG. 8—*Energy absorbed in fracturing a mild steel as a function of test temperature* [25].

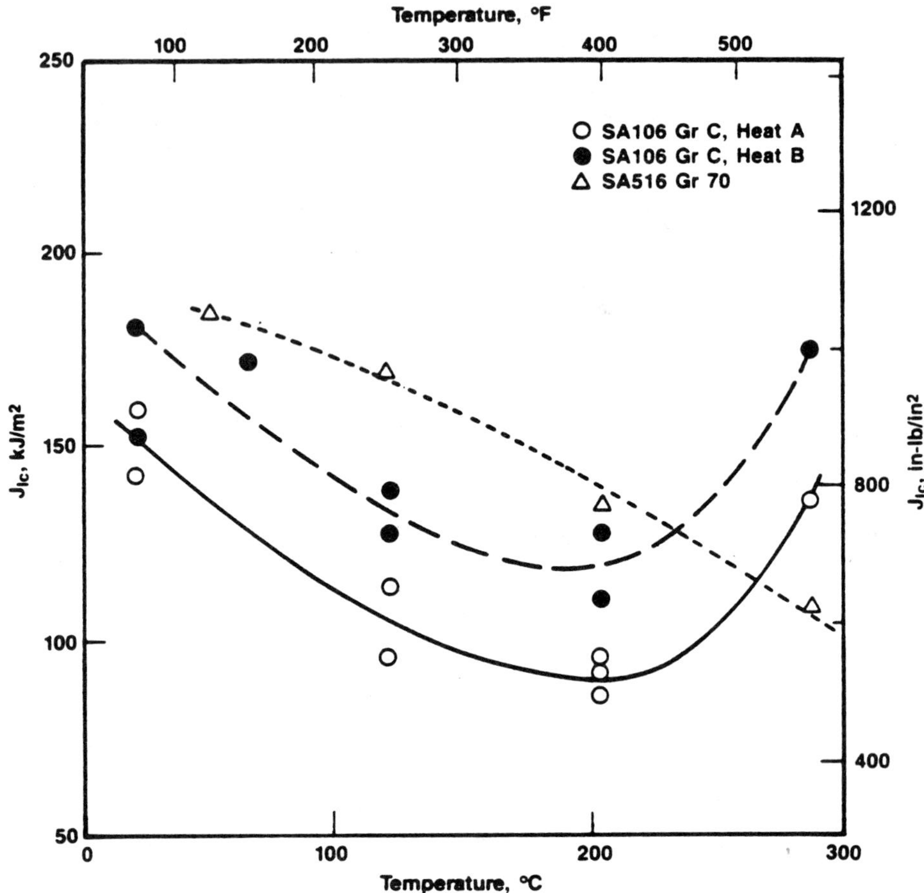

FIG. 9—J_{Ic} *versus temperature for several carbon steels* [26].

J_{Ic} and *J-R* curves for SA 106 Grade B pipes of several diameters were lower at 200°C (392°F) than at 20°C (68°F), both in base metal and in weld metal.

Occurrence of Rapid, Unstable Crack Growth at LWR Temperatures

Another feature of the fracture behavior of carbon steels at moderately elevated temperatures which may be a manifestation of DSA is the occurrence of bursts of rapid, unstable crack growth. Such behavior has been observed frequently in 288°C (550°F) tests conducted at Battelle, both in small compact-tension specimens and in full-scale pipe tests. A satisfactory explanation of this phenomenon is not yet available. Additional details of some of Battelle's experiments in which unstable crack growth was observed are given in the next section.

Results of Experiments on Nuclear Piping Steels

In work being performed at Battelle and several other laboratories for the U.S. NRC's Degraded Piping Program, pipe fracture experiments and laboratory specimen tests are being conducted on austenitic as well as ferritic piping steels [7]. The pipe fracture experiments are generally conducted at 288°C (550°F). The laboratory specimen tests include Charpy, tensile, and fracture toughness. Observations and trends from some of the Battelle tests relative to dynamic strain aging are discussed below.

Observations in Laboratory Specimen Testing

It was noted earlier in this paper that a simple method for assessing a steel's susceptibility to DSA was to examine the ratio S_{200}/S_{22}. In the Degraded Piping Program at Battelle, no strength data were obtained at 200°C, thereby precluding calculation of the above ratio. Data were obtained, however, that permitted calculation of the ratios S_{149}/S_{22} and S_{288}/S_{22}. These ratios are summarized in Table 1 for several carbon steel pipe materials. Table 1 also includes information about the occurrence of serrations on the stress-strain curve at the two temperatures. Such serrations are indicative of DSA; their absence, however, is no guarantee that DSA is not occurring.

Of the eight steels included in Table 1, six clearly are susceptible to DSA; the six had ratios of S_{288}/S_{22} that ranged from 1.16 to 1.24, and they exhibited serrated stress-strain curves at 149°C (300°F). The two remaining steels, while probably also susceptible to DSA, are less susceptible than the other six, as evidenced by lower ratios of S_{288}/S_{22} and absence of serrations at 149°C (300°F).

Earlier in this paper it was noted that unstable crack jumps have occurred in some of the carbon steel compact specimens tested at 288°C (550°F) in the Degraded Piping Program and that this occurrence might be related to DSA. To illustrate unstable cracking, load-displacement curves for compact-specimen tests are shown in Fig. 10 for one of the pipe steels included in Table 1. Figure 10 is for Steel DP2-F26 (ASTM Specification A 155, made from ASTM A516, Grade 70 carbon steel), which was determined to be strongly susceptible to DSA from the results given in Table 1; the load-displacement curve shows four distinct, sudden load drops, indicative of crack jumps, in a test conducted at 288°C (550°F). In a companion test at 149°C (300°F), the load-displacement curve gave evidence of four or five relatively small load drops, plus evidence of serrated yielding.

For comparison purposes, load-displacement curves are shown in Figure 11 for Steel DP2-

TABLE 1—*Results of tensile tests on ferritic pipe materials tested in degraded piping program.*

Steel Ident. No.	Steel Type	Results of Quasi-Static Tensile Tests[a]		Serrations	
		S_{149}/S_{22}	S_{288}/S_{22}	At 149°C	At 288°C
DP2-F29	A106B	1.16	1.18	Yes	No
DP2-F45	A106B	1.18	1.16	Yes	No
DP2-F13	A106B	1.18	1.16	Yes	No
DP2-F30	A106B	1.21	1.19	Yes	No
DP2-F9	A333, Gr. 6	1.07	1.16	Yes	No
DP2-F11	A333, Gr. 6	N.D.	0.98	No[b]	No
DP2-F34	A516, Gr. 70	0.89	0.99	No	No
DP2-F26	A516, Gr. 70	1.20	1.24	Yes	No

[a] S_{149}/S_{22} and S_{288}/S_{22} are ratios of the tensile strengths at the temperatures (°C) indicated by the subscripts.

[b] Although no tensile data were available at 149°C (300°F), compact specimen tests at that temperature exhibited no serrations on the load-displacement curve.

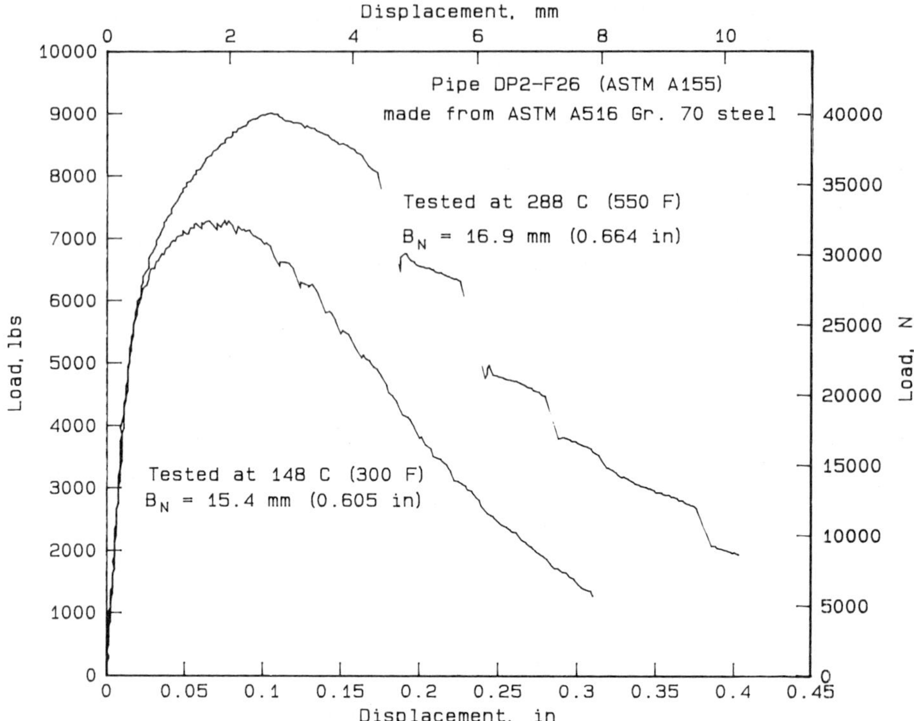

FIG. 10—*Load-displacement records for carbon steel compact specimens to illustrate bursts of unstable crack growth.*

F11, an ASTM A333 Grade 6 steel which, as indicated in Table 1, was less susceptible to DSA than was Steel DP2-F26. For this less susceptible steel, neither a C(T) test at 288°C (550°F) nor at 149°C (300°F) showed evidence of unstable cracking or of serrations.

Full-Scale Pipe Test Results

During the course of conducting circumferential cracked pipe fracture experiments at 288°C (550°F) in the Degraded Piping Program, unstable crack jumps were observed in several ferritic steels.

Figure 12 shows the results from a four-point bend test on a 711 mm (28 in.) diameter through-wall circumferentially-cracked pipe at 288°C (550°F). The pipe was made to ASTM Specification A 155 from ASTM A516 Grade 70 ferritic steel (DP2-F26) which displayed crack instabilities in compact-specimen tests at 288°C (550°F), as shown in Fig. 10. Note in Fig. 12 that there were numerous sudden load drops. These drops corresponded to crack jumps of 6 mm (0.25 in.) to 100 mm (4 in.) in length, as determined using the d-c electric potential method [28]. The fracture surface was a 45-deg slant fracture through the thickness, both during stable ductile tearing and during the crack instability. Notice in Fig. 12 that the crack jumps started prior to the specimen reaching maximum load.

Figure 13 shows a similar test record for a circumferential complex-cracked A106 Grade B pipe bending experiment at 288°C (550°F). In compact-specimen tests at 288°C (550°F)

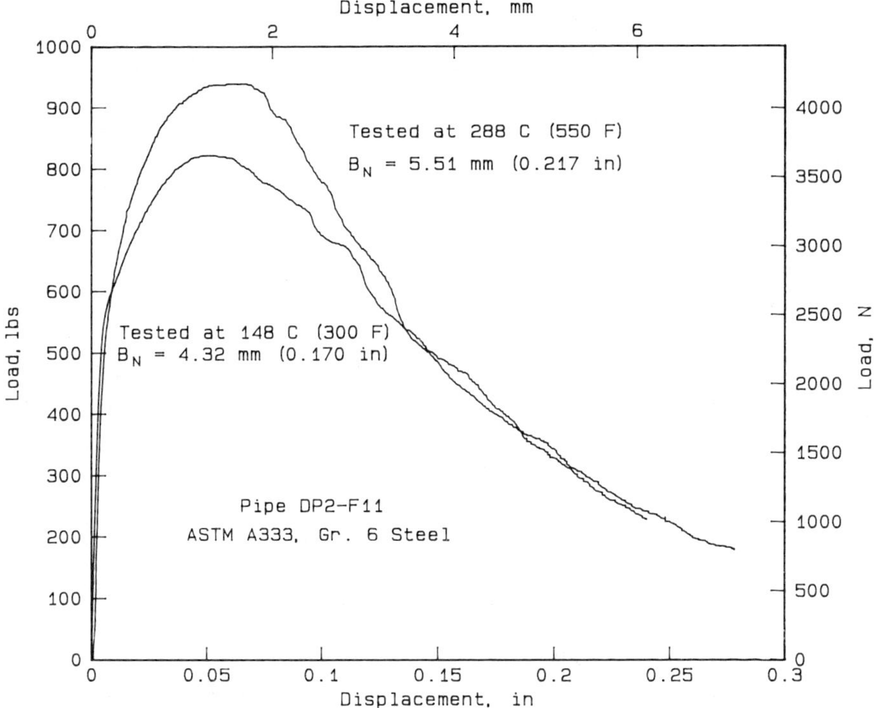

FIG. 11—*Load-displacement records for carbon steel compact specimens which showed only stable, ductile tearing.*

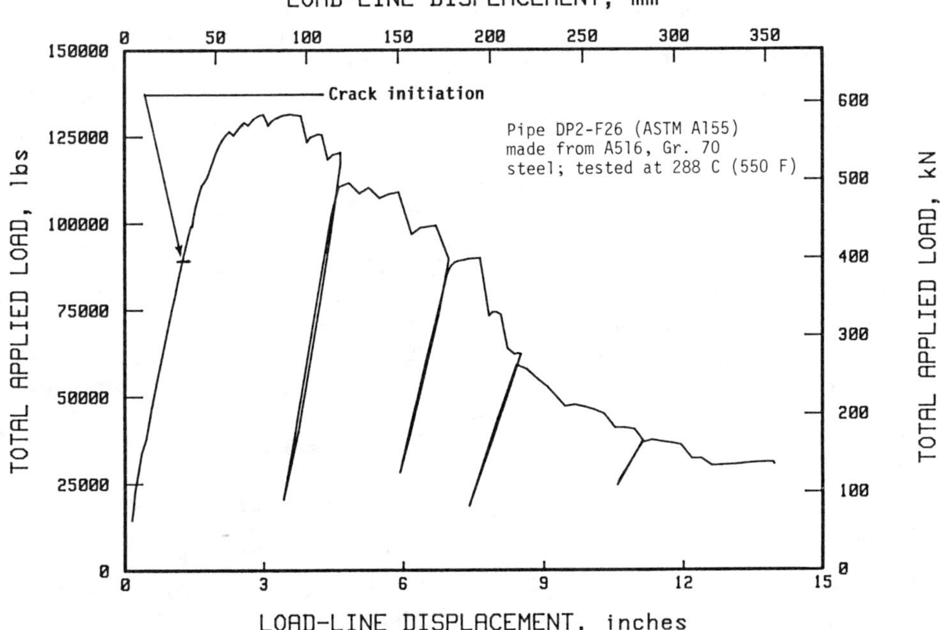

FIG. 12—*Load-displacement record from a full-scale pipe test to illustrate crack instabilities at 288°C (550°F).*

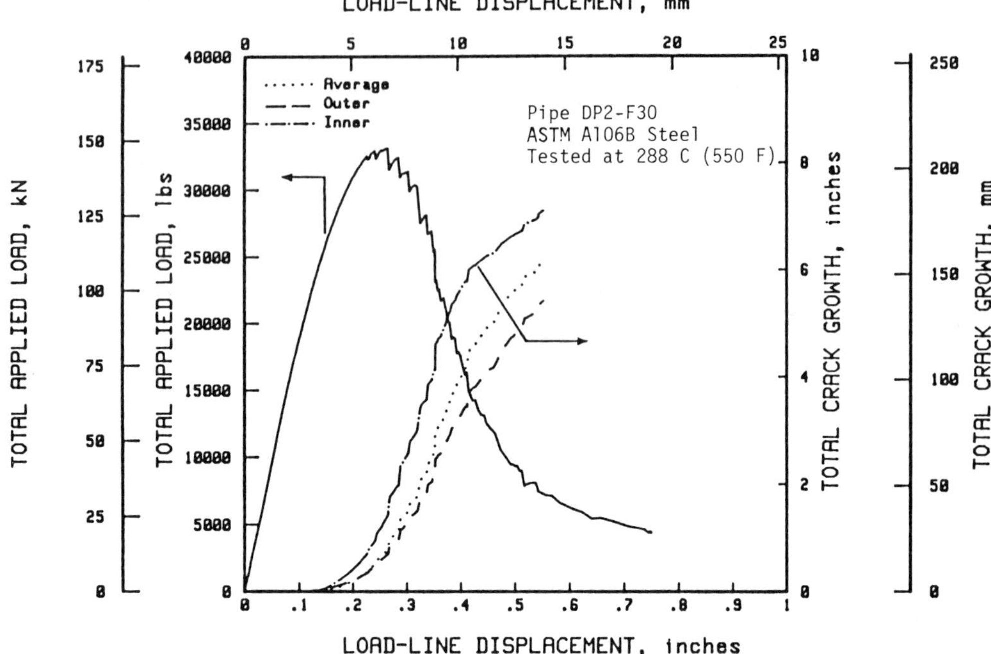

FIG. 13—*Load and total crack growth versus displacement for a full-scale pipe test (complex crack) at 288°C (550°F).*

this material (DP2-F30) also exhibited unstable cracking. As indicated in Table 1, the steel was strongly susceptible to DSA in tensile tests.

A third test record is shown in Fig. 14; in this case the test pipe was 152 mm (6 in.) diameter A106 Grade B pipe with a shielded metal arc weld. A circumferential through-wall crack was in the center of the weld. After a small amount of ductile tearing, where the crack turned 45-deg through the thickness and then reached the toe of the weld, a series of approximately 20 small crack jumps occurred.

An example of a major crack instability was reported by the David Taylor Research Center (DTRC) [29] for a four-point bending experiment on 203 mm (8 in.) diameter A106 Grade B pipe at 288°C (550°F). A through-wall circumferential crack was located in the center of a gas tungsten-arc weld. In this case, once the maximum load was reached, there was a major instability and the load dropped to about half of the load at the start of the instability. The crack jump was approximately a quarter of the pipe circumference.

The results of the pipe experiments show that significant crack instabilities can occur in ferritic nuclear pipe base metals and welds at light-water reactor operating temperatures.

Causes of Unstable Cracking

A satisfactory explanation for unstable cracking has not yet been advanced. Candidate explanations include excessive machine compliance and the occurrence of dynamic strain aging.

With respect to the first candidate, it is well established that crack instabilities are favored in both laboratory specimens and pipes by steep unloading slopes in the load/displacement curve beyond the maximum load point, and by highly compliant loading systems. In each of the tests depicted in Figs. 10 to 14, the load-train compliance was substantially less than that believed necessary to cause crack instabilities, based on an analysis described in Ref *30*. Also, the pipe test at DTRC, described previously, in which the crack jumped approximately one quarter of the pipe circumference, was stated to be a low-compliance test [*29*].

Additional evidence opposing excessive compliance as an explanation for the onset of unstable cracking includes:

- For the pipe tests depicted in Figs. 12 and 13, the first crack jump occurred on the rising part of the load/displacement curve (prior to maximum load); machine compliance can cause a ductile-fracture instability only under falling load in displacement-controlled testing conditions.
- For the C(T) specimen tests in Fig. 10, the upper curve, characterized by relatively large crack jumps, has a much lower unloading slope just prior to the crack jumps than does the lower curve, which shows only small instabilities. Had the instabilities been caused by excessive machine compliance, an opposite result would have been expected.

On the basis of the results shown in Figs. 10 and 11, it is tempting to conclude that the susceptibility to unstable cracking in moderately elevated temperature tests is directly related to DSA susceptibility. As was noted earlier, the steel depicted in Fig. 10, which exhibited crack jumps, was shown in tensile tests to be strongly susceptible to DSA; the steel in Fig. 11, on the other hand, which exhibited no crack instabilities, was less susceptible to DSA on the basis of tensile tests. Those results suggest a connection between DSA and crack jumps.

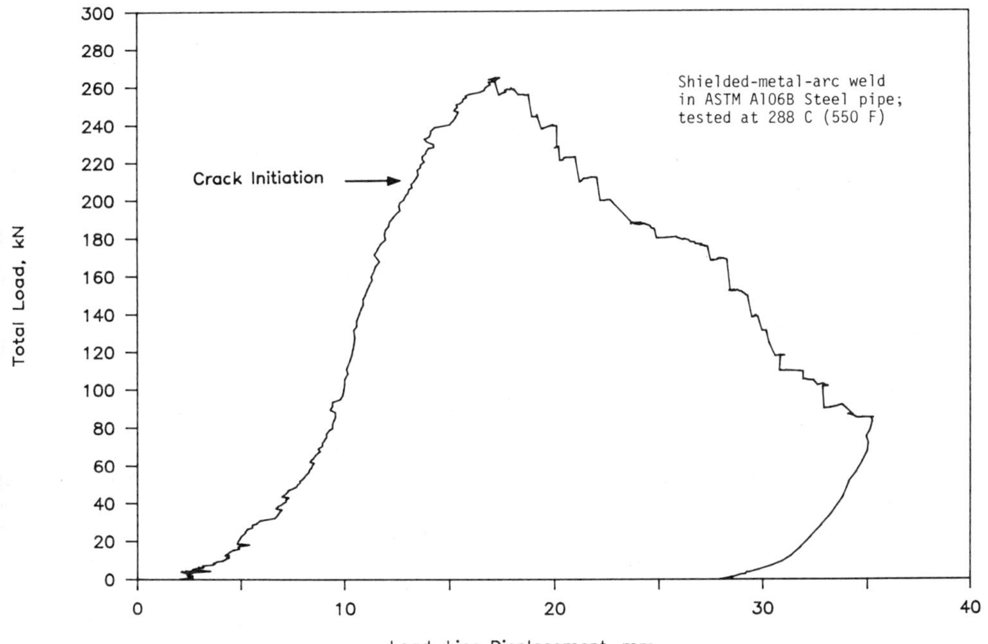

FIG. 14—*Load-displacement record for a full-scale pipe test (through-wall circumferential flaw) of shielded metal arc weld) at 288°C (550°F).*

Other data exist, however, which do not support a direct link between crack instabilities and DSA. Specifically, crack jumps are observed only rarely in 149°C (300°F) C(T) tests of steels that exhibit pronounced crack instabilities in 288°C (550°F) tests, even though both temperatures are well within the DSA temperature range at quasi-static testing rates. Furthermore, some of the pipe steels which have shown a marked susceptibility to DSA in tensile tests have not displayed unstable cracking in compact-specimen tests and vice versa. For example, each of the four A106B steels included in Table 1 were found to be susceptible to DSA in tensile tests, yet only two of them (DP2-F29 and DP2-F30) showed unstable cracking in 288°C (550°F) C(T) tests. Conversely, Steel DP2-F34, an A516 Grade 70 steel, displayed unstable cracking in C(T) tests at 288°C (550°F), even though it was considerably less susceptible to DSA than any of the four A106B steels on the basis of tensile tests.

In addition to a lack of consistent results supporting a direct link between DSA and crack instabilities, no satisfactory explanation has been offered of how DSA can cause crack jumps. It can be conjectured that crack jumps are analogous to serrated yielding which sometimes accompanies DSA and which results from a sudden tearing away of dislocations from their pinning atmospheres of nitrogen and carbon atoms. Crack jumps might be likened also to the occurrence of an upper yield point in steels susceptible to strain aging; at the point of yielding, the load falls sharply and localized straining occurs (Luders bands) with little or no strain hardening (Fig. 1). At the tip of a growing crack, a similar lack of strain hardening could produce highly localized strain and promote crack extension with little energy absorption. However, no physical evidence is available to substantiate that view.

Calculation of J-R Curves in the Presence of Unstable Cracking

Bursts of unstable crack growth can cause difficulties in calculating J-R curves using the procedure set forth in ASTM E 1152-87. One difficulty relates to the theoretical basis for calculating J-R curves; since that basis is stable, continuous, ductile tearing in a homogeneous material, it is perhaps incorrect to calculate J beyond the instability point. Nonetheless, it is important for engineering purposes that an approximate J-R curve be calculated for materials that display unstable crack growth. In attempting to perform such a calculation, a choice must be made between several possible calculation methods, as discussed in the following paragraphs.

Figure 15 helps to illustrate the problem for a single crack instability. The solid line, *ABC*, represents a recorder trace of the event. Since the instability event often is faster than the recording equipment used in quasi-static testing, the actual load-displacement relation might have followed Curve *ADC*. Furthermore, the actual load drop during the instability should depend on load-train compliance; for example, a very stiff load train could result in a near-zero load at Point *D*. Because of the uncertainty in the actual curve during the crack jump and its dependence on compliance, there might be some justification for simply performing a linear interpolation from Point *A* to Point *C*. Curve *AC* clearly would yield higher J values than would Curves *ABC* or *ADC* because of greater area under the curve.

Once a decision is reached regarding which load-displacement curve is to be used, it must also be decided whether to use only data from the points shown in Fig. 15 (Points *A, B, C, D*) or to use an arbitrary number of additional intermediate points. That decision is important because the displacement interval can have an effect on the calculated J-R curve.

Figure 16 shows J-R curves at 288°C (550°F) calculated by the procedures in ASTM E 1152 for an A516 Grade 70 steel (Pipe DP2-F26) whose load-displacement record was shown in Fig. 10. The specimen exhibited four crack jumps, calculated to be 1.96, 1.47, 1.22, and 0.66 mm (0.077, 0.058, 0.048, and 0.026 in.). With reference to Fig. 15, the calculations were performed using Curve *ABC*, without using any data between Points *A* and *B*. Note

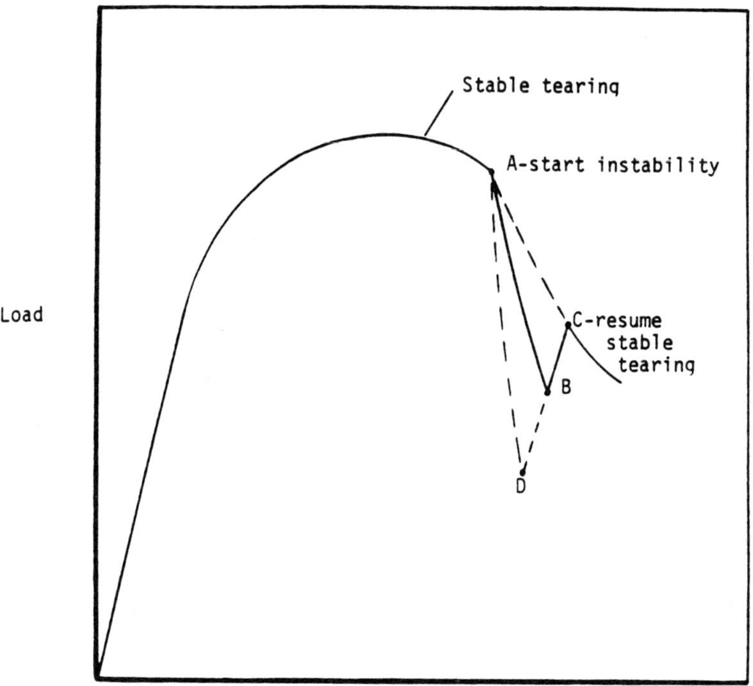

FIG. 15—*Schematic illustration of one burst of unstable crack growth in a fracture toughness test.*

that three of the four regions of unstable crack extension are evident on the *J-R* curve. Also apparent is the marked change in slope of the curves at the first two instabilities.

Two observations are pertinent regarding *J-R* curve calculations in the presence of crack jumps: (1) if crack jumps are small, such as shown by the lower curve in Fig. 10, any of the calculation methods described above will yield approximately the same results, and (2) since *J* by the method of ASTM E 1152 is a cumulative parameter (i.e., $J_{i+1} = J_i + $ a crack growth term), the *J* value just past the onset of an instability is close to the prior value for stable ductile tearing, even though there may be a large instantaneous change in the fracture toughness. Thus the *J-R* curve calculation method tends to mask local regions of unusually low or high toughness.

Clearly, these results suggest that further work is necessary to develop suitable methods for computing *J-R* curves in specimens exhibiting unstable crack extension. A theoretically proper calculation procedure may require using a viscoplastic finite element analysis with a fracture parameter such as T_p^* [*31*].

Discussion

From the evidence available, it is clear that many carbon steel pipes and pressure vessels used in nuclear plants are susceptible to dynamic strain aging. This susceptibility arises from small quantities of uncombined nitrogen and carbon in the steel interacting with dislocations. The degree of susceptibility cannot be predicted readily from chemical analysis results.

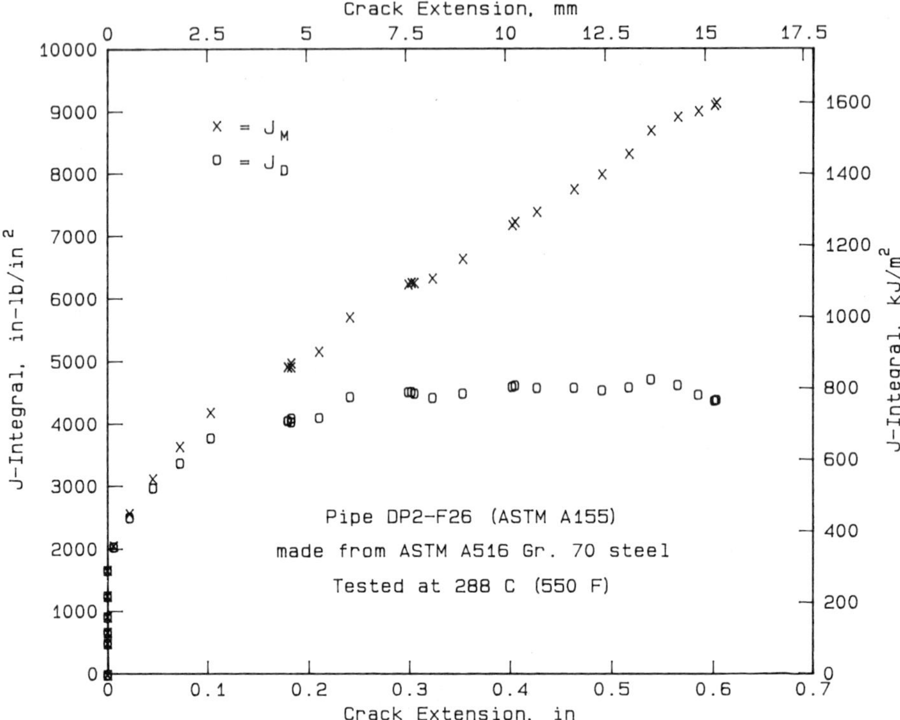

FIG. 16—J *resistance curves for 288°C (550°F) compact-specimen test shown in Fig. 10.*

However, simple quasi-static tensile tests conducted at two temperatures—room temperature and about 200°C (390°F)—can be helpful in assessing the relative susceptibility of a steel; higher ratios of tensile strength at 200°C (390°F) to tensile strength at room temperature indicate higher susceptibility to DSA.

Although DSA can be beneficial to a component from the standpoint of tensile strength, fatigue resistance, and creep resistance, evidence is accumulating which indicates that DSA also presents some decidedly negative aspects to safe performance:

1. If a steel component susceptible to DSA is subjected to plastic strain at temperatures in the DSA range, it will have its ductile-to-brittle fracture transition temperature raised significantly. This behavior can be particularly insidious in causing local embrittlement at a defect whose tip region is plastically strained at an elevated temperature, as could accompany repeated load cycles or introduction of a weld close to the defect. Subsequent stressing near room temperature might cause a crack to initiate in the embrittled notch-tip region.

2. In a cracked steel component operating at LWR temperatures, susceptibility to DSA will make it easier to initiate crack growth (low J_{Ic}) and to achieve crack propagation (low T_{mat}).

3. Although an adequate link with DSA is lacking, cracked steel components susceptible to DSA and operating at LWR temperatures may display sizable crack jumps in addition to stable tearing. In a reactor cooling pipe, unstable crack growth could create a serious leak even in the absence of a guillotine break.

Of the three negative aspects of DSA listed here, only the first has received much attention, primarily in the 1960s and 1970s and involving structures other than nuclear pipes and pressure vessels. More recently, a limited amount of attention has been given to the second and third negative characteristics of DSA listed above in programs studying the integrity of cracked nuclear piping. However, those programs, while revealing that many carbon steel pipes are susceptible to DSA, have not been directed toward gaining an improved understanding of the phenomenon or toward developing methods for minimizing its harmful effects.

Even less attention has been given to potential DSA effects in nuclear pressure vessels. There, it is commonly assumed that operating in the upper-shelf toughness region guarantees safety, so long as the absorbed energy is at least 68 J (50 ft-lb). That assumption may not be correct for certain irradiated welds that have relatively low (but still adequate) upper-shelf toughness values as a result of irradiation. The possibility exists that the ductile fracture resistance is lower at the LWR operating temperature than at the temperature corresponding to the first attainment of the upper shelf.

In view of the fact that LWR pipes and pressure vessels operate within the temperature range of DSA, it would appear worthwhile to gain improved understanding of DSA effects on fracture resistance of carbon steels used in those applications. Areas of particular importance include: (1) effects of temperature and strain rate, (2) explanation of unstable cracking, and (3) steel compositions and thermal treatments to minimize susceptibility to DSA. In addition, work is needed to develop suitable methods for computing *J-R* curves in specimens that exhibit bursts of unstable crack extension.

Acknowledgments

The authors acknowledge the capable assistance of numerous co-workers at Battelle. Paul Held, Paul Mincer, and Earl Swetnam conducted laboratory specimen tests; Greg Kramer and Paul Scott were responsible for conducting the full-scale pipe experiments. Thomas Groeneveld provided helpful discussions and a careful review of the manuscript.

Thanks go also to the U.S. Nuclear Regulatory Commission for its support of work in the Degraded Piping Program.

References

[*1*] Baird, J. D., "Strain Aging of Steel—A Critical Review," *Iron and Steel,* Sept. 1963, pp. 450–457.
[*2*] Baird, J. D., "The Effects of Strain-Ageing Due to Interstitial Solutes on the Mechanical Properties of Metals," *Metallurgical Reviews,* Vol. 16, Feb. 1971, pp. 1–18.
[*3*] Ho, E. T. C. and Lepik, O. E., "Effects of Strain Aging and Related Phenomena Upon the Toughness of Carbon and Carbon-Manganese Weld Metals—A Literature Review," Ontario Hydro Research Division, Report 86-323-K, 14 Jan. 1987.
[*4*] Chakravartty, J. K., Wadekar, S. L., Sinha, T. K., and Asundi, M. K., "Dynamic Strain Ageing of A203D Nuclear Structural Steel," *Journal of Nuclear Materials,* Vol. 119, 1983, pp. 51–58.
[*5*] Formby, C. L. and Charnock, W., "The Effect of Dynamic Strain Ageing Embrittlement on Pressure Vessel Integrity," in *Practical Application of Fracture Mechanics to Pressure-Vessel Technology,* Inst. of Mech. Engineers, London, 1971, Paper C1/71, pp. 1–7.
[*6*] Salma, M. M., "An Assessment of Strain-Aging Effect on Weldable Offshore Structural Steel," *Journal of Petroleum Technology,* Vol. 36, No. 1, Jan. 1984, pp. 141–146.
[*7*] Wilkowski, G. M. and others, "Degraded Piping Program-Phase II," NUREG/CR-4082, Vol. 4, Sept. 1986, pp. 3-22 to 3-26.
[*8*] Keh, A. S., Nakada, Y., and Leslie, W. C., "Dynamic Strain Aging in Iron and Steel," in *Dislocation Dynamics,* A. R. Rosenfield and others, Eds., McGraw-Hill, New York, 1968, pp. 381–408.
[*9*] Manjoine, M. J., "Influence of Rate of Strain and Temperature on Yield Stresses of Mild Steel," *Journal of Applied Mechanics,* Dec. 1944, pp. A211–A218.

[10] Van den Beukel, A., "On the Mechanism of Serrated Yielding and Dynamic Strain Aging," *Acta Metallurgica,* Vol. 28, No. 7, July 1980, pp. 965–969.

[11] Van den Beukel, A., and Kocks, U. F., "The Strain Dependence of Static and Dynamic Strain Aging," *Acta Metallurgica,* Vol. 30, No. 5, May 1982, pp. 1027–1034.

[12] Hume-Rothery, W. H. and Raynor, G. V., *The Structure of Metals and Alloys,* The Institute of Metals, London, 1954, pp 58–59.

[13] Leslie, W. C. and Rickett, R. L., "Influence of Aluminum and Silicon Deoxidation on the Strain Aging of Low-Carbon Steels," *Transactions of AIME,* Aug. 1983, pp. 1021–1031.

[14] Wilson, D. V. and Russell, B., "The Contribution of Precipitation to Strain Aging in Low Carbon Steels," *Acta Metallurgica,* Vol. 8, 1960, pp. 468–479.

[15] Li, C. C. and Leslie, W. C., "Effects of Dynamic Strain Aging on the Subsequent Mechanical Properties of Carbon Steels," *Metallurgical Transactions A,* Vol. 9A, Dec. 1978, pp. 1765–1775.

[16] Kenyon, R. L. and Burns, R. S., "Testing Sheets for Blue Brittleness and Stability Against Changes Due to Aging," *Proceedings of ASTM,* Vol. 34, 1934, pp. 48–58.

[17] Zhemchuzhnikov, G. V. and Givenko, V. S., "The Strain Aging and Brittle Fracture of Steel," *Automatic Welding,* Vol. 86, No. 1964, pp. 7–12.

[18] Burdekin, F. M., "Effects of Thermal Straining During Welding on the Fracture Toughness of a Mild Steel," *British Welding Journal,* Feb. 1967, pp. 81–83.

[19] Burdekin, F. M., "Initiation of Brittle Fracture in Structural Steels," *British Welding Journal,* Dec. 1967, pp. 649–659.

[20] Dolby, R. E., Egan, G. R., Dawes, M. G., Saunders, G. G., and Archer, G. L. "Brittle Fracture Initiation in Welded Low Strength Steels," in *Practical Fracture Mechanics for Structural Steel,* M. O. Dobson, Ed., United Kingdom Atomic Energy Authority, 1969, pp. G1–G18.

[21] Dechaene, R., Soete, W., and Vinckier, A., "Low-Stress Brittle Fracture in Mild Steel," Ship Structure Committee Report SSC-158, 30 Aug. 1963.

[22] Wells, A. A., "Influence of Residual Stresses and Metallurgical Changes on Low-Stress Brittle Fracture in Welded Steel Plates," *Welding Research Supplement,* April 1961, pp. 182s–192s.

[23] Dawes, M. G., "Thermal Cycles and Straining Effects in a Multipass Butt Weld,," *British Welding Journal,* Nov. 1968, pp. 563–570.

[24] Dolby, R. E. and Saunders, G. G., "Sub-Critical HAZ Fracture Toughness of C:Mn Steels," *Metal Construction and British Welding Journal,* May 1972, pp. 185–190.

[25] Maurer, E. and Mailander, R., "The Problem of Blue Brittleness," *Stahl u Eisen,* Vol. 45, Part I, 1925, pp. 895–981.

[26] Miglin, M. T., Van Der Sluys, W. A., Futato, R. J., and Domian, H. A., "Effects of Strain Aging in the Unloading Compliance *J* Test," in *Elastic-Plastic Fracture Test Methods: The User's Experience, ASTM STP 856,* E. T. Wessel and F. J. Loss, Eds., American Society for Testing and Materials, Philadelphia, 1985, pp. 150–165.

[27] Mukherjee, B., "The *J*-Resistance Curve Leak-Before-Break Test Program on Material for the Darlington Nuclear Generating Station," *International Journal of Pressure Vessels and Piping,* Vol. 31, 1988, pp. 363–385.

[28] Schwalbe, K. and Hellman, D., "Application of the Electric Potential Method to Crack Length Measurements Using Johnson's Formula," *Journal of Testing and Evaluation,* Vol. 9, No. 3, May 1981, pp. 218–221.

[29] Hays, R., David Taylor Research Center, private communication.

[30] Wilkowski, G. M. and others, "Degraded Piping Program—Phase II," NUREG/CR-4082, Vol. 3, March 1986, Appendix B.

[31] Nakagaki, M., Marschall, C. W., and Brust, F. W., "An Analysis of a Crack in Welds by Using Near-Field As Well As Far-Field Crack Parameters," to be published in *Transactions of Offshore Mechanics and Arctic Engineering.*

Takashi Miyata,[1] *Akio Otsuka,*[2] *Masahiko Mitsubayashi,*[3]
Toshiaki Haze,[4] *and Syuji Aihara*[5]

Prediction of Fracture Toughness by Local Fracture Criterion

REFERENCE: Miyata, T., Otsuka, A., Mitsubayashi, M., Haze, T., and Aihara, S., **"Prediction of Fracture Toughness by Local Fracture Criterion,"** *Fracture Mechanics: Twenty-First Symposium, ASTM STP 1074,* J. P. Gudas, J. A. Joyce, and E. M. Hackett, Eds., American Society for Testing and Materials, Philadelphia, 1990, pp. 361–377.

ABSTRACT: Incorporation of micromechanistic criteria of failure to analytical or numerical crack tip stress and strain solutions, known as the RKR (Ritchie, Knott, and Rice) model, derives a certain relation between the fracture toughness and flow/fracture properties of materials. Based on the above analytical prediction, correlations between the fracture toughness and yield stress, cleavage fracture stress, and critical plastic strain of the materials have been investigated for cleavage and ductile fracture. Nine types of low carbon structural steel were tested that had various microstructures with yield strengths of 280 to 1110 MPa. Good correlations according to analytical predictions have been obtained for both cleavage and ductile fracture, and the possibility of quantitative prediction of the fracture toughness from a conventional round bar tensile test can be shown. Effects of metallurgical and mechanical factors on the fracture toughness, and temperature dependence of the fracture toughness, can be explained from their effects on the flow and fracture properties.

KEY WORDS: fracture toughness, local fracture criterion, RKR model, cleavage fracture, ductile fracture, low carbon steels, cleavage fracture stress, critical plastic strain

Fracture toughness of a material is determined by fracture characteristics corresponding to a certain failure mode and plastic flow properties which determine the state of plastic stress and strain at the crack tip. Various attempts have been made to relate the microscopic fracture conditions to the local mechanical fracture criteria at the crack tip region, with the objective of a description of macroscopic fracture toughness, as is reviewed by Pineau [1]. In low strength steels for welded structures, fracture occurs by a transgranular cleavage mechanism at a low temperature range, whereas at a higher temperature range fracture occurs by a mechanism of microvoid coalescene.

The micromechanism of the cleavage fracture for mild steels has been investigated, and the cleavage fracture in such steels obeys the critical tensile stress criterion [2]. Ritchie, Knott, and Rice [3] (hereafter referred to as RKR) have related the local fracture criterion to the macroscopic fracture toughness of mild steel by examining the temperature dependence of K_{Ic} over the low temperature cleavage fracture range. Cleavage fracture from a crack

---

[1] Associate Professor, Department of Materials Science and Engineering, Nagoya University, Nagoya, Japan.
[2] Professor, Department of Materials Science and Engineering, Nagoya University, Nagoya, Japan.
[3] Graduate Student, presently Associate Engineer, Toyota Motor Corporation, Toyota, Japan.
[4] Senior Researcher, R&D Laboratories II, Nippon Steel Corporation, Sagamihara, Japan.
[5] Research Engineer, R&D Laboratories II, Nippon Steel Corporation, Sagamihara, Japan.

tip has been assumed to be initiated when the tensile stress exceeds the critical stress over a certain microstructurally determined "characteristic distance X_o," ahead of the crack tip. Applicability of the RKR model to low alloy bainitic steels has been shown by Parks [4] and by Ritchie et al. [5]. Stochastic modeling for the cleavage fracture, which assumes that probabilistic distribution of the crack nucleus is related to the carbide particle distribution and also gives a certain description of the fracture toughness, has been proposed by Curry and Knott [6], Beremin [7], and others [8–10].

On the other hand, for the microvoid coalescence type of ductile fracture, a similar local criterion approach can be applicable as shown by Sailors [11], Hahn and Rosenfield [12], Mackenzie et al. [13], Pineau [1], and Ritchie et al. [5]. Most of them have used the Rice and Johnson [14] model or McMeeking's FEM solution [15] for the stress and strain analysis of a blunted crack. As a ductile fracture criterion, strain criterion as a function of stress triaxiality, firstly proposed by McClintock [16] and developed by Hancock and Mackenzie [17], has been generally used.

When we discuss the effects of metallurgical factors, temperature, strain rate, irradiation, etc., on the fracture toughness, it is convenient to express the fracture toughness in terms of flow and fracture properties of the material, such as the yield stress and the cleavage fracture stress or the critical plastic strain, since these are able to be obtained by conventional tensile tests. If successful correlation between these can be derived, the effects of various factors on the fracture toughness and the transition temperature can be quantitatively discussed through their effects on the flow and fracture properties.

The purpose of this work is to confirm experimentally the methodology of the RKR model to correlate the toughness with the flow and fracture properties for both the cleavage and ductile fracture, and to discuss the possibility of simplification of the RKR model and prediction of the toughness from the conventional tensile tests. In particular, the significance of the characteristic distance in the cleavage fracture, which is considered to be a function of the microstructural unit or those distribution [18], has been investigated for several steels that have various microstructures (i.e., fine grained ferrite, coarse grained bainitic, and martensitic).

Although the stochastic local criterion approach [6–10] might be preferable for analysis of cleavage fracture, data on the statistical material constants are insufficient at the moment for the quantitative analysis. Application of the stochastic models to the present experimental results is currently being developed.

Model of Local Fracture Criterion

In order to correlate the fracture toughness and the flow/fracture properties of the materials, the HRR [19,20] small-scale yielding solution for the plane strain stress state around a stationary crack tip has been used. For a material that obeys the Ramberg-Osgood stress-strain law as $\bar{\varepsilon}_p/\varepsilon_{ys} = (\bar{\sigma}/\sigma_{ys})^n$, stress and strain for the distance r ahead of the crack tip are given by

$$\sigma_{ij} = \sigma_{ys}(J/I_n\varepsilon_{ys}\sigma_{ys}r)^{1/(n+1)}\tilde{\sigma}_{ij}(\theta,n) \tag{1a}$$

and

$$\varepsilon_{ij} = \varepsilon_{ys}(J/I_n\varepsilon_{ys}\sigma_{ys}r)^{n/(n+1)}\tilde{\varepsilon}_{ij}(\theta,n) \tag{1b}$$

where σ_{ys} and ε_{ys} are yield stress and strain, respectively, J is the J contour integral, I_n is a numerical constant weakly dependent on n, and $\tilde{\sigma}_{ij}(\theta,n)$ and $\tilde{\varepsilon}_{ij}(\theta,n)$ are angular factors.

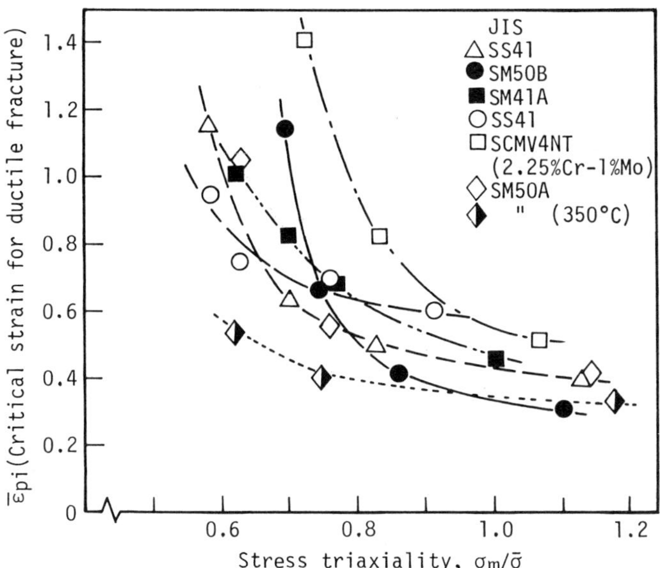

FIG. 1—*Dependence of critical strain at ductile crack initiation on stress triaxiality* [23].

The RKR local fracture criterion that $J = J_c$ when the cleavage fracture stress, σ_c, is exceeded over a microstructurally significant characteristic distance, X_o, gives an expression of the cleavage fracture toughness defined by the stress intensity factor [21]:

$$K_c = C(n) \sqrt{X_o}\, \sigma_c\, (\sigma_c/\sigma_{ys})^{(n-1)/2} \qquad (2)$$

For the dimpled ductile fracture, a similar expression can be obtained. Hancock and Mackenzie [17] have shown that plastic strain, $\bar{\varepsilon}_{pi}$, at the microvoid coalescence strongly depends on stress triaxiality (i.e., the ratio of mean stress, σ_m, to equivalent flow stress, $\bar{\sigma}$), as predicted by the void growth model [16,22]. Figure 1 shows the dependence of $\bar{\varepsilon}_{pi}$ on the stress triaxiality, $\sigma_m/\bar{\sigma}$, for various low carbon steels obtained by some of the authors [23]. Mackenzie et al. [13] have proposed a criterion for ductile fracture from a crack tip, where $\bar{\varepsilon}_{pi}$ is locally exceeded over some distance ahead of the crack tip. Because of the strong dependence of $\bar{\varepsilon}_{pi}$ on the stress triaxiality and the variance of the stress triaxiality depending on the distance from the crack tip, application of the local criterion to ductile fracture seems to be more complicated. However, as shown in Fig. 1, the value of $\bar{\varepsilon}_{pi}$ decreases sharply with an increase in stress triaxiality but takes an almost constant value for a further increase in stress triaxiality. Therefore the constant strain criterion for a high triaxial stress state can be approximately adopted. Considering the variance of $\sigma_m/\bar{\sigma}$ around the crack tip, we assume a more simplified criterion as $\bar{\varepsilon}_p = \bar{\varepsilon}_{pi}$ at $r = \lambda$. Hence the fracture toughness for the onset of ductile fracture is given by [24]

$$J_i = D_j\, \lambda\, \varepsilon_{ys}\, \sigma_{ys}\, (\bar{\varepsilon}_{pi}/\varepsilon_{ys})^{(n+1)/n} \qquad (3)$$

In commercial structural steels, the strain hardening coefficient n takes a value generally larger than 5, so that $n/(n + 1)$ may be regarded as unity:

$$J_i = D_j\, \lambda\, \sigma_{ys}\, \bar{\varepsilon}_{pi} \qquad (4a)$$

or

$$K_i = D(n) \sqrt{\lambda \, \sigma_{ys} \, \bar{\varepsilon}_{pi}} \tag{4b}$$

Experimental Procedure

Materials

The materials used in this work were three vacuum-melted, hot-rolled, and normalized steels, one high carbon (0.15C) steel, and two low carbon (0.08C) steels with/without an addition of 0.025Nb; their chemical compositions are given in Table 1. These steels (in as-received condition, hereafter referred to as T3 materials) were austenized at 1400°C, followed by controlled cooling from 800 to 500°C in 2 s for the T1 process and 40 s for the T2 process. For the low carbon steel with niobium, T2SR steel that was tempered at 600°C for 3 h then air-cooled for stress relief was prepared instead of T1 material. In a normalized condition (T3) the microstructure is fine ferrite with small patches of pearlite, whereas microstructures of T2 and T1 materials are primarily coarse grained upper bainite and coarse grained martensite, respectively, which would be expected to be in the heat-affected zone of welds. Typical microstructures are shown in Fig. 2. Mechanical properties of the materials are given in Table 2. All test specimens were oriented so that the tensile or longitudinal axis coincided with the rolling direction of the plate.

Evaluation of Critical Fracture Stress and Strain

Cleavage fracture stress for each material was determined from unnotched smooth and circumferentially notched (1mmR) round bar specimens broken at −196°C. Specimen configurations are shown in Fig. 3. Stress distributions in the minimum cross section for the smooth specimens were calculated by Bridgman's formula. For the notched specimens, axisymmetric finite element analysis was performed using the stress-strain constitutive relations of the materials at various temperatures. The maximum tensile stress at catastrophic cleavage fracture initiation was taken as the cleavage fracture stress, σ_c, in both specimens. The values of σ_c have often been evaluated from the slip line theory or finite element analysis to V-notched bend bar. However, the ambiguity inherent with the assumption of the plane stress or plane strain condition can not be avoided in such two-dimensional analyses. Application of axisymmetric specimens and their analysis might better obtain more precise fracture stress.

Ductile fracture initiation strain, $\bar{\varepsilon}_{pi}$, was also obtained from the round bar tensile specimens at room temperature. The smooth and notched specimens were loaded with continuous monitering for changes in the minimum diameter. Macroscopic ductile crack initiation at the center of the minimum cross section was defined as the point when the load dropped sharply prior to final failure of the specimen. Effective plastic strain at that point was eval-

TABLE 1—*Chemical composition of steels.*

Steel	C	Si	Mn	P	S	Al[a]	N[b]	Nb
A (High carbon)	0.147	0.214	1.482	0.0059	0.0037	0.028	0.0029	<0.005
B (Low carbon)	0.080	0.195	1.364	0.0048	0.0034	0.025	0.0033	<0.005
C (Low C & Nb)	0.078	0.196	1.347	0.0051	0.0030	0.024	0.0024	0.024

[a] Sol. Al.
[b] Total N.

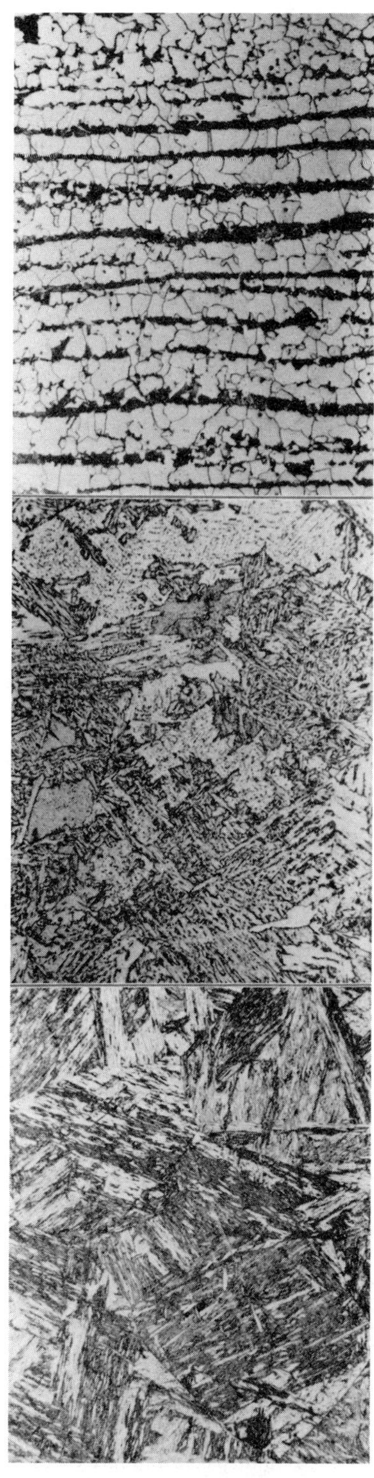

FIG. 2—*Typical microstructures of steels (Steel A, high carbon steel).*

TABLE 2—*Mechanical properties of steels and test results on ductile fracture.*

Steel	Thermal Cycle	Yield Strength (σ_{ys}), N/mm²	Tensile Strength (σ_u), N/mm²	Reduction of Area (ϕ), %	Critical Plastic Strain $(\bar{\varepsilon}_{pi})$ Smooth	Notched	Fracture Toughness (J_i), kN/m
A (High C)	T1	1108	1335	35	0.30	. . .	. . .
	T2	482	619	76	1.04	0.36	167
	T3	326	494	78	1.18	0.70	226
B (Low C)	T1	935	1086	25	0.29	. . .	. . .
	T2	412	538	77	1.17	0.50	221
	T3	279	432	82	1.52	0.94	. . .
C (Low C & Nb)	T2	481	588	78	0.89	0.42	196
	T2SR	489	601	73	1.14	0.41	. . .
	T3	301	459	84	1.48	0.99	. . .

uated as $\bar{\varepsilon}_{pi}$ from the changes of the minimum diameter in both the smooth and the notched specimens.

Fracture Toughness Testing

Three-point-bend specimens with a thickness of 10 mm were used for the fracture toughness test (Fig. 3c). Tests were carried out at various temperatures. The values of fracture

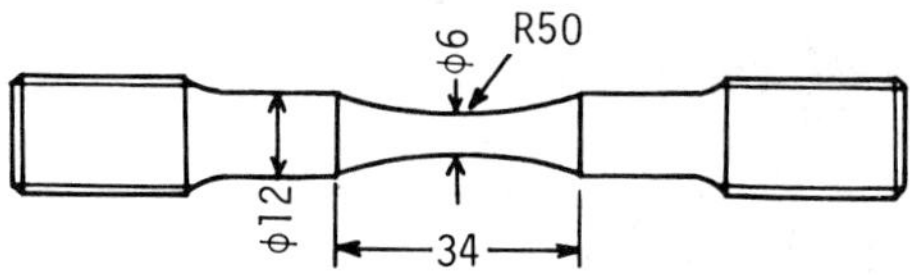

(a) Smooth round bar specimen.

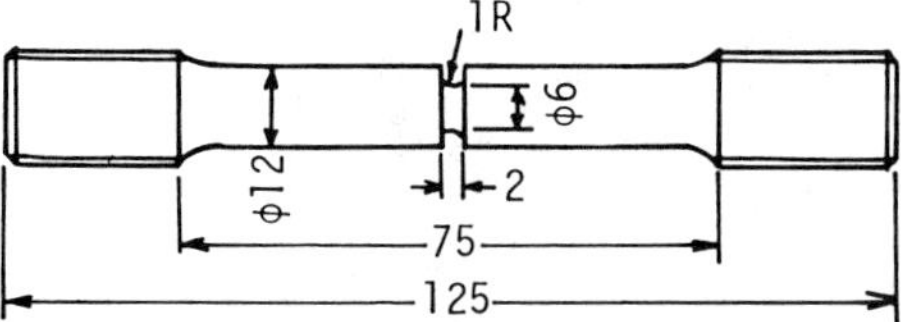

(b) 1mmR notched round bar specimen.

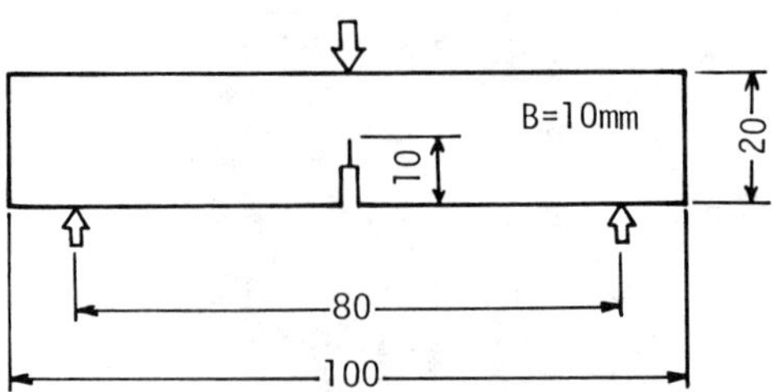

(c) Three point bend specimen.

FIG. 3—*Specimens (dimensions in millimetres).*

toughness were evaluated at the point when the catastrophic cleavage fracture occurred; thus in the transition temperature range the data on the ductile crack preceding the cleavage fracture were involved. The toughness was evaluated in terms of J integral from the measurement of load-displacement relation, except for the values at $-196°C$ of which toughness was defined by the stress intensity factor. Ductile fracture toughness, J_i, for ductile crack initiation from the precrack tip was evaluated at room temperature for some of the steels used in the present work in accordance with ASTM E 813 using the multi-specimen R-curve procedure. The values of cleavage fracture toughness, J_c, were converted to K_c values for later discussions, using the relation $J = K^2(1 - v^2)/E$, where v = Poisson's ratio.

Results and Discussion

The results are presented in two parts. The first part discusses results on the flow and fracture properties, the second discusses the relation between those properties and the fracture toughness.

Variations of the yield stress, σ_{ys} (lower yield stress for T3, 0.2% proof stress for others), with temperature are shown in Fig. 4. The values of the cleavage fracture stress, σ_c, at $-196°C$ for the smooth and notched specimens are listed in Table 3, together with the yield stress at the same temperature. The cleavage fracture stress takes a higher value with the increase of the yield stress, though the high strength material shows low cleavage fracture toughness and high transition temperature. Discussions in detail on the cleavage fracture stress from the metallurgical point of view have been presented elsewhere [25]. It should be noticed that the effect of notch or stress gradient on σ_c can be observed except in T3 steels. The reason for these differences between the smooth and the notched specimens might be attributed to the probabilistic nature of the cleavage fracture [7]. In the smooth specimens the high stressed region is so large compared with that of the notched specimens that the probability of fracture becomes large, which in turn leads to lower fracture stress. Figure 5 shows the Weibull plot of σ_c in the smooth and notched specimens for a mild steel [26].

When the fracture probability in a uniform stress field is given by the Weibull distribution with the shape parameter m and the scale parameter σ_o, the probability of failure for the

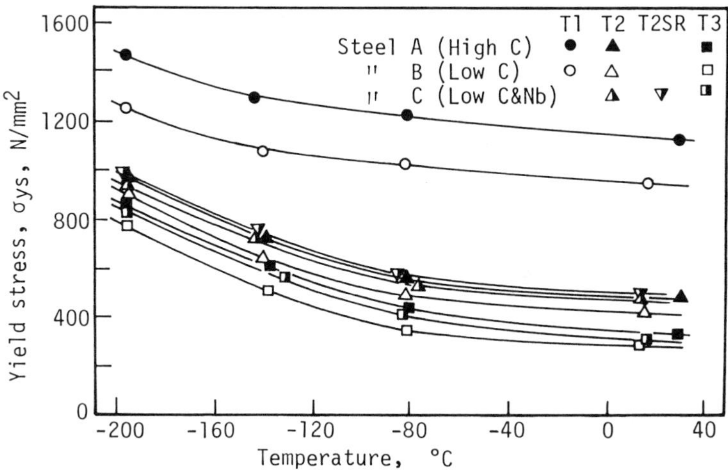

FIG. 4—*Temperature dependence of yield stress of materials.*

TABLE 3—*Cleavage fracture stress and yield stress at* $-196°C$ *and microstructural/fractographic parameters.*

Steel	Thermal Cycle	Ferrite Grain or Packet Size (d), μm	Cleavage Facet Size (d_c), μm	Yield Stress (σ_{ys}), N/mm^2	Cleavage Fracture Stress (σ_c), N/mm^2	
					Smooth	Notched
A (High C)	T1	47	35	1453	1850	2390
	T2	34	143	948	1240	1720
	T3	11	8.4	838	1490	1500
B (Low C)	T1	38	36	1236	1590	2060
	T2	33	137	889	1020	1540
	T3	15	8.1	760	1350	1380
C (Low C & Nb)	T2	31	104	911	1140	1520
	T2SR	30	140	980	1090	1640
	T3	6.4	6.8	818	1540	1500

stress gradient region is given by [7]

$$P = 1 - \exp\left[-\int_V \left(\frac{\sigma_i}{\sigma_o}\right)^m dV/V_o \right] \tag{5}$$

where V_o is the statistical unit volume. The probabilistic distribution of σ_c in the notched specimens can be calculated by the FEM according to Eq 5. Material constants were determined from the results of the smooth specimens, and the integration was performed for the

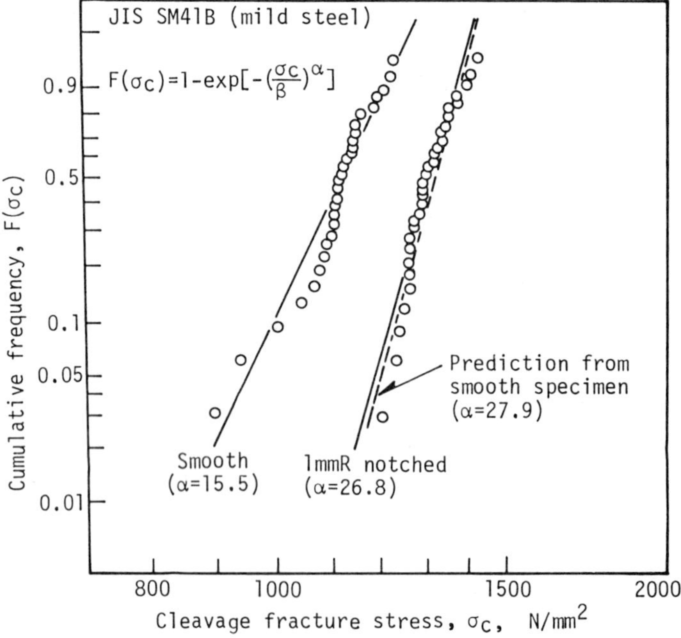

FIG. 5—*Weibull distributions of cleavage fracture stress in smooth and 1mmR notched round bar specimens* [26].

plastically deformed zone [26]. Quite good agreement with the experimental results was obtained (Fig. 5). This indicates the validity of the weakest link analogy for the cleavage fracture of steels. The effect of the stress gradient, namely the statistical size effect, seems to strongly depend on the materials. The ratio of σ_c of the notched specimens to that of the smooth specimens differs in each material and is related to the cleavage facet size as shown in Fig. 6, using data from some of the authors. Cleavage facet size was measured by computer-aided image analysis of the fracture surface, and the average values of measured 400 to 1000 facets are listed in Table 3. Present results suggest the significance of statistical analysis of the cleavage fracture, because the scatter of σ_c must be related to the probabilistic distribution of metallurgical factors such as the grain or carbide particle size [6,9]. Since such an effect as notch on σ_c seems to become a relatively small variance of notch acuity [27], which might be explained by the statistical analysis, σ_c of the notched specimen is adopted to correlate with the fracture toughness in the present work.

The critical plastic strain values, $\bar{\varepsilon}_{pi}$, at the ductile crack initiation in the smooth and notched specimens are shown in Table 2, except for those in the notched specimens of T1 steels in which ductile fracture did not occur even above room temperature. Unique and strict dependence of $\bar{\varepsilon}_{pi}$ on the yield stress (room temperature) is observed. This might be caused from the stress dependency of microvoid nucleation [1], since there is no difference in the state of nonmetallic inclusions and dimple sizes in the present materials.

Variations of the cleavage fracture toughness, K_c, with temperature are shown in Figs. 7 to 9. The materials having higher yield stress show consistently lower fracture toughness and higher transition temperature, independent of microstructures.

If the cleavage fracture stress, σ_c, can be assumed to be independent of temperature, the local criterion approach expressed by Eq 2 indicates that the variation of K_c with temperature is primarily determined by the variation of the yield stress. The relation between the fracture toughness normalized by the reference toughness at a specified temperature (T_o) and the yield stress normalized as well is given as

$$K_{c,T}/K_{c,T_o} = (\sigma_{ys,T_o}/\sigma_{ys,T})^{(n-1)/2} \tag{6}$$

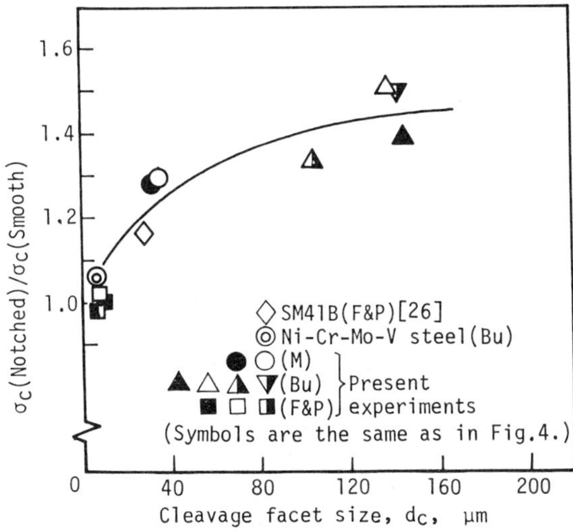

FIG. 6—*Relation between the ratio of cleavage fracture stress of notched specimen and that of smooth specimen and cleavage facet size.*

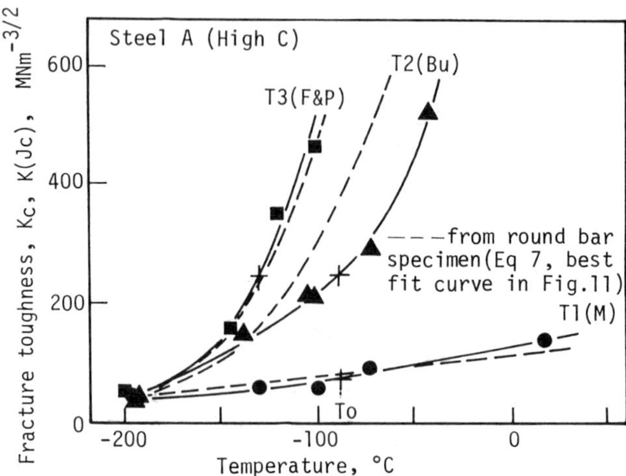

FIG. 7—*Temperature dependence of fracture toughness for Steel A (high carbon).*

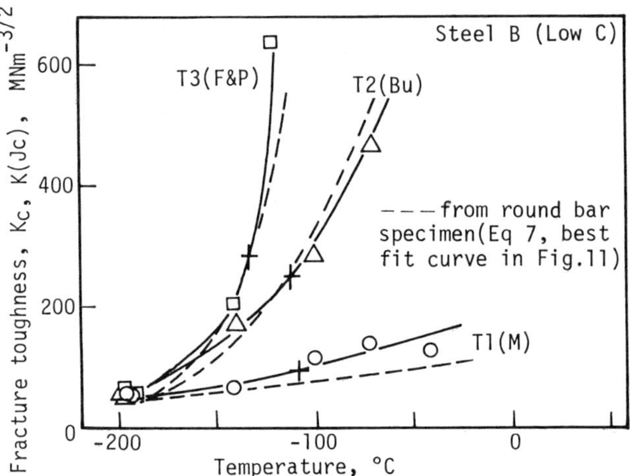

FIG. 8—*Temperature dependence of fracture toughness for Steel B (low carbon).*

Taking the reference temperature T_o as the medium of transition (Figs. 7 to 9), experimental results are shown in Fig. 10. A unique relation, independent of the materials, can be observed, though the strain hardening coefficient n must vary with the material and the temperature. These results imply the validity of the stress criterion independent of temperature and the RKR model incorporated with the HRR solution. Although the n value, in order to fit the data, takes a different value ($n = 7$), a similar result has been obtained in mild steels where the effect of the strain rate was examined [28]. Analysis of the data obtained by Lin et al. [9] for several mild steels also shows a unique relationship but a different n value ($n = 5$ to 7) with the present results. From these results it may be said that simplification of Eq 6 by setting a constant value of n for a range of strength levels of steels will be possible.

Since the temperature dependence of the fracture toughness can be well explained by the local stress criterion approach, correlation between σ_c and K_c on the basis of Eq 2 was investigated. From Fig. 10, the value of n being taken as 10, values of $\sigma_c(\sigma_c/\sigma_{ys})^{(n-1)/2}$ (referred to as the cleavage fracture parameter) were calculated from the σ_{ys} at the same temperature with the fracture toughness tests and the σ_c of the notched specimens at $-196°C$. Figure 11 shows the relation between K_c and the cleavage fracture parameter. Good correlation, independent of materials, can be observed, though Eq 2 predicts different proportional constants involving the characteristic distance, X_o:

$$K_c \, (\text{MN m}^{-3/2}) = 1.6 \cdot 10^{-3}\sigma_c(\sigma_c/\sigma_{ys})^{(n-1)/2} \, (\text{N/mm}^2) \tag{7}$$

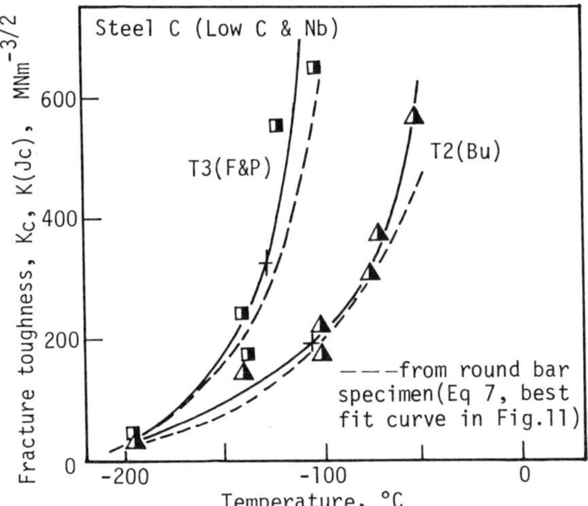

FIG. 9—*Temperature dependence of fracture toughness for Steel C (low C & Nb).*

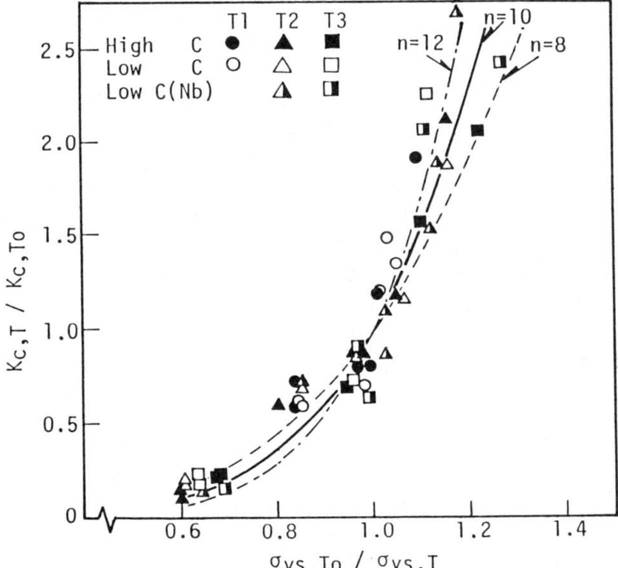

FIG. 10—*Relation between normalized fracture toughness and yield stress.*

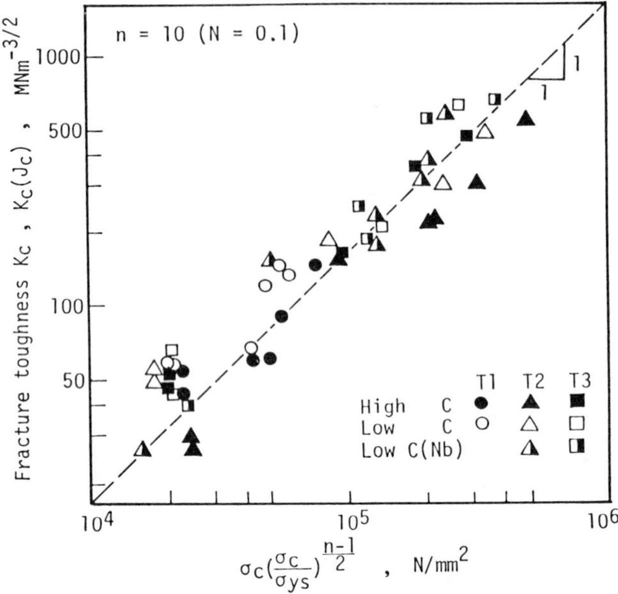

FIG. 11—*Relation between fracture toughness and cleavage fracture parameter.*

Predicted values of K_c using Eq 7 from the properties of the smooth and notched specimens are shown in Figs. 7 to 9 by a broken line.

Although the physical meaning of X_o has not yet been clarified [6], at least the cleavage facet size is quite different in the materials tested, as shown in Table 3. Facet size in T3 steels having fine ferrite microstructure corresponds to ferrite grain size, and those in T1 and T3 steels seem to be related to lath packet size and prior austenite grain size, respectively [25].

Fracture origins of the cleavage fracture on the fracture surface were observed by SEM. Some of the fractographs for the high carbon material are shown in Fig. 12. Figure 13 shows the distance (X_f) from the crack tip to the point of fracture origin for each material, taking the horizontal coordinate as the facet size. Although difficult to discuss in detail because of large scatter and little data, no remarkable difference in X_f which might be concerned with X_o can be observed in spite of a large difference in the cleavage facet size. In the case of mild steels, the variation of X_o with ferrite grain size is small and becomes almost constant for fine grain [29]. Whether this is due to the small variation in X_o or to the RKR model itself is unclear, but a proportional constant involving X_o is likely to have little influence on the relation between K_c and the cleavage fracture parameter, insofar as we discuss fracture toughness in the sense of accuracy shown in Fig. 11.

Furthermore, for the discussion of applicability or generality of the correlation given in Eq 7, the effect of size on fracture toughness must be clarified. Three-point-bend specimens with a thickness of 10 mm were used in the present experiments. Most of the toughness values do not satisfy the K_{Ic} requirement in ASTM E 399, whereas some of the toughness evaluated in terms of J satisfy the plane strain requirement for J_{Ic}, B (specimen thickness) $\geq 25\ J_{Ic}/\sigma_{\text{flow}}$, prescribed in ASTM E 813. Discussions on the size effect on cleavage fracture toughness continue. There are two different ideas. One is a classical mechanical explanation in which the size effect is considered to be caused by the change in stress triaxility, because the cleavage fracture is controlled by the maximum tensile stress. The other is a statistical approach based on the weakest link analogy [30]. If the size effect does not depend on materials or strength

FIG. 12—*Fractographs of cleavage fracture around the fracture origin (Steel A at −196°C).*

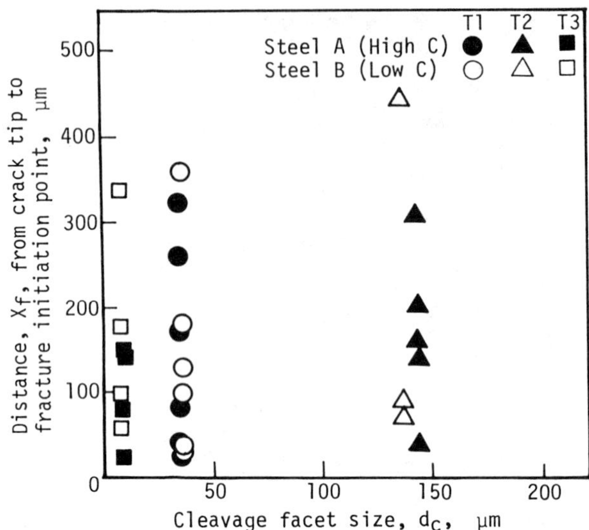

FIG. 13—*Relation between distance from crack tip to fracture initiation point, X_f, and cleavage facet size.*

as derived from statistical theory, the result in Fig. 11 itself has the generality, which shows the correlation of toughness for the 10 mm thick specimen. Wallin [8] has theoretically shown that the probabilistic distribution of the cleavage fracture toughness obeys the Weibull distribution with a constant shape parameter ($\alpha = 4$). That leads to the following relation between toughness for different thicknesses [7,8]:

$$(K_{c,B})^4 \cdot B = (K_{c,B_o})^4 \cdot B_o \tag{8}$$

The statistical approach based on the two-parameter Weibull distribution does not give any lower saturation in the size effect. Then, we assume that the plane strain fracture toughness can be obtained in the specimen with the thickness of

$$B_o = 2.5(K_{Ic}/\sigma_{ys})^2 \tag{9}$$

Using Eqs 8 and 9, K_c values in Fig. 11 were converted to K_{Ic} values; the correlation with the cleavage fracture parameter is shown in Fig. 14. Scatter in the correlation seems to be smaller, but the linear proportional relation (Slope 1 in log-log diagram) in K_{Ic} and the cleavage fracture parameter as predicted by Eq 2 have been lost.

On the other hand, the ductile fracture toughness, J_i or J_{Ic}, is presumed by Eq 4 to be proportional to the product of σ_{ys} and $\bar{\varepsilon}_{pi}$, if the state of the nonmetallic inclusions, which decides λ, is not changed. In the present steels, significant differences in the state of the inclusions and also in the dimple size on the fracture surface were not observed. Figure 15 shows the relation between the J_i (approximately, can be regarded as J_{Ic} according to ASTM E 813) and $\bar{\varepsilon}_{pi} \cdot \sigma_{ys}$. Data for Cr-Mo bainitic steel and a low carbon structural steel in an investigation of the effect of temperature [31] are also shown. Linear proportional relation can be observed for each group of materials, as predicted by the local criterion approach. The structural steel, SM50A, in Fig. 15 has large inclusions and dimple size, and the Cr-Mo steel has a similar situation regarding inclusions and dimple size with the present materials. These results show

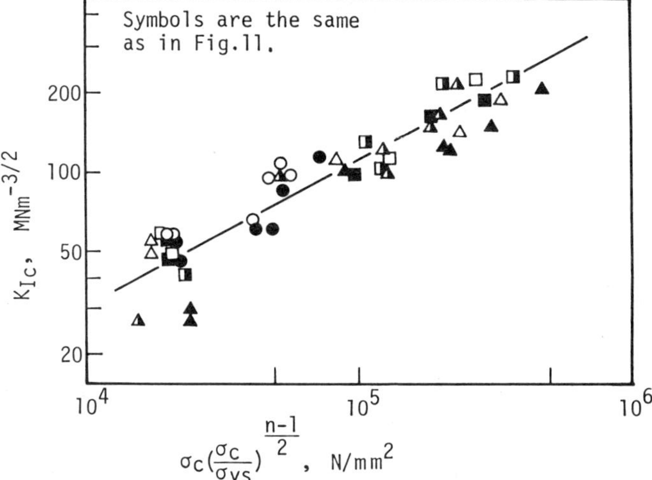

FIG. 14—*Relation between the predicted values of plane strain fracture toughness and cleavage fracture parameter.*

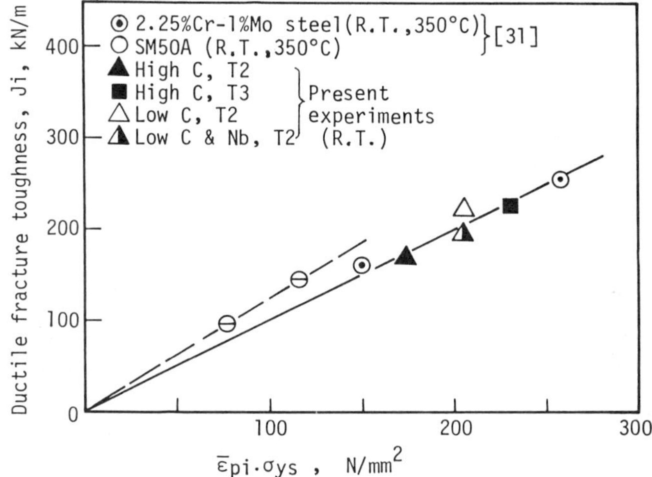

FIG. 15—*Relation between ductile fracture toughness, J_i, and $\bar{\varepsilon}_{pi} \cdot \sigma_{ys}$, where $\bar{\varepsilon}_{pi}$ is critical strain in notched specimen.*

the validity of the simplified strain criterion on the ductile fracture at the crack tip, and that the effect of temperature, strain rate, heat treatment, etc., on J_{Ic} can be discussed through their effects on the σ_{ys} and $\bar{\varepsilon}_{pi}$. Quantitative metallographic and fractographic analyses will be needed for the argument of proportional constants involving λ and for the prediction of the ductile fracture toughness.

From the experimental results and discussions above, it can be conclusively said that the local criterion approach, such as the RKR model, is a valuable methodology for analyzing the fracture toughness of materials. It makes quantitative evaluation of the influence of various factors on the fracture toughness possible. Additionally, prediction of toughness from conventional tensile tests is possible, provided that some problems can be solved. Concerning the cleavage fracture, the values of σ_c have been shown to be closely related to the yield stress at $-196°C$ [25,32]. If such a relation is available, a more simplified prediction will be possible.

Conclusions

Several conclusions were drawn from this work:

1. The cleavage fracture stress, σ_c, is influenced by the sample volume. The difference in σ_c between the smooth and notched round bar specimens is closely related to the cleavage facet size.

2. Temperature dependence of the cleavage fracture toughness is expressed by a function of the yield stress of the materials. This result shows that the maximum stress criterion, independent of the temperature, is valid for the steels having not only the ferrite but also the bainite and martensite microstructure.

3. Good correlation between the cleavage fracture toughness and the cleavage fracture parameter, $\sigma_c(\sigma_c/\sigma_{ys})^{(n-1)/2}$, can be observed independently of the microstructures and the grain or lath packet size, as predicted by the local criterion approach. The influence of the characteristic distance seems to be negligible as a first approximation for the prediction of toughness.

4. Linear proportional relation between the ductile fracture toughness and the product of the yield stress and critical strain in notched specimens can be observed, provided the state of inclusions is not changed.

Acknowledgments

The first author acknowledges support of this work by the Japanese Ministry of Education, Scientific Research Grant 61460206. The authors wish to thank M. Koike, Nippon Steel Corporation, for his stimulating comments and advice.

References

[1] Pineau, A. in *Advances in Fracture Research, Proceedings of ICF 5*, Vol. 2, 1981, pp. 553–577.
[2] Knott, J. F., *Journal of the Iron and Steel Institute*, Vol. 204, 1966, pp. 104–111.
[3] Ritchie, R. O., Knott, J. F., and Rice, J. R., *Journal of the Mechanics and Physics of Solids*, Vol. 21, 1973, pp. 395–410.
[4] Parks, D. M., *Journal of Engineering Materials and Technology, Transactions of ASME, Series H*, Vol. 98, 1976, pp. 30–35.
[5] Ritchie, R. O., Server, W. L., and Wullaert, R. A., *Metallurgical Transactions A*, Vol. 10A, 1979, pp. 1557–1570.
[6] Curry, D. A. and Knott, J. F., *Metal Science*, Vol. 13, 1979, pp. 341–345.
[7] Beremin, F. M., *Metallurgical Transactions A*, Vol. 14A, 1983, pp. 2277–2287.
[8] Wallin, K., *Engineering Fracture Mechanics*, Vol. 19, 1984, pp. 1085–1093.
[9] Lin, T., Evance, A. G., and Ritchie, R. O., *Metallurgical Transactions A*, Vol. 18A, 1987, pp. 641–651.
[10] Anderson, T. and Stienstra, D., *Journal of Testing and Evaluation*, Vol. 17, 1989, pp. 46–53.
[11] Sailors, R. H. in *Properties Related to Fracture Toughness, ASTM STP 605*, American Society for Testing and Materials, Philadelphia, 1976, pp. 34–61.
[12] Hahn, G. T. and Rosenfield, A. R., *Metallurgical Transactions A*, Vol. 6A, 1975, pp. 653–668.
[13] Mackenzie, A. C., Hancock, J. W., and Brown, D. K., *Engineering Fracture Mechanics*, Vol. 9, 1977, pp. 167–188.
[14] Rice, J. R. and Johnson, M. A. in *Inelastic Behavior of Solids*, M. F. Kanninen, W. G. Adler, A. R. Rosenfield, and R. I. Jaffe, Eds., McGraw-Hill, New York, 1970, pp. 641–672.

[15] McMeeking, R. M., *Journal of the Mechanics and Physics of Solids,* Vol. 25, 1977, pp. 357–381.
[16] McClintock, F. A., *Journal of Applied Mechanics Transactions of ASME, Series E,* Vol. 34, 1968, pp. 363–371.
[17] Hancock, J. W. and Mackenzie, A. C., *Journal of the Mechanics and Physics of Solids,* Vol. 24, 1976, pp. 147–169.
[18] Curry, D. A., *Metal Science,* Vol. 14, 1980, pp. 78–80.
[19] Hutchinson, J. W., *Journal of the Mechanics and Physics of Solids,* Vol. 16, 1968, pp. 13–31.
[20] Rice, J. R. and Rosengren, G. R., *Journal of the Mechanics and Physics of Solids,* Vol. 16, 1968, pp. 1–12.
[21] Curry, D. A., *Nature,* Vol. 276, 1978, p. 50.
[22] Rice, J. R. and Tracey, D. M., *Journal of the Mechanics and Physics of Solids,* Vol. 17, 1969, pp. 201–217.
[23] Otsuka, A., Miyata, T., Sakurai, T., and Iida, H., *Journal of the Society of Materials Science,* Japan, Vol. 34, 1985, pp. 622–626.
[24] Lee, S., Majno, L., and Asaro, R. J., *Metallurgical Transactions A,* Vol. 16A, 1985, pp. 1633–1648.
[25] Miyata, T., Yang, R. C., Otsuka, A., Haze, T., and Aihara, S. in *Advances in Fracture Research, Proceedings of ICF7,* Vol. 4, 1989, pp. 2563–2572.
[26] Miyata, T., Otsuka, A., and Katayama, T., *Journal of the Society of Materials Science,* Japan, Vol. 37, 1988, pp. 61–66.
[27] Miyata, T., Otsuka, A., Mitsubayashi, M., Haze, T., and Aihara, S., *Journal of the Society of Materials Science,* Japan, Vol. 37, 1988, pp. 897–903.
[28] Curry, D. A., *Materials Science and Technology,* Vol. 43, 1980, pp. 135–144.
[29] Curry, D. A. and Knott, J. F., *Metal Science,* Vol. 10, 1976, pp. 1–6.
[30] Landes, J. D. and Shaffer, D. H. in *Fracture Mechanics (Twelfth Conference), ASTM STP 700,* American Society for Testing and Materials, Philadelphia, 1980, pp. 368–382.
[31] Otsuka, A., Miyata, T., Tohgo, K., and Iida, H., *Journal of the Society of Materials Science,* Japan, Vol. 33, 1984, pp. 544–549.
[32] Curry, D. A., *Metal Science,* Vol. 18, 1984, pp. 67–76.

A. Saxena,[1] *D. C. Daly,*[2] *H. A. Ernst,*[1] *and K. Banerji*[1]

Microscopic Aspects of Ductile Tearing Resistance in AISI Type 303 Stainless Steel

REFERENCE: Saxena, A., Daly, D. C., Ernst, H. A., and Banerji, K., **"Microscopic Aspects of Ductile Tearing Resistance in AISI Type 303 Stainless Steel,"** *Fracture Mechanics: Twenty-First Symposium, ASTM STP 1074,* J. P. Gudas, J. A. Joyce, and E. M. Hackett, Eds., American Society for Testing and Materials, Philadelphia, 1990, pp. 378–395.

ABSTRACT: The initiation toughness (J_{Ic}) and the resistance to stable crack growth (J_R-curve) of AISI Type 303 stainless steel were characterized using side-grooved compact-type specimens. The test material was produced by a hot extrusion process which resulted in long cylindrical MnS inclusions aligned in the longitudinal direction. In addition, several interrupted tests were conducted in which CT specimens were loaded monotonically to different levels of stable crack growth. These specimens were sectioned at mid-thickness for detailed metallography to observe the microscopic aspects of the ductile crack growth process and the evolution of damage ahead of the crack tip. Following is a summary of the results and the conclusions from the study.

Stable crack growth or ductile tearing occurs by nucleation, growth, and coalescence of voids which nucleate at the inclusions. Void growth appears to be most dominant during stable crack growth. At J_{Ic} the void closest to the initial crack tip coalesces with the crack. During this period, growth of voids further away from the crack tip also occurs. Thus the increment in J required to continue the crack growth process decreases progressively until a steady-state dJ_R/da is obtained.

Based on the observations of this study and the void growth model of Rice and Tracey, a simple model relating the microstructural parameters such as inclusion spacing and size to the stable crack growth resistance of the material is proposed. The model is capable of predicting the J_R-curve. The predicted J_R-curve for the test material agreed with the observed behavior.

KEY WORDS: J-integral, R-curve, crack, fracture, inclusions, microstructure, stainless steel, ductile tearing, void coalescence

Background

Materials such as stainless steels are used in structural applications because of their high resistance to fracture. Therefore fracture in components of stainless steel is accompanied by large amounts of plastic deformation and stable crack growth prior to instability. To develop a sound methodology for fracture prediction in these components, the following ingredients are necessary: (1) the ability to analyze the fracture test specimens and the components using mechanics principles which properly account for large scale plastic deformation, which is accomplished by the use of the J-integral approach [*1,5*], and (2) the ability to understand the micro-mechanisms of ductile fracture.

The latter, which is the subject of this paper, is particularly important if test results from

---

[1] Professor of Materials Engineering, Associate Professor of Mechanical Engineering, and Post-Doctoral Research Fellow, respectively, Mechanical Properties Research Laboratory, Georgia Institute of Technology, Atlanta, GA 30332-0405.
[2] Formerly Graduate Research Assistant, School of Materials Engineering, Georgia Institute of Technology, Atlanta, GA 30332-0245.

small laboratory specimens are to be used to accurately predict the behavior of large structures. Typically, the data developed on small laboratory specimens are restricted to small amounts of stable crack growth under J-dominated conditions, while large structures of stainless steel are capable of sustaining much larger amounts of stable crack growth prior to fracture. For predicting the behavior of the component, it is then necessary to extrapolate the J_R-curve developed from the test specimen. Therefore a model for ductile tearing which will allow for accurate extrapolation of the J_R-curve to larger amounts of stable crack extension is needed.

The objective of this study is to observe the gradual progression of ductile fracture from initiation to instability with particular emphasis on the intermediate step of stable crack growth in an AISI Type 303 stainless steel. Existing models are examined in light of the observations. Considerable use of nonlinear fracture mechanics tests and analytical procedures is made to analyze the test data. In the subsequent discussion, a critique of the pertinent previous work in the area is first provided.

Micro-Mechanisms and Mechanics of Ductile Crack Initiation and Growth

Since the pioneering work of Cox and Low [6], it is well accepted that ductile fracture is a process which proceeds by the nucleation, growth, and coalescence of microvoids. This results in a dimpled or fibrous type fracture. Microvoids can nucleate by either of two processes: (1) separation of the particle/matrix interface, which typically occurs at large particles such as MnS inclusions, or (2) cracking of brittle second-phase particles. Further, by studying the fracture behavior of uniaxial notched and smooth specimens, Cox and Low concluded that void nucleation is a stress-controlled process which is perhaps facilitated by the presence of a triaxial state-of-stress such as at the tips of notches and cracks.

Since second-phase particles are so important in the ductile fracture initiation process due to their role in void nucleation, it is expected that the shape, size, and spacing of inclusions will be important in determining the material's resistance to ductile fracture. Thus one of the objectives of this study is to examine the influence of inclusions on the J_R-curve of Type 303 stainless steel. However, to completely understand and model ductile fracture, it is necessary to also develop an understanding of the void growth and coalescence process. Thus in this work all three stages are considered.

Voids have been observed to grow significantly only under plane strain conditions. Rice and Tracey [7] have derived a model which leads to an exponential growth for an isolated spherical void under the influence of a triaxial stress state in a nonhardening material:

$$\frac{dR}{R} = 0.322 \, d\epsilon \, \exp\left(\frac{1.5\sigma_{\mathrm{m}}}{\overline{\sigma}}\right) \tag{1}$$

where ϵ is strain, R is the radius of the void, σ_{m} is the maximum stress, and $\overline{\sigma}$ is the applied effective stress. McClintock [8] has also reported a similar exponential growth for a cylindrical void growing under a triaxial field in a strain hardening material.

The above laws for void growth were used together with a fracture criterion to predict crack initiation. McClintock used his model with the assumption that fracture occurs after repeated coalescence of the voids with the crack tip. It resulted in a critical crack tip opening displacement. The fracture strain for plane strain conditions is then given by

$$\epsilon_{\mathrm{f}} = \frac{\ln \dfrac{\lambda}{L}(1-N)}{\sinh(1-N)(\sigma_a^\infty + \sigma_b^\infty)/(2\overline{\sigma}/3)} \tag{2}$$

where λ is the particle spacing, $\overline{L}$ is the particle diameter, σ_a^∞ and σ_b^∞ are the remote stresses and $\overline{\sigma}$ the effective stress, and N is the strain hardening exponent. Although the model tends to overestimate the fracture toughness, it is attractive because the concepts of void growth and material purity $\lambda/\overline{L}$ are incorporated into the fracture process. Rice and Tracey [7] used their model with the assumption that fracture was considered to occur when the ligament between the crack and the void was half the void radius, and the crack tip opening displacement (CTOD) at coalescence is considered to be critical. Assuming a spherical void of initial radius R_0 and inclusion spacing X_0, the empirical formula of the model is

$$\frac{\delta_t}{X_0} \left[F(\psi) - (F(0)) \right] \frac{R_0}{R_x + R_y} = \frac{R_0}{R_x} \tag{3}$$

where R_x and R_y are the radius of the void along the x and y axes, δ_t is the crack tip opening displacement at fracture initiation, and $F(\psi)$ is a function that takes the angular arrangement of the voids with respect to the x-axis. The model is more commonly written as

$$\delta_t = 2.7 \, X_0 \tag{4}$$

Following Rice's analysis, many models [9–16] have been based on the critical strain in a complex stress field associated with the onset of crack growth. Attempts to determine the critical strain include the strain to fracture in a tension specimen [9–11], or the strain associated with a unique plastic zone size [12,13], or the strain associated with a critical CTOD [14,15].

The above models assume that fracture occurs at the onset of crack growth when the critical condition is met and have been partially successful in predicting fracture toughness in specific types of steels, for specific heat treatments and fracture mechanisms. Also, some of them involve parameters that are not easily determined.

Most ductile materials also exhibit stable crack growth after initiation. This process is characterized by a resistance curve, a plot of a suitable parameter versus crack extension. Several models have also been proposed to characterize this process. These models typically include the feature that plasticity is an an irreversible process that cannot be suitably modeled by nonlinear elasticity when crack growth occurs. These classes of models are discussed below and are more pertinent to our work.

McClintock [8] proposed a model based on the assumption that crack growth occurs by steps of lengths, ρ, every time that a critical strain is attained a distance ρ ahead of the crack tip. He also assumed that the strain at a given point ahead of a growing crack is less than what one would calculate using the reversible formula. In that way, more deformation has to be applied to reach the critical strain, yielding a rising resistance curve.

Rice et al. [16,17] performed finite element analysis of a growing crack under Mode I loading and small-scale yielding conditions for a nonhardening material. They found an expression for the rate of increase of the opening of the crack, as a function of relative distance r from the crack tip:

$$\dot{\delta} = \frac{\alpha}{\sigma_0} \dot{J} + \frac{\beta \sigma_0}{E} \ln \frac{R}{r} \dot{a} \tag{5}$$

where the overdots denote rates, α and β are material constants possibly depending on the amount of deformation, and R is associated with the maximum length of the plastic zone. Rice et al. also proposed that crack growth proceeds in such a way as to maintain a constant opening, δ_c, a microstructural distance, r_c, behind the tip. The resulting equation for J is

$$\frac{\delta_c}{r_c} = \frac{\alpha_{sy}}{\sigma_0} \frac{dJ}{da} + \frac{\beta\sigma_0}{E} \ln\left(\frac{eR}{r_c}\right) \tag{6}$$

where e is the base of the natural logarithm. They were also able to replace the microstructure parameters β_c and r_c by the reference value of J at initiation, J_{Ic}, and the value of the tearing modulus $(T = (E/\sigma_0^2)(dJ/da))$ at initiation under small-scale yielding conditions, T_0:

$$T = \frac{E}{\sigma_0^2} \frac{dJ}{da} = \frac{\alpha_{sy}}{\alpha} T_0 - \frac{\beta}{\alpha_{sy}} \ln\left(\frac{R}{R_{Ic}}\right) \tag{7}$$

where α_{sy} and α represent the value of α under small-scale yielding and current yielding levels, respectively, and R_{Ic} and R represent the value of R at initiation under small-scale yielding conditions $R_{Ic} = 0.2\, JE/\sigma_0^2$ and at the current state of deformation, respectively. As suggested by the above equations, Suresh and Vasudevan have experimentally shown [18] that it is possible to have materials with comparable crack initiation properties but vastly different tearing moduli.

Ritchie and Thompson [19] proposed a model based on Rice's analysis. They defined a microroughness parameter M, which can be determined experimentally, and is essentially equal to δ_c/r_c from Eq 6.

In summary, several models have been proposed to predict ductile crack initiation and growth. Some of them were partially successful in their predictions under some conditions, but their accuracy and generality are limited.

Experimental Procedures

Testing Material and Specimens

The test material used in the investigation was AISI Type 303 stainless steel of nominal composition (weight percent) 0.15 C, 2.0 Mn, 0.02 P, 0.15 S, 1.00 Si, 17.0 Cr, 8.0 Ni, and 0.6 Mo. The material was supplied in the form of 50.8 mm (2 in.) diameter extruded bar. This material was heat treated at 760°C (1400°F) for 72 h to produce an equiaxed grain structure. (The tensile specimens were machined in the longitudinal (L) and in the radial (R) orientations.) Compact-type (CT) specimens were machined from the bar in the R-L orientation per ASTM E 8. These specimens were of standard proportions (ASTM E 1152) with a width $W = 36.58$ mm (1.44 in.) and a nominal thickness of 18.29 mm (0.72 in.). All CT specimens were side grooved 25% (total).

The microstructure of the test material, particularly as viewed in the plane of interest, is shown in Fig. 1. Grain size, inclusion size, and inclusion spacing were measured. The procedure for determining these parameters and the results are described later. The inclusions are elongated along the extrusion direction as expected.

The tensile properties of the material are shown in Table 1 for the L and R orientations. The yield and ultimate strengths are marginally higher in the L direction as compared with those for the R direction. However, the percent elongation in the L direction is significantly higher than the elongation in the R direction, which is expected from the orientation of the inclusions. The true stress-strain (σ-ϵ) behavior up to 10% strain was fitted to the Ramberg-Osgood type relationship:

$$\epsilon/\epsilon_0 = \sigma/E + \alpha[\sigma/\sigma_0]^n \tag{8}$$

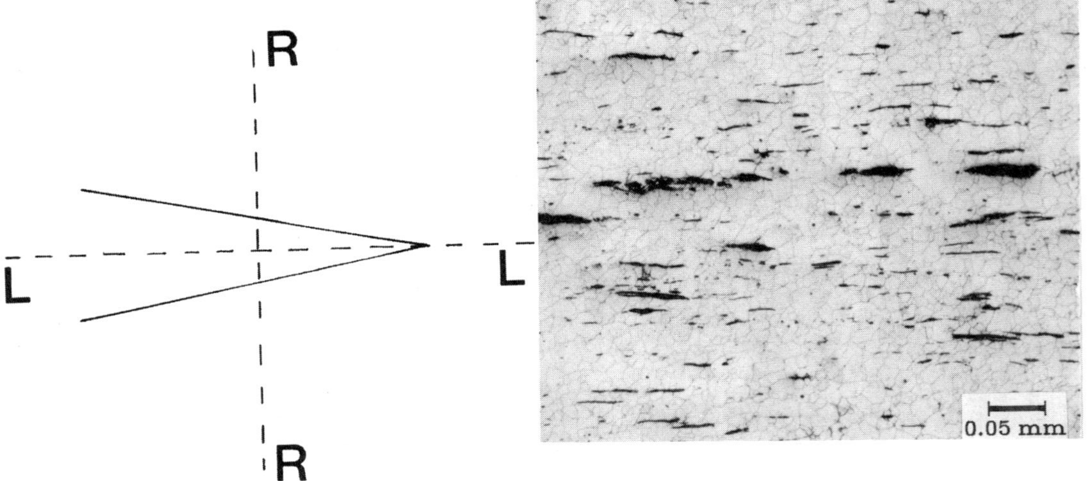

FIG. 1—*Crack-microstructure orientation.*

TABLE 1—*Tensile properties of test material.*

| Orientation | 0.2% Yield Strength | | Ultimate Strength | | Percent Elongation | α | n | ϵ_0 |
	ksi	MPa	ksi	MPa				
R	49.0	338	97.2	670	27.6	16.5	2.26	0.00175
L	52.6	363	106.4	734	62.4	16.0	2.043	0.00175

where σ_0 and ϵ_0 are the 0.2% yield stress and the yield point strain, respectively, and α and n are regression constants. The values of α and n for the two orientations are also listed in Table 1.

Characterization of J_R-Curve

J-resistance curves were established using the single-specimen unloading compliance technique developed by Clarke et al. [20]. Four CT specimens were loaded monotonically and intermittently unloaded several times during the test. The deflections were measured with a clip gage attached to the specimen. The unloading slopes were used to determine the crack size at each of those instants using the crack length versus compliance relationships [21].

In our tests, deflection measurements were made at points other than the load line at a c/W value of 0.293, where c is the distance between the point of deflection measurement and the load line. Load-line deflections were obtained by noting that the axis of rotation is located at a distance of one third the remaining ligament from the crack tip under plastic conditions following the suggestion by Landes [22]. Thus the measured deflections were transformed to load-line by the equation

$$\frac{V_{LL}}{V_c} = \frac{a + 0.33b}{a + 0.33b + c} \tag{9}$$

where b is the uncracked ligament, and V_{LL} and V_c are the load-line deflection and the deflection at the measurement point, respectively.

The method used for obtaining the J_R-curve followed ASTM E 1152.

Metallography and Fractography of Tested Specimens

Several fracture tests were interrupted at different amounts of crack extension and subsequently unloaded. The purpose of these tests was to observe crack tip micromechanisms which accompany stable crack growth at different stages of the process. A considerable amount of metallography was conducted on these specimens which were interrupted at J values ranging from slightly larger than J_{Ic} to those corresponding to substantial amounts of crack extension.

The interrupted fracture toughness test specimens were sectioned through the center of the thickness. One half specimen was mechanically polished in the crack tip region and prepared for metallography. The second half was broken open under cyclic load conditions to measure the amount of stable crack extension and observe the crack profile. Quantitative image analysis of the inclusions was conducted as per procedures outlined by Underwood [23]. The common measurements from which the quantities of interest can be obtained are the point fraction of inclusions, P_p, and the number of inclusions intersected per unit length of test lines oriented parallel to the longitudinal direction, N_L.

The volume fraction of inclusions, V_v, is simply equal to the point fraction, P_p (i.e., proportions of inclusions hit by a probe of an array of test points):

$$V_v = P_p \qquad (10)$$

The mean length of elongated inclusions, $\overline{L}$, is given by

$$\overline{L} = \frac{V_v}{N_L} \qquad (11)$$

where V_v is the volume fraction and N_L is the lineal density.

The mean center to center distance between the inclusions in the longitudinal direction, λ, is given by the empirical relationship

$$\lambda = \frac{1 - V_v}{N_L} + \overline{L} \qquad (12)$$

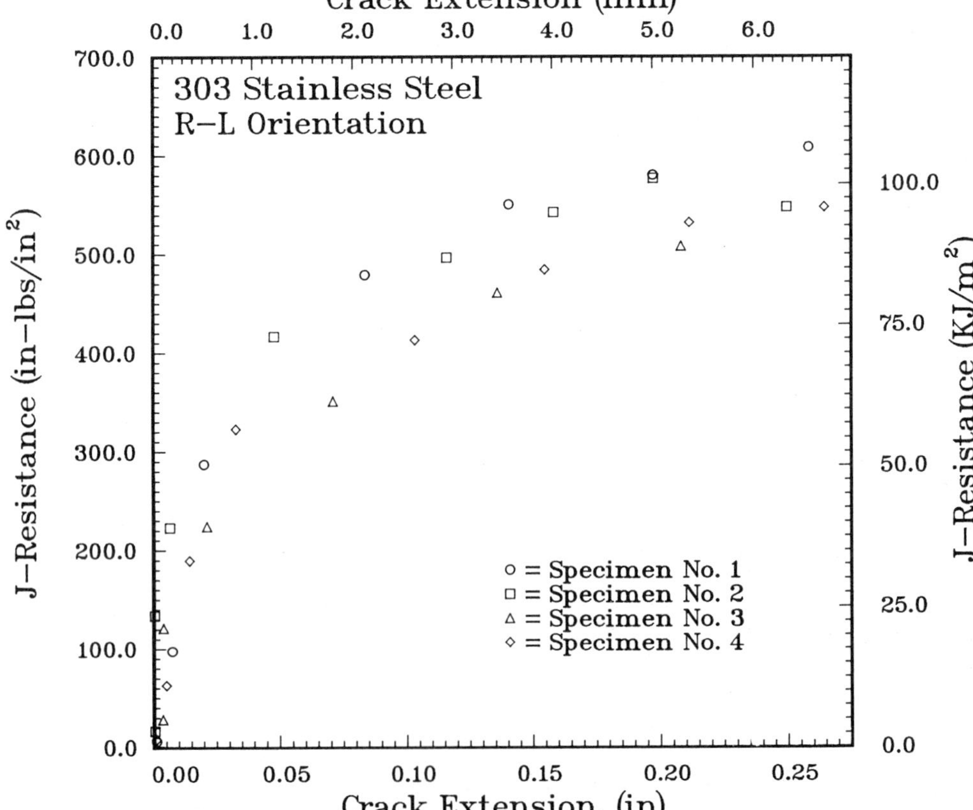

FIG. 2—*J-Resistance curve of AISI 303 stainless steel in the R-L orientation.*

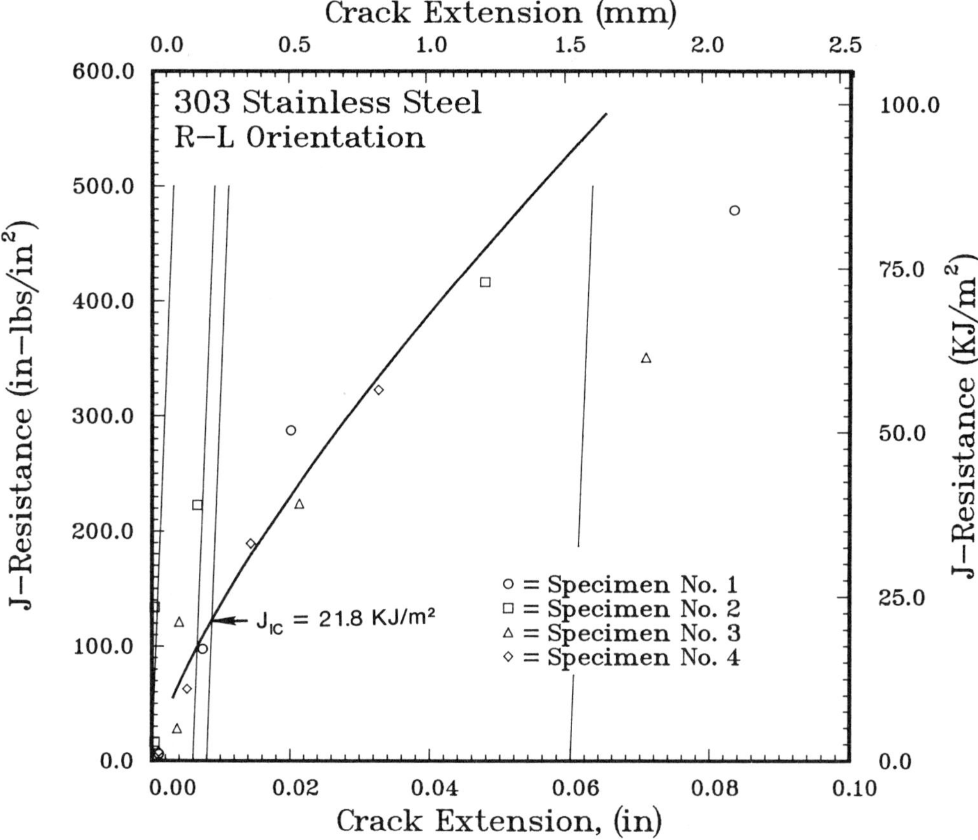

FIG. 3—*Determination of* J$_{lc}$ *for AISI 303 stainless steel.*

The fracture surfaces of specimens used in the determination of J_R-curves were subjected to fractographic analysis using scanning electron microscopy (SEM).

Results and Discussion

The quantitative microstructural results were calculated as outlined in the experimental procedure. The volume fraction of the inclusions, V_v, was 0.0611 and the average grain size was 0.071 mm. The average inclusion size, $\overline{L}$, and spacing, λ, were 0.0175 mm and 0.285 mm, respectively.

Fracture Toughness Test Results

The full *J*-resistance data for the R-L orientation is shown in Fig. 2. The data show crack extension up to 7 mm, which is greater than 10% of the initial remaining ligament. Since *J*-controlled conditions cannot be assured for crack growth beyond 10%, a portion of the *J*-resistance curve is replotted in Fig. 3 to determine the J_{lc} value. The crack blunting line and the appropriate exclusion lines are also plotted and a J_{lc} value of 21.87 kJ/m² (125 in.-lb/ in.²) is determined. The significant crack extension (approximately 7 mm) obtained during testing necessitated the use of modified *J* as defined by Ernst [*24*] to characterize crack

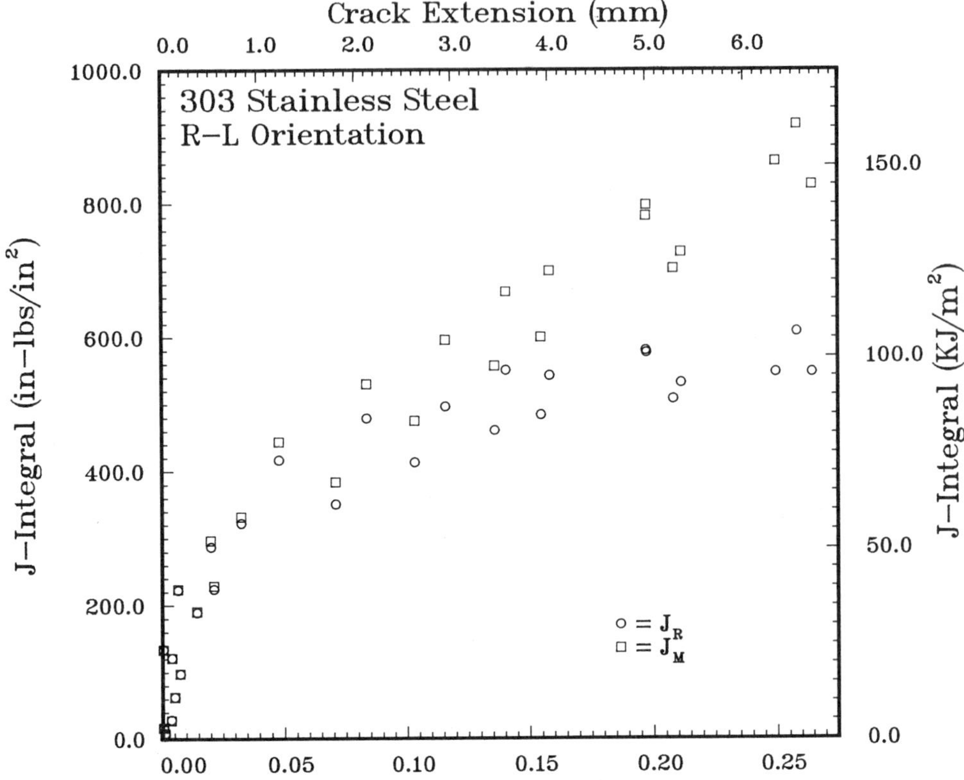

FIG. 4—*Comparison between deformation J and modified J resistance curves.*

growth beyond the *J*-controlled regime. Figure 4 is a plot of modified *J* versus crack extension for crack growth up to 30% of the remaining ligament.

Characterization of Fracture Morphology

The microscopic features of ductile crack growth in the various interrupted fracture tests are shown in Figs. 5 to 7. The three stages of fracture consisting of nucleation, growth, and coalescence were observed for the resulting fibrous fracture mechanism. As one would expect, void nucleation occurs by the separation of the inclusion-matrix interface. The nucleation of voids was not restricted only to the region in front of the crack tip where the large triaxial stress fields exist. The micrographs clearly show nucleation of voids occurring at inclusions throughout the sample, even behind the precrack which is a region subjected to plane stress. This observation tends to support the Cox and Low theory that void nucleation occurs at low stress levels and is not restricted to regions under triaxial state-of-stress.

In contrast to void nucleation, void growth seems to occur directly ahead of the crack tip where the triaxial stress state is expected. This is in agreement with the prediction from the

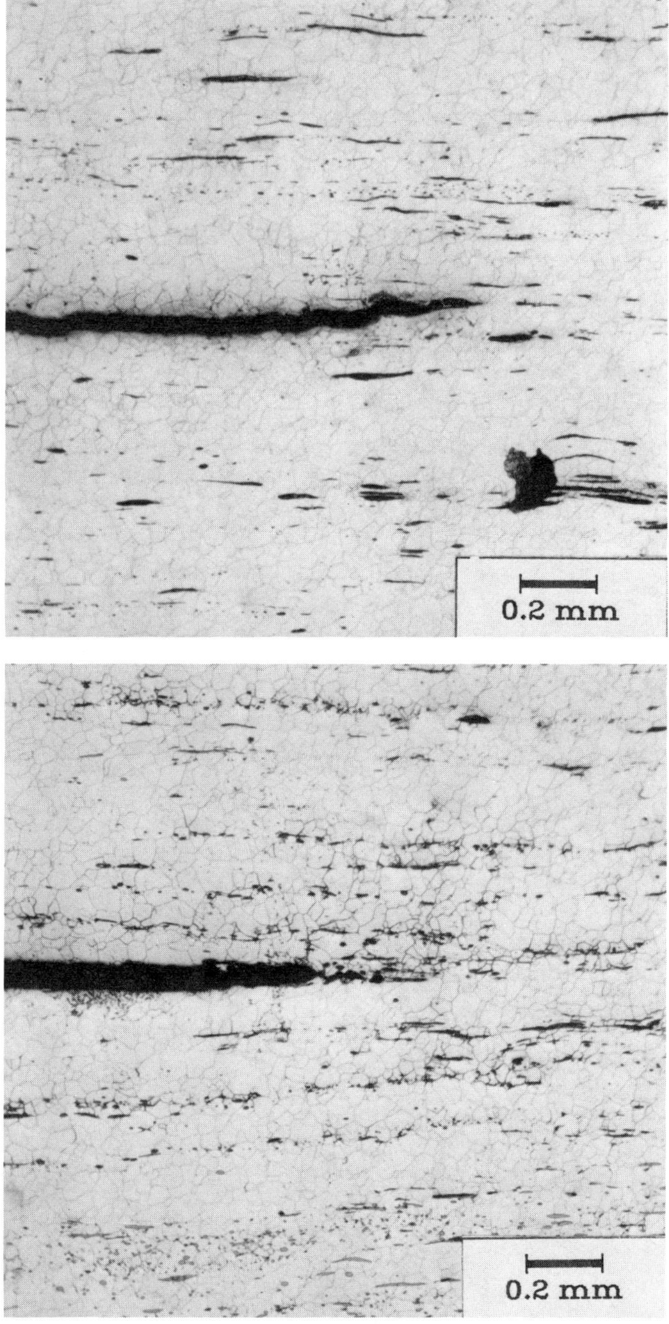

FIG. 5—*Micrographs showing the crack tip regions at* J = 55 kJ/m² *(top) and* J = 43 kJ/m² *(bottom).*

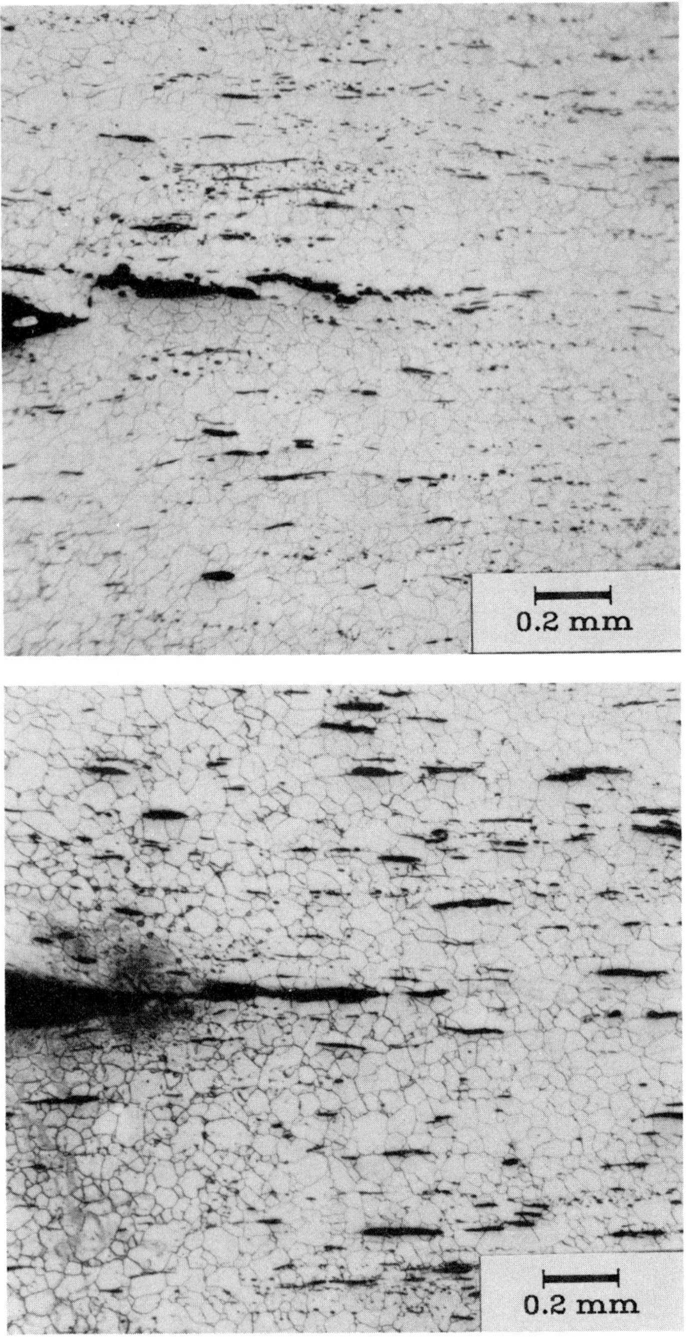

FIG. 6—*Micrographs showing early crack growth at* J = 70 kJ/m² *(top) and* J = 65 kJ/m² *(bottom).*

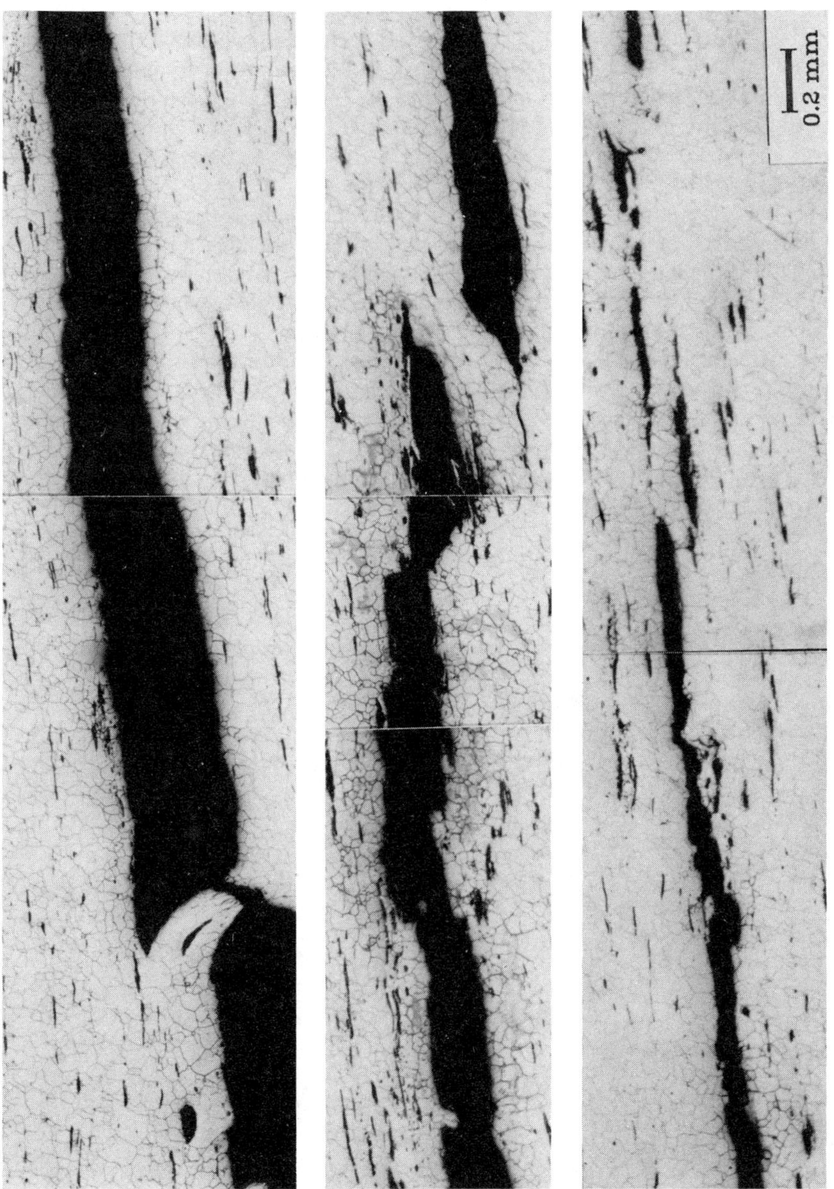

FIG. 7—Montage showing crack growth at J = 105 kJ/m².

Rice and Tracey [7] model which suggests an exponential increase in $dR/d\epsilon$ with stress amplification due to the triaxial state-of stress. Also, void growth is longitudinal rather than spherical, suggesting the final void shape is influenced not only by the state of stress but also by the shape of the inclusions. The grown void appears to coalesce with the main crack once the distance between the crack tip and the void is approximately one to two grain sizes. The rupture of the ligament between the crack tip and the nearest void is known to occur by shear localization [6,19,26] caused by either slip band formation or by nucleation of voids in the vicinity of the smaller carbides at larger stresses. The elongated dimples on the fracture surfaces of the specimens (Fig. 8) and also the micrographs of Fig. 7 showing the ligament which exists between adjacent voids support the former as the mechanism of void coalescence.

Generally, the large elongated inclusions have been thought to be detrimental to crack growth resistance and smaller inclusions are more desirable for improved fracture toughness [19]. It is apparent the large reductions in the diameter of the bar obtained by extrusion have resulted in inclusions which are elongated and separated by large distances. Despite the large inclusion sizes the resistance to ductile tearing was high. This may be due to the large separation between inclusions (0.285 mm). This will be discussed further in the model development section. It also appears from closely examining Fig. 5 that the fracture initiation, J_{Ic}, occurs at the point of coalescence between the precrack and the first adjacent void. This observation is important for developing new models and evaluating existing models for correlating microstructural features with crack initiation toughness.

The micrographs in Fig. 7 show the morphology of extensive crack growth beyond the J-controlled region. Some models (ASTM E 1152) have proposed a relationship between the tearing modulus, T [26], and J_{Ic} based on Rice's crack growth criterion for fully plastic con-

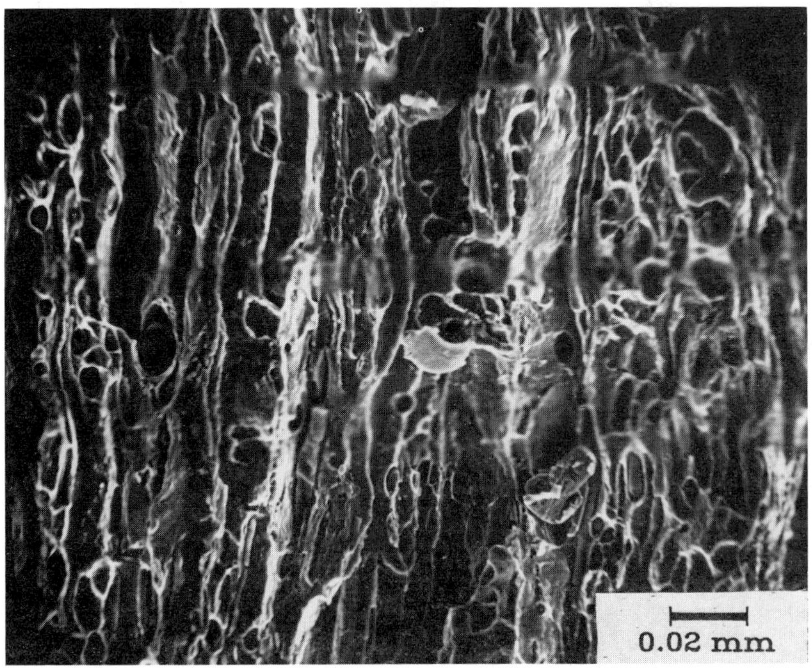

FIG. 8—*Typical SEM fractograph.*

ditions. It is important to note from the micrographs that tearing is affected by void growth and coalescence and therefore these two stages of fracture are expected to dominate the tearing resistance. However, since J_{Ic} is identified with the growth and coalescence of the void associated with the first inclusion, a relationship between J_{Ic} and the tearing modulus, which is perhaps dependent on other factors as well, is implied by our results. If Rice's criterion is applied to the microvoid coalescence mechanism, there should be repeated blunting and coalescence of the crack with voids adjacent to the crack tip as the crack propagates. Then, it would be expected that the crack tip opening displacement should be about the same at crack initiation (Fig. 5) and after substantial crack extension (Figs. 6 and 7). This is in fact observed. However, the incremental strain and J values required for the coalescence of voids subsequent to the first one are smaller. The decreasing slope of the J_R-curve (Figs. 3 and 4) can be attributed to this process. In other words, the smaller increments of J required for continued crack growth are the result of void growth not being limited to the void located immediately in the vicinity of the crack tip but growth of voids located at larger distances from the crack tip. Then, as the crack grows the voids ahead of the crack tip require smaller increments in strain to reach the critical void size and coalesce with the crack. This observation is used to develop a quantitative analysis for tearing resistance at the early stages of crack growth. Such a model leads to the prediction of the complete J_R-curve. This model is described next.

A Model for Predicting Crack Initiation Toughness and Tearing

Since void growth occurs predominantly ahead of the crack tip, the normal strain in the y-direction, ϵ_{yy} (henceforth referred as ϵ), can be deduced from the void growth equations of Rice and Tracey [7] and related to the Hutchinson, Rice, and Rosengren (HRR) singularity [2,3] to derive a model (Fig. 9) for the fracture initiation toughness and the crack growth resistance curve.

The formulation of the proposed model is as follows. A one-dimensional array of voids of the size equaling the average inclusion size is assumed to be present in front of the crack tip along the x-axis (Fig. 9). Since void nucleation was observed to have occurred at very low loads, this step in the fracture process is assumed to not contribute significantly to the fracture toughness. While the void nearest to the crack tip grows in size and coalesces with the crack tip at J_{Ic}, the second, third, and subsequent voids also grow in size. Subsequent crack advance takes place by further growth of what was originally the second void and its coalescence with the crack tip. This results in further growth of the other voids. This process can continue as long as the specimen/component stays stable. The model is analogous to models that have been proposed in creep fracture for crack advance by growth and coalescence of creep cavities [26,27].

In the J-controlled regime, Hutchinson and Paris [28] have derived the relationship between the increment in strain, $d\epsilon$, and the increment in J, dJ. For J-controlled conditions if the nonproportional term can be neglected, the resulting equation will be

$$\frac{d\epsilon}{\epsilon} = \frac{n}{n+1} \frac{dJ}{J} \tag{13}$$

which can be integrated and written as

$$\int_0^{\epsilon^c} d\epsilon = \int_0^{J_{Ic}} \frac{n}{n+1} \frac{\epsilon}{J} dJ \tag{14}$$

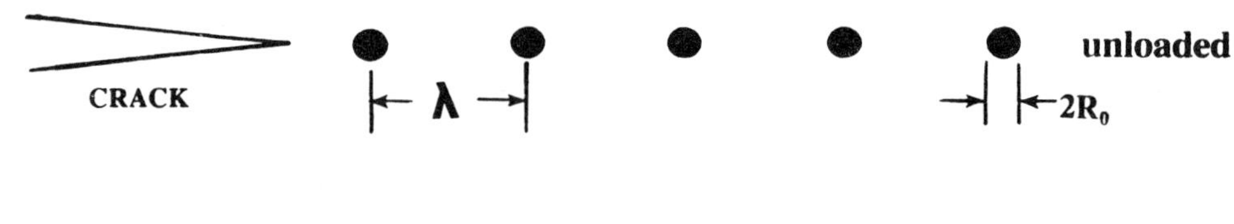

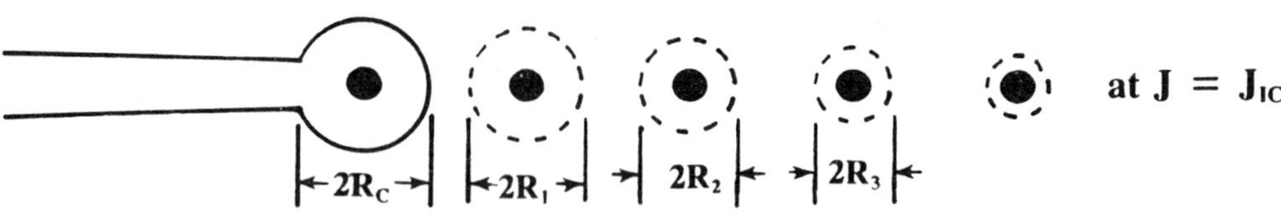

FIG. 9—*Model development.*

The HRR equation for strain is written as

$$\epsilon = \alpha\epsilon_0 \left[\frac{J}{I_n\sigma_0\epsilon_0\alpha r} \right]^{n/n+1} \hat{\epsilon}\,(\theta) \tag{15}$$

Substituting Eq 15 into Eq 14, we get the following result:

$$\epsilon^c = \int_0^{J_{Ic}} \frac{n}{n+1} \alpha\epsilon_0 \left(\frac{J}{\alpha\epsilon_0\sigma_0 I_n r} \right)^{n/n+1} \hat{\epsilon}\,(\theta)\, \frac{dJ}{J} \tag{16}$$

where ϵ^c is the strain necessary to grow the first void to the critical void size, r = distance from the crack tip which at the first void $= \lambda$ and at the second void $= 2\lambda$ and so on, and I_n and $\hat{\epsilon}(\theta)$ are the HRR field quantities listed in tables by Shih [29]. Solving for J_{Ic} in Eq 16 gives

$$J_{Ic} = \left(\frac{\epsilon^c}{\alpha\hat{\epsilon}_0\epsilon(\theta)} \right)^{n/n+1} (\alpha\epsilon_0\sigma_0 I_n\lambda) \tag{17}$$

From Rice and Tracey's model for void growth, we can deduce the plane strain fracture strain by integrating Eq 1:

$$\epsilon^c = \frac{\ln\,(R_c/R_0)}{0.322\,\exp\,(1.5\sigma_m/\bar{\sigma})} \tag{18}$$

On substituting Eq 18 into Eq 17, an expression for fracture toughness initiation is derived as follows:

$$J_{Ic} = \frac{\ln\,(R_c/R_0)}{0.332\,\alpha\epsilon_0\hat{\epsilon}(\theta)\,\exp\,(1.5\sigma/\bar{\sigma}_m)} [\alpha\epsilon_0\sigma_0 I_n\lambda] \tag{19}$$

A model for the initial tearing resistance can be derived by determining the size of the second void in Fig. 9 at the point of coalescence between the first void and the crack tip. The increment in the J value necessary for the second void to grow to the critical void size is then determined and the initial tearing resistance calculated.

The radius of the second void R_1 at the point of coalescence between the first void and crack tip can be determined from the equation

$$\Delta\epsilon_1 = \alpha\epsilon_0\epsilon(\theta)\left(\frac{J_{\mathrm{Ic}}}{2\alpha\epsilon_0\sigma_0 I_n\lambda}\right)^{n/n+1} = \frac{\ln(R_1/R_0)}{0.322\exp(1.5\sigma_{\mathrm{m}}/\bar{\sigma})} \tag{20}$$

where $\Delta\epsilon_1$ is the incremental strain acting on the second void as the first void grew and coalesced with the crack tip. Then, the increment in J value necessary as the second void grew to the critical size and coalesced with the crack tip, J_1, can be determined from the equation

$$\Delta\epsilon_1^c = \alpha\epsilon_0\epsilon(\theta)\left(\frac{1}{\alpha\epsilon_0\sigma_0 I_n\lambda}\right)(J_1^{n/(n+1)} - J_{\mathrm{Ic}}^{n/(n+1)})$$

$$= \frac{\ln(R_c/R_1)}{0.322\exp(1.5\sigma_{\mathrm{m}}/\sigma)} \tag{21}$$

where $\Delta\epsilon_1^c$ is the incremental strain necessary to grow the second void to the critical void size. The analysis can be extended to the other voids ahead of the crack tip, and the increments in J required for continuing crack growth can be determined. Thus a complete J-resistance curve can be generated. These calculations were performed for the test material and the details are described below.

In order to use Eq 19 to estimate J_{Ic}, the values of R_c, R_0, α, ϵ_0, $\hat{\epsilon}(\theta)$, and $\sigma_{\mathrm{m}}/\bar{\sigma}$ are needed. We rely on McMeeking's finited element analysis [30] for the value of stress amplication $(\sigma_{\mathrm{m}}/\bar{\sigma})$ at the crack tip for $n = 2$ $(N = 0.5)$; it is approximately six. McMeeking's analysis was extrapolated to determine the value at $n = 2$. R_0 is taken as half of the average inclusion size 8.75 μm; α, ϵ_0, and σ_0 are obtained from the tensile test results for the R orientation; and I_n and $\hat{\epsilon}(\theta)$ are taken from Ref 29 as 5.94 and 0.00175, respectively. The value of R_c which is the critical void size at coalescence is unknown. In our work, which was primarily aimed at studying the stable crack growth regime, we estimated R_c by using the measured value of J_{Ic} in Eq 19. Subsequently, we used this value of R_c (13.6 μm) in calculating the J_R-curve. Calculations of the J_R-curve were performed for large amounts of crack extension following the procedure outlined in the previous paragraph. The predicted and the measured J_R-curves are plotted in Fig. 10. The excellent agreement between predictions from the simple model and measurements is very encouraging.

To the best of our knowledge, the proposed model is the first of its kind that relates the void growth process to the crack extension at various J levels and proceeds to predict the J_R-curve based on such a mechanism. It also shows that the dJ_R/da decreases immediately following the J_{Ic} point and then settles to an approximately constant value. Such a trend is in agreement with the available crack growth resistance data. The model accurately predicts the observed trend in crack growth resistance curve with respect to inclusion size and spacing. This is a significant factor in judging the appropriateness of the model.

The proposed model uses the J-controlled strain relationships in the crack tip regime. Since the small strain deformation theory of plasticity on which J is based is not applicable in the process zone near the crack tip, it is important to address this point. The process zone size at J_{Ic} is estimated to be on the order of the CTOD, which is approximately J_{Ic}/σ_s, where σ_s is the flow stress that is the average of the yield and tensile strengths. The CTOD was calculated to be on the order of 0.06 mm for the R-L orientation. The inclusion distances

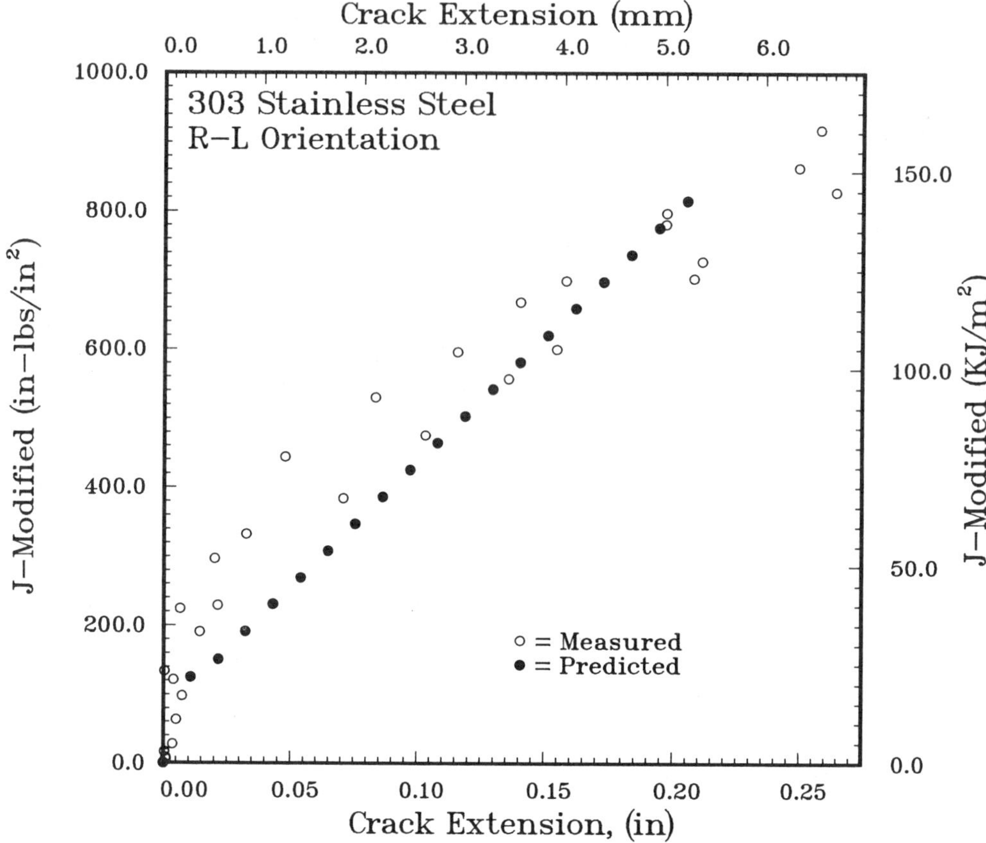

FIG. 10—*Comparison between predicted and measured J-resistance curves.*

are of the order of 0.285 mm for the R-L orientation. Hence the location of the nearest void is clearly in the strain field dominated by J and the use of the HRR strain fields to estimate void growth is justified.

There is also some uncertainty about the stress amplification factor value of six, since it was extrapolated from McMeeking's finite element results. However, even a 20% error in the estimation of the factor will not alter the conclusions of our study.

Overall, the proposed model is unique and appears to rationalize several important experimental observations of this study and is therefore encouraging. It should be further evaluated for other material systems. Improvements in the model by coupling damage and deformation may also be useful in assessing the generality of the model.

It is important to point out that several simplifying assumptions regarding the distribution in shape, size, and spacing of the inclusions were made during the modeling. Also, a linear array of inclusions in front of the crack tip was assumed rather than the complex distribution which actually exists. It should be of interest to evaluate the sensitivity of the predictions to variations in these parameters. These are some areas of future work.

Summary and Conclusions

Compact specimen tests to characterize the J_R-curve of AISI Type 303 stainless steel were conducted. Several tests were interrupted at different amounts of crack extension to observe the microscopic aspects of ductile tearing. The following conclusions can be drawn from the results:

1. The ductile tearing in Type 303 stainless steel appears to be a process governed by nucleation, growth, and coalescence of microvoids forming at second-phase particles.
2. Critical void size to particle size ratio and inclusion spacing are the most relevant microstructural parameters which govern the resistance to ductile tearing.
3. A simple micromechanics model proposed appears to predict the ductile tearing behavior of Type 303 stainless steel reasonably well.

References

[1] Rice, J. R., *Transactions of ASME, Journal of Applied Mechanics,* Vol. 35, 1968, pp. 79–86.
[2] Rice, J. R. and Rosengren, G., *Journal of the Mechanics and Physics of Solids,* Vol. 16, 1968, pp. 1–12.
[3] Hutchinson, J. W., *Journal of the Mechanics and Physics of Solids,* Vol. 25, 1968, pp. 337–347.
[4] Begley, J. A. and Landes, J. D. in *Fracture Toughness, ASTM STP 514,* American Society for Testing and Materials, Philadelphia, 1972, pp. 1–23.
[5] Landes, J. D. and Begley, J. A. in *Fracture Toughness, ASTM STP 514,* American Society for Testing and Materials, Philadelphia, 1972, pp. 24–39.
[6] Cox, T. B. and Low, J. R., Jr., *Transactions of American Society of Metals and Metallurgical Society of AIME,* Vol. 5, 1974, pp. 1457–1470.
[7] Rice, J. R. and Tracey, D. M., *Journal of the Mechanics and Physics of Solids,* Vol. 217, 1969, pp. 201–217.
[8] McClintock, F. A., *Journal of Fracture Mechanics,* Vol. 4, 1968, pp. 101–130.
[9] Hahn, G. T. and Rosenfield, A. R. in *Applications Related Phenomena in Titanium Alloys, ASTM STP 432,* American Society for Testing and Materials, Philadelphia, 1968.
[10] Slatcher, S. and Knott, J. F., *Advance in Fracture Mechanics,* D. Francois, Ed., Pergamon Press, Oxford, 1971.
[11] Chen, C. Q. and Knott, J. F., *Metal Science,* Vol. 15, 1981, pp. 357–364.
[12] Schwalbe, K. H., *Engineering Fracture Mechanics,* Vol. 6, 1974, pp. 415–434.
[13] Schwalbe, K. H., *Engineering Fracture Mechanics,* Vol. 9, 1977, pp. 795–832.
[14] Pandey, R. K. and Banerjee, S., *Engineering Fracture Mechanics,* Vol. 10, 1978, pp. 817–829.
[15] Bates, R. C. in *Fracture: Interactions of Microstructure, Mechanisms and Mechanics,* J. Wells and J. D. Landes, Eds. TMS of AIME, 1984, pp. 255–283.
[16] Rice, J. R. and Sorensen, E. P., *Journal of the Mechanics and Physics of Solids,* Vol. 79, 1978, pp. 163–186.
[17] Rice, J. R., Drugan, W. J., and Sham, T. L. in *Fracture Mechanics: Twelfth Symposium, ASTM STP 700,* American Society for Testing and Materials, Philadelphia, 1980, pp. 189–227.
[18] Suresh, S. and Vasudevan, *Material Science and Engineering,* Vol. 79, 1986, pp. 183–190.
[19] Ritchie, R. O. and Thompson, A. W., *Metallurgical Transactions A,* Vol. 16A, 1985, pp. 233–247.
[20] Clarke, G. A., Andrews, W. R., Paris, P. C., and Schmidt, D. W. in *Mechanics of Crack Growth, ASTM STP 590,* American Society for Testing and Materials, Philadelphia, 1976, pp. 27–41.
[21] Saxena, A. and Hudak, J. S., *International Journal of Fracture,* Vol. 14, 1978, pp. 453–460.
[22] Landes, J. D., *International Journal of Fracture,* Vol. 16, 1980, pp. R183–R186.
[23] Underwood, E. E., *Quantitative Stereology,* Addison-Wesley, Reading, Mass., 1970.
[24] Ernst, H. A. in *Elastic-Plastic Fracture, Second Symposium, Vol. 1—Inelastic Crack Analysis, ASTM STP 803,* C. F. Shih and J. P. Gudas, Eds., American Society for Testing and Materials, Philadelphia, 1983, pp. I-191–I-213.
[25] Birkie, A. J., Wei, R. P., and Pellissier, G. E., *Transactions of the American Society for Metals,* Vol. 59, 1969, pp. 900–981.
[26] Wilkinson, D. and Vitek, V., *Acta Metallurgica,* Vol. 30, 1982, pp. 1723–1732.
[27] Bassani, J. L. and Vitek, V. in *Proceedings,* 9th National Congress of Applied Mechanics, Symposium on Non-Linear Fracture Mechanics, 1982, pp. 127–133.
[28] Hutchinson J. W., and Paris, P. C. in *Elastic-Plastic Fracture, ASTM STP 668,* American Society for Testing and Materials, Philadelphia, 1979, pp. 47–62.
[29] Shih, C. F., "Tables of Hutchinson-Rice-Rosengren Singular Field Quantities," Materials Research Laboratory Report, MRL E-147, Brown University, Providence, R.I., 1983.
[30] McMeeking, R. M., *Journal of the Mechanics and Physics of Solids,* Vol. 25, 1977, pp. 357–381.

J. Zeman,[1] *S. Rolc,*[1] *J. Buchar,*[1] *and J. Pokluda*[2]

Microstructure and Fracture Toughness of Cast and Forged Ultra-High-Strength, Low-Alloy (UHSLA) Steels

REFERENCE: Zeman, J., Rolc, S., Buchar, J., and Pokluda, J., **"Microstructure and Fracture Toughness of Cast and Forged Ultra-High-Strength, Low-Alloy (UHSLA) Steels"** *Fracture Mechanics: Twenty-First Symposium, ASTM STP 1074,* J. P. Gudas, J. A. Joyce, and E. M. Hackett, Eds., American Society for Testing and Materials, Philadelphia, 1990, pp. 396–418.

ABSTRACT: This paper describes the fracture toughness and microstructure relations for ultra-high-strength, low-alloy (UHSLA) steels corresponding to AISI 4340 composition. The steels were either cast or forged, heat treated at austenitizing temperatures from 840 to 1100°C, oil quenched and tempered at 180 and 650°C. Fracture toughness, tensile, and Charpy impact toughness tests, and testing of apparent fracture toughness under the influence of notch root radius, were performed simultaneously. Investigation of microstructures and corresponding mechanical properties, as well as fractographic examinations and evaluation of chemical heterogeneity, made it possible to elucidate the alteration of fracture behavior.

It was found that cast low-tempered steel has higher fracture toughness (K_{Ic}) than forged steels, though its conventional mechanical properties are lower. Increasing K_{Ic} caused by the higher austenitizing temperature of forged steel was also observed. The coarsened prior austenite grains and decreased grain boundary cohesive strength are mainly responsible for the intergranular crack initiation and the related modification of the crack path, the consequence of which may be a change in the effective stress intensity factor.

The proposed model, based on assumed average probable crack path angle deflection controlled by the grain boundary, has yielded satisfactory agreement with experimental results. The mechanical properties of the more heterogeneous cast steel can be elucidated in the same sense. It has been concluded that there exists some uncertainty in the determination of "valid" plane strain fracture toughness, particularly in the case of ultra-high-strength steels, along with their more pronounced macro/microstructure heterogeneity. Due to this effect, the recommended procedure of K_{Ic} measurement can provide undesirable overestimation of the material quality being tested.

KEY WORDS: ultra-high-strength steel, fracture toughness, impact toughness, microstructure, intergranular crack, macrostructure heterogeneity, effective stress intensity factor

There are many engineering applications of forged, ultra-high-strength, low-alloy (UHSLA) structural steels where they could be replaced by cast steels, and, in turn, where the use of UHSLA cast steels in the production of machine parts, especially those with a more complicated shape, can be an effective cost-saving measure. On the other hand, taking full advantage of the strength properties of cast UHSLA steels depends upon their appropriate fracture resistance in service conditions. One of the reasons why cast UHSLA steels have not been widely used is their low toughness. The same could be argued, however, concerning forged UHSLA steels used for dynamically loaded machine parts.

---

[1] Institute of Physical Metallurgy, Czechoslovak Academy of Sciences, Brno, Czechoslovakia.
[2] Technical University, Faculty of Engineering, Brno, Czechoslovakia.

In recent investigations many efforts have been made to increase the toughness of UHSLA steels and to find the testing procedures best able to give a representative estimation of the steel properties being investigated. The close relations of mechanical and microstructural characteristics are commonly accepted. Unfortunately, the number of methods that can describe the microstructure quantitatively is limited; therefore many microstructural features can still be elucidated only qualitatively. A more favorable situation occurs in mechanical testing, where standard methods are available.

The shortage of comprehensive and mutually comparable microstructural characteristics, when the martensitic or a more complex tempered microstructure of structural steel has been achieved, makes the interpretation of final properties particularly uncertain. As an example, for more than 15 years there has been discussion about the relation of fracture toughness and Charpy impact energy under high austenitizing temperatures up to 1200°C [1–4]. Note, however, that the same could be said for complex properties of UHSLA cast steels and their comparison with the properties of forged steels of nearly the same chemical composition. The few papers dealing with the complex properties of UHSLA steels [5,6] have presented experimental evidence for (1) their superior fracture toughness (K_{Ic}) in connection with very high austenitizing temperatures and (2) higher K_{Ic} values of cast steel compared with forged steel. This trend, however, does not correspond to that of Charpy impact toughness. It seems that this improvement in fracture toughness does not reflect real material properties, and it would hardly be utilized generally in practice.

This paper reports on the relations between microstructure and mechanical properties, including the fracture micromechanism of forged and cast UHSLA steels.

Experimental Procedure

The chemical compositions of the forged and cast steels, which are nearly the same as that of AISI 4340 steel, are listed in Table 1. The specimens of forged steel having commercial quality were cut from a 19-mm-diameter soft-annealed rolled rod. The cast steel was induction air melted with final desoxidation by ferrosilicon and aluminum. The semi-finished products for further processing of specimens were cast using precise shell molds. Before the final machining of both steels to desired shapes and sizes, full annealing for 3 ks at 870°C and then air cooling were applied. To obtain a set of different microstructures, the steels were heat treated including austenitizing for 1.8 ks at 840°C up to 1100°C and oil quenched and tempered for 8.7 ks at 180°C and/or 650°C.

The microstructures were studied by optical metallography techniques and by transmission electron microscopy (TEM). X-ray diffraction was used in order to quantify matrix microstrains and to retain austenite volume contents. Details of the latter method are reviewed elsewhere [7,8].

The content and distribution of nonmetallic inclusions were estimated using an image analyzer and appropriate elaboration of results by statistical methods. Quantitative measurement of chemical heterogeneity in correspondence with microstructure heterogeneity of both steels under consideration was carried out.

TABLE 1—*Chemical composition.*

	Composition (wt%)									
Steel	C	Mn	Si	P	S	Cr	Ni	Mo	V	Al
Cast	0.38	0.68	0.23	0.023	0.015	0.99	1.73	0.23	0.02	0.075
Forged	0.37	0.55	0.25	0.021	0.017	1.08	1.84	0.19	0.01	0.023

For this purpose, microprobe analysis of the main alloying elements (chromium, nickel, molybdenum, manganese) was conducted. The concentration distribution of the aforementioned elements along a line in an interval of approximately 2 mm was recorded by a step scan of 3.14 μm and a measuring time of 10 s. The line orientation was perpendicular to the axis of tensile test specimens (i.e., perpendicular to the forging direction). It was assumed that the line allocation in one third of the cast specimen cross section gave the best representation of macrostructure heterogeneity characteristics. The parallel registration of the concentrations of all the elements made it possible to follow quantitatively the chemical heterogeneity, the expression of which represented by an index of heterogeneity (I_i) can be defined as

$$I_i = \frac{4\sigma_i}{\overline{C}_i}$$

where σ_i is the standard deviation of the ith element concentration, and $\overline{C}_i$ is the average concentration in the observed microstructure range.

Mechanical properties in tensile test (6-mm-diameter round bars), Charpy V impact energy, and fracture toughness (K_{Ic} or J_{Ic}) at temperatures ranging from -190 to $20°C$ were calculated in accordance with ASTM E 8 M-85, E 399-83, and E 813-81, respectively.

For specimens subjected to three-point static bending, variations of notch radius from 0.1 to 0.5 mm were also used.

The shape and dimension of K_{Ic} specimens (longitudinal rolling direction in the case of forged steel) and those of blunted ones with different notch radius are shown in Fig. 1.

Results

Microstructure Properties

The main differences in macrostructure properties resulting from the production of both steels are shown in Fig. 2. Characteristic bands along the rolling direction in the forged steel and the dendritic macrostructure of the cast steel are clearly demonstrated. The macrostructural features, as expected, are in mutual correspondence with the chemical heterogeneity.

The distribution of nonmetallic inclusions obtained by image analysis and by statistical elaboration of the data is shown in Fig. 3. It has been shown that the relative number of all inclusions was 14 mm^{-2} and 33.6 mm^{-2} for forged and cast steels, respectively. The position of maximum inclusions incidence rate with respect to that of the cast steel is shifted sub-

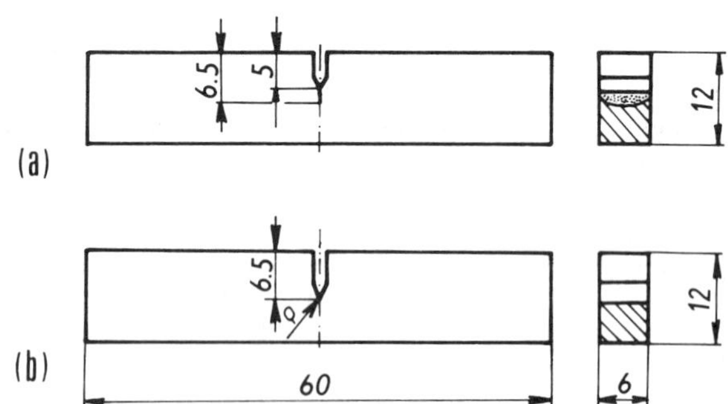

FIG. 1—*Test specimens of* (a) *fracture toughness* K_{Ic} (K_{Jc}) *and* (b) *apparent fracture toughness* K_Q^ρ.

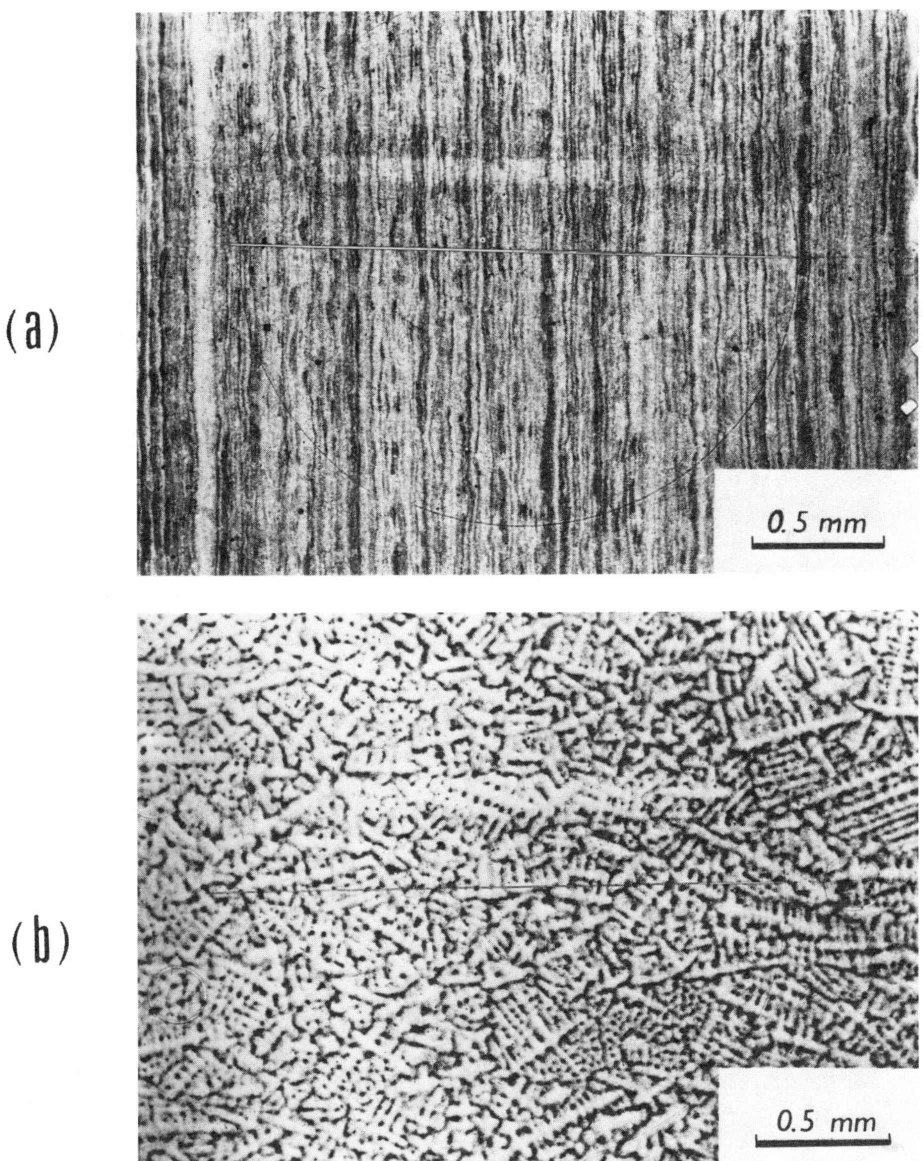

FIG. 2—*Characteristic microstructure heterogeneity of* (a) *forged steel and* (b) *cast steel.*

stantially towards the lower occupied area (i.e., lower inclusions dimension in the case of forged steel). Therefore higher purity of forged steel would be postulated than that of cast steel. It must be pointed out, however, that the histograms in Fig. 3 are somewhat obscure, particularly in their left descending part, because of the limited resolving power of the optical system used. Therefore inclusions less than about 1.5 μm, although able to play a role in the micromechanics of fracture processes, could not be included in full. It is worthy to note here,

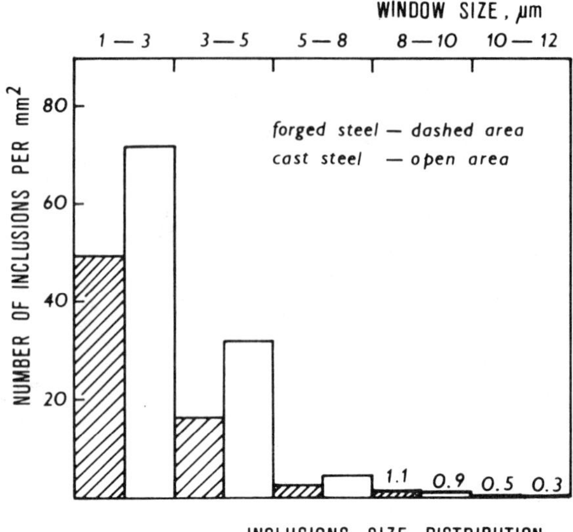

FIG. 3—*Size distribution of inclusions.*

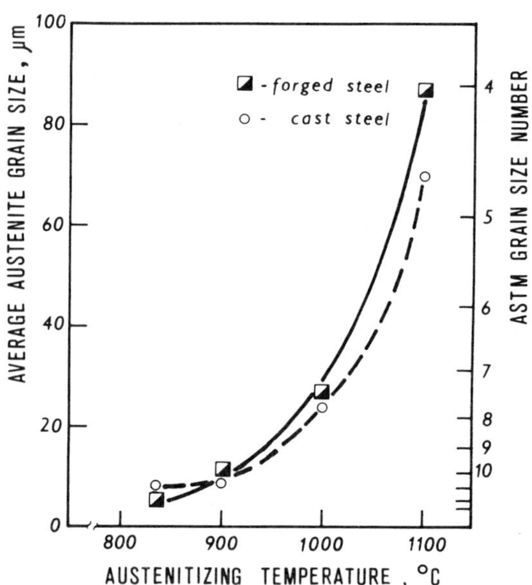

FIG. 4—*Prior austenite grain size versus austenitizing temperatures.*

too, that this shortcoming in evaluation of inclusions distribution is commonly entirely omitted.

Concerning the quality of inclusions, in the case of forged steel there are mainly sulfides in the form of narrow strings elongated along the rolling bands and complex oxides as well. On the other hand, in the cast steel isolated microshrinkages as well as pores were located on dendrite boundaries; they are also included in the histogram in Fig. 3, namely in its right part. Nevertheless, the inclusion distribution is undoubtedly in correspondence with the macrostructure of a particular steel.

Austenite grain size and austenitizing temperature relations are outlined in Fig. 4. The characteristic curves of the forged and cast steels differ a little but, in more detailed observation, the local variation of austenite grain size corresponding to the macrostructure heterogeneity occurred mainly with the characteristic rolling bands of forged steel. Grain size in closely neighboring bands at the lowest austenitizing temperatures used was assumed to be changed by a factor of about 1.5. This slightly decreased with increasing austenitizing temperature. The chemical heterogeneity represented by the segregation index (I_i) is presented in Table 2. It can be seen that the difference in the segregation of chromium and molybdenum between the forged and cast steels is statistically significant as opposed to that of manganese and nickel. The maximum and minimum positions of the peaks of all recorded elements were in mutual correspondence. From Table 2 it can be also deduced that the segregation index decreases following the higher austenitizing temperature, though the average distance between concentration peaks was a little changed. It could be concluded, therefore, that high austenitizing temperature will contribute to substantial improvement of matrix chemical homogeneity.

The quenched and low tempered microstructure consists of martensite plates and very fine dispersion of ϵ_{hcp} carbides and M_3C carbides. For the same heat treatment, if the differences in heterogeneity are omitted, there are no significant changes in matrix microstructures of both steels studied. It is evident that, due to an increasing austenitizing temperature, the final martensite microstructure will change as a consequence of austenite grain size growth, although the related size of martensite plates is not expressed with a simple linear function [9].

Results of X-ray and electron diffraction analysis of quenched and low tempered microstructures have shown relative increases of volume content of ϵ_{hcp} carbides compared to that of M_3C carbides when the austenitizing temperature was increased. The reasons for such behavior are complex, including variations of martensite transformation kinetics, athermal self-tempering intensity, size and distribution of martensite plates, their dislocation substructure characteristics, etc. Nevertheless, it could be concluded that low temperature tempering processes after high austenitizing temperature were realized more intensively.

The volume content of retained austenite in low-tempered conditions (i.e., 180°C) of both steels determined by X-ray analysis ranged from 2 to 4 vol%. A somewhat higher level of retained austenite microstrain after austenitizing at 1100°C, deduced from X-ray diffraction line broadening analysis, unlike that austenitized at 840°C, has indicated that austenite is distributed more frequently in the form of thin films along boundaries of martensite particles than in the form of small islets squeezed among them.

After tempering at 650°C, the reflecting microstructure of the steels was sorbitic without distinctive features between forged and cast steels.

TABLE 2—*Chemical heterogeneity.*

Steel	Austenitizing Temperature, °C	Segregation Index (I_i) of the Elements			
		Cr	Mn	Mo	Ni
Forged	840	0.385	0.564	1.189	0.330
	1100	0.259	0.428	1.030	0.237
Cast	840	0.555	0.485	1.783	0.339
	1100	0.457	0.359	1.414	0.236

TABLE 3—*Mechanical properties.*

Mechanical Properties	Austenitizing Temperature, °C					Note
	840	870	900	1000	1100	
Forged Steel						
σ_y, MPa	1530		1515	1437	1377	
σ_u, MPa	1977		1882	1839	1811	
σ_{Fr}, MPa	2720		2685	2642	2296	
RA, %	44.0		47.0	46.2	32.2	
Cast Steel						
σ_y, MPa	1512	1510	1513	1497	1419	
σ_u, MPa	1522	. . .	. . .	. . .	. . .	$\sigma_{Fr} \leq \sigma_u$
σ_{Fr}, MPa	1522	1511	1513	1538	1673	
RA, %	. . .	. . .	. . .	. . .	. . .	RA $\leq$ 2%

Mechanical Properties

The results of tensile tests carried out at ambient temperature following an influence of austenitizing temperature are summarized in Table 3. The data obtained in the range of test temperature down to $-190°C$ are presented in Fig. 5. From Table 3 it can be seen that all properties—the 0.2% offset yield stress (σ_y), ultimate tensile stress (σ_u), fracture stress (σ_{Fr}), and reduction of area (RA)—of forged steel slightly decrease with increasing austenitizing temperature. Analogical properties of cast steel, however, except for σ_y, were invalid due to the severe detrimental effects of casting defects, like the shrinkages mentioned above, regardless of their local occurrence. As shown in Fig. 5, the same effects have appeared in current tests at decreasing temperatures where the results of forged and cast steels tempered at 180 and at 650°C are reported. Therefore the results of the tensile tests have given convincing evidence for the properties of cast steel being inferior to those of forged steel.

The Charpy impact toughness (CVN) and the plane strain fracture toughness (K_{Ic}) relationships with increasing austenitizing temperature are illustrated in Fig. 6. The same properties test temperatures from 20°C down to $-190°C$ are shown in Fig. 7. As indicated in previous reports [10,11], converse relations between CVN and K_{Ic} results are clearly shown here. On the other hand, both properties of steel tempered at 650°C, mentioned above, are in mutual conformity (i.e., CVN and K_{Ic} values are at a substantially lower level for cast steel at all test temperatures).

The apparent fracture toughnesses (K_Q^ρ) reflecting the influence of notch radius, shown in Fig. 8, were fitted with the linear relationship $K_Q^\rho = A + B \cdot \rho^{1/2}$ using the error mean squared method. The results of the K_{Ic} measurements are depicted here, too. The values of the A and B parameters of the above expression and the corresponding CVN and K_{Ic} values, respectively, are given in Table 4. It is noteworthy that reciprocal differences between individual parameters A and B, as well as CVN and K_{Ic} values, have shown statistical significance. On the other hand, experimental results concerning the K_Q^ρ versus $\rho^{1/2}$ dependence have revealed a higher dispersion, particularly for notch radii less than 0.15 mm, so that an estimation of the so-called "limit radius ρ_0", reported by Ritchie et al. [10], was obscured. The fit of analogical data of cast steel austenitized at 1100°C was impossible in all the ranges of notch radii given. But it is evident that the latter data are the lowest ones overall.

Fracture Micromechanism

Two methods of fracture micromechanism investigation were used: (1) scanning electron microscopy (SEM) of fracture surface observation, and (2) crack-path microstructure inter-

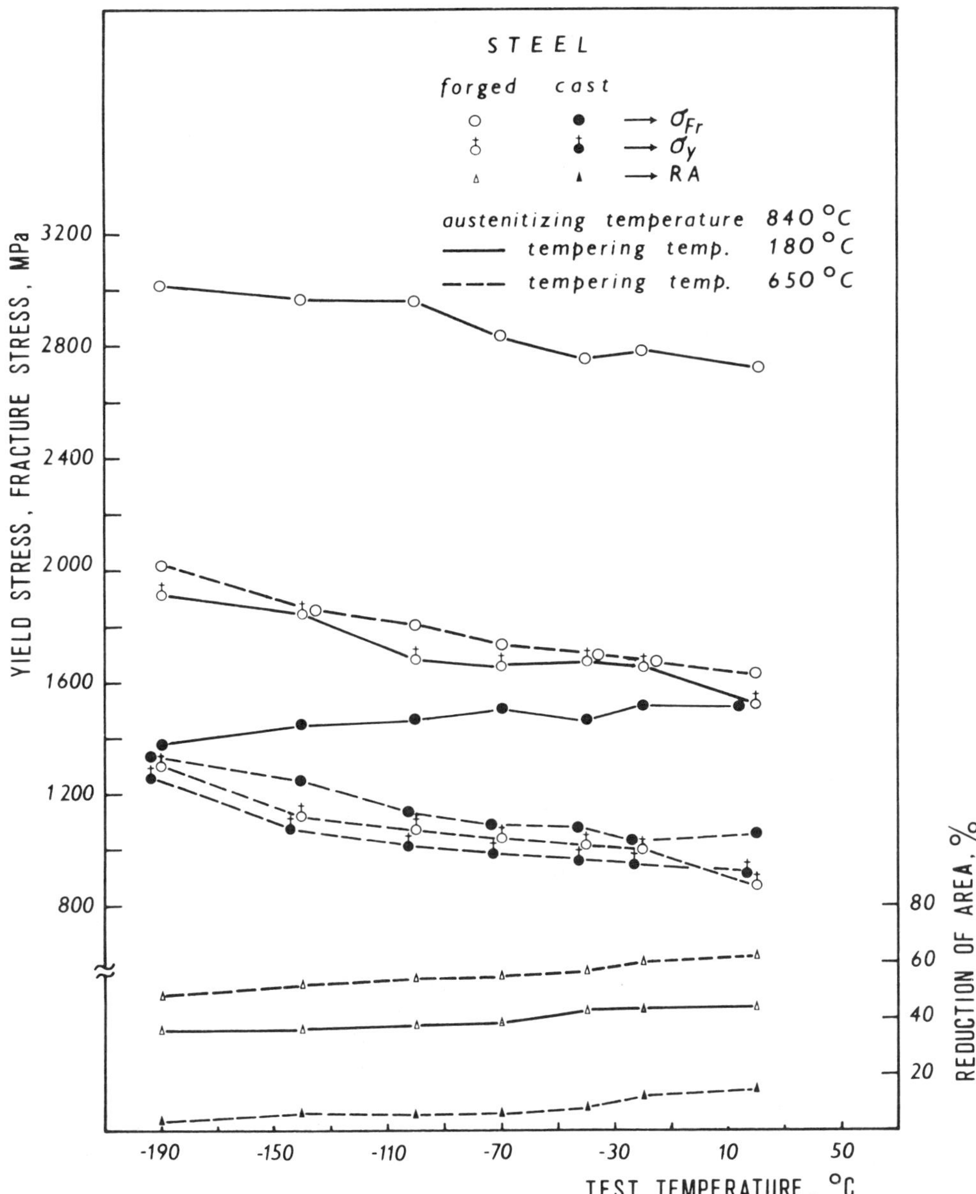

FIG. 5—*Mechanical properties at decreasing test temperatures.*

actions using parallel sections (i.e., perpendicular to a notch) of fractured specimens. Fracture surface micromorphologies of K_{Ic} specimens austenitized at 840°C and low tempered at 180°C (tested at ambient temperature) have shown transgranular void coalescence with correspondence in size and distribution of carbides.

On the other hand, specimens exhibiting coarse austenite grains obtained at higher austenitizing temperatures have contained a number of intergranular facets on their fracture surfaces as the consequence of major initiation and crack growth micromechanism. At the same time, the roughness of fracture surfaces was markedly greater.

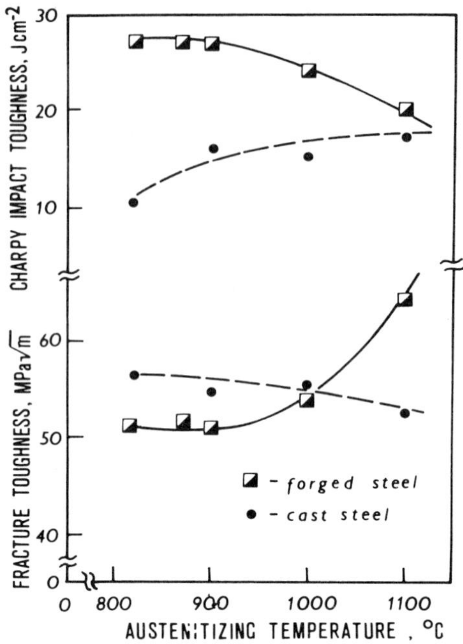

FIG. 6—*Fracture toughness* K_{Ic} *and Charpy impact toughness versus austenitizing temperature.*

Comparing fracture morphology changes between forged and cast steels, we find that they have resulted mainly from the local occurrence of cast defects distributed on boundaries of dendrites, inclusion distribution, and their interaction either with starting fatigue cracks or with propagation cracks. Photographs of characteristic morphologies are shown in Fig. 9.

There was a smaller change in fracture surface micromorphology of individual sets of specimens with decreasing test temperature up to −70°C. Testing at lower temperature, however, has resulted in more pronounced transition to a quasi-cleavage morphology that has reflected an influence of the matrix martensite structure and distribution of tempered carbides.

The specimens tempered at 650°C fractured by the void coalescence mechanism, and the variation between cast and forged steels seemed to alter with inclusions and flaw distribution. It was caused mainly by those of greater size or their local accumulation.

The fracture processes mentioned above were associated with some plastic deformation which manifested itself by blunting and stretch zone formation at the very tip of sharp fatigue cracks. For specimens with originally rounded notch root (i.e., CVN specimens and those with varying radius), the shear fracture initiation which was followed by shear bands along logarithmic spirals (i.e., according to slip line field theory) was more characteristic. The near notch tip shear zone width may be a linear function of the notch root radius [*12*]. Owing to the high scatter of our data, it was impossible to verify this hypothesis.

A comprehensive picture has been offered, in addition, by a more detailed inspection of crack path profiles in parallel sections along specimens. The crack path crossing the forging bands of steel austenitized at 840 and 1100°C can be seen in Fig. 10. While the crack path shown for 840°C is straight, that for 1100°C shows a tendency for kinking and branching or for both phenomena, even in the part of the fatigue crack path. Observation of the crack path and its relation to the relevant microstructure in its immediate vicinity has revealed

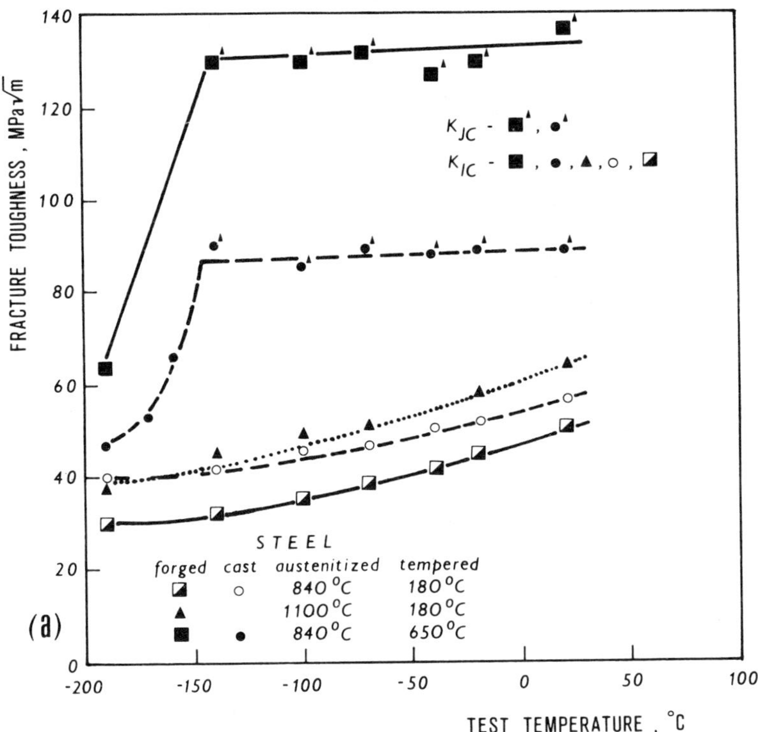

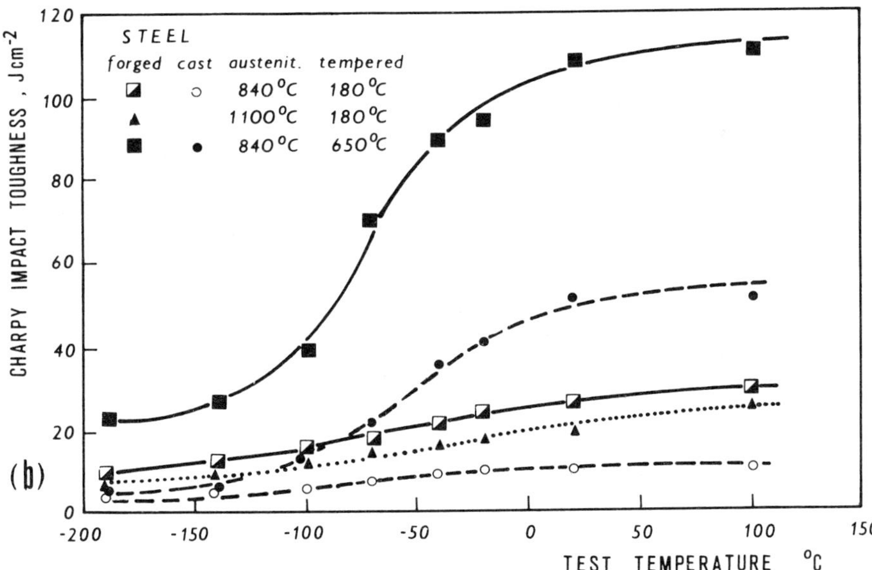

FIG. 7—*Relationships of* (a) *fracture toughness and* (b) *Charpy impact toughness versus test temperature.*

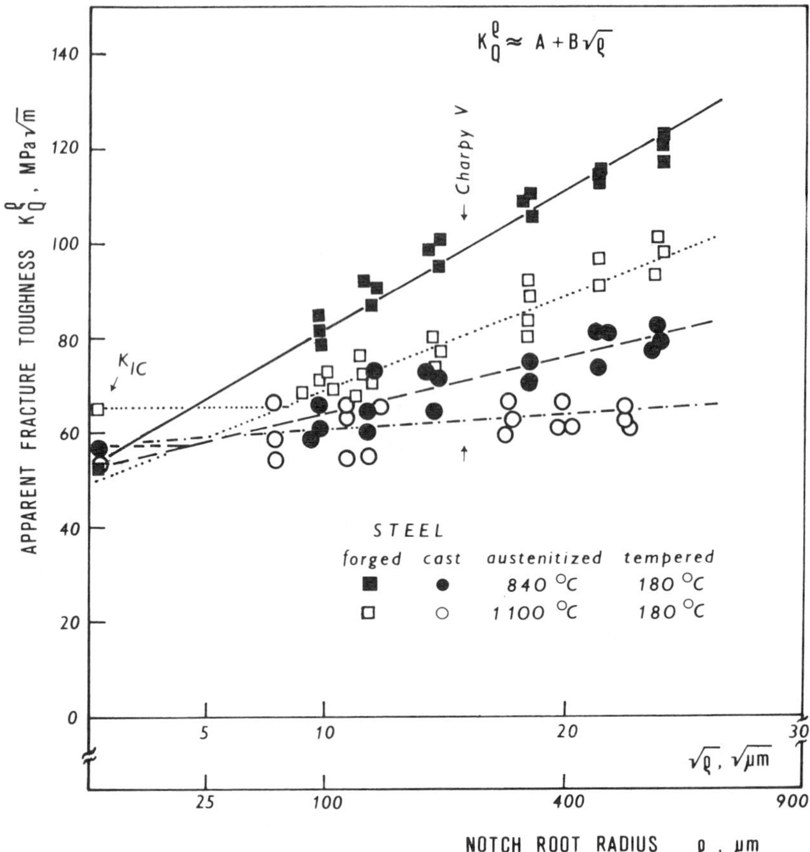

FIG. 8—*Apparent fracture toughness versus notch root radius.*

that the crack profile is wavelike and modulated by forging bands and, in finer scale, also by austenite boundaries (Fig. 11.).

For specimens austenitized at a high enough temperature, the crack path, as concerns crack kinking and branching, was controlled by austenite grain boundaries. It is suggested that this behavior is due to a decrease of grain boundary cohesion strength as a consequence of a more severe effect of impurity segregation. Though there were no means to verify this, it is assumed more probable in analogy with the well known influence of sulfur.

Discussion

The discrepancy in fracture toughness evaluation and Charpy V-notch test, or results of tensile tests of UHSLA steels austenitized at increasing temperatures, have been reported in previous papers [2,4]. The most comprehensive discussion is by Ritchie et al. [10].

The microstructure changes due to the influence of high austenitizing temperature, which have been studied earlier, cannot elucidate the experimental evidence for the behavior of UHSLA steels mentioned above. In addition, there are no meaningful reasons to assume that microstructures of cast steel with regard to all aspects described previously would have

TABLE 4—*Apparent fracture toughness* K_Q^ρ.

Austenitizing Temperature, °C	$K_Q^\rho = A + B \cdot \rho^{1/2}$		Charpy Impact (CVN), $J\ cm^{-2}$	Fracture Toughness (K_{Ic}), $MPa\ m^{1/2}$
	$A,$ $MPa\ m^{1/2}$	$B,$ MPa		
	Forged Steel			
840	58.4 ± 1.8	2434 ± 101	27 ± 1	51.8 ± 1.5
1100	48.8 ± 2.5	2010 ± 144	19.5 ± 2	65.8 ± 2.4
	Cast Steel			
840	51.4 ± 3.3	1266 ± 186	11 ± 1.2	56.5 ± 2.7
1100	. . .	. . .	17 ± 2.4	53.4 ± 1.5

a noticeable advantage over those of forged steel as the reported valid fracture toughness (K_{Ic}) indicates in the works of Floreen [5] and Wei-Di Cao et al. [6]. A markedly higher resistance to environmentally assisted fracture was also reported in both of those papers.

Considering the results obtained in this work, we can suppose that the following microstructure aspects might be beneficial for higher initiation resistance of sharp cracks inside a coarser structure with larger prior austenite grain size resulting from high austenitization temperature:

1. Lower yield strength in virtue of a more intensive self-tempering effect.
2. More intensive low tempering processes that cause the *a priori* sharp crack tip to blunt before a subsequent crack initiation takes place.
3. Appropriate distribution of retained austenite.
4. Increase in chemical homogeneity.

On the other hand, factors like a lower boundary cohesion strength of coarser prior austenite grains, which are prone to enhance impurity segregation effects, and related coarser plates of martensite will probably promote a facility for crack initiation. In the event of comparison between the forged and cast steels the detrimental effect of a higher amount of non-metallic inclusions and other cast defects in the latter case must be noted. There also remain questions as to how the difference in sampling volume of individual tests, along with the changing sensitivity of eligible potential fracture sites to initiate a critical event due to heat treatment used, would influence the final measured values, even when the differences in some of mechanical properties are small in absolute scale. Nevertheless, it was useful to estimate the particular correlations, namely those between fracture toughness and tensile fracture strain, Charpy V impact toughness, and prior austenite grain size. The corresponding relationships are given in Fig. 12, where the experimental results of works quoted herein are also outlined.

The data representing the properties of UHSLA steel under the effect of the equivalent heat treatments shown in Fig. 12 are mutually significant correlations. With regard to the model of strain-controlled fracture mechanism reported in Refs *13* and *14*, which is usually expressed as

$$K_{Ic}^2 / \sigma_y E \approx \epsilon^* \ell^*$$

where ϵ^* is critical fracture strain, ℓ^* is critical microstructural characteristic distance, and E and σ_y have the common meaning, one can suppose that the slope of individual relations

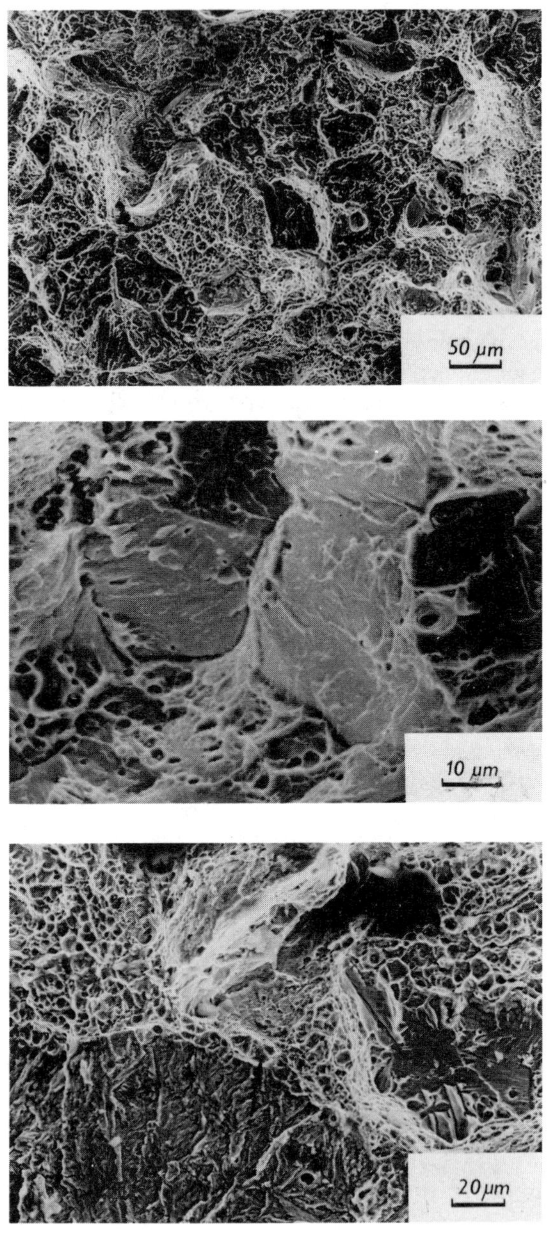

(a)

FIG. 9—*Fracture surface morphology of* K_{Ic} *test specimens austenitized at 1100°C and tempered at 180°C. (a) Forged steel* (from top to bottom): *survey of fracture surface, more detail of intergranular facets, fracture initiation zone. (b) Cast steel: fracture initiation zone including*

shown in Fig. 12*a* might characterize the balance among critical fracture events in the given microstructure. It is also seen that the individual slopes follow approximately the changes in prior austenite grain size.

The collection of results in Fig. 12*b* supports the agreement among experimental evidence for a positive influence of fracture toughness on increasing prior austenite grain size. The problems of K_{Ic} versus grain size relation, recently reviewed by Armstrong [15], particularly for the trend mentioned in this work, can be solved when the dependence of grain size (or, more generally, a critical distance ℓ^*) and the extent of real plastic zone size (r_p) in a near crack tip region are considered. It was also shown [15] that the positive increase in K_{Ic} values with an increase in ℓ^* can be predicted when the ℓ^*/r_p ratio has a value large enough ($\ell^*/r_p \geqslant 1$). This is the very case for UHSLA steels, provided that the grain size or other relevant distances result from micro/macrostructure characteristics.

With the exception of coarse grain size the previously described structural defects of both steels under consideration are also available and might be included into the distribution of all possible defects when they are, with high enough probability, present in the sample volume given. If the properties of cast and forged steels shown in Fig. 12 with a nearly equivalent austenite grain size are compared, it seems more probable that these are differentiated mainly by macrostructure heterogeneity directly connected with inclusion distribution and other structural weak points [16]. In this case fractographic investigations give indications of an enhanced tendency to branching and kinking in an earlier stage of crack growth. Some characteristic cases are shown in Fig. 13. The question arises as to whether the assumed single load Mode I only is operating and whether the linear-elastic fracture mechanics approach can be still applied. These effects will probably be more stressed when the micro/macrostructure becomes coarser, as caused by higher austenitization in the case of forged steel, or when the origin of macrostructure heterogeneity is inherent in the cast process like in the cast steel studied here.

The apparent fracture toughness K_Q^ρ versus $\rho^{1/2}$ dependence resulting from the three-point slow bend test like that used in this work (Fig. 8) has been reported in previous papers [17–19]. In view of recently proposed interpretations, the slope of linear approximation in a limited range of a given notch root radius small enough is believed to be proportional to the

(b)

interaction with microshrinkage. The fatigue crack tip front is arrowed in direction of crack propagation.

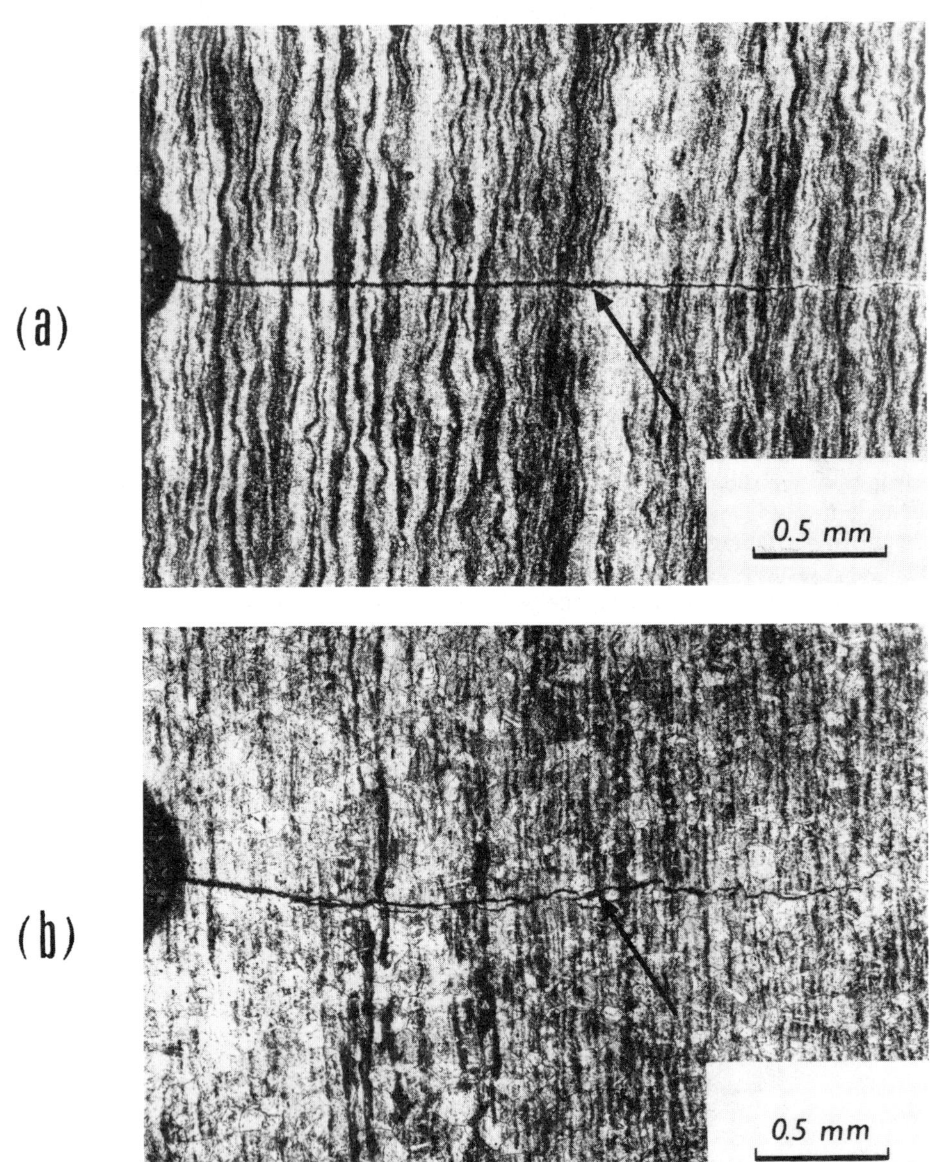

FIG 10—*Crack path profile of fracture toughness specimens. (a) Austenitizing at 840°C. (b) Austenitizing at 1100°C. The fatigue crack tip is arrowed.*

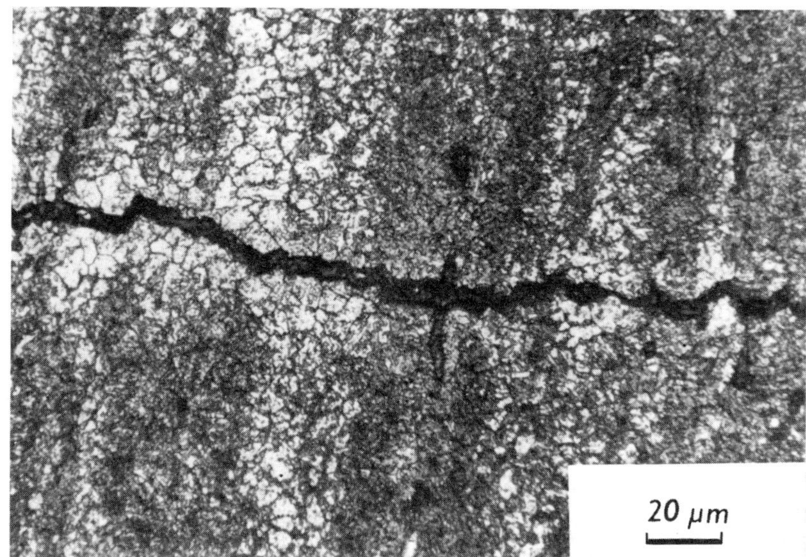

FIG. 11—*Relation between crack path and microstructure of forged steel austenitized at 840°C.*

critical fracture stress in the case of a stress-controlled mechanism and to the critical fracture strain in the case of a strain-controlled mechanism. There is a difficulty, however, in deciding to what extent the transition between the two principal mechanisms has taken place in order to prefer a particular one. Moreover, the ductile-brittle transition behavior and its notch root radius dependence must be taken into account. These could also be the reasons why the K_Q^ρ versus $\rho^{1/2}$ relationship makes more obscure the above approximation when it is extended towards the larger notch radius or when the tests were conducted only within the ductile-brittle transition range. Nevertheless, it seems from the results presented in Table 4 that there is a close relation between slope of K_Q^ρ versus $\rho^{1/2}$ relations and CVN values obtained at equal test temperature. This is in agreement with corresponding relevant changes in tensile fracture strains. The significant correlations between K_{Ic} and CVN values for the characteristic state of materials measured in the range of valid K_{Ic} test temperatures and fitted by commonly used approximations are obvious from Fig. 12c.

As reported earlier [20,21], it is thought that the physical meaning of characteristic distance pertinent to the unique fracture mechanism of a real material results from the complex statistical competition for available eligible sites.

The intergranular fracture mechanisms were more pronounced in all specimens tested, particularly in those of forged steel, because of high austenitizing temperature coarsening of prior austenite grain size and suggested drop of grain boundary cohesion strength. To gain a more objective estimation of the influence of crack path branching or kinking on resulting K_{Ic} values, a simple model was proposed. The model is based on real behavior of orthotropic materials as it was discussed and summarized in Ref 22. Discussions on this topic, particularly the crack branching effect, reviewed in Refs 23 and 24, were also considered. According to Suresh and Shih [25], changes in crack path may be induced by multiaxial far-field stresses, interaction of the crack tip with microstructure heterogeneities, abrupt load excursion, or the embrittling effect of an environment. It is felt here that these same factors can be effectively operated upon in the present case, too, and their beneficial effect on improving the fracture toughness can be therefore expected.

In the event of a crack path along the grain boundary with random orientation, schematically shown in Fig. 14, the average probable angle $\bar{\alpha}$ of a crack declined from the main crack plane can be expressed as

$$\bar{\alpha} = 0.5^n \sum_{x=0}^{n} \binom{n}{x} \arctan \frac{|2x - n|}{\sqrt{3}\, n} \qquad (1)$$

$$\bar{\alpha} = 0.5^n \sum_{x=0}^{n} \binom{n}{x} \arctan \frac{\sqrt{3}\, |2x - n|}{3n - 2} \qquad (2)$$

According to the schema in Fig. 14, Eqs 1 and 2 reflect the situation when the crack arrests in points marked 3, 5, 7 or in those marked 2, 4, 6 respectively. Here x is the number of

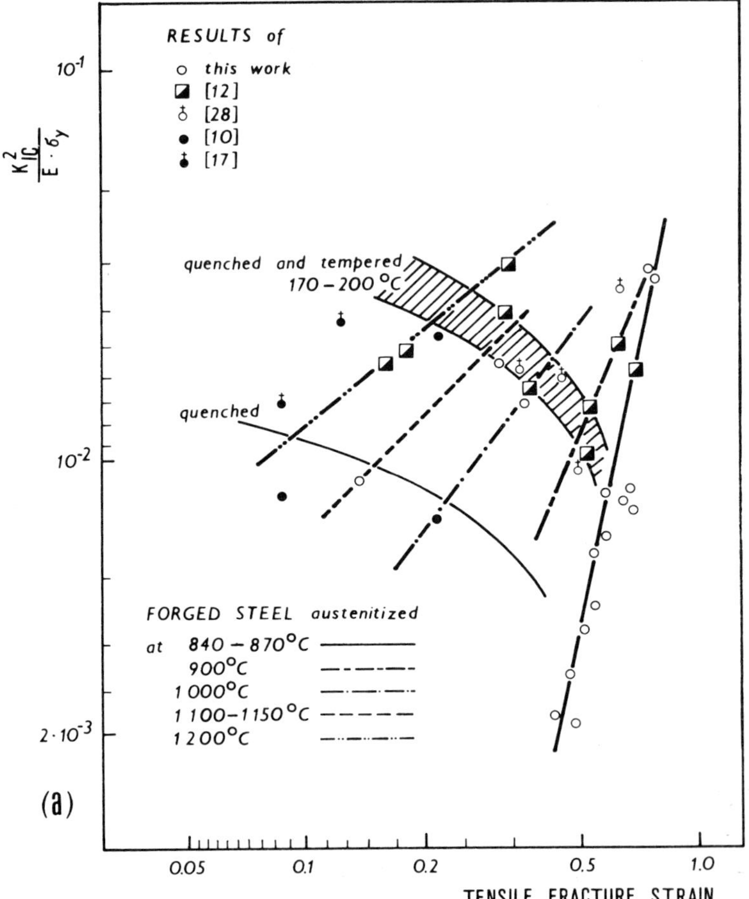

FIG. 12—*Relations between (a) tensile fracture strain and strain-controlled fracture mechanism; (b) K$_{Ic}$ versus average austenite grain size, (c) K$_{Ic}$ versus CVN toughness.*

individual declined segments above the main crack plane, and n represents the number of total declined segments within the plastic zone. In this evaluation the size of the plastic zone

$$r_\mathrm{p} \approx \tfrac{1}{10}(K_\mathrm{Ic}/\sigma_\mathrm{y})^2$$

was assumed. The detailed derivation of Eqs 1 and 2 is shown in the Appendix. Since the parameter n can be expressed as

$$n = (r_\mathrm{p}/d)$$

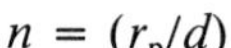

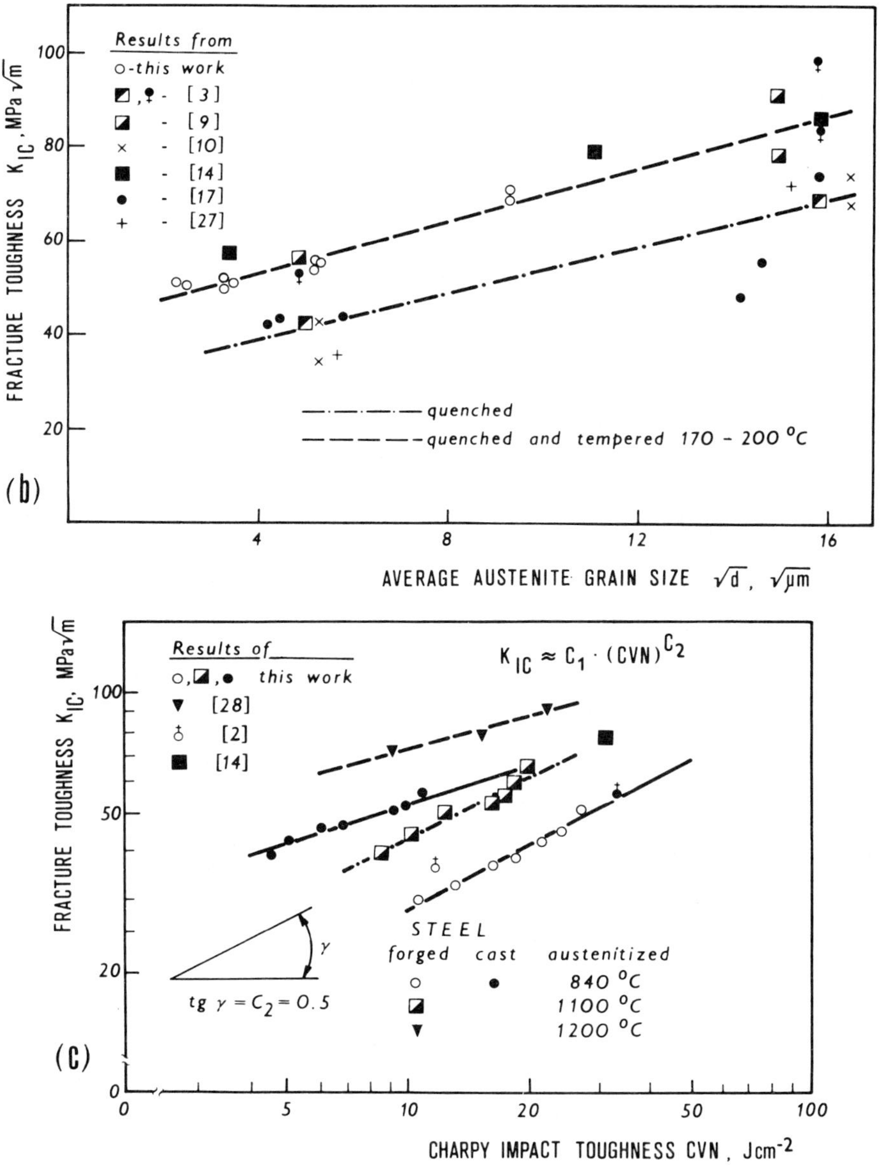

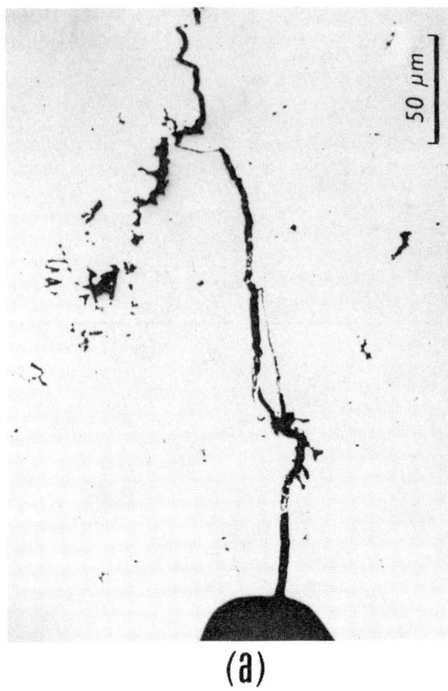

(a)

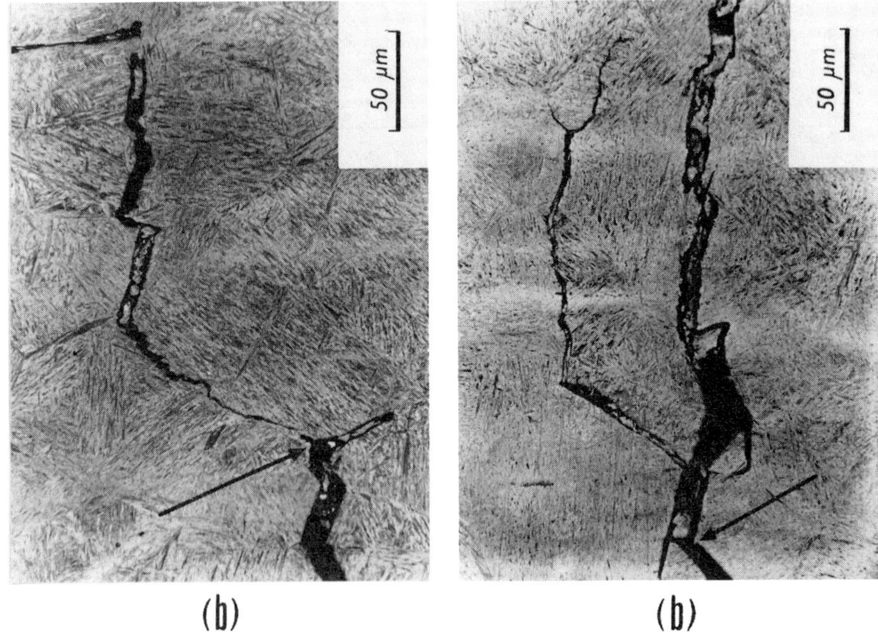

(b)   (b)

FIG. 13—*Examples of branching or kinking in the near fatigue crack tip:* (a) *cast steel,* (b) *forged steel, both austenitized at 1100°C.*

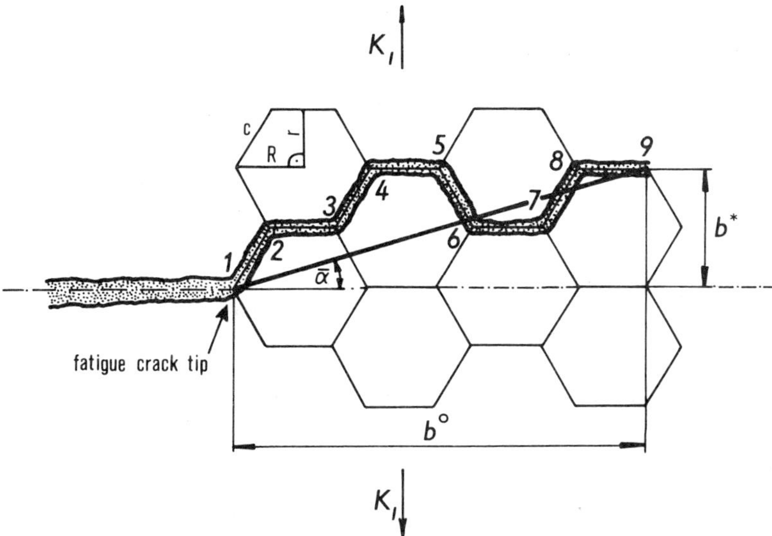

FIG. 14—*Model of crack path controlled by grain boundary decohesion.*

where d is average austenite grain size, the ratio of K_{eff}/K_I versus r_p/d may be calculated using the derivations given in Refs *12* and *15*, considering namely that

$$K_{eff} = (k_1^2 + k_2^2)^{1/2}$$

where k_1 and k_2 are local stress intensity factors at the tip of the declined crack.

The final relationship between K_{eff}/K_I and r_p/d is shown in Fig. 15. This indicates that, if the r_p/d ratio is lower than about two or the angle $\bar{\alpha}$ is higher than about 15°, the ratio $K_{eff}/$

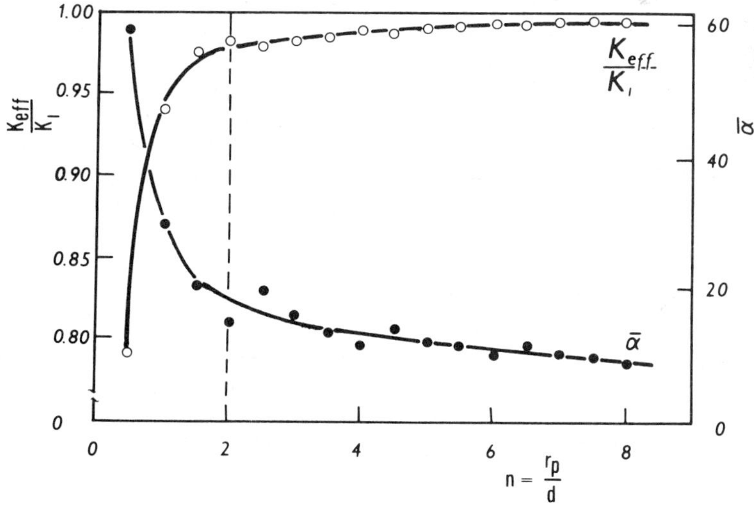

FIG. 15—*Dependence of* K_{eff}/K_I *ratio on the number of grains or average angle of declined crack path inside the plastic zone.*

K_1 is drastically decreasing. In such a case, the K_{Ic} value obtained by standard procedures cannot still represent the real material property and an appropriate correction is necessary [23]. The results offered by the proposed model would be used, for example, for an interpretation of the reversed Hall-Petch relation reported previously [15]. The differentiation in sensitivity of particular mechanical tests to reflect some real macro/microstructural variation would be considered in this sense, too.

Conclusions

The evaluation of experimental results concerning the macro/microstructure and mechanical properties relationships made it possible to estimate the influence of high-temperature austenitizing on behavior of forged and cast UHSLA steels. It was shown that:

1. The observed microstructural differences with respect to their reciprocal competition in fracture process initiation cannot elucidate the observed increase of fracture toughness K_{Ic} in full.

2. The same can be said in the case of compared mechanical properties of forged and cast steels, where the latter, except for fracture toughness K_{Ic}, has demonstrated inferior properties than the former.

3. The results have given indications of significant influence of crack branching or kinking in the early stage of the fracture initiation process. Their effect on resulting K_{Ic} values determined in agreement with recommended procedure is expected to be positive, but real effective stress intensity factor (K_{eff}) following a mixed-mode I + II loading is still at substantially lower values.

4. This is also the case for the forged and cast steels investigated in this work, particularly when the level of macro/microheterogeneity is increased in the course of technology production or following heat treatment. In this sense, the decreased cohesion strength of previous austenite grains modifying the crack initiation path can be understood as a severe local heterogeneity.

5. The proposed simple model of deflected crack path influence on the K_{eff}/K_I ratio with respect to the assumed plastic zone size (r_p) in front of a sharp fatigue crack tip has indicated that, when r_p covers less than two grain sizes (d), the K_{eff}/K_I ratio can significantly decrease. Hence it is more probable that this effect would also be satisfied if, instead of grain size, the characteristic distance among more pronounced microstructural defects is considered.

6. It seems, however, that for K_{Ic} determination of heterogeneous UHSLA steels, careful elaboration of complex mechanical properties can lead to final nonvalid K_{Ic} values, irrespective of the fact that all recommended requirements of a particular K_{Ic} test are satisfied.

APPENDIX

An advanced crack declining along the grain boundary according to the plane of the main crack is, in a more simplified scheme, shown in Fig. 14. The probability (p) that the crack along its path will turn out above or below the magistral plane and reach the points marked 1, 3, 5, etc., can be assumed to be equal, so $p \approx 0.5$. Thus the crack displacement can be described using the binomial probability function:

$$P(x,n) = \binom{n}{x} p^x (1 - p)^{n-x} \tag{3}$$

where x is the number of declined segments (i.e., grain boundary facets above the main plane), and n represents the number of all declined segments (i.e., $n = x + \bar{x}$, where $\bar{x}$ is the number of segments below the magistral plane). Then, for $p = 0.5$, Eq 3 can be written as

$$P(x,n) = 0.5^n \binom{n}{x} \tag{4}$$

Since, with regard to Fig. 14,

$$C = R = 2/r\sqrt{3}, \qquad b^* = r(x - \bar{x}), \qquad b^0 = Rn\,3/2$$

it is obvious that the instantaneous angle α of declined crack can be expressed by

$$\alpha(x,n) = \text{arctg}\,(b^*/b^0) = \text{arctg}\,\frac{|2x - n|}{n\sqrt{3}} \tag{5}$$

The average $\bar{\alpha}$ follows

$$\bar{\alpha} = \sum_{x=0}^{n} P(x,n)\alpha(x,n) \tag{6}$$

Substituting Eqs 4 and 5 into Eq 6 gives

$$\bar{\alpha} = 0.5^n \sum_{x=0}^{n} \binom{n}{x} \text{arctg}\,\frac{|2x - n|}{\sqrt{3}\,n} \tag{7}$$

Equation 7 is satisfied if the crack is finished at points 3, 5, 7, etc. (Fig. 14) that are equivalent to $n = 1, 2, 3$, etc. Since it is assumed that the described crack propagation is realized inside the plastic zone, the dimension of which is r_p, the number of all declined segments n is equal to r_p/d, where d represents average grain size. When the number of grains covered by plastic zone is incomplete (i.e, the crack is finished at points 2, 4, 6), then the argument in Eq 7 will be exchanged for

$$\theta = \frac{\sqrt{3}\,|2x - n|}{(3n - 2)} \tag{8}$$

The evaluation of $\bar{a}$ versus n dependence (Fig. 15) was done for $n = 0.5, 1.0, 1.5$, etc.

Acknowledgments

The authors wish to acknowledge the support of Professor P. Ryš during the course of this work. Sincere thanks are expressed to J. Škarek and S. Korčák for many valuable discussions.

References

[1] Zackay, V. F., Parker, E. R., Goodsby, R. D., and Wood, W. E., *Nature of Physics Science,* Vol. 236, 1972, pp. 108–109.
[2] Ritchie, R. O. and Horn, R. M., *Metallurgical Transactions,* Vol. 9A, No. 3, March 1978, pp. 331–341.
[3] Datta, K. P. and Wood, W. E., *Journal of Testing and Evaluation,* Vol. 9, No. 2, March 1981, pp. 111—117.

[4] Wei-Di Cao and Xiao-Ping Lu, *Metallurgical Transactions,* Vol. 18A, No. 10, Sept. 1987, pp. 1569–1585.

[5] Floreen, S., *Journal of Engineering Materials and Technology,* Vol. 99, Jan. 1977, pp. 70–75.

[6] Wei-Di Cao, Xiao-Ping Lu, and Jia-Jun Lin, *Materials Science and Engineering,* Vol. 63, 1984, pp. 165–187.

[7] Zeman, J., Rolc, S., and Škarek, J., *Non-Destructive Testing,* Vol. 4, Pergamon Press, London, 1988, p. 1963.

[8] Rothman, R. L., and Cohen, J. B., *Advances in X-Ray Analysis,* Vol. 12, 1969, pp. 205–235.

[9] Guimaraes, R. and Savedra, A., *Metallurgical Transactions,* Vol. 16A, No. 3, March 1985, pp. 329–336.

[10] Ritchie, R. O., Francis, B., and Server, W. L., *Metallurgical Transactions,* Vol. 7A, June 1976, pp. 831–838.

[11] Ritchie, R. O. and Knott, J. F., *Metallurgical Transactions,* Vol. 5, No. 3, March 1974, pp. 782–788.

[12] Fujita, H., Kamiya, I., and Tanaka, M., *Journal of Materials Science,* Vol. 17, 1982, pp. 3541–3548.

[13] Ritchie, R. O., Server, W. L., and Wullaert, R. A., *Metallurgical Transactions,* Vol. 10A, Oct. 1979, pp. 1557–1570.

[14] Lee, S., Majno, L., and Asaro, R. J., *Metallurgical Transactions,* Vol. 16A, Sept. 1985, pp. 1633–1648.

[15] Armstrong, R. W., *Engineering Fracture Mechanics,* Vol. 28, No. 5/6, 1987, pp. 529–538.

[16] Lin, T., Evans, A. G., and Ritchie, R. O., *Acta Metallurgica,* Vol. 34, No. 11, 1986, pp. 2205–2216.

[17] Firrao, D., Roberti, R., and Silva, G., in *Proceedings,* 4th European Conference on Fracture, Leoben, Austria, 22–24 Sept. 1982, pp. 727–735.

[18] Yoda, M., *Engineering Fracture Mechanics,* Vol. 26, No. 3, 1987, pp. 425–431.

[19] Firrao, D., Begley, J. A., Silva, G., Roberti, R., and De Benedeti, B., *Metallurgical Transactions,* Vol. 13A, No. 6, 1982, pp. 1003–1013.

[20] Curry, D. A., *Materials Science and Engineering,* Vol. 85, 1987, pp. 53–58.

[21] Tsann Lin, Evans, A. G., and Ritchie, R. O., *Acta Metallurgica,* Vol. 34, No. 11, 1986, pp. 2205–2216.

[22] Lewis, G. in *Proceedings,* International Conference on the Mechanism of Fracture, American Society for Metals, Salt Lake City, Utah, 2–6 Dec. 1985, pp. 365–373.

[23] Evans, A. G. and Cannon, R. M., *Acta Metallurgica,* Vol. 34, No. 5, 1986, pp. 761–800.

[24] Suresh, S. and Vasudevan, A. K., *Materials Science and Engineering,* Vol. 79, 1986, pp. 183–190.

[25] Suresh, S. and Shih, C. F., *International Journal of Fracture,* Vol. 30, 1986, pp. 237–259.

[26] Zhoo Yishu, *Engineering Fracture Mechanics,* Vol. 26, No. 5, 1987, pp. 683–689.

[27] Lai, G. Y. and Wood, W. E., *Metallurgical Transactions,* Vol. 5, No. 6, 1974, pp. 1663–1670.

[28] Padmanabhan, R. and Wood, W. E. in *Proceedings,* 4th European Conference on Fracture, Leoben, Austria, 22–24 Sept. 1982, pp. 264–271.

Computational Mechanics

K. S. Kim,[1] R. H. Van Stone,[1] J. H. Laflen,[1] and T. W. Orange[2]

Simulation of Crack Growth and Crack Closure under Large Cyclic Plasticity

REFERENCE: Kim, K. S., Van Stone, R. H., Laflen, J. H., and Orange, T. W., **"Simulation of Crack Growth and Crack Closure under Large Cyclic Plasticity,"** *Fracture Mechanics: Twenty-First Symposium, ASTM STP 1074,* J. P. Gudas, J. A. Joyce, and E. M. Hackett, Eds., American Society for Testing and Materials, Philadelphia, 1990, pp. 421–447.

ABSTRACT: The propagation of a crack during cyclic loading with large plasticity was simulated using a finite element method. The numerical procedure included the determination of crack closing and opening by examining the contact condition of the crack surfaces in the plasticity iteration. The method was applied to the growth of cracks in Alloy 718 single edge notch specimens at 538°C subject to $R_\epsilon = -1$ strain cycling under both nominally elastic and elastic-plastic conditions. Experimentally measured remote displacements were used to establish the boundary conditions for the finite element analysis. The measured load and crack mouth opening displacement were used to verify the results. The results show, among other things, that the crack closure and opening stresses become more compressive as the loading becomes more plastic, and that, under a given control condition, the closure and opening stresses become less compressive as the crack propagates.

KEY WORDS: crack closure, elastic-plastic finite element method, crack growth simulation, fatigue crack growth, elevated temperature, nickel-base superalloy, Alloy 718

Hot section gas turbine components experience complex thermal-mechanical fatigue (TMF) cycling. The growth of cracks under small-scale yielding conditions can be predicted to a reasonable accuracy by linear elastic fracture mechanics (LEFM). While this approach is sufficient for most engineering applications, there are circumstances which lead to crack propagation in much larger plastic fields. This situation is more likely to happen at elevated temperatures due to reduced strength of materials. It is known that LEFM does not provide adequate prediction of the crack growth behavior for large-scale yielding. Path-independent (P-I) integrals which were proposed to overcome the limitations of the *J*-integral may provide an approach to modeling the crack growth in these circumstances and also for nonisothermal conditions. These P-I integrals were recently reviewed by Kim and Orange [*1*]. Subsequently, a combined analytical and experimental investigation was initiated to evaluate the ability of P-I integrals to correlate the crack growth data under various cyclic conditions including elastic-plastic and TMF loading. The major step in the evaluation and development of such a methodology is the accurate determination of the stress and strain field of a cracked body during crack propagation. The effects of the plastic wake, which gives rise to crack closure under cyclic loading, must also be considered in the analysis.

This paper describes a procedure of simulating crack growth using an elastic-plastic finite element method which can predict the closing and opening of a crack under cyclic loading.

---

[1] GE Aircraft Engines, Mail Drop A333, Cincinnati, OH 45215.
[2] NASA Lewis Research Center, Cleveland, OH 44135.

The finite element method was applied to model the growth of cracks in Alloy 718 button-head single edge notch specimens under displacement control. The range of test conditions included nominally elastic and elastic-plastic conditions. The level of plastic deformation which occurred in these tests was much higher than those in most published papers of crack closure analysis. Hence, there were more difficulties in the numerical procedure. These test conditions were chosen in consideration of the ultimate goal of the present program. Remote displacements measured during the experiments were used to establish the boundary conditions for the finite element analysis. The results of analysis were verified by comparing the predicted and experimentally measured values of load and crack mouth opening displacement (CMOD). The present paper limits its scope to investigation of the deformation behavior of the specimen in the crack propagation process. The question on the utility of P-I integrals for crack growth prediction will be addressed in a forthcoming paper.

Since the work of Elber [2], there has been a considerable amount of research performed to characterize crack closure mechanisms and behaviors. Suresh and Ritchie [3] have reviewed the experimental closure literature and shown that the closure mechanisms could be characterized as being induced by plasticity, fracture surface roughness, or a layer of corrosion or oxide debris. It would be anticipated that plasticity-induced closure would dominate the other mechanisms for the large plasticity conditions studied in the present investigation. Other recent investigations also suggest that most of the closure which occurs in Alloy 718, the nickel-base superalloy modeled in this investigation, is induced by plasticity even in the near-threshold regime [4,5].

The application of a finite element method to analyze plasticity-induced crack closure usually involves an enormous amount of computation due to the fact that the equilibrium must be achieved such that both the material nonlinear behavior and the contact condition of the crack surface are satisfied. As a result, there have been relatively few investigations performing this type of simulation. Newman [6] investigated the effects of mesh size, load ratio (R), and the sequence of loading on the crack closing and opening stresses under nominally elastic loading in a center cracked plate. His results were consistent with Elber's experimental data [2] and rationalized crack retardation and acceleration behavior. No fracture criterion was used in this study, and the crack extension was made at the maximum load in each cycle by releasing the crack tip node. In a later investigation, Newman [7] used the crack tip strain as the crack growth criterion and examined the effects of mesh size, critical strain, hardening rule, and specimen type on the crack growth behavior.

Ohji et al. [8], Ogura and Ohji [9], and Nakagaki and Atluri [10] have also performed similar analyses under nominally elastic loading conditions. Lalor and Sehitoglu [11] performed finite element analyses of a crack emanating from a circular hole in a panel. The loads applied in this model were much higher than those used by other investigators [6–10], resulting in a larger amount of plasticity. Similar work was done by McClung [12]. He also correlated the crack growth data with closure-modified parameters such as ΔK_{eff} and ΔJ_{eff}.

Subsequent sections of this paper will describe the crack growth experiments, the finite element technique, the results of finite element analyses, and comparison of the analytical and experimental results.

Crack Growth Experiments

This section of the paper will describe the material, test specimen, and experimental techniques used in the displacement-controlled crack growth rate tests. These have been described in detail elsewhere [14] and will be summarized here.

Material

The γ''-strengthened nickel base superalloy Alloy 718 was used in these experiments. Alloy 718 plate with a nominal thickness of 22.2 mm was procured in a mill-annealed condition. The plates were solution treated at 968°C for 1 h and cooled to room temperature. They received a duplex aging treatment where they were aged at 718°C for 8 h, then cooled at 56°C/h to 621°C, where they were held for 8 h. The microstructure of this plate had a duplex grain size with some large, elongated grains, surrounded by smaller, more equiaxed grains. The larger grains are most likely warm-worked but unrecrystallized, while the small, uniform grains have been recrystallized during the rolling of the plate. These types of duplex microstructures are typical of rolled superalloys.

Cyclic Properties

The cyclic stress-strain response of Alloy 718 was determined using a buttonhead specimen with a circular cross section having a nominal diameter of 6.4 mm and a nominal gage length of 25.4 mm. The axial orientation of the specimens was parallel to the longitudinal direction in the Alloy 718 plates. All tests were performed in closed-loop equipment under strain control. The strain was measured using an elevated-temperature, water-cooled extensometer with a gage length of 12.7 mm.

Six cyclic stress-strain tests were performed at 538°C with an R_ϵ ratio (maximum strain/minimum strain) of -1. These comprised a matrix of three strain ranges by two strain rates. Strain ranges of 1.15, 1.70, and 3.50% were employed at strain rates of 0.02 and 0.50%/s. The specimen with the highest strain range and strain rate failed prematurely due to the inability of the testing machine's function generator to respond to the serrated yielding which occurs at the higher strain levels in Alloy 718 at this temperature. The results of this specimen were not used to determine the cyclic stress-strain curves.

Alloy 718 experiences significant cyclic softening early in the tests [*15*], but the softening was completed after approximately 20% of the fatigue life [*14*]. The cyclic stress-strain response showed very little change between 20 and 80% of the fatigue lives, so the stress-strain hysteresis loops from cycles corresponding to approximately 20, 50, and 80% of the fatigue lives were used to construct the cyclic stress-strain curves. Figure 1 shows the variation of the alternating stress with cycle for an Alloy 718 cyclic test performed at 538°C. The data from the lower two strain ranges are similar, but the data from the 3.5% strain range tests have a higher stress for a given value of strain. The data from these tests are shown in Fig. 2 as small dots. No appreciable strain rate sensitivity was observed. The data from the tests at the lower two strain ranges were used to collapse the data to a Ramberg-Osgood relationship with the additional restriction that the curve had to pass through the tip of the stable hysteresis loop obtained in the 3.5% strain range test. This curve was then converted to a piecewise linear stress-strain curve for use as input to the finite element analysis. The resultant cyclic stress-strain curve is shown in Fig. 2 as a solid line for alternating strains of up to 2%. The curve was extended up to strains of 8% using the Ramberg-Osgood constants for the cyclic stress-strain curve. At high strain levels, the extrapolation of the cyclic stress-strain curve exceeded the tensile stress-strain curve, so the tensile data were used above this crossover point.

Crack Growth Tests

The specimen used to evaluate the crack growth response of Alloy 718 was a buttonhead single edge notch (SEN) specimen (Fig. 3). This specimen has a thickness of 2.54 mm and a

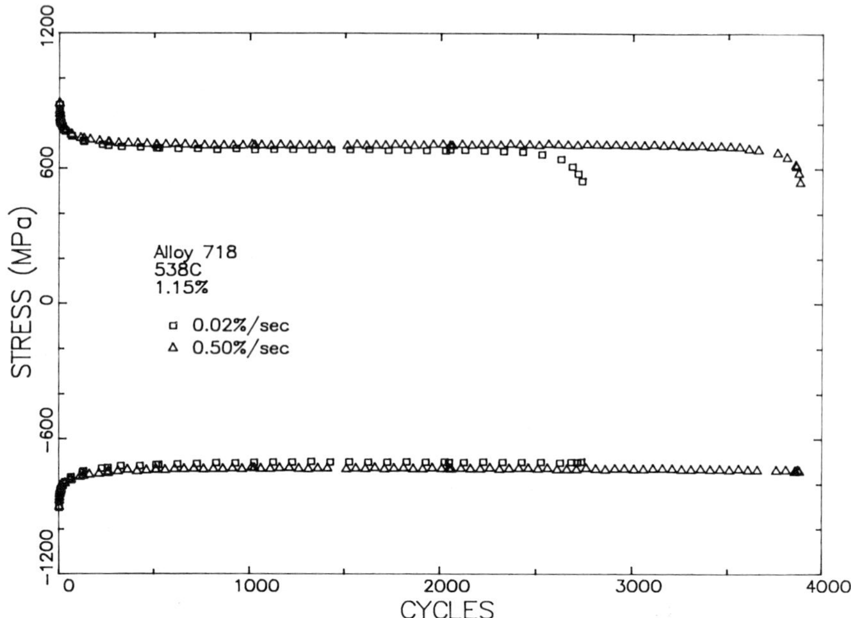

FIG. 1—*Variation of maximum and minimum stress measured during fully reversed ($R_\epsilon = -1$) Alloy 718 fatigue tests performed at 538°C with a strain range of 1.15% and two strain rates.*

width of 10.2 mm. Pin-loaded SEN specimens with a similar gage section geometry have previously been used to determine the linear elastic fracture mechanics crack growth properties of a variety of superalloys over a range from room temperature to 980°C [4,5]. The modification of the specimen from pin to buttonhead loading was made to accommodate cyclic loading with compressive loads. The gage length of this specimen was shortened to avoid buckling. The specimen was oriented so that the axial direction was parallel to the rolling direction of the plate and the crack propagated in a transverse direction.

Three-dimensional elastic and two-dimensional elastic-plastic finite element analyses were performed on the whole specimen to examine the boundary conditions necessary to perform the finite element simulation of crack growth. The analytical results showed that there was a linear variation in axial displacement across the end of the gage section [*14*]. This was confirmed in experiments which measured the axial displacements at three locations across the width of the gage length. The slope of this gradient was dependent on both strain range and crack size. As a result of these studies, the linear displacement gradient was used as the input boundary conditions of the finite element model.

The values of this gradient were experimentally determined by performing the tests with the three extensometers as shown schematically in Fig. 4. The specimen displacement was controlled using an extensometer with a gage length of 12.7 mm mounted at the center of the 10.2 mm wide gage section. A CMOD gage with a gage length of 0.76 mm was positioned across the mouth of the notch. A third extensometer was positioned on the back face of the specimen (opposite the notch) so that it would not collide with the other extensometers. The controlling and back face extensometers had the same gage lengths and established the linear displacement gradient 6.35 mm above and below the plane of the crack.

The crack length was monitored using a direct-current, on-off potential drop technique

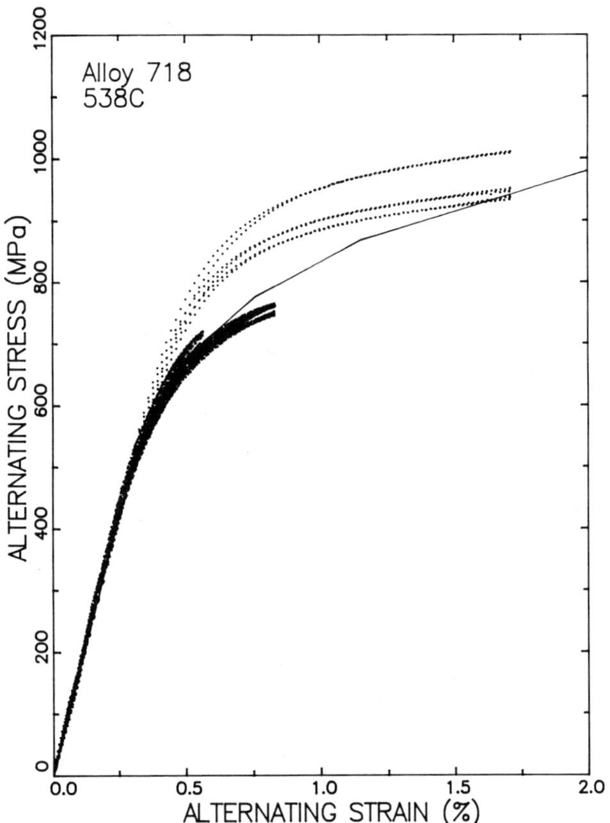

FIG. 2—*Comparison of the variation of alternating stress with alternating strain measured in Alloy 718 fatigue tests performed with three strain ranges at 538°C and* $R_\epsilon = -1$ (small dots) *with the multi-linear curve used to represent the cyclic stress-strain curve.*

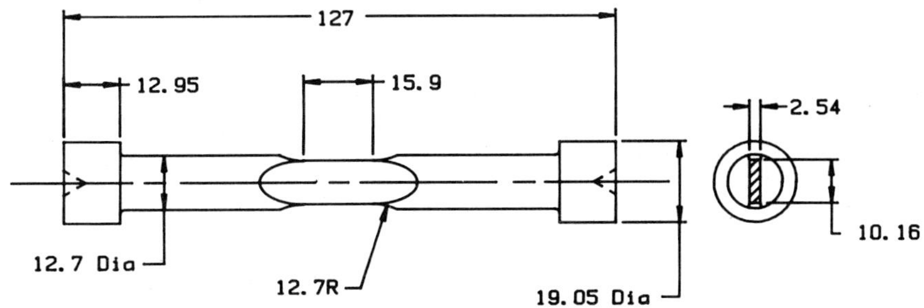

FIG. 3—*Drawing of buttonhead single edge notch (SEN) crack growth specimen (dimensions in millimetres).*

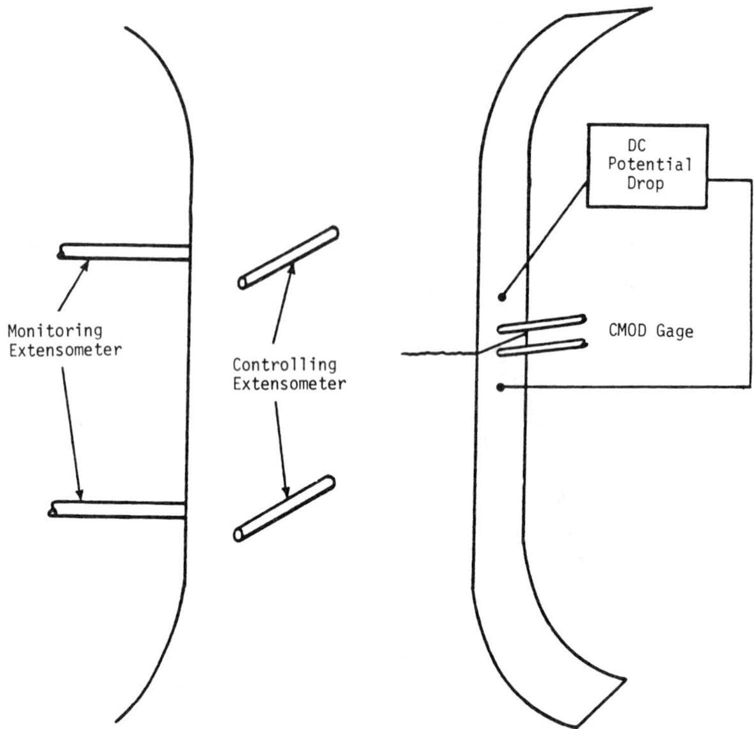

FIG. 4—*Schematic drawing of SEN displacement control crack growth test method.*

used extensively in linear elastic fracture mechanics characterization of materials with a pin-loaded SEN specimen [4,5]. The apparatus was originally developed by Gangloff [16] and adapted to the SEN geometry by Wilcox and Henry [17]. The potential probe spacing was approximately 1.52 mm so that the potential probes did not interfere with the CMOD gage. The potential is averaged over a predetermined number of cycles and converted to crack length using the potential solution developed by Johnson [18]. A detailed description of the data analysis procedure was given by Krueger [19]. This technique permitted the tests to be performed without the requirement for visual monitoring of the crack length.

These tests were further automated using an ETS test control and data acquisition system to monitor load and displacements of the three extensometers. The load-displacement hysteresis loops were recorded at preselected cycles. The data were stored on floppy disks for subsequent data analysis.

The rest of this section describes the results of the tests performed at 538°C with a mean strain of zero ($R_\epsilon = -1$). Crack growth tests were performed with total strain ranges of 0.50, 1.15, and 1.70%. These strain levels correspond to plastic strain ranges of 0.01, 0.34, and 0.75% in an uncracked specimen and thus cover the range from nominally elastic to elastic-plastic conditions. Figure 5 shows the crack lengths as measured with a d-c potential drop technique plotted as a function of the number of cycles. The duplicate tests show the excellent reproducibility of the test method. The crack growth rates were analyzed using linear elastic fracture mechanics and P-I integrals. The results will be reported elsewhere.

It should be noted that for strain ranges in excess of 0.75% the crack originally grew normal to the loading axis and then started to form shear lips at the intersection of the crack and the specimen surface, and finally grew in a totally shear mode.

Crack closure was detected using the change in the slope of the load-CMOD curve at the point where the crack opens or closes. In nominally elastic tests, the loading and unloading curves are virtually linear with load, and crack closure can be detected by subtracting an apparent elastic slope from the measured displacements and using an offset value. The use of this technique becomes more difficult for these hysteresis loops because in the higher strain range tests there was not a distinct linear region due to extensive plasticity. The technique used in this study to determine closure was to search for a change in slope rather than finding the offset value from the nominally elastic behavior. The change in slope corresponds to a discontinuity in the second derivative of load with respect to CMOD. A seven-point sliding polynomial technique [20] was used to calculate the second derivative for both increasing and decreasing load portions of each available load-CMOD hysteresis loop. The points of the crack closure (decreasing load) and crack opening (increasing load) were determined to occur when the absolute value of the second derivative reached a maximum value. This technique accurately determined the closure/opening points in the available load-CMOD loops with readily apparent cusps. By knowing these locations, the values of the specimen displacements at the controlling and back face extensometer locations when crack opening and closure occurred were also known.

The analysis of the load-displacement data will be illustrated using the results from a buttonhead SEN specimen tested at 538°C with a strain range of 1.15% and R_ϵ of -1. The results of this test are shown in Figs. 6 to 9. Figure 6 shows the variation of maximum stress ($+$), minimum stress (x), closure stress (square), and opening stress (triangle) with cycle number for this specimen. Here, the stress is the load divided by the uncracked cross-sectional area. At the beginning of the test the CMOD loop does not show much cusping and it is difficult to reliably detect the presence of closure as indicated by the large amount of scatter in the opening and closure loads. As the crack grows, the cusping increases and the detection of closure becomes much more reproducible. Figures 7 to 9 show the variation in

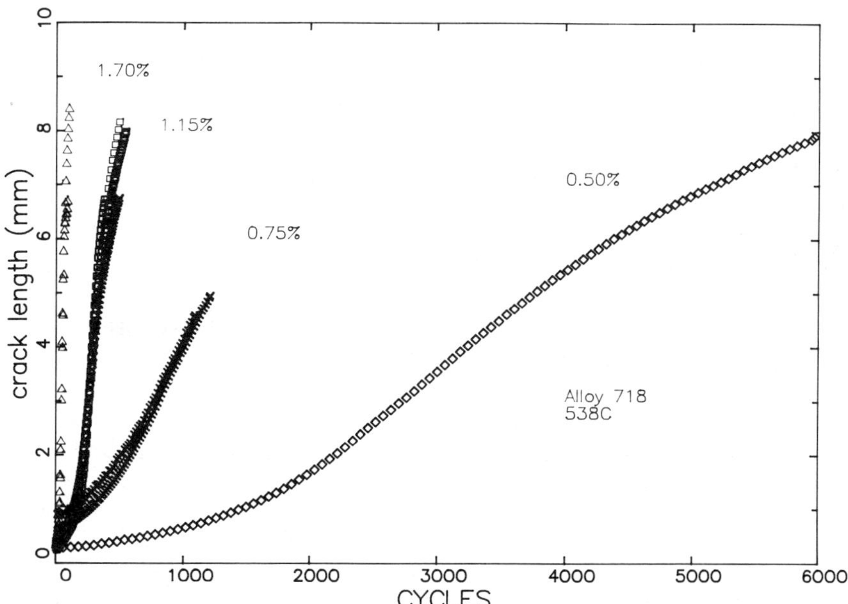

FIG. 5—*Variation of crack length with cycles in strain-controlled Alloy 718 crack growth tests performed at 538°C with* $R_\epsilon = -1$.

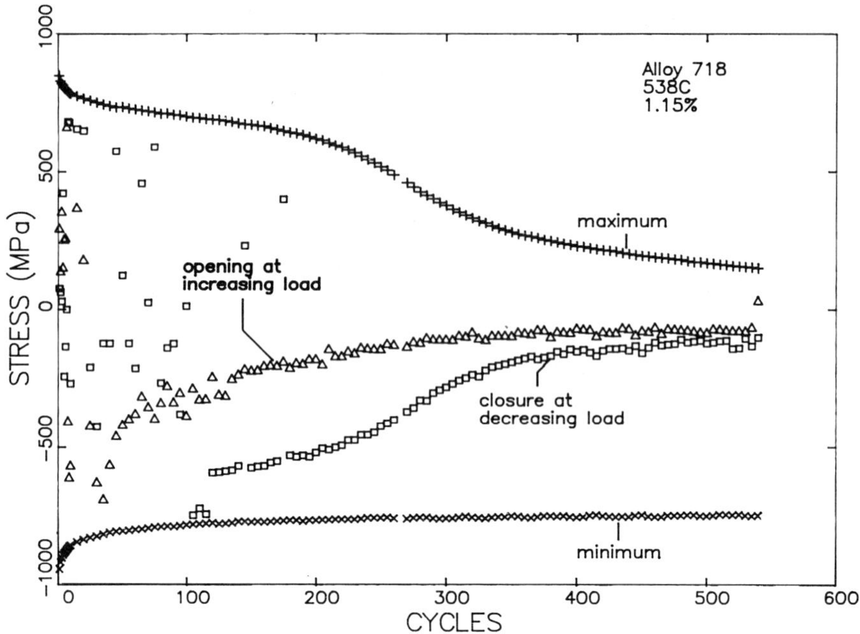

FIG. 6—*Variation of maximum, minimum, crack closing, and opening stresses with cycle measured during an Alloy 718 crack growth test performed at 538°C with a strain range of 1.15% and $R_\epsilon = -1$.*

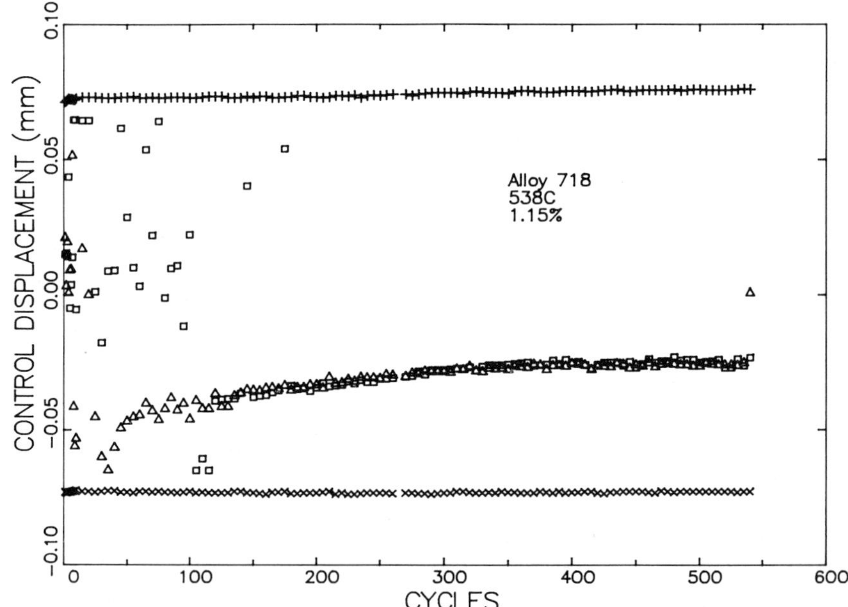

FIG. 7—*Variation of control extensometer displacement at maximum load, minimum load, crack closing, and crack opening with cycle measured during an Alloy 718 crack growth test performed at 538°C with a strain range of 1.15% and $R_\epsilon = -1$.*

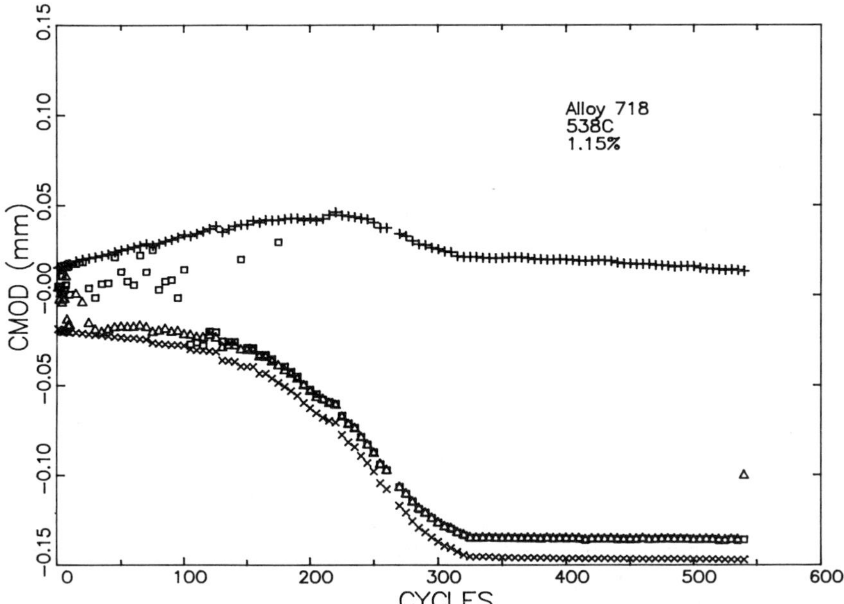

FIG. 8—*Variation of crack mouth opening displacement (CMOD) at maximum load, minimum load, crack closing, and crack opening with cycle measured during an Alloy 718 crack growth test performed at 538°C with a strain range of 1.15% and $R_\epsilon = -1$.*

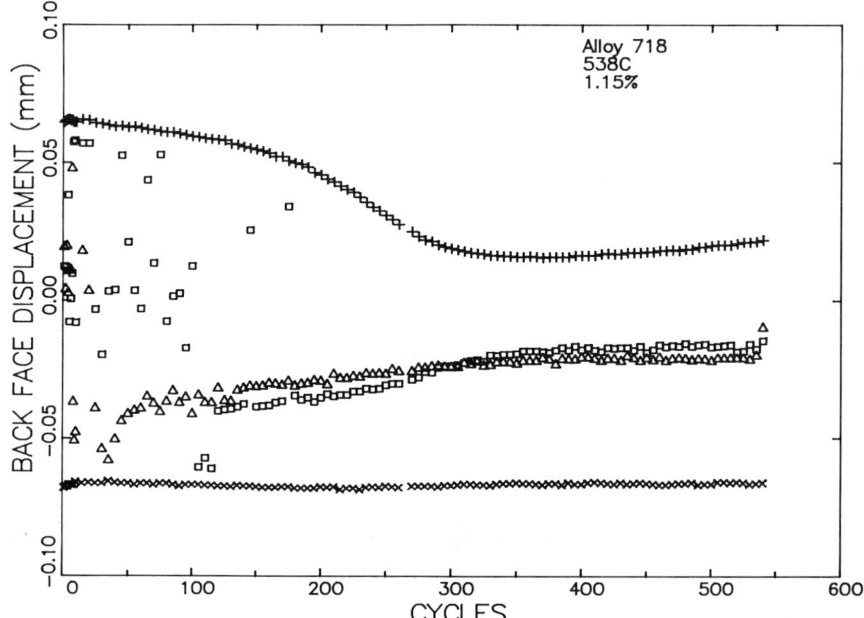

FIG. 9—*Variation of back face displacement at maximum load, minimum load, crack closing, and crack opening with cycle measured during an Alloy 718 crack growth test performed at 538°C with a strain range of 1.15% and $R_\epsilon = -1$.*

the control, crack mouth, and back face deflections at maximum load, minimum load, crack closure, and crack opening as a function of cycle number in the same test. These figures show that the specimen displacements are almost identical at the crack opening and closing events. The values of the control and back face extensometers shown in Figs. 7 and 9 were used to set the boundary conditions of the finite element model. Figure 8 shows a sharp drop in both of the CMOD values. Comparison of the CMOD histories of a large number of specimens suggests that this shift is an experimental artifact induced by the shear crack growth. This resulted in the shift of the mean value of the CMOD, so the experimental CMOD data will be compared with the numerical results on the basis of the CMOD range. This artifact did not diminish the ability to detect the positions within the cycle where crack closure and opening occurred.

Finite Element Method

The simulation of plasticity-induced closure requires the development of elastic-plastic contact capabilities in the finite element code. This section describes the finite element method used in this study to model crack growth including crack closure.

The classical incremental theory of plasticity which utilizes the Prandtl-Reuss flow rule, von Mises yield criteria, and the kinematic hardening rule in the strain space is the basis of the constitutive model employed in this method. The Besseling's subvolume method [13] is used within this constitutive framework.

The numerical scheme of the elastic-plastic finite element analysis used herein is the initial strain method analogous to the method used by Newman [6]. For a given load increment the finite element equation of equilibrium at the elemental level can be written as

$$[K]\{\Delta u\} = \{\Delta f\} \tag{1}$$

$$\{\Delta f\} = \{\Delta f^{a}\} + \{\Delta f^{\theta}\} + \{\Delta f^{P}\} \tag{2}$$

where $\{\Delta u\}$ is the increment of displacement, $\{\Delta f\}$ is the total increment of the load, and $\{\Delta f^{a}\}$ is the increment of the applied load. The stiffness $[K]$, the increment of thermal load $\{\Delta f^{\theta}\}$, and the increment of plastic load $\{\Delta f^{P}\}$ are given by

$$[K] = \int_{V} [B]^{T}[C][B]dV \tag{3}$$

$$\{\Delta f^{\theta}\} = \int_{V} [B]^{T}[C]\{\Delta\epsilon^{\theta}\}dV \tag{4}$$

$$\{\Delta f^{P}\} = \int_{V} [B]^{T}[C]\{\Delta\epsilon^{P}\}dV \tag{5}$$

where the $[B]$ matrix and the elastic properties matrix $[C]$ are as found in standard textbooks and hence omitted here. The plastic strain and the thermal strain are

$$\{\Delta\epsilon^{P}\} = \{\Delta\epsilon^{P}_{11}, \Delta\epsilon^{P}_{22}, \Delta\epsilon^{P}_{33}, \Delta\gamma^{P}_{12}\}^{T} \tag{6}$$

$$\{\Delta\epsilon^{\theta}\} = \alpha\Delta\theta\{1,1,1,0\}^{T} \tag{7}$$

The prescribed displacement boundary conditions are enforced using penalty constants along the diagonal of the assembled stiffness matrix and replacing the corresponding entries of the force vector by Δu prescribed times the penalty constant.

For analysis of the contact condition of the crack surface, gap elements, which are actually springs with either very high moduli or zero moduli depending on whether the gap is zero or positive, are introduced on the crack plane. The high modulus must be taken several orders of ten higher than the highest entry in the stiffness matrix. In this analysis it was uniformly set to 7×10^{11} MPa. It is noted that gap elements are also located ahead of the crack tip for use in the propagation of the crack. A gap element connects a node on the crack surface to a nonstructural node placed on the line of symmetry. If the gap becomes negative in the analysis, the high modulus is used for the spring, whereas if the force in the spring becomes tensile while in contact, the modulus is set to zero. Thus, if contact occurs, the high modulus enters into the diagonal of the assembled stiffness for the y-directional degree of freedom of the crack surface node of the gap element, and the force on the right-hand side is set to zero. The original stiffness and the force vector corresponding to the degree of freedom are saved and later restored to their places if the gap opens up. The analysis procedure for an isothermal loading step is given below:

1. Read the boundary conditions for the loading step.
2. Construct the stiffness (for the first load case) or retrieve the original stiffness saved in a file.
3. Modify the stiffness and the force vector for displacement boundary conditions and for gap element contact conditions.
4. Solve for the increment of displacement for the incremental elastic force.
5. Compute (or revise) the plastic force and obtain the increment of displacement due to plasticity.
6. Update the total displacement increment and check the contact conditions of gap elements. If there are changes, repeat Steps 3 to 5 with modified contact conditions.
7. Check the plasticity convergence criteria. If not converged, go back to Step 5.
8. If plastic convergence were achieved and there were no changes in the contact condition, update the field data and write the results in the output file.
9. Go back to Step 1 for the next loading step.

The stiffness must be computed in each loading step for TMF cycling where the material properties change with temperature. Notice that the contact conditions of gap elements are examined and revised in the plastic iteration process. This procedure will permit precise assessment of the contact conditions even when the loading step is taken as large. For the convergence criteria, the increment of the effective plastic strain defined by

$$\Delta\epsilon^{p}_{\text{eff}} = (\tfrac{2}{3}\, \Delta\epsilon^{p}_{ij}\, \Delta\epsilon^{p}_{ij})^{1/2} \tag{8}$$

was used. It was assumed that the convergence requirement was met if either the absolute value of the change of the effective plastic strain or the percentage change in the two consecutive iterations is smaller than the value given in the input for all the elements in the model.

Finite Element Simulation of Crack Growth and Crack Closure

Finite Element Model

The two-dimensional finite element model of the SEN gage length used in this study is shown in Fig. 10. Due to the symmetry of the geometry and loading conditions, only half of

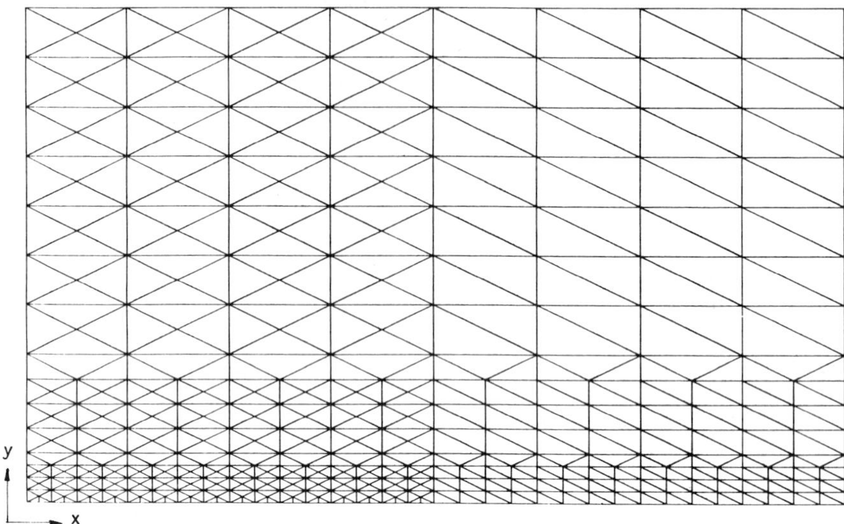

FIG. 10—*Finite element model of gage section of SEN specimen.*

the gage length was included in the model. The model consists of 688 constant strain triangular elements and 421 nodes, 33 of which are the nonstructural nodes for the gap elements located on the line of symmetry. The model is 10.16 mm wide and 6.35 mm high. As discussed earlier, the specimen gage length is rather short to avoid buckling, so the finite element mesh is shorter than other SEN specimen models appearing in the literature.

The element size along the crack tip path is certainly a factor which influences the closing and opening behavior of the crack. Obviously, finer crack tip elements permit higher residual plastic deformation on the new crack surface as the crack is extended. This will influence the closing stress as Newman [6] reported for $R = 0$ loading.

The size of the mesh along the crack plane in Fig. 10 is approximately 0.15 mm. This size is close to the intermediate size of the three mesh sizes studied by Newman [6]. However, the ratio of the mesh size to the specimen width is much larger for the present mesh. In other papers [8,9,11,12], smaller ratios were used for crack closure analysis. The mesh size effect is thought to vary according to load ratio and magnitude of loads. Newman [6] showed that the opening stress was identical for the three meshes he used for $R = 0$ loading with a maximum stress equal to the half of the yield stress, but that it varied at lower stresses. Recently, McClung [12] suggested the use of mesh sizes less than 5% of the plastic zone size at the maximum load for accurate determination of the crack closure behavior. He assumed that the crack length is incremented by one element each time. In this study, the crack length is incremented by two elements in each cycle to save computational time. Therefore the mesh size should be smaller than 2.5% of the plastic zone size. The plastic zones for the three control strain ranges were determined in this study and are reported with the results of finite element analyses. The results showed that, for strain ranges of 1.15% and 1.7%, the whole crack ligament was in the plastic zone at most of the crack lengths considered in the analysis. The ratio of the crack length increment to the plastic zone size ranged between 3.2% and 4.5% in these cases, thus satisfying McClung's criterion. This criterion was not satisfied for the case of 0.5% strain range where the ratio exceeded 14.5%. Nevertheless, the results of this analysis appear to be in good agreement with experimental data, and we believe that the mesh is also acceptable for this case.

Analysis Procedure

The finite element analysis was carried out for three specimens with total strain ranges of 0.50, 1.15, and 1.70% as measured by the controlling extensometer.

The analysis was two-tiered. The first part was crack growth simulation where the crack tip advanced into the material incrementally in each cycle. The purpose of this analysis was to produce a plastic wake along the newly created crack surface. This analysis was called a *crack growth simulation.*

The second type of analysis was performed with multiple load cases per loading cycle at constant crack lengths. The purpose of this type of analysis was to perform a more detailed simulation of crack closing and opening. This analysis was called a *crack closure analysis.*

In both analyses the state of stress was assumed to be plane stress. In view of the highly plastic deformation under the loading in consideration, the plane stress state is believed to be much closer to the real test condition than the plane strain state.

Crack Growth Simulation

The depth of the EDM starter notch in the SEN test specimens was approximately 0.3 mm. Thus the initial position of crack tip was positioned at the third node from the left on the crack plane in the model, which results in an initial crack length of 0.3175 mm. The material at this stage is assumed to be stress-free. In this analysis, each cycle had three load cases as shown in Fig. 11: (1) loading up to the tensile peak in a single step, (2) releasing two nodes at the crack tip resulting in a crack length increase of 0.3175 mm, and (3) loading down to the compressive peak in a single step. The process was repeated until the crack length reached 2.54 mm. At this point another cycle of loading was applied and the analysis terminated. This procedure required 22 load cases. The loading steps in this analysis may appear to be too large to adequately characterize the nonproportionality of the crack tip field and the residual plastic deformation wake left behind as the crack grows. The simultaneous release of two nodes does not allow the second node to experience the full severity of the crack tip field. Some preliminary studies were made to investigate the effects of smaller load steps and the release of one rather than two nodes at a time. It was found that the changes were rather small for much increased computational efforts as the procedure became more refined. For instance, the differences in the maximum load, minimum load, and CMOD for the 1.15% strain range specimen with a 2.54 mm crack length between a single node and two node release simulations were within 4%. The effects of additional loading steps were also small as will be discussed when presenting the results of crack closure analyses. In this connection, it is worth noting the work of Krieg and Krieg [20] and Schreyer et al. [21], who investigated the numerical accuracy of an elastic-plastic analysis for various algorithms when one large loading step is taken along a wide range of nonproportional loading paths. For the numerical scheme implemented in the present study, which is essentially the elastic predictor-radial corrector method, and for the loading conditions analyzed, it is expected from these papers that satisfactory results will be obtained without subdividing the loading steps. Furthermore, considering errors which could be attributed to other factors, such as discretization and the constitutive relation itself, the procedure to be taken here is well justified.

It is also noted that the deformation field may vary somewhat depending on where the nodes are released in a cycle. The determination of such a position, however, requires advanced knowledge of the fracture criteria, which, in turn, are determined from the mechanical field data. The simple approach of node release taken in this work was assumed

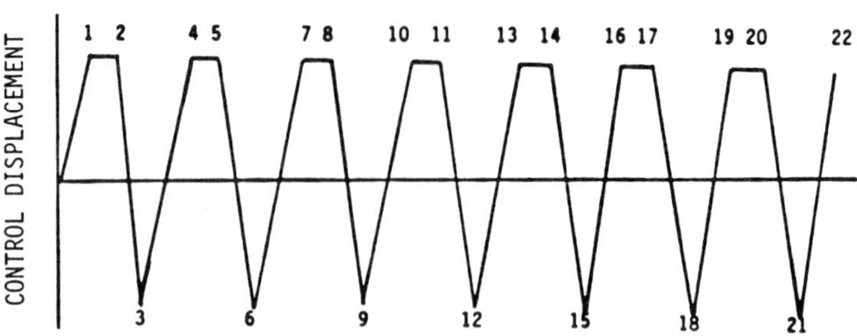

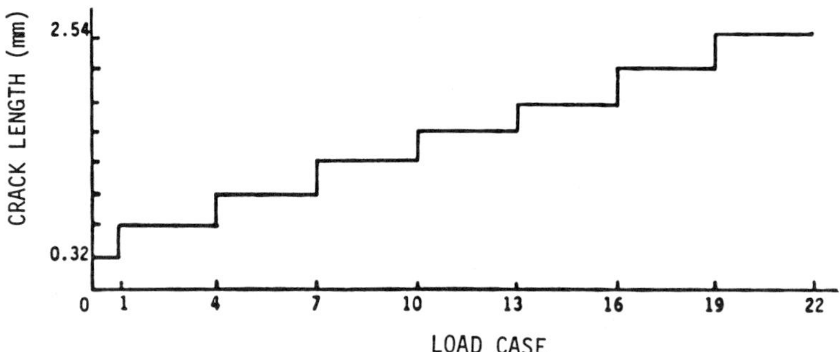

FIG. 11—*Schematic representation of variation of control displacements and crack lengths used during the finite element crack growth simulation.*

to provide a reasonable data base for evaluation of fracture criteria. The results of evaluation of some criteria will be published in a separate paper as mentioned previously.

Crack Closure Analysis

This analysis can be performed at crack lengths associated with any of the cycles appearing in Fig. 11. Due to the cost involved in the analysis we have considered only two crack lengths here, 0.95 mm and 2.54 mm. The same crack lengths were used for all three specimens. These crack lengths were selected to analyze situations with significant differences in crack length, crack growth rate, and extent of deformation. For both crack lengths, the fatigue cycle was divided into ten loading steps. The loading steps are indicated by solid triangles in Fig. 12 for the 2.54 mm crack. The ten load cases include the tensile peak, the compressive peak, the experimentally measured crack closing, and opening points on the stress-CMOD hysteresis loop. Four load cases were positioned close to the crack closing and opening points. These load cases were selected to examine the closing and opening behavior, but not necessarily to precisely identify the closing and opening stresses. This could be accomplished by the addition of more load cases.

The crack closure analysis was initiated using the results from the crack growth simulation run which corresponded to the appropriate crack length (load case 7 or 22). The starting point of the analysis is the load case where the material has undergone a whole cycle after node release and not the load case immediately following node release. This will most likely result in a more accurate determination of the closure stress because the segment of crack surface most recently created by node release has to be subjected to the contact pressure of crack closing during one simulated fatigue cycle. In most previous investigations, the closure stresses were determined using small unloading steps immediately after the crack tip node was released at the upper peak. The opening stress is nevertheless unaffected, since the reverse deformation on the contact area of the crack at the lower peak has already been accounted for.

Boundary Conditions

As discussed previously, the boundary data obtained on the SEN specimen were the remote vertical displacements measured on a plane 6.35 mm above and below the plane of the crack. The displacement data from the control and backface extensometers were linearly interpolated and extrapolated to determine the boundary conditions at the nodes on the boundary. The lateral boundary conditions on the remote boundary are unknown. However,

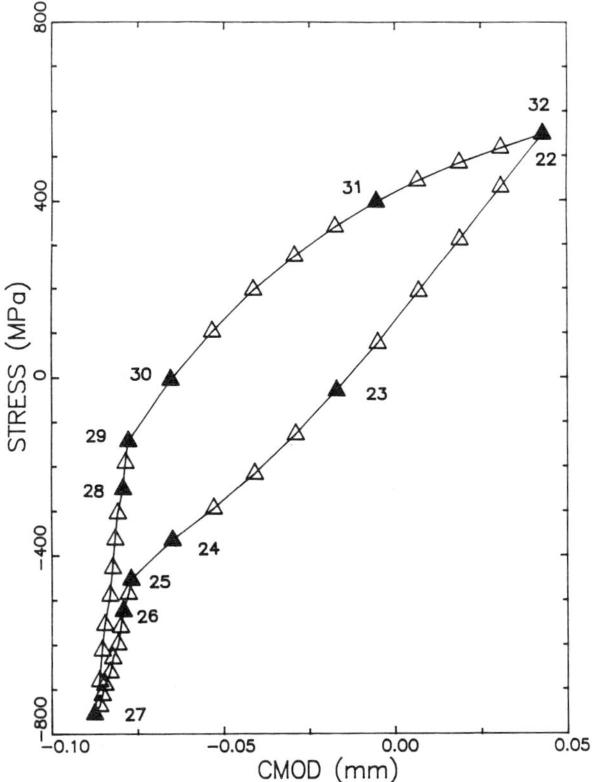

FIG. 12—*Stress-CMOD hysteresis loop determined at a crack length of 2.54 mm during an Alloy 718 crack growth test performed at 538°C with a strain range of 1.15% and $R_\epsilon = -1$. The closed points and corresponding numbers indicate the load cases used in the crack closing and opening analyses.*

previous analyses using the results of a full specimen elastic-plastic analysis and zero lateral displacements yielded insignificant changes in the stress intensity factor. Consequently, the lateral displacements at the upper boundary nodes were set to zero. This will give rise to inaccurate shear stress distribution along the boundary, but these stresses are small and the effects will be diminishing with distance from the boundary.

The hysteresis loop data of the specimen were recorded every few cycles during the crack growth test. As a result, these data were not available at precise crack lengths corresponding to the finite element node positions on the crack plane. The displacement and load data for a given crack length were determined by linear interpolation of the data at crack lengths which bracket the desired crack length. Figure 12 shows the 40 points thus determined for a stress-CMOD hysteresis loop for the 2.54 mm crack.

Results of Analysis

Crack Growth Simulation

Crack growth simulation of the Alloy 718 SEN tests with three different strain ranges (0.50, 1.15, and 1.70%) were performed using the procedure just described. Figures 13 and 14 show typical examples of the crack surface profile as predicted during the crack growth simulation of the test with the intermediate strain range. The material just behind the crack tip has a cusped morphology due to the residual plastic deformation as the nodes are released (Fig. 13). The cusp is obliterated as the material experiences load reversals (Fig. 14) and the crack completely closes during the compressive loading. Notice the difference of the crack surface

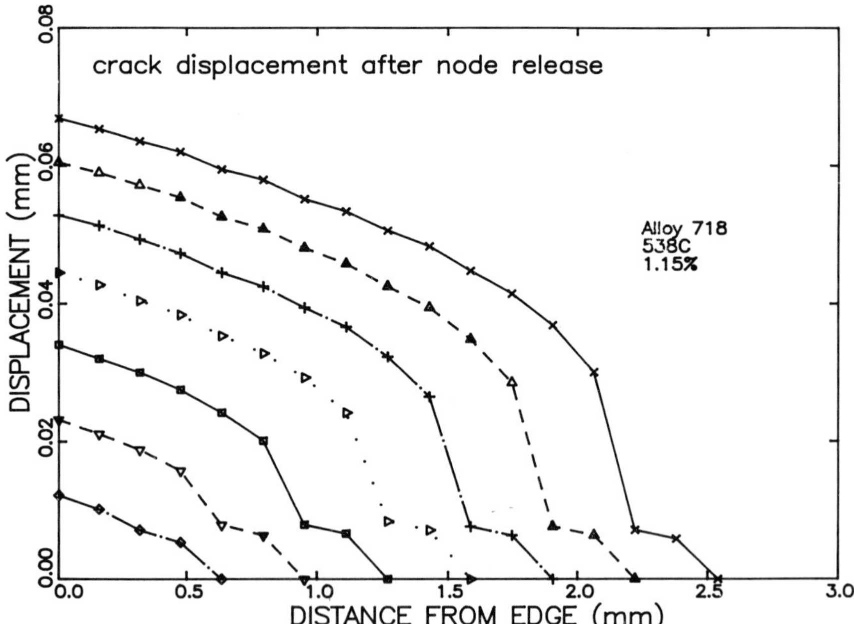

FIG. 13—*Calculated crack surface profile of an Alloy 718 SEN specimen at 538°C with a strain range of 1.15% and $R_\epsilon = -1$ determined after node release during a crack growth simulation.*

profile from the case of $R = 0$ loading for which the crack surface remains cusped even after reloading [6].

The plastic zones are plotted in Fig. 15 for the intermediate strain range. The plastic zone became smaller because of load drop, as the crack grew. The plastic zone was oriented backward from the crack tip and spread over the whole or most of the uncracked ligament at the crack lengths analyzed. For the 1.7% strain range, the plastic zone spread over the whole specimen and diminished somewhat as the crack propagated. For the 0.5% strain range (nominally elastic case), the initial plastic zone was small but grew as the crack propagated. The shape of the plastic zone was similar to oval and inclined forward as usually observed in a small-scale yielding case. It rotated a little backward and grew wider with crack propagation.

The average maximum tensile and minimum compressive stresses, defined by the total load predicted from the crack growth simulation divided by the cross section area, are plotted in Fig. 16 (lines) as a function of crack length for the three strain ranges. The experimental results (symbols) are also shown. The correlation seems to be good for all three test conditions. The stress at the tensile peak decreases as the crack becomes longer. This is caused by the increasing contribution of the specimen compliance as the crack propagates under displacement control. The degree of stress drop increased with the amount of associated plasticity. The minimum compressive stress remains relatively constant, because at the minimum load the crack is totally closed and the specimen behaves like an uncracked specimen. Notice also in Fig. 16 that the stress level at the peaks varies from 75% to 185% of the yield stress (450 MPa) depending on the control strain range and crack length.

The computed CMOD range (lines) was compared with the experimental results (symbols)

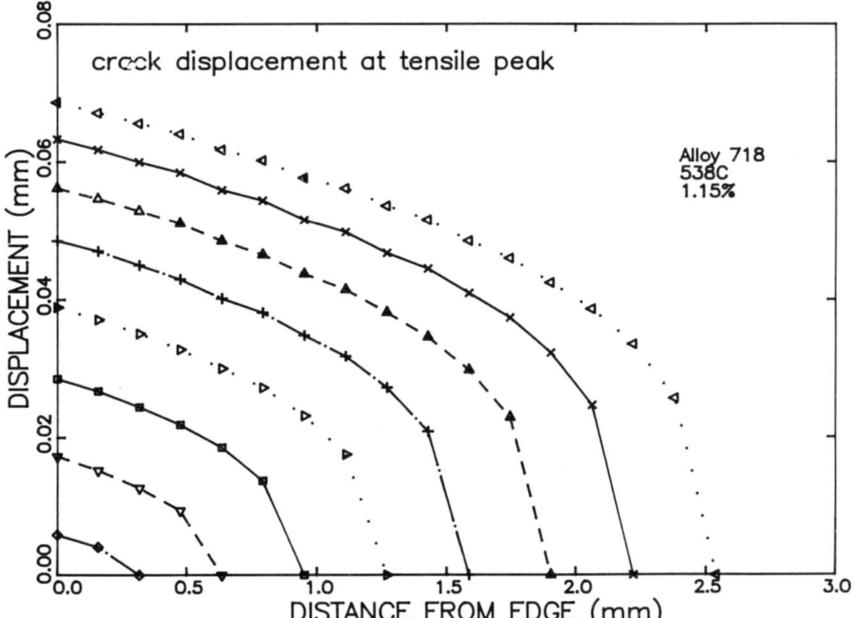

FIG. 14—*Calculated crack surface profile of an Alloy 718 SEN specimen at 538°C with a strain range of 1.15% and $R_\epsilon = -1$ determined at maximum load conditions during a crack growth simulation.*

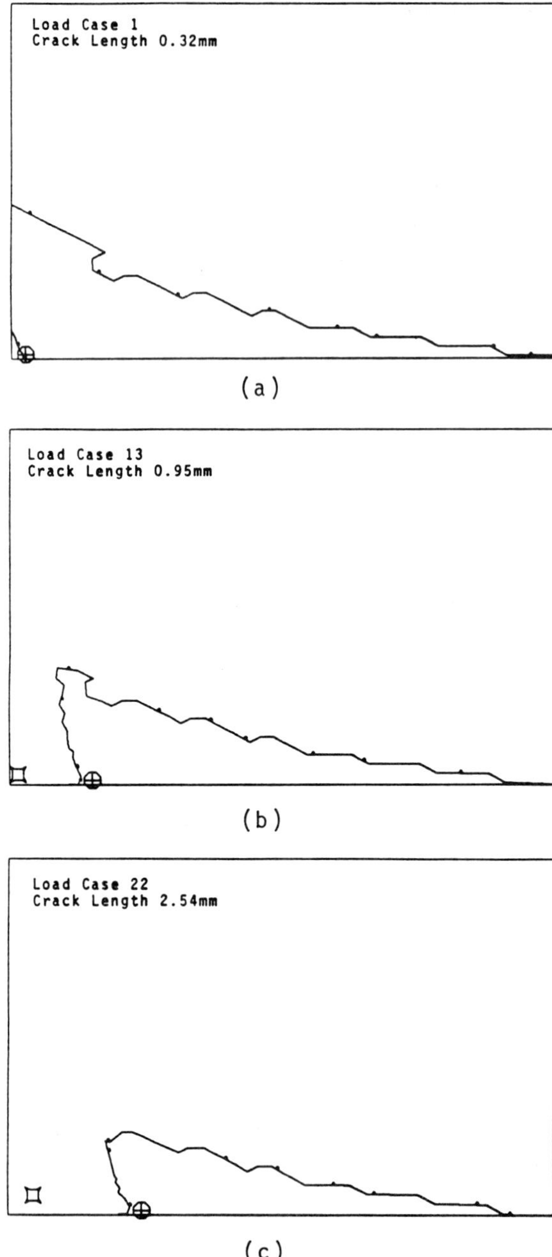

FIG. 15—*Plastic zones at the tensile peaks for 538°C, strain range of 1.15%, $R_\epsilon = -1$; crack lengths:* (a) *0.32 mm,* (b) *0.95 mm,* (c) *2.54 mm.*

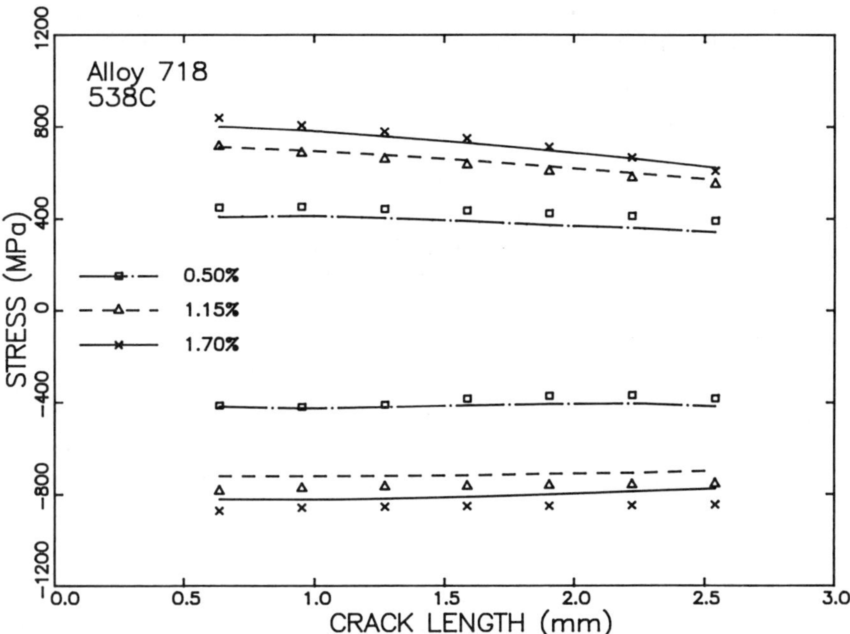

FIG. 16—*Comparison of predicted and experimentally measured values of maximum and minimum cross section stresses with crack length in Alloy 718 crack growth tests at 538°C and $R_\epsilon = -1$ having three different strain ranges.*

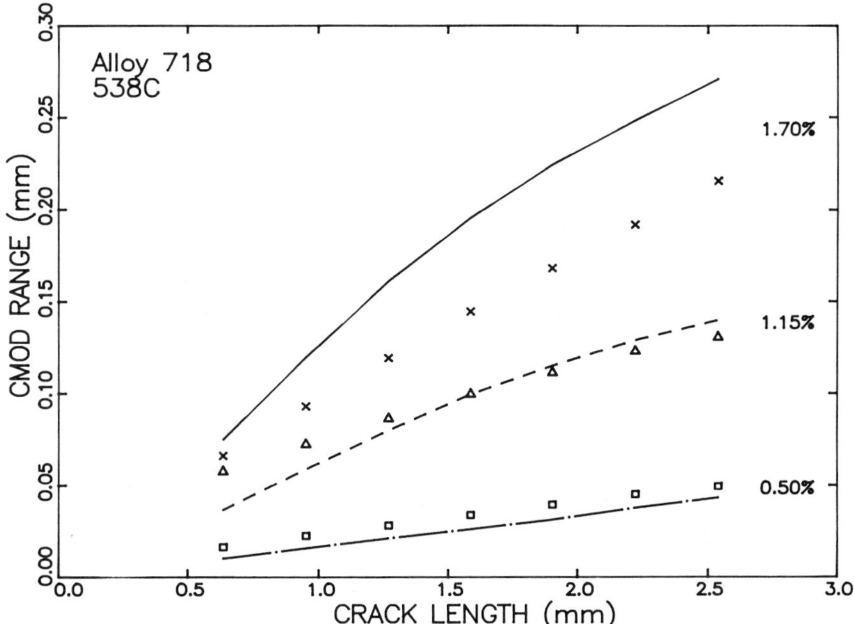

FIG. 17—*Comparison of predicted and experimentally measured CMOD range with crack length in Alloy 718 crack growth tests at 538°C and $R_\epsilon = -1$ having three different strain ranges.*

in Fig. 17 for the three test conditions. The correlation was again good for the lower strain range specimens, but the quality of the correlation diminishes with increasing strain range.

Crack Closure Analysis

Crack closure analyses were performed for crack lengths of 0.95 mm and 2.54 mm for all three strain ranges. Figure 18 shows the crack surface profiles for the load cases in the analysis of a 2.54 mm long crack cycled with a 1.15% strain range. The surface profiles at load cases 26, 27, and 28 are given at the lowest part of the figure. These load cases are not marked separately because of their near-zero displacements. The crack was still open at load case 25, which corresponds to the experimentally detected point of crack closure. The first contact was seen at the mouth of the crack in load case 26. This is an interesting observation because crack closure in the experiments was monitored by the displacement at the crack mouth (CMOD). In the analysis of other strain ranges, however, the crack mouth and the near-crack-tip area were simultaneously in contact in the loading step where the first contact appeared. The crack was closed in load case 27 except at the point just behind the crack tip, while it was totally closed in the crack growth simulation run. The near-crack-tip opening was, however, extremely small. The crack was partially closed in load case 28 and was fully open in load case 29. The crack surface profile of load case 32, the tensile peak, coincided precisely with that of load case 22, which was not shown in Fig. 18. Notice in the figure that the crack surface is slightly wavy due to double-node release in the crack growth simulation. The lower peaks in the wave represent the nodes where the crack tip was positioned in the crack growth simulation run. It does not seem that these small ripples affect the closing and

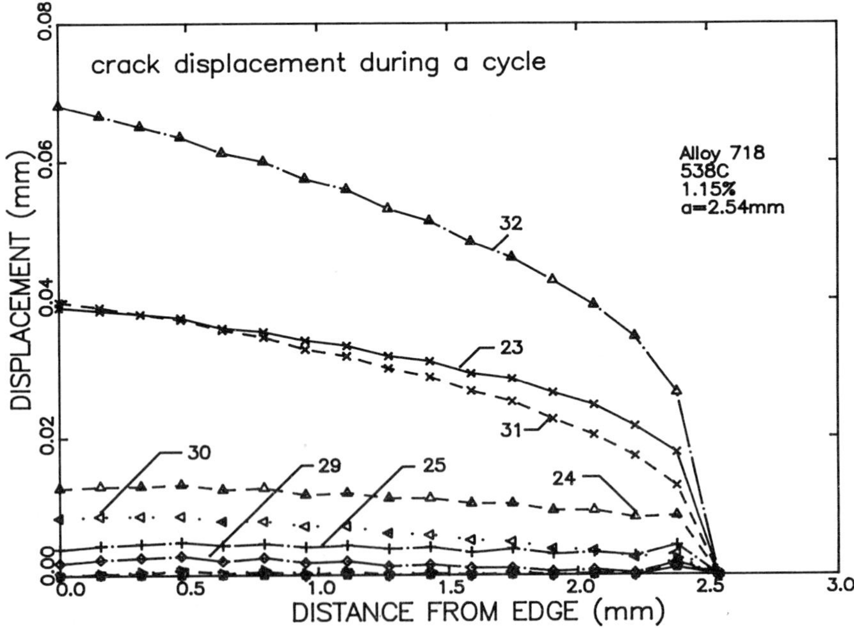

FIG. 18—*Calculated crack surface profile determined during a hysteresis loop analysis of an Alloy 718 SEN specimen at 538°C with a 2.54 mm long crack, a strain range of 1.15%, and* $R_\epsilon = -1$.

opening stresses significantly nor does the irregularity of contact pressure distort the overall crack surface profile to a meaningful extent.

The plastic zones at the tensile peak, the experimental closing and opening points, and the compressive peak are presented in Fig. 19 for the intermediate strain range. Notice that the plastic zone at the closing point is significant while it disappeared at the opening point. This indicates that a considerable reverse plastic deformation occurs around the crack tip before the crack starts to close in the reverse loading, and that the stresses at the compressive peak are much relieved at the opening point. Notice also that the plastic zone is stretched over the crack surface at the compressive peak, meaning that the crack surface has undergone reverse plastic deformation due to contact pressure. For the 1.7% strain range, the plastic zone covered much of the area of the specimen for the four load cases shown in Fig. 19, but to a somewhat less extent for the crack opening point. For the case of 0.5% strain range, the plastic zone at the compressive peak was much smaller than that of the tensile peak and was oriented slightly backward from the crack tip. The plastic zone was very small at the closing point and disappeared at the opening point. The reverse plastic zone on the crack surface at the compressive peak was unidentifiable. It is also noted that the plastic zones at load cases 22 and 32 were virtually identical for all three strain ranges (compare load case 22 in Fig. 15 with load case 32 in Fig. 19).

The average stress versus CMOD hysteresis loops for the three specimens are shown in Figs. 20, 21, and 22. The experimental CMOD data were shifted so that the minimum CMOD measurement and predicted value were equal to compensate for the drift of the CMOD measurements as described previously. The stress-CMOD hysteresis loops predicted with the finite element method agree closely with the experimentally measured loops despite various approximations made in the analysis. One can observe the following from Figs. 20 to 22:

• The area of the average stress–CMOD hysteresis loop increases as the loading becomes more nonlinear, which occurs with increasing displacement ranges and increasing crack length.

• The displacements at the crack closing and opening points are nearly identical for a given hysteresis loop.

• The closing stress and the opening stress are compressive for all loading conditions evaluated. The closing stress and the opening stress are approximately identical for the smallest displacement range, which had an almost elastic response. As the displacement range and the level of net section plasticity increases, the closing and opening stresses become more compressive and, within a given hysteresis loop, the closing stress becomes more compressive than the opening stress.

• As the crack propagates, the closing and opening stresses become less compressive. There were no significant changes in the minimum compressive stress. As a result, the portion of the hysteresis loop where the crack was closed became relatively longer with increasing crack length.

The third observation agrees well with the results of other researchers [11,12]. The predicted values of stress and CMOD are nearly identical for load cases 22 and 32 (before and after the multiple load case hysteresis cycle). This provides some substantiation of reliability of numerical results obtained by taking large load steps in the crack growth simulation.

The stress range between the crack opening point and the tensile peak becomes larger as the displacement range increases. This implies that the effect of crack closure on the crack growth rate becomes less significant with increasing plasticity for $R_\epsilon = -1$ loading.

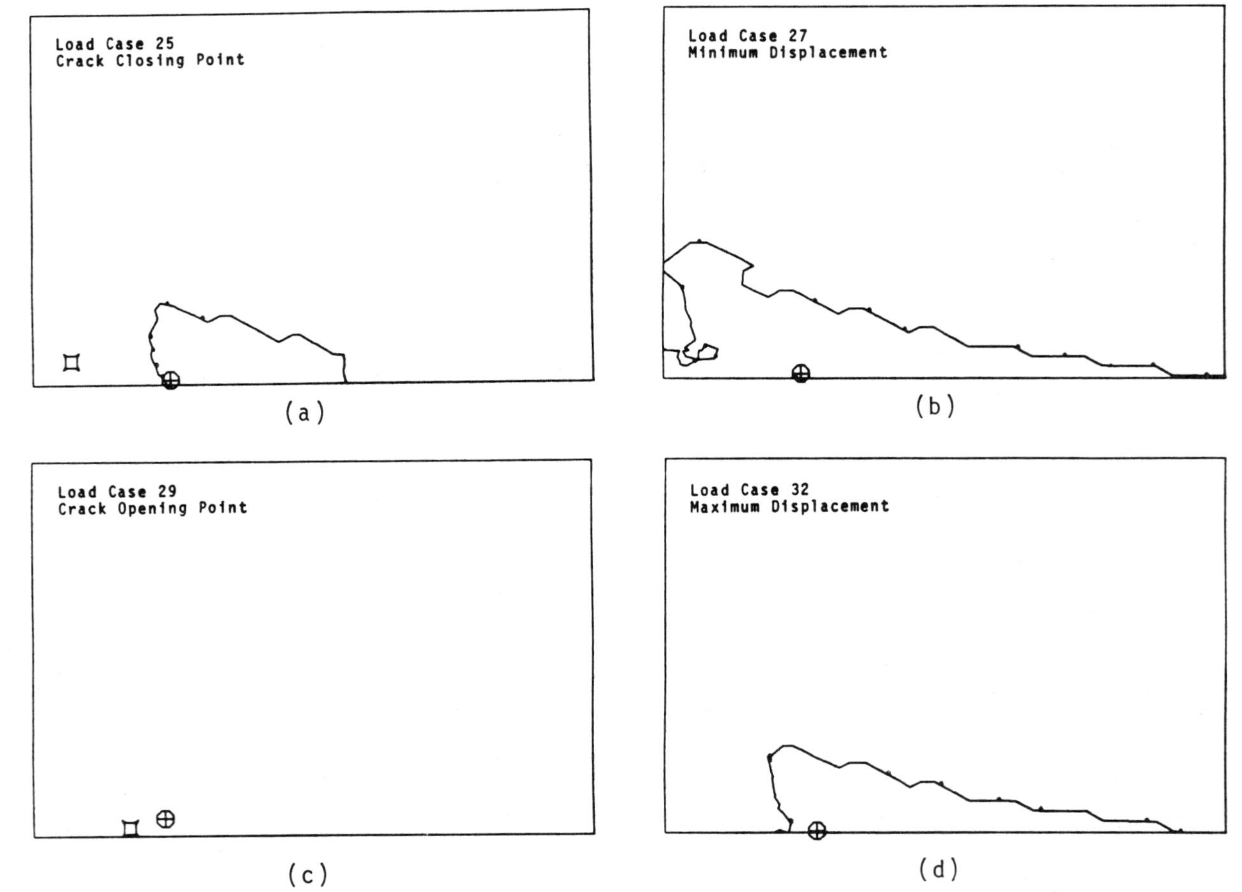

FIG. 19—*Plastic zones at (a) the closing point, (b) the compressive peak, (c) the opening point, and (d) the tensile peak for 538°C, strain range of 1.15%, $R_\epsilon = -1$ and crack length of 2.54 mm.*

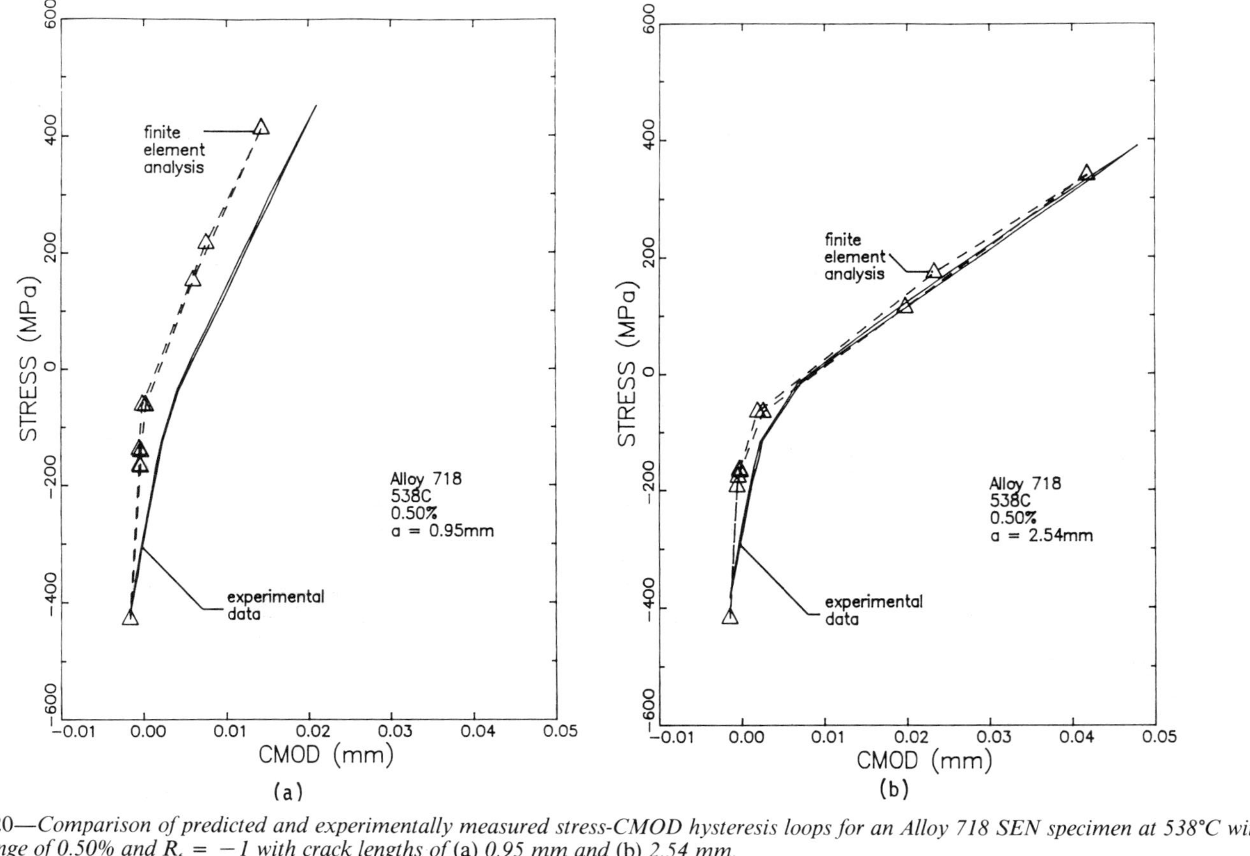

FIG. 20—*Comparison of predicted and experimentally measured stress-CMOD hysteresis loops for an Alloy 718 SEN specimen at 538°C with a strain range of 0.50% and $R_\epsilon = -1$ with crack lengths of (a) 0.95 mm and (b) 2.54 mm.*

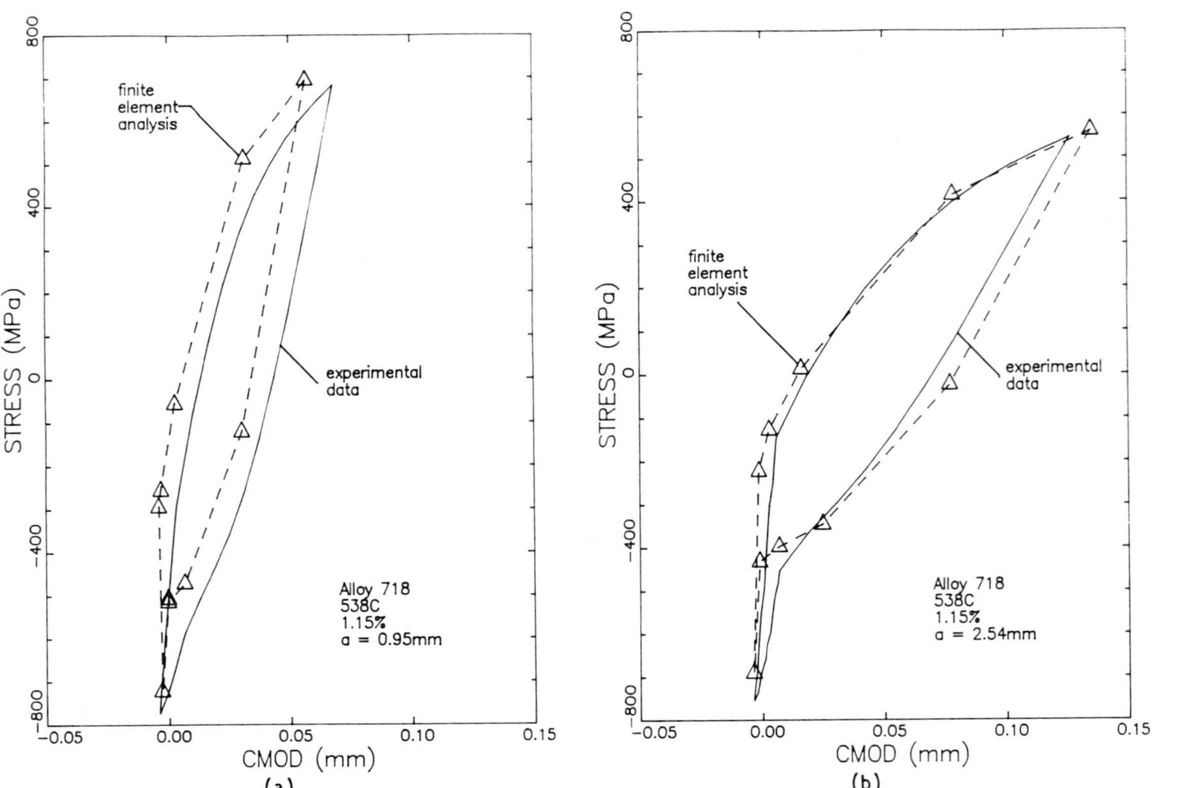

FIG. 21—*Comparison of predicted and experimentally measured stress-CMOD hysteresis loops for an Alloy 718 SEN specimen at 538°C with a strain range of 1.15% and $R_\epsilon = -1$ with crack lengths of (a) 0.95 mm and (b) 2.54 mm.*

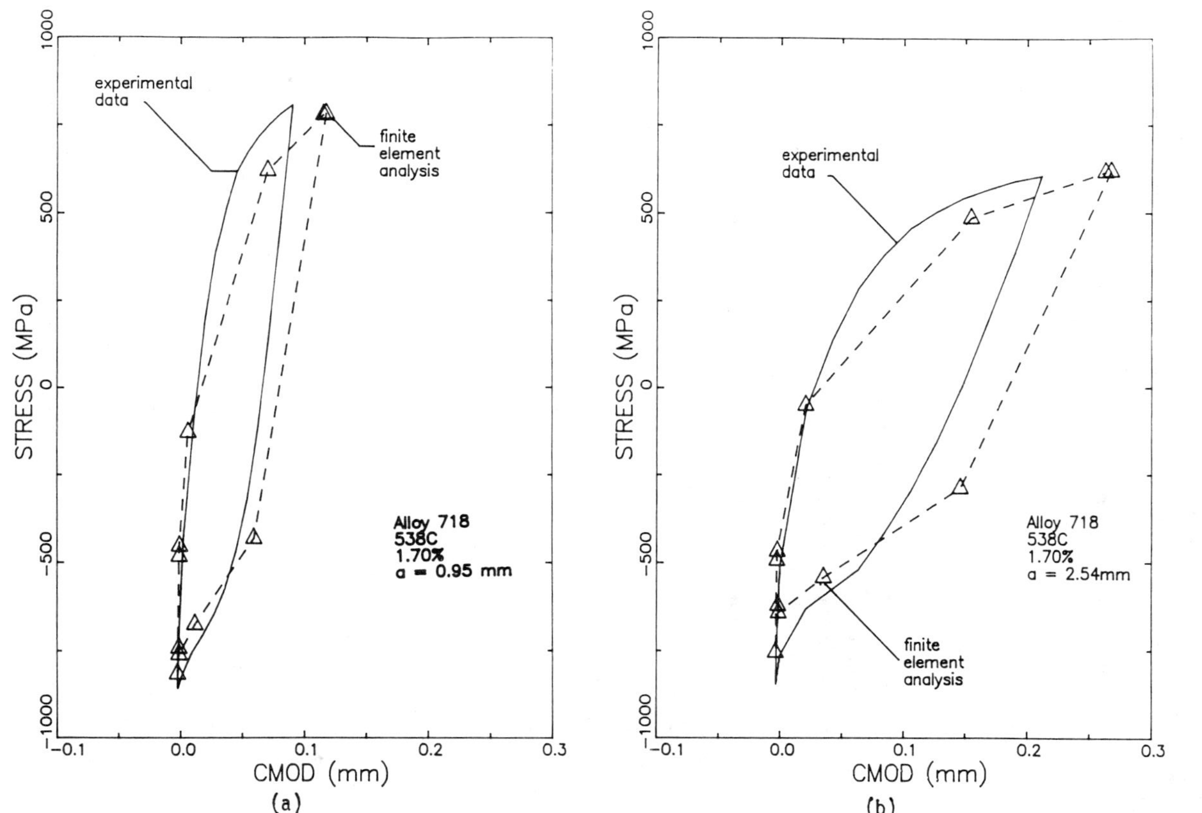

FIG. 22—*Comparison of predicted and experimentally measured stress-CMOD hysteresis loops for an Alloy 718 SEN specimen at 538°C with a strain range of 1.70% and $R_\epsilon = -1$ with crack lengths of (a) 0.95 mm and (b) 2.54 mm.*

The test data (Fig. 6) show that the crack closure and opening loads increase and converge as the crack propagates. Since the maximum load decreases with increasing crack length in these strain control tests, the range of load over which the crack is open decreases drastically as the crack becomes longer. This results in increasing the ratio of crack opening load range to the total load range as the crack propagates.

Agreement between the predicted and observed hysteresis loops of the stresses and displacements at the positions of the control and backface extensometers was quite good. The only major discrepancy between the experiment and analysis was the prediction of cusps at the closing and opening points; in the experimental data, the cusps were only slightly visible in the control displacement loops and virtually nonexistent in the backface displacement loops. It is conjectured that the plane stress assumption in the analysis is related to this deviation.

Concluding Comments

This is the first study to perform a combined experimental and analytical evaluation of crack closure. The deformation associated with crack growth in a single edge notch specimen subject to nominally elastic and elastic-plastic loading at an elevated temperature was analyzed. A finite element procedure was developed to model the crack growth and crack closure where the contact condition of the crack surface is examined in each plasticity iteration. This technique allows determination of the crack closure accurately without regard to the loading step size, provided that there is not excessive nonproportional loading. This situation is relevant to most standard test specimens and loading conditions. The experimentally measured remote displacements were used to develop remote boundary conditions for the analysis, and the load and CMOD were used to validate the results of analysis.

Cyclic crack growth was simulated using a single load step from peak to peak and advancing the crack by releasing nodes at the tensile peaks. Two nodes were released at each tensile peak rather than a single node as performed in other investigations. This created a minor disturbance in the distribution of residual plasticity along the crack, but analysis of the load versus CMOD hysteresis loop showed good agreement between predicted and experimentally measured crack closure and opening loads. Good agreement was also found between the experimentally observed and analytically predicted load and CMOD values in the crack growth simulation.

The results of load versus CMOD hysteresis loop analyses show excellent agreement with the experimental data. It was observed that increasing plasticity resulted in crack opening and closing at more compressive stresses for the $R_t = -1$ loading used in this study. This implies that the role of the crack closure on the crack growth becomes less significant with increasing net section plasticity. It was also shown that the influence of crack closure becomes more significant as the crack propagates under a given value of remote displacement range.

A salient feature of the two-tiered analysis approach was that the crack closure analysis (refined loading step run) can be performed at selected values of crack length by restarting the crack growth simulation run (coarse loading step run) rather than by evaluating all the cycles using small load increments. Consequently, there can be a significant reduction in computational effort and cost if the closure and opening loads are not required for each cycle. The automated data acquisition procedure was invaluable in generating sufficient data to analyze any portion of the test.

Acknowledgments

The development of crack closure analysis capability in the finite element code was supported by the Independent Research and Development Program of GE Aircraft Engines. The crack growth experiments and simulation were performed under funding provided by the NASA Lewis Research Center under Contract NAS3-23940. The authors are also indebted to E. D. Deaton and D. Miller, who conducted the crack growth experiments.

References

[1] Kim, K. S. and Orange, T. W., "A Review of Path-Independent Integrals in Elastic-Plastic Fracture Mechanics," in *Fracture Mechanics: Eighteenth Symposium, ASTM STP 945,* American Society for Testing and Materials, Philadelphia, 1988, pp. 713–729.

[2] Elber, W., "The Significance of Fatigue Crack Closure," in *Damage Tolerance in Aircraft Structures, ASTM STP 486,* American Society for Testing and Materials, Philadelphia, 1971, pp. 230–242.

[3] Suresh, S. and Ritchie, R. O., "Near-Threshold Fatigue Crack Propagation: A Perspective on the Role of Crack Closure," in *Fatigue Crack Growth Threshold Concepts,* D. Davidson and S. Suresh, Eds., The Metallurgical Society of AIME, 1984.

[4] Kreuger, D. D., Antolovich, S. D., and Van Stone, R. H., *Metallurgical Transactions,* Vol. 18A, 1987, pp. 1431–1449.

[5] Van Stone, R. H. and Krueger, D. D. in *Fracture Mechanics: Nineteenth Symposium, ASTM STP 969,* American Society for Testing and Materials, Philadelphia, 1988, pp. 883–906.

[6] Newman, J. C., Jr., "A Finite-Element Analysis of Fatigue Crack Closure," in *Mechanics of Crack Growth, ASTM STP 590,* American Society for Testing and Materials, Philadelphia, 1976, pp. 281–301.

[7] Newman, J. C., Jr., "Finite-Element Analysis of Crack Growth under Monotonic and Cyclic Loading," in *Cyclic Stress-Strain and Plastic Deformation Aspects of Fatigue Crack Growth, ASTM STP 637,* American Society for Testing and Materials, Philadelphia, 1977, pp. 56–80.

[8] Ohji, K., Ogura, K., and Ohkubo, Y., "Cyclic Analysis of a Propagating Crack and its Correlation with Fatigue Crack Growth," *Engineering Fracture Mechanics,* Vol. 7, 1975, pp. 457–464.

[9] Ogura, K. and Ohji, K., "FEM Analysis of Crack Closure and Delay Effect in Fatigue Crack Growth under Variable Amplitude Loading," *Engineering Fracture Mechanics,* Vol. 9, 1977, pp. 471–480.

[10] Nakagaki, M. and Atluri, S. N., "Elastic-Plastic Analysis of Fatigue Crack Closure in Modes I and II," *AIAA Journal,* Vol. 18, No. 9, 1980, pp. 1110–1117.

[11] Lalor, P. L. and Sehitoglu, H., "Fatigue Crack Closure Outside a Small-Scale Yielding Regime," in *Mechanics of Fatigue Crack Closure, ASTM STP 982,* American Society for Testing and Materials, Philadelphia, 1987, pp. 342–360.

[12] McClung, R. C., "Fatigue Crack Closure and Crack Growth Outside the Small Scale Yielding Regime," Ph.D. thesis, Department of Mechanical and Industrial Engineering, University of Illinois at Urbana-Champaign, 1987.

[13] Besseling, J. F., "A Theory of Plastic Flow for Anistropic Hardening in Plastic Deformation of an Initially Isotropic Material," Report S 410, National Aeronautical Research Institute, Amsterdam, 1953.

[14] Kim, K. S., Van Stone, R. H., Malik, S. N. and Laflen, J. H., "Elevated Temperature Crack Growth," Final Report, NASA Contract NAS3-23940, General Electric Aircraft Engines, 1988.

[15] Cook, T. S. in *Mechanical Testing for Deformation Model Development, ASTM STP 765,* American Society for Testing and Materials, Philadelphia, 1982, pp. 269–283.

[16] Gangloff, R. P., *Fatigue of Engineering Materials and Structures,* Vol. 4, 1982, pp. 15–31.

[17] Wilcox, J. R. and Henry, M. F., General Electric Corporate Research and Development, Schenectady, N.Y., 1982, unpublished research.

[18] Johnson, H. H., *Materials Research and Standards,* Vol. 5, 1965, pp. 442–445.

[19] Krueger, D. D., "Effects of Grain Size and Precipitate Size on the Fatigue Crack Growth Behavior of Alloy 718 at 427°C," M.S. thesis, University of Cincinnati, Ohio, 1984.

[20] "Standard Test Method For Measurement of Fatigue Crack Growth Rates," E 647-86a, *1987 Annual Book of ASTM Standards,* Vol. 03.01, pp. 899–926; also available separately.

[21] Krieg, R. D. and Krieg, D. B., "Accuracies of Numerical Solution Methods for the Elastic-Perfectly Plastic Model," *Journal of Pressure Vessel Technology,* Nov. 1977, pp. 510–515.

[22] Schreyer, H. L., Kulak, R. F., and Kramer, J. M., "Accurate Numerical Solutions for Elastic-Plastic Models," *Journal of Pressure Vessel Technology,* Aug. 1979, pp. 226–234.

F. W. Brust,[1] M. Nakagaki,[2] and P. Gilles[3]

Comparison of Elastic-Plastic Fracture Mechanics Techniques

REFERENCE: Brust, F. W., Nakagaki, M., and Gilles, P., **"Comparison of Elastic-Plastic Fracture Mechanics Techniques,"** *Fracture Mechanics: Twenty-First Symposium, ASTM STP 1074,* J. P. Gudas, J. A. Joyce, and E. M. Hackett, Eds., American Society for Testing and Materials, Philadelphia, 1990, pp. 448–469.

ABSTRACT: The first part of this paper reviews the adequacy of simple estimation procedures for predicting the crack growth and failure behavior of cracked structures. These simple methods include: J-estimation schemes (where five different schemes are presented including two new methods), the R6 approach, and the V_R-curve technique. The second part of the paper discusses methods for predicting elastic-plastic fracture under conditions where the simple schemes are not applicable. This includes a discussion of integral parameters, mainly focusing on the T^*-integral, and the local (or damage) based approaches. These two approaches are required where the degradation near a crack tip depends on the history of damage accumulation.

KEY WORDS: integral parameters, T^*, J-integral, estimation schemes, R6, V_R-curve, local approach, damage, crack growth, elastic-plastic, fracture, history dependence, cracked pipe

This paper has two objectives: (1) to present a review and theoretical discussion of available nonlinear fracture mechanics parameters, and (2) discuss practical applications of these various parameters which reveal their respective strengths and weaknesses.

The numerous elastic-plastic fracture mechanics parameters which have appeared over the years will be discussed. The parameters which are discussed in detail are: J-integral estimation techniques, the CEGB R6 method, the V_R-curve approach, the numerous crack tip integral parameters which have appeared in recent years including T^* and $\hat{J}$, and the various local approaches based upon damage mechanics. This review will investigate the theoretical basis and corresponding limitations of the methods. The ease with which each method may be used will also be considered. It is noted that, while the other methods may be subject to less theoretical restrictions, J-estimation techniques, the R6 method, and the V_R-curve approach are the only current approaches which do not require detailed numerical analyses.

The first portion of this paper discusses simple engineering methods to predict elastic-plastic crack growth behavior of structures subjected to overloads. Included is a discussion of five different J-estimation methods for through-wall cracked pipes, including two new methods. This first portion also discusses the R6 and V_R-curve approaches and compares some predictions with experimental data.

The second part of the paper discusses fracture methods applicable to history-dependent loadings which lead to history-dependent crack tip damage accumulation. Here, integral parameters and local approaches are discussed and example predictions provided.

---

[1] Battelle Columbus Division, Columbus, OH 43201.
[2] Kyushu Institute of Technology, Iizuka-City, Japan.
[3] Framatome, Paris, France.

J-Estimation Techniques

Elastic-plastic crack growth and instability predictions based on the *J*-integral have proven to be very valuable for assessing the behavior of ductile cracked structures subjected to overloads. This method consists of first developing a material *J*-resistance curve from a laboratory specimen. Next, for the service structure whose crack growth performance is of concern, an analysis is performed assuming that the laboratory specimen resistance curve governs the service crack growth behavior. The analysis may be performed by calculating *J* via finite element analysis or from a *J*-estimation scheme. *J*-estimation scheme techniques have proven valuable in giving timely, cost effective, and conservative predictions of ductile crack growth behavior in service structures subjected to overloads. It should be noted that ductile fracture mechanics methods based upon *J* cannot assure accurate or conservative predictions for history-dependent loading, as discussed later. A rather complete discussion of the phenomenological theory of nonlinear fracture mechanics based on *J*-integral theory is given by Hutchinson [1].

This section discusses the philosophy behind the development of five *J*-estimation techniques for through-wall cracked pipes subjected to bending loads. Comparisons between prediction and theory are also provided.

Generally, in the development of a *J*-estimation scheme it is assumed that the generalized displacements and the *J*-integral may be divided into their elastic and plastic components; i.e.,

$$\delta_T^c = \delta_e^c + \delta_p^c \tag{1}$$

and

$$J = J_e + J_p \tag{2}$$

where δ_T^c, δ_e^c and δ_p^c are, respectively, the total, elastic, and plastic generalized load point displacements due to the presence of the crack. J_e and J_p are elastic and plastic components of *J*. In the elastic range, the displacement and the load are uniquely related. In addition, assuming that the deformation theory of plasticity holds, a unique relationship exists between generalized load (*P*) and the plastic component of the generalized displacement. That is,[4]

$$\delta_e^c = \delta_e^c(P) \tag{3}$$

$$\delta_p^c = \delta_p^c(P) \tag{4}$$

Equation 2 may also be written as

$$J = \frac{K_I^2}{E'} + J_p \tag{5}$$

where

$E' = E$ for plane stress, or
   $= E/(1 - \nu^2)$ for plane strain,
$K_I =$ (known) elastic stress intensity factor,
$E =$ elastic modulus, and
$\nu =$ Poisson's ratio.

---

[4] This form applies for a given cracked structure and material properties.

In addition:

$$J_{\mathrm{p}} = \int_0^{\delta_{\mathrm{p}}^c} \left(\frac{\partial P}{\partial a}\right)\bigg|_{\delta_{\mathrm{p}}^c} d\delta_{\mathrm{p}}^c \tag{6}$$

Hence J-estimation schemes simply reduce to determining the relationship between δ_{p}^c and P in Eq 4 as a function of crack size. Once this relationship is determined, Eq 6 can be used to find J_{p} and Eq 5 to find J.

For circumferentially through-wall cracked (TWC) pipes in bending, the generalized load is the bending moment and the generalized displacement is the corresponding load-point rotation. Currently five different J-estimation schemes exist for TWC pipes. Therefore the differences between the schemes amount to how Eq 4 is developed and, correspondingly, how Eq 6 is evaluated.

GE/EPRI Scheme

The GE/EPRI scheme [2] uses the finite element method to establish Eq 4. In addition, rather than using Eq 6 to determine J_{p}, the virtual crack extension method is used. The GE/EPRI method also used the finite element method to develop the elastic solutions for TWC pipes.

Paris/Tada Method

The Paris/Tada method interpolates between the known linear elastic and known rigid plastic solutions. Equation 4 is determined by applying an Irwin plastic zone correction to the elastic solution in a novel way. This technique is described elsewhere [3].

LBB.NRC Scheme

The Paris/Tada method depends on material properties through the flow stress and does not explicitly account for material strain hardening. The LBB.NRC technique modifies the Paris/Tada relation for Eq 4 by accounting for material strain hardening. This method is fully described elsewhere [4].

LBB.GE and LBB.ENG Methods

The remainder of this section describes two alternative J-estimation procedures referred to as the LBB.GE and LBB.ENG methods. The LBB.GE method uses the numerical results of the GE/EPRI Handbook [2] to establish Eq 4, and then uses Eq 6 directly to determine J_{p}. The LBB.ENG uses an engineering estimate of the reduced pipe compliance due to crack tip plasticity to establish Eq 4.

To illustrate that J-estimation schemes may be developed in a rather simple fashion, we briefly outline the development of these two methods before providing sample predictions. Both of these methods are more fully discussed elsewhere [5]. The elastic components of these solutions are based upon the work of Sanders [6], and results are compiled in equation form in [5]. Here we discuss the development of the plastic component of the solution.

A Ramberg-Osgood stress-strain relation is assumed to adequately describe the material constitutive relations. In applying the Ramberg-Osgood relation to the cracked pipe problem, it is necessary to relate the stresses to displacements. Ilyushin [7] showed that the field solutions to the boundary value problem involving a monotonically increasing load or displacement type parameter is "proportional"; that is, deformation-theory plasticity is assumed to be valid.

Moreover, the corresponding load-displacement relationship is proportional:

$$\phi_p^c = K' \alpha \left(\frac{\sigma}{\sigma_f}\right)^{n-1} \phi_e^c \tag{7}$$

where ϕ_p^c and ϕ_e^c are, respectively, the plastic and elastic kink angle components at the load point, K' is a constant,[5] α and n are Ramberg-Osgood coefficients, and σ_f is the flow stress, defined as one half the yield plus ultimate stress.

For the LBB.GE method, one may easily show that K' from Eq 7 is

$$K' = K^n/4(h_2/V_1) \tag{8}$$

$$K = \frac{\pi}{4 \, (\cos \theta/2 \, - \, 1/2 \, \sin \theta)} \tag{9}$$

where h_2 and V_1 are as tabulated in [2], and θ is one half the crack angle.

For the LBB.ENG method, the actual circumferentially through-wall cracked pipe is replaced by a pipe with a reduced thickness, t_e, extending over a short distance near the center of the pipe. The reduced section, t_e, is defined by setting the limit moment for the cracked pipe to be equivalent to the limit moment for the reduced section pipe. This assumption leads to K' from Eq 7:

$$K' = (t/t_e)^{n-1} \left(\frac{\pi}{4\hat{K}}\right)^n \tag{10}$$

$$\hat{K} = \int_0^{\pi/2} (\sin \theta)^{n+1/n} \, d\theta = \frac{\sqrt{\pi}}{2} \, \frac{\Gamma(1 + 1/2n)}{\Gamma(3/2 + 1/2n)} \tag{11}$$

where the gamma function is accurately described by Sterling's formula, and the t_e values are as given in [5]. Hence, with K' described by Eq 8 or 10 inserted into Eq 7, and this result inserted to Eq 6, the plastic component of J is established, and the estimation scheme is complete. Other simple estimation schemes which also do not require numerical solution are given in [8] for elasticity and [9,10] for ductile fracture in laboratory specimens.

J-Estimation Scheme Predictions

Sample predictions using all five J-estimation techniques for through-wall cracked pipes are presented in Figs. 1 to 3. In all cases, J-resistance curves were developed from small compact tension specimens cut from pipe. Because crack growth in the pipes was much greater than that for the compact specimens, the resistance curves were extrapolated. Details of extrapolation procedures, material properties, and so on are provided in [5] and the references discussed therein. In addition, [5] provides detailed analyses for a number of different stainless, carbon, and welded pipes.

---

[5] Since σ is proportional to M, Eq 7 gives the moment-rotation relationship for a cracked body governed by deformation theory plasticity.

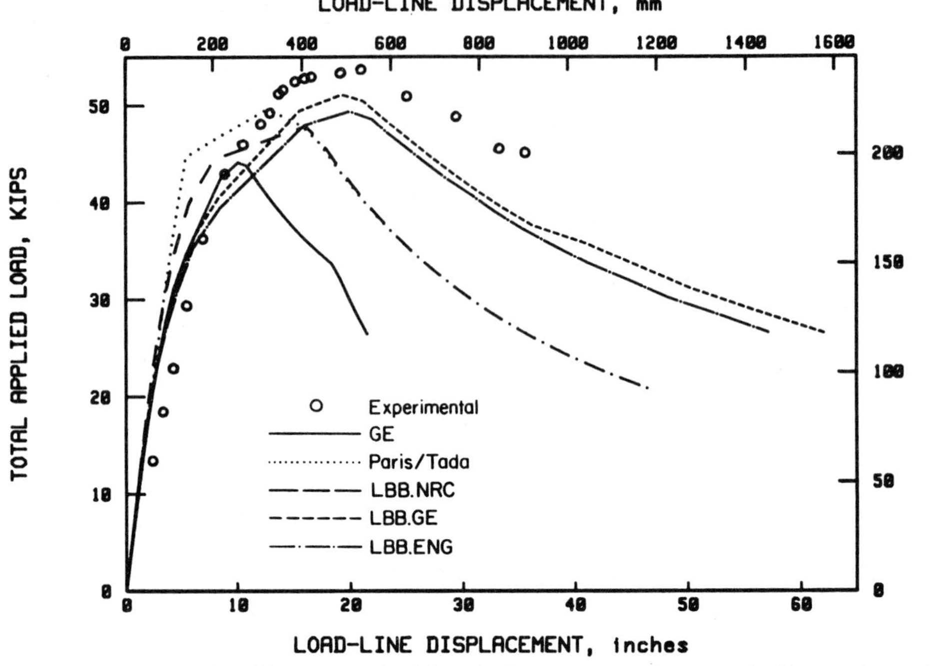

FIG. 1—*Predicted load versus load-line displacement record compared with experimental data from 102-mm (4-in.)-diameter Schedule 80 SA-376 TP304 stainless steel pipe tested at room temperature.*

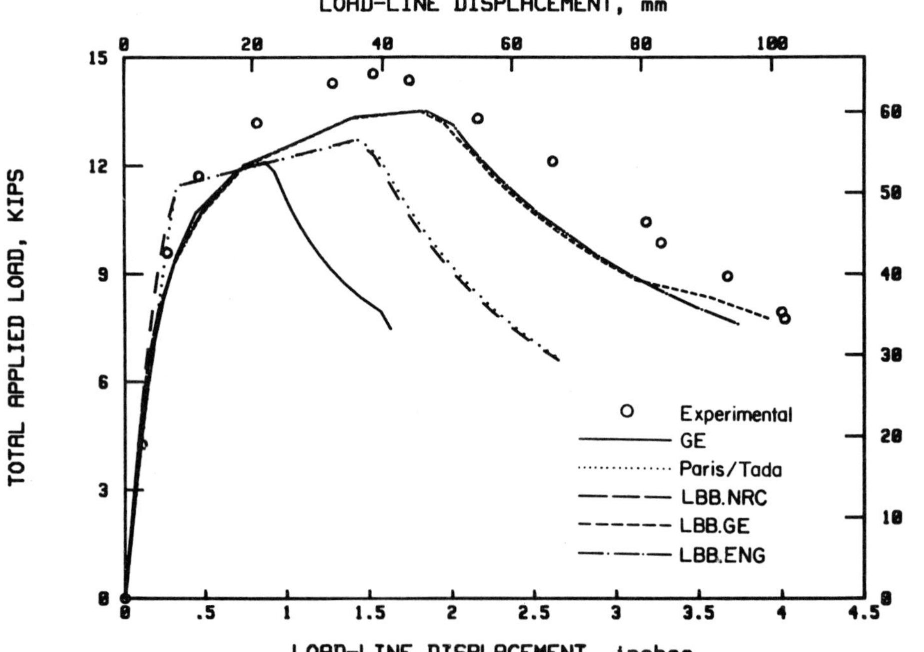

FIG. 2—*Predicted load versus load-line displacement record compared with experimental data from a 406-mm (16-in.)-diameter Schedule 100 SA-358 TP304 stainless steel pipe tested at room temperature.*

The results shown in Figs. 1 to 3, along with numerous analyses presented in [5] and [11], reveal that the simple *J*-estimation scheme analyses give conservative predictions of loads in through-wall cracked pipes in almost all cases examined to date. This is despite the fact that the predictions were made in crack growth ranges that are far beyond the limits for valid *J*-controlled growth established by Hutchinson and Paris [12].

Double-Criteria Approaches (DPFAD and R6 Rule)

The double-criteria approaches examined in this section, the CEGB procedure (or R6 rule [13,14]) and the Ductile Fracture Assessment Diagram [15], rely on the same foundations.

The R6 rule is an attempt to establish an EPFM criterion combining two different failure mechanisms: brittle fracture and plastic collapse. These two types of failure are predicted in these approaches by the following commonly used criteria:

1. The stress intensity factor (SIF) K should not exceed a critical value, i.e., the fracture toughness K_{Ic} which is a material characteristic depending on environmental conditions (temperature, irradiation, etc.).

2. The load must be less than the limit load of the flawed structure. This load corresponds to collapse; buckling phenomena are excluded.

Theoretical Foundations

Dowling and Townley [16] introduced the two-criteria approach, stating that structural failure occurs for the lower of either the critical load calculated by LEFM theory or the ulti-

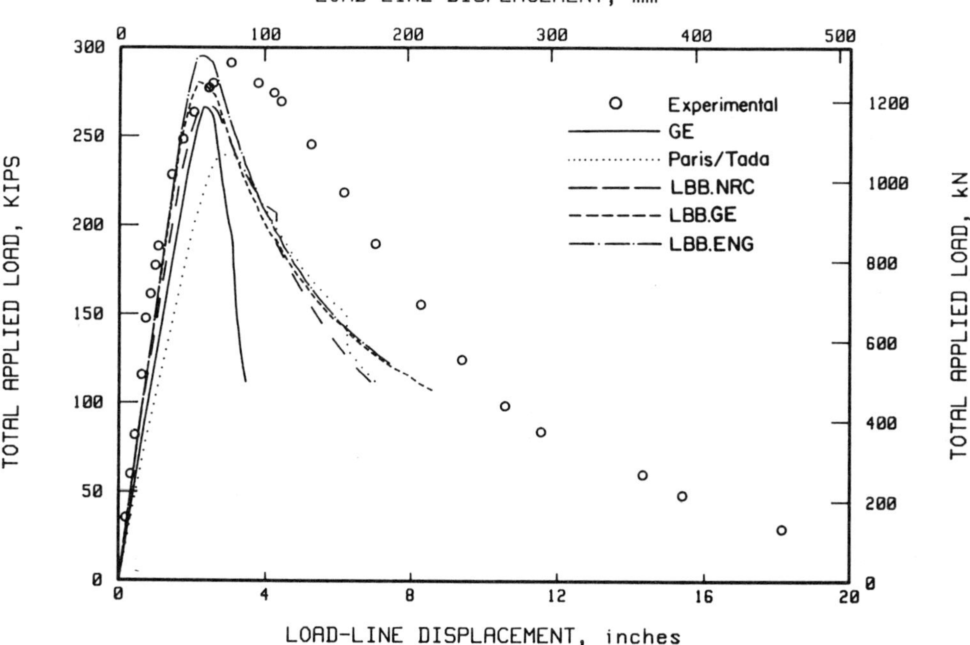

FIG. 3—*Predicted load versus load-line displacement record compared with experimental data from a 1067-mm (42-in.)-diameter, 15.9-mm (0.625-in.)-thick API 5LX65 carbon steel pipe tested at ambient temperature.*

mate collapse load. On the basis of experiments conducted in CT specimens and cracked vessels they supported this statement and evidenced an overestimate of failure load in the transition region between brittle fracture and collapse. Using the post-yield fracture mechanics formula of Heald et al. [*17*], they provided the following general failure criterion working in LEFM and collapse regions as well in transition regions:

$$\frac{L_F}{L_U} = \frac{2}{\pi} \cos^{-1} \left[\exp\left(- \frac{\pi^2 L_K^2}{8 L_U^2} \right) \right]$$

(12)

where L_F, L_U, and L_K represent respectively the failure load, the collapse load, and the failure load in LEFM theory. It may be useful to point out the limitations of this formula.

The formula is derived from the Dugdale-Barrenblat model which develops a plane stress solution in perfect plasticity ($\sigma \leq \sigma_0$) for an infinite centrally cracked plate subjected to a remote uniform stress distribution. This solution gives the opening at the crack tip:

$$\Delta_p = - \frac{4}{\pi\mu(1 + \nu)} \sigma_0 \, a \, \log \left[\cos\left(\frac{\pi}{2} \frac{\sigma}{\sigma_0} \right) \right]$$

(13)

where a is the half crack length, μ is the shear modulus, ν is Poisson's ratio, and σ_0 represents the yield stress.

For a small-scale yielding (i.e., σ/σ_0 small):

$$\Delta_p \rightarrow \Delta_{ssy} = \frac{4}{\pi\mu(1 + \nu)} \sigma_0 \, a \left[\frac{1}{2} \left(\frac{\pi}{2} \frac{\sigma}{\sigma_0} \right)^2 \right]$$

(14)

Recalling that

$$\Delta_{ssy} = \frac{K^2}{E\sigma_0}$$

(15)

one may define an equivalent SIF K_J from

$$\Delta_p = \frac{K_J^2}{E\sigma_0}$$

(16)

and write the ratio of these two CTOD as

$$\frac{\Delta_{ssy}}{\Delta_p} = \frac{K^2}{K_J^2} = -\frac{1}{2} \left(\frac{\pi}{2} \frac{\sigma}{\sigma_0} \right)^2 \frac{1}{\log\left(\cos \frac{\pi}{2} \frac{\sigma}{\sigma_0} \right)}$$

(17)

If $\sigma/\sigma_0 \rightarrow 0$, this ratio goes to one, and the brittle fracture criterion $K = K_{Ic}$ may be rewritten as

$$\frac{K^2}{K_{Ic}^2} = -\frac{1}{2} \left(\frac{\pi}{2} \frac{\sigma}{\sigma_0} \right)^2 \frac{1}{\log\left(\cos \frac{\pi}{2} \frac{\sigma}{\sigma_0} \right)}$$

(18)

This is the formula that Heald et al. [17] and then Dowling and Townley [16] proposed to extend to post-yield fracture, the right-hand side of Eq 7 being understood as a function reducing the toughness from K_{Ic}^2 to 0 as σ comes close to σ_0. The first R6 procedure (R6.1) uses this criterion [13].

Revised R6 Formulations

The aforementioned assumptions do not extend very well to three-dimensional cracked structures, and the physical basis for this two-criteria approach is very narrow. However, the simplicity of the procedure, which does not require any elastic-plastic finite element computations, led the CEGB to develop the so-called R6 rule from the original report number [13]. Other authors proposed similar approaches, deriving the criterion from EPFM analyses or experiments. Reference 18 shows an interesting comparison between a number of these models.

The latest revision of the R6 rule, R6.3 [14], retains the two following criteria:

$$K_r \equiv \frac{K}{K_{Ic}} \leq f(L_r) \tag{19}$$

$$L_r \equiv \frac{P^p}{P_y} \leq L_r^{max} \tag{20}$$

These conditions are represented on the failure assessment diagram (FAD) schematically drawn in Fig. 4.

K is the sum of K^p and K^s corresponding respectively to the loads giving rise to primary and secondary stresses. A plastic zone size correction is applied to K^s in order to take into account interactions between the two types of stresses.

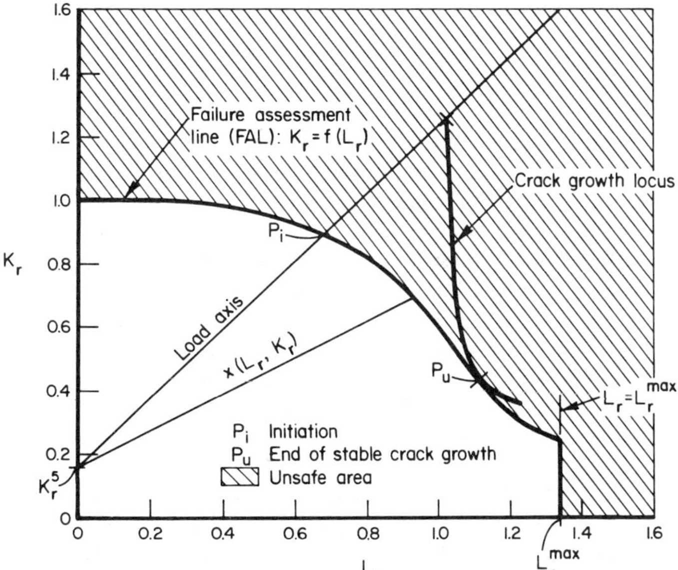

FIG. 4—*Failure assessment diagram (FAD).*

The L_r parameter is the ratio of the *primary* load to the *plastic yield load* (P_y) of the flawed structure. In earlier versions, as in Dowling and Townley [16], the S_r ratio of primary load to plastic collapse load is used instead of L_r. L_r^{max} corresponds to plastic collapse. The limit value of K_r is a function of L_r which depends on the material and the geometry.

This illustrates the empirical nature of the R6 rule and has led the CEGB to define three options for the failure assessment line (FAL) described by $K_r = f(L_r)$. These options allow the user to take into account, with increasing precision, work hardening and geometrical effects. Each of these options, which will not be shown here, essentially allows a different form for $f(L_r)$ given in Eq 19. See [14] and [19] for details.

For a given configuration (material, geometry, and type of loading) the R6 procedure allows one to check if the cracked structure is safe or not, to find the initiation load for a given loading path, and to evaluate the amount of stable crack growth.

As long as the assessment point K_r,L_r is inside the domain bounded by the axes of the FAD and the FAL (Fig. 4), the integrity of the structure is ensured. For a given cracked structure, the crack initiates when the assessment point reaches the FAL. This requires one to compute, for each loading step, primary and secondary stresses and the related K_r and S_r. In the case of a proportional loading the assessment point moves on a straight line which allows one to easily determine the initiation state.

Extensive comparisons with experimental data and elasto-plastic computations have been conducted to ensure the validity of the FAL [14]. R6 and other closely related double-criteria approaches present a considerable advantage for engineering analyses: they use only limit moment formulas and known elastic solutions for the cracked structure. This technique also leads to a visual representation of the type of failure that may occur in the structure and of the margins of safety. However, the user must be aware that a number of options are provided, and results may be sensitive to the choice of $f(L_r)$ or the limit load formula. Bloom's procedure [15] and R6, Rev. 3, Option 2 are the most reliable formulations.

Three main concerns associated with the R6-type approaches may be summarized as: (1) displacement solutions are not provided, (2) the separation between secondary and primary stresses is difficult in structures subjected to complex loading, and (3) the limit load formulas are often difficult to derive precisely.

V_R-*Curve Method*

Newman and co-workers [20–22] have developed a crack tip opening displacement resistance curve technique which appears promising as a simple engineering approach for predicting elastic-plastic crack growth behavior of flawed components. This technique is based on the Dugdale model. A V_R-resistance curve is developed, which depends only upon crack growth in the same fashion as a J-resistance curve, and is assumed to be a material property. To predict the crack growth and failure condition for a cracked structure of interest, one only needs to determine the crack driving force CTOD from a Dugdale-type analysis and follow the known resistance curve until failure is predicted.

This method is quite simple since only elastic finite element analyses need be compiled for the cracked structure of interest in order to determine the driving force crack tip opening displacement. From the results presented in [20–22], the V_R-curve appears to be independent of specimen configuration; that is, compact tension (CT), middle crack tension (MCT), and single edge notch (SEN) specimens produce identical V_R-curves. While we present no comparisons here between J-estimation techniques, the R6 method, and the V_R-curve approach, this new approach appears quite attractive from an applications viewpoint. This method, to our knowledge, has not been applied to extremely ductile stainless steels.

History-Dependent Crack Growth

The methods discussed earlier are engineering approaches. Numerous articles have been written over the years spelling out the theoretical concerns regarding these engineering approaches. However, a large data base now exists, and continues to grow, which clearly illustrates that these methods give reasonable predictions of the failure load for a cracked structure compared with experimental data. Moreover, engineering methods can be developed to ensure that predictive errors occur on the conservative side. These methods are easy to apply, give fast answers, and do not require a numerical or fracture mechanics expert to apply them. However, caution should be used when applying these methods in situations other than for assessing the effect of a single overload situation on a cracked structure.

In some situations the loading may be such that the accumulation of damage near the crack, which contributes to crack growth and eventual failure, may depend on load history. Cracked components which experience time-dependent straining, low cycle fatigue, proof testing, etc., may require monitoring of the crack tip damage accumulation during the load history to admit accurate crack growth and failure predictions. The foregoing engineering approaches, as they currently stand, cannot monitor load history dependent damage accumulation adequately.

The latter portion of this paper discusses two techniques for predicting load history dependent damage accumulation due to the presence of a crack: integral parameters and the local approach.

Integral Parameters

A number of integral parameters have been defined in recent years for application to nonlinear fracture mechanics. Blackburn [23] defined an integral, J^*, and later provided applications [24,25]. The $\hat{J}$-integral of Kishimoto, Aoki, and Sakata [26] has been applied in the context of nonlinear fracture in [27], for example. Other integral parameters have been suggested by McClintock [28] and Watanabe [29,30]. The ΔT^* and T^* integrals [31,32] have also been studied in the context of nonlinear fracture [32–34].

A review of these integrals, along with a number of other integral parameters, has recently been provided by Kim and Orange [35]. Recent studies of the performance of the above five integrals in the context of thermal fracture [36] and creep [37] are soon to appear. Because of the authors' previous experience, the following paragraphs provide a discussion of the T^*-integral parameter and the corresponding philosophy behind its use. References 36 and 37 provide discussion of the other integral parameters.

The T^*-*Integral*

Consider Fig. 5a. Here we have a two-dimensional volume of material with an embedded crack of initial size a_0. This volume of material is bounded by a fixed material curve, Γ_f, and by a curve, Γ_ε, which encircles the current crack tip and moves with the crack tip having constant size and shape, with speed $\dot{a}$. Note that because Γ_ε moves, different material particles are within the volume $V - V_\varepsilon$ as a function of time. Willis [38] showed, from simple energy balance considerations, and for arbitrary constitutive relations (neglecting heat flux and inertial effects and for Mode I crack growth):

$$\int_{\Gamma_f} (\sigma_{ij}\dot{u}_i)n_j ds - \frac{D}{Dt}\int_{V-V_\varepsilon} W\, dV = \int_{\Gamma_\varepsilon} [W\dot{a} + \sigma_{ij}\dot{u}_i]\, n_j\, dS \qquad (21)$$

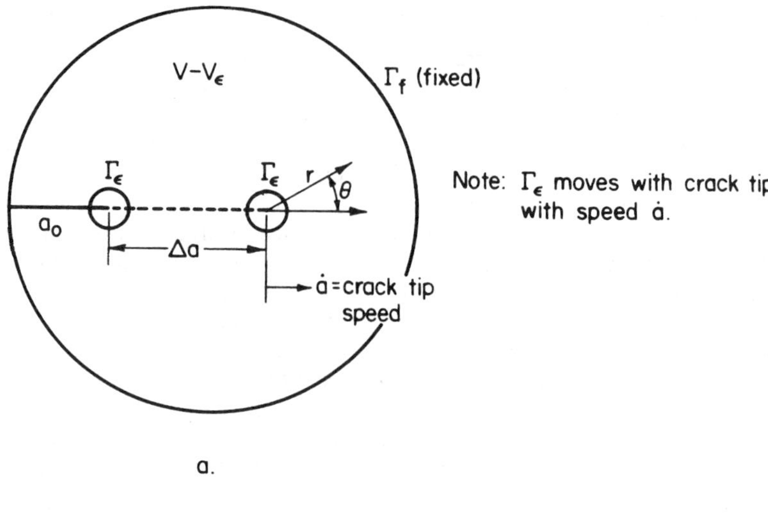

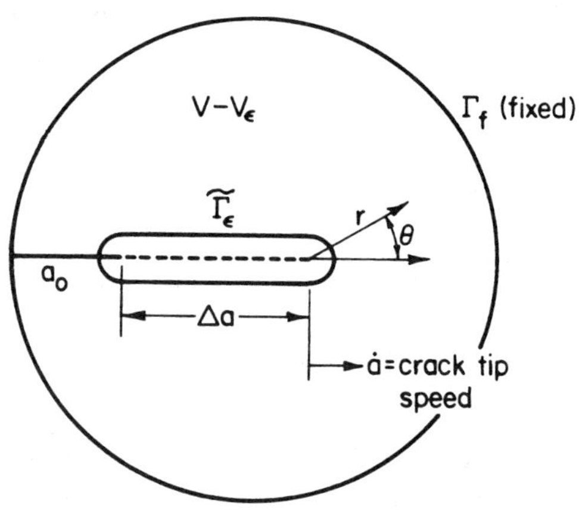

FIG. 5—*Integral path geometry.* (a) *Circular path.* (b) *Dugdale path.*

The first term on the extreme left-hand side of Eq 21 may be interpreted as the rate of energy input to $V - V_\varepsilon$, the second term represents the rate of change of energy within $V - V_\varepsilon$, while the term on the right-hand side represents the energy flux through Γ_ε. Hence, one may write for the energy lost through Γ_ε:

$$\dot{E} = \int_{\Gamma_\varepsilon} (\dot{a} W n_1 + \sigma_{ij}\dot{u}_i n_j)\, dS \tag{22}$$

Or, for the energy flux through Γ_ε per unit crack growth,

$$\dot{E}/\dot{a} = \int_{\Gamma_\varepsilon} \left(W n_1 + \frac{\sigma_{ij}\dot{u}_i n_j}{\dot{a}} \right) dS \tag{23}$$

If ε is taken to be small so as to be within a separable asymptotic crack field, one may write

$$\dot{u}_i \,\tilde{=}\, -\dot{a}u_{i,1} \tag{24}$$

Equation 23 may thus be rewritten as

$$\dot{E}/\dot{a} = \int_{\Gamma_\varepsilon} (n_1 W - n_j\sigma_{ij}u_{i,1})\, dS \tag{25}$$

For a circular path within a separable asymptotic field one may easily show (see, for example, Moran and Shih [39]) that $\dot{E}/\dot{a}$ in Eq 25 is independent of the Γ_ε path chosen. Hence, one may write

$$G = \dot{E}/\dot{a} = \lim_{\varepsilon \to 0} \int_{\Gamma_\varepsilon} (n_1 W - n_j\sigma_{ij}u_{i,1})\, dS \tag{26}$$

and Eq 26 represents the energy release rate to the crack tip per unit crack advance.

Let us now completely repeat the arguments leading to Eq 26, except now Γ_ε is defined as in Fig. 5b, and we refer to it as $\tilde{\Gamma}_\varepsilon$ here. As seen in Fig. 5b, $\tilde{\Gamma}_\varepsilon$ is circular at the initial crack size, a_0, and extends in a Dugdale fashion as the crack grows. In this case, the arguments of Eqs 21 to 24 lead to a different Eq 25 as

$$T^* \equiv \frac{\dot{E}}{\dot{a}} = \int_{\tilde{\Gamma}_\varepsilon} (n_1 W - n_j\sigma_{ij}u_{i,1})\, dS \tag{27}$$

Hence T^* and all the other integrals we discussed earlier are meant to be calculated along a "Dugdale" path. Only when T^* is calculated along a circular path such as in Fig. 5a does T^* become equivalent to G. Recall that the developments leading to Eqs 25 to 27 are valid for arbitrary constitutive relations.

When using Eq 27 calculated along a Dugdale path as in Fig. 5b, the major contribution to T^* comes from integrating Eq 27 along the circular portion of the path; i.e.,

$$T^* \cong \int_{-\pi/2}^{\pi/2} (n_1 W - n_j\sigma_{ij}u_{i,1})\, \varepsilon d\theta \tag{28}$$

as one might expect by noting that $n_1 = 0$ and $t_2 \,\tilde{=}\, 0$ along the long straight line portion of $\tilde{\Gamma}_\varepsilon$, and observing the terms within Eq 27. This was also verified numerically [32]. Hence we interpret T^* evaluated along a Dugdale path as the energy flow into a finite sized process zone near the current crack tip. $\tilde{\Gamma}_\varepsilon$ must be finite sized. Hence T^* is a function of ε. Implications of this ε dependence of T^* will be discussed shortly.

There is an important practical reason for requiring T^* to be evaluated on a Dugdale-type path. Consider crack growth to be modeled by a node release technique in a finite element model. If we consider Fig. 5a to be a finite element mesh, at the current crack tip, $a_0 + \Delta a = a$, two elements model the complete angular variation of the crack tip fields from 0 to π, with the coordinate system as in Fig. 5a or 5b. It is clear that this discretization is grossly inadequate and that the calculated stresses, strains, etc., in the growing wake region are incorrect. If one attempts to evaluate T^* along a circular path which passes through this wake, large errors are introduced into T^* (i.e., in effect, T^* cannot be calculated). Moreover, if one uses a moving crack tip mesh to model crack growth, the errors associated with mesh shifting and remeshing lead to similar errors if Γ_ε is circular [32].

A final comment about the choice of a Dugdale path is in order. If the stress state saturates to a finite value during crack growth in an elastic-plastic body, the energy release rate, G, from Eq 26 tends to zero [40]. Indeed, finite element modeling of elastic-plastic crack growth in strain hardening materials reveals that G is a function of crack growth increment size [41]. However, T^* evaluated on a Dugdale path is mesh size independent, as it should be. Figure 6, taken from [32], shows that T^* converges to a finite value, independent of mesh size. This is a very important feature of the Dugdale path. For this same mesh size increment [32], the energy release rate was strongly dependent on mesh size.

Importance of History Dependence

Finally, an example illustrating the importance of considering history dependence in fracture mechanics is presented. The example is taken from [32] and deals with stable elastic-plastic time-independent fracture. An ASTM A508 steel compact tension specimen was loaded beyond crack initiation, then completely unloaded and reloaded. Figure 7 shows that the crack growth reinitiation after unloading was accurately predicted using the T^*-parameter, which inherently accounts for the history effects of previous crack growth and inelastic deformation. The analyses based on J or CTOA (crack tip opening angle) parameters yielded nonconservative predictions.

This study is one of the few we are aware of that examines the history dependence of crack growth. For time-dependent deformation this problem has not been addressed. There is evidence that history effects may be more pronounced for creep crack growth compared with time-independent plasticity.

There is still much work needed before an integral parameter, whether T^* or any other, can be used with confidence for arbitrarily cracked structures. Indeed, issues such as what

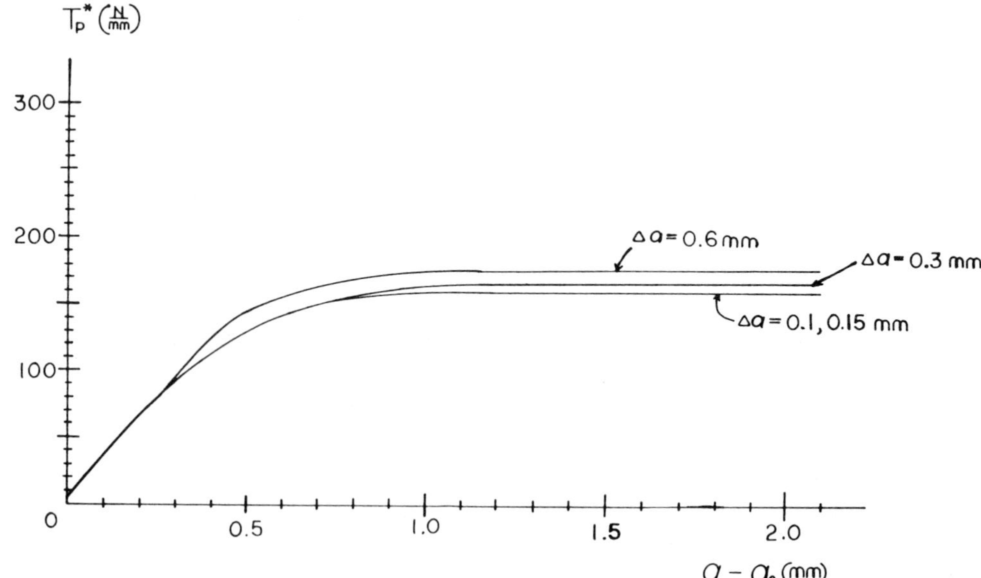

FIG. 6—*T^* resistance curves for different near field mesh refinement* [32].

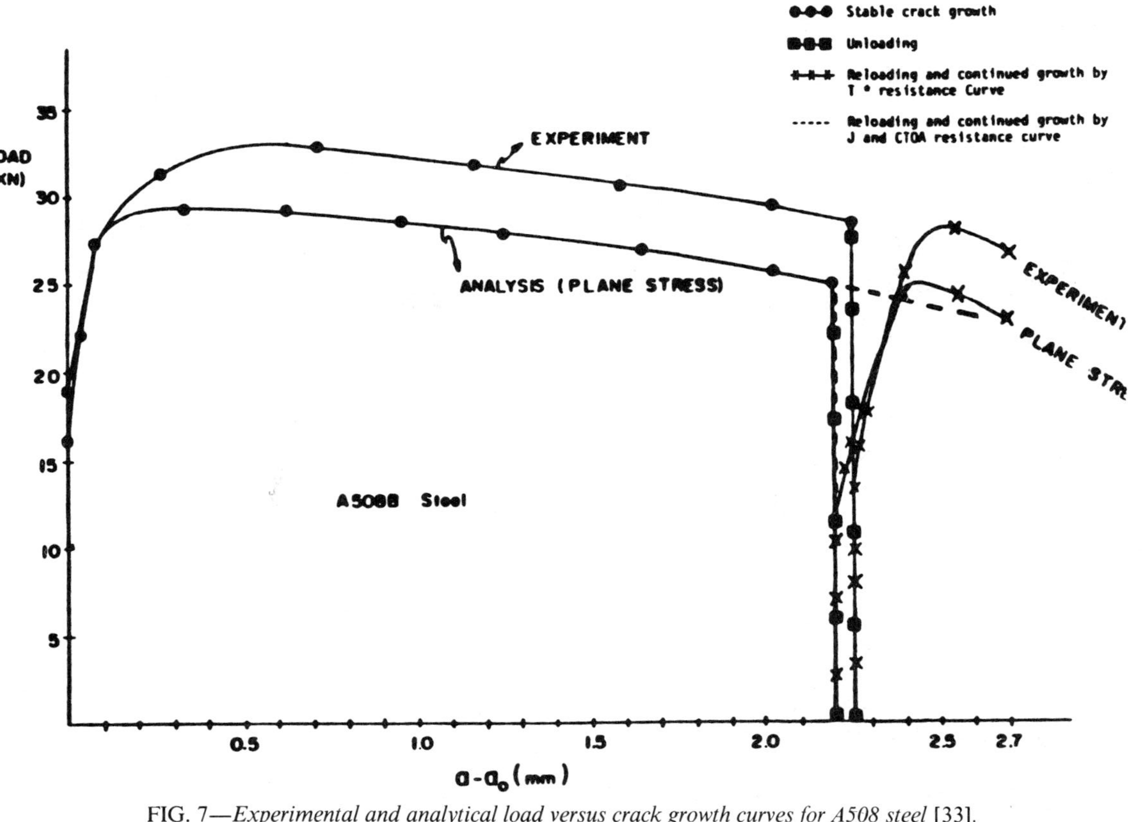

FIG. 7—*Experimental and analytical load versus crack growth curves for A508 steel* [33].

size ε should one choose when going from one cracked structure, where the T^* resistance curve was developed, to a real structure which may be much larger must be addressed. This issue and many more are dealt with in some detail elsewhere [42].

Local Approaches in Elastic-Plastic Fracture

Local approaches predict fracture at a point or in a small volume of the solid through a criterion involving local stresses and/or strains. The main feature of these approaches is that they take into account the microscopic breaking processes such as void nucleation and growth or microcracking, although staying in a framework of continuum mechanics. This is achieved through a damage function and the following three-step procedure:

1. The physical micromechanisms leading to fracture for a given material for given environmental conditions (temperature, humidity, radioactivity) are studied experimentally.
2. Based on these observations, a macroscopic model is established which describes the damage mechanisms inside a small volume (the scale depends on the material—e.g., for a metal the order of magnitude is 0.1 mm) as a function of local stresses and eventually strains.
3. The local criteria derived for each type of failure are introduced either in a post-processor of the stress and strain algorithm or directly embedded within the constitutive equations.

Local approaches were initially developed for fatigue initiation and creep crack growth [43,44], then extended during the 1970s to fatigue and ductile fracture by F. A. Leckie in Great Britain, J. Lemaitre and J. L. Chaboche in France, J. Hult in Sweden, ... These approaches are now flourishing as shown in a review article [45].

Classification of Local Approaches Used in Elastic-Plastic Fracture

Before describing in more detail local approaches related to elastic-plastic fracture, we elaborate on the basic problems to be solved using such models. The different scales at which micromechanisms take place and continuum mechanics apply are to be defined for each material. This is achieved through experimental observations and must be included in the model by defining a size characterizing the "small volume" where damage is estimated or by checking the convergence of the computations. The relative importance of each mechanism involved in the type of damage considered must be carefully analyzed in order to obtain a simple model representative of the main damage process and also take into account secondary fracture processes through some adjustments. Overly complex models require difficult experimental technique and may imply very delicate material characterizations.

For structures where fatigue is not a concern and for temperature ranges and materials where creep may be considered negligible, elastic-plastic fracture is mainly due to transgranular fracture, either via cleavage fracture or ductile rupture. In ferritic steels intergranular brittle fracture may exist as well as delamination for interface cracks, but most research work ignore these phenomena. The local modelization of cleavage or/and ductile mode of fracture may be classified in three categories: implicit and explicit formulations of damage and continuum damage mechanics.

In *implicit formulations of damage,* the process of microfracture is not described but represented by the evolution of a single parameter. The criterion in this case is given by equating the scalar damage parameter to a critical value. Cleavage is usually described by such a criterion [46]:

$$\text{maximal principal stress} = \sigma_c \tag{29}$$

But in the presence of high stress gradients in the front of a crack, this criterion is applied at a characteristic distance from the crack tip [47]. The critical stress is greater than yield stress because plastic flow is a necessary condition for cleavage fracture and is greater for smaller grain size. This approach is hampered by the scatter exhibited by experimental results and by the difficulty in physically justifying the notion of characteristic distance. The Weibull statistical approach may be used to handle the scatter ([47]; [48], Chapter VI).

A totally different approach using analytical fracture mechanics is presented elsewhere [49] for brittle dynamic fracture in Mode III. This theory shows that the height of the damaged layer surrounding the crack is a characteristic of the structure depending on fracture stress, crack velocity, and loading condition.

In the explicit formulations of damage, the micromechanisms are described by the criterion equations. Such a formulation has been derived for crack initiation and stable crack growth in ductile fracture. Again, the main difficulty stems from the very steep stress and strain gradients near the crack tip. A classic way to handle this is to average stress and strain values over a so-called process zone whose size is to be related to the spacing between inclusions. But, as explained in [47], this spacing depends on the particle's type and shape. Moreover, stress triaxiality which may strongly affect the critical material parameter is not taken into account.

With these considerations in mind, several researchers have attempted to analyze ductile fracture which is characterized by a dimpled fracture appearance. Three stages have been evidenced: void nucleation, growth, and coalescence. In an effort to be brief, we present here three examples of criteria and two models describing the whole process. More details can be found elsewhere [48–52].

Void nucleation in A508 Class 3 steel used for nuclear power plant vessels is described in [53] by the equation

$$\sigma_1 + k(\sigma_{eq} - \sigma_y) = \sigma_c \tag{30}$$

where σ_1 is the maximal principal stress, σ_{eq} is the von Mises stress, σ_y is the yield stress, σ_c is a critical stress, and k is a parameter depending on particle shape and direction of loading.

For void growth prediction the most successful models [54,55] are expressed through Rice and Tracey's formula [55] modified to account for strain hardening:

$$\ln\left(\frac{R}{R_0}\right)_c = \int_1^2 A \exp\left(B \frac{\sigma_m}{\sigma_{eq}}\right) d\varepsilon_{eq}^P \tag{31}$$

where R is the void radius, R_0 is its initial value, state 1 corresponds respectively to void formation and state 2 to fracture, σ_m/σ_{eq} is the ratio of the first two stress tensor invariants, and ε_{eq}^P is the von Mises equivalent plastic strain. The constant A typically varies between 0.18 and 0.7 and $B \simeq 1.5$. Experimental results show the validity of such a relationship ([48], p. 265) and confirm the value of B. The factor A depends strongly on the material and may be related to the initial distribution of particles ([47]; [48], p. 276). This critical void growth rate $(R/R_0)_c$ can be considered as a material constant for a given loading direction if the stress triaxiality is not too low [48].

The third stage, void coalescence, is difficult to understand in spite of the large amount of experimental work conducted in the past decade (see [47] for a review). Coalescence may be interpreted as the result of a local necking and may be described by a plastic deformation model. More successful approaches include this effect in constitutive equations (see next section and [55–57]), but the decrease in the distance between voids must be expressed by modifying the previous void growth criterion:

$$\frac{R\ell_0}{R_0\ell} = \frac{R}{R_0} \times \min \exp \varepsilon = \text{critical value} \tag{32}$$

In this formula the distance between voids is meant to vary with the mean strain in the damaged small volume ([58]; [48], p. 264).

In practical applications the first stage is usually neglected because the plastic strains creating voids are small in comparison to those occurring at fracture, and the model based on Eq 32 has been applied successfully to crack initiation and stable growth in Mode I for planar cracked geometries [58,59]. In these finite element analyses the "small volume" is taken as the first element in front of the crack. The size of this element ($\simeq 0.2$ mm) is a material characteristic chosen by convergence analysis [58] or related to the number of particles per unit volume [59]. The critical value $(R\ell_0/R_0\ell)_c$, which is close to 2, is obtained from tests on notched axisymmetrical bars in tension. Once this value is reached in the crack tip element, relaxation of nodes in several steps is applied through this element. This procedure is repeated as the load is increased; hence the load versus crack growth curve is predicted. Good agreement between analytical and experimental results on CT and axisymmetrical specimens has been obtained (see Fig. 8 taken from [48], p. 268), even though crack growth is slightly underestimated for axisymmetric specimens. The maximum load computed in plane strain is 10% less than the experimental one.

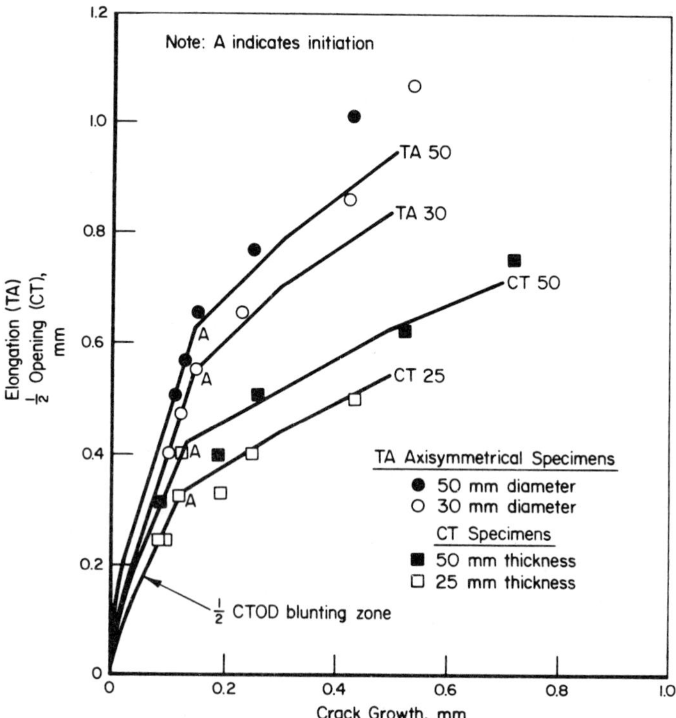

FIG. 8—*Comparison of computed and measured crack growth for four specimens taken from A508 Class 3 steel at 100°C in the longitudinal direction* [48].

This model has been used to carry out more complex problems for strains evolving under nonproportional loading [60], mixed mode (I and II) fracture, and fracture in the ductile/brittle transition region (see [48], p. 269 for references).

Continuum damage mechanics approaches constitute the third type of local approach. Following the earliest work of Kachanov [44], several scientists such as Lemaitre, Leckie, Hult, Chaboche, Murakami, and Krajinovic (see [61] for references) described damage phenomena in the general frame of thermodynamics of irreversible processes. This is a macroscopic approach based on defining internal variables to account for dissipation due to damage, formulation of constitutive equations governing the evolution of these variables, and derivation of mechanical equations for the damaged material.

We shall limit the following to a brief description of Lemaitre's [62,63] and Rousselier's [64] formulations of ductile fracture damage. In this case we have two internal variables: the strain hardening variable, p, and the damage variable, D. σ_{eq} designates the von Mises stress.

Lemaitre's model has the following features:

- It is based on an effective stress introduced by Kachanov [44]:

$$\overline{\sigma} = \sigma/(1 - D) \tag{33}$$

D varies from 0 to 1. This definition is completed by an equivalence principle which states that the damaged material behavior is described by the nondamaged material constitutive laws where σ is replaced by $\overline{\sigma}$.

- Elastic strain and damage are coupled in the thermodynamic potential equation

$$\phi = \phi_e(\varepsilon^e, D) + \phi_p(p) \tag{34}$$

- The plastic potential is expressed as

$$F = \overline{\sigma}_{eq} + P(p) + F_D(y) \tag{35}$$

and

$$y = -\frac{1}{2}\overline{\sigma} : \epsilon^e \tag{36}$$

- The damage evolution law

$$\left(\dot{D} = -\dot{p}\frac{\partial F_D}{\partial y} \right) \tag{37}$$

is given by a power law:

$$\dot{D} = \left(\frac{-y}{S_0} \right)^{s_0} \tag{38}$$

where S_0 and s_0 are material constants.

- The fracture criterion is $D = D_c$.

In Lemaitre's approach a critical value of damage has to be measured. The crack initiates when $D = D_c$ at the gauss points of the finite element mesh of a characteristic size. The crack is a series of elements in which critical damage is reached.

In a later paper [45] Lemaitre explains a preference for an energy criterion, but the need for a sufficiently refined mesh and then for a convergence study remains.

This model may be modified in order to take into account void nucleation.

Rousselier's approach is derived assuming isotropic damage and relies on five hypotheses:

- Classical thermodynamics is applied.
- Internal variables p and D are scalars.
- Damage is uncoupled; the thermodynamical potential is such that

$$\phi = \phi_e(\varepsilon^e) + \phi_p(p) + \phi_D(D) \tag{39}$$

- The plastic potential is given by

$$F = \frac{\sigma_{eq}}{\rho(D)} + P(p) + A \cdot \sigma_a \cdot f \cdot \exp\left(\frac{\sigma_1}{\rho\sigma_a}\right) \tag{40}$$

where ρ is the density of the damaged structure, P is the variable associated with the accumulated plastic strain p, σ_1 is the maximum principal stress, and f is the initial void volume fraction. A, σ_a, and f are material characteristics. The last term of Eq 40 has been derived from the main conservation law, the normality rule, and a fifth hypothesis on void growth.

In this model, as opposed to Lemaitre's, damage is coupled with strain through the plastic potential F and with the density ρ, but not with elastic strain. No equivalence principle is needed, and the damage evolution law is based on physical considerations.

Also with this approach, the crack tip elements have a characteristic size ℓ_c and are considered as totally damaged when the stresses collapse to zero at the gauss points. The author proposes a very convenient way to find ℓ_c: In a tensile test of an uncracked notched bar the sharp decrease of the slope of the load displacement curve corresponds to damage initiation, and in the numerical simulation the slope beyond this point is directly related to ℓ_c. Good agreement between these measured and computed slopes ensures the adequacy of the mesh.

Conclusion

Fracture mechanics methodology, we believe, should be as simple as possible to apply, while providing accurate, yet conservative, failure predictions. The reason for this is clear: Often engineers who are required to apply fracture mechanics methodology routinely do not have the time or the expertise to apply extremely complicated methods. However, the limitations of the simple approaches must be understood.

Herein, we have discussed the simple engineering methods: J-integral estimation schemes, the failure assessment diagram (R6) approach, and the V_R-resistance curve method. A large experimental data base exists which clearly shows that, for numerous different types of cracked structures, the J-integral and R6 methods provide accurate failure predictions. (The data base for the V_R-curve approach is still emerging.) Despite numerous theoretical concerns in the literature regarding these methods, they give good predictions for overload situations.

For situations where strain rate effects are important, or where load history affects the damage accumulation and hence crack growth and failure response, the simple schemes are inadequate. For these situations, the more complicated integral parameter and local approaches should be used. Integral parameter approaches, while still in the early stages of development, are easier to apply than local approaches, since material property requirements are much simpler and, at least in the case of T^*, are closely related to the J-integral approach.

On the other hand, local approaches are more developed in their theoretical foundation compared to integral approaches. These two detailed approaches need to be utilized to assist in the development of simple estimation schemes. Finally, it should be remarked that integral parameter approaches should perhaps be included under the category of a local approach.

References

[1] Hutchinson, J. W., "Fundamentals of the Phenomenological Theory of Nonlinear Fracture Mechanics," *Journal of Applied Mechanics,* Vol. 50, 1983, pp. 1042–1051.

[2] Kumar, V. et al., "Advances in Elastic-Plastic Fracture Analysis," EPRI Final Report NP-3607, Electric Power Research Institute, Palo Alto, Calif., Aug. 1984.

[3] Paris, P. C. and Tada, H., "The Application of Fracture Proof Design Methods Using Tearing Instability Theory to Nuclear Piping Postulating Circumferential Through Wall Cracks," NUREG/CR-3464, Sept. 1983.

[4] Klecker, R., Brust, F. W., and Wilkowski, G., "NRC Leak-Before-Break (LBB.NRC) Analysis Method for Circumferentially Through-Wall Cracked Pipes under Axial Plus Bending Loads," NUREG/CR-4572, May 1986.

[5] Brust, F. W., "Approximate Methods for Fracture Analyses of Through-Wall Cracked Pipes," NUREG/CR-4853, Feb. 1987.

[6] Sanders, J. L. R., "Circumferential Through-Crack in a Cylindrical Shell under Combined Bending and Tension," *Journal of Applied Mechanics,* Vol. 50, March 1983, p. 221.

[7] Ilyushin, A. A., "The Theory of Small Elastic-Plastic Deformations," *Prikladnaia Matematika I Mekhanika,* Vol. 10, 1946, p. 347.

[8] Herrmann, G. and Sosa, H., "On Bars with Cracks," *Engineering Fracture Mechanics,* Vol. 24, No. 6, 1986, pp. 889–894.

[9] Abeyaratne, R., "On the Estimation of Energy Release Rates," *ASME Journal of Applied Mechanics,* Vol. 50, 1983, pp. 19–23.

[10] Yang, J. S., "Estimation of Energy Release Rates: An Alternative to Interpolation," *ASME Journal of Applied Mechanics,* Vol. 52, March 1985, pp. 25–26.

[11] Scott, P. and Brust, F. W., "An Experimental and Analytical Assessment of Circumferential Through-Wall Cracked Pipes under Pure Bending," NUREG/CR-4574, Sept. 1986.

[12] Hutchinson, J. W. and Paris, P. C. in *Elastic-Plastic Fracture, ASTM STP 668,* American Society for Testing and Materials, Philadelphia, 1979, pp. 37–64.

[13] Harrison, R. P., Loosemore, K., and Milne, I., "Assessment of the Integrity of Structures Containing Defects," CEGB Report R/H/R6, 1976.

[14] Milne, I., Ainsworth, R. A., Dowling, A. R., and Stewart, A. T., "Assessment of the Integrity of Structures Containing Defects," CEGB Report R/H/R6, Rev. 3, 1986.

[15] Bloom, J. M. and Malik, S. N., "A Procedure for the Assessment of the Integrity of Nuclear Pressure Vessels and Piping Containing Defects," EPRI Report NP-2431, Research Project R37-2, June 1982.

[16] Dowling, A. R. and Townley, C. H. A., "The Effect of Defects on Structural Failure: A Two-Criteria Approach," *International Journal of Pressure Vessels and Piping,* Vol. 3, 1975, p. 77.

[17] Heald, P. T., Spink, G. M., and Worthington, P. J., "Post Yield Fracture Mechanics," *Materials Science and Engineering,* Vol. 10, 1972, p. 129.

[18] Taupin, Ph., Ferry, F., Bhandari, S., and Pellissier-Tandon, A., "Validation of Engineering Ductile Fracture Analysis Methods," in *Fatigue and Fracture Assessment by Analysis and Testing,* PVP, Vol. 103, 1986, p. 9.

[19] Ainsworth, R. A., "The Assessment of Defects in Structures of Strain Hardening Materials," *Engineering Fracture Mechanics,* Vol. 19, 1984, p. 633.

[20] Newman, J. C., "Prediction of Stable Crack Growth and Instability Using the V_R-Curve Method," in *Elastic-Plastic Fracture Mechanics Methodology, ASTM STP 896,* American Society for Testing and Materials, Philadelphia, 1986, pp. 139–166.

[21] Newman, J. C., "Evaluation of the Crack-Tip-Opening-Displacement (V_R) Resistance Curve Method," Workshop on CTOD Methodology, Geesthacht, W. Germany, 23–25 April 1985.

[22] Newman, J. C., McNeill, S. R., and Sutton, M. A., "Application of the V_R Resistance Curve Method to Fracture of Various Crack Configurations," in *Fracture Mechanics: Eighteenth Symposium, ASTM STP 945,* American Society for Testing and Materials, Philadelphia, 1988, pp. 103–117.

[23] Blackburn, W. S., "Path Independent Integrals to Predict Onset of Crack Instability in an Elastic Plastic Material," *International Journal of Fracture Mechanics,* Vol. 8, 1972, pp. 343–346.

[24] Batte, A. D., Blackburn, W. S., Elsender, A., Hellen, T. K., and Jackson, A. D., "A Comparison of the J^* Integral with Other Methods of Post Yield Fracture Mechanics," *International Journal of Fracture Mechanics,* Vol. 21, 1983, pp. 49–66.

[25] Blackburn, W. S., "Contour Integrals Around Crack Tips for Reversed Loading," *International Journal of Fracture Mechanics,* Vol. 8, 1985, pp. R73–R78.

[26] Kishimoto, K., Aoki, S., and Sakata, M., "On the Path-Independent Integral-$\hat{J}$," *Engineering Fracture Mechanics,* Vol. 13, 1980, pp. 841–850.

[27] Aoki, S., Kishimoto, K., and Sakata, M., "Elastic-Plastic Analysis of Crack in Thermally-Loaded Structures," *Engineering Fracture Mechanics,* Vol. 16, 1982, pp. 405–413.

[28] McClintock, F. A. in *Fracture 3,* H. Liebowitz, Ed., Academic Press, New York, 1971.

[29] Watanabe, K., "The Conservation Law Related to Path Independent Integral and Expression of Crack Energy Density by Path Independent Integral," *Bulletin of JSME,* Vol. 28, No. 235, Jan. 1985.

[30] Watanabe, K. and Tsugai, "Evaluation of Crack Energy Density by Using ϵ_J-Integral in Case of Elastic-Plastic and Creep Crack," *Bulletin of JSME,* Vol. 29, No. 248, Feb. 1986.

[31] Atluri, S. N., Nishioka, T., and Nakagaki, M., "Incremental Path-Independent Integrals in Inelastic and Dynamic Fracture Mechanics," *Engineering Fracture Mechanics,* Vol. 20, No. 2, 1984, pp. 209–244.

[32] Brust, F. W., Nishioka, T., Nakagaki, M., and Atluri, S. N., "Further Studies on Elastic-Plastic Stable Fracture Utilizing the T^* Integral," *Engineering Fracture Mechanics,* Vol. 22, No. 6, 1985, pp. 1079–1103.

[33] Brust, F. W., McGowan, J. J., and Atluri, S. N., "A Combined Numerical/Experimental Study of Ductile Crack Growth after a Large Unloading Using T^*, J, and CTOA Criteria," *Engineering Fracture Mechanics,* Vol. 23, No. 3, 1986, pp. 537–550.

[34] Brust, F. W. and Atluri, S. N., "Studies on Creep Crack Growth Using the T^* Integral," *Engineering Fracture Mechanics,* Vol. 23, No. 3, 1986, pp. 551–574.

[35] Kim, K. S. and Orange, T. W., "A Review of Path-Independent Integrals in Elastic-Plastic Fracture Mechanics," in *Fracture Mechanics: Eighteenth Symposium, ASTM STP 945,* American Society for Testing and Materials, Philadelphia, 1988, pp. 713–729.

[36] Brust, F. W., Nakagaki, M., and Springfield, C., "Integral Parameters For Thermal Fracture," submitted to *Engineering Fracture Mechanics.*

[37] Brust, F. W., Nakagaki, M., and Miyazaki, N., "Integral Parameters for Creep Fracture with Applications," to be submitted.

[38] Willis, J. R., "Equations of Motion for Propagating Cracks," in *The Mechanics and Physics of Fracture,* The Metal Society, 1975, pp. 57–67.

[39] Moran, B. and Shih, C. F., "Crack Tip and Associated Domain Integrals from Momentum and Energy Balance," *Engineering Fracture Mechanics,* Vol. 27, No. 6, 1987, pp. 615–642.

[40] Rice, J. R., "The Mechanics of Quasi-Static Crack Growth," in *Proceedings,* 8th U.S. National Congress of Applied Mechanics, R. E. Kelly, Ed., Western Periodicals, 1979, pp. 191–216.

[41] Kfouri, A. P. and Rice, J. R., "Elastic/Plastic Separation Energy Rate for Crack Advance in Finite Growth Steps," in *Fracture 1977,* D. M. R. Taplin, Ed., Vol. 1, University of Waterloo Press, 1977, pp. 43–59.

[42] Brust, F. W., Atluri, S. N., Nakagaki, M., and Nishioka, T., "The T^*-Integral: Theory, Analysis, and Application," to be submitted.

[43] Marco, S. M. and Starkey, W. L., "A Concept of Fatigue Damage," *Transactions of the American Society of Mechanical Engineers,* Vol. 76, 1954, pp. 627–632.

[44] Kachanov, L. M., "Time of the Rupture Process under Creep Conditions," *Izv. Akad. Nauk. SSR otd. Teckh. Nauk.,* No. 8, 1958, pp. 26–31.

[45] Lemaitre, J., "Local Approach of Fracture," *Engineering Fracture Mechanics,* Vol. 25, Nos. 5/6, 1986, pp. 523–537.

[46] Knott, J. F., *Journal of the Iron and Steel Institute,* Vol. 204, 1966, pp. 104–111.

[47] Pineau, A., "Review of Fracture Micromechanisms and a Local Approach to Predicting Crack Resistance in Low Strength Steels," in *Proceedings,* ICF5, D. Francois, Ed., 1981, pp. 553–577.

[48] Mudry, F. in *Proceedings,* International Seminar on Local Approach of Fracture, 3–5 June 1986, EDF Research Center, "Les Renardières" 77250 Moret/Loing (France), Chapter VI; see also *Nuclear Engineering and Design,* Vol. 105, No. 1, Dec. 1987.

[49] Bui, H. D. and Ehrlacher, A., "Propagation of Damage in Elastic and Plastic Solids," in *Proceedings,* ICF5, Vol. 2—Advances in Fracture Research, D. Francois, Ed., Pergamon Press, Oxford, 1981, pp. 533–551.

[50] Beremin, F. M., "Experimental and Numerical Study of the Different Stages in Ductile Rupture:

Application to Crack Initiation and Stable Crack Growth," in *Three Dimensional Constitutive Relations and Ductile Fracture,* S. Nemat-Nasser, Ed., North Holland, London, 1981, pp. 181–201.

[*51*] Rosenfield, A. R., *Metallurgical Reviews,* Vol. 52, 1968, p. 29.

[*52*] McClintock, F. A., *Fracture,* Academic Press, New York, 1971, p. 47.

[*53*] Beremin, F. M., *Metallurgical Transactions,* Vol. 12A, 1981, p. 723.

[*54*] McClintock, F. A., *International Journal of Fracture Mechanics,* Vol. 4, 1968, p. 101.

[*55*] Rice, J. R. and Tracey, D. M., *Journal of the Mechanics and Physics of Solids,* Vol. 17, 1969, p. 201.

[*56*] Gurson, A. L., *Journal of Engineering Materials and Technology,* Vol. 99, 1977, p. 2.

[*57*] Rousselier, G., "Contribution à l'étude de la rupture des métaux dans le domaine de l'élastoplasticité," Thèse Docteur-ès-Sciences, Ecole Polytechnique, 1979.

[*58*] Yamamoto, M., *International Journal of Fracture,* Vol. 14, 1978, pp. 347–364.

[*59*] D'Escatha, Y. and Devaux, J. C., "Numerical Study of Initiation, Stable Crack Growth and Maximum Load, with a Ductile Fracture Criterion Based on the Growth of Holes," in *Elastic-Plastic Fracture, ASTM STP 668,* American Society for Testing and Materials, Philadelphia, 1979, pp. 229–248.

[*60*] Devaux, J. C., Rousselier, G., Mudry, F., and Pineau, A., "An Experimental Program for the Validation of Local Ductile Fracture Criteria Using Axisymmetrically Cracked Bars and Compact Tension Specimens," *Engineering Fracture Mechanics,* Vol. 21, No. 2, 1985, pp. 273–283.

[*61*] Marini, B., Mudry, F., and Pineau, A., "Ductile Rupture of A508 Cl 3 Steel under Nonradial Loading," *Engineering Fracture Mechanics,* Vol. 22, No. 3, 1985, pp. 375–386.

[*62*] Lemaitre, J., "How to Use Damage Mechanics," *Nuclear Engineering and Design,* 1984, pp. 233–245.

[*63*] Lemaitre, J., "A Continuous Damage Mechanics Model for Ductile Fracture," *Journal of Engineering Materials and Technology, Transactions of ASME,* Vol. 107, Jan. 1985, pp. 83–89.

[*64*] Rousselier, G., "Constitutive Relations Including Ductile Fracture Damage—Application to Cracked Bodies," in *Proceedings,* ICF5, Vol. 2—Advances in Fracture Research, D. Francois, Ed., Pergamon Press, Oxford, 1981, pp. 803–807.

K. N. Shivakumar[1] and I. S. Raju[1]

Treatment of Singularities in a Middle-Crack Tension Specimen

REFERENCE: Shivakumar, K. N. and Raju, I. S., **"Treatment of Singularities in a Middle-Crack Tension Specimen,"** *Fracture Mechanics: Twenty-First Symposium, ASTM STP 1074,* J. P. Gudas, J. A. Joyce, and E. M. Hackett, Eds., American Society for Testing and Materials, Philadelphia, 1990, pp. 470–489.

ABSTRACT: A three-dimensional finite-element analysis of a middle-crack tension (M-T) specimen subjected to Mode I loading was performed to study the stress singularity along the crack front. The specimen was modeled using 20-node isoparametric elements with collapsed non-singular elements at the crack front. The displacements and stresses from the analysis were used to estimate the power of singularities, by a log-log regression analysis, along the crack front. Analyses showed that finite-sized cracked bodies have two singular stress fields of the form $\sigma = C_0(\theta,z)r^{-1/2} + D_0(\theta,\phi)R^{\lambda_\sigma}$. The first term is the cylindrical singularity with the power $-\frac{1}{2}$ and is dominant over the middle 96% of the crack front and becomes nearly zero at the free surface. The second term is a vertex singularity that exists at the intersection of the crack front and the free surface. Because of two singular stress fields near the free surface and the classical square root singularity elsewhere, the strain energy release rate (G) appears to be an appropriate parameter all along the crack front.

KEY WORDS: two-dimensional analysis, three-dimensional analysis, finite-element method, crack, cylindrical singularity, polar singularity, vertex singularity, strain energy release rate, boundary layer, middle-crack tension specimen

Nomenclature

a	Crack length, m
E	Young's modulus, GPa
F_j	y-direction force at jth node, N
G	Strain energy-release rate, J/m^2
G_{P_ϵ}	Plane-strain strain-energy-release rate, J/m^2
H	Height of specimen, m
R, θ, ϕ	Spherical coordinates
r, θ, z	Cylindrical coordinates
S	Remote tension stress, GPa
t	Specimen thickness, m
t_i	ith layer thickness, m
u, v, w	Displacements in x-, y-, and z-directions, m
v_j	v-displacement at jth node, m
x, y, z	Cartesian coordinates, m
W	Half-width of specimen, m

---

[1] Analytical Services and Materials Inc., Hampton, VA 23666.

Δ	Finite-element length behind and ahead of crack front, m
λ_σ	Power of stress singularity
λ_v	Power of displacement field
ν	Poisson's ratio
σ_y	Normal stress in y-direction, GPa

Introduction

A crack front intersecting the free surface in a three-dimensional (3-D) body has been a subject of interest for the last two decades [1–12]. Hartranft and Sih [2] and Sih [3] considered a through-the-thickness crack intersecting normal to the free surfaces in a finite size plate. They argued that the classical two-dimensional (2-D) singular (square root) stress and displacement fields are not valid at the free surface because such a field would violate the physical deformation characteristics of the body, namely the nonzero deformation at the free surface. They defined a boundary-layer thickness β near the free surface by an empirical equation $\beta = t/(4 + 16t/a)$ in which the square-root singularity is not valid. In this equation, t is the thickness of the plate and $2a$ is the crack length.

In 1977, Benthem [4,5] showed that if separation of variables is valid at the intersection of the crack front and the free surface, then a vertex singularity exists. The power of singularity is weaker than the classical value of -0.5. He analyzed a problem of a quarter-plane crack in a semi-infinite half-space subjected to symmetric loading. He expressed the near field stresses by a vertex singularity, $\sigma = E(\theta,\phi)R^{\lambda_\sigma}$, and the displacement $v = F(\theta,\phi)R^{\lambda_v}$. Both stress and displacement are in the spherical coordinate system, R, θ, and ϕ, with the intersection of crack plane and free surface as the origin. Benthem showed that the value of λ_σ varies between -0.5 to -0.332 for Poisson's ratios between 0.0 and 0.5, respectively. About the same time, Bazant and Estenssoro [6,7] presented identical results through a 3-D finite-element analysis.

Later efforts, both analytical [8] and experimental [9], focused on confirming results obtained by Benthem [4,5] and Bazant and Estenssoro [6,7]. Smith and his co-workers [9–11] used stress-freezing photoelastic experiments to measure the power of the singularity along the crack front. They assumed that the near field stress and displacements are represented by an extended classical 2-D singularity equation. (In this paper such a singularity is referred to as the *cylindrical singularity*.) By invoking the separation of variables assumption they calculated the stress singularity from the power of displacement field equation $v = B_0 r^{\lambda_v}$. Their predicted free-surface singularity for Poisson's ratio ν of 0.40 agreed reasonably well with Benthem's results. Furthermore, the measured power of the singularity monotonically changed from -0.5 at the interior to -0.37 at the free surface. The transition region near the free surface was about 20% of the specimen thickness.

One common assumption in all previous studies was that the space variables (r,θ) are separable in the near field stress and displacement equations. The validity of this assumption itself is not known. Furthermore, the variation of λ_σ and λ_v within the boundary-layer region has not been analytically investigated for finite-size cracked bodies. The objectives of the present paper are to evaluate the stress field singularity along the crack front, to assess the validity of separation of variables for a finite size cracked body, and to compare the calculated power of singularity with the results in the literature.

A middle-crack tension (M-T) specimen (Fig. 1) with homogeneous, isotropic, and elastic properties was considered. Finite-element analyses using 20-node isoparametric elements were performed. The M-T specimen was analyzed with several values of Poisson's ratios. In each case, the displacement and stress fields around the crack front were evaluated. The

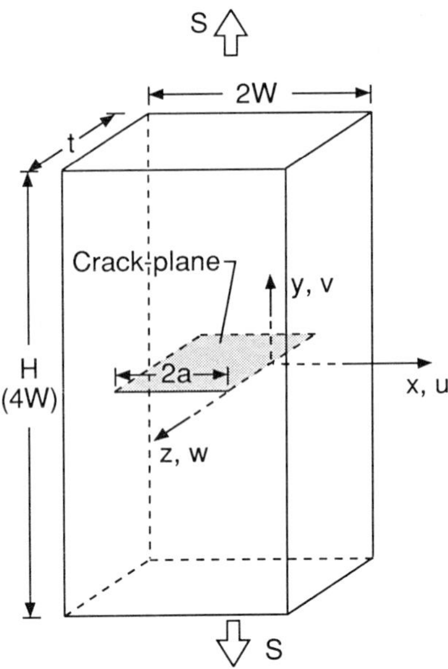

FIG. 1—*Middle-crack tension specimen configuration and loading.*

power of the singularity of the Mode I stress field was evaluated from the finite-element results, and the value at the free surface is compared with Benthem's [4,5] results.

The stress-intensity factor (K) is based on an assumed square-root singularity and hence may not be valid in the boundary-layer region. On the other hand, the strain-energy release rate (G), which requires no such assumption, is an appropriate alternative fracture parameter. Folias [12] showed that the strain energy around the crack front is finite for a range of λ_σ ($-2.5 \le \lambda_\sigma \le -0.5$). Therefore the strain-energy-release rate is used to characterize the severity of the crack in the boundary-layer region, where the stress singularity is complex. While several methods exist to compute G, a simple and an accurate method based on the virtual crack closure technique (VCCT) developed by Shivakumar et al. [13] is used here. The G distribution was also computed assuming that the classical square-root singularity is valid *all along* the crack front. The accuracy of this assumption is assessed.

Treatment of Singularities in Cracked Bodies

In this section, the mathematical forms of 2-D and 3-D singular stress and the corresponding displacement fields near a crack front are discussed.

Two-Dimensional Problems

Classical solutions to the 2-D crack problems were given by Westergaard [17] and Williams [18]. The near field stress singularity was found to be of square root r type, where r is the radius vector in the polar coordinate system. The near field stresses and displacements are expressed in a variable separable form as

$$\sigma = A(\theta)r^{\lambda_\sigma}$$

and

$$v = B(\theta)r^{\lambda_v}$$

with $\lambda_\sigma = -0.5$ and $\lambda_v = 0.5$. The terms A and B are functions of θ only. This type of stress singularity is referred to as a *polar singularity*.

Three-Dimensional Problems

For 3-D crack problems, two forms of singularities have been postulated: the cylindrical singularity, which is an extension of polar singularity to 3-D problems, and the vertex singularity, which is assumed to exist at the intersection of the crack front and the free surface [4–7]. This section discusses these two singularities.

Cylindrical Singularity—This is a direct extension of the 2-D polar singularity of the near field stresses along the crack front. Figure 2 shows the cylindrical coordinate system r, θ, z with the crack front along the z-axis. The near field stresses and displacements at a point P are assumed to be

$$\sigma = A_0(\theta,z)r^{\lambda_\sigma} \tag{1}$$

and

$$v = B_0(\theta,z)r^{\lambda_v} \tag{2}$$

where λ_σ is the power of the stress singularity and λ_v is the power of the displacement field. A_0 and B_0 are functions of only θ and z. Because a cylindrical coordinate system with the crack front as the z-axis is used in Eqs 1 and 2, this singularity is termed a *cylindrical singularity*. If λ_σ and λ_v are independent of θ for all values of z and $\lambda_\sigma = \lambda_v - 1$, the cylindrical singularity equations are valid for the complete crack front. Equations 1 and 2 reduce to the classical square-root singularity if $\lambda_\sigma = -0.5$ and $\lambda_v = 0.5$. However, this can not be valid at the free surface $z = t/2$ [3].

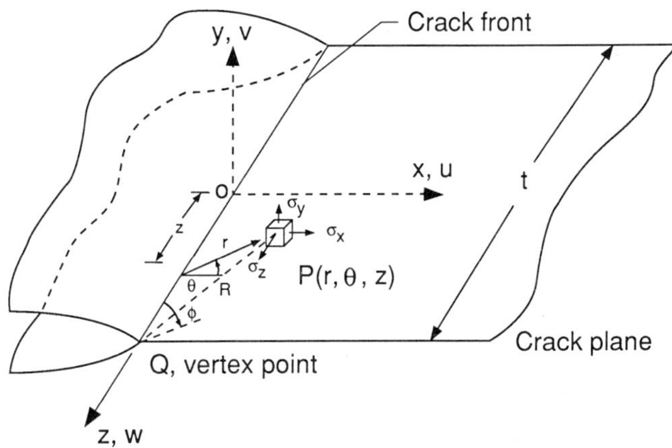

FIG. 2—*Coordinate systems and normal stresses at a point* P.

Vertex Singularity—As previously explained, Benthem postulated the *vertex singularity* at the intersection of the crack front and the free surface (point Q in Fig. 2) in the form

$$\sigma = E(\theta,\phi)R^{\lambda_\sigma} \tag{3}$$

and

$$v = F(\theta,\phi)R^{\lambda_v} \tag{4}$$

where R is the distance between the vertex point Q and the point P. The terms E and F are independent of R and functions of θ and ϕ. The exponents λ_σ and λ_v are constants and are functions of only Poisson's ratio. Furthermore, Eqs 3 and 4 assume separation of variables; hence $\lambda_\sigma = \lambda_v - 1$.

Both the cylindrical and vertex forms of the singularities are based on the fundamental assumption that the separation of space variables is applicable to the stress and displacement fields near the crack front. However, the validity of this assumption is not proved. In the next section, a new look at the 3-D problem based on superposition is used to hypothesize the singular stress fields.

Superposition Analysis

The complete stress and displacement fields in a linear elastic cracked body like the M-T specimen can be hypothesized as a sum of two solutions: a plane-strain solution and a 3-D solution where surface traction are prescribed on $z = \pm t/2$ planes (Fig. 3). The plane-strain solution requires the displacements w to be zero on $z = \pm t/2$ planes. This constraint gives rise to a traction σ_z on these planes. Therefore the negative of the surface traction calculated for the plane-strain problem needs to be imposed on the second problem (Fig. 3c) to satisfy the stress-free conditions on $z = \pm t/2$ in the original 3-D problem. For convenience of presentation, the second problem is referred to as the *residual problem*. This superposition analysis is schematically shown in Fig. 3. Therefore the near field stress for the original 3-D

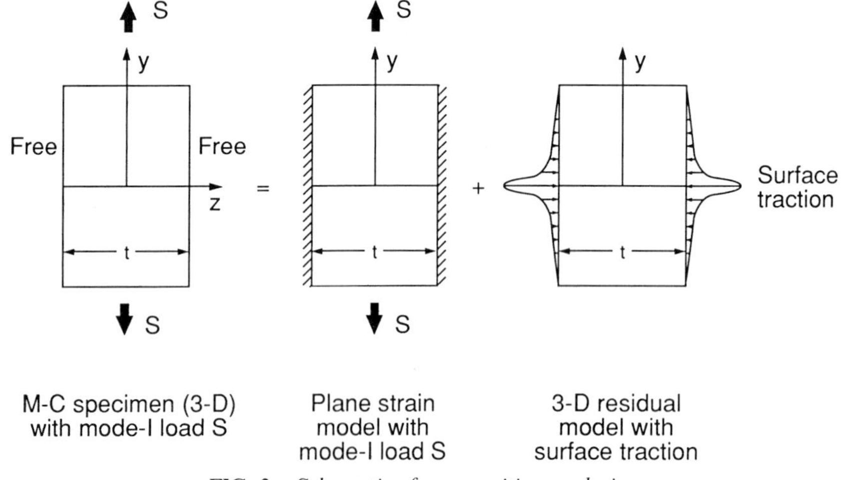

FIG. 3—*Schematic of superposition analysis.*

problem can be expressed as

$$\sigma = C_0'(\theta)r^{-1/2} + T'(r,\theta,z) \tag{5}$$

where the first term is the solution to the plane-strain problem, with a square root r type singularity. The second term is a solution to the residual problem, which could also represent a singular field with a power different from -0.5 [3]. Therefore a 3-D cracked body, in general, has two singular stress fields. Since the residual problem is also a 3-D problem, its solution should also have two singular stress fields: square-root singularity and an unknown singular field. Therefore $T'(r,\theta,z)$ is expressed as

$$T'(r,\theta,z) = C_0''(\theta,z)r^{-1/2} + T(r,\theta,z) \tag{6}$$

Combining Eqs 5 and 6, the general solution to the near field stresses for the original 3-D cracked body is

$$\sigma = C_0(\theta,z)r^{-1/2} + T(r,\theta,z) \tag{7}$$

The first term in Eq 5 is the square-root singular field and the second term is the general singularity, which needs to be established. The term C_0, function of the polar angle θ and z, defines the magnitude of the square root singular stress field. For very thick specimens the interior develops a plane strain stress state; hence in the interior only the square root singular stress field exists. Several investigators [14–16], including the present authors, have found in their 3-D analysis of cracked bodies that plane-strain constraint exists for most of the crack front. (In the present analysis, the constraint was evaluated by comparing the $\sigma_z/(\sigma_x + \sigma_y)$ value with the Poisson's ratio ν of the material. For plane-strain condition both are equal. The constraint factor $\sigma_z/(\sigma_x + \sigma_y)$ was more than 0.9ν in the region enveloping 0.1 mm behind, 0.6 mm ahead, and interior 95% length of the crack front. This factor reduced monotonically to nearly zero in the boundary layer region. Therefore $T(r,\theta,z)$ must become zero in the interior.

The second term, $T(r,\theta,z)$, can be expressed either as a cylindrical or a vertex singularity using a cylindrical or spherical coordinate system, respectively. Hence Eq 7 is rewritten as

$$\sigma = C_0(\theta,z)r^{-1/2} + D_0(\theta,z)r^{\lambda_\sigma} \tag{8}$$

and

$$\sigma = C_0(\theta,z)r^{-1/2} + E_0(\theta,\phi)R^{\lambda_\sigma} \tag{9}$$

where λ_σ is the unknown power of singularity, D_0 is a function of θ and z, and E_0 is a function of θ and ϕ. At the free surface ($z = \pm t/2$), $R = r$; hence Eq 9 reduces to Eq 8. Equations 8 and 9 assume the variable separable form is valid. If this assumption is not valid, the finite element results would show a dependency on θ. At $z = t/2$, the value of λ_σ must be other than -0.5 to satisfy the compatibility requirement suggested by Sih [3]. It is clear from Eqs 8 and 9 that the plane strain solution is a special case of the 3-D solution wherein the coefficient D_0 (or E_0) is zero and C_0 is constant for all values of z.

The displacement field for the cracked body can be expressed, similarly, in two-term form as

$$v = F_0(\theta,z)r^{1/2} + G_0(\theta,z)r^{\lambda_v} \tag{10}$$

and

$$v = F_0(\theta,z)r^{1/2} + H_0(\theta,\phi)R^{\lambda_v} \tag{11}$$

The terms F_0 and G_0 are functions of θ and z; H_0 is a function of θ and ϕ. The exponent λ_v is the power of the displacement field. As already explained, in the interior of the crack front C_0 and F_0 are dominant and the other coefficients are either zero or have negligible influence. But in the neighborhood of the free surface ($z = t/2$), the coefficients of the second singularity become dominant. For the displacements in Eqs 10 and 11 to correspond to the stress field in Eqs 8 and 9, λ_σ should be equal to $\lambda_v - 1$. Strains (and hence stresses) are the first-order derivatives of the displacements. If λ_σ is not equal to $\lambda_v - 1$, then the variable separable form is not valid in Eqs 8 to 11. Furthermore, if Eqs 8 and 9 are the correct form for stresses, then Eqs 10 and 11 are not the form for displacements and *vice versa*. In the next section, evaluation of the second singularity is discussed.

Evaluation of Second Singular Stress Field

Utilizing stresses and displacements from an accurate numerical solution (such as a finite-element analysis), the coefficients C_0, D_0, and E_0 and exponent λ_σ from the stress data (Eqs 8 and 9) and F_0, G_0, H_0 and λ_v from the displacement data (Eqs 10 and 11) can, in principle, be evaluated along the crack front. Each of Eqs 8 to 11 have two constants and the exponent (λ_σ or λ_v) is complex. Direct evaluation of two constants and the exponent is very difficult. Several numerical procedures (Newton-Raphson and Levenberg-Marquardt methods) were attempted but none proved reliable. However, the evaluation of one constant and one exponent can be performed very easily and accurately by a log-log fit to the data. Therefore Eqs 8 to 11 were examined to reduce the problem to one coefficient and one unknown exponent; that is, the stress and displacements are represented by

$$\sigma = C_b r^{\lambda_\sigma} \tag{12}$$

and

$$v = G_b r^{\lambda_v} \tag{13}$$

For the purpose of discussion, the crack front is considered to be divided into three regions: the interior region, the boundary layer region, and the free surface location (region). The interior region is the region where the stress field is dominated by the square-root singularity (the first term in Eqs 8 to 11). The boundary-layer region is the region near the free surface and is beyond the square-root singularity dominant region. The free-surface ($z = 0.5t$) is the special location in the boundary-layer region. Here Eq 9 reduces to Eq 8 and Eq 11 reduces to Eq 10, because $R = r$. As mentioned previously, in the interior region the contribution of the second singularity to total stresses and displacements is nearly zero. Therefore in the interior region Eqs 12 and 13 are valid. The region over which the finite-element results fit to Eqs 12 and 13 give $\lambda_\sigma = -0.5$ and $\lambda_v = 0.5$ establishes the square-root singular region. To establish the type of stress singularity field in the boundary-layer region and at the free surface, consider the following three hypotheses:

(*a*) $C_0(\theta,z)$ is dominant and constant all along the crack front; that is, $C_0(\theta,z) = C_0(\theta,0)$, for $0 \leq z \leq t/2$.

(b) $C_0(\theta,z)$ is dominant in the interior of the crack front, varies along the z-axis, and at the free surface ($z = t/2$) C_0 is zero or very small compared with the second term in Eq 8 or 9.

(c) $C_0r^{-1/2}$ is of the same order of magnitude as $D_0r^{\lambda_\sigma}$ or E_0R^{λ}.

If condition (a) is true, then the constants C_0 and F_0 can be evaluated using stress data at the midplane ($z = 0$). Eqs 8 and 9 can now be rewritten as

$$\sigma - C_0(\theta,0)r^{-1/2} = D_0r^{\lambda_\sigma}$$

and

$$\sigma - C_0(\theta,0)r^{-1/2} = E_0R^{\lambda_\sigma} \tag{14}$$

Similarly, for displacements:

$$v - F_0(\theta,z)r^{1/2} = G_0r^{\lambda_v}$$

and

$$v - F_0(\theta,z)r^{1/2} = H_0R^{\lambda_v} \tag{15}$$

The left-hand side values of Eqs 14 and 15 are known, stresses σ and displacements v from the finite-element analysis and constants C_0 and F_0 from the finite-element stress and displacement fit to Eqs 12 and 13, respectively, at $z = 0$ plane. For various values of r a fit can be made to determine the constants D_0, E_0, G_0, and H_0 and the exponents λ_σ and λ_v. If the computed λ_σ and λ_v are the same for various angles θ and satisfy the condition $\lambda_\sigma = \lambda_v - 1$, then λ_σ and λ_v are unique and the variable separable form for stress and displacement field is valid. Otherwise, either condition (a) or the variable separable form for the second singularity field is not valid.

If condition (b) is true, then at the free surface ($z = t/2$) Eqs 8 and 10 reduce to Eqs 12 and 13, respectively, as

$$\sigma = D_0r^{\lambda_\sigma} \tag{16}$$

and

$$v = G_0r^{\lambda_v} \tag{17}$$

The exponents λ_σ and λ_v can easily be calculated from finite-element stresses and displacements using Eqs 16 and 17, respectively. If λ_σ and λ_v show independence with θ, and $\lambda_\sigma = \lambda_v - 1$, then the variable separable assumption and condition (b) are valid. Otherwise, either condition (b) or the variable separable assumption is not valid. Within the boundary-layer region (second region), Eqs 16 and 17 are not correct representation of stresses (Eqs 8 and 9) and displacements (Eqs 10 and 11) because both the terms are dominant. Therefore, even if λ_σ and λ_v are independent of θ at $z = t/2$, they will show dependence on θ at $z < t/2$. If such a condition exists, then condition (b) is still valid.

If condition (c) is true, the stresses and displacements cannot be represented in a form like the one-term Eqs 12 and 13. Therefore numerical fit to Eqs 12 and 13 will yield meaningless values for λ_σ and λ_v. Also, even if the variable separable form for stresses and displacements is valid, λ_σ and λ_v would show dependence on θ.

In summary, two fits of the form of Eqs 14 to 17 are needed to determine the form of the singular stress in the boundary-layer region and at the free-surface location. If the fit to Eqs 14 and 15 produces λ_σ and λ_v values independent of θ and $\lambda_\sigma = \lambda_v - 1$, then condition (a) is valid and the second singularity is given by λ_σ. If the fit to Eqs 16 and 17 produces λ_σ and λ_v values independent of θ, satisfy the condition $\lambda_\sigma = \lambda_v - 1$ at $z = t/2$, and depend on θ within the boundary layer region ($z < t/2$), then condition (b) exists and the second singularity is given by λ_σ of Eq 16. On the other hand, if λ_σ and λ_v do not show independence with θ at $z = t/2$ and in the boundary-layer region, then condition (c) exists.

Finite-Element Analysis

Specimen Configuration

Figure 1 shows a middle-crack tension (M-T) specimen with loading. The specimen half-width W is 50 mm, height H is 200 mm, and the half-crack length a is 25 mm. The thickness t of the specimen is 25 mm. The material is assumed to be homogeneous and isotropic with elastic modulus $E = 70$ GPa and Poisson's ratio $v = 0.3$.

The M-T specimen is loaded by imposing a uniform v-displacement at $y = \pm H/2$ planes. The equivalent remote stress is S. Since the height H of the specimen was twice the width and eight times the half-crack length a, the remote displacements or stress would produce similar stress distributions at the crack front. Exploiting the symmetry in the specimen geometry and loading, only one eighth of the specimen was modeled in the finite-element analysis.

Finite-Element Model

Figure 4 shows the finite-element idealization of one eighth of the M-T specimen. Twenty-node isoparametric elements were used everywhere except along the crack front, where pen-

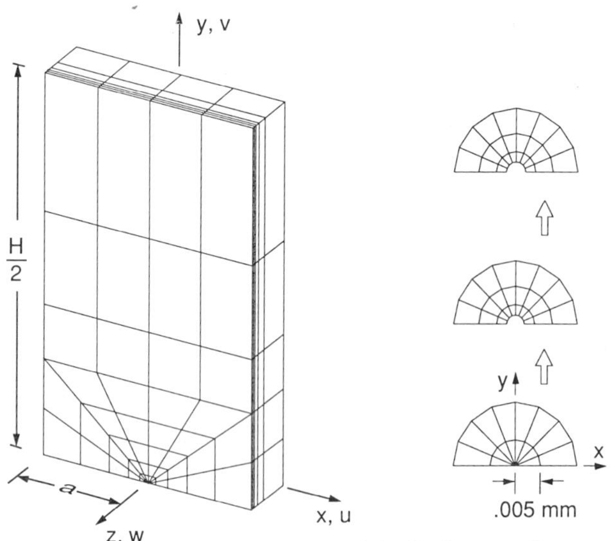

FIG. 4—*Finite-element idealization of one eighth of M-T specimen.*

TABLE 1—*Layer thickness in different finite-element models.*

Layer No.	Layer thickness, t_i/t			
	6-Layer Model	7-Layer Model	8-Layer Model	9-Layer Model
1	0.304	0.304	0.304	0.304
2	0.100	0.100	0.100	0.100
3	0.0492	0.0492	0.0492	0.0492
4	0.0300	0.0300	0.0300	0.0300
5	0.0128	0.0120	0.0116	0.0114
6	0.004	0.0036	0.0036	0.0036
7	...	0.0012	0.0012	0.0012
8	...	...	0.0004	0.0004
9	...	...	...	0.0002

tahedron elements (15 nodes) were used. The insert in Fig. 4 describes idealization near the crack front in the x-y plane. In pentahedron (collapsed) elements, the intermediate nodes were retained at the mid-sides of the element edges. Note that a square-root singularity of the stress and displacement fields would have been reproduced if the mid-side nodes were moved to the quarter points [19–21]. This was not done because it would have biased the stress field singularity at the crack front. Instead, the singularity was calculated from the computed stresses ahead of the crack front.

The 3-D finite-element model was developed using a 2-D idealization of the $z = 0$ plane and then translating that plane in the z-direction. Adequacy of the inplane (x-y) and through-thickness (z-direction) modeling was established by a 2-D plane-strain analysis and a convergence study on the 3-D models. Details of these analyses are presented in the Appendix. The model selected for the present study had nine layers of unequal thicknesses (Table 1) with 5348 nodes, 1116 elements, and 16 044 degrees of freedom. The layer thickness near the free surface was 0.005 mm, and the smallest element size in the x-y plane was also 0.005 mm (see insert in Fig. 4). A local least-square smoothing procedure [22] was used to extrapolate the Gauss point stresses to the element nodes.

Results and Discussion

In this section, a plane-strain analysis of the M-T specimen was performed to reproduce some of the established results in 2-D crack problems. The nodal displacements and stresses near the crack front were used to evaluate power of singularity through log-log least square regression analysis. Details of the regression analysis are presented in the Appendix. Secondly, 3-D stress and displacement fields near the crack front for the M-T specimen are presented. The power of singularities along the crack front of the specimen are then evaluated. Finally, the effect of multiple singularity of the stress field near the free surface on strain-energy-release rate G is assessed.

Two-Dimensional Plane-Strain Analysis

The objective of the plane-strain analysis is to verify that the finite element model is capable of reproducing the classical 2-D singular stress and displacement fields near the crack tip. The plane-strain results were obtained from the same 3-D, F-E model shown in the Fig. 4

by specifying $w = 0$ on $z = t/2$ plane. The Mode I stress σ_y and the opening displacement v were used to calculate the exponents in the stress and displacement equations. The power of singularity λ_σ was calculated by fitting a straight line to log r versus log σ_y results (referred to as the log-log fit). The procedure was repeated along radial lines at intervals of 22.5° for $0 \leq \theta \leq 135°$. A similar straight line fit (log r versus log v) procedure was also used for v-displacement to evaluate the exponent λ_v. The procedure was repeated along each radial line at $22.5° \leq \theta \leq 180.0°$ in intervals of 22.5°. As shown in Fig. 5, consistent values of λ_σ of about -0.497 were obtained for $0 \leq \theta \leq 67.5°$. The λ_v from this fit was 0.505 at $\theta = 180°$ and varied less than 1% for θ between 112.5 and 180°.

The fit to the σ_y stresses for $\theta \geq 90°$ gave λ_σ values that deviated by more than 2% from the classical value -0.5. Similarly, for $\theta \leq 90°$ the fit to the v-displacements gave λ_σ values that deviated by more than 2% from 0.5. The variations in λ_σ values for $\theta \geq 90°$ (λ_v values for $\theta \leq 90°$) is expected, because the magnitude of σ_y and v decreases and eventually approaches zero at $\theta = 180°$ and $\theta = 0°$, respectively. Hence the log-log fit is expected to yield accurate results only in the regions where σ_y and v are predominant. For this reason these values are not shown in Fig. 5.

Figure 5 shows that the exponents λ_σ and λ_v are independent of the polar angle θ and they satisfy the condition $\lambda_\sigma = \lambda_v - 1$. Thus the plane-strain finite element model reproduced accurately the square-root singularity of the stress field and variable separable form of the 2-D singular stress and displacement fields at the crack tip.

The 2-D analysis suggests that if the exponents λ_σ and λ_v are invariant with respect to θ, then the log-log fit would yield nearly identical slopes at the different θ = constant lines. Hereafter, the invariance of λ_σ and λ_v with respect to θ will be evaluated by comparing the values at $\theta = 0$ and 45° for λ_σ and at $\theta = 180°$ and 135° for λ_v. If the exponents are nearly identical, the invariance with respect to θ is assumed to exist. The invariance suggests that a variable separable form for stresses and displacement fields in polar coordinates is valid.

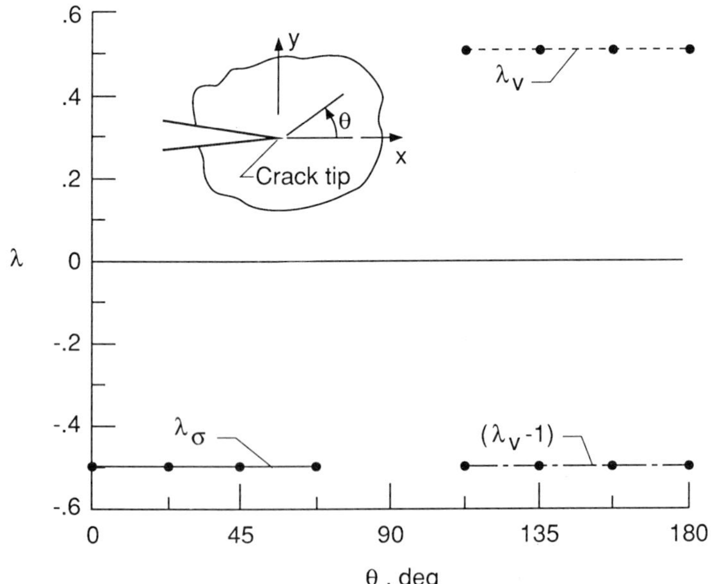

FIG. 5—*Variation of λ_σ and λ_v for various angles θ for the plane-strain problem.*

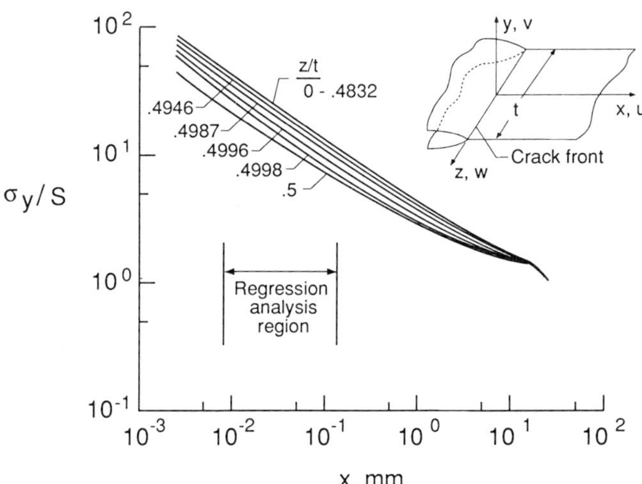

FIG. 6—*Normalized σ_y distribution ahead of the crack front at various z = constant stations from the mid-plane (v = 0.3).*

Stress and Displacement Distributions in the M-T Specimen

Figure 6 shows the distribution of stress σ_y ahead of the crack front at various z = constant planes. Note the log-log scale in the figure. The stress is normalized by the remote stress S at $y = \pm H/2$. The figure shows that both slope and magnitude of σ_y are nearly identical for $0 \leq z/t \leq 0.48$. For $z/t > 0.48$, slopes appear to change and the magnitude of the stress shows a drop compared to the interior ($z/t = 0$) values. At the free boundary $z = t/2$, the slope is smaller than at the mid-plane ($z = 0$). The variation of slope (power of the singularity λ_σ) in the z-direction will be discussed later.

Figure 7 shows the distribution of opening displacement v behind the crack front at various z = constant planes. Note that $S = 57$ N/mm² for the displacement plot. Slopes of displacements curves increased towards the free surface. Furthermore, opening displacement decreased as the free surface is approached [14–16]. The variation of the slopes (exponent λ_v) of these curves in the z-direction will be discussed later.

Comparison of v-displacement at the $z = 0$ and $t/2$ at the center-plane ($x = -25$ mm) of the specimen showed that the free surface ($z = t/2$) displacement is about 4.5% higher than at the mid-plane ($z = 0$). This trend is contrary to the distribution near the crack front and common knowledge. However, for the bend specimen the v-displacement was nearly constant at $x = -25$ mm. Therefore care must be taken in measurement and interpretation of surface displacements.

Singular Stress Fields in the M-T Specimen

As previously mentioned, two types of singular stress fields (Eqs 8 and 9) and the corresponding displacement fields (Eqs 10 and 11) are possible for a 3-D cracked body. Furthermore, three regions along the crack front, interior, boundary layer, and free surface, and three possible solutions to near field singular stresses were identified. Here, the possibility of conditions (a) and (b) are assessed. If neither (a) nor (b) is possible, then the condition (c) defines the near field singular stresses in a 3-D cracked body.

Condition (a)—In this definition, the coefficient C_0 of square-root singularity is constant

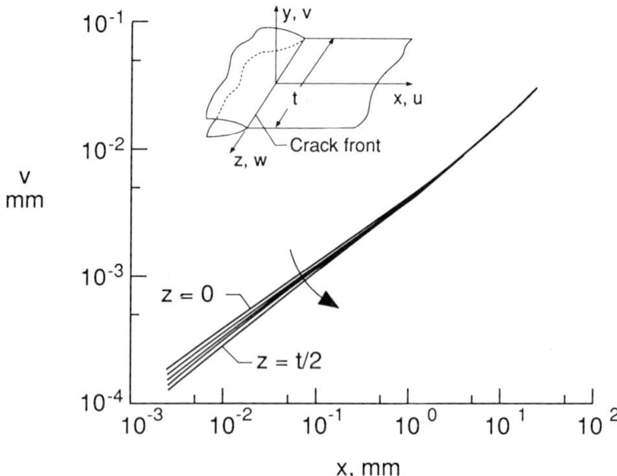

FIG. 7—*Opening displacement distribution behind the crack front at various z = constant stations from the mid-plane (v = 0.3).*

all along the crack front. If this is true, Eqs 14 and 15 fit to finite-element stresses and displacements, respectively, at $z = t/2$ should define the second singular field. The free surface stresses and displacements were subtracted by their respective mid-plane values. Resulting stresses and displacements were used to evaluate λ_σ and λ_v by log-log regression analysis at various θ = constant lines. Table 2 shows that computed values of λ_σ (at $\theta = 0$ and 45°) and λ_v (at $\theta = 135$ and 180°) are dependent on θ and also λ_σ is not equal to $\lambda_v - 1$. Dependence of λ_σ and λ_v on θ shows that either this condition or the variable separable form for the residual stresses and displacements is not valid.

Condition (b)—In this definition, the coefficient of the square-root singularity is dominant in the interior of the crack front and is zero or negligibly small at the free surface. As mentioned previously, the coefficient of the second singularity is negligible or zero in the interior of the crack front and becomes dominant at the free surface. Therefore simplified Eqs 16 and 17 are a valid representation for all z, except in the boundary-layer region. This type of fit should also define the region of dominance of each of the terms in Eqs 8 to 11. First, to verify the validity of separation of variables, a log-log least square regression was performed at midplane ($z = 0$) and at the free surface ($z = t/2$) for $\theta = 0$ and 45° for stress σ_y and $\theta = 135$ and 180° for v-displacement. Table 3 presents λ_σ and λ_v at the mid-plane and at free surface. At $z = 0$ plane, power of singularity λ_σ is nearly identical to the 2-D classical value of -0.5 and is the same for $\theta = 0$ and 45°. The corresponding v-displacement slope λ_v is 0.5 and is also independent of θ. These results satisfy the relationship $\lambda_\sigma = \lambda_v - 1$ both at the midplane and free surface. Therefore separation of variables is valid both at the midplane

TABLE 2—*Variation of λ_σ and λ_v in θ-direction at free-surface for condition* (a).

θ, deg.	λ_σ	λ_v
0	-0.534	. . .
45	-0.590	. . .
135	. . .	0.330
180	. . .	0.300

and at the free surface of the specimen. At the free surface, the power of singularity is weaker ($\lambda_\sigma = -0.451$) than the square-root singularity and agreed very well with Benthem's solution [4,5]. Actually, Benthem assumed separations of variable and proved that the vertex singularity exists at the vertex point Q. Therefore, proving the validity of variable separable form for stress and displacements shows the possible existence of vertex singularity at the free surface. However, at the free surface, $R = r$, the vertex singularity reduces to cylindrical singularity. Therefore one would argue that the second singularity could be a cylindrical singularity.

Table 4 shows the comparison of λ_σ calculated at $\theta = 0$ and $(\lambda_v - 1)$ calculated at $\theta = 180°$ with Benthem's solutions [4,5] for various Poisson's ratios ($v = 0.0$ to 0.45). The present λ_σ and $(\lambda_v - 1)$ agreed very well with Benthem's vertex singularity results. Therefore the second singularity may be a vertex singularity and the coefficients of the square-root singular term (first term in Eqs 8 and 11) at the free surface are negligibly small or identically equal to zero in comparison to the second term.

Figure 8 shows the variation of λ_σ and λ_v along the crack front. Both λ_σ and λ_v were evaluated by fitting Eqs 12 and 13 to finite-element stresses and displacements. In this figure only the results for $z/t > 0.45$ are shown. In the region $0 \leq z/t \leq 0.48$, both λ_σ and λ_v are constant and equal to the classical 2-D values ($\lambda_\sigma = -0.5$ and $\lambda_v = 0.5$). This indicates that over the interior, 96% of the crack front, the power of singularity is -0.5 and the influence of the second singularity is negligibly small or zero. Furthermore, at $z/t = 0.48$ calculated λ_σ at $\theta = 0$ and 45° and λ_v at $\theta = 135$ and 180° did not show the dependency on θ. Both λ_σ and λ_v start deviating from the 2-D values for $z/t > 0.48$ and become dependent on θ (except at $z = t/2$). This indicates that both singualr fields in Eqs 8 to 11 are becoming equally dominant; hence the one-term fit based on Eqs 12 and 13 is not valid in this region. Therefore

TABLE 3—*Power of stress* (λ_σ) *and displacement* (λ_v) *fields at mid-plane and free surface of M-T specimen* ($v = 0.3$).

θ, deg.	Mid-Plane ($z = 0$)		Free Surface ($z = t/2$)	
	λ_σ	λ_v	λ_σ	λ_v
	-0.495	. . .	-0.451	. . .
45	-0.494	. . .	-0.442	. . .
135	. . .	0.510	. . .	0.556
180	. . .	0.505	. . .	0.548

TABLE 4—*Comparison of* λ_σ *and* λ_v *calculated from the present analysis with Benthem's solution (free surface).*

Poisson's Ratio (v)	Benthem [4,5] (λ_σ)	Present Results	
		λ_σ	$\lambda_v - 1$
0.0	-0.500	-0.497	-0.497
0.30	-0.452	-0.451	-0.452
0.40	-0.414	-0.407	-0.417
0.45	. . .	-0.356	-0.391
0.50	-0.332	. . .	. . .

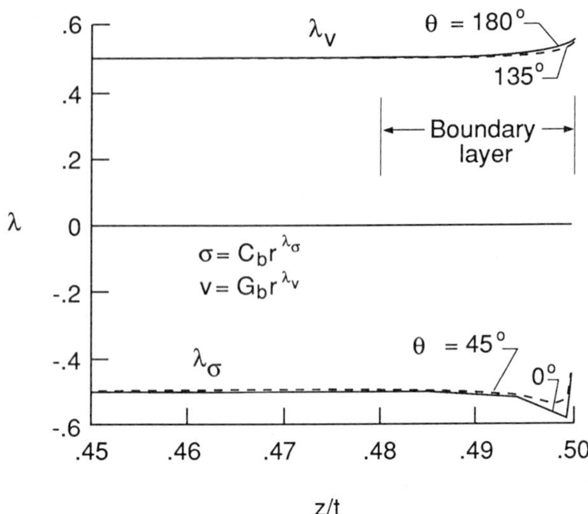

FIG. 8—*Variation of λ_σ and λ_v along the crack front.*

the question of type of the second singularity (vertex or cylindrical) is still not completely resolved unless a reliable procedure to determine two constants and an exponent is available.

In summary, a 3-D cracked body has two singularities: (1) the square-root cylindrical singularity and (2) the vertex or cylindrical singularity. Comparison of present results with Benthem's [4,5] and Bazant and Estenssoro's [6,7] results appear to show that the second singularity is a vertex singularity. The square-root singularity dominates in the middle 96% of the crack front and its coefficient is nearly zero at the free surface. The influence of the second singularity becomes dominant near the free-surface region ($0.48 \leq z/t \leq 0.5$).

Strain-Energy Release Rate Distribution

It is apparent from the previous section that two stress singularities exist within the boundary-layer region. The effect of ignoring the vertex singularity and assuming the square-root singularity in $0.48 \leq z/T \leq 0.50$ region to calculate the strain-energy-release rate G is assessed in this section.

Figure 9 shows a comparison of normalized G calculated from the 3-D VCCT [13] and COD [19–21] method for a M-T specimen. G was normalized by the plane-strain value ($G_{p_\epsilon} = 1.462 \times 10^{-3}$ N/mm) for an applied remote tension stress $S = 1$ N/mm^2. The results for the COD method were obtained by using square root singular elements (collapsed, quarter-point 20-node element) around the crack front. Circular symbols are G calculated from the COD method using the plane-strain assumption along the crack front:

$$G = [(2\pi E)/\{16(1 - \nu^2)\}] (V_\Delta/2)^2 (1/\Delta) \tag{18}$$

where V_Δ is the total opening displacement measured at Δ behind the crack front at any z = constant plane. At the free boundary ($z/T = 0.5$), G was calculated also from the plane-stress assumption by replacing $(1 - \nu^2)$ in Eq 18 with unity. The G value from the plane-strain COD method agreed very well with the 3-D VCCT method for $0 \leq z/T \leq 0.49$ and reasonably well for $0 \leq z/T \leq 0.4996$ (Table 5). At the free boundary, plane-stress approximation gave better agreement with VCCT, as shown by the square symbol in Fig. 9 (Table

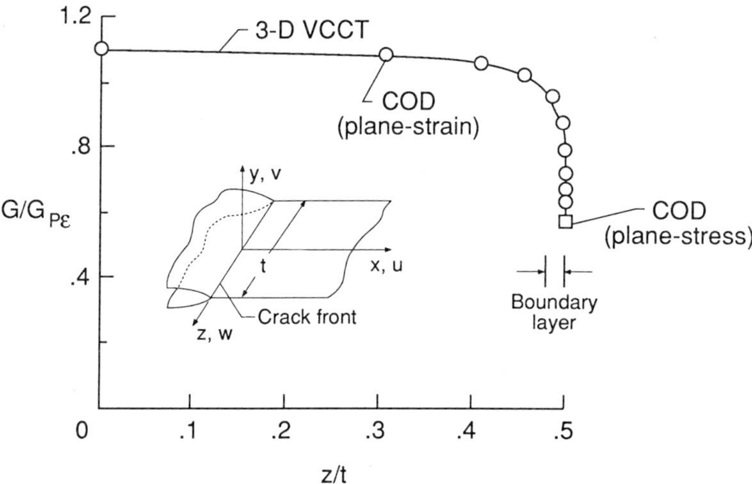

FIG. 9—*Variation of normalized strain-energy-release rate along the crack front.*

5). Note that the COD method assumed the square-root singularity ($\lambda_\sigma = -0.5$) all along the crack front, even though it is not truly valid in the boundary-layer region ($0.48 \leq z/t \leq 0.5$). Therefore the assumption of square-root singularity all along the crack front has very little effect on G, provided the plane-stress approximation at the free-surface and plane-strain approximation in the remaining length of the crack front are used.

Concluding Remarks

A three-dimensional finite-element analysis of the middle-crack tension (M-T) specimen was performed. Twenty-node isoparametric elements were used in the model. The displacements and stresses from the analysis were used to estimate the power of singularities along the crack front. A log-log regression analysis was performed to calculate the power of stress singularity. The finite-element model and the regression analysis procedure were verified by analyzing a two-dimensional (2-D) plane-strain problem. Finite-element results reproduced the classical singular stress field and the corresponding displacement field.

In the three-dimensional M-T specimen, two singularities exist along the crack front and

TABLE 5—*Comparison of G from COD method with 3-D VCCT in the boundary layer region.*

| | $G/G_{p\epsilon}$ | | |
| | COD Method | | |
z/t	Plane Strain	Plane Stress	3-D VCCT
0.4964	0.857	0.800	0.861
0.4988	0.782	0.712	0.781
0.4996	0.720	0.655	0.704
0.4999	0.675	0.615	0.608
0.5000	0.658	0.600	. . .

the near field stresses have the form $\sigma = C_0 r^{-1/2} + D_0 R^{\lambda_\sigma}$. The first term is the cylindrical singularity with the power of -0.5, which is the same as the classical 2-D value. The second term may be a vertex or cylindrical singularity. Comparison of the present results with the literature suggests that the second singularity is a vertex singularity with the power λ_σ and located at the intersection of the crack front and the free surface. The cylindrical singularity dominates over the interior 96% of the crack front, while the second singularity dominates in the boundary layer (2% of the specimen thickness from the free surface). The coefficient of the cylindrical singularity becomes nearly zero at the free surface, whereas the coefficient of the second singularity is nearly zero outside the boundary layer and largest at the free-surface.

The calculated power of the vertex singularity λ_σ at the free surface agreed very well with Benthem's results for all Poisson's ratios. The separation of variables assumption in stress and displacement fields was found to be valid at the free surface and at the middle 96% of the crack front. Hence this assumption may be valid in the boundary layer (intermediate) region as well.

Because two singular stress fields exist within the boundary-layer region, the strain-energy-release rate G is an appropriate parameter to define the severity of the crack all along the crack front. For practical applications, the assumption of a square root singular stress field all along the crack front has little effect on the accuracy of the calculated G.

APPENDIX

Evaluation of Finite-Element Model

The accuracy of the 9-layer, 3-D, finite-element model used in the present study is evaluated here. First, the adequacy of the mesh refinement in the x-y plane (refer to Fig. 4) was verified by a 2-D analysis. The thickness or z-direction refinement in the 3-D model was evaluated by the successive subdivision of the specimen thickness near the free surface (a high stress gradient region).

Two-Dimensional Evaluation

A 2-D plane-strain analysis of the middle-crack specimen (Fig. 1) was conducted using 8-node isoparametric elements. The finite-element mesh, at z = constant plane, shown in the Fig. 4 was used. Two analyses were conducted: one using collapsed singular elements [*16–18*] and the other using collapsed non-singular elements. The classical square-root singularity at the crack tip was achieved by moving the mid-side nodes to the quarter points [*16–18*].

Figure 10 shows the σ_y distribution ahead of the crack tip on the $y = 0$ plane. Results from the singular and non-singular elements were identical beyond the first three nodal values (i.e., beyond the first element at the crack tip). A straight line fit to log r and log σ_y stress data was made to the region beyond three nodes (one element) from the crack tip and extending to about 0.15 mm. This region included 10 nodal stress values. A least-square regression gave a slope of -0.497 with the correlation factor greater than 0.995. This value is in excellent agreement with the classical 2-D value of -0.50. The regression analysis was repeated by selecting several segments over the region 0.005 to 0.2 mm. The calculated slopes were found to be consistent with the above value and the variation was less than about 1%.

The stress-intensity factor (K) for both singular and non-singular element analyses was calculated using the virtual crack-closure technique [*24*] and the equation $K = \sqrt{GE/(1 - \nu^2)}$. Calculated values of K/S for singular and non-singular element analyses were 1.192 and 1.178, respectively, which are 0.5% higher and 0.7% lower, respectively, than Isida's solution [*23*] for this problem.

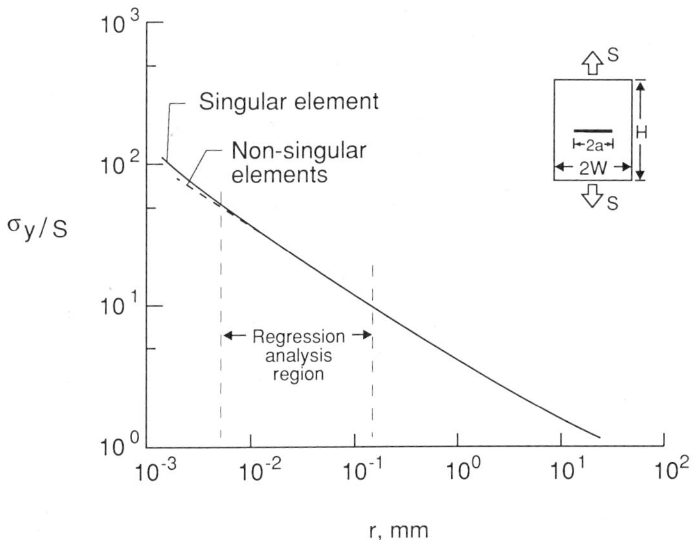

FIG. 10—*Distribution of* σ_y *ahead of the crack front calculated using singular and non-singular elements.*

Three-Dimensional Evaluation

The 3-D finite-element model was developed by translating the 2-D mesh in the *z*-direction. Four different mesh idealizations consisting of 6, 7, 8, and 9 layers were considered. Layer thicknesses in each of the four models are given in Table 1. Layer thicknesses were smaller near the free surface to account for the high stress gradient there. The ninth layer

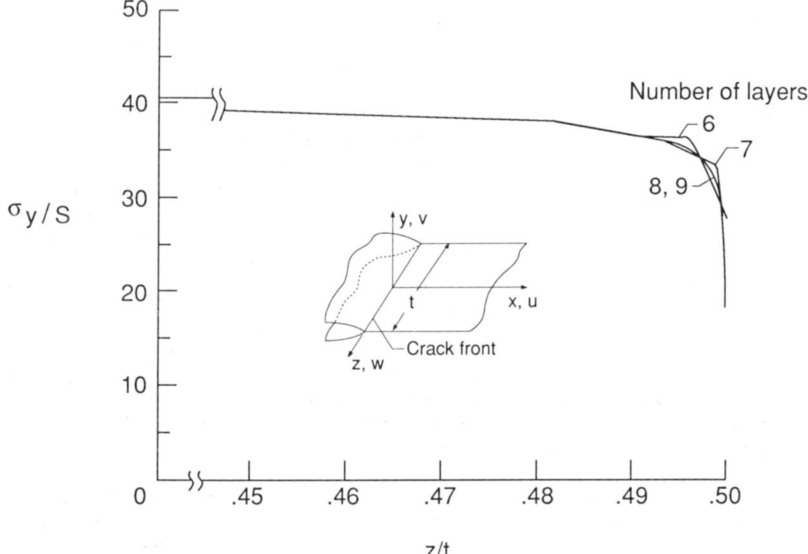

FIG. 11—*Convergence of* σ_y *stress with finite-element mesh refinement in thickness (z) direction.*

thickness for the 9-layer model was about $0.002t$ or 0.005 mm, which is the same as the crack-tip element size (0.005 mm) in the x-y plane.

Figure 11 shows the variation of normalized opening mode stress (σ_y/S) in the z-direction at $x = 0.01$ mm (very near the crack front). All four models gave nearly identical results except in 2% of the thickness near the free surface. However, 8- and 9-layer models showed very good agreement through the complete thickness of the specimen.

Results presented here suggest that the 9-layer model is sufficiently fine to produce accurate results; hence this model was used in the rest of this study.

Acknowledgments

The authors are deeply indebted to Dr. J. H. Crews, Jr., for many inspiring discussions and for drawing attention to the superposition principle. The authors take this opportunity to thank him. This work was performed at the Langley Research Center as part of NASA Contracts NAS1-18256 and NAS1-18599.

References

[1] Sih, G. C., Williams, M. L., and Swedlow, J. L., "Three-Dimensional Stress Distribution Near a Sharp Crack in a Plate of Finite Thickness," AFML-TR-66-242, Air Force Materials Laboratory, Wright-Patterson Air Force Base, 1966.

[2] Hartranft, R. J. and Sih, G. C., "An Approximate Three-Dimensional Theory of Plates with Application to Crack Problems," *International Journal of Engineering Science,* Vol. 8, 1970, pp. 711–729.

[3] Sih, G. C., "A Review of the Three-Dimensional Stress Problem for a Cracked Plate," *International Journal of Fracture Mechanics,* Vol. 7, No. 1, 1971, pp. 39–61.

[4] Benthem, J. P., "State of Stress at the Vortex of a Quarter-Infinite Crack in a Half-Space," *International Journal of Solids and Structures,* Vol. 13, 1977, pp. 479–492.

[5] Benthem, J. P., "The Quarter-Infinite Crack in a Half-Space: Alternate and Additional Solutions," *International Journal of Solids and Structures,* Vol. 16, 1980, pp. 119–130.

[6] Bazant, Z. P. and Estenssoro, L. F., "General Numerical Method for Three-Dimensional Singularities in Cracked or Notched Elastic Solids," *Advances in Research on the Strength and Fracture of Materials,* D. M. R. Taplin, Ed., Vol. 3a, Pergamon Press, Oxford, 1977, pp. 371–385.

[7] Bazant, Z. P. and Estenssoro, L. F., "Surface Singularity and Crack Propagation," *International Journal of Solids and Structures,* Vol. 15, 1979, pp. 405–426.

[8] Solecki, J. S. and Swedlow, J. L., "On the Three-Dimensional Implications of LEFM: Finite Element Analysis of Straight and Curved Through-Cracks in a Plate," in *Fracture Mechanics: Sixteenth Symposium, ASTM STP 868,* M. F. Kanninen and A. T. Hopper, Eds., American Society for Testing and Materials, Philadelphia, 1985, pp. 535–553.

[9] Smith, C. W., Epstein, J. S., and Olaosebikan, O., "Boundary Layer Effects in Cracked Bodies: An Engineering Assessment," in *Fracture Mechanics: Seventeenth Volume, ASTM STP 905,* J. H. Underwood, R. Chait, C. W. Smith, D. P. Wilhem, W. A. Andrews, and J. C. Newman, Eds., American Society for Testing and Materials, Philadelphia, 1986, pp. 775–788.

[10] Smith, C. W., Epstein, J. S., and Olaosebikan, O., *Advances in Aerospace Structures, Materials and Dynamics,* ASME-Aero Div.-06, 1983, pp. 119–126.

[11] Smith, C. W. and Epstein, J. S., "Measurement of Three Dimensional Effects in Cracked Bodies," in *Proceedings,* 5th International Congress on Experimental Stress Analysis, June 1984.

[12] Folias, E. S., "On the Three-Dimensional Theory of Cracked Plates," *Journal of Applied Mechanics, Transactions of ASME,* Vol. 42, Sept. 1975, pp. 663–674.

[13] Shivakumar, K. N., Tan, P. W., and Newman, J. C., Jr., "A Virtual Crack-Closure Technique for Calculating Stress-Intensity Factors for Cracked Three-Dimensional Bodies, *International Journal of Fracture,* Vol. 36, March 1988, pp. R43–R50.

[14] Raju, I. S., Shivakumar, K. N., and Crews, J. H., Jr., "Three-Dimensional Elastic Analysis of a Composite Double-Cantilever-Beam Specimen," AIAA Paper No. AIAA-87-0864, AIAA/ASME/ASCE/AHS 28th Structures, Structural Dynamics, and Materials Conference, Monterey, Calif., April 1987.

[15] Raju, I. S. and Newman, J. C., Jr., "Three-Dimensional Finite Element Analysis of Finite-Thickness Fracture Specimens," NASA TN D-8414, 1977.

[*16*] Atluri, S. N., Kathiresan, K., and Kobayashi, A. S., "Three-Dimensional Linear Elastic Fracture Mechanics Analysis by a Displacement Hybrid Finite Element Model," Paper No. L-7/3, Trans. 3rd Conf. on Structural Mechanics in Reactor Technology, London, Sept. 1975.

[*17*] Westergaard, H. M., "Bearing Pressures and Cracks," *Journal of Applied Mechanics,* Vol. 61, 1939, pp. A49–A53.

[*18*] Williams, M. L., "On the Stress Distribution at the Base of a Stationary Crack," *Journal of Applied Mechanics,* Vol. 24, 1957, pp. 109–204.

[*19*] Barsoum, R. S., "On the Use of Isoparametric Finite Elements in Linear Fracture Mechanics," *International Journal of Numerical Methods in Engineering,* Vol. 10, 1976, pp. 25–37.

[*20*] Henshell, R. D. and Shaw, K. G., "Crack Tip Elements Are Necessary," *International Journal of Numerical Methods in Engineering,* Vol. 9, 1975, pp. 496–507.

[*21*] Hibbit, H. D., "Some Properties of Singular Isoparametric Elements," *International Journal of Numerical Methods in Engineering,* Vol. 11, 1977, pp. 180–184.

[*22*] Hinton, E., Scott, F. C., and Ricketts, R. E., "Local Least Square Stress Smoothing for Parabolic Isoparametric Elements," *International Journal of Numerical Methods in Engineering,* Vol. 9, 1975, pp. 235–256.

[*23*] Tada, H., Paris, P. C., and Irwin, G. R., *The Stress Analysis of Cracks Handbook,* Del Research Corp., Hellertown, Pa., 1975.

[*24*] Rybicki, E. F. and Kanninen, M. F., "A Finite Element Calculation of Stress Intensity Factors by Modified Crack Closure Integral," *Engineering Fracture Mechanics Journal,* Vol. 9, 1977, pp. 931–938.

[*25*] Irwin, G. R., "Analysis of Stresses and Strains Near the End of a Crack Traversing a Plate," *Journal of Applied Mechanics,* Vol. 24, 1957, pp. 361–364.

M. Oore[1]

Assessment of Influence Function for Elliptical Cracks Subjected to Uniform Tension and to Pure Bending

REFERENCE: Oore, M., "Assessment of Influence Function for Elliptical Cracks Subjected to Uniform Tension and to Pure Bending," *Fracture Mechanics: Twenty-First Symposium, ASTM STP 1074,* J. P. Gudas, J. A. Joyce, and E. M. Hackett, Eds., American Society for Testing and Materials, Philadelphia, 1990, pp. 490–508.

ABSTRACT: A root mean square (RMS) type stress intensity factor, $\overline{K}$, which is defined and used in [1] and [2] for crack growth and fracture calculations, is investigated in this study. Whereas the common $\overline{K}$ studies use approximate numerical integration of the influence function to analyze various configurations, in the present study the $\overline{K}$ definition is used analytically to derive accurate $\overline{K}$ values for various elliptical cracks subjected to uniform tensile stress and also to pure bending. The significance of these $\overline{K}$ results are explored via comparison with the exact classical maximum and minimum stress intensity factors along the front of the corresponding elliptical cracks.

KEY WORDS: stress intensity factor, elliptical crack, influence function, fatigue crack growth, strain energy release, weight function

Among the various methods for calculating stress intensity factors for three-dimensional (3-D) crack geometries, one which receives considerable attention and appears to be widely used is the Influence Function presented by Cruse and Besuner [1,2]. This method, which is based on the work in [3], does not give the variation of the stress intensity factor K along the crack front of a 3-D crack but rather an averaged value, $\overline{K}$, depending on the assumed direction of propagation. For crack front extension in one degree of freedom i, the Influence Function is intended to calculate a $\overline{K}_i$ value equal to

$$\overline{K}_i = \left[\frac{1}{\delta A_i} \int \int_{\delta A_i} K^2(s)\, dA_i \right]^{1/2} \tag{1}$$

where δA_i is the change in crack area due to the front extension (in the degree of freedom i), $K(s)$ is the local K at any point along the crack front s, and dA_i is local crack area increase at a point on s.

In the existing literature, Eq 1 is used for definition (not for computation) because in the practical cases where $\overline{K}_i$ is sought, $K(s)$ is usually not known along the crack front. In order to calculate $\overline{K}_i$, Cruse and Besuner derived influence functions $F_i(x;y)$ (weight functions) which depend only on the particular geometry, not on the loading. They demonstrated that $\overline{K}_i$ can be obtained by integrating over the crack area (A) the product of $F_i(x;y)$ and the

---

[1] Chief, Aeronautical Support Engineering, I.M.P. Aerospace Division, Halifax, Nova Scotia, Canada B3L 4T1.

normal stress $\sigma_z(x;y)$ in the uncracked material as follows:

$$\overline{K}_i = \int\int_{(A)} F_i(x;y)\sigma_z(x;y)\, dA \tag{2}$$

This procedure, like any other method based on numerical integration of weight functions with any given stress field over the crack area, is usually more economical than finite element (FE) analysis.

To demonstrate the accuracy of the procedure, Besuner applied it to an embedded elliptical crack (Fig. 1) in a cyclic stress field [2]. By using $\overline{K}_x$ and $\overline{K}_y$ in the major and minor axis directions respectively (Fig. 2) in conjunction with a fatigue crack growth formula, Besuner predicted the number of cycles required for that crack to grow (assuming elliptical growth pattern) until it reaches a critical size (i.e., at which fracture occurs). Repeating the fatigue crack growth calculations but using the exact solution for K_x and K_y and assuming elliptical crack growth pattern gave similar cyclic life [2]. Besuner demonstrated this agreement for several aspect ratios of the initial elliptical crack and for uniform as well as parabolic stress distributions.

Since that time, this influence function concept has been adapted to analyze a variety of crack configurations: a semi-elliptical surface crack in a half-infinite solid, a quarter-elliptical corner crack in a quarter-infinite solid, a quarter-circular crack in the crotch of a vessel nozzle junction, and others [4,5]. However, no direct comparison of the $\overline{K}$ values with the corresponding classical K solutions is provided. The intent of the present study is to derive such a direct comparison in order to assess the potential accuracy of predictions by the Influence Function—in particular, to identify whether the use of $\overline{K}$ could produce significantly unconservative results.

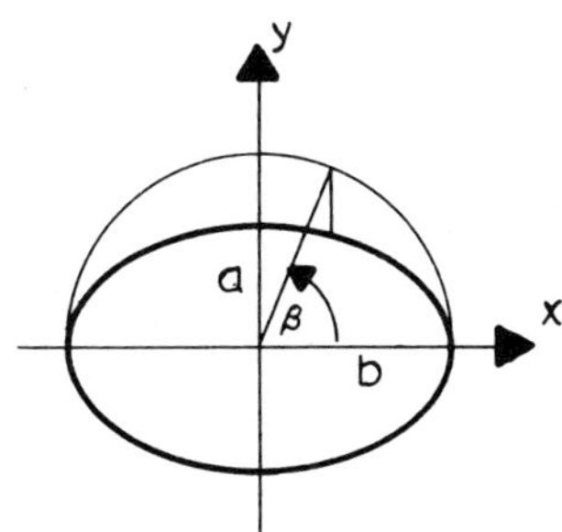

FIG. 1—*Elliptical crack.*

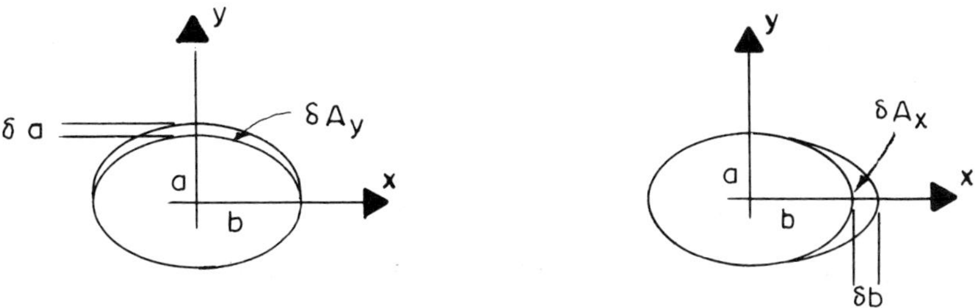

FIG. 2—*Elliptical crack extensions in two degrees of freedom.*

Summary

The $\overline{K}$ method is widely used for crack growth and fracture calculations. Various investigators who seem to favor this method calculate the $\overline{K}$ values by some approximate techniques for selected crack configurations. They then demonstrate the $\overline{K}$ proximity to the local K values derived from FE (or other numerical) analysis or from experimental crack growth data. The latter agreement is regarded by some as a substantiation of $\overline{K}$ for general use. Only satisfactory agreement has been published to date; no significant discrepancy has been reported to the present author's knowledge.

The objective of this study is to derive an exact comparison between accurately calculated $\overline{K}$ values and a known closed-form K solution. For that purpose the well-known configuration of embedded elliptical crack is selected as the most suitable choice for comparison. The $\overline{K}$ functions are then derived analytically and calculated to an accuracy of 0.01% for a family of elliptical cracks of various aspect ratios. Comparing these $\overline{K}$ results with the corresponding classical point-wise K solution shows that a significant discrepancy does exist for slender elliptical cracks. This suggests that the $\overline{K}$ method should be used with caution when applied to new geometry loading configurations. This discrepancy may also invalidate the $\overline{K}$ value (despite its being related to energy change) as a quantity of fundamental significance for evaluating onset of fracture in elliptical as well as other 3-D crack configurations.

It will be noted that in the present study, $\overline{K}$ is calculated directly, using Eq 1, for embedded elliptical cracks and then ratioed by the exact maximum and minimum K along the front. The use of the definition of $\overline{K}$ (Eq 1) for computation is intended to avoid possible inaccuracies of the weight function expression $F_i(x;y)$ in Eq 2, which is commonly used by others via approximate numerical procedure.

Application of Influence Function to Elliptical Cracks

For an embedded elliptical crack (Fig. 1), four directions of propagation are identified (four degrees of freedom): $+x$, $-x$, $+y$, and $-y$. Due to symmetry, only the positive directions are considered (Fig. 2). The general $\overline{K}$ expression per Eq 1 in terms of the parametric angle β is

$$\overline{K} = \left[\frac{1}{\delta A} \int \int_{\delta A} K^2(\beta)\, dA \right]^{1/2} \tag{3}$$

The $\overline{K}$ in the y direction is

$$\overline{K}_y = \left[\frac{1}{\delta A_y} \int \int_{\delta A y} K^2(\beta)\, dA \right]^{1/2} \tag{4a}$$

and in the x direction is

$$\overline{K}_x = \left[\frac{1}{\delta A_x} \int \int_{\delta A x} K^2(\beta)\, dA \right]^{1/2} \tag{4b}$$

where δA_x and δA_y are crack area changes due to growth δb and δa of the major and minor axes of the ellipse respectively.

In order to perform the integration in Eqs 3 and 4, dA is expressed as

$$dA = (d\delta)(ds) \tag{5}$$

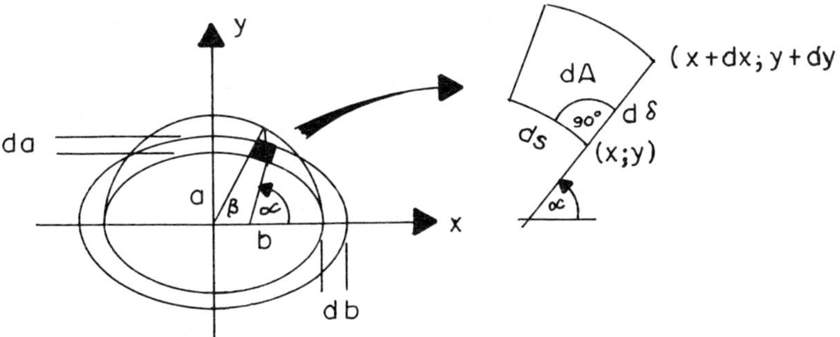

FIG. 3—*Simultaneous front extension in two degrees of freedom of elliptical crack.*

where ds is an infinitesimal length of crack front, and $d\delta$ is an increment of crack front extension perpendicular to the crack front (Fig. 3).

To use Eq 5 in the integration in Eqs 3 and 4, ds and $d\delta$ must be expressed in terms of the parametric angle β.

The expression for ds is derived below.

Consider the ellipse equation defined by

$$x = b \cos \beta \tag{6}$$

$$y = a \sin \beta \tag{7}$$

By differentiating Eqs 6 and 7, we obtain

$$dx = -b \sin \beta \, d\beta \tag{8}$$

$$dy = a \cos \beta \, d\beta \tag{9}$$

Substituting Eqs 8 and 9 in

$$ds = \sqrt{dx^2 + dy^2} \tag{10}$$

gives

$$ds = b \left[\sin^2 \beta + \left(\frac{a}{b}\right)^2 \cos^2 \beta \right]^{1/2} d\beta \tag{11}$$

The expression for $d\delta$ in terms of the parametric angle β and the growth of the axes da and db is derived below.

The ellipse equation is given by

$$\frac{x^2}{b^2} + \frac{y^2}{a^2} = 1 \tag{12}$$

Considering the growing crack to maintain an elliptical shape, the parameters in the ellipse equation, which are time-dependent $x(t)$, $y(t)$, $a(t)$, and $b(t)$, are related by

$$\frac{x^2(t)}{b^2(t)} + \frac{y^2(t)}{a^2(t)} = 1 \tag{13}$$

In the following, it will be designated that $x(t) = x$, $y(t) = y$, $a(t) = a$, and $b(t) = b$, and Eq 13 will take the form of Eq 12. Differentiating the ellipse equation (12) with respect to t gives

$$\frac{\partial}{\partial t}\left(\frac{x^2}{b^2} + \frac{y^2}{a^2}\right) = \frac{\partial}{\partial t}(1) = 0 \tag{14}$$

Performing the differentiation on the left-hand side of Eq 14 gives

$$\frac{2x\dfrac{\partial x}{\partial t}b^2 - 2x^2 b\dfrac{\partial b}{\partial t}}{b^4} + \frac{2y\dfrac{\partial y}{\partial t}a^2 - 2y^2 a\dfrac{\partial a}{\partial t}}{a^4} = 0 \tag{15}$$

Consider the propagating crack front in the first quadrant ($x \geq 0$, $y \geq 0$) as shown in Fig. 3 (due to symmetry, this applies also to the crack front in the other quadrants). It can be seen that if we consider the increments dx and dy as the point $(x;y)$ propagates along the normal extension $d\delta$ at this point on the front, then

$$dx = d\delta \cos \alpha \tag{16}$$

$$dy = d\delta \sin \alpha \tag{17}$$

where α is the angle that the normal to the front (at this point) makes with the x axis. Considering the increment of growth as a function of the growth rate gives

$$dx = \frac{\partial x}{\partial t} dt \tag{18}$$

$$dy = \frac{\partial y}{\partial t} dt \tag{19}$$

$$d\delta = \frac{\partial \delta}{\partial t} dt \tag{20}$$

Substituting Eqs 18 and 20 in Eq 16, and Eqs 19 and 20 in Eq 17, gives the following growth rates relations:

$$\frac{\partial x}{\partial t} = \frac{\partial \delta}{\partial t} \cos \alpha \tag{21}$$

$$\frac{\partial y}{\partial t} = \frac{\partial \delta}{\partial t} \sin \alpha \tag{22}$$

Substituting Eqs 21 and 22 in Eq 15, rearranging and isolating $\partial \delta / \partial t$ gives

$$\frac{\partial \delta}{\partial t} = \frac{\dfrac{1}{b}\left(\dfrac{x}{b}\right)^2 \dfrac{\partial b}{\partial t} + \dfrac{1}{a}\left(\dfrac{y}{a}\right)^2 \dfrac{\partial a}{\partial t}}{\dfrac{1}{b}\left(\dfrac{x}{b}\right)\cos \alpha + \dfrac{1}{a}\left(\dfrac{y}{a}\right)\sin \alpha} \tag{23}$$

To replace α with the parametric angle β, the following mathematical manipulation is used. Differentiating the ellipse equation with respect to x and isolating dy/dx gives

$$\frac{dy}{dx} = -\frac{x}{y}\frac{a^2}{b^2} \tag{24}$$

which is the slope of the tangent at that point. The slope of the normal at that point is tan α, and by utilizing Eq 24 it can be expressed as

$$\tan \alpha = -\frac{1}{\dfrac{dy}{dx}} = \frac{y}{x}\frac{b^2}{a^2} \tag{25}$$

Substituting Eqs 6 and 7 in Eq 25 gives

$$\tan \alpha = \frac{b}{a}\frac{\sin \beta}{\cos \beta} \tag{26}$$

By substituting Eq 26 in known trigonometrical formulae for sin α and cos α, we obtain

$$\sin \alpha = \frac{\tan \alpha}{\sqrt{1 + \tan^2 \alpha}} = \frac{\sin \beta}{\sqrt{\sin^2 \beta + \left(\dfrac{a}{b}\right)^2 \cos^2 \beta}} \tag{27}$$

$$\cos \alpha = \frac{1}{\sqrt{1 + \tan^2 \alpha}} = \frac{\cos \beta}{\sqrt{\sin^2 \beta + \left(\dfrac{a}{b}\right)^2 \cos^2 \beta}} \tag{28}$$

Substituting Eqs 6, 7, 27, and 28 in Eq 23 and rearranging gives

$$\frac{\partial \delta}{\partial t} = \frac{\dfrac{\partial a}{\partial t}\sin^2 \beta + \dfrac{\partial b}{\partial t}\dfrac{a}{b}\cos^2 \beta}{\left[\sin^2 \beta + \left(\dfrac{a}{b}\right)^2 \cos^2 \beta\right]^{1/2}} \tag{29}$$

To obtain an expression for $d\delta$, Eq 29 is substituted in Eq 20 giving

$$d\delta = \frac{\dfrac{\partial a}{\partial t}dt \sin^2 \beta + \dfrac{\partial b}{\partial t}dt \dfrac{a}{b}\cos^2 \beta}{\left[\sin^2 \beta + \left(\dfrac{a}{b}\right)^2 \cos^2 \beta\right]^{1/2}} \tag{30}$$

Since

$$da = \frac{\partial a}{\partial t}dt \tag{31}$$

and

$$db = \frac{\partial b}{\partial t} dt \tag{32}$$

Eq 30 becomes

$$d\delta = \frac{da \sin^2 \beta + db \frac{a}{b} \cos^2 \beta}{\left[\sin^2 \beta + \left(\frac{a}{b} \right)^2 \cos^2 \beta \right]^{1/2}} \tag{33}$$

Substituting Eqs 11 and 33 in Eq 5 gives

$$dA = b \sin^2 \beta \, d\beta \, da + a \cos^2 \beta \, d\beta \, db \tag{34}$$

For small increments δa and δb of the ellipse axes due to growth in the y and x directions, the increment δA of the area along a quarter of the ellipse perimeter is obtained by integrating Eq 34 as follows:

$$\delta A = \int \int dA = \int_{a=a}^{a+\delta a} \int_{\beta=0}^{\pi/2} b \sin^2 \beta \, da \, d\beta + \int_{b=b}^{b+\delta b} \int_{\beta=0}^{\pi/2} a \cos^2 \beta \, db \, d\beta \tag{35}$$

Due to the small magnitude of δa and δb it can be assumed that a and b are constants during the integration. Integrating and substituting the limits in Eq 35 gives

$$\delta A = \frac{\pi}{4} (b\delta a + a\delta b) \tag{36}$$

This expression for the incremental area along a quarter of the perimeter agrees with the expression that can be easily derived from the formula of the ellipse area and thus confirms the validity of Eq 34. Note that, due to symmetry, the investigation of one quarter of the ellipse front is sufficient.

Substituting Eqs 34 and 36 in Eq 3 gives

$$\overline{K} = \left\{ \frac{1}{\frac{\pi}{4} (b\delta a + a\delta b)} \int \int_{(\delta A)} K^2 (\beta)[b \sin^2 \beta \, d\beta \, da + a \cos^2 \beta \, d\beta \, db] \right\}^{1/2} \tag{37}$$

Introducing the limits of integrations in Eq 37 gives

$$\overline{K} = \left\{ \frac{4}{\pi(b\delta a + a\delta b)} \left[\int_{a=a}^{a+\delta a} \int_{\beta=0}^{\pi/2} K^2(\beta)b \sin^2 \beta \, d\beta \, da \right. \right.$$
$$\left. \left. + \int_{b=b}^{b+\delta b} \int_{\beta=0}^{\pi/2} K^2(\beta)a \cos^2 \beta \, d\beta \, db \right] \right\}^{1/2} \tag{38}$$

Since δa and δb are small, a and b will be considered constants during the integration. Performing the integration once, for da and for db gives

$$\overline{K} = \left[\frac{\delta a \int_{\beta=0}^{\pi/2} K^2(\beta) b \sin^2 \beta \, d\beta + \delta b \int_{\beta=0}^{\pi/2} K^2(\beta) a \cos^2 \beta \, d\beta}{\frac{\pi}{4}(b\delta a + a\delta b)} \right]^{1/2} \tag{39}$$

for crack growth in the y and x directions simultaneously.

To obtain $\overline{K}_y$, we have to consider crack growth in the y direction only (i.e., to substitute $\delta b = 0$ in Eq 39), giving

$$\overline{K}_y = \left[\frac{4}{\pi} \int_{\beta=0}^{\pi/2} K^2(\beta) \sin^2 \beta \, d\beta \right]^{1/2} \tag{40}$$

To obtain $\overline{K}_x$, we have to consider crack growth in the x direction only (i.e., to substitute $\delta a = 0$ in Eq 39), giving

$$\overline{K}_x = \left[\frac{4}{\pi} \int_{\beta=0}^{\pi/2} K^2(\beta) \cos^2 \beta \, d\beta \right]^{1/2} \tag{41}$$

Investigation of the Influence Function for Elliptical Cracks Subjected to Uniform Stress

For an embedded elliptical crack in an infinite solid subjected to a uniform normal stress σ, the stress intensity factor $K(\beta)$ along the crack front is given [6] by:

$$K(\beta) = \frac{\sigma \sqrt{\pi a}}{E(k)} \left(\sin^2 \beta + \frac{a^2}{b^2} \cos^2 \beta \right)^{1/4} \tag{42}$$

$E(k)$ is a complete elliptic integral of the second kind given by

$$E(k) = \int_0^{\pi/2} (1 - k^2 \sin^2 \phi)^{1/2} \, d\phi \tag{42a}$$

where

$$k^2 = 1 - \frac{a^2}{b^2} \tag{42b}$$

Substituting Eq 42 in Eq 40 and Eq 42 in Eq 41 gives the following expressions for $\overline{K}_y$ and $\overline{K}_x$ respectively for uniform stress field:

$$\overline{K}_y = \frac{2\sigma \sqrt{a}}{E(k)} \left[\int_0^{\pi/2} \left(\sin^2 \beta + \frac{a^2}{b^2} \cos^2 \beta \right)^{1/2} \sin^2 \beta \, d\beta \right]^{1/2} \tag{43}$$

$$\overline{K}_x = \frac{2\sigma \sqrt{a}}{E(k)} \left[\int_0^{\pi/2} \left(\sin^2 \beta + \frac{a^2}{b^2} \cos^2 \beta \right)^{1/2} \cos^2 \beta \, d\beta \right]^{1/2} \tag{44}$$

The exact stress intensity factors at the end of the minor and major axes of the ellipse are obtained from Eq 42 as

$$K_{a(\beta=\pi/2)} = \frac{\sigma \sqrt{\pi a}}{E(k)} \tag{45}$$

and

$$K_{b(\beta=0)} = \frac{\sigma \sqrt{\pi a}}{E(k)} \sqrt{\frac{a}{b}} \tag{46}$$

K_a and K_b are in fact respectively the maximum and minimum stress intensity factors along the crack front. For comparison of $\overline{K}_y$ and $\overline{K}_x$ with the exact stress intensity factors in the two axis directions, $\overline{K}_y$ from Eq 43 is ratioed by K_a from Eq 45 and $\overline{K}_x$ from Eq 44 is ratioed by K_b from Eq 46 giving

$$\frac{\overline{K}_y}{K_a} = \frac{2}{\sqrt{\pi}} \left[\int_0^{\pi/2} \left(\sin^2 \beta + \frac{a^2}{b^2} \cos^2 \beta \right)^{1/2} \sin^2 \beta \, d\beta \right]^{1/2} \tag{47}$$

and

$$\frac{\overline{K}_x}{K_b} = \frac{2}{\sqrt{\pi}} \sqrt{\frac{b}{a}} \left[\int_0^{\pi/2} \left(\sin^2 \beta + \frac{a^2}{b^2} \cos^2 \beta \right)^{1/2} \cos^2 \beta \, d\beta \right]^{1/2} \tag{48}$$

The above ratios of $\overline{K}_y/K_a$ and $\overline{K}_x/K_b$ for uniform stress loading were calculated by numerical integration of Eqs 47 and 48 for various values of a/b. A numerical integration subroutine with accuracy better than 0.01% (on the pocket calculator HP 15C) was used and gave the results shown in Table 1. It is seen from Table 1 that for $a/b = 1$ both $\overline{K}_y/K_a$ and $\overline{K}_x/K_b$ are equal to 1. This is expected since a circular crack ($a/b = 1$) subjected to uniform stress field has a uniform K along the front and an RMS averaging will give the same K value. As the aspect ratio a/b decreases towards zero, the ratio $\overline{K}_y/K_a$ is decreasing too, indicating that $\overline{K}_y$ is underestimating somewhat K_a. For aspect ratios a/b between 0.5 and 1.0 the underestimation is less than 5.5%. For slender ellipses with aspect ratios between 0.05 and 0.1 the underestimation is approximately 8%. These underestimations are not very significant but could have an effect in some calculations, especially if an exponential function of $\overline{K}_y$ is used. The ratio $\overline{K}_x/K_b$, on the other hand, is increasing significantly above one as the aspect ratio a/b is decreasing towards zero. This indicates that $\overline{K}_x$ is always an overestimate of K_b in an elliptical crack subjected to a uniform stress.

TABLE 1—*Comparison of K values for uniform stress.*

a/b	0.05	0.1	0.2	0.4	0.6	0.8	0.1
$\overline{K}_y/K_a$	0.9216	0.923	0.926	0.938	0.955	0.976	1.00
$\overline{K}_x/K_b$	2.933	2.104	1.55	1.21	1.090	1.033	1.00

Investigation of the Influence Function for Embedded Elliptical Cracks Subjected to Pure Bending

For an embedded elliptical crack in an infinite solid and subjected to a linearly varying stress field σ of the form

$$\sigma = \sigma_0 \left(1 + \frac{y}{a} \right) \tag{49}$$

a $K(\beta)$ solution along the front is given [7] by

$$K(\beta) = \frac{\sigma_0 \sqrt{\pi a}}{E(k)} \left(\sin^2 \beta + \frac{a^2}{b^2} \cos^2 \beta \right)^{1/4} \left[1 + \frac{k^2 E(k) \sin \beta}{(1 + k^2)E(k) - k'^2 K(k)} \right] \tag{50}$$

where $E(k)$ and k are specified along with Eq 42 and

$$k'^2 = 1 - k^2 \tag{51}$$

$$K(k) = \int_0^{\pi/2} \frac{d\phi}{(1 - k^2 \sin^2 \phi)^{1/2}} \tag{52}$$

$K(\beta)$ in Eq 50 is in fact a superposition of K solutions for uniform stress field; $\sigma = \sigma_0$ and pure bending; $\sigma = \sigma_0(y/a)$. By subtracting $K(\beta)$ for uniform stress (Eq 42) from Eq 50, the $K(\beta)$ solution for pure bending is obtained as

$$K(\beta) = \frac{\sigma_0 \sqrt{\pi a}\, k^2}{(1 + k^2)E(k) - k'^2 K(k)} \sin \beta \left(\sin^2 \beta + \frac{a^2}{b^2} \cos^2 \beta \right)^{1/4} \tag{53}$$

Note that the expression in Eq 53 does not consider crack closing. Substituting Eq 53 in Eq 40 and Eq 53 in Eq 41 gives the following expressions for $\overline{K}_y$ and $\overline{K}_x$ respectively for pure bending:

$$\overline{K}_y = \frac{2\sigma_0 \sqrt{a}\, k^2}{(1 + k^2)E(k) - k'^2\, K(k)} \left[\int_0^{\pi/2} \left(\sin^2 \beta + \frac{a^2}{b^2} \cos^2 \beta \right)^{1/2} \sin^4 \beta\, d\beta \right]^{1/2} \tag{54}$$

$$\overline{K}_x = \frac{2\sigma_0 \sqrt{a}\, k^2}{(1 + k^2)E(k) - k'^2\, K(k)} \left[\int_0^{\pi/2} \left(\sin^2 \beta + \frac{a^2}{b^2} \cos^2 \beta \right)^{1/2} \sin^2 \beta \cos^2 \beta\, d\beta \right]^{1/2} \tag{55}$$

The exact stress intensity factors at the ends of the minor and major axes of the ellipse are obtained from Eq 53 as

$$K_{a(\beta = \pi/2)} = \frac{\sigma_0 \sqrt{\pi a}\, k^2}{(1 + k^2)E(k) - k'^2 K(k)} \tag{56}$$

and

$$K_{b(\beta = 0)} = 0 \tag{57}$$

K_a is the maximum stress intensity factor along the crack front. For comparison of $\overline{K}_y$ with the exact stress intensity factor K_a in that direction, $\overline{K}_y$ from Eq 54 is ratioed by K_a from Eq 56. Since K_b is equal to zero, $\overline{K}_x$ will also be ratioed by K_a from Eq 56. These ratios are

$$\frac{\overline{K}_y}{K_a} = \frac{2}{\sqrt{\pi}} \left[\int_0^{\pi/2} \left(\sin^2 \beta + \frac{a^2}{b^2} \cos^2 \beta \right)^{1/2} \sin^4 \beta \, d\beta \right]^{1/2} \tag{58}$$

and

$$\frac{\overline{K}_x}{K_a} = \frac{2}{\sqrt{\pi}} \left[\int_0^{\pi/2} \left(\sin^2 \beta + \frac{a^2}{b^2} \cos^2 \beta \right)^{1/4} \sin^2 \beta \cos^2 \beta \, d\beta \right]^{1/2} \tag{59}$$

The above expressions of $\overline{K}_y/K_a$ and $\overline{K}_x/K_a$ for pure bending were computed by numerical integration of Eqs 58 and 59 for various values of a/b. A numerical integration subroutine with accuracy better than 0.01% was used and gave the results shown in Table 2. It is seen from Table 2 that $\overline{K}_y$ is always less than K_a. This is in fact expected even for $a/b = 1$ because the non-uniform stress loading is causing a non-uniform $K(\beta)$ distribution even for a circular crack. Therefore the RMS average stress intensity factor, $\overline{K}_y$, is less than the maximum K_a. For $a/b = 1$ the underestimation is 13% and it grows larger for lower values of a/b. For a slender ellipse with an aspect ratio of $a/b = 0.1$ the underestimation is 17.5%. Although pure bending may not be the actual type of loading in practice (more probably a combination of tension and bending), the bending results indicate the potential difference between $\overline{K}$ and the exact K. It is also realized from Table 2 that the $\overline{K}_x$ values are significant (about one half of $\overline{K}_y$) and are therefore conservative in comparison with the exact K_b which is zero for pure bending. In summary, the $\overline{K}_y/K_a$ and $\overline{K}_x/K_b$ values for both uniform tension and bending loadings are expected to deviate from one, especially for slender elliptical cracks (i.e., when $a/b \simeq 0$).

Discussion

The Influence Function procedure calculates $\overline{K}_y$ and uses it for fracture evaluation by comparison with K_c (which is obtained from two-dimensional testing). The fact that $\overline{K}_y$ is lower than the local maximum K for both uniform and bending stress loadings (as quantified in this study) can be significant in fracture evaluation if the onset of fracture depends on the maximum local K along the front rather than on an averaged value. Hence, the use of $\overline{K}_y$ as a criterion for fracture needs to be verified by proper testing before it can be used for fracture evaluation. It was reported [8] that fracture testing of corner cracks in nozzles showed that fracture occurred when the local $K_{\max} \approx K_c$. Hence, a maximum K along the front, rather than an averaged value, may need to be used for fracture evaluation of such a configuration.

The $\overline{K}_x$ by the Influence Function is higher than K_b for both the uniform and the bending

TABLE 2—*Comparison of* K *values for pure bending.*

a/b	0.05	0.1	0.2	0.4	0.5	0.6	0.8	1.0
$\overline{K}_y/K_a$	0.824	0.825	0.826	0.832	0.836	0.841	0.852	0.866
$\overline{K}_x/K_a$	0.412	0.413	0.418	0.432	0.442	0.452	0.475	0.5

stress loadings. This will not have a direct effect on fracture evaluation, because the fracture is assessed by the larger value of $\overline{K}$ which is $\overline{K}_y$ for the subject configuration and loading. However, $\overline{K}_x$ will affect fatigue crack growth calculations by Influence Function. The use of $\overline{K}_x$ will give a faster crack growth than the use of K_b in that direction. On the other hand, the use of $\overline{K}_y$ in the calculation of the simultaneous fatigue crack growth in the y direction will give slower crack growth than the use of K_a. Therefore the crack shapes (aspect ratio) predicted by the Influence Function procedure will be different (i.e., more slender than those calculated using the local maximum and minimum K values in the two directions) (for the conditions for exact elliptical growth pattern see [9]). It is reported [2] that the number of uniform stress cycles calculated to grow an elliptical crack to a critical size using the Influence Function procedure is close to that calculated using the exact K_a and K_b solutions. This suggests that the overestimation of crack growth by Influence Function in the x direction and its underestimation of growth in the y direction averaged out. This means that after N cycles, the Influence Function gave a crack shape with $\overline{K}_y(N)$ which is close to the $K_a(N)$ of the crack that was calculated to grow through the same number of cycles but using the classical K_a and K_b. However, if fracture is to occur after a number of cycles which is not large, the results by the Influence Function could be unconservative because the crack may not have time to grow sufficiently to match $\overline{K}_y$ to K_a. (Note that after a large number of cycles the shape of the crack approaches a circular shape for which $\overline{K}_y = K_a$.)

The question arises as to what is the justification for using $\overline{K}_y$ and $\overline{K}_x$ for fatigue crack growth and for fracture evaluation. Should they be regarded merely as approximation for the maximum and minimum K along the crack front, or do they represent some established physical quantities that validate the general use of $\overline{K}_y$ and $\overline{K}_x$ in fracture mechanics? Another possibility is that $\overline{K}_i$ is a quantity that has empirical validity for specific configurations, that can be substantiated by experimental data.

To explore the meaning of $\overline{K}$, consider Eq 1 for $\overline{K}_i$ which assumes crack growth in one degree of freedom (i). K^2 is proportional to the energy release G per crack area extension as given for plane strain by the equation

$$G = \frac{K^2}{E/(1 - \nu^2)} = \frac{K^2}{H} \tag{60}$$

where $E/(1 - \nu^2)$ is the appropriate elastic modulus. It follows that the quantity $K^2(s)\, dA_i$ is proportional to the energy released locally at point s due to crack extension in degree of freedom i. Hence, the double integral in Eq 1 is proportional to the total energy released due to increase δA_i in crack area, and $\overline{K}_i^2$ is therefore equivalent to the total energy released per crack area increase due to crack growth in the i direction only. It stems from the above that $\overline{K}_y^2$ (Eq 40) represents the energy released per crack area increase due to elliptical crack growth in the y direction only ($\delta b = 0$), whereas $\overline{K}_x^2$ (Eq 41) represents the energy released per area increase due to crack growth in the x direction only ($\delta a = 0$). However, if the crack grows in the x and y directions simultaneously (as usually happens in practice), then $\overline{K}_x^2$ and $\overline{K}_y^2$ may lose their meaning as discrete energy release rate because δa and δb are both non-zero. In such a simultaneous crack growth state, the total energy released per crack area increase, will be a single value represented by $\overline{K}^2$ in Eq 39, which is different from $\overline{K}_y^2$ or $\overline{K}_x^2$ or $\overline{K}_y^2 + \overline{K}_x^2$.

It will be noted, however, that the total energy released δE can be shown to be equal to

$$\delta E = \frac{\overline{K}^2\, \delta A}{H} = \frac{\overline{K}_x^2\, \delta A_x}{H} + \frac{\overline{K}_y^2\, \delta A_y}{H} \tag{60a}$$

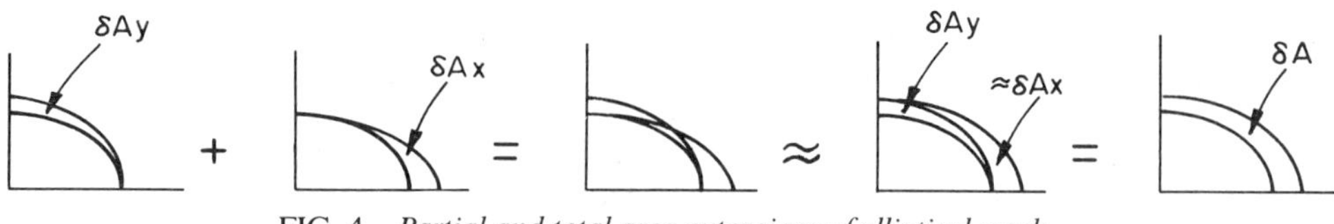

FIG. 4—*Partial and total area extensions of elliptical crack.*

and that the total area increase δA due to simultaneous growth of the axes in x and y (Fig. 4) is given by

$$\delta A = \delta A_x + \delta A_y \tag{60b}$$

as derived in Eqs 77, 79, and 63 below.

The use of $\overline{K}_y$ and $\overline{K}_x$ to calculate fatigue crack growth for simultaneous extension in y and x directions is shown to give good predictions in some cases [5] where there was a very small growth in one direction. It is not likely to give good prediction in other cases [10] where the highest K near the surface is required in order to give good crack growth predictions, as quantified in [11]. Note that also during stable crack growth an elliptical crack can be expected to grow in the x and y directions simultaneously; thus the above discussion is applicable also for this case.

In general, the merit of $\overline{K}_y$ and $\overline{K}_x$ is mainly in being an economical approximation for K_{max} and K_{min}. In cases where experimental data show high correlation with the $\overline{K}_i$ values, the RMS Influence Function method can be regarded as having empirical validity, limited to these cases.

It will be pointed out that although the assessment in this study is for uniform tensile stress and for bending, the $\overline{K}_{max}/K_{max}$ ratio for a loading combination of uniform stress and bending is expected to be bounded between the values obtained for the two separate load cases. On the other hand, for higher polynomial stress distribution, the underestimation of K_{max} by $\overline{K}_{max}$ could be larger than the values obtained herein for bending stress. This is because the averaged stress intensity factor, $\overline{K}_i$, is lower (relative to K_{max}) if the K variation is stronger along the front. If the polynomial stress field is such that it produces a more uniform K distribution along the crack front (as the parabola stress distribution in [2]), then $\overline{K}_{max}$ tends to get closer to $\overline{K}_{max}$ and the life predictions by $\overline{K}_{max}$ will be closer as well. For surface cracks where the calculated K on the free surface is lower than the K near the surface, the $\overline{K}_{surface}$ may be almost equal to the calculated K on the surface. For semi-elliptical surface cracks, with the major axis along the surface, the proximity to the surface tends to increase the stress intensity factor in the region near the tip of the major axis. This can result in a reduced variation of K_{min} to K_{max} for certain load cases, and thus the underestimation of K_{max} by $\overline{K}_{max}$ could be very small. This geometry is outside the scope of the present study.

General Investigation of Relationship Between Energy Release Rate and $\overline{K}_i$ Due to Crack Extension in Various Degree of Freedom

Consider a crack shape defined by a function of n degrees of freedom $q_1 \ldots q_i \ldots q_n$. The area A of such a crack is a function of the degrees of freedom; hence

$$A = A(q_1 \ldots q_i \ldots q_n) \tag{61}$$

The change in area δA_i due to infinitesimal change dq_i in degree of freedom q_i is given by

$$\delta A_i = \frac{\partial A}{\partial q_i} dq_i \tag{62}$$

The total change in crack area, δA, due to sequential or simultaneous changes in all the degrees of freedom is, according to the differential theorem, given by

$$\delta A = \sum_{i=1}^{n} \frac{\partial A}{\partial q_i} dq_i = \sum_{i=1}^{n} \delta A_i \tag{63}$$

The elastic strain energy released due to the extension dq_i of q_i is

$$\delta E_i = \frac{\partial E}{\partial q_i} dq_i \tag{64}$$

The total energy released, δE, due to sequential or simultaneous increments in all degrees of freedom is

$$\delta E = \sum_{i=1}^{n} \frac{\partial E}{\partial q_i} dq_i \tag{65}$$

Since the energy can be regarded as a function of the crack area, and the crack area is a function of the degrees of freedom, we can write

$$\frac{\partial E}{\partial q_i} = \frac{\partial E}{\partial A_i} \frac{\partial A}{\partial q_i} \tag{66}$$

Substituting Eq 66 in Eq 64 gives

$$\delta E_i = \frac{\partial E}{\partial A_i} \frac{\partial A}{\partial q_i} dq_i \tag{67}$$

Substituting Eq 62 in Eq 67 gives

$$\delta E_i = \frac{\partial E}{\partial A_i} \delta A_i \tag{68}$$

Note that the quantity $\partial E/\partial A_i$ is the energy released per crack area increase, commonly designated by G.

From Eq 67 we can see that the total energy released due to increments dq_i in all degrees of freedom can be expressed as

$$\delta E = \sum_{i=1}^{n} \frac{\partial E}{\partial A_i} \frac{\partial A}{\partial q_i} dq_i \tag{69}$$

Substituting Eq 62 in Eq 69 gives

$$\delta E = \sum_{i=1}^{n} \frac{\partial E}{\partial A_i} \delta A_i \tag{70}$$

for sequential or simultaneous extension of the degrees of freedom.

Consider the energy released due to extension in a degree of freedom by integration of the energy released locally along the crack front s. The energy $dE(s)$ released per infinitesimal crack area increase $dA(s)$ at a local point (s) is known to be related to the local stress intensity factor $K(s)$ at that point by Eq 60:

$$\frac{dE(s)}{dA(s)} = \frac{K^2(s)}{H} \tag{71}$$

where H is an appropriate elastic modulus. For an isotropic material in plane strain $H = E/(1 - v^2)$. From Eq 71 we obtain

$$dE(s) = \frac{1}{H} K^2(s) \, dA(s) \tag{72}$$

By applying integration to Eq 72 over crack area increase δA_i (this results from an increase of dq_i) we obtain the energy δE_i released due to the corresponding increase δA_i in crack area as

$$\delta E_i = \int\int_{\delta A_i} \frac{1}{H} K^2(s) \, dA(s) \tag{73}$$

Substituting Eq 68 in Eq 73 and rearranging gives the energy released per crack area increase (due to dq_i) to be

$$\frac{\partial E}{\partial A_i} = \frac{1}{H \delta A_i} \int\int_{\delta A_i} K^2(s) \, dA(s) \tag{74}$$

The right-hand side of Eq 74 corresponds to the definition of $\overline{K}_i$ by the Influence Function method per Eq 1. Thus

$$\frac{\partial E}{\partial A_i} = \frac{1}{H} \overline{K}_i^2 \tag{75}$$

Applying δA_i to both sides of Eq 75 gives

$$\frac{\partial E}{\partial A_i} \delta A_i = \frac{1}{H} \overline{K}_i^2 \, \delta A_i \tag{76}$$

Summing each side of Eq 76 and using Eq 70 gives

$$\delta E = \sum_{i=1}^{n} \frac{\partial E}{\partial A_i} \delta A_i = \sum_{i=1}^{n} \frac{1}{H} \overline{K}_i^2 \, \delta A_i \tag{77}$$

as the total energy released due to sequential or simultaneous crack growth in all degrees of freedom.

By applying integration to both sides of Eq 72 over total crack area increase δA (due to extension of all degrees of freedom) we obtain

$$\delta E = \frac{1}{H} \int \int_{\delta A} K^2(s)\, dA(s) \tag{78}$$

Note that the right-hand side of Eq 78 represents the general $\overline{K}$ per Eq 3 for simultaneous extension of degrees of freedom. Substituting Eq 3 in Eq 78 gives

$$\delta E = \frac{1}{H} \overline{K}^2\, \delta A \tag{79}$$

By equating the expressions for the total energy released of Eq 77 and Eq 79 we obtain

$$\sum_{i=1}^{n} \overline{K}_i^2\, \delta A_i = \overline{K}^2\, \delta A \tag{80}$$

By substituting Eq 63 in Eq 80 we obtain

$$\sum_{i=1}^{n} \overline{K}_i^2\, \delta A_i = \overline{K}^2 \sum_{i=1}^{n} \delta A_i \tag{81}$$

Equation 81 provides the relationship among the $\overline{K}_i$ values and the "total" average $\overline{K}$ value, and is based on the derivation that shows that each side of Eq 81 represents the total energy released due to simultaneous crack extension in all the degrees of freedom.

Investigation of the Relationship Between $\overline{K}_x$, $\overline{K}_y$ and General $\overline{K}$ for the Special Case of Embedded Elliptical Cracks

It will be checked below whether the solutions obtained for $\overline{K}$, $\overline{K}_y$ and $\overline{K}_x$ in Eqs 39, 40 and 41 respectively satisfy the general relationship that has been derived and presented in Eq 81.

From Eq 36 we obtain

$$\delta A_y = \frac{\pi}{4}\, b\, \delta a \tag{82}$$

$$\delta A_x = \frac{\pi}{4}\, a\, \delta b \tag{83}$$

For an ellipse with two degrees of freedom a and b in the y and x directions respectively, the left-hand side of Eq 81 becomes

$$\sum_{i=1}^{n} \overline{K}_i^2\, \delta A_i = \overline{K}_y^2\, \delta A_y + \overline{K}_x^2\, \delta A_x \tag{84}$$

Substituting Eqs 40, 41, 82, and 83 in the right-hand side of Eq 84 gives for the elliptical crack:

$$\sum_{i=1}^{n} \overline{K}_i^2 \, \delta A_i = b \, \delta a \int_0^{\pi/2} K^2(\beta) \sin^2 \beta \, d\beta + a \, \delta b \int_0^{\pi/2} K^2(\beta) \cos^2 \beta \, d\beta \qquad (85)$$

The right-hand side of Eq 81 will be, for the present case of an ellipse, equal to

$$\overline{K}^2 \sum_{i=1}^{n} \delta A_i = \overline{K}^2 \, (\delta A_y + \delta A_x) \qquad (86)$$

Substituting Eqs 39, 82, and 83 in Eq 86 gives for the elliptical crack:

$$\overline{K}^2 \sum_{i=1}^{n} \delta A_i = b \, \delta a \int_0^{\pi/2} \overline{K}^2 (\beta) \sin^2 \beta \, d\beta + a \, \delta b \int_0^{\pi/2} \overline{K}^2 (\beta) \cos^2 \beta \, d\beta \qquad (87)$$

From the equality of the right-hand sides in Eqs 85 and 87 it is realized that the solutions obtained for the elliptical crack in Eqs 39, 40, and 41 satisfy the general equation (Eq 81). By substituting Eqs 84 and 86 in Eq 81 we obtain for the elliptical crack:

$$\overline{K}_y^2 \, \delta A_y + \overline{K}_x^2 \, \delta A_x = \overline{K}^2 \, (\delta A_y + \delta A_x) \qquad (88)$$

From Eq 88 we obtain the relationship

$$\overline{K}^2 = \overline{K}_y^2 \frac{\delta A_y}{\delta A_y + \delta A_x} + \overline{K}_x^2 \frac{\delta A_x}{\delta A_y + \delta A_x} \qquad (89)$$

As observed from Eq 63, or from substitution of Eqs 82 and 83 in Eq 36, we obtain for the elliptical crack:

$$\delta A = \delta A_y + \delta A_x \qquad (90)$$

Substituting Eq 90 in Eq 89 gives

$$\overline{K}^2 = \overline{K}_y^2 \frac{\delta A_y}{\delta A} + \overline{K}_x^2 \frac{\delta A_x}{\delta A} \qquad (91)$$

to be the relationship among $\overline{K}$, $\overline{K}_y$ and $\overline{K}_x$ in terms of the total and partial changes in crack area.

Conclusion

For embedded elliptical cracks subjected to uniform stress field, the RMS stress intensity factors, $\overline{K}_y$, which result from the basic definition of the Influence Function (Eq 1), underestimate the classical maximum stress intensity factor by up to 8%. If the elliptical crack aspect ratio is larger than 0.5, the underestimation will be less than 5.5%. For embedded elliptical cracks subjected to pure bending, the above RMS stress intensity factor can be as much as 18% below the classical K_{max}. On the other hand, the $\overline{K}_x$ results overestimate the classical minimum K for uniform stress field and also for pure bending.

Fatigue crack growth analysis using $\overline{K}_y$ and $\overline{K}_x$ for elliptical cracks can give crack shape patterns and a number of cycles to failure different from those predicted by using the classical K_a and K_b solutions. The discrepancy in the results of cycles to failure may be significant if failure is to occur after a small amount of crack growth, but tends to become insignificant if the cyclic life is long enough to allow substantial crack growth before reaching a critical size.

It is also concluded that the total strain energy released due to crack growth of simultaneous extension in several degrees of freedom is represented by $\Sigma \overline{K}_i^2 \, \delta A_i$.

References

[1] Cruse, T. A. and Besuner, P. M., "Residual Life Prediction for Surface Cracks in Complex Structural Details," *Journal of Aircraft,* Vol. 12, No. 4, 1975, pp. 369–375.

[2] Besuner, P. M., "Residual Life Estimates for Structures with Partial Thickness Cracks," in *Mechanics of Crack Growth, ASTM STP 590,* 1976, pp. 403–419.

[3] Rice, J. R., "Some Remarks on Elastic Crack-Tip Stress Yields," *International Journal of Solids and Structures,* Vol. 8, 1972, American Society for Testing and Materials, Philadelphia, pp. 751–758.

[4] Mattheck, C., Morawietz, P., and Munz, D., "Stress Intensity Factor at the Surface and at the Deepest Point of a Semi-Elliptical Surface Crack in Plates under Stress Gradients," *International Journal of Fracture,* Vol. 23, 1983, pp. 201–212.

[5] Perez, R., Ray, S. K., and Grandt, A. F., "Application of a Weight Function Method to Predict the Fatigue Life of Corner Cracked Holes Loaded in Bending," *Engineering Fracture Mechanics,* Vol. 28, No. 3, 1987, pp. 283–291.

[6] Irwin, G. R., "The Crack Extension Force for a Part-Through Crack in a Plate," *Transactions of ASME, Journal of Applied Mechanics,* 1962, pp. 651–654.

[7] Shah, R. C. and Kobayashi, A. S., "Stress Intensity Factor for an Elliptical Crack Under Arbitrary Normal Loading," *Engineering Fracture Mechanics,* Vol. 3, 1971, pp. 71–96.

[8] Broekhoven, M. J. G., "Fatigue and Fracture Behaviour of Cracks at Nozzle Corners: Comparison of Theoretical Predictions with Experimental Data," Report MMPP—157, Delft University of Technology, The Netherlands, July 1976.

[9] Oore, M., "Fatigue Crack Growth with Varying Stress Intensity Factor Along the Front," *Predictive Capabilities in Environmentally Assisted Cracking—PVP,* Vol. 99, 1985, pp. 341–352.

[10] Schijve, J., "Comparison Between Empirical and Calculated Stress Intensity Factors of Hole Edge Cracks," *Engineering Fracture Mechanics,* Vol. 22, No. 1, 1985, pp. 49–58.

[11] Oore, M., "Stress Intensity Factor of Three-Dimensional Cracks at the Edge of a Hole," in *Proceedings,* 16th Congress of the International Council of the Aeronautical Sciences, 1988, pp. 1451–1460.

Walter H. Gerstle[1] and Joao Elias Abdalla, Jr.[1]

Finite Element Meshing Criteria for Crack Problems

REFERENCE: Gerstle, W. H. and Abdalla, J. E., Jr., **"Finite Element Meshing Criteria for Crack Problems,"** *Fracture Mechanics: Twenty-First Symposium, ASTM STP 1074,* J. P. Gudas, J. A. Joyce, and E. M. Hackett, Eds., American Society for Testing and Materials, Philadelphia, 1990, pp. 509–521.

ABSTRACT: Rules for the meshing of finite element problems with elastic cracks have never been established. Therefore finite element results for crack problems presented in the literature are of uncertain accuracy. This paper establishes the finite element meshing criteria, based upon a "least dimension" concept, necessary to calculate mixed-mode stress-intensity factors in linear elastic crack problems to a pre-specified accuracy.

Finite element convergence studies, using meshes which consist of eight-noded quadrilateral elements and corresponding six-noded quarter-point singularity elements, are presented for several benchmark problems with known stress-intensity factor solutions. It is assumed that the stress-intensity factors are extracted from the analysis using the displacement correlation method, a common practice in the analysis of mixed-mode problems.

The convergence studies give the analyst valuable information about meshing requirements for arbitrary crack problems. In performing a convergence study for such analyses, three key meshing parameters associated with a given crack tip must be considered: size of the crack tip elements, number of crack tip elements surrounding the crack tip, and size of regular elements near the crack tip. Recommendations for each of these three parameters to achieve an efficient mesh and a prespecified accuracy are presented.

KEY WORDS: finite element, crack, singularity, convergence study, stress-intensity factor, linear elastic fracture mechanics

Several finite element meshing criteria for mixed-mode crack problems have been proposed by investigators during the last ten years [*1-4*]. Problems with various crack geometries were studied by these researchers. The methodology adopted by each investigator differed significantly, and the meshing criteria established in each work are different. Further, the stress-intensity factor accuracy considered adequate in some of the work [*1,3,4*] does not seem to be sufficient for many engineering problems. It is believed that, for the finite element method to be useful to a fracture mechanician, it must calculate stress-intensity factors to arbitrarily pre-specified accuracy. Therefore finite element meshing techniques to achieve arbitrary accuracy for crack problems are established in this paper.

Several plane strain crack problems with known stress-intensity factor solutions are analyzed extensively, using different meshes. The results of these convergence studies can be used to establish finite element meshing criteria to be applied to any crack problem. Both pure Mode I and pure Mode II problems are studied.

The displacement correlation method [*1*] of stress-intensity factor extraction is used. This

---

[1] Assistant Professor and Graduate Student, respectively, Department of Civil Engineering, University of New Mexico, Albuquerque, NM 87131.

is a common method for mixed-mode crack analysis, being simple to implement. Energy methods for stress-intensity factor extraction are not considered in this paper.

Finally, to show that stress-intensity factors for arbitrary problems are calculated to within the predicted accuracy for a given mesh configuration, several example problems are analyzed and the results presented.

Theoretical Finite Element Crack Tip Meshing Criteria

Least Dimension

Any crack tip has an associated least dimension (LD) [2,5]. This quantity must be clearly understood (as illustrated in Fig. 1) in order to successfully analyze a crack problem. The LD will be used to calculate an approximate radius surrounding the crack tip within which the

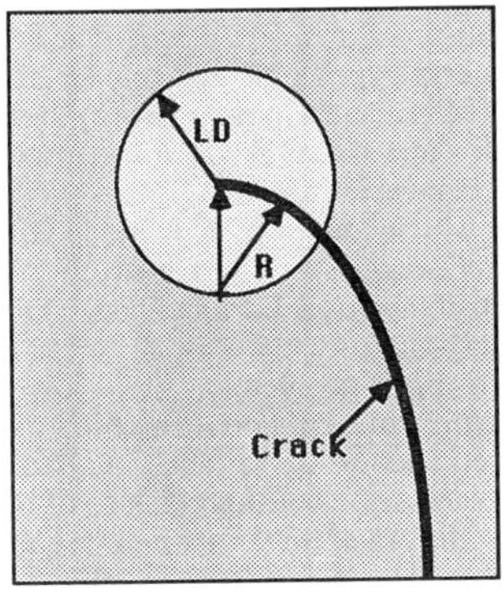

(a) Least Dimension (LD) of this Crack is governed by the Crack Curvature.

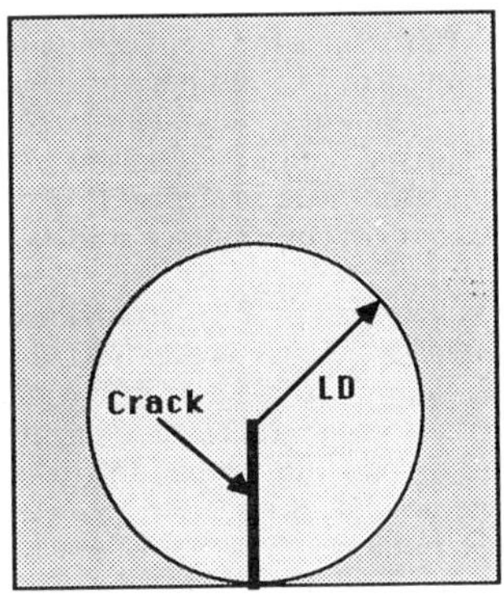

(b) Least Dimension (LD) of this Crack is governed by the Crack Length.

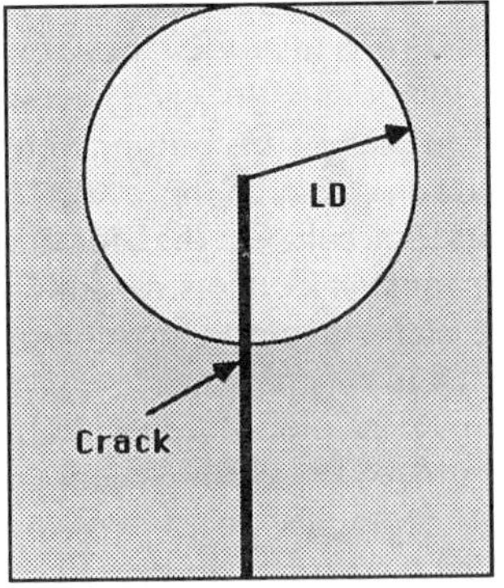

(c) Least Dimension (LD) of this Crack is governed by the Ligament Size.

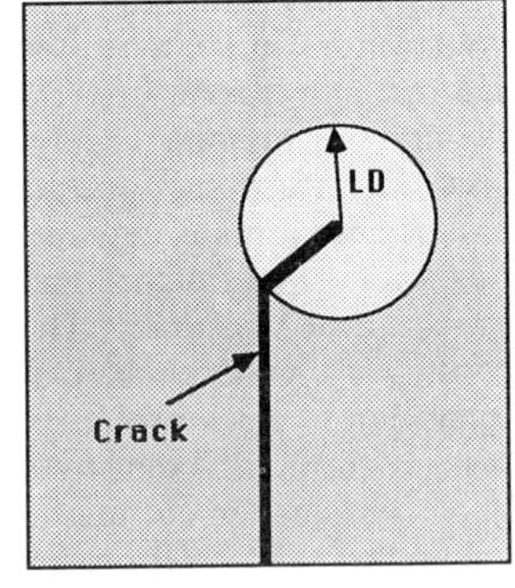

(d) Least Dimension (LD) of this Crack is governed by a Kink in the Crack.

FIG. 1—*The least dimension (LD) associated with a crack tip may be governed by characteristics such as* (a) *crack curvature,* (b) *crack length,* (c) *ligament size, and* (d) *a kink in the crack.*

singular stress field can be guaranteed to dominate the solution. A working definition of the least dimension is presented below.

Linear elastic fracture mechanics makes the assumption that, "close" to the crack tip, the stress field in a problem with finite dimensions is the same as the singular stress field near a crack tip in an infinite plate. If the crack has some radius of curvature close to the crack tip, then the singularity can dominate the solution only within a region of radius which is some small percentage of the radius of curvature of the crack. Therefore the crack radius of curvature will be one of the quantities which must be considered in computing the crack tip LD (Fig. 1*a*).

Other quantities must also be used to determine the LD. For example, if any geometrical boundary or material discontinuity occurs near the crack tip, this discontinuity causes a disturbance in the stress field. Therefore the distance from the crack tip to the nearest discontinuity will be used to determine the LD for a given crack, as shown in several cases in Fig. 1. Often, the crack length itself may be the quantity which governs the LD (Fig. 1*b*) or the size of the ligament may govern the LD (Fig. 1*c*). In some instances (Fig. 1*d*) the LD associated with a crack tip will be limited by the distance to a discrete kink in the crack.

The applied loading can also affect the size of the region surrounding the crack tip in which the singular term in the stress field dominates the solution. For instance, if a point load is applied near the crack tip, this applied load will distort the stress field locally. Therefore the region in which the singular terms dominate the stress field cannot contain any point loads. Accordingly, the LD may be determined by the applied loading. For example, in the case of a continuously varying applied body force near the crack tip, or a surface traction applied to the crack face, the size of the singular region will be affected. Similarly, a continuously varying material thickness or continuously varying material properties near the crack tip will affect the size of the region within which the singular $r^{-1/2}$ terms dominate the stress field solution.

With the foregoing considerations in mind, the LD is the minimum of the radius of crack curvature, the distance to the nearest material discontinuity or geometrical discontinuity (other than the crack surfaces themselves), and the distance to the nearest applied point load. Also, for cases where continuously varying body forces or surface tractions are applied to the region close to the crack tip, the LD must be selected to be small enough so that no quantity changes significantly within the region of radius of LD. The LD is therefore defined as the radius of the largest circle, centered at the crack tip, within which no geometrical or applied load quantity varies, as shown by the various examples in Fig. 1.

Note that after a finite element analysis of a particular linear elastic crack problem has been performed, the numerically calculated stress field near the crack tip should evidence the necessary $r^{-1/2}$ radial distribution of stress. *A posteriori* comparison of the numerically calculated stresses with the required $r^{-1/2}$ stress distribution will show whether the initially assumed LD was valid.

The singular terms of the stress field dominate the solution within a small percentage of the LD from the crack tip [5]. The singular elements should therefore have a size which is some small percentage of the LD. Decreasing the singular element size below this size will not have a large effect upon the solution error, since these elements can represent the $r^{-1/2}$ singular radial distribution of stress exactly.

Meshing Criteria

The stress field within a small percentage of the LD from the crack tip of a linear elastic crack is dominated by a $r^{-1/2}$ singularity [6,7]. This singularity can be exactly modeled with finite elements by using quadratic triangular elements whose mid-side nodes have been shifted to the quarter-points of the element sides radiating from the crack tip [8,9]. These elements model a displacement variation in the radial direction r as

$$u_i(r) = A_i + B_i \sqrt{r} + C_i r \tag{1}$$

and a stress variation in the radial direction of

$$\sigma_{ij}(r) = \frac{D_{ij}}{\sqrt{r}} + E_{ij} \tag{2}$$

where r is the distance from the crack tip in a given direction, and A_i, B_i, C_i, D_{ij}, and E_{ij} are constants which depend only on the direction of r

The displacement and stress variations in the η direction (parallel to the element side opposite the crack tip node) are quadratic and linear respectively and are not altered by the shift of mid-side nodes to the quarter points [8,9].

Close to a crack tip, the singularity dominates the solution. Therefore, as the singular elements are made smaller, they do a better job of modeling the stress field. However, because the singular term dominates the solution close to the crack tip, the singular elements need not be made smaller than this critical size. But two additional considerations must also be addressed: (1) how many singular elements are required around the crack tip, and (2) what size must the elements surrounding the singular elements be to obtain the stress-intensity factors to within a given accuracy.

These considerations can be addressed by considering a mixed-mode crack in an infinite plate, where the stress solution is known exactly, and is given by [6,7]

$$\sigma_{xx} = \frac{1}{\sqrt{2\pi r}} \left\{ K_I \cos \frac{\theta}{2} \left[1 - \sin \frac{\theta}{2} \sin \frac{3\theta}{2} \right] + K_{II} \sin \frac{\theta}{2} \left[2 + \cos \frac{\theta}{2} \cos \frac{3\theta}{2} \right] \right\} \tag{3}$$

$$\sigma_{yy} = \frac{1}{\sqrt{2\pi r}} \left\{ K_I \cos \frac{\theta}{2} \left[1 + \sin \frac{\theta}{2} \sin \frac{3\theta}{2} \right] + K_{II} \sin \frac{\theta}{2} \cos \frac{\theta}{2} \cos \frac{3\theta}{2} \right\} \tag{4}$$

$$\sigma_{xy} = \frac{1}{\sqrt{2\pi r}} \left\{ K_I \sin \frac{\theta}{2} \cos \frac{\theta}{2} \cos \frac{3\theta}{2} + K_{II} \cos \frac{\theta}{2} \left[1 - \sin \frac{\theta}{2} \sin \frac{3\theta}{2} \right] \right\} \tag{5}$$

where the coordinates r, θ, x, and y are shown in Fig. 2; and K_I and K_{II} are the Mode I and Mode II stress-intensity factors respectively.

The crack tip meshing criteria can be established by examining the theoretical (known, except for three as yet undetermined constants: K_I, K_{II} and K_{III}) stress field near the crack tip, and comparing its form to the form of the stress field permitted by the finite elements in the crack tip region. In the next three sections, first singular element size requirements, then tangential meshing requirements, and finally radial meshing requirements near a crack tip are discussed.

Singular Element Size (L)—The singular element size, L, required for a required accuracy, is problem dependent. Problems with a high amount of bending in the ligament ahead of the crack tip will typically have a smaller region surrounding the crack tip within which the $r^{-1/2}$ singularity dominates the solution. Our analyses indicate that for problems with little bending in the ligament the singularity dominates the solution, with an error of less than 1%, within a region of LD/10, while for problems with a high amount of ligament bending, the singularity dominates the solution, to within 1%, within a region of LD/20.

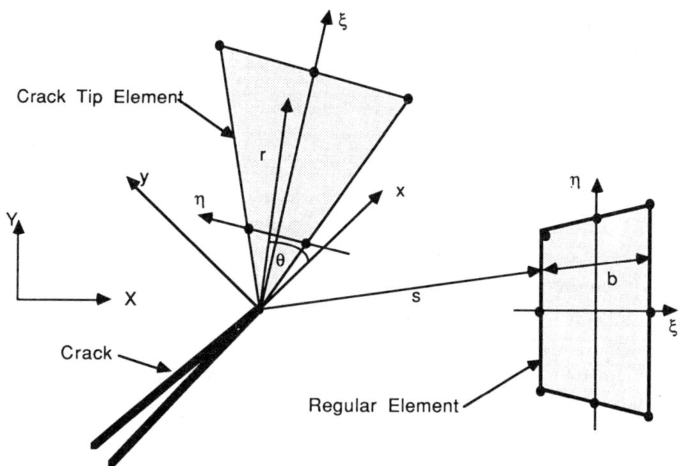

FIG. 2— *Nomenclature for crack tip and finite elements in the region of the crack tip.*

Number of Singular Crack Tip Elements (N)—The σ_{ij} components of stress in a straight-sided regular quadratic element vary linearly in the η direction and are given by

$$\sigma_{ij} = A_{ij} + B_{ij}\eta \tag{6}$$

where A_{ij} and B_{ij} are constants which depend only on radius r from the crack tip (Fig. 2). The tangential direction does not correspond exactly with the η direction of a crack tip element. However, at least for singular elements which subtend no more than 45 deg, with sides radiating from the crack tip of approximately equal length, and which are straight-sided, the tangential direction is closely approximated by the η direction.

The theoretical tangential variation in each component of stress, as shown in Eqs 3 to 5, is composed of trigonometric functions of $\theta/2$ and $3\theta/2$.

Figure 3 shows how four equi-angular singular elements would model this stress field,

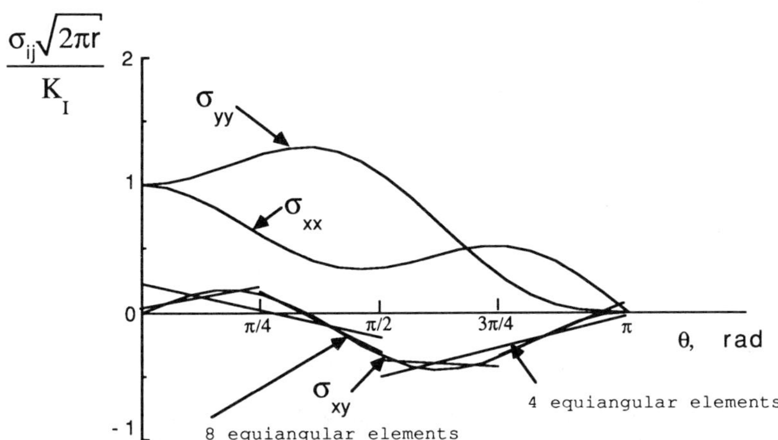

FIG. 3—*Normalized stress versus theta for a near-crack tip stress field. A quadratic fintie element approximates the theoretical stress field with piecewise linear segments.*

assuming that the theoretical stresses are matched exactly at the second-order Gauss point locations within each element. As shown, eight such elements model the theoretical distribution of stresses much more accurately. As more elements are placed around the crack tip, the ability of these elements to represent the theoretical stress field increases. Note that with four elements surrounding the crack tip, the maximum error in representing tangential stresses is on the order of 100%, while with eight elements surrounding the crack tip, the maximum error in stress representation drops to less than 20%.

Radial Meshing Criteria (b/s)—A similar analysis may be performed to determine the maximum error in stress modeling as a function of element size in the radial direction. Figure 4 shows the normalized near-tip stress variation in the radial direction as a function of *r*. Assume that a regular element is placed a distance *s* from the crack tip, and that the element dimension in the radial direction is *b*, as shown in Fig. 2. Again assuming that the element models the stress correctly at the Gauss point locations [*12*], situated a distance *a* from the element edges, the error in stress, as a function of *b/s*, is plotted in Fig. 4. This

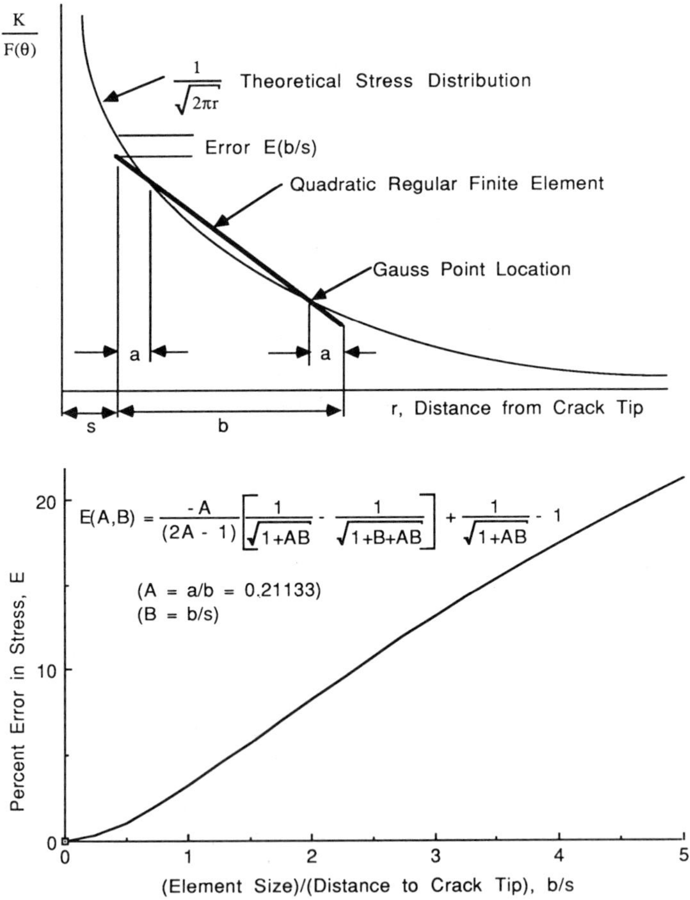

$$E(A,B) = \frac{-A}{(2A-1)}\left[\frac{1}{\sqrt{1+AB}} - \frac{1}{\sqrt{1+B+AB}}\right] + \frac{1}{\sqrt{1+AB}} - 1$$

$$(A = a/b = 0.21133)$$
$$(B = b/s)$$

FIG. 4—*Normalized stress versus* r *for a near-crack tip stress field. A quadratic regular finite element of size* b *and distance* s *from the crack tip approximates the theoretical stress field with piecewise linear segments. The error* E *is a function of* b/s.

figure shows that for b/s equal to 1.0, the error E is about 5%; while for b/s smaller than 0.5, the error E drops to about 2%.

Summary of Crack Tip Meshing Criteria—To achieve 1% maximum expected error in any stress component due to any mixed-mode crack problem, one would expect that approximately sixteen equi-angular singular elements would be required ($N = 16$). Additionally, for this level of desired accuracy, the size b of a regular element near the crack tip should in no case exceed about 0.25 times its clear distance s from the crack tip, as shown in Fig. 2 ($b/s = 0.25$). Finally, the singular elements must be small enough to be completely contained within the singularity-dominated region in the vicinity of the crack tip. For 1% accuracy, the size of these singular elements should be about 1/20 of the LD associated with this crack tip ($L/\text{LD} = 1/20$).

The rest of the mesh must be refined sufficiently to model the non-singular stress field away from the crack tip.

A well designed crack tip mesh should have regular elements with approximately equal sizes in the r and θ directions, because the curvature of the stress field is approximately equal in both directions. Therefore b/s and N required for a given accuracy are not independent. For the regular elements surrounding the crack tip to have approximately the same dimension in the radial and tangential directions, N should be approximately $2\pi/(b/s)$.

Because the rate of convergence of displacements for a quadratic finite element is $0(\text{h}^3)$, one would expect that the rate of convergence of stresses would be $0(\text{h}^2)$, and that the stress-intensity factors would converge at the same rate as the stresses, $0(\text{h}^2)$. Therefore, if a mesh conforming to the recommendations given above produces approximately 1% error in stresses, then a mesh with elements of size α times as large as the elements in this mesh would have an expected error in stress-intensity factors on the order of α^2 times 1%. These theoretical guidelines must be confirmed by convergence studies, which is the topic of the next section.

Finite Element Crack Tip Meshing Convergence Studies

The structure chosen for this convergence study is a square plate of 40.6 cm (16.0 in.) side length, 1.0 in. (2.54 cm) thickness, and a central horizontal crack of 20.3 cm (8 in.) length, as shown in Fig. 5. The loading cases for both pure Mode I and pure Mode II problems are shown in the figure. Advantage of symmetry is taken so that only one quarter of the plate is analyzed. Plane strain conditions are assumed. The applied stress is equal to 1.0 ksi (6.895 MPa) for both Mode I and Mode II loading conditions. The finite element meshes consist of eight-noded quadrilaterals (Q8) and six-noded triangles (T6). The latter are the singular quarter-point elements which model the stress singularity near the crack tip. A 2×2 Gauss integration scheme is used for all elements. The stress-intensity factors are computed by the displacement correlation method [1], using displacement components computed at the crack-surface nodes of the singular elements.

Many meshes were analyzed during these convergence studies. However, to not overwhelm the reader with a large quantity of figures and numbers, only nine of the typical finite element meshes (Fig. 6a), along with analyses results (Fig. 6b), are presented. These meshes are well suited to illustrate the general behavior observed.

The results obtained in the convergence studies described above, along with the theoretical understanding of finite element behavior, allow for modification of the theoretical finite element meshing criteria presented above. To achieve approximately 1% solution accuracy, the crack tip singular elements must have a size approximately equal to one tenth of the LD ($L/\text{LD} = 0.10$), there must be at least 16 equi-angular singular elements placed around the crack tip node ($N = 16$), and any non-singular element in the mesh must have a size less than one

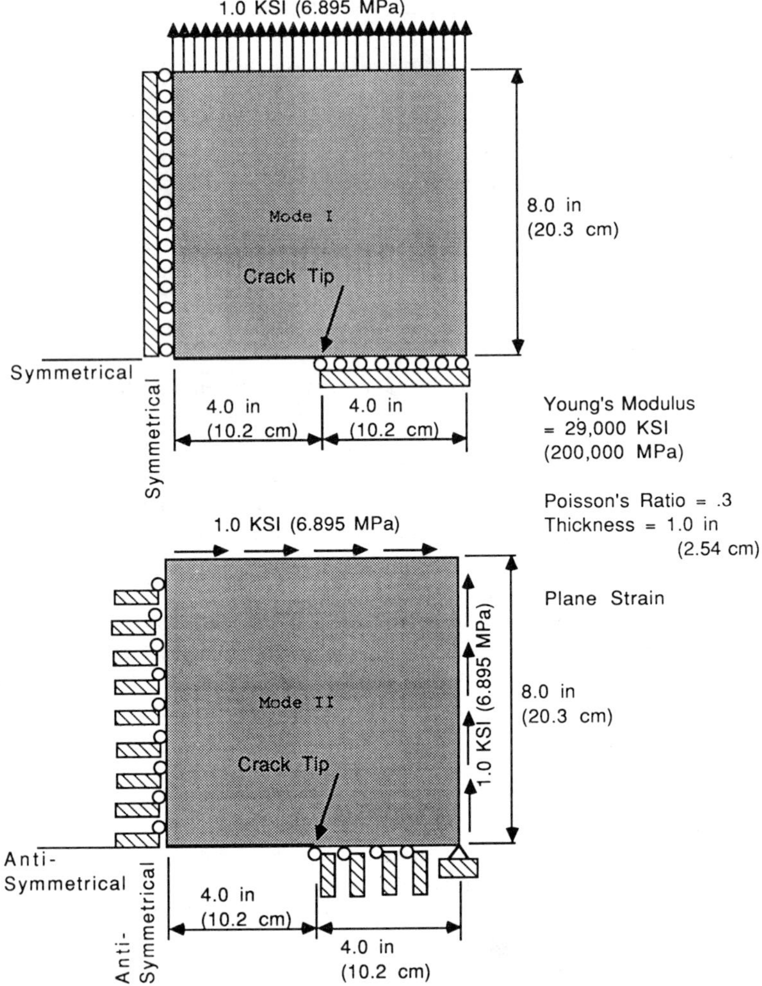

FIG. 5—*Modeling assumptions for the center-cracked finite plate used in the convergence study.*

half ($b/s = 0.5$) its clear distance to the crack tip. Additionally, regions remote from the crack tip must be appropriately refined.

Figure 7 summarizes the meshing criteria necessary to achieve a prespecified accuracy in stress-intensity factors for a crack problem, and it can be used as a tool by the finite element analyst for the generation of crack tip meshes.

Sample Problems

As sample problems to test the validity of the proposed crack tip meshing criteria, a center-cracked plate and an edge-cracked plate are subjected to the five different loading conditions shown in Fig. 8. The meshes used to analyze these problems are shown in Fig. 9. Where possible, advantage of symmetry is taken. The results of these analyses are shown in Fig. 10.

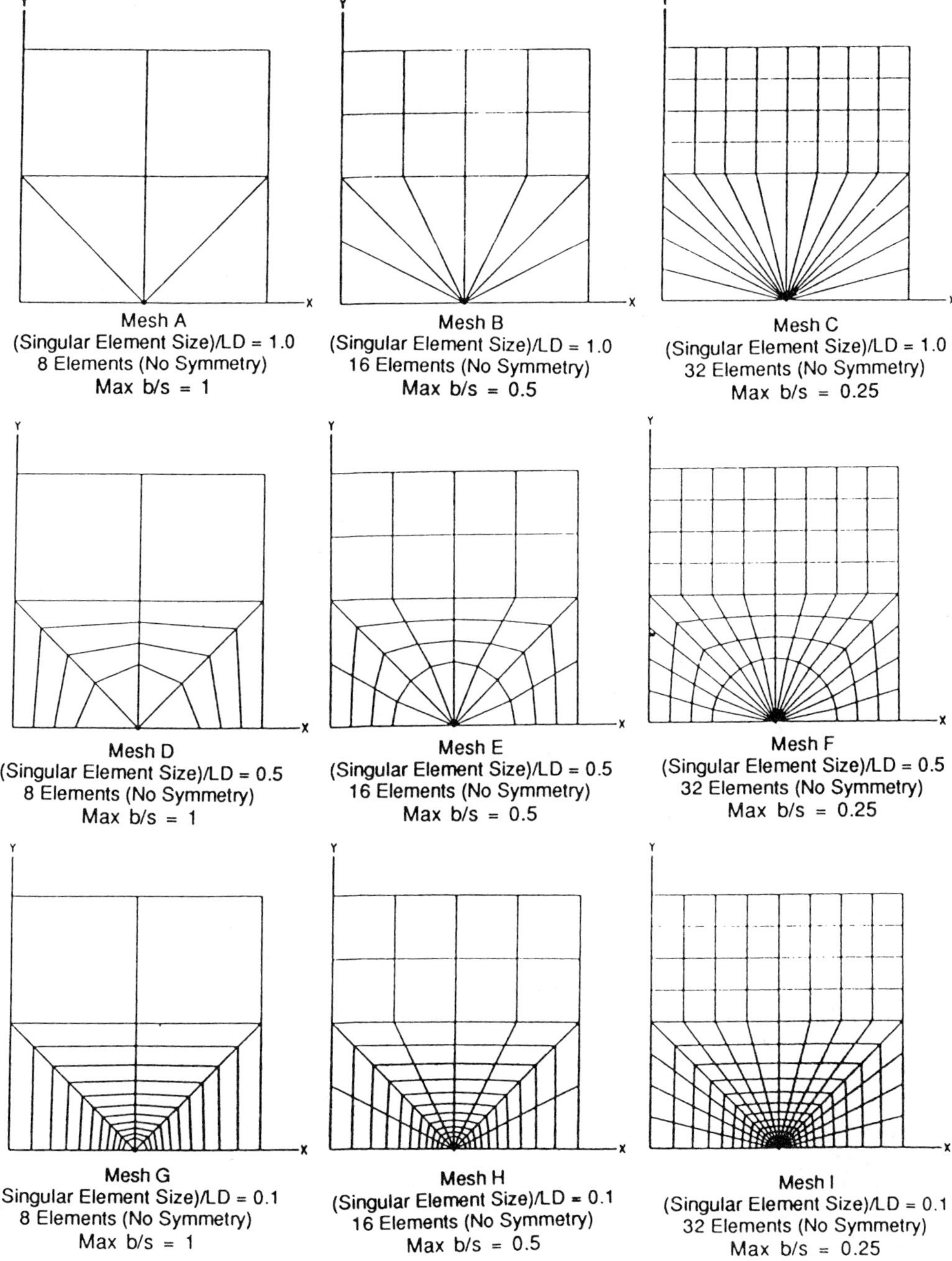

FIG. 6a—*Finite element meshes used in pure Mode I and pure Mode II convergence studies.*

		Number of Singular Elements (no Symmetry)			
		8	16	32	
(Singular Element Size)/LD	1.0	5.6025 / 61.565 (A) / (+18.8%)	5.7123 / 62.771 (B) / (+21.2%)	5.7306 / 62.972 (C) / (+21.5%)	KSI-$\sqrt{\text{in}}$ K_I / MPa$\sqrt{\text{cm}}$ / Error
	0.50	4.8978 / 53.821 (D) / (+3.8%)	5.0206 / 55.170 (E) / (+6.5%)	5.0415 / 55.400 (F) / (+6.9%)	KSI-$\sqrt{\text{in}}$ K_I / MPa$\sqrt{\text{cm}}$ / Error
	0.10	4.6532 / 51.133 (G) / (-0.9%)	4.7184 / 51.850 (H) / (+0.1%)	4.7696 / 52.412 (I) / (+1.2%)	KSI-$\sqrt{\text{in}}$ K_I / MPa$\sqrt{\text{cm}}$ / Error

Mode I Convergence Study Results. Exact K_I Solution assumed 4.72 KSI-$\sqrt{\text{in}}$ (51.8 MPa$\sqrt{\text{cm}}$) Accurate to within One Percent. (13)

		Number of Singular Elements (no Symmetry)			
		8	16	32	
(Singular Element Size)/LD	1.0	4.6992 / 51.639 (A) / (+9.3%)	4.8161 / 52.923 (B) / (+12.0%)	4.8163 / 52.925 (C) / (+12.0%)	KSI-$\sqrt{\text{in}}$ K_{II} / MPa$\sqrt{\text{cm}}$ / Error
	0.50	4.3486 / 47.786 (D) / (+1.1%)	4.5035 / 49.488 (E) / (+4.8%)	4.5347 / 49.831 (F) / (+5.5%)	KSI-$\sqrt{\text{in}}$ K_{II} / MPa$\sqrt{\text{cm}}$ / Error
	0.10	4.1259 / 45.339 (G) / (-4.0%)	4.2947 / 47.194 (H) / (-0.1%)	4.3303 / 47.585 (I) / (+0.7%)	KSI-$\sqrt{\text{in}}$ K_{II} / MPa$\sqrt{\text{cm}}$ / Error

Mode II Convergence Study Results. Exact K_{II} Solution assumed 4.30 KSI-$\sqrt{\text{in}}$ (47.2 MPa$\sqrt{\text{cm}}$) Accurate to within One Percent.

FIG. 6b—*Results of the convergence studies shown in Fig. 6a.*

Using the crack tip meshing criteria established in this paper, and presented in Fig. 7, one would expect Mesh 3 to provide stress-intensity factor accuracy of approximately 4%. As shown in Fig. 10, most of the results are between 1 and 2% accurate, except for problem *e*, which was accurate to within only 4.7%.

Mesh 2 has one half as many crack tip elements as Mesh 3, and they are five times as large as the crack tip elements in Mesh 3. The *b/s* ratios are approximately twice those in Mesh 3. From Fig. 7, one would expect the accuracy to be about 15% for this mesh. The actual errors vary between 7% for Problem *a* to −14% for Problem *e*

Mesh 1 has one quarter as many crack tip elements as Mesh 3, and they are ten times as large. The *b/s* ratio is approximately one. From Fig. 7, one would expect the accuracy to be about 20% for this mesh. The actual errors vary between −21% for Problem *b* to 19% for Problem *d*.

Conclusions

Based upon theoretical considerations and upon the results of a convergence study, Fig. 7 summarizes the meshing criteria developed in this research. The relationships between least

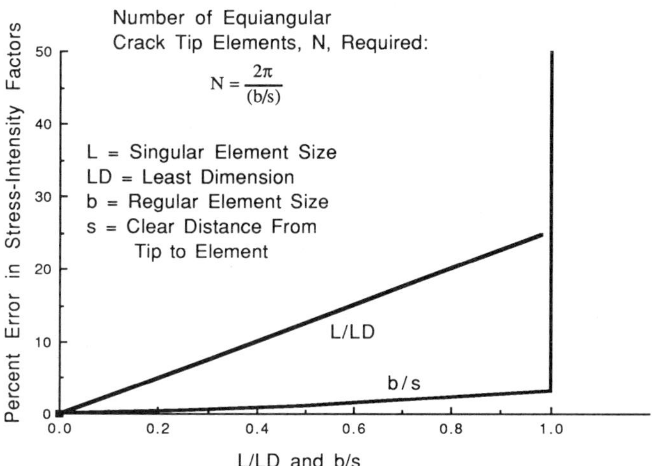

FIG. 7—*For a required percent accuracy in stress-intensity factors this graph may be used to predict the required meshing characteristics:* L/LD, b/s, *and* N.

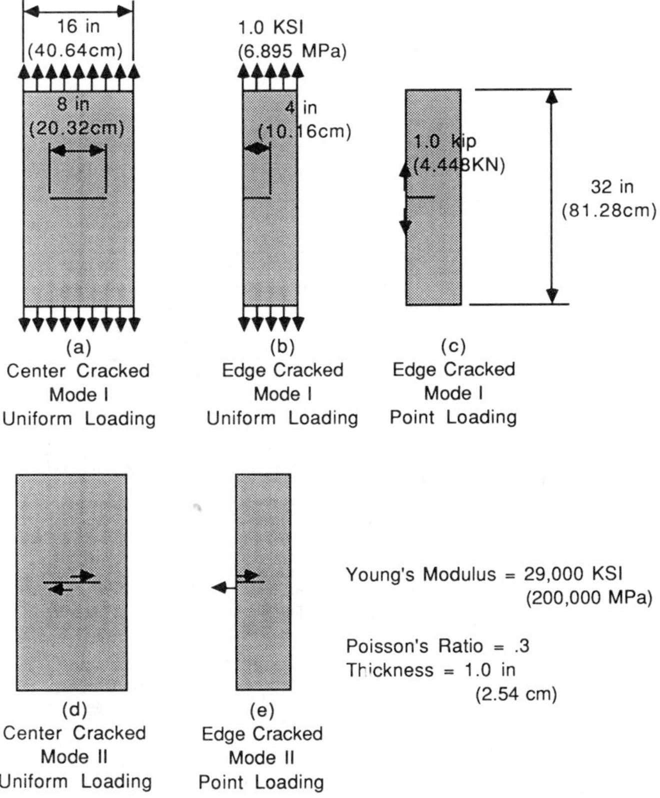

FIG. 8—*Geometry and boundary conditions assumed in the sample problems. The dimensions are typical for all sample analyses.*

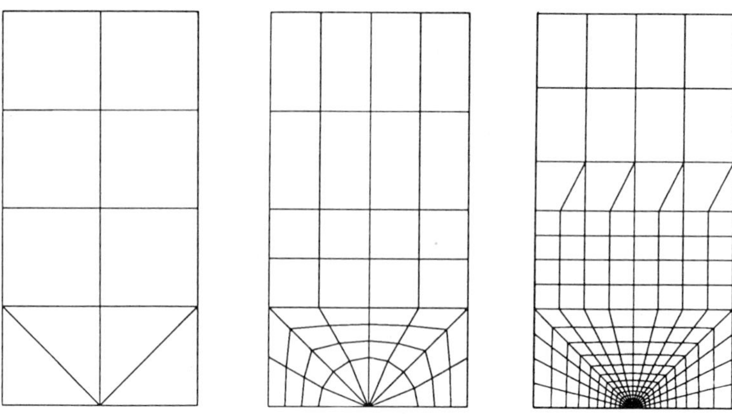

FIG. 9—*Finite element meshes used to analyze the problems shown in Fig. 8. Symmetry used wherever possible.*

Problem (shown in Fig. 8)		Mesh Type (shown in Fig. 9) 1	2	3	Exact (+ or - 1%) (13)
	a	4.969 KSI 54.60 MPa	4.502 KSI 49.48 MPa	4.262 KSI 46.84 MPa	4.207 KSI 46.23 MPa
		18%	7.0%	1.3%	Error
	b	7.936 KSI 87.21 MPa	9.375 KSI 103.0 MPa	9.859 KSI 108.4 MPa	10.03 KSI 110.2 MPa
		-21%	-6.5%	-1.7%	Error
	c	1.710 KSI 18.79 MPa	2.718 KSI 29.87 MPa	3.166 KSI 34.79 MPa	NA
		NA	NA	NA	Error
	d	0.4604 KSI 5.059 MPa	0.3637 KSI 3.997 MPa	0.3910 KSI 4.296 MPa	0.3870 KSI 4.252 MPA
		19%	-6.0%	1.0%	Error
	e	0.7960 KSI 8.7471 MPa	0.7525 KSI 8.269 MPa	0.8373 KSI 9.201 MPa	0.8797 KSI 9.667 MPa
		-9.5%	-14%	-4.7%	Error

FIG. 10—*Stress-intensity factors calculated in sample analyses. The exact solutions are also reported [13]. Percent error between numerical result and exact solution is tabulated.*

dimension (LD), singular element size (L), regular element size (s), and number of singular elements (N) surrounding the crack tip to produce well-balanced crack tip meshes have been presented. The analyst now has some guidelines for the development of reasonably efficient crack tip finite element meshes. In addition, Fig. 7 presents the meshing requirements needed to achieve a prespecified stress-intensity factor accuracy.

While these recommendations are not guaranteed to produce the predicted accuracy in every case, 15 sample problems have been presented in Figs. 8 to 10. Comparison of the results of these sample problems with the predicted accuracy presented in Fig. 7 shows that the recommendations are reasonable for a wide variety of problems.

Acknowledgments

Ferhat Akgul, a masters student in the Department of Civil Engineering at the University of New Mexico, performed many analyses included in this paper.

References

[1] Ingraffea, A. R., "Numerical Modelling of Crack Propagation," in *Rock Fracture Mechanics,* Springer-Verlag, Wien-New York, 1983, pp. 151–208.

[2] Gerstle, W. H., "Finite and Boundary Element Modeling of Crack Propagation in Two- and Three-Dimensions using Interactive Computer Graphics," Department of Structural Engineering Report 85-8, Program of Computer Graphics and Department of Structural Engineering, Cornell University, Ithaca, N.Y., 1985.

[3] Saouma, V. E. and Schwemmer, D., "Numerical Evaluation of the Quarter-Point Crack Tip Element," *International Journal of Numerical Methods in Engineering,* Vol. 20, 1984, pp. 1629–1641.

[4] Sabouni, A-R. and Elzanaty, A., "Finite Element Idealization of Mixed-Mode Fracture," Report 83-9, Department of Structural Engineering, School of Civil and Environmental Engineering, Cornell University, Ithaca, N.Y., 1983.

[5] Abdalla, J. E. Filho and Gerstle, W. H., "A Finite Element for Arbitrarily Precise Determination of Stress Intensity Factors," Bureau of Engineering Report CE-84(88), College of Engineering, University of New Mexico, Albuquerque, 1988.

[6] Rolfe, S. T. and Barsom, J. M., *Fracture and Fatigue Control in Structures—Applications of Fracture Mechanics,* Prentice-Hall, Englewood Cliffs, N.J., 1977.

[7] Irwin, G. R., "Analysis of Stresses and Strains Near the End of a Crack Traversing a Plate," *Transactions of ASME, Journal of Applied Mechanics,* Vol. 24, 1957, pp. 361–364.

[8] Barsoum, R. S., "On the Use of Isoparametric Finite Elements in Linear Fracture Mechanics," *International Journal of Numerical Methods in Engineering,* Vol. 10, 1976, pp. 25–37.

[9] Barsoum, R. S., "Triangular Quarter-Point Elements as Elastic and Perfectly-Plastic Crack Tip Elements," *International Journal of Numerical Methods in Engineering,* Vol. 11, 1977, pp. 85–98.

[10] Parks, D. M., "A Stiffness Derivative Finite Element Technique for Determination of Crack Tip Stress Intensity Factor," *International Journal of Fracture Mechanics,* Vol. 10, 1974, pp. 487–503.

[11] McGuire, W. and Gallagher, R. H., *Matrix Structural Analysis,* Wiley, New York, 1979.

[12] Huebner, K. H. and Thornton, E. A., "The Finite Element Method for Engineers," 2nd ed., Wiley, New York, 1982.

[13] Tada, H., Paris, P., and Irwin, G., *The Stress Analysis of Cracks Handbook,* Del Research Corporation, Hellertown, Pa., 1973.

Fracture Mechanics Applications

G. G. Chell[1]

Application of the CEGB Failure Assessment Procedure, R6, to Surface Flaws

REFERENCE: Chell, G. G., "**Application of the CEGB Failure Assessment Procedure, R6, to Surface Flaws,**" *Fracture Mechanics: Twenty-First Symposium, ASTM STP 1074,* J. P. Gudas, J. A. Joyce, and E. M. Hackett, Eds., American Society for Testing and Materials, Philadelphia, 1990, pp. 525–544.

ABSTRACT: The application of the R6 failure assessment procedure to surface flaws requires the definition of a plastic yield load that characterizes the crack tip deformation at both the deepest and surface points on the defect. In this paper the results of finite element computations of the J-integral are used to identify the most appropriate yield loads. This is achieved by assuming that the Option 2 failure assessment diagram (FAD), which is based on a reference stress approximation to J, provides a good approximation to the Option 3 FAD, which is derived from the computed J values. The finite element results for surface flaws in plates and cylinders are used in the investigation. It is demonstrated that global yield loads, which characterize the overall plastic deformation of the cracked structure, produce a better agreement between the R6 procedure, and the finite element solutions, than do local yield loads, which are usually chosen to characterize the plastic deformation at a specific location on the crack front.

KEY WORDS: R6 procedure, surface flaws, failure assessment diagram, J-integral, plastic yield load, cracked structures, elastic-plastic fracture, failure analysis

The R6 procedure [1] has been developed by the Central Electricity Generating Board in the United Kingdom in order to provide a relatively simple diagrammatic method of assessing the integrity of cracked structures. It is based on the concept of a failure assessment diagram (FAD) consisting of two axes, L_r and K_r, and a failure assessment curve (FAC) (Fig. 1). The approach has been extensively validated using elastic-plastic computations based on the J-integral and the results of structural and laboratory fracture testing [1]. However, the validation for surface flaws is less extensive than that for fully extended defects, partly because of the computational and testing complications associated with semi-elliptical cracks, and partly because of difficulties in the definition of the parameter L_r.

The parameter K_r is determined using the linear elastic stress intensity factor (SIF). This can now be accurately calculated for semi-elliptical defects as a function of position along the crack front. The parameter L_r is evaluated using an appropriate plastic yield load; it is not clear *a priori* whether the relevant yield load should characterize the overall plastic deformation of the cracked structure (global yield or net section yield) or the plastic deformation local to a point on the defect (local yield).

These aspects have been studied by Miller [2] using the results of published elastic-plastic computations that employed the finite element method or a line spring model. He concluded

---

[1] Materials Branch, Central Electricity Research Laboratories, Kelvin Avenue, Leatherhead, Surrey KT22 7SE, U.K.

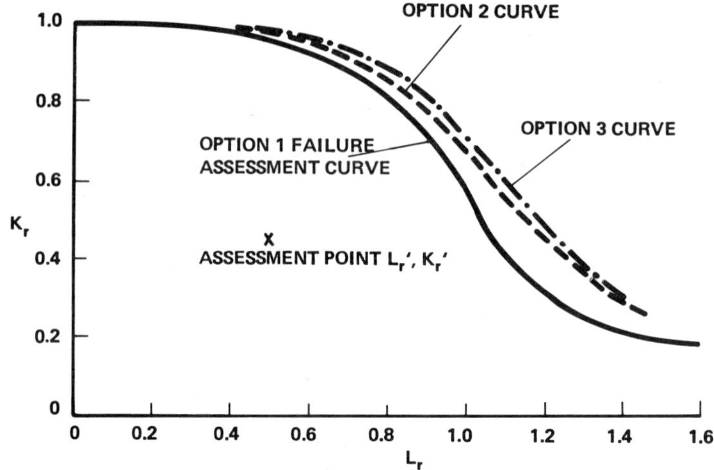

FIG. 1—*Failure assessment diagram showing typical failure curves.*

that, in general, a global yield load gave better agreement between the predictions of the R6 method and the computed results than did a local yield load.

This paper follows closely the work of Miller [2] and extends his investigations to data that have only recently become available in the open literature. Furthermore, whereas Miller restricted his calculations to assessing the effects of plasticity at the deepest point on the crack front, the present work also considers the deformation of the crack where it intersects the free surface. Assessments made with respect to both the deepest and surface points on a defect form an integral part of many failure analyses (e.g., in establishing safety cases based upon leak-before-break).

Failure Assessment Diagram

The FAD may be used in assessments involving primary and secondary loads in order to determine limiting defect sizes and limiting load conditions. The limiting conditions may be calculated for surface flaws with respect to the fracture toughness at the initiation of crack extension, the toughness after a certain amount of ductile tearing, or ductile instability based on the full toughness resistance curve. The details of the procedures to be followed in performing these kind of analyses are given in the R6 document [1] and are not addressed further in this paper. Here it is noted that the analyses involve calculating failure assessment points, K_r' and L_r', which are defined as

$$K_r' = K/K_c$$

$$L_r' = \sigma_r/\sigma_y = L/L_1$$

where K is the applied SIF, K_c the appropriate fracture toughness, σ_r the reference stress corresponding to the applied load L, σ_y the yield stress, and L_1 the appropriate plastic yield load (either global or local in the present context). In general:

$$\sigma_r = \sigma/f(a/t, a/c, \phi) \tag{1}$$

where

$$f(a/t, a/c, \phi) = L_1/L_y \tag{2}$$

and σ is the applied stress, and L_y is the load required to cause general yielding in the uncracked component. For surface flaws the collapse function f depends on the ratio of crack depth, a, to section thickness, t; the aspect ratio a/c, where c is half the surface length of the defect; and possibly the position on the crack front, ϕ (Fig. 2). An appropriate value of the yield function f is essential for a failure analysis based on R6.

In the R6 method the integrity of a cracked structure is judged by plotting the assessment point on the FAD and comparing its position against the FAC. The structure is safe if the point falls inside the curve, but safety can not be assured if it falls outside of it (Fig. 1).

In general, the FAC depends on material deformation characteristics, crack and structural geometry and the form of the applied loading. To help avoid the complexities of a rigorous elastic-plastic failure analysis the R6 procedure allows three forms of the FAC corresponding to three levels of sophistication. In the first, Option 1, a general FAC is defined by the equation

$$K_r = (1 - 0.14L_r^2)[0.3 + 0.7 \exp(-0.65L_r^6)] \tag{3}$$

which represents a lower-bound curve independent of material and structure. Here L_r and K_r are the values of L_r' and K_r' respectively that define the failure curve. In the second, Option 2, a material specific FAC is generated using the equation

$$K_r = \left[\frac{E\epsilon_r}{\sigma_r} + \frac{1}{2} L_r^2/(E\epsilon_r/\sigma_r) \right]^{-1/2} = F(L_r\sigma_y) \tag{4}$$

where ϵ_r is the reference strain corresponding to the reference stress, $\sigma_r = L_r\sigma_y$, and is related to it through the uniaxial stress-strain curve of the material, and E is Young's modulus [1]. Although in Eq 4 the value of K_r for a given value of L_r is dependent on the form of the material stress-strain curve, for convenience only the dependence on $\sigma_r = L_r\sigma_y$ is shown in the function F. In the plastic limit, $E\epsilon_r/\sigma_y \gg 1$ and

$$K_r = (E\epsilon_r/\sigma_r)^{-1/2} \tag{5}$$

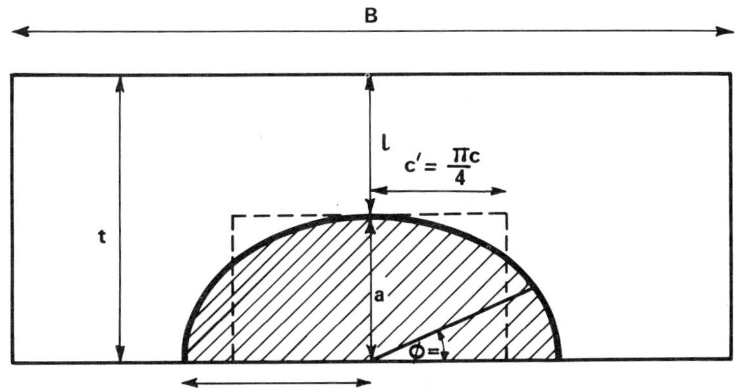

FIG. 2—*Surface flaw and dimensions.*

In the third option the FAC is specific to the actual structure and material and is derived from an elastic-plastic determination of the J-integral using the equation

$$K_r = (J_e/J)^{1/2} \tag{6}$$

where $J_e = K^2/E'$ is the elastic value of J, $E' = E/(1 - \nu^2)$ for plane strain and $E' = E$ for plane stress deformation, and ν is Poisson's ratio [1]. The J-integral varies with the applied stress, σ, and is also dependent on the geometry and material flow properties. However, for convenience, the right-hand side of Eq 6 is written principally as a function, F', of L_r using the definition

$$L_r = \sigma/\sigma_y f \tag{7}$$

Thus

$$K_r = F'(L_r \sigma_y f) \tag{8}$$

The Option 3 FAC is explicitly dependent on the choice of the yield function f, whereas the Option 2 curve is not (cf. Eqs 4 and 8). This fact provides a means of estimating the yield function f from numerical computations of the J-integral, provided it is assumed that with the appropriate choice of f that the Option 2 curve will be a good approximation to the Option 3 curve for a given material, especially at large values of L_r (cf. the curves in Fig. 1).

Estimating the Plastic Yield Function

The function f may be determined graphically or algebraically depending on whether the computed J solutions are presented numerically or in an analytical form similar to that used in the engineering approach developed at General Electric under various EPRI projects [3–5]. In the former case the value of f in Eq 8 is varied until the Option 3 curve closely agrees with the Option 2 curve (see Fig. 3 where f' represents an initial guess for f).

If the J solutions are expressed as in the engineering approach, then for a fully plastic material with a power law relationship between stress (σ') and strain (ϵ') of the form

$$\epsilon'/\epsilon_0 = \alpha(\sigma'/\sigma_0)^n \tag{9}$$

the J-integral may be written as

$$J = \alpha\sigma_0\epsilon_0 a(1 - a^*)h_1(a^*,n)(\sigma/\sigma_0)^{n+1} \tag{10}$$

where α is a constant, $\epsilon_0 = \sigma_0/E$ and σ_0 is a material constant (approximately equal to the yield stress in many cases), and $a^* = a/t$. The function $h_1(a^*,n)$ is dependent on the structural geometry, crack shape, and strain hardening exponent n.

The fully plastic expression for J represented by the Option 2 curve can be obtained by comparing Eqs 5 and 6 as

$$J = J_e \cdot E\epsilon_r/\sigma_r$$

or, using the stress-strain relationship of Eq 9,

$$J = J_e\alpha(\sigma_r/\sigma_0)^{n-1} \tag{11}$$

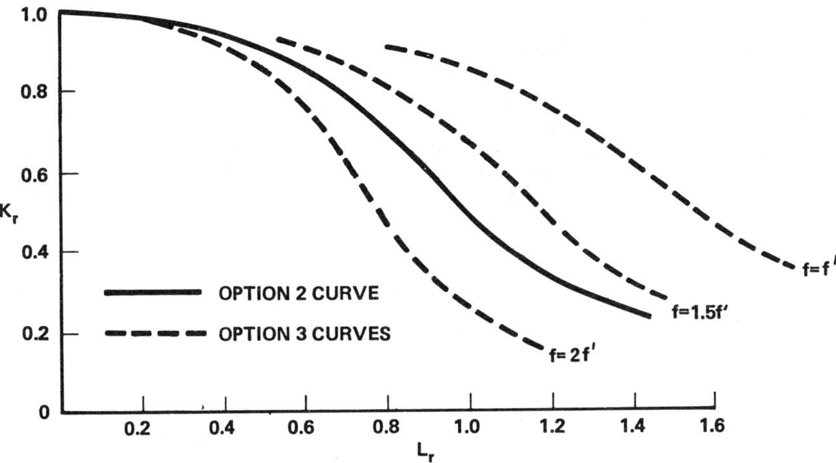

FIG. 3—*Failure assesssment curves.*

where $J = J_e$ for $n = 1$ and $\alpha = 1$. From Eq 10 with $n = 1$:

$$J_e = \sigma_o\epsilon_o a(1 - a^*)h_1(a^*,1)(\sigma/\sigma_o)^2 \tag{12}$$

Hence Eq 10 can be written as

$$J = J_e \cdot \alpha \cdot \frac{h_1(a^*,n)}{h_1(a^*,1)} \cdot (\sigma/\sigma_o)^{n-1} \tag{13}$$

Writing $\sigma_r = \sigma/f$ and comparing Eqs 11 and 13 leads to the following expression for f:

$$f = \left[\frac{h_1(a^*,1)}{h_1(a^*,n)} \right]^{1/(n-1)} \tag{14}$$

The functions $h_1(a^*,n)$, or similar functions, for various geometries and n values have been tabulated, for example, in Refs *3* to *5*.

Surface Flawed Plate in Tension

Plastic Yield Functions

The simplest choice of yield function f is that based on the reduction in load bearing area due to the presence of the defect. This gives the global yield function corresponding to fixed grip loading:

$$f^g = 1 - \frac{\pi ac}{2Bt} \tag{15}$$

where B is the breadth of the plate (Fig. 2). It is possible to construct a local yield function for this geometry by using the approximation shown schematically in Fig. 4a. In this figure the plate is divided into two unnotched sections and a central notched part whose breadth is

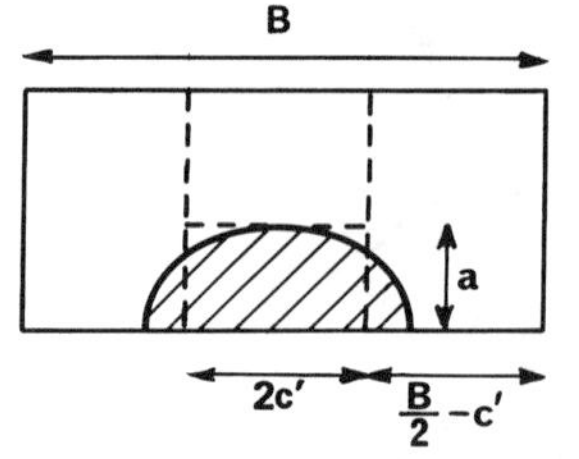
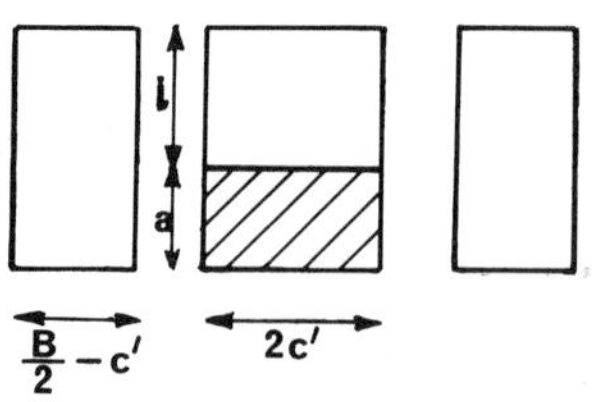

(a) SINGLE EDGE CRACKED PLATE, LOCAL YIELD

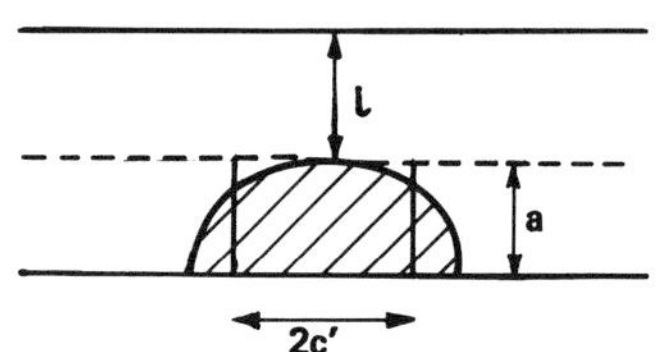
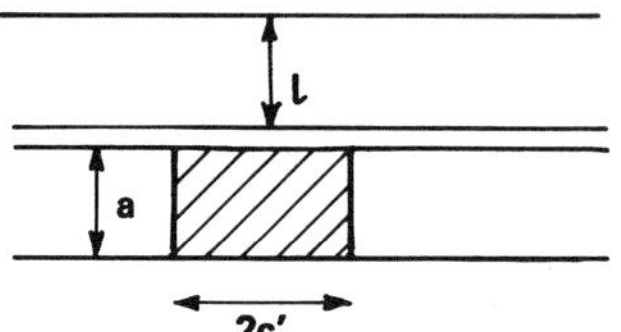

(b) AXIAL SURFACE FLAW, GLOBAL YIELD

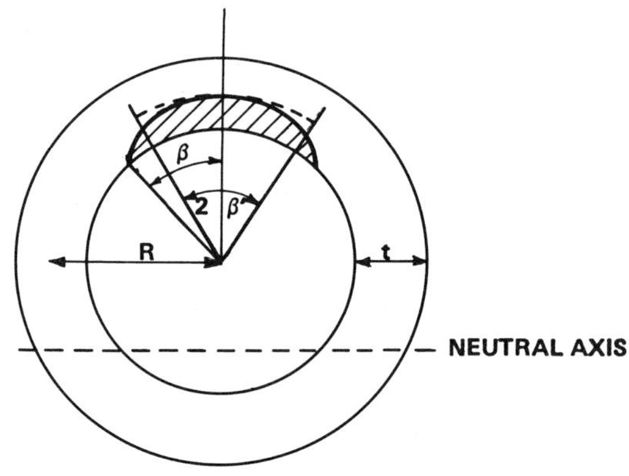

(c) CIRCUMFERENTIAL SURFACE FLAW, GLOBAL YIELD

FIG. 4—*Determining plastic yield functions. (a) Single edge cracked plate, local yield. (b) Axial surface flaw, global yield. (c) Circumferential surface flaw, global yield.*

$2c'$, where c' ($= \pi c/4$, if the defect is semi-elliptical) is chosen so that the area of the original surface flaw is equal to that of the rectangular defect in the central section. An expression for f can then be derived by calculating the plastic yield loads for each section individually and adding these to define the yield load for the original cracked plate.

There are two possible yield solutions for the central section depending on whether it is assumed to deform in plane stress or plane strain. The plane stress solution is [6]

$$f' = [(1 - a^*)^2 + a^{*2}]^{1/2} - a^* \qquad (16)$$

The plane strain solution for $a^* \leq 0.545$ is [7]

$$f' = 1 - a^* - 1.232\, a^{*2} + a^{*3} \tag{17a}$$

and for $a^* > 0.545$ it is [6]

$$f' = 1.702\,\{[(0.794 - X)^2 + 0.5876X^2]^{1/2} - [0.794 - X]\} \tag{17b}$$

where $X = 1 - a^*$.

The local yield function is thus

$$f^\ell = 1 - \frac{\pi c}{2B} + \frac{\pi c}{2B} f'(a^*) \tag{18}$$

where $f'(a^*)$ is set equal to Eqs 16 or 17 depending on whether plane stress or strain is assumed. The value of $L_r = \sigma/\sigma_y f$, where f adopts the appropriate value of f^g or f^ℓ.

Comparison of Option 2 and Option 3 Curves

Finite element elastic-plastic computations to determine J for surface flaws in plates have been performed by Hodulak and Stöckl [8] and Yagawa et al. [9]. The former used a piecewise linear stress-strain curve and calculated J at the deepest and surface crack positions on a flaw of depth $a = 13.3$ mm, surface length $2c = 40$ mm, in a plate where the thickness $t = 20$ mm and the breadth $B = 120$ mm. The yield stress was 542 MPa. The results transformed into an Option 3 FAC using Eq 6 are shown in Fig. 5 together with the Option 2 FAC derived using Eq 4. The abrupt changes in the Option 2 curve reflect the piecewise linear nature of the stress-strain curve. The Option 3 curves refer to global and local yield loads, the latter only for the plane stress result, as this, and the curve for plane strain, are very similar in this case.

It can be seen from Fig. 5 that the Option 3 curves for the deepest and surface points are similar, and that overall the global yield function provides the best fit to the Option 2 curve, although the resulting curve is slightly optimistic (falls inside the Option 2 curve) at large L_r values. In this regime the local collapse function gives the better correlation with the Option 2 curve.

The results of Yagawa et al. [9] are given in terms of the function $h_1(a^*,n)$ for the deepest point on the crack front and were derived for a power law stress-strain curve. The plate dimensions were $B/t = 30$, the surface lengths of the defects were fixed at $c/t = 7.5$, and the crack depths were $a/t = 0.3, 0.7,$ and 0.9. The values of the function f derived using Eq 14 are shown in Table 1 together with the values of the global and local yield functions obtained from Eqs 15 to 17.

The value of n is usually in the range 5 to 10 for austenitic and ferritic steels; hence the results of Yagawa et al. [9] for $n = 5$ are of particular interest. Overall the global yield functions give the better agreement with the derived result for this value of n.

Axially Oriented Surface Flaw in a Pressurized Cylinder

Plastic Yield Functions

Following Miller [10] a lower bound to the global yield solution may be derived using the construction shown schematically in Fig. 4b. The cylinder is split into a defect-free cylinder

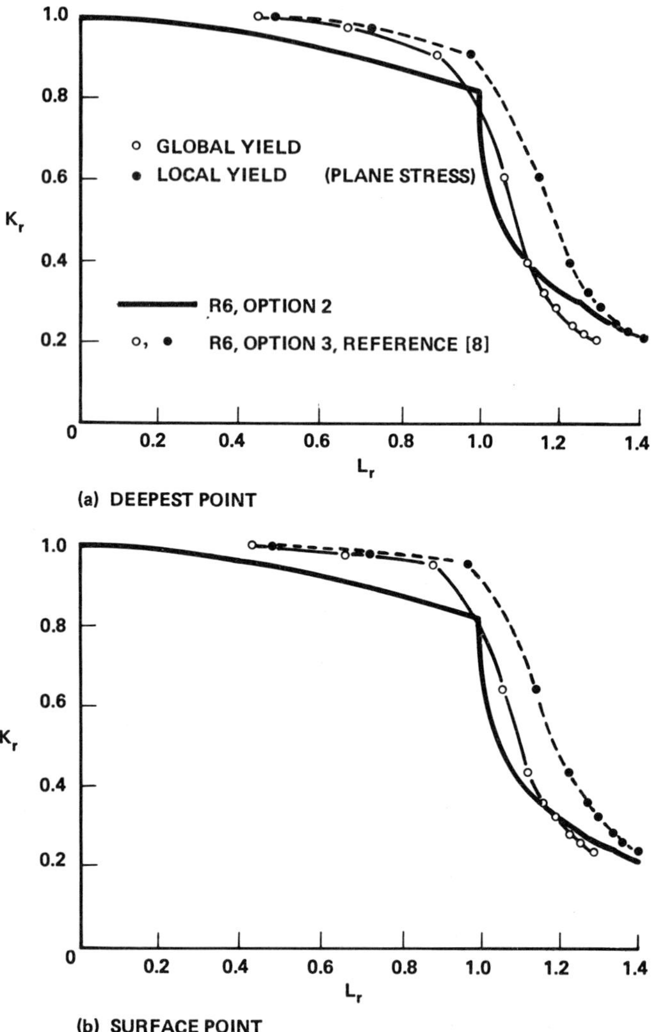

(a) DEEPEST POINT

(b) SURFACE POINT

FIG. 5—*FAD for a surface flaw in a plate subjected to uniform tenstion* (a/t $= 0.665$, a/c $= 0.665$). (a) *Deepest point.* (b) *Surface point.*

of thickness $\ell = t - a$ and a cylinder of thickness a which contains a through-wall crack of length $2c'$. The plastic yield pressure is then taken as the sum of the pressures required to yield each of the cylinders. The result is

$$f^{g} = 1 - a^{*} + a^{*}f'(c'^{2}/Ra) \qquad (19)$$

where the function f' is taken from Kiefner et al. [*11*] as

$$f'(x) = (1 + 1.61x)^{-1/2} \qquad (20)$$

TABLE 1—*Plastic yield functions* (f) *for surface flaw in a plate subjected to tensile loading.*

a/t	a/c	n	f^1	f^2	f^3	f^4
0.3	0.04	2	0.67			
		3	0.72	0.88	0.79	0.86
		4	0.75			
		5	0.77			
0.7	0.09	2	0.63			
		3	0.68	0.73	0.63	0.64
		4	0.71			
		5	0.74			
0.9	0.12	2	0.65			
		3	0.67	0.65	0.61	0.61
		4	0.69			
		5	0.70			

[1] Derived from Ref 9.
[2] Global, reduction in area (Eq 15).
[3] Local, plane stress (Eq 16).
[4] Local, plane strain (Eq 17).

A local yield pressure is also defined by Kiefner et al. [11] as the empirical equation

$$f^\ell = (1 - a^*)/[1 - a^* f'(c'^2/Rt)] \tag{21}$$

where f' is the same function as in Eq 20.

The value of L_r equals $RP/t\sigma_y f$ for vessels with mean radius R and internal pressure P.

Comparison of Option 2 and Option 3 Curves

There have been several elastic-plastic J computations performed on axially cracked cylinders. The results of Aurich et al. [12] and Brocks and Noach [13] are shown in Fig. 6 as Option 3 curves for both the deepest and surface points. The calculations were performed on a cylinder of mean radius $R = 1400$ and wall thickness $t = 140$ mm containing a defect of depth $a/t = 0.59$ and aspect ratio $a/c = 0.346$. A piecewise linear fit to the stress-strain curve of the material was used. (In the present calculations this has been fitted by a power law curve to good accuracy out to a strain of 5%, which covers the part of the curve of interest.)

The results of two sets of computations are shown in Fig. 6. In one the cylinder was open and not subjected to any axial stressing, and the crack faces were not loaded by the internal pressure. In the other, the cylinder was closed and the crack faces loaded. As can be seen from Fig. 6 there was little difference between the Option 3 curves derived for these two cases. The Option 2 curve shown in Fig. 6 was determined using the power law curve fit to the stress-strain curve.

It is interesting to note from Fig. 6b that the effects of crack tip plasticity on the surface point of the defect are predicted initially to reduce the crack tip loading at that position with respect to the results of linear elasticity (i.e., K_r is greater than 1). Due to this fact the Option 2 curve of R6 is overly pessimistic compared with the Option 3 results for the surface point.

There is little difference between the Option 3 curves based on either the global or local yield function, but of the two the global solution produces the better agreement between the Option 2 and Option 3 curves.

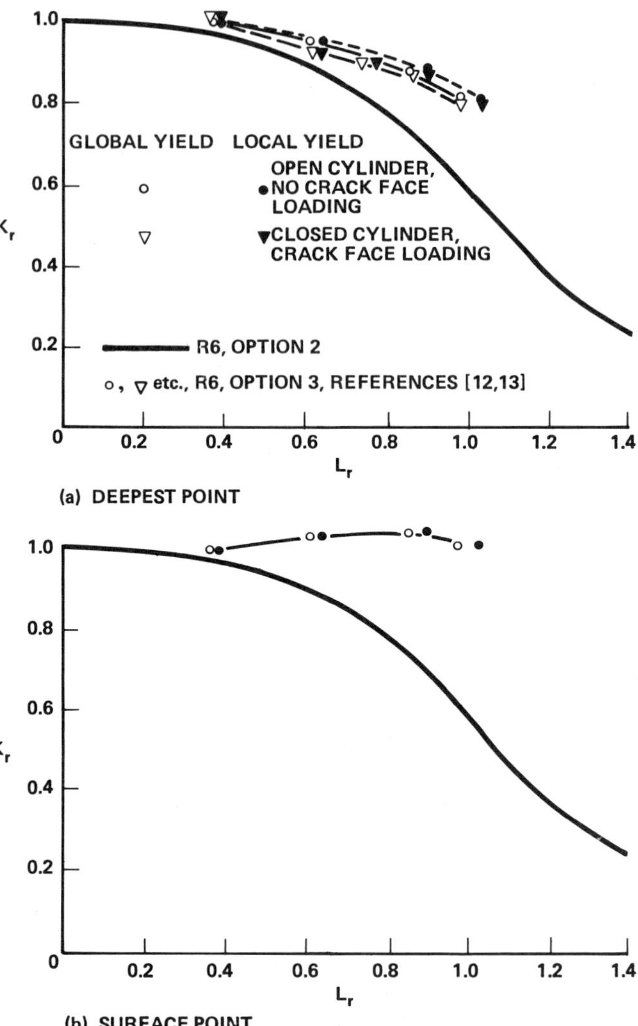

FIG. 6—*FAD for a surface, axially oriented internal flaw in a pressurized cylinder* (R/t = 10, a/t = 0.59, a/c = 0.346). (a) *Deepest point.* (b) *Surface point.*

Wilkening et al. [*14*] have computed J values for axial defects in pressurized cylinders where $R = 2.4$ m and $t = 0.229$ m using a Ramberg-Osgood representation of the stress-strain curve. The Option 3 curves for the deepest and surface points on the flaw, together with the dimensions of the defects they considered, are shown in Fig. 7. Also shown is the Option 2 curve derived from the stress-strain curve. The Option 3 results for the surface point on the defect show K_r values greater than 1 for small values of L_r, as was also the case in Fig. 6*b*.

The Option 3 curves derived assuming either global or local yield show significant differences for the deepest crack. In this case, and for all the other defects considered by Wilkening et al., the use of the global yield function produces the best correlation between the Option 2 and Option 3 curves.

Bryan et al. [15] and Bass et al. [16] have computed J solutions for surface flaws in thick cylinders ($R = 495.3$ mm, $t = 152.4$ mm) subjected to internal pressure in order to study their fracture behavior under conditions similar to those encountered in full-scale reactor pressure vessels. Some of the results of Bryan et al. [15] for a defect of depth $a/t = 0.67$ and

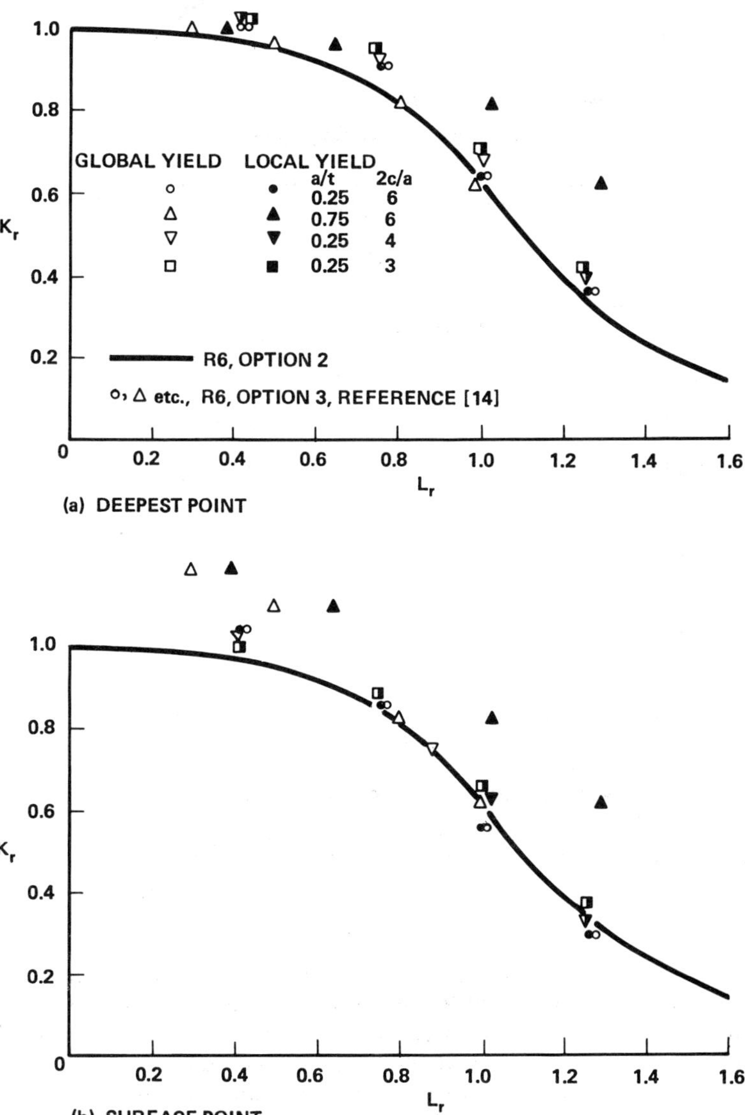

FIG. 7—*FAD for a surface, axially oriented internal flaw in a pressurized cylindrical vessel* ($R/t = 10.5$). (a) *Deepest point.* (b) *Surface point.*

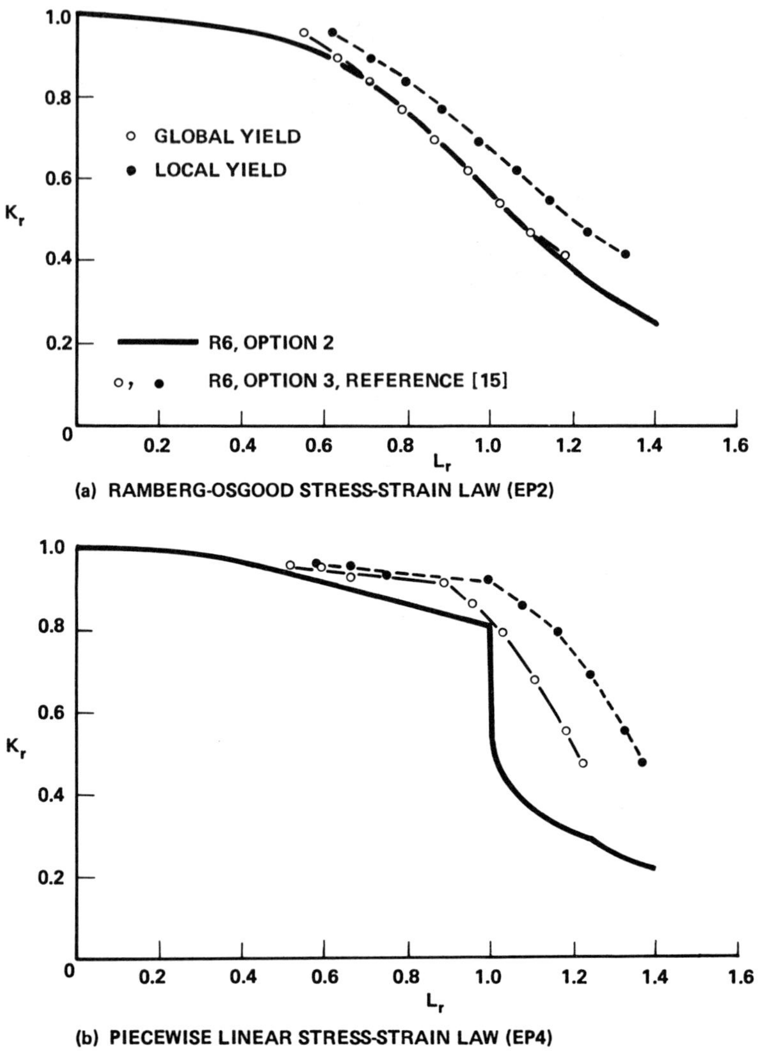

(a) RAMBERG-OSGOOD STRESS-STRAIN LAW (EP2)

(b) PIECEWISE LINEAR STRESS-STRAIN LAW (EP4)

FIG. 8—*FAD for a surface, axially oriented internal flaw in a pressurized cylinder* (R/t = 2.75, a/t = 0.67, a/c = 0.67, ϕ = 34.62°). *(a) Ramberg-Osgood stress-strain law (EP2). (b) Piecewise linear stress-strain law (EP4).*

aspect ratio a/c = 0.67 are shown expressed in the form of Option 3 curves in Fig. 8 for Ramberg-Osgood (case number V8EP2) and piecewise linear (V8EP4) representations of the stress-strain curve. These curves are based on J values for the position ϕ = 34.62° on the crack front (Fig. 2), the position at which J had its maximum value. Also shown in Fig. 8 are the Option 2 curves. In these cases the global yield function (Eq 19) clearly produces the better agreement between the Option 2 and Option 3 curves.

The results of similar computations by Bass et al. [*16*] are shown in Figs. 9 and 10 for the piecewise linear curve, V8EP4, and crack depths of a/t = 0.59 and a/t = 0.656, both with aspect ratios of 0.67. In these cases the deepest (ϕ = 90°) and surface (ϕ = 0°) point J values were given. It can be seen that in both instances the global yield function provides the better correlation between the Option 2 and Option 3 curves.

Circumferential Cracked Cylinders Subjected to Tension

Plastic Yield Functions

A global yield function may be obtained by balancing external forces and moments against the internally induced forces due to yielding [10] to give

$$f^g = f'(a^*, \beta') \tag{22}$$

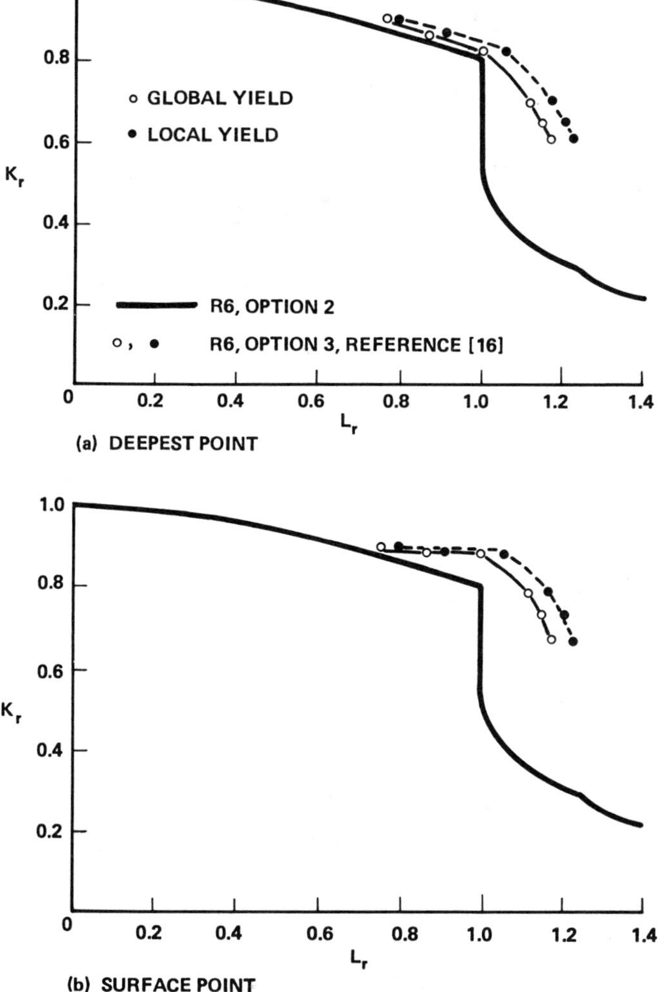

FIG. 9—*FAD for a surface, axially oriented internal flaw in a pressurized cylinder (R/t = 2.75, a/t = 0.5, a/c = 0.67). (a) Deepest point. (b) Surface point.*

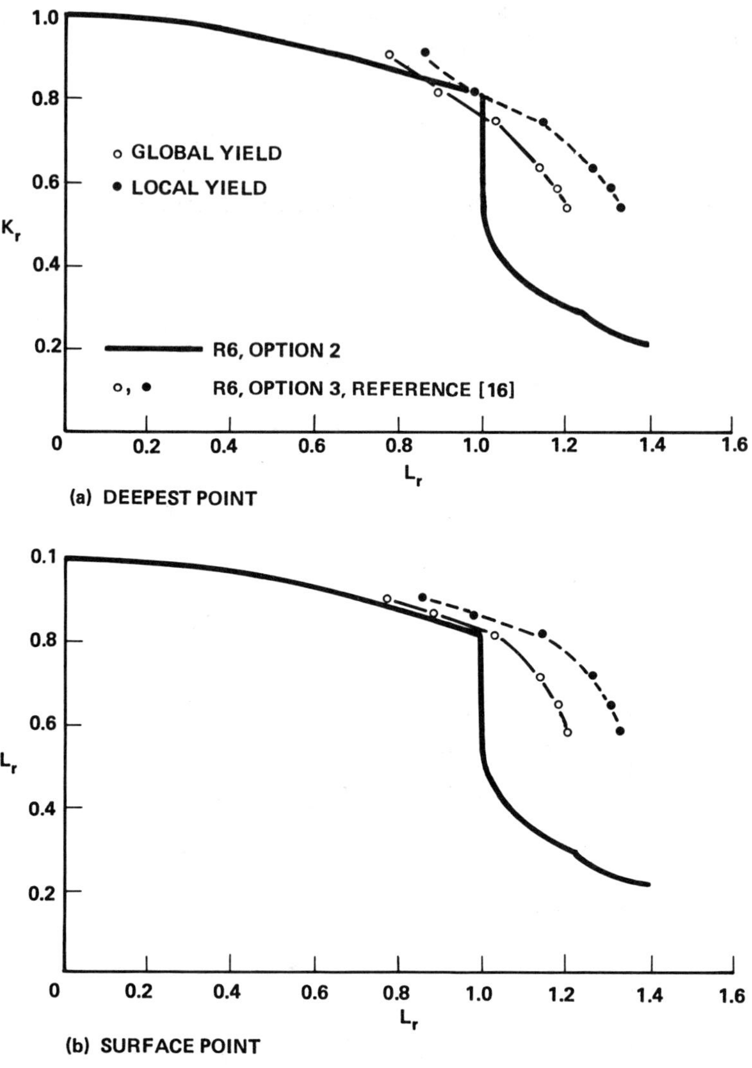

FIG. 10—*FAD for a surface, axially oriented internal flaw in a pressurized cylinder* (R/t) = 2.75, a/t = 0.656, a/c = 0.67). (a) *Deepest point.* (b) *Surface point.*

where $\beta' = \pi\beta/4$, β is half the angle subtended by the semi-elliptical defect (Fig. 4c), and

$$f'(a^*,\beta) = 1 - \{\beta a^* + 2\sin^{-1}(a^* \sin \beta/2)\}/\pi \tag{23}$$

Thin shell theory is assumed in deriving this expression.

A local yield function may be estimated by analogy with Eq 21 as

$$f^\ell = (1 - a^*)/[1 - a^* f'(1,\beta')] \tag{24}$$

where f' is the same function as Eq 23 with $a/t = 1$. L_r is simply given by $\sigma/\sigma_y f$.

Comparison of Option 2 and Option 3 Curves

Both Yagawa et al. [9] and Kumar and German [5] have computed the function $h_1(a^*,n)$ at the deepest point on the crack for this geometry using power law stress-strain relationships. In the former the yield function f may be determined from Eq 14, whereas in the latter this equation has to be modified slightly because of the different way loads are normalized to read

$$f = \left[\frac{A_{nc}}{A}\right] \left[\frac{h_1(a^*,1)}{h_1(a^*,n)}\right]^{1/(n-1)} \tag{25}$$

where A and A_{nc} are the cross-sectional areas of the uncracked and cracked cylinders respectively, and

$$\frac{A_{nc}}{A} = [(R_o^2 - R_c^2) + (1 - \beta/\pi)(R_c^2 - R_i^2)]/(R_o^2 - R_i^2)$$

where R_o and R_i are the outer and inner radii respectively, and $R_c = R_i + a$. The function $h_1(a/t,1)$ is given by

$$h_1(a^*,1) = \left[\frac{A_{nc}}{A}\right]^2 \frac{(1 - \nu^2)\pi F^2}{(1 - a^*)}$$

where F is defined as in Table 3.4 of Ref 5. The defect shape of the crack analyzed by Kumar and German was rectangular; hence β' should be replaced by β in Eqs 22 and 24.

The functions f derived from Eqs 14 and 25 are shown in Tables 2 and 3 respectively, together with the values of f determined from Eqs 22 and 24. Also given in the tables are the relative dimensions of the cylinders and cracks. Overall the global yield function provides better agreement with the computed values than does the local yield for n values of 5 and 10, which span the range of values characteristic of steels.

TABLE 2—*Plastic yield functions* (f) *for circumferentially oriented flaw in a cylinder (R/t = 10.9) subjected to tension.*

a/t	2β	n	f^1	f^2	f^3
0.3	60°	2	0.80		
		3	0.83	0.94	0.92
		4	0.86	0.92^4	
		5	0.87		
0.7	60°	2	0.70		
		3	0.75	0.85	0.67
		4	0.78	0.81^4	
		5	0.80		
0.9	60°	2	0.72		
		3	0.75	0.81	0.34
		4	0.77	0.76^4	
		5	0.79		

[1] Derived from Ref 9.
[2] Global yield (Eq 22).
[3] Local yield (Eq 24).
[4] Global yield using Eq 22 with $\beta' = \beta$.

TABLE 3—*Plastic yield functions* (f) *for circumferentially oriented rectangular flaw in a cylinder (R/t = 10) subjected to tension.*

a/t	2β	n	f^1	f^2	f^3
0.5	45°	2	0.69		
		5	0.77	0.90	0.83
		10	0.82		
	90°	2	0.65		
		5	0.70	0.80	0.72
		10	0.75		
	180°	2	0.63		
		5	0.65	0.65	0.58
		10	0.67		
0.75	45°	2	0.57		
		5	0.73	0.85	0.62
		10	0.80		
	90°	2	0.48		
		5	0.63	0.70	0.46
		10	0.70		
	180°	2	0.39		
		5	0.51	0.47	0.32
		10	0.59		

[1] Derived from Ref 5.
[2] Global yield (Eq 22).
[3] Local yield (Eq 24).

Circumferential Cracked Cylinders Subjected to Bending

Plastic Yield Functions

The global yield function derived from balancing moments is [*10*]

$$f^g = f'(a^*,\beta) \tag{26}$$

where β is the half angle subtended by a rectangular flaw and

$$f'(a^*,\beta) = \frac{4}{\pi} \{\cos (a^*\beta/2) - a^* \sin \beta/2\} \tag{27}$$

A local yield function may be obtained from

$$f^\ell = (1 - a^*)/[1 - a^*f'(1,\beta)] \tag{28}$$

where the function f' is given by Eq 27. L_r is calculated as $\sigma_b/\sigma_y f$, where σ_b is the bending stress given by $4RM/\pi(R_o^4 - R_i^4)$ and M is the applied moment.

Comparison of Option 2 and Option 3 Curves

Values of the functions $h_1(a/t,n)$ for power law hardening materials are given by Kumar and German [*5*] for the deepest point on the defect. From these the functions f may be calculated using the equation

$$f = \left[\frac{A_{nc}}{A}\right] \frac{1}{D} \left[\frac{h_1(a^*,1)}{h_1(a^*,n)}\right]^{1/(n-1)} \tag{29}$$

where

$$D = \pi(R_o^4 - R_i^4)/8Rt(R_o^2 + R_i^2)$$

and

$$h_1(a^*,1) = \left[\frac{A_{nc}}{A}\right]^2 \frac{1}{D^2} \frac{(1 - \nu^2)\pi F^2}{(1 - a^*)}$$

and the values of F are as given in Table 3.4 of Ref 5.

The results of calculating f from Eq 29 and Eqs 26 and 28 are shown in Table 4. Again the global yield function is in better agreement with the computed values for $n = 5$ and 10 than is the local yield function.

Review of Circumferential Cracked Cylinders

The values of the yield functions obtained from global and local collapse considerations are plotted in Fig. 11 against the values derived from the finite element computations for power law stress-strain exponents of 5 and 10. The results for tension and bending are given. This figure demonstrates more clearly than the values in Tables 2, 3, and 4 the good correlation between the f values derived from the global yield functions and those derived from the finite element results. Ideally the values shown in Fig. 11 should fall on the 1:1 line.

Discussion

From the results presented in the previous section it is clear that the R6 procedure may be applied to the deepest and surface points on surface flaws, provided that the plastic yield function, f, which is defined in Eq 2, is known. The present work has closely followed the method used by Miller [2] in order to try and identify the function f most appropriate to an R6 analysis. The results of both studies indicate that yield functions derived from global yield

TABLE 4—*Plastic yield functions* (f) *for circumferentially oriented rectangular flaw in a cylinder* (R/t = 10) *subjected to bending.*

a/t	2β	n	f^1	f^2	f^3
0.5	90°	2	0.89		
		5	0.95	1.05	0.79
		10	1.01		
0.75	90°	2	0.69		
		5	0.88	0.94	0.55
		10	0.97		

[1] Derived from Ref 5.
[2] Global yield (Eq 26).
[3] Local yield (Eq 28).

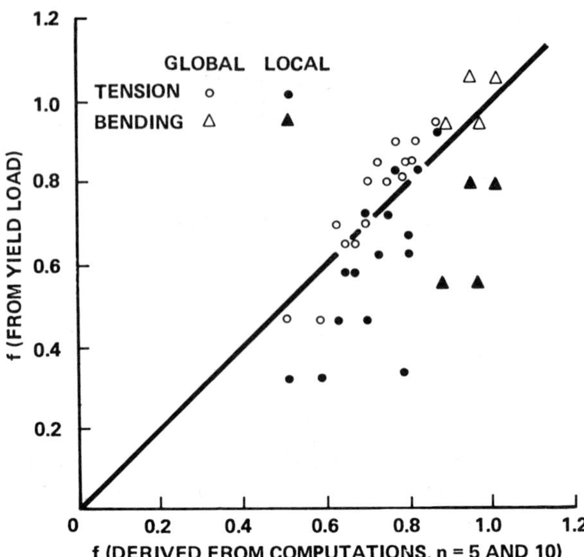

FIG. 11—*Yield functions for circumferential flaws in cylinders.*

loads rather than local yield loads provide the best agreement between R6 and computed *J* values.

The method used assumes that the Option 2 curve will be a good approximation to the Option 3 curve. The Option 2 curve is based on a reference stress formulation for the *J*-integral derived by Ainsworth [*17*] and is dependent on the materials unaxial stress-strain deformation and independent of the geometry of the cracked structure. Thus the yield function which results in the best agreement between the Option 2 and Option 3 curves will not necessarily represent the plastic yield load of the structure. Instead it represents, in general, an empirical yield load which will produce good agreement between the R6 procedure and an elastic-plastic computation of the *J*-integral.

Because of this fact the use of the derived yield functions in an R6 analysis needs to be further qualified by the necessity to establish the accuracy of the numerical methods used in the *J* computations. Recently Miller and Ainsworth [*18*] have investigated the accuracy of finite element solutions using a consistency check based on a theorem of nesting surfaces introduced by Calladine and Drucker [*19*]. They applied the check to computed *J* results for a number of standard specimen geometries and found that most of the finite element results passed the check, although a few gave errors of up to 15% on load. However, Miller and Ainsworth [*18*] concluded that reference stress estimates of *J* (i.e., Option 2 curves) tend to conservatively reproduce the finite element results. Although this conclusion is also borne out by the present study, a similar consistency check on the accuracy of the finite element results for surface flaws would provide an extra level of assurance regarding the appropriateness of the yield functions to an R6 analysis.

Most of the surface flaws considered have been semi-elliptical and, for the purposes of determining the plastic yield functions, they have been re-characterized as rectangular defects of the same area. A more pessimistic representation of the flaws is to re-characterize them as rectangular defects of surface length $2c$ rather than $2c'$. This would result in only a small change in the yield functions for the range of defects analyzed in the present work, as is illustrated in Table 2 where the values of the global yield functions calculated according to this re-characterization are shown. The differences resulting from the two re-characterization procedures would be more significant for defects with very small aspect ratios.

Conclusions

The following conclusions may be drawn concerning the application of the R6 procedure to surface flaws based on analysis of published elastic-plastic finite element computations of the J-integral:

1. The R6 procedure may be used to assess both the deepest and surface points on a surface flaw, provided the appropriate plastic yield load is used.

2. The same plastic yield load may be used to conservatively assess all points on the crack front, provided the stress intensity factor appropriate to each position is used in the R6 analysis.

3. Overall a global yield load gave the better agreement between Option 2 failure curves based on a reference stress formulation for J and Option 3 curves derived from the computed J results. This conclusion is in agreement with Miller [2]. The global yield functions used herein to determine the plastic yield loads via Eq 2 are:

(i) Eq 15—for surface flawed plates in tension.

(ii) Eq 19—for axially oriented internal flaws in cylinders subjected to internal pressure.

(iii) Eq 22—for circumferentially oriented internal flaws in cylinders subjected to an axial tensile loading.

(iv) Eq 26—for circumferentially oriented internal flaws in cylinders subjected to external bending.

4. The use of local yield loads in an R6 analysis is conservative.

Acknowledgment

This work was performed at the Central Electricity Research Laboratories and is published by permission of the Central Electricity Generating Board.

References

[1] Milne, I., Ainsworth, R. A., Dowling, A. R., and Stewart, A. T., "Assessment of the Integrity of Structures Containing Defects," Report R/H/R6-Rev. 3, CEGB, London, 1986, also *International Journal of Pressure Vessels and Piping,* Vol. 32, 1988, p. 196.

[2] Miller, A. G., "J Estimation for Surface Defects," Report TPRD/B/0811/R86, CEGB, Berkeley Nuclear Laboratories, Gloucestershire, U.K., 1986.

[3] Kumar, V., German, M. D., and Shih, C. F., "An Engineering Approach for Elastic-Plastic Fracture Analysis," Report EPRI NP-1931, General Electric Company, Schenectady, N.Y., 1981.

[4] Kumar, V., German, M. D., Wilkening, W. W., Andrews, W. R., de Lorenzi, H. G., and Mowbray, D. F., "Advances in Elastic-Plastic Fracture Analysis," Report EPRI NP-3607, General Electric Company, Schenectady, N.Y., 1984.

[5] Kumar, V. and German, M.D., "Elastic-Plastic Fracture Analysis of Through-Wall and Surface Flaws in Cylinders," Report EPRI NP-5596, General Electric Company, Schenectady, N.Y., 1988.

[6] Ewing, D. F. J. and Richards, C. E., *Journal of the Mechanics and Physics of Solids,* Vol. 22, 1974, pp. 27–36.

[7] Ewing, D. J. F., "Plastic Collapse Loads and S_r Ratios for Edge Cracked Plates under Tension and Bending," Report TPRD/L/2362/N82, CEGB, Leatherhead, Surrey, U.K., 1982.

[8] Hodulak, L. and Stöckl, H., "Bestimmung von J-Risswiderstandskurven an Platten mit Oberflächenrissen," Report W11/85, Fraunhofer-Institut der Werkstoffmechanik, Freiburg, Germany, 1985.

[9] Yagawa, G., Ueda, A., and Takahashi, Y., "Surface Crack Behaviour in Ductile Materials," in *Proceedings,* 8th SMIRT Conference, Brussels, Paper G6/2, 1985.

[10] Miller, A. G., "Review of Limit Loads of Structures Containing Defects," 3rd ed., Report TPRD/B/0098/N82, CEGB, Berkeley Nuclear Laboratories, Gloucestershire, U.K., 1987, also *International Journal of Pressure Vessels and Piping,* Vol. 32, 1988, pp. 197–327.

[11] Kiefner, J. F., Maxey, W. A., Eiber, R. J., and Duffey, A. R. in *Progress in Flaw Growth and Fracture Toughness Testing, ASTM STP 536,* American Society for Testing and Materials, Philadelphia, 1973, pp. 461–481.

[12] Aurich, D., Brocks, W., Noack, H.-D. and Veith, H. in *Fracture Mechanics: Sixteenth Symposium, ASTM STP 868,* American Society for Testing and Materials, Philadelphia, 1985, pp. 617–631.

[13] Brocks, W. and Noack, H.-D., "Elastic-Plastic *J* Analysis for an Inner Surface Flaw in a Pressure Vessel," in *Proceedings,* 6th European Conference on Fracture, EMAS, Amsterdam, H. C. van Elst and A. Bakker, Eds., 1986, pp. 617–630.

[14] Wilkening, W. W., de Lorenzi, H. G., and Barishpolsky, M., *Journal of Pressure Vessel Technology,* Vol. 106, 1984, pp. 247–254.

[15] Bryan, R. H., Bass, B. R., Bryson, J. W., and Merkle, J. G., "Experimental Investigation of Tearing Behaviour of a Flaw in a Thick Pressure Vessel," in *Light Water Reactor Structural Integrity,* Proceedings of 3rd International Seminar on Assuring Structural Integrity of Steel Reactor Pressure Boundary Components, K. E. Stahlkopf and L. E. Steele, Eds., Elsevier Applied Science, 1983, pp. 175–209.

[16] Bass, B. R., Bryan, R. H., Bryson, J. W., and Merkle, J. G., *Journal of Pressure Vessel Technology,* Vol. 104, 1982, pp. 308–316.

[17] Ainsworth, R. A., *Engineering Fracture Mechanics,* Vol. 19, 1984, pp. 633–642.

[18] Miller, A. G. and Ainsworth, R. A., "Consistency of Numerical Results for Power Law Hardening Materials and the Accuracy of the Reference Stress Approximation for *J*," Report TPRD/B/1005/R87, CEGB, Berkeley Nuclear Laboratories, Gloucestershire, U.K., 1987 (to be published in *Engineering Fracture Mechanics*).

[19] Calladine, C. R. and Drucker, D. C., *Journal of Mechanical Engineering Science,* Vol. 4, 1962, pp. 1–11.

Thomas W. Orange[1]

Method and Models for *R*-Curve Instability Calculations

REFERENCE: Orange, T. W., "**Method and Models for *R*-Curve Instability Calculations,**" *Fracture Mechanics: Twenty-First Symposium, ASTM STP 1074,* J. P. Gudas, J. A. Joyce, and E. M. Hackett, Eds., American Society for Testing and Materials, Philadelphia, 1990, pp. 545–559.

ABSTRACT: This paper presents a simple method for performing elastic *R*-curve instability calculations. For a single material-structure combination, the calculations can be done on some pocket calculators. On microcomputers and larger, it enables the development of a comprehensive program having libraries of driving-force equations for different configurations and *R*-curve model equations for different materials. The paper also presents several model equations for fitting to experimental *R*-curve data, both linear elastic and elasto-plastic. The models are fit to data from the literature to demonstrate their viability.

KEY WORDS: *R*-curve, *J-R* curves, instability, elastic-plastic fracture, fracture tests, data reduction, mathematical models

Nomenclature

a	Crack length
A, B, C, D, F, H	Empirical coefficients in Eqs 3 to 7
G	Strain energy release rate
J	Nonlinear crack parameter
K	Stress intensity factor
L, M, N, P	Empirical coefficients in Eqs 8 to 11
W	Width of specimen (or structure)
Y	Stress intensity calibration factor
Δ	Crack extension (effective or physical, as noted)
σ_f	Flow stress

Subscripts

o	At initial (unloaded) condition
c	Critical, at the instability point
R	Related to the material's resistance

Specimen Notation

M(T)	Center-crack tension
CS	Rectangular compact
CLWL	Crack-line wedge loaded

[1] Research Engineer, NASA Lewis Research Center, Cleveland, OH 44135.

Introduction

The *R*-curve is one of the most powerful concepts available to the fracture analyst. It is probably the best phenomenological description of the monotonic fracture process available, but it has not received the acclaim and widespread use that it deserves. Perhaps the instability calculations are thought to be too involved or tedious. The literature contains very little information on instability calculations and nothing recent. Creager [1] presented a good graphical method. However, it involves overlay transparencies, is labor-intensive, and may lack precision. In a predictive round-robin program [2], seven participants used the *R*-curve method. Only the present author used anything more elaborate than trial-and-error to perform the instability calculations.

The method used is really quite simple. For a single material-structure combination, the calculation can be done on some pocket calculators. On microcomputers and larger, a comprehensive program having libraries of driving-force equations and *R*-curve model equations is possible.

This paper will describe the method of performing the instability calculation. It will also present several model equations for fitting to experimental *R*-curve data, both linear-elastic (K, G) and elastoplastic (J).

Instability Calculation Method

The method was presented previously [3], but will be repeated here. The instability condition requires that both the magnitudes and the slopes of the crack-driving force (G or K) curve and the material resistance curve (G_R) be equal at the point of instability. So we must solve two simultaneous equations. After writing the equations in a general form and doing a little algebra, we can write the instability condition as

$$0 = \frac{G_R(\Delta_c)}{G_R'(\Delta_c)} - \frac{a_o + \Delta_c}{1 + 2\alpha_c} \tag{1}$$

where

$$G_R'(\Delta) = dG_R(\Delta)/d\Delta$$

and

$$\alpha = (a/Y)(dY/da)$$

For elastic *R*-curves Δ is the effective crack extension, a_o is the initial crack length, Y is the stress intensity calibration factor, and the subscript c means "evaluated at the instability point." (If you prefer to work with the stress intensity factor K, substitute $K/2K'$ for G/G'.) Now if we put in the expressions for G_R, G_R', and α, then, for prescribed values of a_o and specimen width W, Δ_c is the least positive root of Eq 1. Any of several numerical methods will solve for the root, even on a programmable hand calculator. Once Δ_c is known, we can calculate G_c and the failure stress. The beauty of this equation is that the first term on the right-hand side is a function of the material's *R*-curve only and the second is a function of the structural geometry only. Equations for the second term can be obtained from standard handbooks and will not be discussed here. Model *R*-curve equations to be used in the first term will be discussed later.

The effectiveness of this method is shown in Appendix 11 of Ref *2*. In that analytical round-robin program, good predictions of fracture strengths were made for simple and complex specimen geometries.

Unfortunately this simple method is not applicable to elastic-plastic (J_R) calculations. As discussed in Sections 2 and 6 of Ref *4*, the problem is much more complicated. Material properties are an integral part of the calculation of the driving force (J), and there appears to be no general way to separate them from the geometrical terms. However, the model equations for *J-R* curves to be discussed later may still prove useful.

Elastic *R*-Curve Models

To do these calculations we need an equation for the *R*-curve. While it would be nice if the equation had some physical significance, it is not absolutely necessary. Until a good theoretical analysis is available, all we need is a continuous equation that approximates reality, is readily differentiable, and fits the data. At present there is no single equation that fits all *R*-curve data. But there are several usable equations, so we have to pick the one that best fits the particular data. Let ('*R*') be a generic term to indicate either G_R or K_R.

Wang and McCabe [5] suggested fitting a polynomial to data in the region of the *R*-curve where one expects instability to occur. One could also describe the entire curve by a low-order (say, cubic) spline function. For either case we have

$$R_0 = A_0 + A_1\Delta + A_2\Delta^2 + A_3\Delta^3 \tag{2}$$

(Note: Different polynomial coefficients are used for each spline segment.)

Broek and Vlieger [6] suggested a power curve of the form

$$R_1 = A\Delta^B \tag{3}$$

where $0 < B < 1$. This particular form works very well for the more ductile materials such as 2000-series aluminums and some stainless steels. It has two unique features. The slope is infinite at $\Delta = 0$, and there is no asymptote. In physical terms these imply there is no crack extension at small loads (which is tenable) and that a material's fracture toughness is limited only by the size of the specimen (which is not). So it may not be a good fundamental model, but it is acceptable for curve fitting.

R-curves for lower-toughness materials are usually asymptotic to a "plateau value" of toughness. Several models fit this description. Bluhm [7] proposed an exponential model:

$$R_2 = R_2^* \cdot (1 - e^{-\Delta/C}) \tag{4}$$

This form has a finite slope at $\Delta = 0$ and is asymptotic to R_2^*. A somewhat similar model, which I will call the hyperbolic model, is

$$R_3 = R_3^*\Delta/(D + \Delta) \tag{5}$$

This is actually an inverted and transposed rectangular hyperbola. Like the exponential model it has a finite slope at $\Delta = 0$ and is asymptotic to R_3^*. Trigonometry suggests two additional asymptotic forms. They are the arctangent function

$$R_4 = R_4^*(2/\pi) \arctan (\pi\Delta/2F) \tag{6}$$

and the hyperbolic tangent

$$R_5 = R_5^* \tanh (\Delta/H) \tag{7}$$

The power model (Eq 3) is simple enough that an illustration is not required. The four asymptotic models (Eqs 4 to 7) are shown schematically in Fig. 1a. The first empirical coefficient (the asymptote) is the "plateau value" of fracture toughness, the second is a characteristic value of crack extension. Note that in all cases the initial slope is equal to the first empirical coefficient divided by the second. That is why the second coefficient was placed in the denominator of the arguments in Eqs 4, 6, and 7 rather than in the numerator. Note also that Eqs 4 to 7 are linear in the first coefficient and nonlinear in the second. Most nonlinear regression routines require initial estimates of the coefficients. These estimates can easily be made from a plot of the raw data using Fig. 1a as a guideline. Furthermore, the fitted coefficients can easily be converted from one system of units to another.

Figure 1b shows the asymptotic models, all with the same asymptote and initial slope. This helps to show the inherent characteristics of each. The hyperbolic tangent model (curve A) has the sharpest "knee," rising quickly to approach the asymptote. The knee of the exponential (B) is not as sharp, but the curve still approaches the asymptote fairly quickly. The arctangent model (C) rises quickly at first but then more slowly. The hyperbolic model (D) has the slowest approach of all. The characteristics of these models should be kept in mind when attempting to fit them to data.

The low-order polynomial (or spline) (Eq 2) might well be the most accurate way to fit a single R-curve. But a comprehensive computer program must store the polynomial coefficients for each spline segment as well as the location of the knots. Interpolation for thickness or temperature effects could be a problem. The other R-curve models, however, only require that we store two constants and one equation code for each thickness and temperature. Interpolation for thickness or temperature should be much simpler.

Elastic-Plastic R-Curve Models

The four asymptotic models for elastic R-curves can be modified to represent elastic-plastic J-R curves by replacing $[R^*]$ with $[R^* + T\Delta]$ and renaming the coefficients. Then Eqs 4 to 7 become

$$R_6 = (R_6 + T_6\Delta)[1 - \exp(-\Delta/L)] \tag{8}$$

$$R_7 = (R_7^* + T_7\Delta)\Delta/(M + \Delta) \tag{9}$$

$$R_8 = (R_8^* + T_8\Delta)(2/\pi) \arctan (\pi\Delta/2N) \tag{10}$$

$$R_9 = (R_9^* + T_9\Delta) \tanh (\Delta/P) \tag{11}$$

These four models are shown schematically in Fig. 2. Here the first coefficient R^* is a reference toughness value, similar to J_{Ic}. The second coefficient, T, is proportional to the tearing modulus. The initial slope is equal to the first coefficient, R^*, divided by the third (L, M, N, or P). Using Fig. 2 as a guideline, all three coefficients can be estimated from a plot

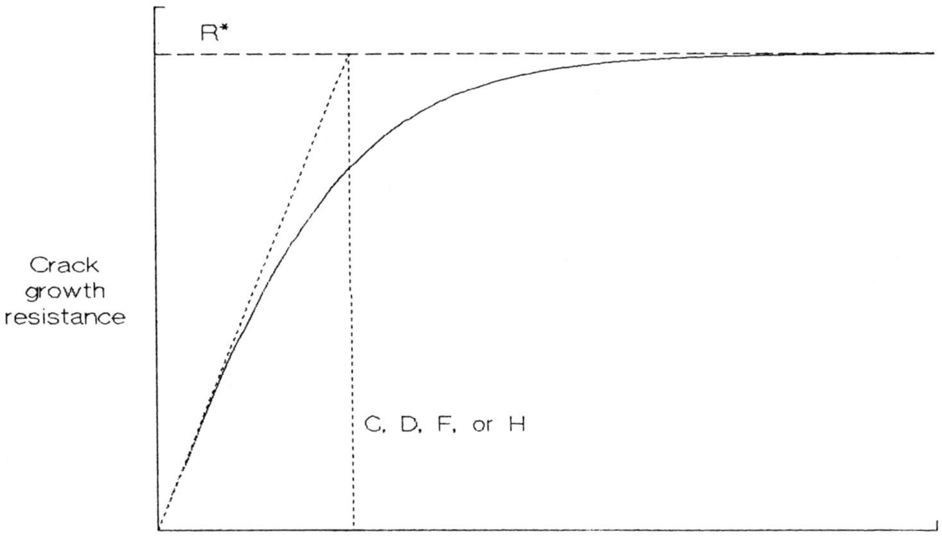

FIG. 1a—*Schematic representation of elastic asymptotic* R-*curves (C, D, F, and H are coefficients in Eqs 4 to 7).*

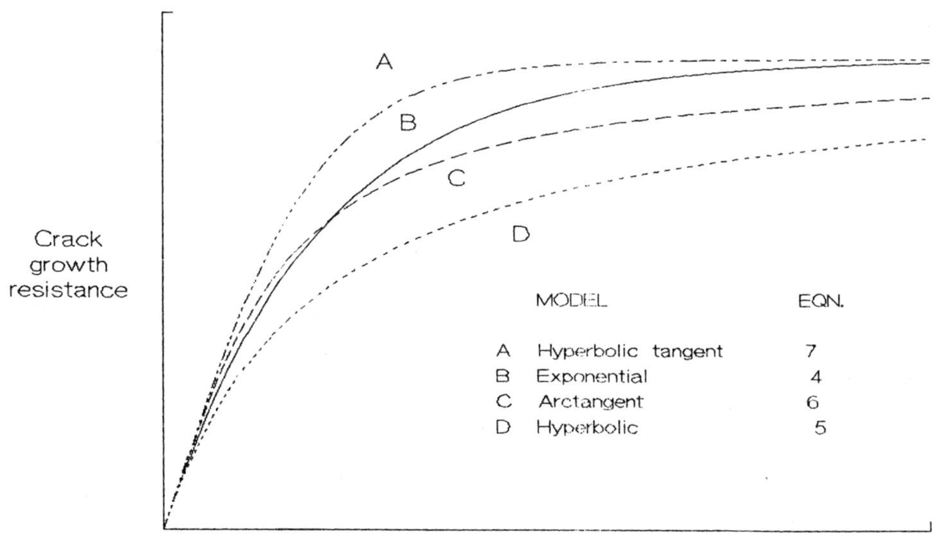

FIG. 1b—*Elastic asymptotic* R-*curves having the same asymptote and initial slope.*

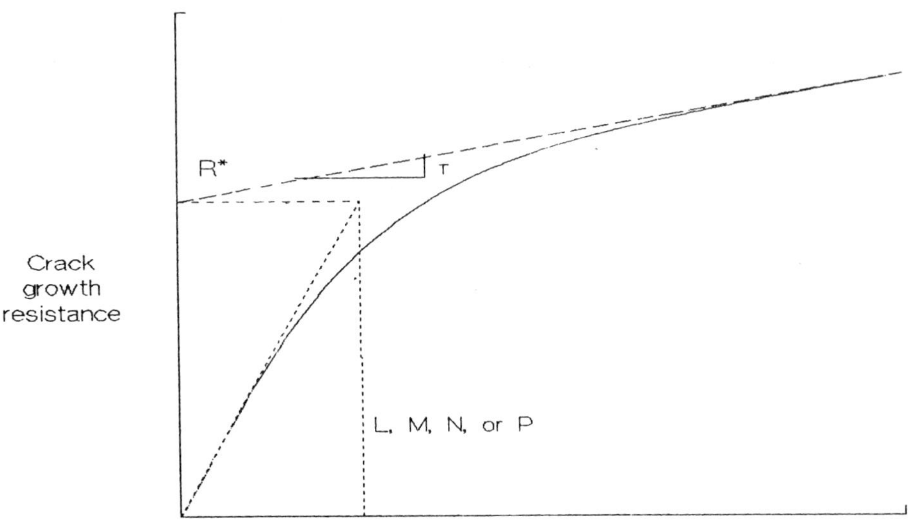

FIG. 2—*Schematic representation of elastic-plastic* R-*curves (L, M, N, or P are coefficients in Eqs 8 to 11).*

of the raw data. However, the construction is a bit more elaborate than for elastic *R*-curves. As before, the fitted coefficients may easily be converted from one system of units to another.

At this point we have a choice. We can prescribe the initial slope as twice the flow stress, which is a customary assumption. Then only the parameters R^* and T need to be determined empirically. Equations 8 to 11 then become

$$R_{6A} = (R^*_{6A} + T_{6A}\Delta)[1 - \exp(-2\sigma_f\Delta/R^*_{6A})] \tag{8a}$$

$$R_{7A} = (R^*_{7A} + T_{7A}\Delta)[\Delta/(R^*_{7A}/2\sigma_f) + \Delta] \tag{9a}$$

$$R_{8A} = (R^*_{8A} + T_{8A}\Delta)(2/\pi)\arctan(\pi\sigma_f\Delta/R^*_{8A}) \tag{10a}$$

$$R_{9A} = (R^*_{9A} + T_{9A}\Delta)\tanh(2\sigma_f\Delta/R^*_{9A}) \tag{11a}$$

On first thought, prescribing the flow stress appears attractive. However, we still really have three parameters in the equation. We are merely fixing one of them. In multiparameter curve fitting, this can sometimes result in poor fits. Remember, too, that the flow stress is not readily measured and is only defined by custom. But both approaches are worthy of investigation. The coefficients of these elastic-plastic equations can also be estimated from a plot of the raw data (Fig. 2) and readily converted from one system of units to another.

Applications to Data

To use this method and these models, we must first obtain the model equation coefficients by fitting to experimental data. This can be done by nonlinear regression analysis. The data

TABLE 1—*Elastic R-curves.*

Fig.	Eq	Fitted Equation
3	3	$K_R = 43.92\Delta^{0.2175}$
4	4	$K_R = 161.3[1 - \exp(-\Delta/10.43)]$
5	5	$K_R = 50.98\Delta/(0.9718 + \Delta)$
6	6	$K_R = 155.0(2/\pi) \arctan [(\pi/2)(\Delta/15.06)]$
7	7	$K_R = 158.6 \tanh (\Delta/13.76)$

TABLE 2—*Elastic-plastic* R-*curves.*

Fig.	Eq	Fitted Equation
8	8	$J_R = (315.1 + 129.6\Delta)[1 - \exp(-\Delta/0.3300)]$
9	9	$J_R = (357.7 + 3.600\Delta)\Delta/(5.172 + \Delta)$
10	10	$J_R = (373.5 + 130.7\Delta)(2/\pi) \arctan [(\pi/2)(\Delta/0.3258)]$
11	11	$J_R = (502.7) + 177.6\Delta) \tanh (\Delta/0.6318)$
12	8a	$J_R = (347.6 + 122.6\Delta)[1 - \exp(-\Delta/0.4237)]$
13	9a	$J_R = (38.87 + 13.00\Delta)\Delta/(0.06645 + \Delta)$
14	10a	$J_R = (396.7 + 124.2\Delta)(2/\pi) \arctan [(\pi/2)(\Delta/0.2493)]$
15	11a	$J_R = (553.5 + 165.2\Delta) \tanh (\Delta/0.8763)$

to follow were all fitted using the program MARQFIT [8] on a microcomputer. If the data used were not available in tabular form, published figures were enlarged and digitized. Fitted values of the coefficients for the data sets presented here are given in Tables 1 and 2. In the discussion to follow, the models of Eqs 8 to 11 will be referred to as "modified" models and those of Eqs 8a to 11a as "special" models.

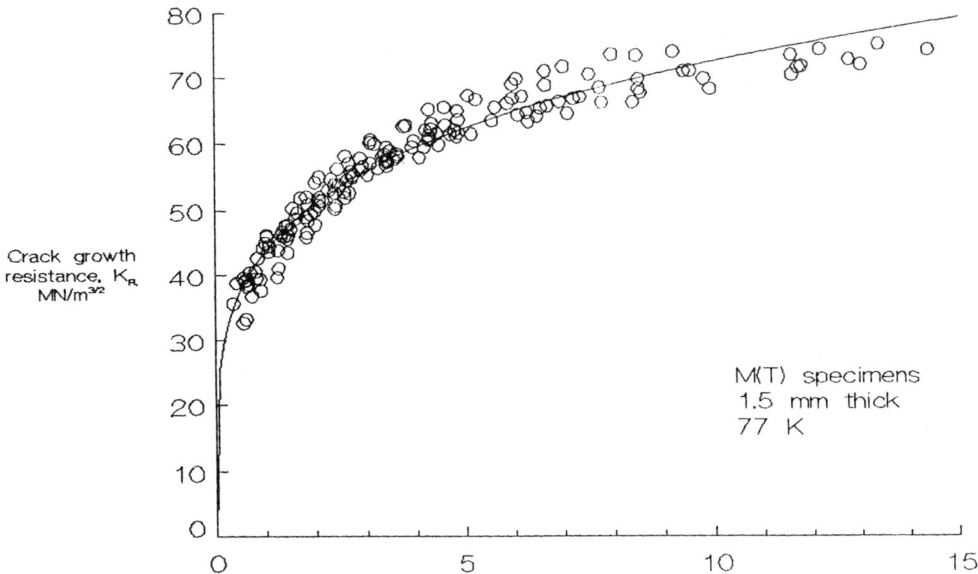

FIG. 3—*Power model (Eq 3) fit to data for 2014-T6 aluminum* [9].

Elastic *R*-Curves

Figure 3 shows the power model (Eq 3) fit to data for 2014-T6 aluminum center-crack specimens tested at 77 K [9]. Figure 4 shows the exponential model (Eq 4) fit to data for 7475-T761 aluminum CLWL specimens [5]. Figure 5 shows the hyperbolic model (Eq 5) fit to data for 7075-T651 aluminum compact specimens [2]. Figure 6 shows the arctangent model (Eq 6) fit to data for 2024-T3 aluminum CLWL specimens [5]. (Tabular data for Figs. 4 and 6 were obtained by private communication from the second author.) Figure 7 shows the hyperbolic tangent model (Eq 7) fit to the same data as Fig. 4.

These figures all show good fits, but these are not necessarily the best fits that could be obtained. All possible combinations of model equations and data sets cannot be shown for space reasons. The combinations that are presented were selected to show that each model is a viable one. That is, there is at least one data set that each model can describe well.

Elastic-Plastic *R*-Curves

The following figures were developed using Eqs 8 to 11. That is, the initial slope (flow stress) was considered a third fitting parameter. Figure 8 shows the modified exponential model (Eq 8) fit to data for A106C steel compact specimens at 135C [10]. Figure 9 shows the modified hyperbolic model (Eq 9) fit to data for 2024-T351 compact specimens [2]. Figure 10 shows the modified arctangent model (Eq 10) fit to data for A533B steel compact specimens at 149C [11]. Figure 11 shows the modified hyperbolic tangent model (Eq 11) fit to data for A106B steel compact specimens [12]. Again the figures presented were chosen to show that each model is a viable one.

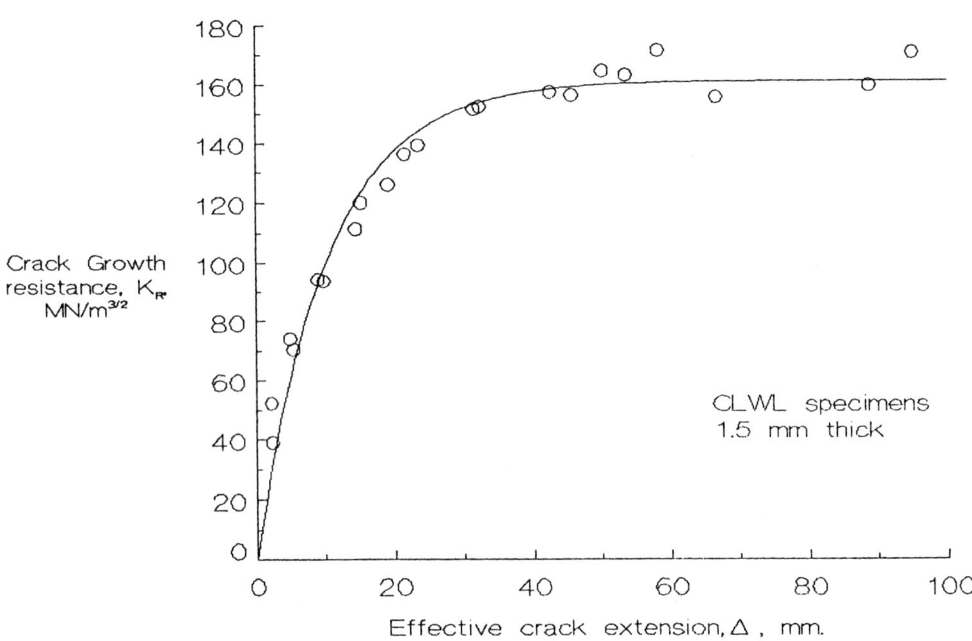

FIG. 4—*Exponential model (Eq 4) fit to data for 7475-T761 aluminum* [5].

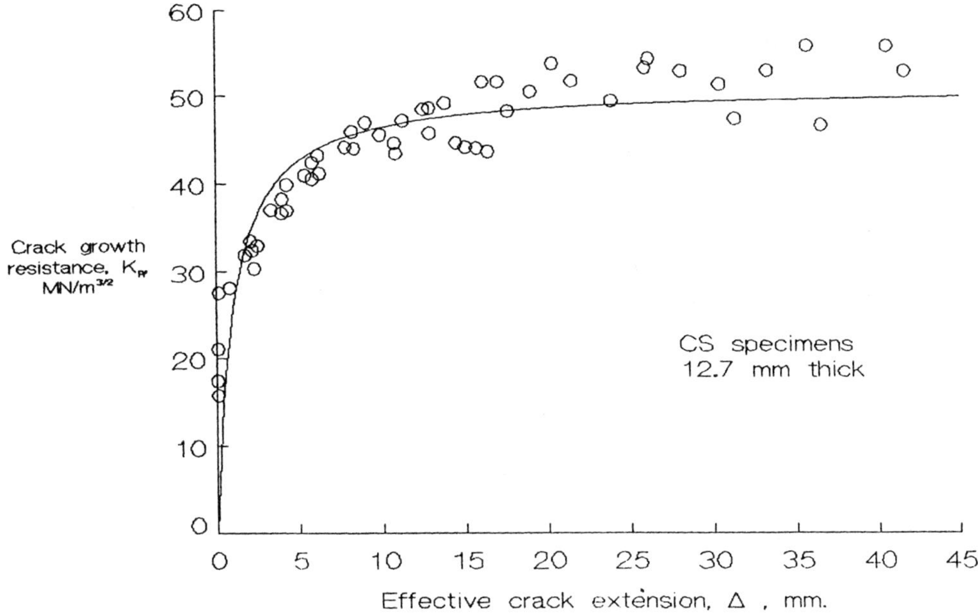

FIG. 5—*Hyperbolic model (Eq 5) fit to data for 7075-T651 aluminum* [2].

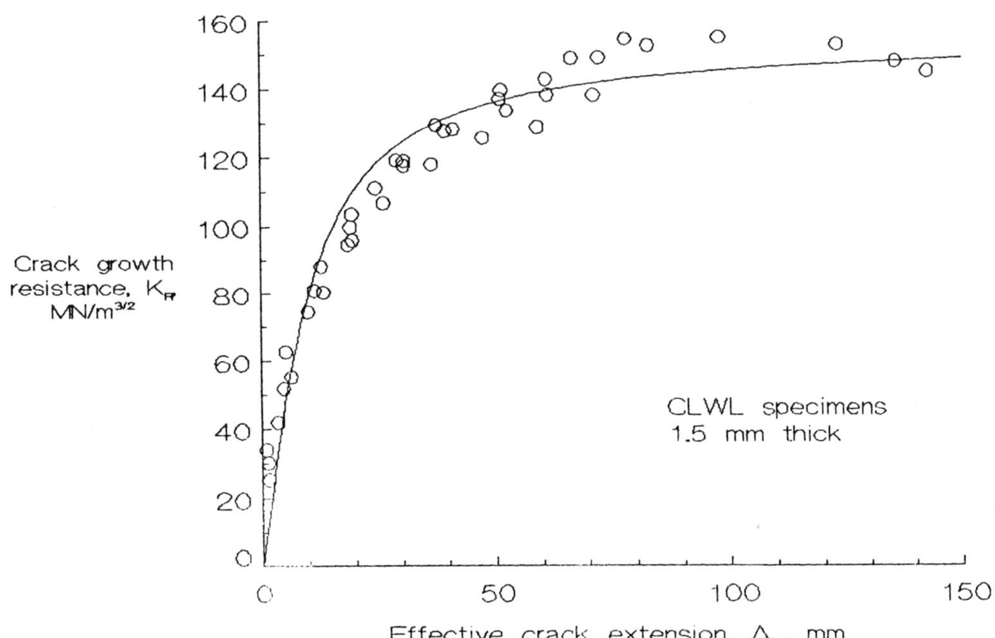

FIG. 6—*Arctangent model (Eq 6) fit to data for 2024-T3 aluminum* [5].

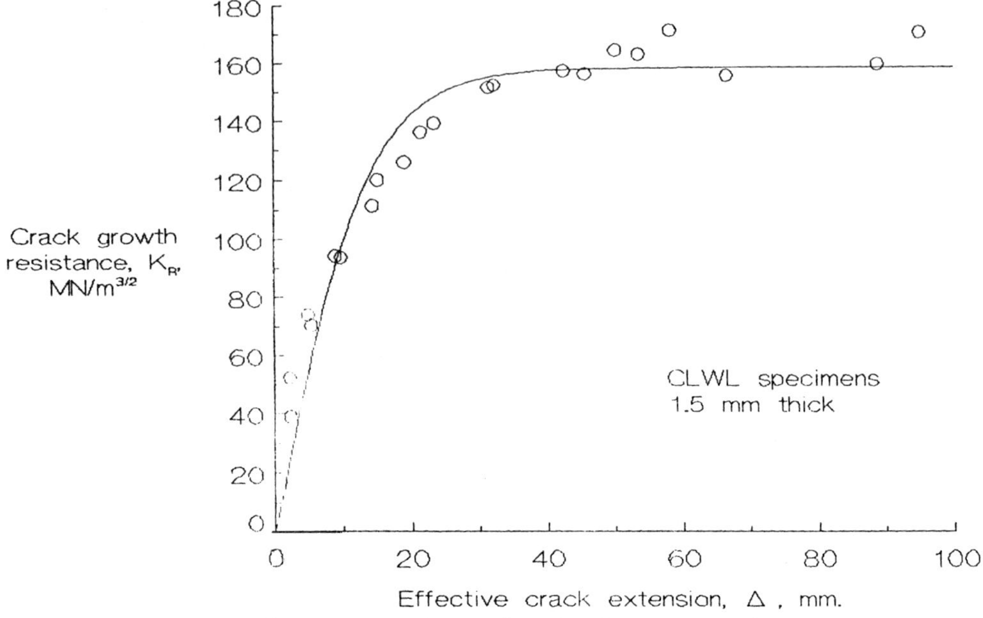

FIG. 7—*Hyperbolic tangent model (Eq 7) fit to data for 7475-T761 aluminum* [5].

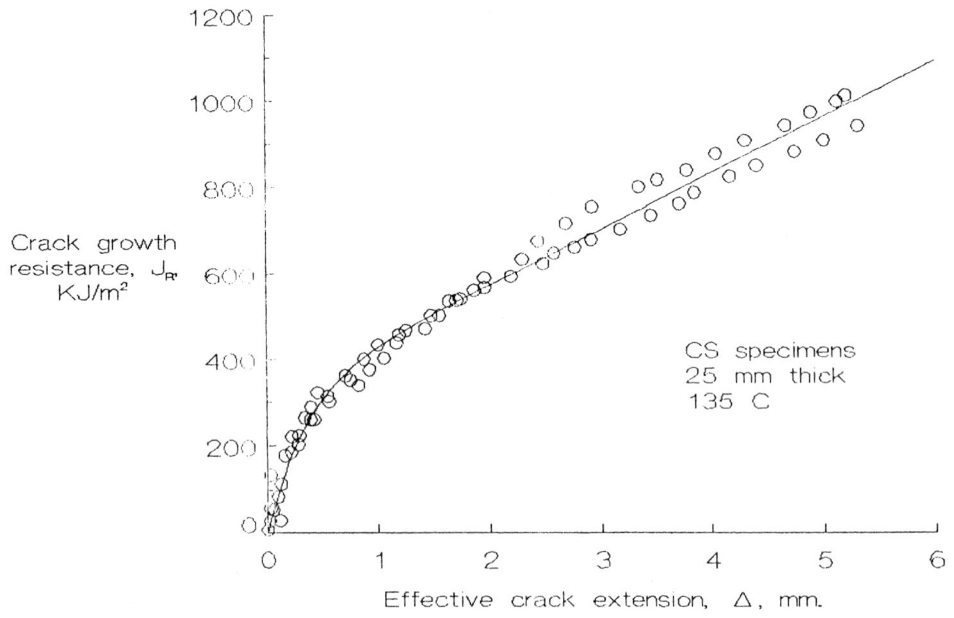

FIG. 8—*Modified exponential model (Eq 8) fit to data for A106C steel* [10].

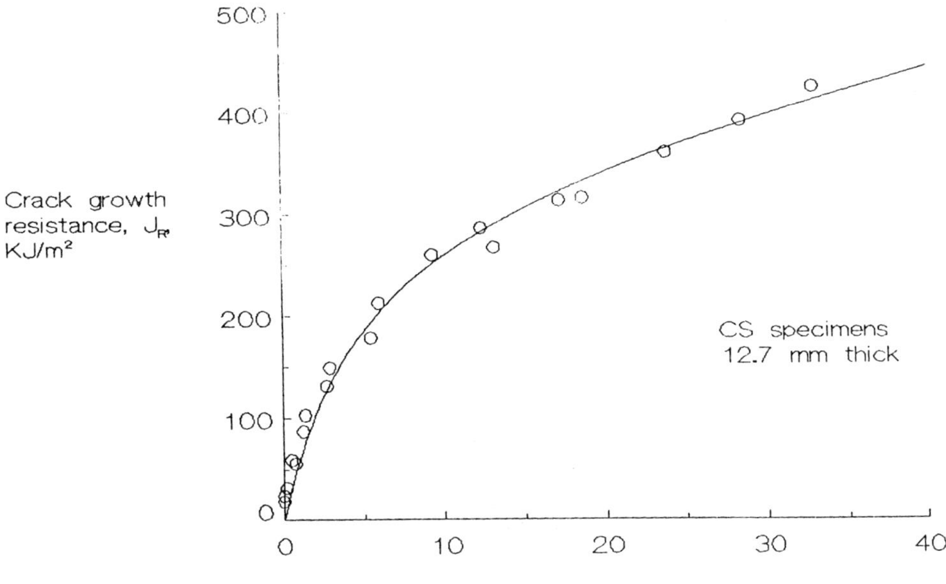

FIG. 9—*Modified hyperbolic model (Eq 9) fit to data for 2024-T351 aluminum* [2].

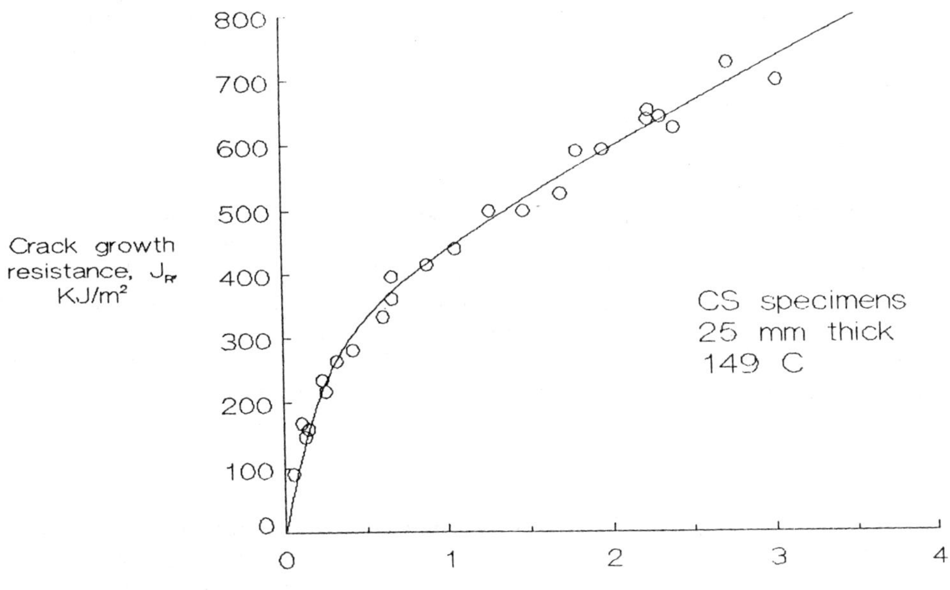

FIG. 10—*Modified arctangent model (Eq 10) fit to data for A533B steel* [11].

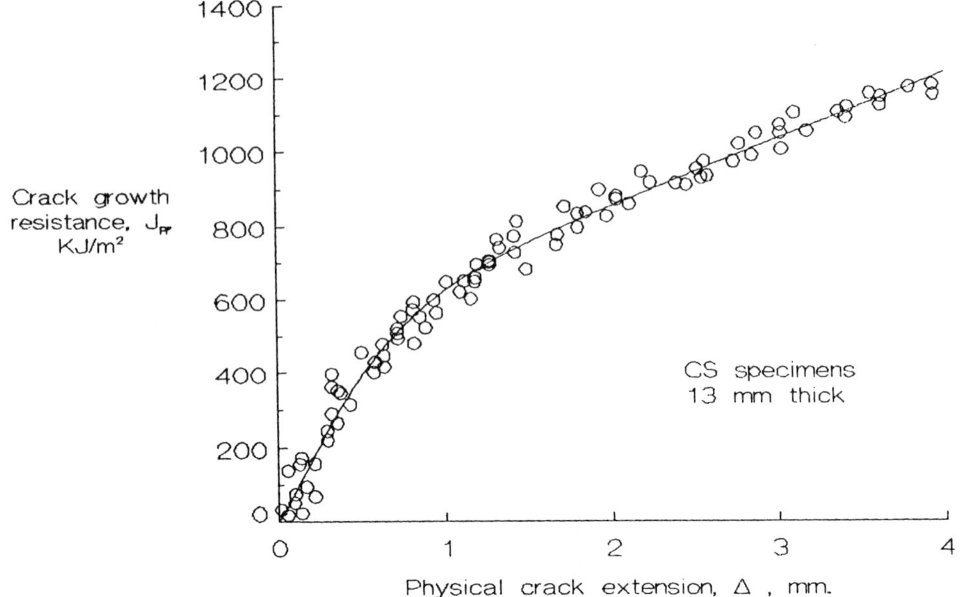

FIG. 11—*Modified hyperbolic tangent model (Eq 11) fit to data for A106B steel* [12].

Elastic-Plastic *R*-curves (Flow Stress Prescribed)

Figures 12 to 15 were developed using Eqs 8*a* to 11*a*. The flow stress was taken as the mean of the reported yield and ultimate strengths. Figure 12 shows the special exponential model (Eq 8*a*) fit to the same data as Fig. 8. Figure 13 shows the special hyperbolic model fit to data for 5456-H117 aluminum compact specimens [*13*]. Figure 14 shows the special arctangent model fit to the same data as Fig. 10. Figure 15 shows the special hyperbolic tangent model fit to the same data as Fig. 11.

Discussion

All possible combinations of model equations and data sets cannot be shown for space reasons. The combinations that are presented were chosen to show that each model is a viable one. And, indeed, each model is a good fit to at least one data set.

For elastic *R*-curves, the eye is usually able to judge whether a power model or an asymptotic model is appropriate. But there is no obvious way to predict which asymptotic model will be the best fit. For the elastic-plastic case, there also is no way to tell in advance whether the flow stress can be prescribed or whether it should be a parameter to be fitted. However, for four of the six sets of *J-R* data, the best fit overall was obtained when the initial slope (flow stress) was fitted rather than prescribed.

Conclusions

The instability calculation method is simple and effective. The *R*-curve model equations presented here are all viable, and at least one of them should fit almost any data. The empirical coefficients are easily estimated from a plot of the raw data and converted from one system of units to another. In combination, the method and the models make powerful tools for fracture analysis.

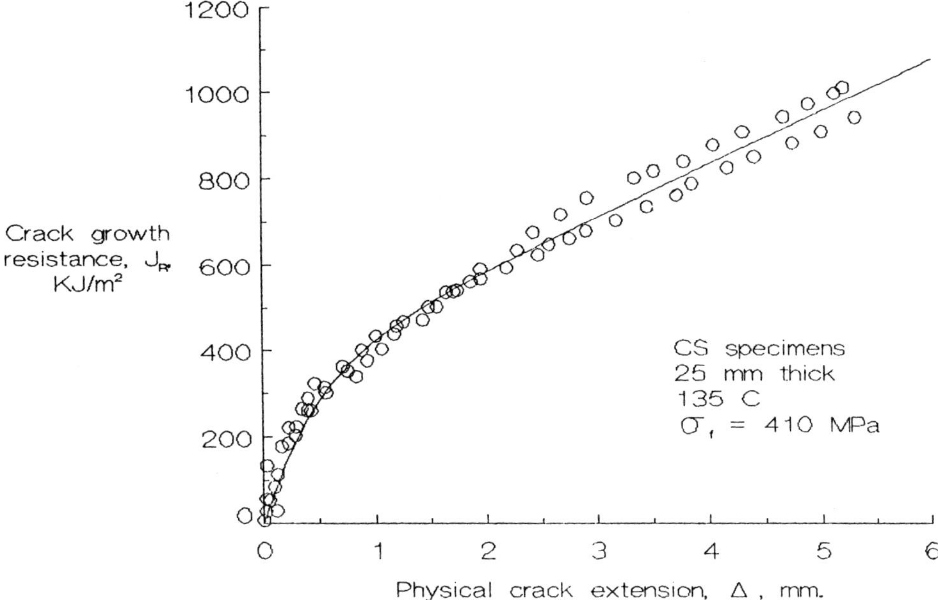

FIG. 12.—*Special exponential model (Eq 8a) fit to data for A106C steel* [10].

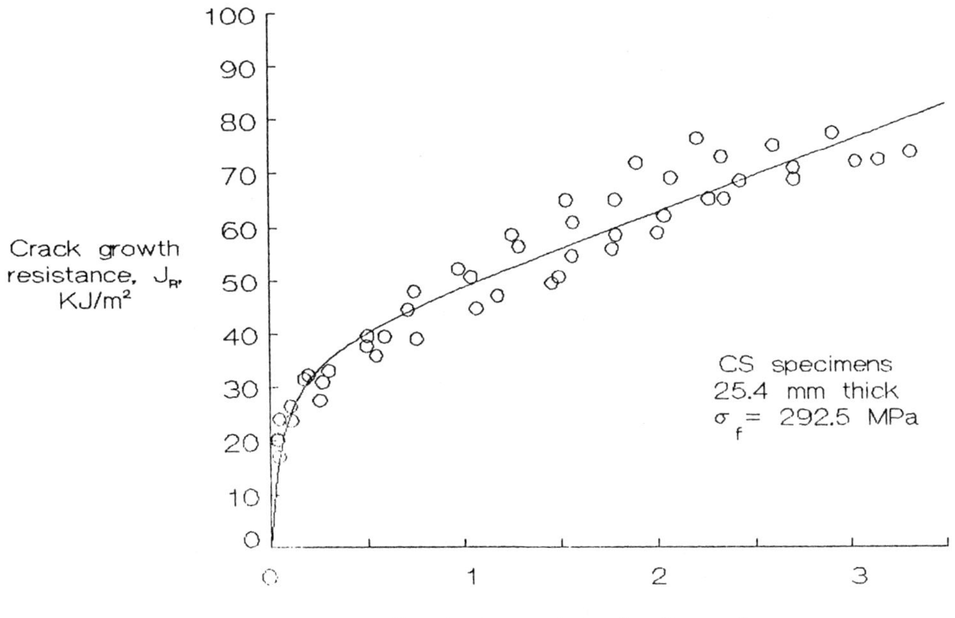

FIG. 13—*Special hyperbolic model (Eq 9a) fit to data for 5456-H117 aluminum* [13].

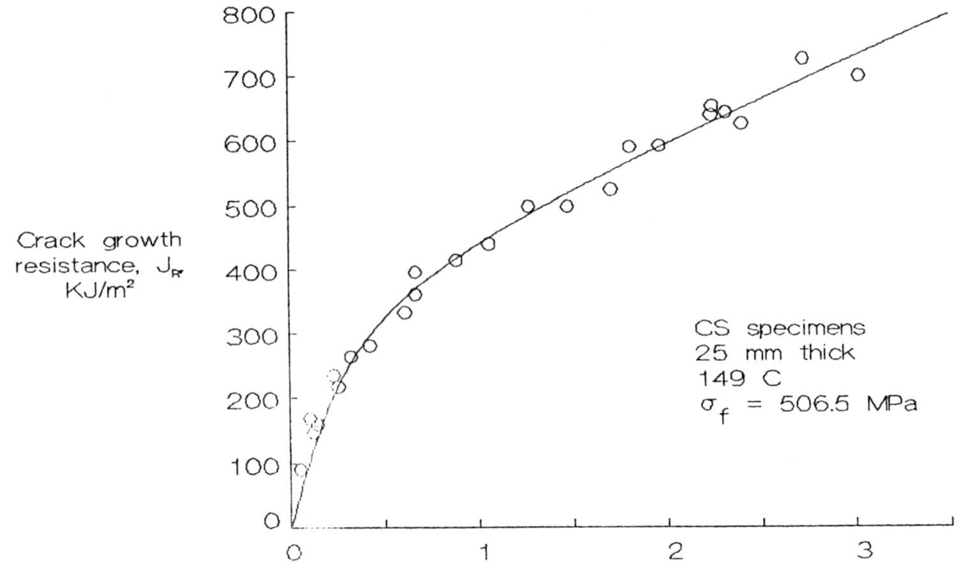

FIG. 14—*Special arctangent model (Eq 10a) fit to data for A533B steel* [11].

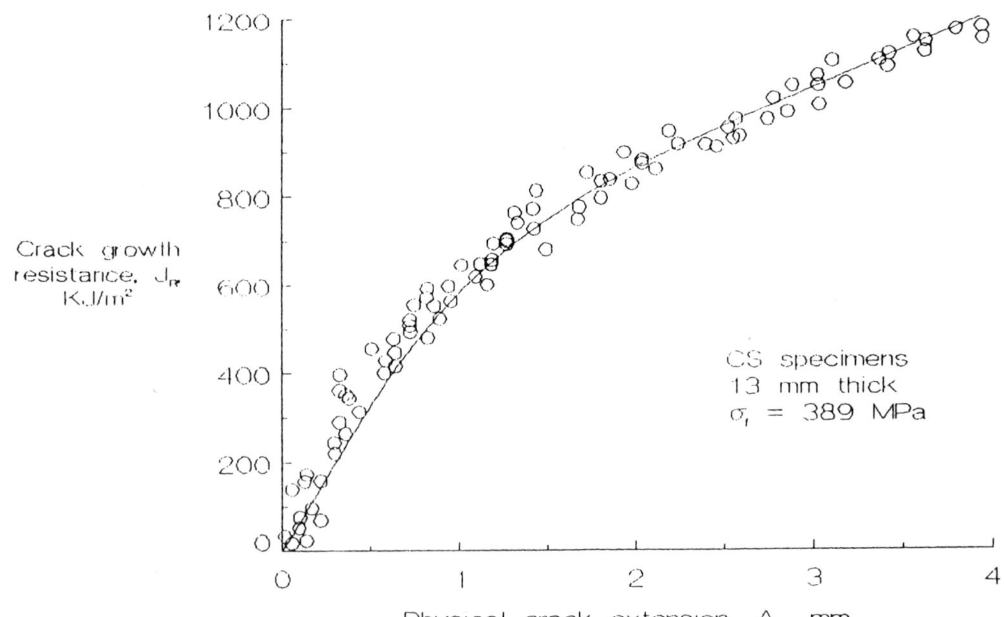

FIG. 15—*Special hyperbolic tangent model (Eq 11a) fit to data for A106B steel* [12].

References

[1] Creager, M., "A Note on the Use of a Simple Technique for Failure Prediction Using Resistance Curves," in *Fracture Toughness Evaluation by R-Curve Methods, ASTM STP 527,* American Society for Testing and Materials, Philadelphia, 1973, pp. 105–112.

[2] Newman, J. C., "An Evaluation of Fracture Analysis Methods," in *Elastic-Plastic Fracture Mechanics Method, ASTM STP 896,* American Society for Testing and Materials, Philadelphia, 1985, pp. 5–96.

[3] Orange, T. W., "Method for Estimating Crack-Extension Resistance Curve from Residual-Strength Data," NASA Technical Paper 1753, Nov. 1980.

[4] Kumar, V., German, M. D., and Shih, C. F., "An Engineering Approach for Elastic-Plastic Fracture Analysis," Report NP-1931, Electric Power Research Institute, Palo Alto, Calif., 1981.

[5] Wang, D. Y. and McCabe, D. E., "Investigation of *R*-Curve Using Comparative Tests with Center-Cracked-Tension and Crack-Line-Wedge-Loaded Specimens," in *Mechanics of Crack Growth, ASTM STP 590,* American Society for Testing and Materials, Philadelphia, 1976, pp. 169–193.

[6] Broek, D. and Vlieger, H., "The Thickness Effect in Plane Stress Fracture Toughness," NLR TR 74032 U, National Aerospace Laboratory (The Netherlands), 1973.

[7] Bluhm, J. I., "Appendix IV-1: Resistance Method," in *Fracture Mechanics of Aircraft Structures,* H. Liebowitz, Ed., AGARDograph No. 176, AGARD, 1973, pp. 74–88.

[8] Schreiner, W., Kramer, M., Krischer, S., and Langsam, Y., "Nonlinear Least-Squares Fitting," *PC Tech Journal,* May 1985, pp. 170–190.

[9] Orange, T. W., "Fracture Toughness of Wide 2014-T6 Aluminum Sheet at −320°F," Technical Note TN D-4017, National Aeronautics and Space Administration, Washington, D.C., 1967.

[10] Sutton, G. E. and Vassilaros, M. G., "Influence of Partial Unloadings Range on the J_I-R Curves of ASTM A106 and 3-Ni Steels," in *Fracture Mechanics: Seventeenth Volume, ASTM STP 905,* J. H. Underwood, R. Chait, C. W. Smith, D. P. Wilhem, W. A. Andrews, and J. C. Newman, Eds., American Society for Testing and Materials, Philadelphia, 1986, pp. 364–378.

[11] Carlson, K. W. and Williams, J. A., "A More Basic Approach to the Analysis of Multiple-Specimen *R*-Curves for Determination of J_c," in *Fracture Mechanics: Thirteenth Conference, ASTM STP 743,* Richard Roberts, Ed., American Society for Testing and Materials, Philadelphia, 1981, pp. 503–524.

[12] Vassilaros, M. G., Hays, R. A., and Gudas, J. P., "*J*-Resistance Curve Analysis for ASTM A106 Steel 8-Inch-Diameter Pipe and Compact Specimens," in *Fracture Mechanics: Seventeenth Volume, ASTM STP 905,* J. H. Underwood, R. Chait, C. W. Smith, D. P. Wilhem, W. A. Andrews, and J. C. Newman, Eds., American Society for Testing and Materials, Philadelphia, 1986, pp. 435–453.

[13] Joyce, J. A. and Vassilaros, M. G., "An Experimental Evaluation of Tearing Instability Using the Compact Specimen," in *Fracture Mechanics: Thirteenth Conference, ASTM STP 743,* Richard Roberts, Ed., American Society for Testing and Materials, Philadelphia, 1981, pp. 525–542.

Fracture Mechanics Testing

G. Marci,[1] D. E. Castro,[1] and V. Bachmann[1]

Closure Measurements via a Generalized Threshold Concept

REFERENCE: Marci, G., Castro, D. E., and Bachmann, V., **"Closure Measurements via a Generalized Threshold Concept,"** *Fracture Mechanics: Twenty-First Symposium, ASTM STP 1074,* J. P. Gudas, J. A. Joyce, and E. M. Hackett, Eds., American Society for Testing and Materials, Philadelphia, 1990, pp. 563–580.

ABSTRACT: Fatigue thresholds have been determined for Inconel 617 and Ti-6Al-4V that are unaffected by any closure phenomenon. These results show that a nonpropagation condition exists between K_{op} and K_{max} of unique amplitude ΔK_T which is unaffected by the position between K_{op} and K_{max}. This experimental fact opens up the possibility of measuring the dividing point K_{op} (fixing the ΔK_{eff}) based on a propagation/nonpropagation criterion. Three such test methods are proposed. Results of these measurements are compared with K_{op} determinations via load-displacement curves made in parallel on Inconel 617. Comparison of both results show that they contradict each other.

It is concluded that the two different experimental procedures determine and characterize two independent phenomena affecting the fatigue crack growth process differently. Characterizing the two different phenomena by K_{op} is incorrect.

KEY WORDS: fatigue threshold, closure, effective stress intensity range, test methods, *R*-dependence, fatigue crack growth curves

Nomenclature

σ_{max}	Maximum stress of a cycle
σ_{min}	Minimum stress of a cycle
K_{max}	Maximum stress-intensity factor of a cycle
K_{min}	Minimum stress-intensity factor of a cycle
$R = \sigma_{min}/\sigma_{max} = K_{min}/K_{max}$	Stress ratio
$\Delta K = K_{max} - K_{min}$	Stress-intensity range
ΔK_{th}	Threshold stress-intensity range, determined as the lower end of a da/dN versus ΔK curve
K_{op}	Stress-intensity factor at which a fatigue crack opens; above K_{op} the crack front experiences tensile loading
$\Delta K_T = K_{max,j} - K_{min,j}$	Maximum stress-intensity range that does not produce fatigue crack growth but is completely above K_{op}
ΔK_T	Fatigue threshold range, free of closure effects
$K_{max,i}$	Stress-intensity factor producing the monotonic plastic zone presently associated with the fatigue crack
$\Delta K_i = K_{max,i} - K_{min,i}$	Stress-intensity range producing the cyclic plastic zone presently associated with the fatigue crack
$\rho = K_{max,j}/K_{max,i}$	Load ratio of the fatigue tolerance range

[1] Branch Head, Graduate Student, and Senior Research Engineer, respectively, Fracture and Mechanical Property Branch, Institute for Materials Research, DFVLR, German Aerospace Research Establishment, P.O. Box 906058, 5000 Köln 90, F.R.G.

Introduction

The lack of understanding of closure and fatigue threshold and their role in the fatigue crack propagation process are presently the obstacles to improved fatigue crack growth life prediction schemes. It has been the opinion of our research group that closure and threshold are Siamese twins and that only the separation of the two can lead to a better understanding of each phenomenon and their roles in the fatigue crack propagation process. Furthermore, better experimental measuring procedures might be developed from the insight gained by such a separation.

In the first part of the research program the fatigue thresholds of Inconel 617 [*1–3*] and Ti-6Al-4V [*3*] were investigated. Parts of the threshold results are repeated here, since novel experimental procedures to measure closure are derived from these results. Tests were made on Inconel 617 with one of these novel test procedures. For comparison, K_{op} values determined via load versus crack mouth opening displacement curve (CMOD) and via load versus back face strain signal curve (BFS) were obtained. As will be shown, the "K_{op}" values obtained from the two conceptually different test procedures contradict each other. Some ideas are presented to reconcile the two different sets of results.

Previous Results on Fatigue Threshold Pertinent to the Present Investigation

The experimental methods shown in Fig. 1 were used to determine nonpropagation conditions for fatigue crack growth. Each of the test methods can be applied over the K_{max} range possible with the subject material. Type I and Type III test methods are used to determine the bounds on nonpropagation fatigue conditions under tension fatigue (i.e., they encompass any nonpropagation condition in the realm of tension fatigue). Type II test method is between Types I and III in that the size of the cyclic plastic zone at a given K_{max} is between that of Types I and III tests at the same K_{max}. Type IV test method is, in respect to the plastic zone condition (monotonic and cyclic), identical to that of Type III for the same K_{max}, but now the K_{max} of the nonpropagation condition is below the K_{max} producing the monotonic plastic zone. The mean K_I of the non-propagation condition was kept at 75% of K_{max} producing the monotonic zone. A simpler definition of the shift between the K_{max} producing the monotonic plastic zone, say $K_{max,i}$, and the K_{max} of the nonpropagation condition, say $K_{max,j}$, is the ratio of the two. The load ratio ρ was defined as

$$\rho = K_{max,j}/K_{max,i} \tag{1}$$

For Type I through III tests the load ratio $\rho = 1.0$; however, for Type IV tests it ranged from $0.77 \leq \rho \leq 0.89$. Implicit in the Type II through IV test methods is the assumption that no additional damage occurs in the plastic zone if the applied stress intensity factor does not exceed the bounds set by $K_{max,i}$ and $K_{min,i}$ producing the plastic zone condition and no crack growth occurs. Detailed experimental descriptions can be found in Refs *1* to *3*. The nonpropagation conditions were established in the range of K_{max} from 10 to 50 MPa $\cdot$ m$^{1/2}$ for Inconel 617 and 10 to 25 MPa $\cdot$ m$^{1/2}$ for Ti-6Al-4V in increments of 5 MPa $\cdot$ m$^{1/2}$. Crack growth under static loads corresponding to $\approx$ 55 MPa $\cdot$ m$^{1/2}$ with Inconel 617 and $\approx$ 28 MPa $\cdot$ m$^{1/2}$ with Ti-6Al-4V make a fatigue threshold irrelevant beyond these stress intensity factors. The nonpropagation conditions ΔK_T are plotted versus K_{max} in Figs. 2 and 3 for Inconel 617 and Ti-6Al-4V respectively. The results can be summarized as follows:

- A threshold exists for any fatigue condition in the tension fatigue regime.
- The magnitude of the threshold decreases with increasing K_{max}; the decrease depends on the material [*3*].

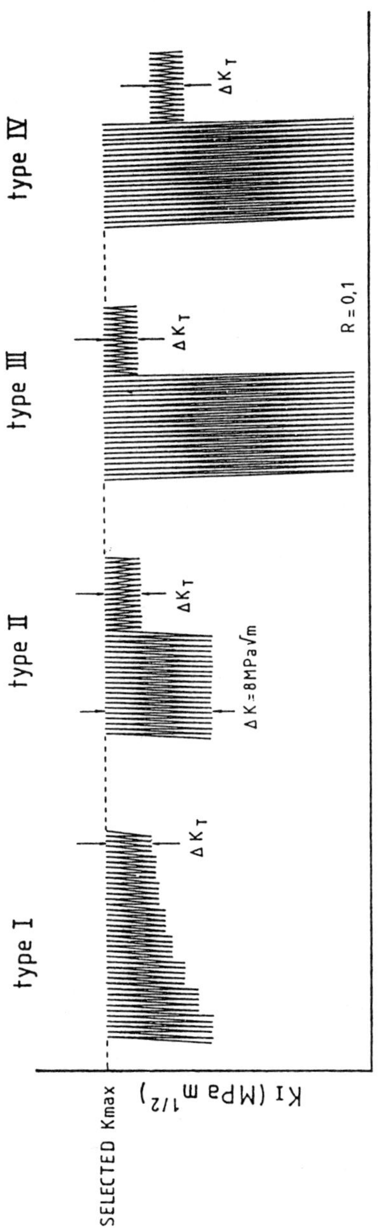

FIG. 1—*Experimental procedures used for threshold determination; constant* K_{max} *(Type I) and "jump-in method" (Types II, III, and IV).*

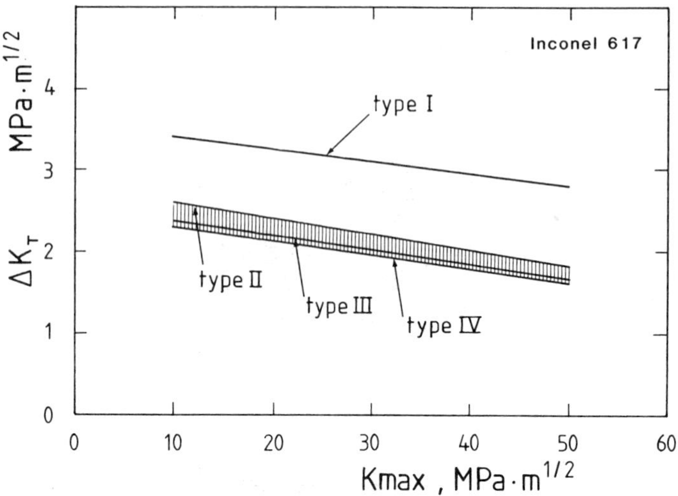

FIG. 2—*Threshold values measured with the four test procedures on Inconel 617 as a function of* K_{max}.

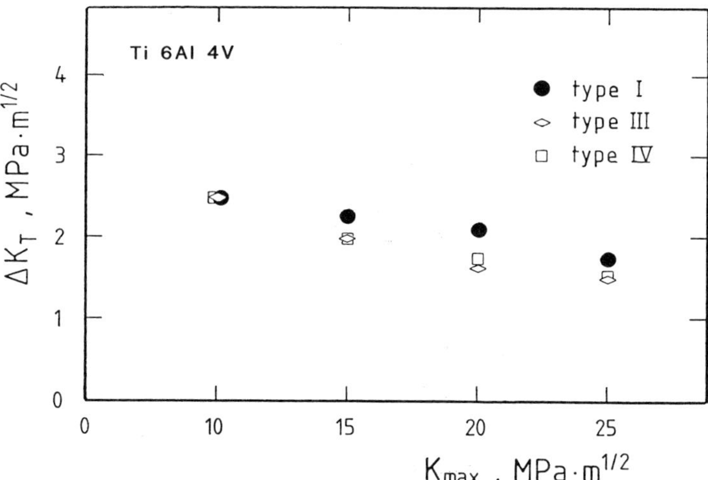

FIG. 3—*Threshold values as a function of* K_{max} *for Ti-6Al-4V alloy.*

- The magnitude of the threshold at a given K_{max}, as measured by Type I test method, is equal to or larger than that measured by the other test methods. The difference is dependent on the material.
- The load ratio ρ has no effect on the threshold.

Basis for Novel Test Methods to Determine K_{op}

The last result above is of particular importance to the present investigation. For Type IV test procedures the nonpropagation condition would be trivial if it were below K_{op}. This was checked by a ΔK_j somewhat larger than the threshold ΔK_T, either by decreasing the respective

K_{min} or by increasing the respective K_{max}. When ΔK_j was made some what larger than ΔK_T, fatigue crack growth commenced at once. Therefore the results concerning the influence of the load ratio ($\rho = 1.0$ for Type III tests and $0.77 \leq \rho \leq 0.89$ for Type IV tests) indicate that a unique threshold exists anywhere between K_{max} and K_{op}. This is graphically shown in Fig. 4.

Elber's equations [4,5]:

$$\frac{da}{dN} = f(\Delta K_{eff}) \tag{2}$$

where f is some function, and

$$\Delta K_{eff} = K_{max} - K_{op} \tag{3a}$$

or, if $K_{min} > K_{op}$,

$$\Delta K_{eff} = K_{max} - K_{min} \tag{3b}$$

state that for crack growth to occur it is necessary that $\Delta K_{eff} > 0$ (from Eq 2), but that any stress intensity factor amplitude above K_{op} results in a $\Delta K_{eff} > 0$ and thereby in fatigue crack propagation (Eqs 3). All results from Types I to IV test methods contradict Eqs 3 in that the growth rate is zero below a given ΔK_T. But a minor modification to Eqs 3 resolves the contradiction without any change of the K_{op} concept:

$$\Delta K_{eff} = (K_{max} - K_{op}) - \Delta K_T \tag{4a}$$

or, if $K_{min} > K_{op}$,

$$\Delta K_{eff} = (K_{max} - K_{min}) - \Delta K_T \tag{4b}$$

According to Eqs 4, there can be stress intensity factor excursions above K_{op} which lead to $\Delta K_{eff} = 0$, namely

$$K_{max} - K_{op} \leq \Delta K_T \tag{5a}$$

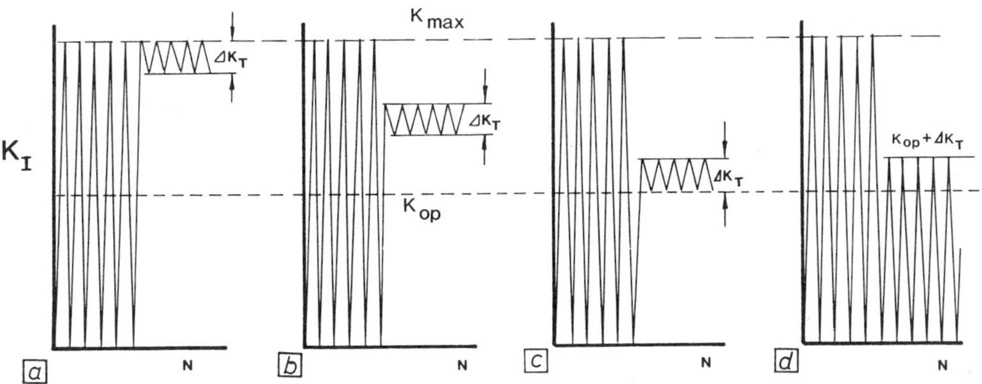

FIG. 4—*Graphical interpretation of the threshold results from Type III and Type IV tests.*

or

$$K_{\max} - K_{\min} \leq \Delta K_T \tag{5b}$$

The existence of ΔK_T by itself allows one to propose experimental test methods with which a K_{op} can be determined, if it exists and has the property attributed to it as being a partition point for ΔK_{eff}. If K_{op} partitions the ΔK applied into two parts, one of which is uniquely related to the fatigue crack growth rates (ΔK_{eff}), then this partitioning point can be determined by the test methods shown in Fig. 5. Any phenomena that reduce ΔK_j to ΔK_T (ΔK_j somewhat larger than ΔK_T) changes a propagation condition into a nonpropagation condition.

With the test method shown in Fig. 5a a certain fatigue cycle with $K_{\max,i}$ and $K_{\min,i}$ should be tested for K_{op}. Since it is the hypothesis to be tested that the part of ΔK below K_{op} does not produce fatigue crack growth (if a point K_{op} exists), $K_{\min,j}$ is kept the same as $K_{\min,i} \cdot K_{\max,j}$ is dropped instantaneously to a level below the expected K_{op}. If crack growth occurs, the plastic zone condition has to be established again by advancing the crack under $K_{\max,i}$ and $K_{\min,i}$. Then $K_{\max,j}$ is dropped below the previous $K_{\max,j}$. If instead no crack growth occurs after the drop of $K_{\max,j}$, $K_{\max,j}$ is increased in small steps, say 0.25 to 0.5 MPa $\cdot$ m$^{1/2}$, until crack growth is detected at a certain $K_{\max,j}$. The previous $K_{\max,j}$, say $\overline{K_{\max,j}}$, is then used and, if ΔK_T is known, K_{op} is determined via Eq 4a:

$$\Delta K_{\mathrm{eff}} = 0 = (\overline{K_{\max,j}} - K_{\mathrm{op}}) - \Delta K_T \tag{6a}$$

$$K_{\mathrm{op}} = \overline{K_{\max,j}} - \Delta K_T \tag{6b}$$

The precision of the measurements depend on the steps between successive $K_{\max,j}$ (besides other experimental factors). This method was used previously to determine K_{op} at $R = 0.05$ on 2024-T3 aluminum [6] over the range of possible $K_{\max}$ values and on Ti-6Al-4V [7]. Pellas et al. [8] and Baudin and Robert [9] determined the $K_{\max,j}$ with the same procedure for 2024-T3 aluminum and determined $\Delta K_{\mathrm{eff}} = K_{\max,i} - K_{\max,j}$ which is identical to that given by Eq 4a.

With the method shown in Fig. 5b [10], K_{op} is determined again for a fatigue condition with $K_{\max,i}$ and $K_{\min,i}$. A ΔK_j somewhat larger than ΔK_T is selected and shifted in small steps

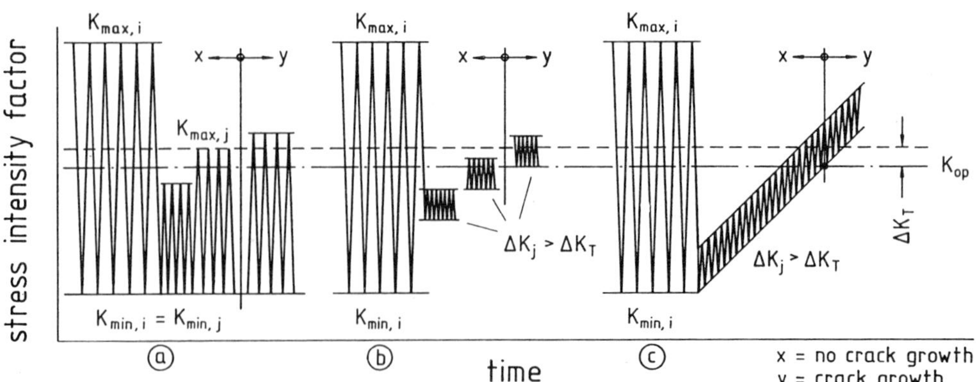

FIG. 5—*Test methods to determine the partition point* K_{op} *in respect to* ΔK_{eff} *based on closure-free fatigue threshold.*

from a level below the expected K_{op} to higher mean K_I levels. The $\overline{K_{max,j}}$ of the last step producing no crack growth is used to determine K_{op}:

$$K_{op} = \overline{K_{max,j}} - \Delta K_T \tag{7}$$

The precision of the K_{op} measurement depends on the size of steps between successive mean K_I levels and the difference between ΔK_j and ΔK_T.

The same procedure presented for the method shown in Fig. 5*b* also applies to the method shown in Fig. 5*c*. However, instead of a stepwise increase of the mean K_I, the mean K_I is increased continuously. The method shown in Fig. 5*c* was used in the present investigation and in Ref *11*.

Materials and Experimental Procedure

The high temperature creep and corrosion resistant alloy Inconel 617 as sheet material was the subject of the present investigation. The percentage weight chemical composition was: 55.2 Ni, 21.95 Cr, 12.0 Co, 8.90 Mo, 0.074 C, 0.0 Mn, 0.16 Fe, 0.004 S, 0.11 Si, 0.012 P, 1.05 Al, and 0.44 Ti; the mechanical properties of the alloy are: 400 MPa yield strength, 890 MPa ultimate strength, 53% elongation, and 48% reduction in area. In addition, the Ti-6Al-4V alloy, which was included in the fatigue threshold investigation and for which K_{op} measurements were made in Ref *11*, was analyzed. The mechanical properties of the Ti-6Al-4V alloy are: 985 MPa yield strength, 1025 MPa ultimate strength, and 14.6% elongation.

The tests on Inconel 617 were performed on CT specimens taken from a plate with a thickness of 17 mm, machined out symmetrically to the mid-plane of the plate with TL-orientation. Similarly, CT specimens were machined out of a Ti-6Al-4V plate with a thickness of 16 mm. The CT specimen dimensions were $W = 50$ mm (in accordance with ASTM E 399 except for thickness $B = 12$ mm), while the threshold investigation was made on CT specimens with $W = 60$ mm and $B = 12.5$ mm.

The tests were carried out in laboratory atmosphere at room temperature on load-controlled servohydraulic testing machines. A sinusoidally changing load with a frequency of 30 to 50 Hz was applied. The fatigue crack propagation was measured by the d-c potential drop method. At a source current of about 18 A, the voltage change per millimetre of crack length was 49 μV. The condition of a nonpropagating fatigue crack was taken as established, when the d-c potential drop method showed no indication of crack propagation during a minimum of $1.5 \cdot 10^6$ cycles. This was always checked by optical crack length measurements with a high-power optical microscope. The d-c potential drop method allowed the detection of crack growth rates down to 10^{-11} m/cycle; this method has been used extensively for similar tests previously. Particular care was used to shield the electrical connections for the potential drop signal against temperature variations. The tests on an individual specimen were done so that the specimen had not experienced a K_{max} higher than that of the present test schedule. This means that during testing of an individual specimen the subject K_{max} was either constant or increased from one test to another.

For Inconel 617 the opening point of the fatigue crack K_{op} during a loading cycle was determined by two distinctly different methods. With the first method it was determined from a load versus crack mouth opening displacement (CMOD) plot and a load versus back face strain (BFS) signal curve. The lowest point of the upper linear branch of the respective curve was taken as K_{op} in accordance with Ref *6*. Usually with R ratios above 0.35 to 0.45, the K_{min} of the respective fatigue cycle is above any K_{op} as measured by CMOD or BFS. In order to gain, if possible, a better understanding of the phenomena involved, the K_{op} was determined in increments of 0.1 from $R = 0.1$ to $R = 0.9$ with CMOD and BFS for three

K_{max}-values (15, 30, and 45 MPa · m$^{1/2}$). This was done by advancing the fatigue crack at a given R-ratio and a given K_{max} until da/dN had stabilized at a certain value. Usually this corresponded to a crack advance of 0.8 to 2 mm (note that on any one specimen K_{max} applied was either constant or increasing). Then the stress intensity factor was lowered to nearly zero and increased again up to K_{max}. For this cycle, between the K_{max} of the respective fatigue cycle and $K_I \approx 0$, the load versus CMOD and load versus BFS were recorded and analyzed as described above.

With the second method, K_{op} was determined with the test shown in Fig. 5c for Inconel 617 at $K_{max,i}$ of 15, 30, and 45 MPa · m$^{1/2}$ and R-ratios from $R = 0.1$ to 0.9 in increments of 0.1. A ΔK_j value was chosen so that $\Delta K_j = 1.1 \times \Delta K_T(I)$, where $\Delta K_T(I)$ is the threshold measured with Type I tests at the respective $K_{max,i}$. The K_{op} value was determined from the test result in accordance with Eq 7.

Results and Discussion

Figure 6 shows the results of the K_{op} determination for Inconel 617 by means of CMOD and BFS. The measured K_{op} has been normalized with respect to K_{max}. The normalized K_{op} (i.e., K_{op}/K_{max}) is plotted versus the R-ratio. Figure 6a shows the results obtained from load versus CMOD curves, while Fig. 6b shows the results obtained from load versus BFS curves. Both results were superposed and represented by an upper and lower bound curve (except two data points) in Fig. 7; the line A-B represents the position of K_{min} at the respective R-ratio. To the right of line A-B in Fig. 7, K_{min} of the respective fatigue cycle is above K_{op} as measured via CMOD or BFS. As can be seen from Figs. 6 and 7, the phenomenon (or phenomena) detected with CMOD or BFS and represented by the respective K_{op} is continuously detected over the range of R-ratios. But to the right of line A-B in Fig. 7 it (they) should not have any influence on fatigue crack growth. The data shown in Fig. 6 are in agreement with the data found in the literature for this type of experimental technique.

The values of K_{op} obtained for Inconel 617 by the experimental method shown in Fig. 5c and Eq 7 are normalized with their corresponding K_{max} and plotted in Fig. 8 versus R-ratio. Döker and Bachmann [11] determined K_{op} with the same method for a Ti-6Al-4V alloy.

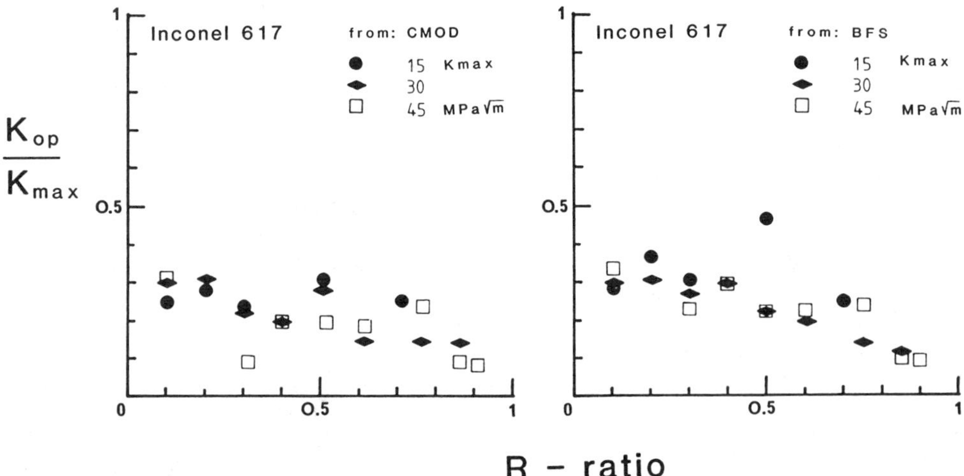

FIG. 6—*Normalized* K$_{op}$/K$_{max}$ *versus* R = K$_{min}$/K$_{max}$ *obtained via load-CMOD curves* (a) *and via load-BFS curves* (b) *for Inconel 617.*

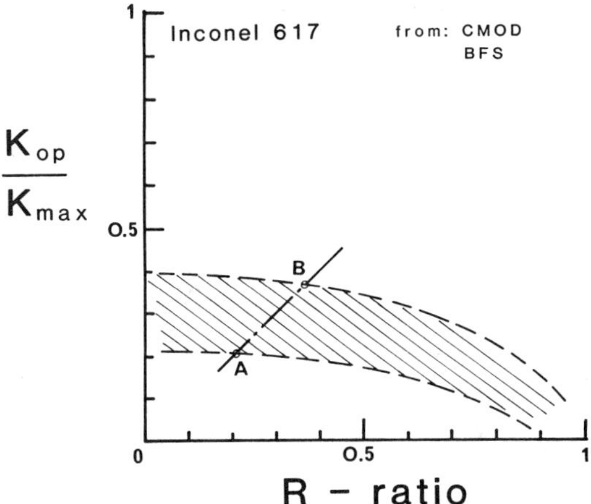

FIG. 7—*Results from Fig. 6 superpositioned; line* A-B *trace of* K_{min} *at the respective* R *ratios.*

Their data were similarly normalized by K_{max} and are replotted as a function of R in Fig. 9. By comparing Fig. 9 with Fig. 8 it can be seen that in both sets of data K_{op}/K_{max} follows similar functional dependence on R. In Fig. 10 both sets of data are compared using two differently shaded areas.

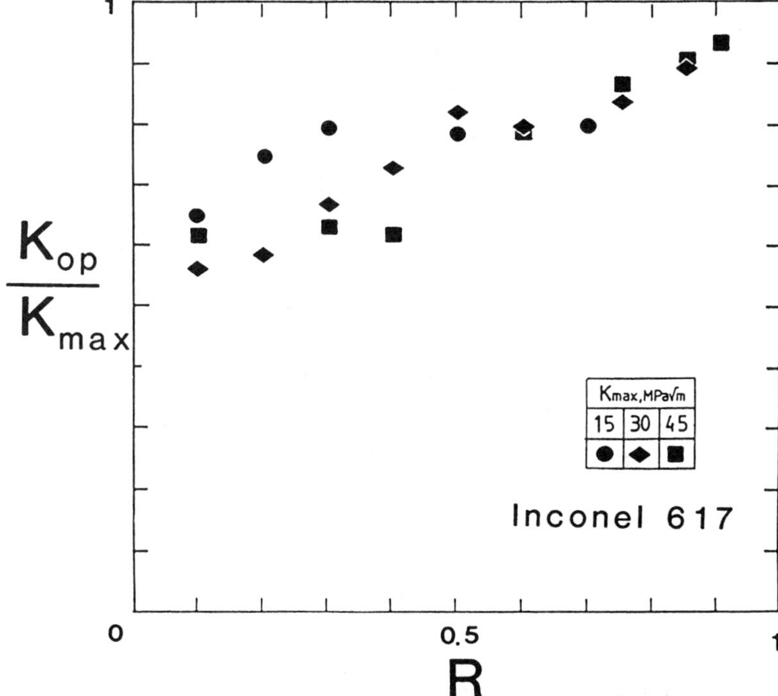

FIG. 8—*Partition point* K_{op} *as measured by the method shown in Fig. 5c for Inconel 617;* K_{op} *normalized by* K_{max}; K_{op}/K_{max} *as function of* R.

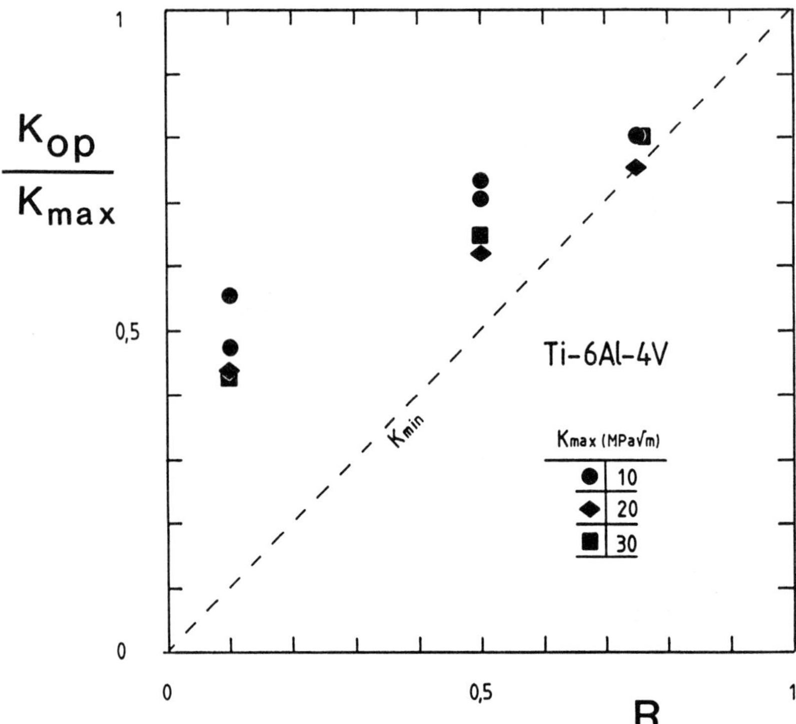

FIG. 9—*Partition point* K_{op} *normalized by* K_{max} *as a function of* R *for Ti-6Al-4V* [11].

Elber's equation [4,5] transposed to

$$K_{op}/K_{max} = 0.5 + 0.1R + 0.4R^2 \qquad (8)$$

and Schijve's equation [12]

$$K_{op}/K_{max} = 0.45 + 0.2R + 0.25R^2 + 0.1R^3 \qquad (9)$$

cannot be compared directly with the results shown in Figs. 8 to 10.

Schijve's equation [9] is empirically based on the fatigue crack growth behavior of 2024-T3 aluminum alloys. This means that the growth rates are correlated to a ΔK_{eff} based on Eqs 3, without consideration of a threshold. Therefore Eq 9 should be written more correctly as

$$K_{max,j}/K_{max,i} = 0.45 + 0.2R + 0.25R^2 + 0.1R^3 \qquad (10)$$

and

$$K_{op} = K_{max,j} - \Delta K_T \qquad (11)$$

The situation with Elber's equation [8] is a little more complex. It is obtained by analysis of load versus COD curves and was found to give a ΔK_{eff} in good agreement with fatigue crack growth rates of 2024-T3 aluminum alloys. If Eq 8 is considered to be empirical, then

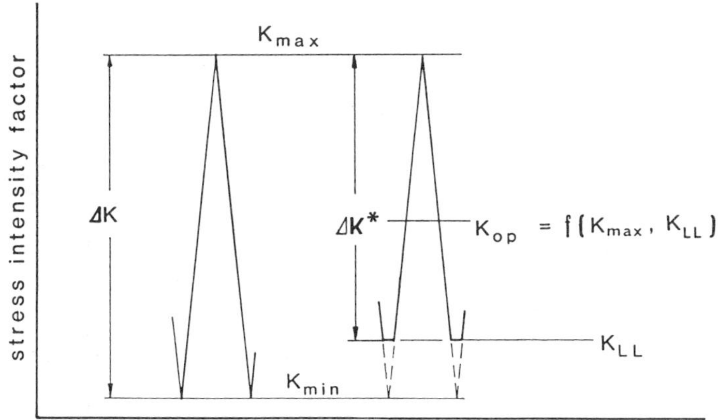

FIG. 11—*Graphical representation of the effect of wake zone and associated phenomena on the applied cycle and its modification.*

ing R, not only ΔK but also ΔK_{eff} is reduced by the influence of the wake zone. One fact concerned us at the time the work reported in Ref *20* was done: namely, the erratic changes of the magnitude by which $K_{\min}$ was modified. This point will be discussed later.

In the meantime, other experimental results in support of a mechanical picture of the wake zone effect (in Fig. 11) became available. Figure 12 shows the fatigue threshold results of Schmidt and Paris [*21*] for 2024-T3 aluminum as a function of R. The data marked by an arrow were replotted [*22*] as a function of $K_{\max}$ in Fig. 13. As can be seen in Fig. 13, all data points in region A in Fig. 12 correspond to a fatigue cycle with $K_{\max} = 3.74$ MPa · m$^{1/2}$ (3.4 ksi $\sqrt{\text{in.}}$). A fatigue threshold without a wake zone influence can be obtained from Fig. 13 as $\Delta K_{\text{T}} = 0.88$ MPa · m$^{1/2}$ (0.8 ksi $\sqrt{\text{in.}}$) for $K_{\max} = 3.74$ MPa · m$^{1/2}$ (3.4 ksi $\sqrt{\text{in.}}$). This value inserted into Eq 5*a* gives a K_{op} (partition point for ΔK_{eff}) of 2.86 MPa · m$^{1/2}$ (2.6 ksi $\sqrt{\text{in.}}$). With the ratio of $K_{\text{op}}/K_{\max} = 2.6/3.4 = 0.76$, it can be seen from Fig. 10 that the $K_{\min}$ expe-

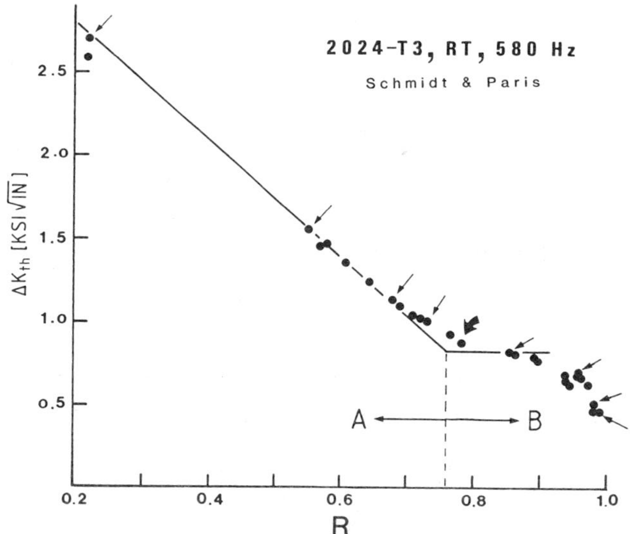

FIG. 12—*Fatigue threshold as a function of* R *for 2024-T3 aluminum alloy of Ref* 21.

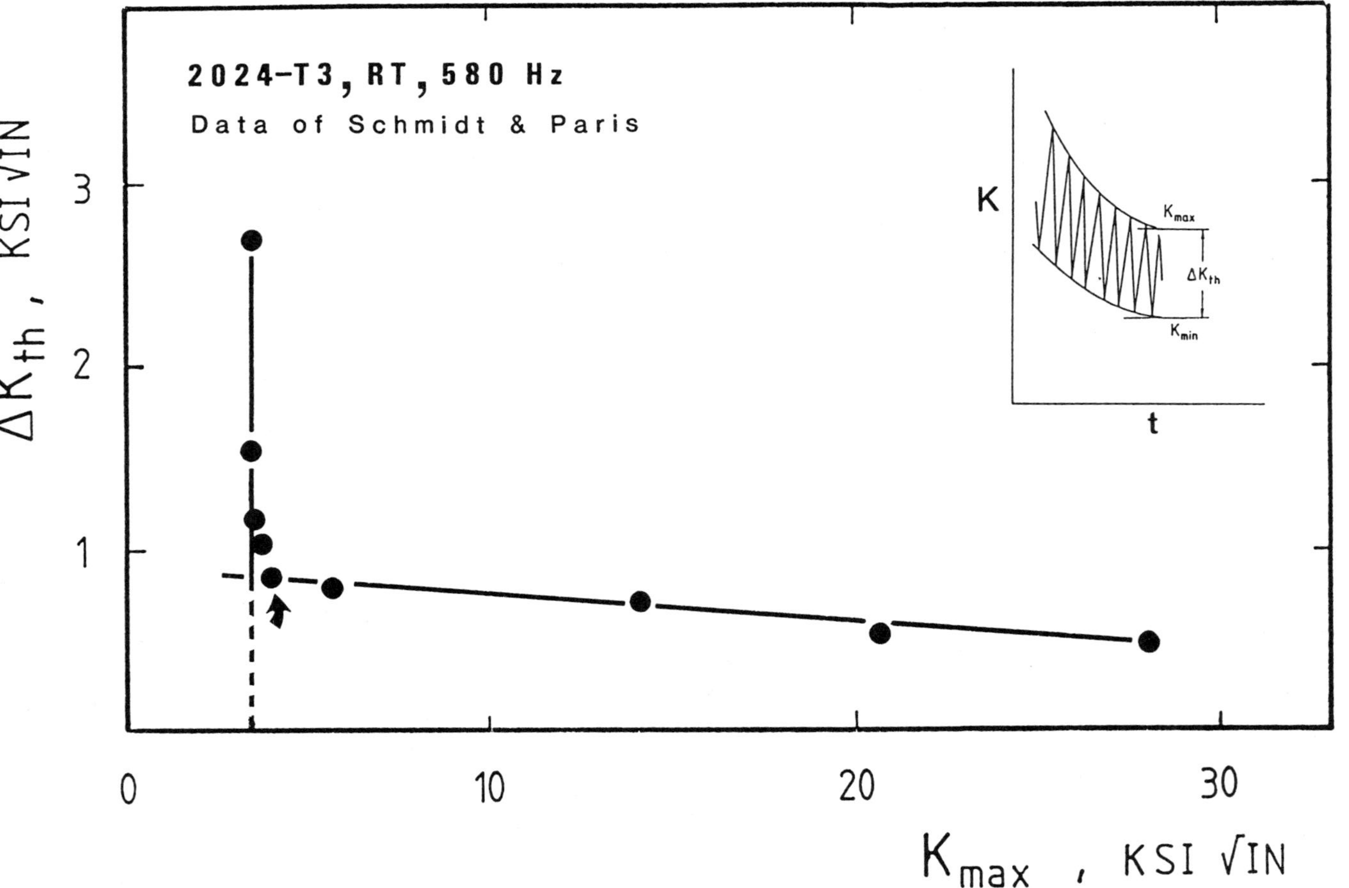

FIG. 13—*Threshold values marked by arrows in Fig. 12 replotted as a function of K_{max}.*

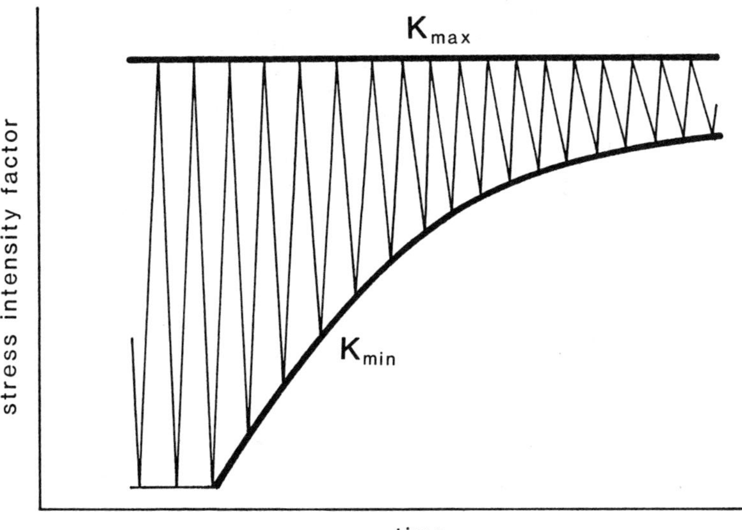

FIG. 14—*Constant* K_{max} *test to obtain* da/dN *versus* ΔK *curves.*

rienced by the crack tip region, namely K_{LL}, must be approximately equal to K_{op}. Reducing the externally applied K_{min} below 2.86 MPa · m$^{1/2}$ (2.6 ksi $\sqrt{in.}$) has no effect on K_{LL} or K_{op} and consequently on ΔK_{eff}. Therefore all data points in region A of Fig. 12 have the same K_{max}, K_{op} and ΔK_T. Only the externally applied K_{min} is different, which is without consequence because of the wedging action of the wake zone.

Hertzberg and co-workers [23–25] used the constant K_{max} procedure (Fig. 14) to determine da/dN as a function of ΔK. They found that the short crack fatigue growth behavior was adequately covered by the da/dN–ΔK curve thus obtained. Figure 15 shows da/dN–ΔK curves obtained in accordance with ASTM E 647 for different R-ratios (lines) and such curves (data points) obtained by the constant K_{max} test for a quenched and tempered steel with German designation 42CrMo4. As can be seen in Fig. 15, the usual R dependence of the fatigue crack growth rates is not visible in the results from constant K_{max} tests. The reason for this is that the small ΔK is always applied with high R-ratios such that K_{min} is above the influence of the wake zone. This influence of the wake zone is even more pronounced with aluminum-lithium alloys due to the deformation by planar slip.

Figure 16 compares again da/dN–ΔK [26] curves obtained in accordance with ASTM E 647 (lines) with a curve obtained by the constant K_{max} test. The growth rates obtained by the constant K_{max} test are approximately equal in the ΔK range between 3 and 8 MPa · m$^{1/2}$. This means that the ΔK_{eff} stays constant and increasing K_{min} (K_{max} stays constant) has no effect on K_{op} in this ΔK range. Therefore the externally applied K_{min} is changing but not the minimum stress intensity factor experienced by the crack tip region K_{LL}.

In Ref 20 it was found that thickness of specimen, K_{max}-level producing the fracture surfaces (see also Ref 27), and material defects can strongly influence the modification of K_{min} toward K_{LL}. The Al-Li alloy shown in Fig. 16 gives an example of the influence of the slip mechanism when compared to other aluminum alloys. Coarse grained materials modify K_{min} much more than fine grained materials, as the well known R-dependence of da/dN–ΔK curves of such materials document. All these factors can contribute to the modification of K_{min} at low R-ratios toward a higher transmitted lower limit of the stress intensity range in

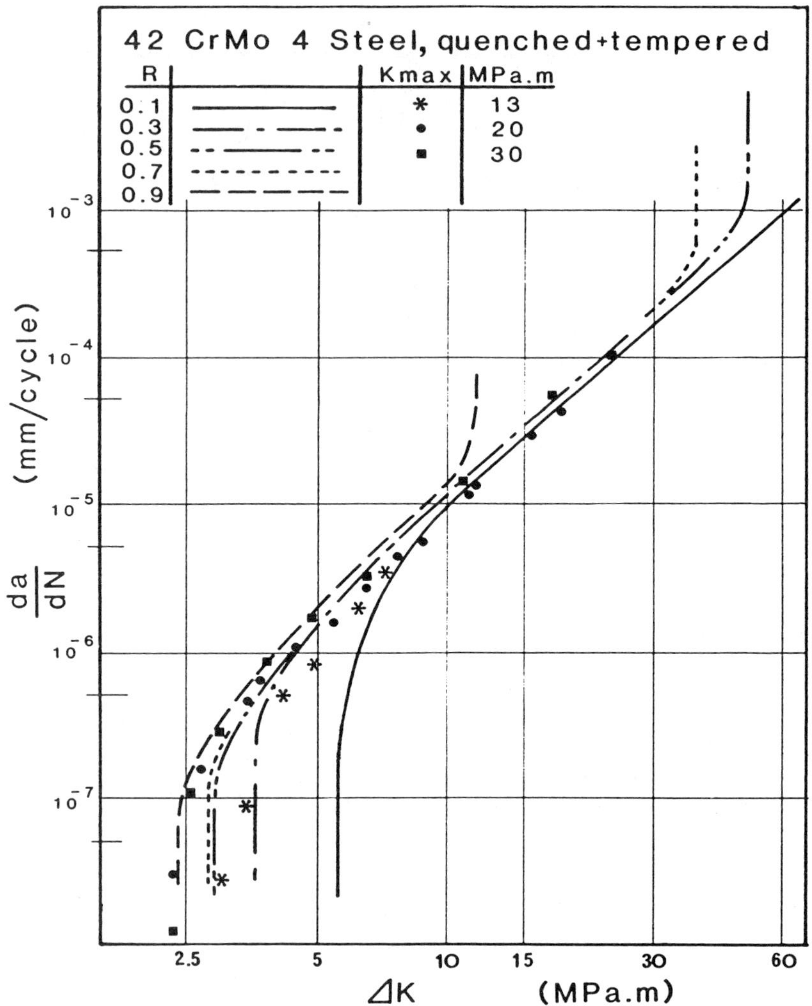

FIG. 15—*Comparison of* da/dN–ΔK *curves obtained in accordance with ASTM E 647 and* K_{max}*–constant tests of a quenched and tempered steel.*

sporadic and unpredictable ways. This makes the use of conventional *da/dN–ΔK* curves as a general design base questionable.

Conclusions

1. Experimental methods are proposed to measure "K_{op}" via a propagation versus non-propagation criterion. The "K_{op}" measured by this technique partitions the ΔK into two parts: in the upper part crack growth commences, while no crack growth can occur in the lower part. This "K_{op}" is the partition point necessary to determine ΔK_{eff}.

2. The "K_{op}" values measured by the novel test method on Inconel 617 and Ti-6Al-4V agree in their R dependence with those obtained empirically from fatigue crack growth rates or from finite element calculations.

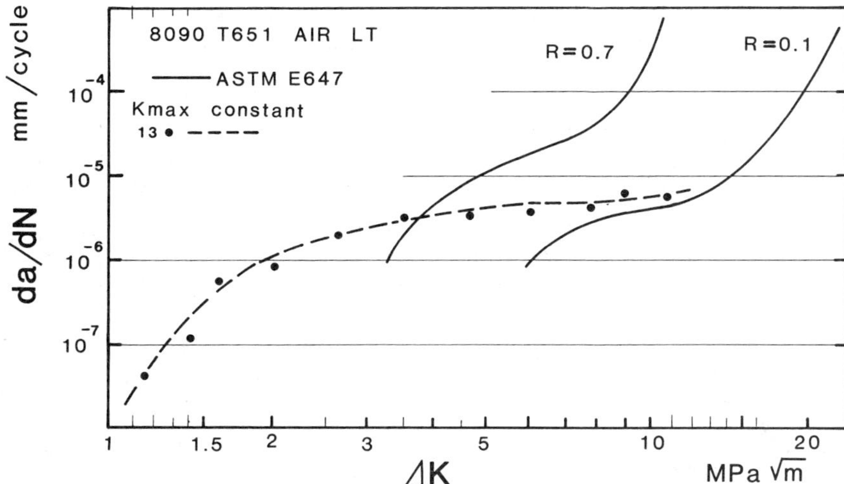

FIG. 16—*Comparison of* da/dN-ΔK *curves obtained in accordance with ASTM E 647* [26] and K$_{max}$-*constant tests for an Al-Li alloy (8090-T651).*

3. The "closure" phenomenon characterized by the lowest point of the upper linear branch of load-displacement curves contradicts the above results and is shown not to affect ΔK_{eff} directly.

4. It is postulated that the "closure" phenomenon as obtained by a load-displacement curve only reflects the wedging action of the wake zone associated with fatigue crack. With low R-ratios the wedging action of the wake zone modifies K_{min} to some higher value K_{LL} experienced by the near-region of the crack front. ΔK experienced by the near-region of the crack front is therefore reduced, the R-ratio and the partition point "K_{op}" are increased. Consequently, ΔK_{eff} is reduced.

Acknowledgments

Financial support of D. E. Castro by DAAD (German Academic Exchange Service) during this investigation is thankfully acknowledged.

References

[1] Castro, D. E., Marci, G., and Munz, D., "A Generalized Concept of a Fatigue Threshold," *Journal of Fatigue Fracture of Engineering Materials and Structures,* Vol. 10, No. 4, 1987, pp. 305–315.

[2] Castro, D. E., Marci, G., and Munz, D., "Threshold and Non-propagation of Fatigue Cracks under Service Loading," in *Fracture Mechanics: 19th Symposium, ASTM STP 969,* T. A. Cruse, Ed., American Society for Testing and Materials, Philadelphia, 1987, pp. 818–829.

[3] Marci, G., Castro, D. E., and Bachmann, V., "Fatigue Crack Propagation Threshold," *Journal of Testing and Evaluation,* Vol. 17, No. 1, Jan. 1987, pp. 28–39.

[4] Elber, W., "Fatigue Crack Propagation under Random Loading: An Analysis Considering Crack Closure," presented to I.C.A.F., Stockholm, Sweden, May 1969.

[5] Elber, W., "The Significance of Fatigue Crack Closure," *Damage Tolerance in Aircraft Structures, ASTM STP 486.* American Society for Testing and Materials, Philadelphia, 1971, pp. 230–242.

[6] Marci, G., "Effect of the Active Plastic Zone on Fatigue Crack Growth Rates," in *Fracture Mechanics (Eleventh Conference), ASTM STP 677,* C. W. Smith, Ed., American Society for Testing and Materials, Philadelphia, 1979, pp. 168–186.

[7] Döker, H., Bachmann, V., and Marci, G., "A Comparison of Different Methods of Determination

of the Threshold for Fatigue Crack Propagation," in *Fatigue Thresholds,* Proc. Int. Conf., Stockholm, Sweden, 3–1 June 1981, J. Bäcklund, A. F. Blom, and C. J. Beevers, Eds., EMAS, Warley, 1982, p. 45.

[8] Pellas, J., Baudin, G., and Robert, M., "Fatigue Crack Growth Model Prediction with Two Coupled Differential Equations," in *Fracture 1977,* Proc. 4th Intern. Conf. on Fracture, D. M. R. Taplin, Ed., Vol. 2, 1977, pp. 1353–1360.

[9] Baudin, G. and Robert, M., "Crack Growth Model for Flight Type Loading," ONERA T.P. No. 1981-44, presented at International Committee on Aeronautical Fatigue (ICAF), Noordwijkerhout (P.B.), 18–21 May 1981.

[10] Marci, G. and Bazant, E., "Über die physikalische Bedeutung des Threshold-Spannungsintensitätsbereiches bei Schwingbelastungen," *Materialprüfung,* Vol. 19, No. 11, 1977, pp. 455–462.

[11] Döker, H. and Bachmann, V., "Determination of Crack Opening Load by Use of Threshold Behavior," in *Mechanics of Fatigue Crack Closure, ASTM STP 982,* J. C. Newman, Jr., and W. Elber, Eds., American Society for Testing and Materials, Philadelphia, 1987, pp. 247–259.

[12] Schijve, J., "The Stress Ratio Effect on Fatigue Crack Growth in 2024-T3 Alclad and Relation to Crack Closure," Delft University of Technology, Department of Aerospace Engineering, Memorandum M-336, Aug. 1979.

[13] De Koning, A. U., "A Simple Crack Closure Model for Prediction of Fatigue Crack Growth Rates under Variable Amplitude Loading," in *Fracture Mechanics: 13th Conference, ASTM STP 743,* Richard Roberts, Ed., American Society for Testing and Materials, Philadelphia, 1981, pp. 63–85.

[14] Newman, J. C., Jr., "A Finite-Element Analysis of Fatigue Crack Closure," in *Mechanics of Crack Growth, ASTM STP 590,* American Society for Testing and Materials, Philadelphia, 1976, pp. 281–301.

[15] Ritchie, R. O. and Suresh, S., "Communications, Some Considerations on Fatigue Crack Closure at Near-Threshold Stress Intensities Due to Fracture Surface Morphology," *Metallurgical Transactions A,* Vol. 13A, 1982, pp. 937–940.

[16] Suresh, S., "Crack Deflection: Implications for the Growth of Long and Short Fatigue Cracks," *Metallurgical Transactions A,* Vol. 14A, 1983, pp. 2375–2385.

[17] Suresh, S., Zamiski, G. F., and Ritchie, R. O., "Oxide-Induced Crack Closure: An Explanation for Near-Threshold Corrosion Fatigue Crack Growth Behavior," *Metallurgical Transactions A,* Vol. 12A, 1981, pp. 1435–1443.

[18] Nordmark, G. E. and Fricke, W. G., "Fatigue Crack Arrest at Low Stress Intensities in a Corrosive Environment," *Journal of Testing and Evaluation,* Vol. 6, No. 5, Sept. 1978, pp. 301–303.

[19] Hertzberg, R. W., Newton, C. H., and Jaccard, R., "Crack Closure: Correlation and Confusion," in *Mechanics of Fatigue Crack Closure, ASTM STP 982,* J. C. Newman, Jr., and W. Elber, Eds., American Society for Testing and Materials, Philadelphia, 1988, pp. 139–148.

[20] Marci, G. and Bachmann, V., "Direct Measurement of ΔK Transmitted to the Active Plastic Zone in Fatigue," in *Fracture Mechanics, 14th Symposium—Vol. II: Testing and Applications, ASTM STP 791,* J. C. Lewis and G. Sines, Eds., American Society for Testing and Materials, Philadelphia, 1983, pp. II-47–II-66.

[21] Schmidt, R. A. and Paris, P. C., "Threshold for Fatigue Crack Propagation and the Effects of Load Ratio and Frequency," in *Progress in Flaw Growth and Fracture Toughness Testing, ASTM STP 536,* American Society for Testing and Materials, Philadelphia, 1973, pp. 79–94.

[22] Döker, H., Bachmann, V., Castro, D. E., and Marci, G., "Schwellwert für Ermüdungsrißausbreitung: Bestimmungsmethoden, Kennwerte, Einflußgrößen," *Zeitschrift für Werkstofftechnik,* No. 18, 1987, pp. 323–329.

[23] Hertzberg, R. W., Herman, W. A., and Ritchie, R. O., "Use of a Constant K_{max} Test Procedure to Predict Small Crack Growth Behavior in 2090-T8E41 Aluminum-Lithium Alloy," *Scripta Metallurgica,* Vol. 21, 1987, pp. 1541–1546.

[24] Herman, W. A., Hertzberg, R. W., Newton, C. H., and Jaccard, R., "A Re-Evaluation of Fatigue Threshold Test Methods," in *Proceedings,* Fatigue '87, Volume II, R. O. Ritchie and E. A. Starke, Jr., Eds., EMAS Ltd., 1987, pp. 819–828.

[25] Herman, W. A., Hertzberg, R. W., and Jaccard, R., "A Simplified Laboratory Approach for the Prediction of Short Crack Behavior in Engineering Structures," *Journal of Fatigue and Fracture of Engineering Materials and Structures,* Vol. 11, No. 4, 1988, pp. 303–320.

[26] Peters, M., Bachmann, V., and Welpmann, K., "Fatigue Crack Propagation Behavior of the Al-Li Alloy 8090 Compared to 2024," in *Proceedings,* 4th International Al-Li Conference, Paris, 1987.

[27] James, M. N. and Knott, J. F., "An Assessment of Crack Closure and the Extent of the Short Crack Regime in QIN (HY80) Steel," *Journal of Fatigue and Fracture of Engineering Materials and Structures,* Vol. 8, No. 2, 1985, pp. 177–191.

C. W. Marschall,[1] *P. R. Held,*[1] *M. P. Landow,*[1] *and P. N. Mincer*[1]

Use of the Direct-Current Electric Potential Method to Monitor Large Amounts of Crack Growth in Highly Ductile Metals

REFERENCE: Marschall, C. W., Held, P. R., Landow, M. P., and Mincer, P. N., **"Use of the Direct-Current Electric Potential Method to Monitor Large Amounts of Crack Growth in Highly Ductile Metals,"** *Fracture Mechanics; Twenty-First Symposium, ASTM STP 1074,* J. P. Gudas, J. A. Joyce, and E. M. Hackett, Eds., American Society for Testing and Materials, Philadelphia, 1990, pp. 581–593.

ABSTRACT: The direct-current electric potential method is receiving increasing attention for monitoring crack extension in *J*-resistance curve testing. Among its advantages over the unloading-compliance method are: (1) no time-consuming unloadings are required, (2) a continuous record of crack extension versus displacement can be obtained, and (3) the method can be used at higher strain rates where unloading compliance cannot be used.

Despite the advantages of the direct-current electric potential method, questions persist regarding its ability to monitor large amounts of crack growth in highly ductile materials where large displacements and large amounts of plastic strain occur. This paper presents details of an experiment conducted on a 3T planform-size compact specimen of 25.4 mm thickness to assess the ability of the direct-current electric potential method to accurately measure crack extension in a highly ductile material. The material selected was Type 304 austenitic stainless steel. It was found that the Johnson expression, often used to calculate crack extension from direct-current electric potential data, significantly underestimated the actual amount of crack extension. However, a simple modification of the Johnson expression resulted in excellent agreement between calculated and measured crack extensions.

KEY WORDS: electric potential, crack growth, austenitic stainless steel, Johnson equation

Direct-current electric potential (d-c EP) measurements are receiving increasing attention for monitoring stable crack extension in laboratory fracture toughness specimens [*1–5*]. The concept is relatively simple. As shown schematically in Fig. 1, leads for supplying a constant direct current and for measuring the potential are attached to a compact specimen at points for which calibration curves have been established. Calibration can be performed by taking potential readings at constant current as the crack (notch) is extended by saw cutting. In an actual test, the d-c EP is monitored along with the load and displacement and, from the calibration curve, crack length can be obtained from the d-c EP value at any point in the test.

Use of the d-c EP method offers several advantages over the unloading-compliance method commonly used in single-specimen *J-R* curve tests. These advantages include: (1) no time-consuming unloadings (or hold times at constant displacement prior to unloading) are required, (2) a continuous record of crack length versus displacement can be obtained,

---

[1] Battelle, Columbus, OH 43201.

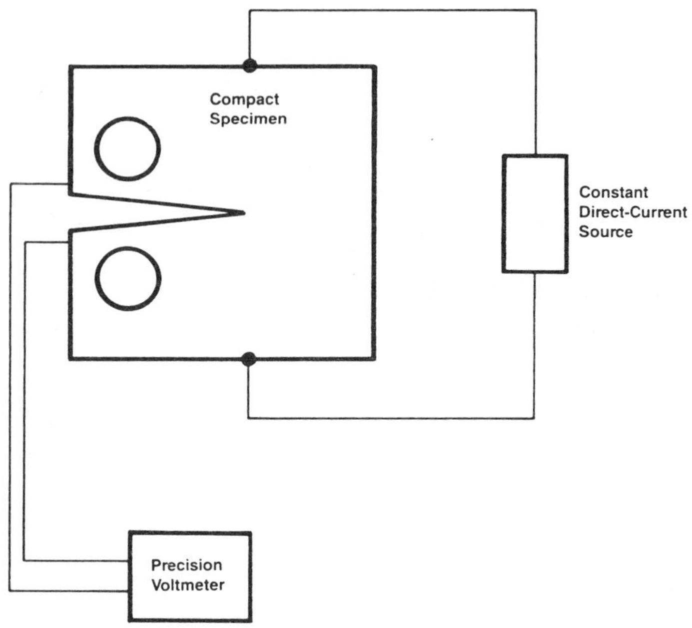

FIG. 1—*Schematic illustration of direct-current electric potential method for monitoring crack extension.*

and (3) the method can be used at higher displacement rates where unloading compliance cannot be used.

The ability of d-c EP data to predict crack growth has been demonstrated to be good where very little plastic deformation accompanies crack extension (fatigue crack growth, for example) or in materials that exhibit relatively low stable-fracture resistance. However, in very tough materials that exhibit extensive plasticity during stable crack growth, such as some of the materials used in nuclear reactor coolant piping, the ability of d-c EP data to accurately predict crack growth is not well established.

The problems anticipated in employing d-c EP on very tough materials are twofold. First, extensive plastic deformation occurs prior to initiation of a stable crack from the existing fatigue precrack. This deformation causes a change in the observed d-c EP value, due to a change in resistivity caused by work hardening of a zone near the crack tip, that is unrelated to crack extension. Thus the crack initiation point cannot be determined merely by noting the load or displacement at which the d-c EP begins to increase. Rather, the initiation point lies somewhere along a rising curve of d-c EP versus displacement or d-c EP versus load. In practice, selection of the appropriate point is not always obvious.

The second problem pertains to crack growth. In addition to the plastic deformation increasing the resistivity of the material, the zone ahead of the crack experiences thickness changes as the crack extends. The material immediately ahead of the crack tip undergoes thinning, sometimes as much as 30 or 40% in very tough materials, while that near the back face undergoes significant thickening. Thinning increases the electric resistance and thickening decreases the electric resistance, leading to changes in d-c EP values that are not related to crack extension.

This investigation was undertaken in an attempt to better define the ability of the d-c EP method to identify the crack initiation point and to accurately measure crack extension in highly ductile materials.

Material

The material used in this study was annealed ASTM A240, Type 304 austenitic stainless steel in the form of 25.4 mm (1 in.) thick plate. Its chemical composition and tensile properties are shown in Table 1.

Experimental Procedure

A compact specimen of 3T plan dimensions ($W = 152.4$ mm [6 in.]) was machined from the 25.4 mm (1 in.) thick plate in the L-T orientation. It was fatigue precracked in accordance with ASTM Test for J_{Ic}, A Measure of Fracture Toughness (E 813-87). The ratio of crack length to specimen width was approximately 0.5. The specimen was tested in displacement control at a low loading rate at room temperature. Load, load-line displacement, and d-c EP were recorded during the test on an X-Y-Y plotter and on a computer floppy disk.

Preparation of the compact specimen for testing is shown in Fig. 2. Leads to supply constant current were attached at points C and C', midway between the load line and the back face. Potential leads were placed at four different locations:

1. A-A', the conventional location.
2. B-B', a location close to the initial crack tip that might provide a better indication of crack initiation than would location A-A'.
3. C-C', a location identical to that of the current leads.
4. D-D', a location at the back face that might provide an indication of back face thickening as the test progresses.

A grid was scribed on the specimen above and below the crack plane (Fig. 2).

Testing of the compact specimen employed partial unloadings at various stages of the test. The purposes of the partial unloadings were to (1) mark the crack front for later measurement of actual crack extension, and (2) permit measurement of thickness changes at grid intersection points at each unloading. The thickness changes at the grid points are not included in this paper. Ten unloadings were used, based on preselected amounts of observed extension of the crack at the surface, as shown in Table 2. Figure 3 shows the location of the crack front after each unloading, as revealed by surface marks from the partial unloadings. A nine-point average crack length was measured for each of the ten unloading points. It should be noted that the fracture surface was nominally flat (i.e., no shear lips were observed), despite the large amount of plastic deformation that accompanied crack growth.

TABLE 1—*Chemical composition and tensile properties of annealed ASTM A240 Type 304 stainless steel plate.*

Chemical Composition (Percent by Weight)								
C	Mn	P	S	Si	Ni	Cr	Cu	Sn
0.048	1.87	0.027	0.005	0.63	8.0	18.4	0.42	0.016
Mo	Al	V	Cb	Zr	Ti	B	Co	W
0.22	0.002	0.09	0.026	0.002	0.003	0.0008	0.12	0.01

Tensile Properties (from Mill Report)					
Yield Strength		Tensile Strength		Elongation in	Area
MPa	ksi	MPa	ksi	2 in., %	Reduction, %
263	38.2	593	86.0	68	75

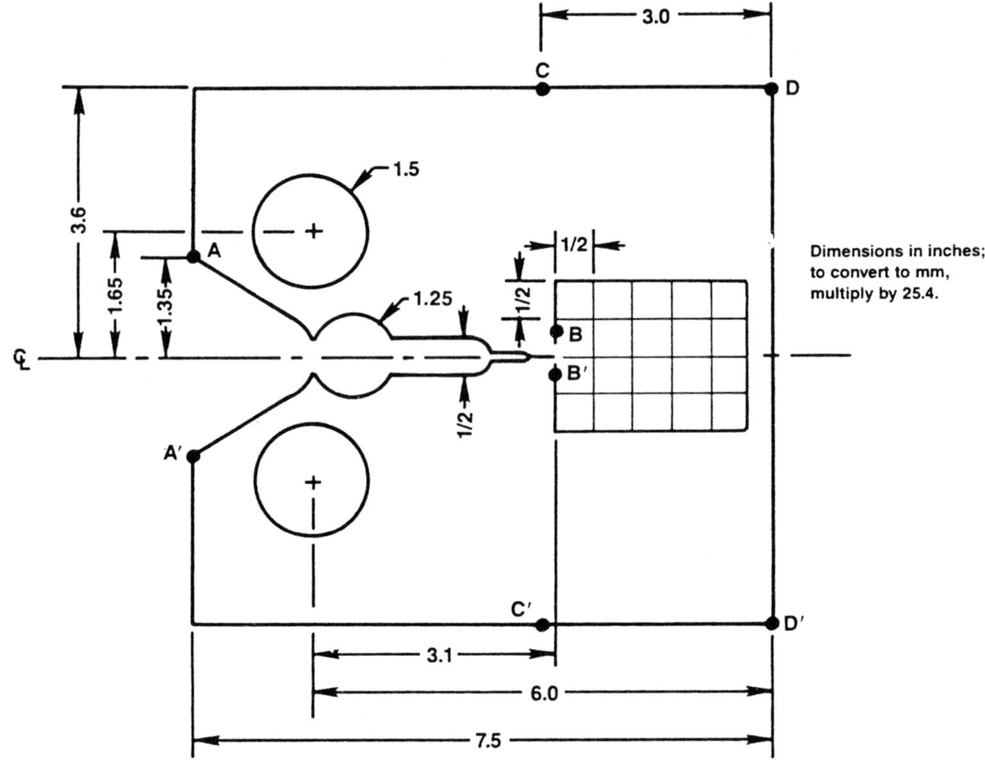

FIG. 2—*Plan view of compact specimen.*

Results

Detecting the Point of Crack Initiation

Figure 4 is the load versus displacement record for the compact specimen, showing the ten unloadings. As was noted earlier, the first unloading occurred prior to any visual evidence of crack extension. Figures 5 to 8 are graphs of electric potential (U), measured at A-A', B-B', C-C', and D-D', respectively, versus load-line displacement. Each of the four curves displays

TABLE 2—*Unloading points in compact specimen tests.*

No.	Approximate Surface Crack Extension, mm (in.)
1	0.0 (0.0) (no initiation)
2	2.5 (0.1)
3	5.1 (0.2)
4	10.2 (0.4)
5	15.2 (0.6)
6	20.3 (0.8)
7	25.4 (1.0)
8	31.5 (1.25)
9	38.1 (1.5)
10	44.5 (1.75)

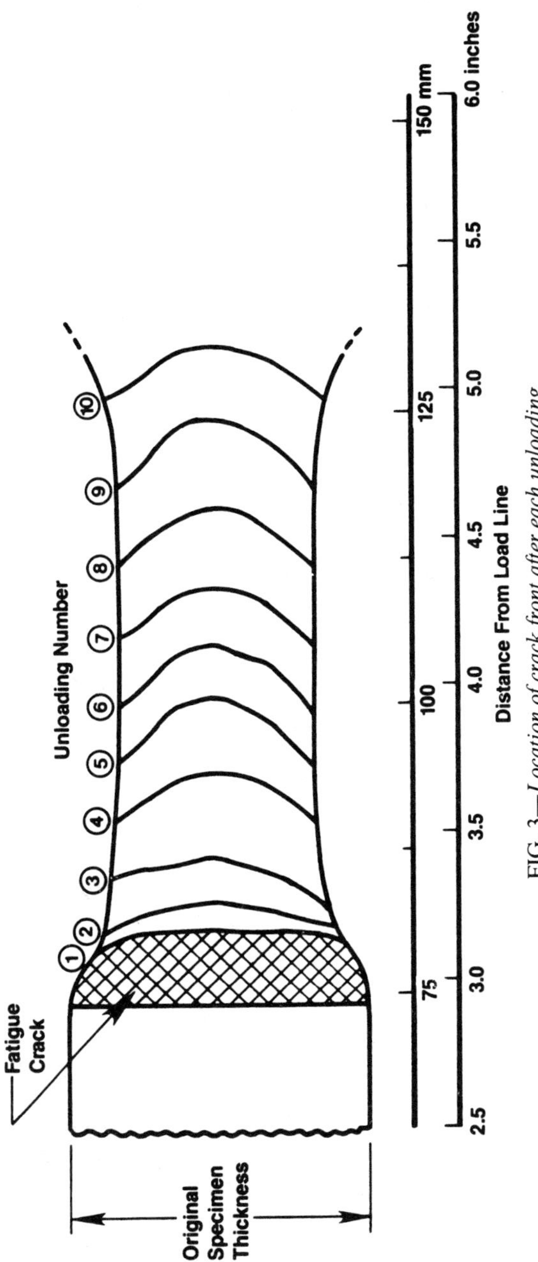

FIG. 3—*Location of crack front after each unloading.*

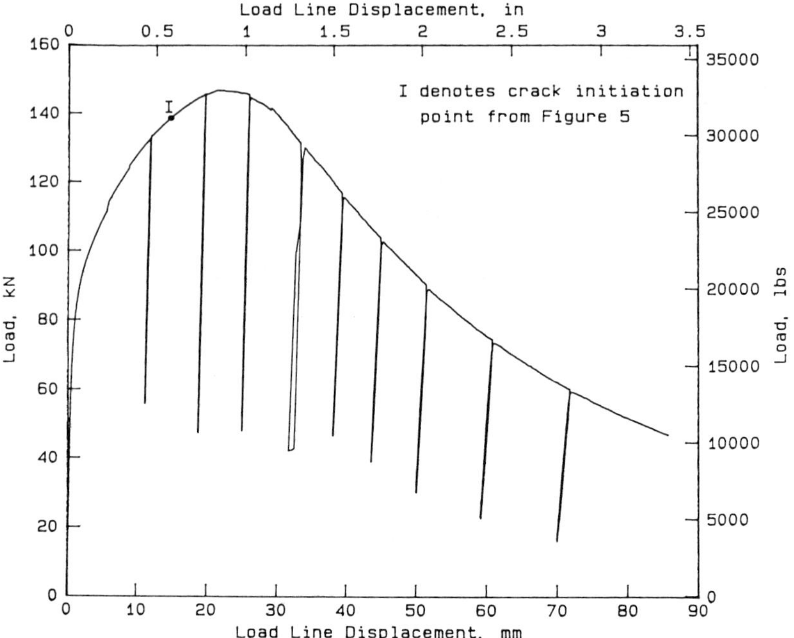

FIG. 4—*Load versus displacement record for compact specimen.*

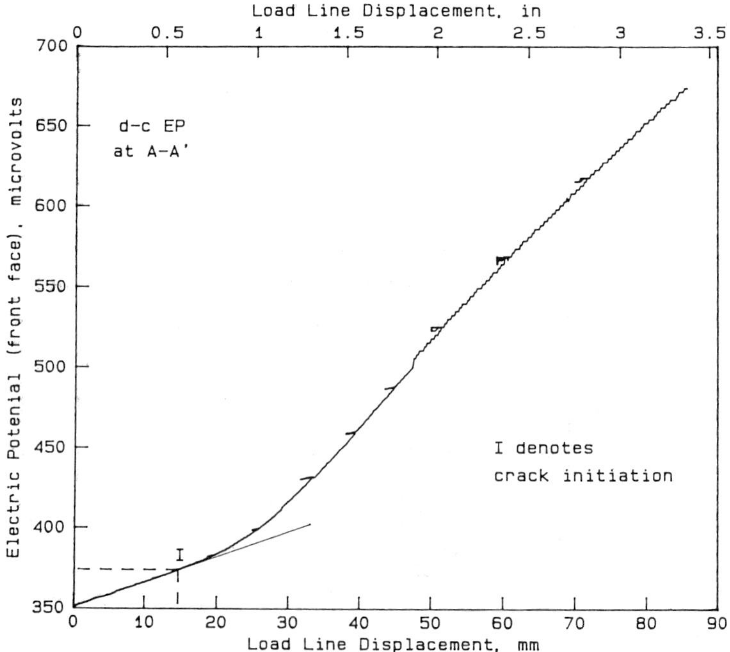

FIG. 5—*Electric potential at* A-A' *versus displacement.*

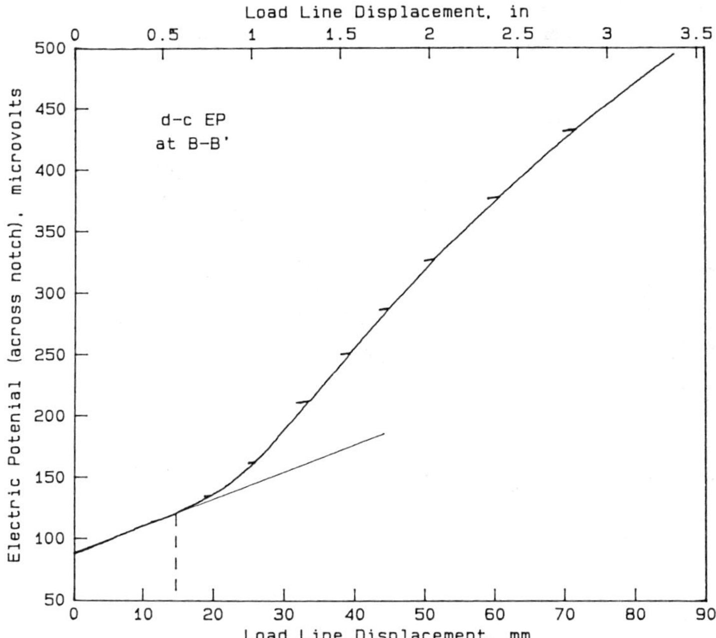

FIG. 6—*Electric potential at* **B-B'** *versus displacement.*

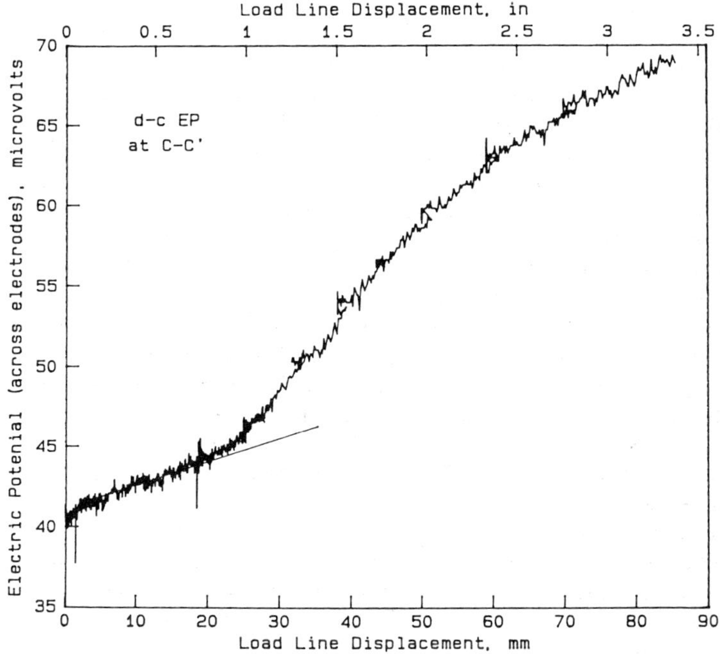

FIG. 7—*Electric potential at* **C-C'** *versus displacement.*

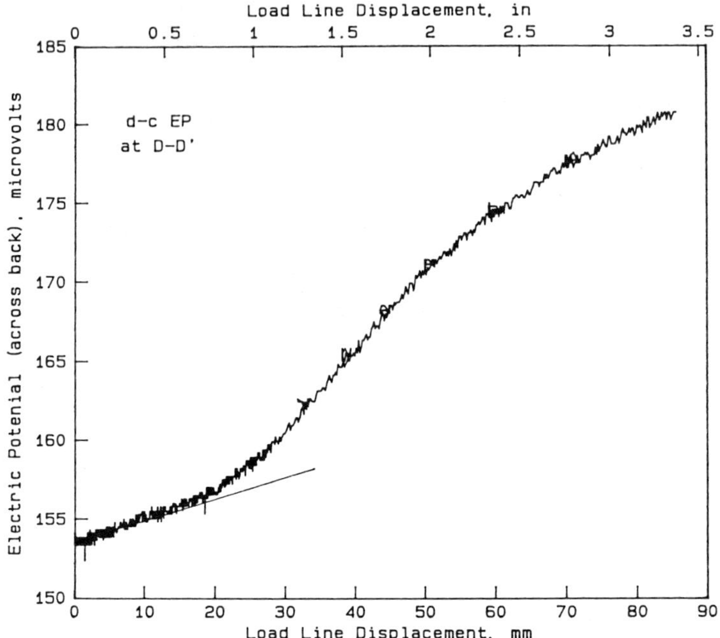

FIG. 8—*Electric potential at* D-D' *versus displacement.*

an initial region of rising potential that is approximately linear with displacement. This increase in potential is the result of crack-tip blunting (i.e., the resistivity of material near the crack tip is increased as a result of plastic deformation occurring there), though there is no firm rationale for the linear relation between potential and displacement.

Referring to Fig. 5, it is widely held that the first deviation from linearity of the potential (measured at *A-A'*) versus displacement curve is indicative of crack initiation [2]. Again, there is no firm basis for this assumption, except that it has been shown experimentally to provide reasonable results and to agree approximately with the unloading compliance method and the multiple specimen method for determining the crack-initiation point [3,4]. The method is not totally satisfactory because of the uncertainties inherent in selecting such a point on a smooth curve and the need to use engineering judgment. Nonetheless, using Fig. 5 to select the initiation point indicates that initiation occurred at a displacement of about 14.2 mm (0.56 in.) and a potential at *A-A'* of about 373 μV. That this procedure provided a reasonable result can be seen in Fig. 4, where the initiation point is marked on the load versus displacement curve. It is seen to lie between the first unloading, where no crack growth was evident, and the second unloading, where crack growth had occurred.

Another method sometimes used to establish the crack initiation point from d-c EP data is the examination of a graph of d-c EP versus load to detect significant slope changes [5]. Such a graph is shown in Fig. 9 for the potential measured at *A-A'*. It is evident from the graph that this method is not useful for defining crack initiation in this particular specimen: the only significant slope changes occur at loads appreciably below the first unloading. Nonetheless, in other experiments conducted at Battelle, particularly on materials less tough than austenitic stainless steels, this method has sometimes been useful in helping to define the crack initiation point.

It was believed that the potential data obtained at *B-B'* might show a more abrupt departure from linearity at crack initiation when plotted against displacement than did the data obtained at *A-A'*, because of the close proximity of *B-B'* to the crack tip. This occurrence would make the crack initiation point easier to define. However, this result was not realized, as indicated in Fig. 6. Nonetheless, the deviation from linearity in Fig. 6 occurs at very nearly the same displacement as in Fig. 5, indicating that both locations are equally capable of defining the point of crack initiation.

Figures 7 and 8 show trends similar to those in Figs. 5 and 6, but are harder to interpret because of noise in the potential record. Using engineering judgment with Figs. 7 and 8, it is possible to select the point of departure from linearity at approximately the same displacement value as in Fig. 5. However, because of the noise, the curves easily could be interpreted as indicating crack initiation at a somewhat larger displacement, perhaps 20 mm (0.8 in.). That value can be seen to be incorrect by noting in Fig. 4 that it lies beyond the second unloading point where significant crack growth already had been observed visually.

Thus the results obtained here confirm that selecting the crack initiation point from the point of departure of linearity of a curve of electric potential (at point *A-A'*) versus displacement provides a reasonable result. However, it is subject to error because of the uncertainties inherent in the method. Contrary to expectations, locating the potential probes closer to the crack tip does not appear to increase the certainty in selection of the crack initiation point. The two other potential-probe locations used here likewise appear to offer no advantages in defining the initiation point.

Calculating Crack Growth from d-c EP Data

With respect to crack growth, three different procedures were applied to calculate crack extension from the potential change measured at *A-A'*. Each was based on the Johnson equa-

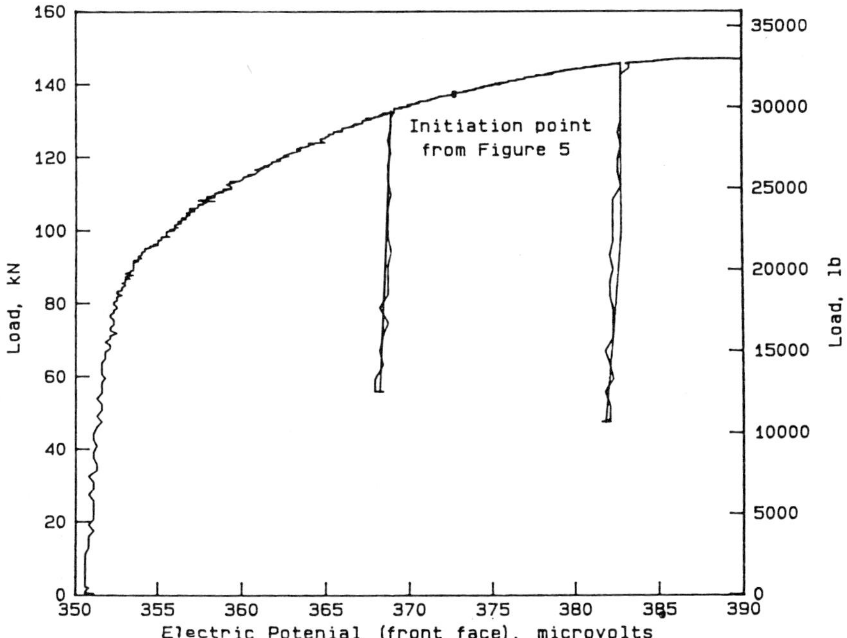

FIG. 9—*Load versus electric potential at* A-A'.

tion [1,6]:

$$\frac{U}{U_0} = \frac{\cosh^{-1}\left[\cosh\left(\pi y/2W\right)/\cos\left(\pi a/2W\right)\right]}{\cosh^{-1}\left[\cosh\left(\pi y/2W\right)/\cos\left(\pi a_0/2W\right)\right]}$$

where U_0 and a_0 are the values of potential and crack length, respectively, at the point of crack initiation; U and a are the actual values of both quantities as the crack extends; y is the half-spacing of the potential probes across the notch mouth; and W is specimen width.

The three methods are described below:

• *Method 1:* Any increase in potential observed after the test began was assumed to indicate crack extension (i.e., the starting potential of 300.6 μV was assumed to be the crack initiation potential, U_0. Crack growth was calculated from U/U_0. Obviously, this method will overestimate crack growth at the beginning of the test, because of the potential increase resulting from plastic deformation near the crack tip.

• *Method 2:* A crack initiation potential (U_0) of 373 μV was selected from Fig. 5. Crack growth beyond initiation was calculated from U/U_0.

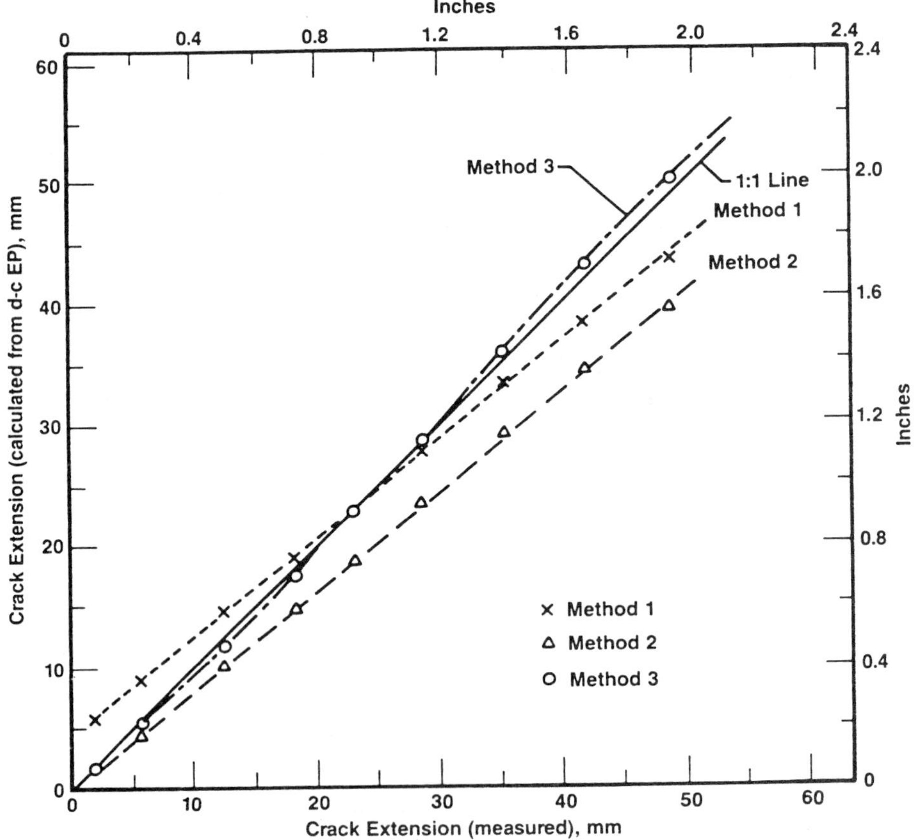

FIG. 10—*Comparison of calculated crack extension with measured crack extension.*

TABLE 3—*Comparison of measured crack lengths with crack lengths calculated from Johnson's equation.*[a]

Unloading No.	Measured Extension, mm (in.)	Calculated Crack Extension, mm (in.), and Percentage Error		
		Method 1	Method 2	Method 3
1	0.000 (0.000)	. . . . . .	0.00 (0.000) 0.0%	0.00 (0.000) 0.0%
2	1.88 (0.074)	5.82 (0.229) +209%	1.42 (0.056) −24.3%	1.83 (0.072) −2.7%
3	5.54 (0.218)	9.02 (0.355) +62.8%	4.60 (0.181) −17.0%	5.33 (0.210) −3.7%
4	12.60 (0.496)	14.58 (0.574) +15.7%	10.19 (0.401) −19.2%	11.96 (0.471) −5.0%
5	18.16 (0.715)	19.08 (0.751) +5.0%	14.73 (0.580) −18.9%	17.58 (0.692) −3.2%
6	22.99 (0.905)	23.22 (0.914) +1.0%	18.92 (0.745) −17.7%	22.91 (0.902) −0.3%
7	28.50 (1.122)	27.86 (1.097) −2.2%	23.67 (0.932) −16.9%	28.91 (1.138) +1.4%
8	35.41 (1.394)	33.40 (1.315) −5.7%	29.39 (1.157) −17.0%	36.17 (1.424) +2.2%
9	41.94 (1.651)	38.53 (1.517) −8.5%	34.70 (1.366) −17.6%	43.33 (1.706) +3.3%
10	48.97 (1.928)	43.59 (1.716) −11.0%	40.01 (1.575) −18.3%	50.37 (1.983) +2.9%

[a] Material: Type 304 stainless steel
Temperature: 22°C (72°F)
Specimen Type: 3T planform-size compact specimen
$\quad\quad\quad\quad B = 25.4$ mm (1 in.)
$\quad\quad\quad\quad W = 152.4$ mm (6 in.)
$\quad\quad\quad\quad a_0 = 80.11$ mm (3.154 in.)
Potential Probe Location: A-A' in Fig. 2
See Text for Description of Methods 1, 2, and 3

• *Method 3:* This method was identical to Method 2 with only one exception—the value of y in the Johnson equation was allowed to increase in proportion to the load-line displacement, rather than to remain at its original value as in Method 2. This modification, which appears to be justified because of the large load-line displacements (up to 86 mm [3.4 in.]) observed, would be expected to increase the crack growth calculated by Method 2.

The results of the three methods for calculating crack extension are shown in Fig. 10. Note that Method 1, as expected, resulted in a large overestimate of crack extension at the start of the test. By the end of the test, however, Method 1 was underestimating crack growth.

Method 2 underestimated crack growth throughout the test. However, the simple modification to Method 2 to obtain Method 3 produced excellent agreement between calculated and measured crack extension. Table 3 provides values of the results along with error percentages for each method. Note that Method 3 yields crack extension values within 5% of the measured result for the entire test. This level of agreement is unexpectedly good in view of the curved crack fronts at each unloading (see Fig. 3).

Based on the results, it appears that a simple modification to the Johnson equation can provide a close estimate of the crack extension in highly ductile materials, even for crack extensions exceeding half of the original uncracked ligament. This finding may be fortuitous, since it is based on only one test, or it may mean that: (*a*) the potential probe spacing should, in fact, be treated as a variable in the Johnson expression; (*b*) once the crack has initiated, blunting makes no further contribution to the measured potential (i.e., any increase in the potential beyond initiation is due to crack extension only and does not include any additional blunting contribution); and (*c*) the thinning of the specimen immediately ahead of the crack and the thickening near the back face, both of which cause electric resistance changes, may be self-compensating.

Discussion

This study has demonstrated that the d-c EP method of estimating crack extension, designed for situations of limited plasticity, can provide surprisingly accurate results in highly ductile materials and for large amounts of crack extension. A simple modification of the Johnson equation, namely, to allow the potential-probe spacing to increase in proportion to the load-line displacement, provided crack growth estimates within 5% of actual values in a Type 304 stainless steel compact specimen for crack growth of $\Delta a/b_0$ ranging from 0.03 to 0.68.

The study also demonstrated that locating a set of potential probes close to the crack tip (position B-B' in Fig. 2) did not simplify detection of the crack initiation point. Nonetheless, use of the potential data from any of the four probe locations investigated provided a reasonable estimate of the crack initiation point. The method for defining that point, namely, noting the point of departure of linearity of a curve of potential versus displacement, is inherently uncertain and requires the use of engineering judgment.

The use of d-c EP measurements, as described here, for detecting crack initiation and for calculating crack growth has been generally successful in tests on various materials conducted at Battelle. Occasionally, crack initiation is found to be unusually difficult to pinpoint because of a very gradual change in the slope of the curve of d-c EP versus displacement prior to maximum load. Final calculated crack extension values generally agree with measured values within about 10%, similar to results reported by Schwalbe et al. [5], though occasionally the agreement is not that good for reasons that are not readily apparent. Extremely good agreement between measured crack growth and crack growth calculated from d-c EP data was found in recent tests conducted at Battelle as part of a US NRC sponsored round-robin program. When potential-probe location A-A' (Fig. 2) was used and the d-c EP data were treated according to Method 3, described earlier, five of six (CT) specimens tested at Battelle showed agreement to within 3% when calculated and measured crack extension were compared. The results are shown in Table 4. In each specimen, the fracture surface was nominally flat because of side grooves. Crack growth as a percentage of the original uncracked ligament ranged from 36 to 50%.

The promising results obtained to date using d-c EP techniques and the wide applicability of those techniques in fracture toughness testing will likely lead additional investigators to use this method. Thus development of appropriate standards is urgently needed.

TABLE 4—*Final crack growth calculated from d-c EP in round-robin program.*[a]

Specimen Identification No.	Material	Crack Extension, mm (in.)		Difference, %
		Measured	Calculated from d-c EP	
1	Al06B steel	8.48 (0.334)	8.28 (0.326)	−2.4
2	Al06B steel	8.33 (0.328)	8.18 (0.322)	−1.8
3	Al06B steel	7.32 (0.288)	7.11 (0.280)	−2.8
4	Aluminum alloy	7.09 (0.279)	6.96 (0.274)	−1.8
5	Aluminum alloy	7.59 (0.299)	7.52 (0.296)	−1.0
6	Aluminum alloy	6.40 (0.252)	5.99 (0.236)	−6.3

[a] Specimen Type = 1T compact
a_0/W = ~0.65
Temperature = 22°C (72°F)

Acknowledgments

The authors acknowledge the financial support of Battelle Columbus Division in conducting the research investigation reported here. The encouragement of Dr. John Holbrook, Manager of the Physical Metallurgy Section, is especially appreciated.

References

[*1*] Schwalbe, K. H. and Hellmann, D., "Application of the Electrical Potential Method of Crack Length Measurements Using Johnson's Formula," *Journal of Testing and Evaluation,* Vol. 9, No. 3, 1981, pp. 218–221.

[*2*] Wilkowski, G. M., Wambaugh, J. O., and Prabhat, K., "Single Specimen *J*-Resistance Curve Evaluations Using the Direct-Current Electric Potential Method and a Computerized Data Acquisition System," in *Fracture Mechanics: Fifteenth Symposium, ASTM STP 833,* R. J. Sanford, Ed., American Society for Testing and Materials, Philadelphia, 1984, pp. 553–576.

[*3*] Hackett, E. M., Kirk, M. T., and Hays, R. A., "An Evaluation of *J-R* Curve Testing of Nuclear Piping Materials Using the Direct Current Potential Drop Technique," David Taylor Naval Ship R&D Center, NUREG/CR-4540, Aug. 1986.

[*4*] Vassilaros, M. G. and Hackett, E. M., "*J*-Integral *R*-Curve Testing of High Strength Steels Utilizing the Direct Current Potential Drop Method," in *Fracture Mechanics: 15th Symposium, ASTM STP 833,* R. J. Sanford, Ed., American Society for Testing and Materials, Philadelphia, 1984, pp. 535–552.

[*5*] Schwalbe, K. H., Hellmann, D., Heerens, J., Knaack, J., and Muller-Roos, J., "Measurement of Stable Crack Growth Including Detection of Initiation of Growth Using the DC Potential Drop and the Partial Unloading Methods," in *Ductile Fracture Test Methods,* Proceedings of CSNI Workshop, Organization for Economic Cooperation and Development, Paris, 1983, pp. 18–53.

[*6*] Johnson, H. H., "Calibrating the Electric Potential Method for Studying Slow Crack Growth," *Materials Research and Standards,* Vol. 5, 1965, pp. 442–445.

Francis I. Baratta,[1] Joseph A. Kapp,[2] and David S. Saunders[3]

Load-Point Compliance for the Arc-Bend/Arc-Support Fracture Toughness Specimen

REFERENCE: Baratta, F. I., Kapp, J. A., and Saunders, D. S., **"Load-Point Compliance for the Arc-Bend/Arc-Support Fracture Toughness Specimen,"** *Fracture Mechanics: Twenty-First Symposium, ASTM STP 1074*, J. P. Gudas, J. A. Joyce, and E. M. Hackett, Eds., American Society for Testing and Materials, Philadelphia, 1990, pp. 594–612.

ABSTRACT: The primary objective of this work is to provide the theoretical computation of load-point compliance for the arc-bend/arc-support specimen and some experimental verification of these predictions. This specimen is a segment of an annular disk loaded in three-point bending with the roller supports positioned at the inside surface.

Load-line compliance was determined principally by utilizing Irwin's equation that relates the compliance rate of change with crack length to the strain energy release rate. The strain energy release rate was determined from the stress intensity solution available in the literature for this sample. Corroboration was also provided by utilizing boundary collocation, boundary integral element (BIE), and finite element (FEM) methods. Representative cases were experimentally tested to confirm theoretical and numerical results. During the tests, it was also found that duplication of idealized support and fixture conditions was extremely important with regard to accurate measurement of load-point compliance.

Two statically equivalent models were used to calculate the load-line compliance using the numerical methods. The first we call Model 1, where the load point is fixed and the roller reactions are applied at the appropriate location. The second we call Model 2, where the load is applied at the load point and the roller reaction points are fixed in the appropriate manner. Although both models are statically equivalent, significantly different load-line compliances are obtained depending upon the model chosen.

Experimental results were also obtained. Specimens were tested with machined slots to simulate a crack and other specimens tested had fatigue cracks. The measured compliances were found to agree with Model 2 results in some cases, with Model 1 in others, and with neither in still others.

KEY WORDS: fracture mechanics, fracture testing, stress intensity factors, numerical stress analysis, compliance

The arc-shaped tension loaded (A(T)) specimen has been accepted as one of the standard fracture toughness samples in ASTM E 399 since 1978. This specimen, which is a single-edge notched and fatigue cracked ring segment loaded in tension can be easily fabricated from hollow thick-walled cylinders. For some applications, such as relatively thin-walled tubing, a bending sample would be more convenient. Two bending sample variations of the arc-shaped sample have been suggested. One of these is the arc-bend chord support sample, where the reaction loads are applied to a chord which is machined in the specimen (Fig. 1*a*).

---

[1] U.S. Army Materials Technology Laboratory, Watertown, MA 02172.
[2] U.S. Army Armament R, D & E Center, Benet Laboratories, Watervliet, NY 12189-4050.
[3] Aeronautical Research Laboratories, Melbourne, Victoria 3001, Australia.

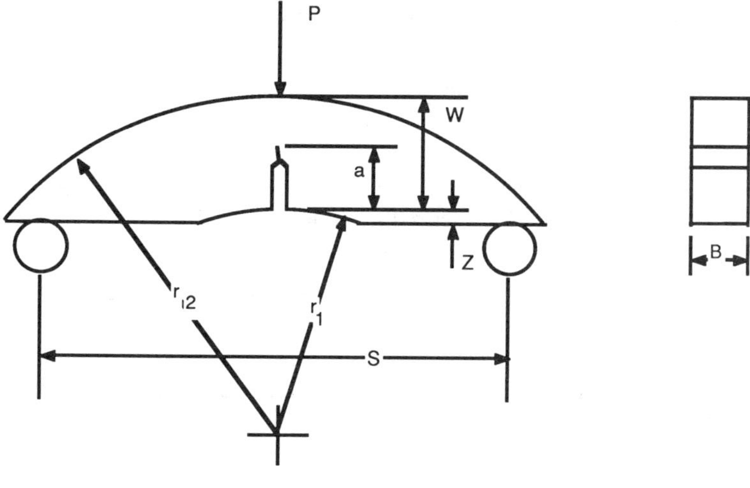

Figure 1a

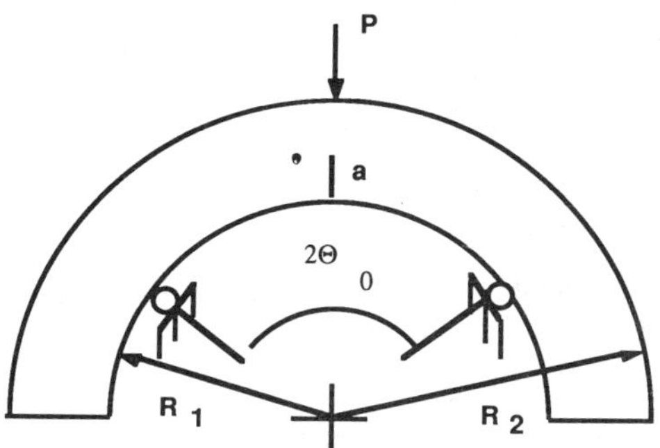

Figure 1b

FIG. 1—*Arc-shaped bending samples.* (a) *Arc-bend chord support.* (b) *Arc-bend/arc-support.*

The second type is the arc-bend/arc-support sample, where the reaction loads are applied at the inside diameter of the sample (Fig. 1b). This paper will deal only with the latter sample.

ASTM Task Group E24.01.05 on Fracture Specimen Design has considered that the arc-bend chord support sample has distinct advantages over the arc-bend/arc-support sample. Thus the arc-bend/arc-support sample has been eliminated from consideration as a standard fracture toughness sample. Nevertheless, useful information has been generated in the analysis of the arc-bend/arc-support sample, such as the development of stress intensity factor and crack mouth opening displacement solutions using boundary collocation [1]. This paper presents the results of further analyses used to calculate and measure the loading compliance of the arc-bend/arc-support sample.

The primary load-line compliance results were obtained in the following manner. It is assumed that the total load-line compliance of the sample (Δ_{tot}) is given by

$$\Delta_{tot} = \Delta_{crack} + \Delta_{no\ crack} \tag{1}$$

where $\Delta_{no\ crack}$ is the compliance of the specimen with no crack present and Δ_{crack} is the additional displacement due to the presence of the crack. Using this formulation, $\Delta_{no\ crack}$ is determined using Winkler's theory as referenced by Jones [2], and Δ_{crack} is found by integrating Irwin's compliance equation [3]:

$$G = \frac{P^2}{2B} \frac{\partial C}{\partial a} \tag{2}$$

where G is the strain energy release rate, C is the compliance of the sample ($C = \Delta_{crack}/P$; P being the load), and a is the crack depth. We determine the energy release rate from [4]

$$G = K^2 E' \tag{3}$$

where K is the stress intensity factor and E' is the elastic modulus (E) for plane stress conditions or ($E' = E/1 - \mu^2$); μ being Poisson's ratio) for plane strain.

In addition, load-line compliance solutions were obtained using boundary collocation (BC), finite elements (FEM), and the boundary integral element method (BIE). Two different modeling methods were used when these numerical techniques were employed. Also, load-line compliance was measured on several specimen geometries.

Load-Line Compliance Using Irwin's Equation

As stated, the load-line compliance can be determined with Eq 1, obtaining Δ_{crack} by integrating Eq 2. The formulation of $\Delta_{no\ crack}$ using Winkler's theory has been accomplished by

TABLE 1—$EB\Delta_{total}/P$ determined using Irwin's equation approach.

					$r_2/r_1 = 1.1$				
$\backslash\theta_0$	11.46	18	27	36	45	54	63	72	81
a/W									
0.0	23.27	85.51	302.7	826.7	2006.	4729.	11770.	35130.	179100.
0.1	25.30	89.90	316.2	854.4	2058.	4829.	11970.	35640.	181200.
0.2	29.35	100.8	343.7	911.2	2167.	5037.	12400.	36680.	185700.
0.3	36.50	120.2	392.8	1013.	2361.	5407.	13160.	38550.	193800.
0.4	48.47	152.5	474.8	1182.	2685.	6025.	14420.	41680.	206700.
0.5	69.02	207.7	614.4	1470.	3234.	7073.	16560.	46970.	229200.
0.6	107.9	311.4	874.7	2007.	4255.	9018.	20530.	56770.	270600.
0.7	195.8	543.2	1454.	3199.	6517.	13320.	29300.	78410.	362200.
0.8	450.9	1212.	3117.	6619.	12989.	25610.	54340.	140200.	587500.

					$r_2/r_1 = 1.15$				
$\backslash\theta_0$	3.5	18	27	36	45	54	63	72	81
a/W									
0.0	13.42	29.31	100.2	268.7	646.2	1517.	3763.	11210.	57100.
0.1	14.63	31.61	106.0	280.8	669.5	1561.	3855.	11440.	58050.
0.2	17.17	36.48	118.6	307.0	719.8	1657.	4052.	11930.	60120.
0.3	21.79	45.36	141.5	354.8	811.9	1834.	4413.	12820.	64420.
0.4	29.63	60.41	180.2	435.7	967.4	2131.	5023.	14330.	70350.
0.5	43.19	86.30	246.3	573.5	1232.	2636.	6058.	16900.	81200.
0.6	68.97	135.1	370.0	830.4	1724.	3572.	7975.	21630.	101300.
0.7	127.6	245.3	646.6	1402.8	2815.	5646.	12210.	32090.	145600.
0.8	301.2	569.0	1453.	3066.	5977.	11640.	24430.	62220.	273300.

$r_2/r_1=1.2$

$\backslash\theta_0$ a/W	13.5	18	27	36	45	54	63	72	81
0.0	7.12	14.69	47.59	124.9	297.1	693.0	1714.	5093.	25890.
0.1	7.84	16.07	51.14	132.3	311.3	720.6	1769.	5232.	26480.
0.2	9.30	18.91	58.58	147.9	341.5	778.8	1888.	5527.	27740.
0.3	11.94	24.06	72.06	176.2	396.3	884.4	2105.	6066.	30030.
0.4	16.43	32.78	94.92	224.2	488.9	1062.	2471.	6974.	33890.
0.5	24.30	47.97	134.3	306.4	647.2	1366.	3095.	8521.	40460.
0.6	39.48	76.88	208.5	460.5	943.0	1934.	4256.	11390.	52640.
0.7	74.43	142.6	375.3	805.0	1601.	3195.	6831.	17740.	79550.
0.8	178.1	335.5	860.4	1802.	3500.	6826.	14240.	35960.	156600.

$r_2/r_1=1.25$

$\backslash\theta_0$ a/W	18	27	36	45	54	63	72	81
0.0	9.00	27.59	70.66	166.0	384.5	946.7	2806.	14240.
0.1	9.92	30.00	75.71	175.8	403.2	985.1	2901.	14640.
0.2	11.82	35.02	86.29	196.3	442.8	1067.	3104.	15510.
0.3	15.22	44.09	105.5	233.6	514.8	1215.	3473.	17080.
0.4	21.02	59.46	138.0	296.6	636.3	1465.	4095.	19730.
0.5	31.13	86.02	193.7	404.4	843.7	1892.	5154.	24240.
0.6	50.57	136.4	298.8	606.4	1231.	2686.	7123.	32630.
0.7	95.24	250.6	535.2	1059.	2096.	4454.	11500	51230.
0.8	227.8	586.1	1226.	2373.	4602.	9570.	24130	150000.

$r_2/r_1=1.5$

$\backslash\theta_0$ a/W	27	36	45	54	63	72	81
0.0	6.500	14.74	32.31	71.72	171.9	500.3	2508
0.1	7.300	16.46	34.59	78.28	185.5	534.4	2654
0.2	8.890	19.94	42.56	91.74	213.5	604.9	2956
0.3	11.71	26.13	54.86	115.8	263.9	731.2	3499
0.4	16.51	36.59	75.58	156.4	348.4	943.1	4409
0.5	24.93	54.79	111.4	226.0	493.2	1305	5957
0.6	41.37	89.77	179.7	357.9	766.4	1986	8855
0.7	79.85	170.5	336.2	657.6	1385	3522	15360
0.8	196.4	412.1	802.0	1545	3210	8033	34420

$r_2/r_1=2.0$

$\backslash\theta_0$ a/W	31.5	36	45	54	63	72	81
0.0	3.33	4.64	9.110	18.69	42.39	118.7	578.4
0.1	3.79	5.32	10.49	21.46	48.18	133.6	643.0
0.2	4.68	6.65	13.23	26.97	59.94	163.5	773.6
0.3	6.24	8.99	18.06	36.68	80.80	216.4	1004
0.4	8.93	12.97	26.22	52.98	115.7	304.6	1386
0.5	13.72	20.02	40.49	81.23	175.7	455.4	2037
0.6	23.32	33.97	68.29	135.6	289.8	741.5	3264
0.7	46.45	67.24	133.7	262.0	551.4	1396	6055
0.8	118.6	170.2	333.7	645.0	1337	3360	14370

$r_2/r_1=2.5$

$\backslash\theta_0$ a/W	36	45	54	63	72	81
0.0	2.88	5.30	10.33	22.46	60.87	289.5
0.1	3.30	6.22	12.20	26.50	71.30	335.4
0.2	4.15	7.99	15.86	34.43	91.87	426.0
0.3	5.61	11.10	22.25	48.24	127.7	583.5
0.4	8.11	16.36	32.95	71.23	186.9	842.8
0.5	12.62	25.65	51.58	110.8	288.0	1282
0.6	21.70	43.94	87.73	186.8	480.3	2111
0.7	43.74	87.47	172.5	362.8	922.5	4001
0.8	112.7	222.0	431.6	896.8	2256	9666

Jones [2] and will not be discussed in detail here. The result for $\Delta_{\text{no crack}}$ is

$$\Delta_{\text{no crack}} = \frac{P}{BE}\left\{\frac{k+1}{8(k-1)}\left[\left(\frac{\theta_0}{\cos^2\theta_0} - \tan\theta_0\right)\bigg/\left(1 - \frac{2(k-1)}{(k+1)\ln(k)}\right)\right.\right.$$
$$\left.\left. + \frac{(2+3\mu)\theta_0}{\cos^2\theta_0} + (4+3\mu)\tan\theta_0\right]\right\} \quad (4)$$

where B is the through-thickness, k is the radius ratio (r_2/r_1), and θ_0 is the angle of support.

To determine the value of Δ_{crack}, we use the stress intensity factor solution of Gross [1] in Eq 3 and substitute the result into Eq 2. Upon rearranging and integrating, the result is

$$\Delta_{\text{crack}} = 2\tan^2\theta_0 \int_0^{a/W} [f(a/W)]^2 \, d(a/W) \quad (5)$$

$$f(a/W) = \frac{KB(W)^{1/2}}{P\tan\theta_0} \quad (6)$$

where K is the stress intensity factor, W is the specimen width, and P is the applied load.

The stress intensity factor solution developed in Ref 1 is a numerical solution and is reported only at discreet values of a/W; thus to evaluate the integral in Eq 5 the numerical solution must be estimated at values of a/W other than those reported. To do this an interpolation equation of the following type was fitted to the K solutions:

$$f(a/W) = \frac{KB(W)^{1/2}}{P\tan\theta_0} = A_1\, e^{\{(A_2(a/W)+A_3(a/W)^2+A_4(a/W)^3\}} \quad (7)$$

where A_1, A_2, A_3, and A_4 are constants determined by least squares fitting to the numerical solution and e is the base of the Napierian logarithm. The integration was performed using Simpson's rule. The results for Δ_{total} are given in Table 1, for a wide range of radius ratios (r_2/r_1), support angles (θ_0) in degrees, and crack depths (a/W).

For completeness, the load-line compliances for the arc-bend/arc-support specimen with external cracks were considered in the same manner as outlined above (i.e., using Irwin's equation approach). A brief discussion of this approach is given in the Appendix.

Load-Line Displacement Solutions with Numerical Methods

For analysis purposes the arc-bend/arc-support sample can be modeled as shown in Fig. 2a, if it is assumed that the roller supports cause a reaction force to act radially as shown. This is probably the case if the displacement of the sample is small. Large displacements will cause the inside surface of the specimen to roll over the reaction rollers. This changes the boundary conditions, making the problem nonlinear. For this study small displacements are assumed and a linear analysis performed. A second model for the specimen is pictured in Fig. 2b. In this case the load is applied as in the actual loading of the specimen and the reactions are at the inside radius of the sample. Both of these models are used to determine the load-line displacements.

Three methods were used to develop the load-line displacements: a finite element method [5], boundary value collocation [6], and the boundary integral element method [7]. The boundary value collocation results here are generated using exactly the same stress function and its derivative as reported in Ref 1. In that earlier study, stress intensity factors and crack

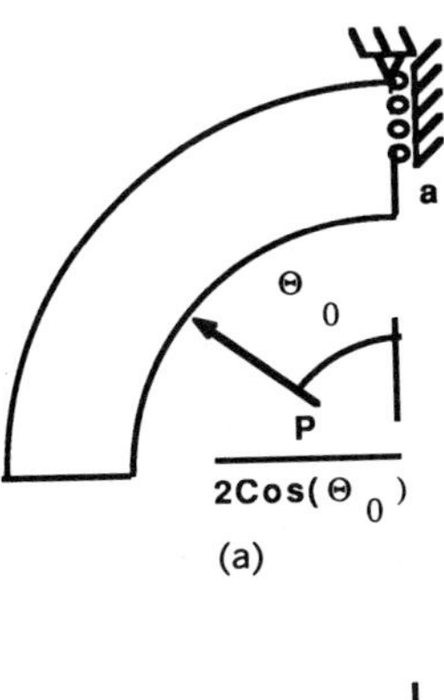

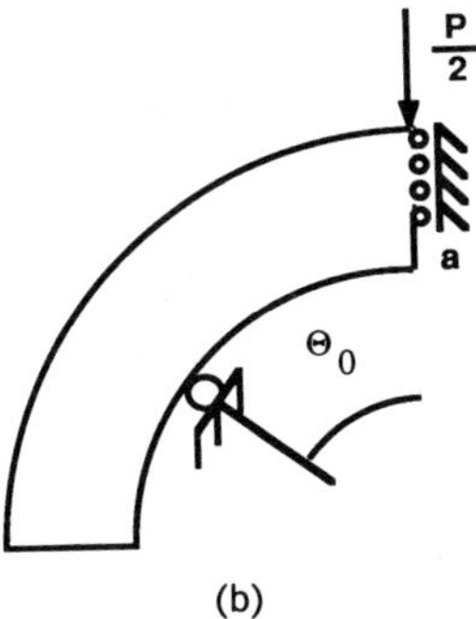

FIG. 2—*Two models used to analyze the arc-bend/arc-support specimen.*

mouth opening displacements only were generated. If only these two parameters are sought, a relatively small boundary is all that is required for the analysis. The stresses that result on a radial plane between the reactions at $\theta = \theta_0$ and $\theta = 0$ can be calculated and applied to that annular segment at the intermediate radial plane. The loading is equivalent to the actual loading and sufficient to determine stress intensity factors and crack mouth opening displacements, but this modeling is not sufficient to determine the load-line displacement. The boundary collocation method used here and in Ref *1* is based on the stress function reported in Ref *6*. With this stress function all displacements are relative to the displacement of the crack tip or it assumes that the crack tip is stationary. In order to determine the load-line displacement with this method the displacement of two points must be known: (1) the displacement of the point of applied external force relative to the crack tip, and (2) the component of displacement in the direction of the applied external force at the roller reaction. In order to accomplish this, the reaction point must be included on the boundary that is analyzed. The boundaries studied here included the reaction point. This is the only difference between this analysis and that reported in Ref *1*.

The finite element method applied here [5] utilizes cubic isoparametric elements and is written so that either model for the specimen can be applied. Grid points at any location can be loaded or constrained. We used this method to model the problem in the two ways described above. In this study, the boundary integral element code used utilized cubic hermation elements [7]. This code was written such that only entire elements could be constrained or loaded. This makes it difficult to model the loading of the arc-bend/arc-support sample by loading at the physical load point and reacting the load at the inside radius (the physical reaction point). These loads and reactions are in the ideal case point loads and reac-

TABLE 2—*Numerical solutions for arc-bend/arc-support sample with $\theta_0 = 36°$.*

$$\frac{KB\sqrt{W}}{P \tan \theta_0}$$

Method	r_2/r_1	a/W 0.3	0.4	0.5	0.6	0.7
Coll[1]	1.1	34.97	45.32	60.37	84.23	130.9
Coll (TS)	1.1	--	--	--	--	--
BIE	1.1	35.26	45.83	61.28	86.70	135.8
FM (ARF)	1.1	34.20	44.66	60.11	85.36	131.8
FM (RS)	1.1	--	--	--	--	--
Coll[1]	1.5	8.632	11.31	15.24	21.82	34.29
Coll (TS)	1.5	8.604	11.25	15.20	21.69	33.65
BIE	1.5	8.634	11.29	15.27	21.90	34.60
FM (ARF)	1.5	8.560	11.26	15.33	22.09	34.77
FM (RS)	1.5	--	9.501	12.90	18.60	--
Coll[1]	2.0	5.309	6.996	9.565	13.88	22.14
Coll (TS)	2.0	5.386	7.024	9.557	13.84	22.04
BIE	2.0	5.270	6.966	9.545	13.86	22.13
FM (ARF)	2.0	5.247	6.976	9.631	14.08	22.46
FM (RS)	2.0	--	--	--	--	--

$$\frac{E'B\Delta_{cm}W}{6M}$$

Method	r_2/r_1	a/W 0.3	0.4	0.5	0.6	0.7
Coll[1]	1.1	2.191	3.714	6.314	10.96	22.40
Coll (TS)	1.1	--	--	--	--	--
BIE	1.1	2.274	3.803	6.431	11.51	23.16
FM (ARF)	1.1	2.126	3.570	6.014	10.60	20.58
FM (RS)	1.1	--	--	--	--	--
Coll[1]	1.5	2.219	3.620	6.194	11.04	22.02
Coll (TS)	1.5	2.177	3.650	6.152	10.97	21.80
BIE	1.5	2.252	3.709	6.223	11.30	22.20
FM (ARF)	1.5	2.184	3.661	5.991	10.49	20.34
FM (RS)	1.5	--	3.598	5.986	10.46	--
Coll[1]	2.0	2.229	3.661	6.115	10.85	21.67
Coll (TS)	2.0	2.275	3.704	6.183	10.88	21.63
BIE	2.0	2.216	3.621	6.054	10.78	21.61
FM (ARF)	2.0	2.226	3.605	5.935	10.34	20.05
FM (RS)	2.0	--	--	--	--	--

tions. It was decided therefore to use only the first modeling described with the boundary integral element method.

Numerical Results

Solutions were obtained for a wide range of specimen geometries and loading conditions. Because the arc-bend/arc-support sample can have many radius ratios (r_2/r_1), several reaction points (θ_0), and several crack depths (a/W), all of these variables were studied. In the interest of brevity only a representative number of these solutions are presented here. These are the cases of $r_1/r_2 = 1.1$, 1.5, and 2.0 at $\theta_0 = 36°$ or 54° and $0.3 \leq a/W \leq 0.7$. These results are shown in Table 2 for $\theta_0 = 36°$ and in Table 3 for $\theta_0 = 54°$. Included in the tables are the solutions for the stress intensity factor (K) and the crack mouth opening displacement (Δ_{cm}) as well as the load-line displacement. The K and Δ_{cm} solutions are presented to give a measure of the accuracy of the numerical solutions.

TABLE 2—*(continued)*

$$\frac{E'B\Delta_{total}}{P}$$

Method			a/W			
	r_2/r_1	0.3	0.4	0.5	0.6	0.7
Irwin's Eq.	1.1	1013.	1182.	1470.	2007.	3199.
Coll (TS)	1.1	--	--	--	--	--
BIE	1.1	924.2	853.0	1073.	1489.	2413.
FM (ARF)	1.1	712.3	832.2	1037.	1412.	2199.
FM (RS)	1.1	--	--	--	--	--
Irwin's Eq.	1.5	26.13	36.59	54.79	89.77	170.5
Coll (TS)	1.5	14.61	20.13	30.54	49.54	90.92
BIE	1.5	14.91	20.63	30.70	50.16	94.21
FM (ARF)	1.5	15.23	20.78	30.32	48.24	87.17
FM (RS)	1.5	--	35.67	53.01	85.65	--
FM (RS,rel)	1.5	--	20.67	30.20	48.02	--
Irwin's Eq.	2.0	8.99	12.97	20.02	33.97	67.24
Coll (TS)	2.0	5.10	6.60	9.42	14.77	27.87
BIE	2.0	4.80	6.33	9.09	14.70	27.60
FM (ARF)	2.0	5.42	6.93	9.58	14.72	26.30
FM (RS)	2.0	--	--	--	--	--

Legend:
Coll[1] - Collocation Ref [1]
Coll(TS) - Collocation This Study
BIE - Boundary Integral Equation
FM(ARF) - Finite Element with Physical Load Point Fixed and Roller Reaction Applied at the Inside Radius
FM(RS) - Finite Element with Load Applied at the Physical Load Point and the Inside Radius

Constrained at θ_0

Irwin's Eq. - Irwin's Equation (N. B. These results are for the plane stress condition)
FM(RS,rel) - Finite Element with Load Applied at the Physical Load Point and the Inside Radius

Constrained at θ_0, relative displacement of the load point with respect to the roller reaction point

Discussion of Numerical Results

Considering the results for $\theta_0 = 36°$, we find that all the K results generated in this study agree very well with the solutions reported in Ref *1*. The maximum difference appears to be about 3.0% in some cases, with most of the solutions being closer if the specimen is modeled by applying the roller reaction as the active load and fixing the load point or the crack tip. If the specimen is modeled by applying the physical load at the physical load point and fixing the inside radius at θ_0 to move only in the radial direction of θ_0 FM(RS), K does not agree at all with the solution reported in Ref *1*. This finite element model predicts that the stress intensity factor should be about 18% less than that reported in Ref *1* and developed here using the other model. Considering the crack mouth opening displacements, all the solutions generated agree with one another quite well, including both finite element solutions, although for deep cracks both finite element solutions seem to predict less displacement than the other methods.

The major discrepancies occur with the load-line displacements. Using Irwin's equation to predict the load-line displacement solution by integrating the K solutions in Ref *1*, a much greater load-line displacement is predicted than if the specimen is modeled with the roller support reaction applied and the physical load point or the crack tip is fixed. The difference can be greater than 50%. Furthermore, all the numerical solutions developed here agree with one another (within about 4%). The interesting finding is that when the specimen is modeled so that the active load is applied at the physical load point and the inside radius is constrained at θ_0 in the direction of θ_0, the finite element solution agrees with Irwin's equation and dis-

TABLE 3—*Numerical solutions for arc-bend/arc-support sample with $\theta_0 = 54°$.*

$$\frac{KB\sqrt{W}}{P \tan \theta_0}$$

Method	r_2/r_1	a/W 0.3	0.4	0.5	0.6	0.7
Coll[1]	1.1	35.28	45.32	60.37	84.70	130.9
Coll (TS)	1.1	--	--	--	--	--
BIE	1.1	35.86	46.62	62.29	88.11	138.1
FM (ARF)	1.1	34.62	45.16	60.67	85.98	132.5
FM (RS)	1.1	--	--	--	--	--
Coll[1]	1.5	8.987	11.71	15.69	22.30	34.82
Coll (TS)	1.5	9.063	11.71	15.70	22.28	34.80
BIE	1.5	9.006	11.73	15.78	22.51	35.42
FM (ARF)	1.5	9.039	11.81	15.94	22.76	35.50
FM (RS)	1.5	--	9.501	13.33	19.06	--
Coll[1]	2.0	5.700	7.432	10.03	14.38	22.68
Coll (TS)	2.0	5.657	7.393	10.01	14.34	22.48
BIE	2.0	5.675	7.418	10.06	14.44	22.85
FM (ARF)	2.0	5.738	7.534	10.24	14.74	23.19
FM (RS)	2.0	--	--	--	--	--

$$\frac{E'B\Delta_{cm}W}{6M}$$

Method	r_2/r_1	a/W 0.3	0.4	0.5	0.6	0.7
Coll[1]	1.1	2.209	3.744	6.363	11.02	22.53
Coll (TS)	1.1	--	--	--	--	--
BIE	1.1	2.310	3.867	6.541	11.71	23.60
FM (ARF)	1.1	2.144	3.602	6.067	10.68	20.70
FM (RS)	1.1	--	--	--	--	--
Coll[1]	1.5	2.302	3.780	6.428	11.38	22.56
Coll (TS)	1.5	2.370	3.866	6.443	11.41	22.58
BIE	1.5	2.343	3.861	6.467	11.50	22.91
FM (ARF)	1.5	2.261	3.738	6.217	10.85	20.92
FM (RS)	1.5	--	3.739	6.212	10.81	--
Coll[1]	2.0	2.389	3.916	6.503	11.43	22.55
Coll (TS)	2.0	2.332	3.833	6.408	11.33	22.36
BIE	2.0	2.383	3.885	6.485	11.41	22.63
FM (ARF)	2.0	2.341	3.809	6.273	10.89	20.93
FM (RS)	2.0	--	--	--	--	--

agrees with the other numerical solutions. An additional solution is presented for the load-line displacements: that the specimen is modeled by application of the load at the physical load point and constraining at θ_0 in the direction of θ_0, but now determining the displacement of the load point relative to the roller reaction point FM(RS,rel). This relative displacement agrees with the solutions generated by fixing the load point and applying the roller reaction as the active load. This observation suggests that the numerical solutions and the Irwin's equation solutions are both correct but solve for a different parameter.

The discrepancy between the two solution methods is unusual and unexpected. The two models for the specimen are statically equivalent (i.e., from a mechanics of materials approach, they should produce the same solutions. This is the case for the two parameters that have been reported in the literature; K and Δ_{cm} are compared with the numerical solutions generated here. This would lead one to believe that the load-line displacements generated with these models should be very accurate. As it turns out they are indeed accurate in

TABLE 3—*(continued)*

$$\frac{E'B\Delta_{total}}{P}$$

Method	r_2/r_1	0.3	0.4	0.5	0.6	0.7
			a/W			
Irwin's Eq.	1.1	5407.	6025.	7073.	9018.	13320.
Coll (TS)	1.1	--	--	--	--	--
BIE	1.1	2656.	2999.	3589.	4701.	7168.
FM (ARF)	1.1	2595.	2914.	3455.	4444.	6516.
FM (RS)	1.1	--	--	--	--	--
Irwin's Eq.	1.5	115.8	156.4	226.0	357.9	657.6
Coll (TS)	1.5	43.28	58.40	86.06	139.3	257.6
BIE	1.5	43.86	60.12	88.23	142.6	262.8
FM (ARF)	1.5	43.79	59.44	86.12	135.6	241.9
FM (RS)	1.5	--	151.4	218.0	341.9	--
FM (RS,rel)	1.5	--	59.26	85.87	135.1	--
Irwin's Eq.	2.0	36.68	52.98	81.23	135.6	262.0
Coll (TS)	2.0	11.08	15.95	23.50	39.39	75.05
BIE	2.0	11.29	15.93	24.14	40.19	76.91
FM (ARF)	2.0	11.69	16.21	24.05	38.92	71.58
FM (RS)	2.0	--	--	--	--	--

Legend:
Coll[1] - Collocation Ref [1]
Coll(TS) - Collocation This Study
BIE - Boundary Integral Equation
FM(ARF) - Finite Element with Physical Load Point Fixed and Roller Reaction Applied at the Inside Radius
FM(RS) - Finite Element with Load Applied at the Physical Load Point and the Inside Radius
Constrained at θ_0
Irwin's Eq. - Irwin's Equation (N. B. These results are for the plane stress condition)
FM(RS,rel) - Finite Element with Load Applied at the Physical Load Point and the Inside Radius
Constrained at θ_0, relative displacement of the load point with respect to the roller reaction point

determining the displacement of the reaction point relative to the load point, but it is clear there is another component of the load-line displacement that is not accounted for. This additional displacement in the direction of the load-line at first suggests that a general translation of the specimen must occur, but of course that is impossible with the static conditions that must apply. Rather, the additional displacement is a consequence of the imposed conditions that the specimen cannot move in the θ_0 direction. This condition can be met with the reaction point moving in the direction of the applied load if there is additional displacement in the direction perpendicular to the load-line direction such that the vector sum of these displacements is zero in the θ_0 direction. Such a constraining condition is easily applied to a computer stress analysis, but may be very difficult to produce in the laboratory. In the laboratory, the conditions for supporting the arc-bend/arc-support sample would be to fix the roller location in directions parallel and perpendicular to the direction of the applied load in some sort of a bearing such that roller rotation is allowed. When the load is applied to maintain the condition that the displacement in the direction of θ_0 is zero, the rollers must move in the direction of the applied load so that the initial point of roller contact remains the only point of contact. Since the rollers would normally be fixed in space, the additional displacement must be accommodated by the specimen moving relative to the rollers. This changes the boundary conditions, making the specimen in actual testing nonlinear in the sense that the boundary conditions are variable (θ_0 changes with the magnitude of applied load and crack length). Of course, if the loads are low and the resulting displacements of the specimen are also small, the laboratory conditions are close to the idealized boundary con-

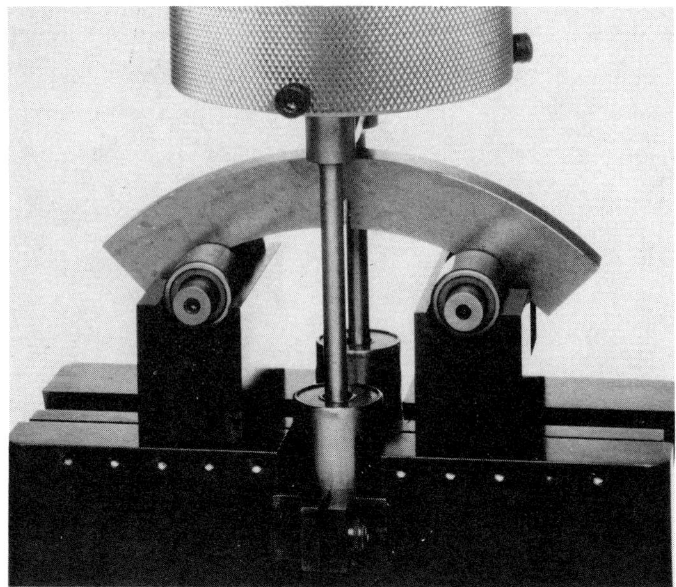

FIG. 3—*Experimental arrangement with rollers fixed in space and supported on bearings.*

ditions outlined. This suggests that if the specimen is to be used on a material with properties such that relatively large displacements will result, the arc-bend/arc-support specimen is of questionable applicability. Such conditions may only be obtained when testing very high strength brittle materials. Use of this specimen for J testing, K-R testing, or even K_{Ic} testing (where as much as 2% crack growth is allowed) may not be prudent due to the large displacements involved.

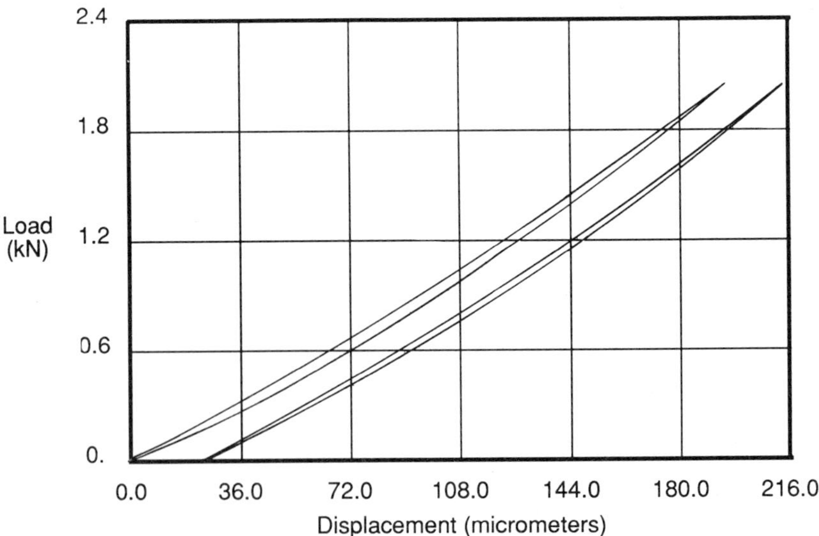

FIG. 4—*Typical load-line displacement versus applied load traces when using the experimental arrangement shown in Fig. 3. The radius ratio was 1.5, θ_0 was 54°, and* a/W *was 0.2.*

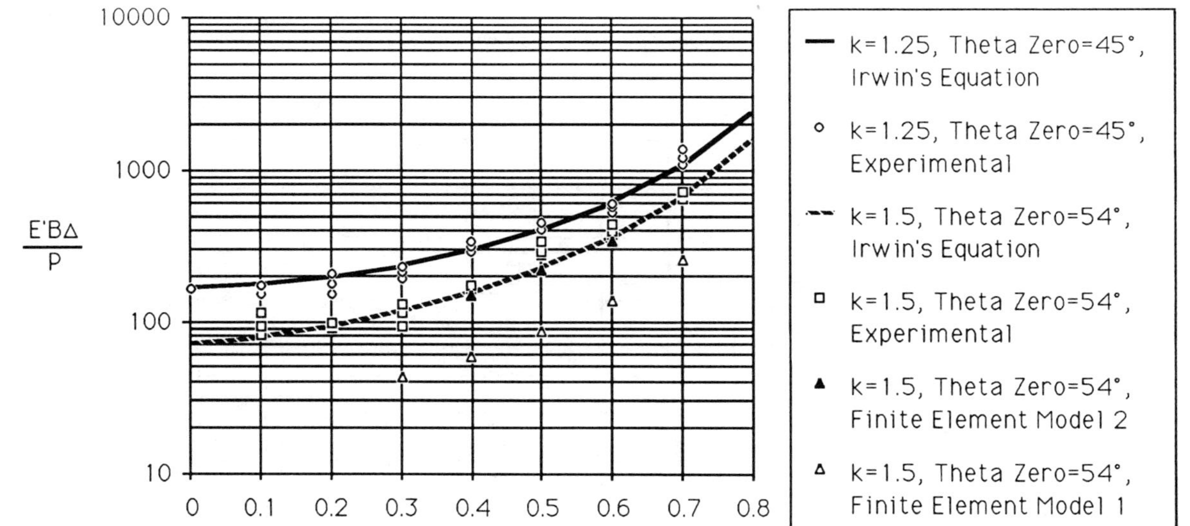

FIG. 5—*Comparison of experimentally measured load-line compliances obtained from the arrangement shown in Fig. 3 with predicted load-line compliances. The cracks are machined in the specimen.*

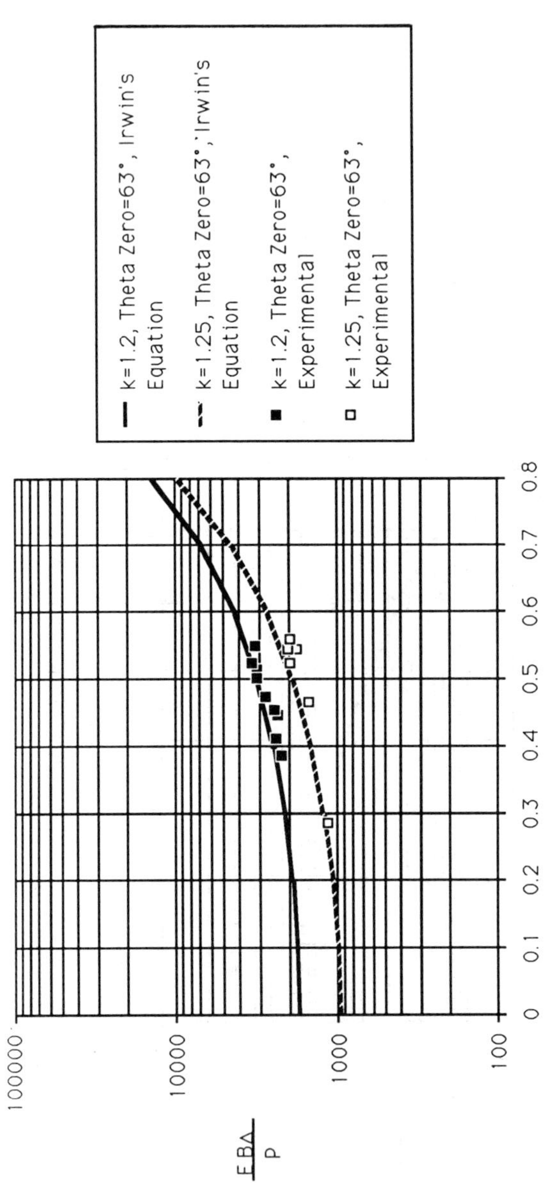

FIG. 6—*Comparison of experimentally measured load-line compliances obtained from the arrangement shown in Fig. 3 with predicted load-line compliances. The cracks are from fatigue precracking.*

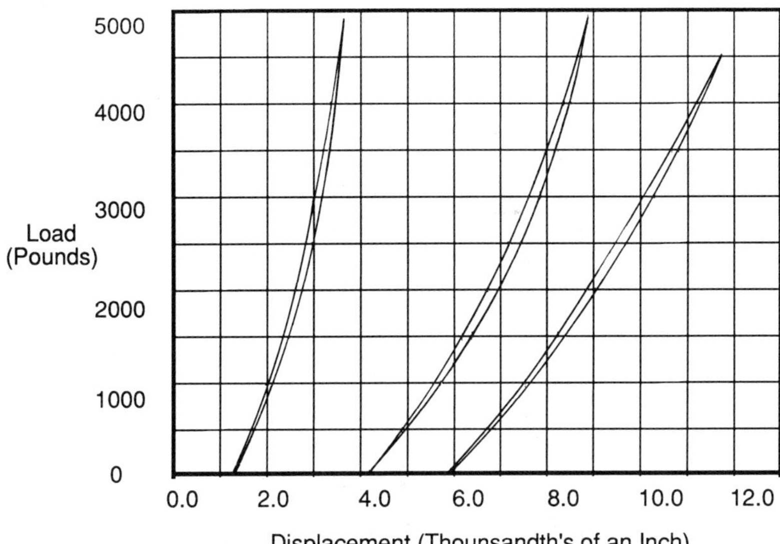

FIG. 7—*Typical load-line displacement versus applied load traces when using the experimental arrangement for three-point bending samples suggested in ASTM E 399. The radius ratio was 2.0, θ_0 was 36°, and the normalized crack lengths were (from left to right) 0.35, 0.45, and 0.55.*

The discrepancy in the K solutions with modeling is also very confusing. As stated the two models for the specimen are statically equivalent. Therefore the stresses produced by either loading condition should be the same. Of the three specimen parameters reported in this paper, K should be the parameter most closely associated with stress. Therefore it would be expected that the parameter most insensitive to modeling should be K. This is apparently not the case. A possible explanation is that the additional displacement introduced by the idealized boundary conditions would cause the resultant bending stress on the uncracked ligament to change. If the stress on the uncracked ligament changes, certainly the stress intensity factor would change.

Experimental Procedure

Experimental measurements of load-line compliances for the arc-bend/arc-support sample in several possible specimen geometries were obtained. Two sets of experiments were run. In the first, as shown in Fig. 3, great pains were taken to position the fixtures of the reaction rollers. The rollers were fixed in space, but were supported in bearings such that free rotation of the supports was allowed. Although not shown in Fig. 3, the columns supporting the rollers and bearings were stiffened during testing in the horizontal direction to minimize lateral deflection of the fixture. In the second study, the rollers were positioned as suggested in ASTM E 399. In this second case, the rollers are fixed so that they cannot move in the direction of the applied load, but are free to move perpendicular to the direction of the applied load. This is another way to accommodate rotation of the roller supports. Also, in the first study cracks were approximated by machined saw cuts or by fatigue cracking, while in the second group of experiments the cracks were produced by fatigue cracking only. The material used in the first study was a brittle martensitic steel whose properties are presented in

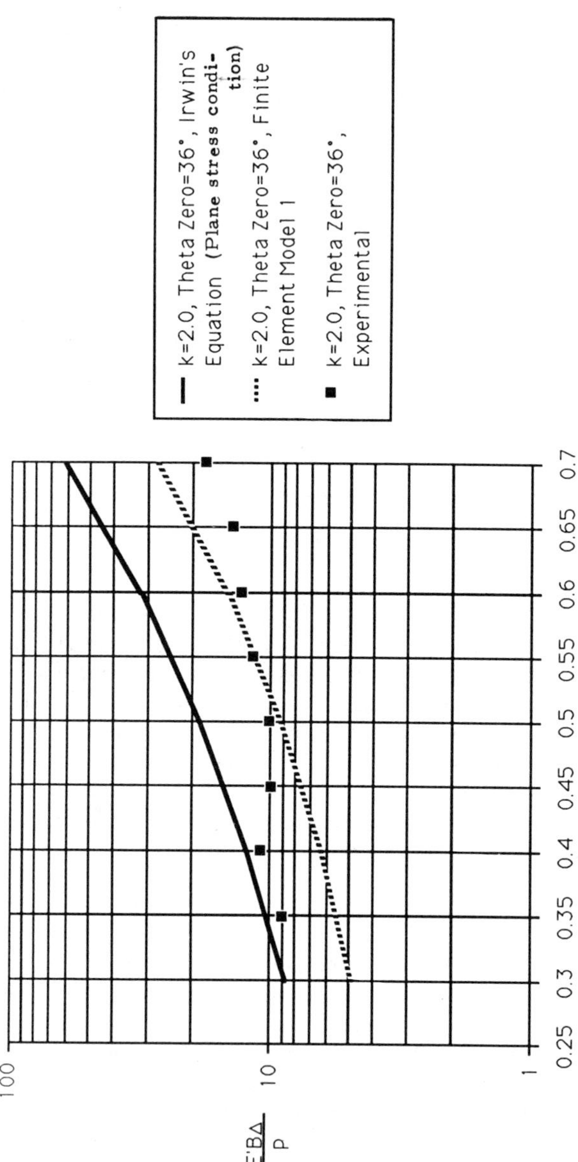

FIG. 8—*Comparison of experimentally measured load-line compliances with prediction using the experimental arrangement for three-point bending samples suggested in ASTM E 399.*

Ref *8*. In the second study the material was ASTM A723, Grade 2, Class 3, pressure vessel steel. The results from the first experimental study are discussed first.

A typical load versus load-line displacement plot from an autographic plotter is shown in Fig. 4 for rollers fixed in space and supported on bearings. As can be seen from the plot, there is some hysteresis and nonlinearity in the load-displacement behavior. For those tests performed with the rollers fixed in space and supported in bearings, the nonlinearity and hysteresis was smoothed by using least squares to determine the compliance. The data shown in Fig. 4 were digitized and the analysis performed.

The results for two specimens over a wide range of crack depths are given in Fig. 5. In this plot k is the radius ratio r_2/r_1. Also, the experimental results are compared with various predictions of the load-line compliance. For each crack depth there are three points plotted in the figure: the highest recorded compliance, the lowest recorded compliance, and the average of all the measurements. Usually there were five loadings and unloadings for each experimental condition. The solid line in Fig. 5 is Irwin's equation prediction of compliance from Table 1 for the case of $k = 1.25$ and $\theta_0 = 45°$. This solution agrees quite well with the experimentally determined compliances at all crack depths studied. The dashed line is Irwin's equation solution for $k = 1.5$ and $\theta_0 = 54°$. The agreement between this prediction and the experimentally determined compliances is also very good except when the normalized crack depth is either 0.1 or 0.5, where the experimental data is 22% and 40% respectively higher than predicted. All possible explanations for this discrepancy were explored without success. Figure 5 also shows the finite element predictions for $k = 1.5$ and $\theta_0 = 54°$ using both models of analysis. It is clear that the predictions using Model 1 (the load point fixed and the reaction load actively applied) do not agree at all with the experimental results, but the results obtained using Model 2 (the load actively applied and the reaction points fixed in the θ_0 direction) do indeed agree with those from the experiment.

A second series of tests was performed using the fixturing shown in Fig. 3 on actual fatigue precracked K_{Ic} samples [9]. The load-line compliance results for those specimens are shown in Fig. 6. Two specimen geometries were studied here, one with $k = 1.20$ and the other with $k = 1.25$. In both of these cases, the agreement with Irwin's equation prediction of compliance was very good.

Another series of tests was performed with a precracked sample of ASTM A723 pressure vessel steel with the condition that $k = 2.0$ and $\theta_0 = 36°$ with the rollers supported as suggested in ASTM E 399. Typical load versus load-line displacement traces obtained from this study are shown in Fig. 7. When these results are compared with the traces given in Fig. 4, it is clear that with these types of supports, more nonlinearity is induced and there is a greater amount of hysteresis behavior. Both of these observations are expected due to increased roller friction in this case. The load-line compliances from this testing were determined graphically by taking the slope of the most linear portion of the load displacement trace. The measured compliances appear in Fig. 8. Also shown in Fig. 8 are Irwin's equation prediction and the finite element prediction using Model 1 (load point fixed and the roller reactions actively applied). The comparisons are quite interesting. When the crack length is small and the specimen is stiff, the measured compliances are in good agreement with Irwin's equation prediction. When the crack gets a little deeper and the specimen gets more compliant, the specimen must move relative to the rollers. Since the amount of relative motion of the specimen with respect to the rollers is restricted because of roller friction, the measured compliances deviate from Irwin's equation prediction and follow the finite element, Model 1 prediction for a time. Finally, when the crack is very deep and the specimen is very compliant, the roller friction severely limits the specimen's compliance. At very deep crack depths, significant plastic deformation at the roller supports was observed.

These experimental results suggest that when small amounts of displacement are expected,

TABLE 4—*EBΔ_{total}/P determined using Irwin's equation approach for an externally cracked sample.*

$r_2/r_1 = 1.1$

$\backslash\theta_0$ a/W	11.46	18	27	36	45	54	63	72	81
0.0	23.32	84.75	303.6	829.1	2012.	4743.	11810.	35240.	179600.
0.1	24.93	89.04	314.3	851.3	2054.	4823.	11970.	35640.	183300.
0.2	28.32	98.12	337.3	898.6	2145.	4996.	12320.	36510.	185000.
0.3	34.45	114.7	379.5	985.9	2311.	5315.	12970.	38120.	191800.
0.4	44.89	143.0	451.3	1134.	2595.	5856.	14080.	40860.	203400.
0.5	62.93	191.6	574.3	1387.	3078.	6779.	15960.	45510.	223100.
0.6	96.92	282.4	802.8	1856.	3973.	8486.	19440.	54210.	259400.
0.7	172.6	483.0	1305.	2884.	5929.	12220.	27010.	72810.	338500.
0.8	387.6	1050.	2719.	5776.	11420.	22690.	48240.	125100.	560200.

$r_2/r_1 = 1.15$

$\backslash\theta_0$ a/W	13.5	18	27	36	45	54	63	72	81
0.0	13.49	29.47	100.8	270.4	650.3	1526.	3788.	11290.	57470.
0.1	14.45	31.29	105.4	280.0	668.7	1561.	3859.	11460.	58220.
0.2	16.49	35.20	115.5	300.9	709.0	1638.	4018.	11860.	59880.
0.3	20.22	42.38	134.1	339.7	783.6	1781.	4311.	12580.	62960.
0.4	26.60	54.58	165.6	405.5	710.2	2023.	4809.	13820.	68170.
0.5	37.55	75.47	219.5	517.3	1125.	2433.	5652.	15900.	76900.
0.6	58.14	114.4	319.2	723.5	1521.	3186.	7200.	19720.	93130.
0.7	104.1	200.5	538.2	1174.	2384.	4824.	10570.	28020.	128100.
0.8	236.6	446.5	1160.	2449.	4819.	9435.	20050.	51330.	226400.

$r_2/r_1 = 1.20$

$\backslash\theta_0$ a/W	13.5	18	27	36	45	54	63	72	81
0.0	7.170	14.81	48.06	126.2	300.3	700.4	1732.	5149.	26180.
0.1	7.710	15.84	50.69	131.7	310.8	720.6	1773.	5251.	26610.
0.2	8.820	17.99	56.28	143.3	333.4	763.9	1802.	5472.	27550.
0.3	10.82	21.89	66.52	164.8	374.9	843.8	2027.	5881.	29280.
0.4	14.24	28.57	83.98	201.5	445.8	980.	2306.	6578.	32230.
0.5	20.21	40.10	114.0	264.1	566.7	1212.	2783.	7763.	37240.
0.6	31.50	61.71	169.7	379.9	789.7	1638.	3657.	9939.	46420.
0.7	56.57	109.2	291.0	630.2	1271.	2556.	5537.	14610.	66080.
0.8	126.7	240.8	625.0	1316.	2586.	5062.	10660.	27350.	119500.

the specimen will deform according to Irwin's equation prediction. In order for this to occur, the specimen must be relatively free to move relative to the rollers. When the rollers restrict this relative motion, the only displacement observed is the motion of the roller support location relative to the load point motion. Finally, if the specimen is very compliant, pin friction can drastically change the compliance response.

Acknowledgments

The authors would like to thank the following personnel for their contributions without which this work would not have been completed: I. Burch, the Materials Research Laboratory, Australia, who conducted the bulk of the experimental measurements reported; L. Spiridigliozzi, Materials Technology Laboratory, who developed the specific interpolating polynomials for the K solutions so that the numerical integration of Irwin's equation could be performed; P. Perrone, Materials Technology Laboratory, who conducted an independent finite element analysis of the arc-bend/arc-support sample; R. Abbott, Benet Laboratories,

TABLE 4—*(continued)*

$r_2/r_1 = 1.25$

θ_0 / a/W	18	27	36	45	54	63	72	81
0.0	9.10	27.47	71.71	168.6	390.6	961.9	2852.	14470.
0.1	9.73	29.62	75.17	175.3	403.4	988.2	2917.	14750.
0.2	11.07	33.16	82.63	189.7	431.3	1046.	3059.	15350.
0.3	13.51	39.65	96.38	216.4	482.7	1152.	3324.	16480.
0.4	17.67	50.7	119.8	246.7	570.1	1333.	3772.	18390.
0.5	24.84	69.6	159.6	338.7	718.4	1639.	4532.	21620.
0.6	38.29	104.2	233.0	480.1	990.2	2200.	5921.	27520.
0.7	68.08	181.4	392.8	780.4	1577.	3412.	8911.	40190.
0.8	152.5	396.4	837.7	1635.	3201.	6762.	17160.	75110.

$r_2/r_1 = 1.50$

θ_0 / a/W	27	36	45	54	63	72	81
0.0	6.71	15.33	33.73	75.09	180.3	525.5	2637.
0.1	7.10	16.17	35.40	78.34	187.0	542.4	2709.
0.2	7.94	18.00	39.02	85.41	201.8	579.5	2868.
0.3	9.48	21.37	45.71	98.52	229.2	648.4	3163.
0.4	12.10	27.10	57.12	120.8	275.9	765.8	3666.
0.5	16.60	36.92	76.56	158.7	355.0	964.4	4514.
0.6	25.01	55.05	112.3	228.0	499.4	1326.	6054.
0.7	43.44	94.40	189.4	376.6	808.5	2099.	9329.
0.8	94.90	203.2	401.	782.9	1651.	4202.	18200.

$r_2/r_1 = 2.00$

θ_0 / a/W	31.5	36	45	54	63	72	81
0.0	3.55	4.99	9.96	20.72	47.45	133.8	655.6
0.1	3.70	5.21	10.40	21.59	49.28	138.4	675.6
0.2	4.01	5.68	11.35	23.50	53.31	148.7	720.1
0.3	4.58	6.52	13.09	27.00	60.77	167.8	803.1
0.4	5.92	7.94	16.03	32.91	73.35	200.0	943.3
0.5	7.15	10.35	20.97	42.84	94.45	253.9	1177.
0.6	10.16	14.80	30.02	60.88	132.6	350.9	1596.
0.7	16.81	24.51	49.51	99.46	213.6	555.7	2476.
0.8	35.64	51.74	103.4	205.2	434.1	1110.	4844.

$r_2/r_1 = 2.50$

θ_0 / a/W	36	45	54	63	72	81
0.0	3.16	5.97	11.90	26.39	72.6	349.6
0.1	3.26	6.17	12.32	27.27	74.9	359.3
0.2	3.47	6.62	13.23	29.23	79.9	381.5
0.3	3.85	7.44	14.91	32.88	89.4	423.2
0.4	4.49	8.81	17.73	39.00	105.2	493.3
0.5	5.57	11.09	22.41	49.14	131.4	609.1
0.6	7.54	15.21	30.81	67.22	177.9	813.6
0.7	11.85	24.08	48.64	105.3	274.9	1239.
0.8	24.16	--	97.70	208.9	536.9	2379.

who performed some of the experimental measurements; and Dr. R. Fenner, Imperial College of Science and Technology, for his many discussions regarding the analysis of this problem.

The authors wish to dedicate this paper to the memory of a former colleague, Dr. John E. Srawley, who made numerous significant contributions to the field of fracture mechanics.

APPENDIX

Load-Line Compliances for Arc-Bend/Arc-Support Samples with External Cracks

An additional sample that has been suggested is the arc-bend/arc-support sample with the crack on the external surface. This specimen is the same basic geometry as shown in Fig. 1b, except that the crack and roller supports are on the outside diameter of the specimen, and the load is applied at mid-span on the inside diameter of the specimen. The stress intensity factor and crack mouth opening displacement solutions for this sample have been developed using the collocation in Ref 1. It is a relatively simple matter to apply Irwin's equation approach to predict the load-line compliance of this sample. In the course of the work presented in the main body of this paper, the analysis of the externally cracked arc-bend/arc-support sample was performed. The equation for $\Delta_{\text{no crack}}$ is easily determined using Winkler's theory; the expression is

$$\Delta_{\text{no crack}} = \frac{P}{BE}\left\{\frac{k+1}{8(k-1)}\left[\left(\frac{\theta_0}{\cos^2\theta_0} - \tan\theta_0\right)\Bigg/\left(1 - \frac{2(k-1)}{(k+1)\ln(k)}\right)\right.\right. $$
$$\left.\left. + \frac{3(2+\mu)\theta_0}{\cos^2\theta_0} + 3\mu\tan\theta_0\right]\right\}$$

The contribution of the crack to the deflection (Δ_{crack}) was determined as described in the main body of the paper (Eqs 5 to 7). The only difference here is that the interpolating polynomials (Eq 7) were fit to the numerical stress intensity factors for externally cracked arc-bend/arc-support samples from Ref 1. The results of the integration are given in Table 4.

References

[1] Gross, B. and Srawley, J. E., "Stress Intensity and Displacement Coefficients for Radially Cracked Ring Segments Subject to Threepoint Bending," NASA TM 83059, March 1983.

[2] Jones, A. T., "Fracture Toughness Testing with Sections of Cylinders," *Engineering Fracture Mechanics,* Vol. 6, 1974, pp. 653–662.

[3] Irwin, G. R. and Kies, J. A., "Fracturing and Fracture Dynamics," *The Welding Journal, Research Supplement,* Vol. 31, 1952, pp. 95s–100s.

[4] Irwin, G. R., "Analysis of Stresses and Strains Near the End of a Crack Transversing a Plate," *Journal of Applied Mechanics, Transaction of ASME,* Vol. 24, 1957, p. 361.

[5] Gifford, I. N., Jr., and Hilton, P. D., "Stress Intensity Factors by Enriched Finite Elements," *Engineering Fracture Mechanics,* Vol. 10, 1978, pp. 485–496.

[6] Gross, B., Roberts, E. J., and Srawley, J. E., *International Journal of Fracture Mechanics,* Vol. 4, No. 3, Sept. 1968, pp. 267–276 (Errata, Vol. 6, 1970, p. 87).

[7] Watson, J. O., "Hermatian Cubic Boundary Elements for Plane Problems in Fracture Mechanics," *Res. Mechanica,* Vol. 10, No. 1, 1982, pp. 23–42.

[8] Saunders, D. S. and Baratta, F. I., "The Measurement of Fracture Toughness of Steels from the M795 and M549 Artillery Projectiles Using Small-Sized Fracture Toughness Specimens," MTL TR 86-22, U.S. Army Materials Technology Laboratory, Watertown, Mass., June 1986.

[9] Baratta, F. I. and Saunders, D. S., "Comparative Evaluation of LAW Motor Bodies of Three Materials," AMMRC TR 85-7, May 1985.

Author Index

Subject Index

W

Warm pre-stressing, 264
Weight function, 490

Y

Yield loads, 525
Yield stress, 361